THE METABOLISM, STRUCTURE, AND FUNCTION OF PLANT LIPIDS

THE METABOLISM, STRUCTURE, AND FUNCTION OF PLANT LIPIDS

Edited by

Paul K. Stumpf
University of California, Davis
Davis, California

J. Brian Mudd
ARCO Plant Cell Research Institute
Dublin, California

and

W. David Nes
U.S. Department of Agriculture
Albany, California

PLENUM PRESS • NEW YORK AND LONDON

Library of Congress Cataloging in Publication Data

International Symposium on Plant Lipids (7th: 1986: University of California, Davis)
The metabolism, structure, and function of plant lipids.

"Proceedings of the Seventh International Symposium on Plant Lipids, held July 27–August 1, 1986, at the University of California, Davis, Davis, California"–T.p. verso.
Includes bibliographies and index.
1. Plant lipids–Congresses. I. Stumpf, Paul K. (Paul Karl), 1919–   . II. Mudd, J. Brian (John Brian), 1929–   . III. Nes, W. David. IV. Title. [DNLM: 1. Lipids–metabolism–congresses. 2. Plants–metabolism–congresses. W3 IN921BU 7th 1986b / QK 898.L56 I61 1986b]
QK898.L56I556 1986   581.19'247   87-1749
ISBN 0-306-42492-4

Proceedings of the Seventh International Symposium on Plant Lipids,
held July 27–August 1, 1986, at the University of California, Davis,
Davis, California

© 1987 Plenum Press, New York
A Division of Plenum Publishing Corporation
233 Spring Street, New York, N.Y. 10013

All rights reserved

No part of this book may be reproduced, stored in a retrieval system, or transmitted in any form or by any means, electronic, mechanical, photocopying, microfilming, recording, or otherwise, without written permission from the Publisher

Printed in the United States of America

DEDICATION

Dr. Andrew A. Benson, professor of biology,
Scripps Institution of Oceanography, University
of California, San Diego.

It is customary in the Proceedings of the Symposia on Plant Lipids to honor one of the major contributors to this field of research.

On this occasion we would like to recognize important discoveries made by

ANDREW ALM BENSON

who is currently Professor at the Scripps Institute of Oceanography, La Jolla, California.

Andy Benson obtained his Ph.D. in organic chemistry from the California Institute of Technology. He subsequently joined the bio-organic chemistry group at the Lawrence Radiation Laboratory of the University of California at Berkeley at what proved to be a very productive time. Together with Calvin and Bassham he was responsible for the complete elucidation of the pathway of carbon dioxide assimilation in photosynthesis.

During the research on photosynthesis it is clear that Andy Benson noticed that in addition to the water soluble compounds labeled after the supply of radioactive carbon dioxide, lipid soluble compounds were also synthesized. When he moved to Pennsylvania State University in 1955, he turned his attention to these compounds. During the six years he spent at Pennsylvania State University Andy and his students made several important discoveries, including the isolation and characterization of both phosphatidylglycerol and sulfoquinovosyldiacylglycerol, two lipids we now know to be of particular significance in the chloroplast. No account of the chemistry and biochemistry of these compounds is complete without discussion of the results from Andy Benson's laboratory.

Work on the sulfoquinovosyldiacylglycerol continued in Andy's laboratories after he moved back to California, but his association with the Scripps Institute of Oceanography led to new discoveries, particularly in the lipids of marine organisms. The finding of arsenic containing lipids has been particularly interesting and impressive.

Andy Benson has made a deep impression on research in plant lipids not only by the contributions from his own laboratory but also by fostering and encouraging the research in other laboratories. We hope that he will continue to inspire us for years to come.

PREFACE

The Seventh International Symposium on the Structure and Function of Plant Lipids took place at the University of California, Davis, California July 27th to August 1st, 1986.

This was the first time the Symposium was held in the United States. The list of previous host cities reads, Norwich, Karlsruhe, Göteborg, Paris, Groningen, Neuchâtel. The addition of Davis to this distinguished list was made by the organizers with the doubts of people who give invitations to parties - will anybody come? In fact 155 participants registered and there were 21 spouses in attendance.

The scientific program was composed of nine sessions: biochemistry of isoprenoids and sterols, function of isoprenoids and sterols, structure and function of lipids, biosynthesis of complex lipids, fatty acid oxygenases and desaturases, medium and long chain fatty acids, interaction of university, government and industrial research, algal lipids, and genetics and biotechnology. In addition to these sessions of plenary lectures, there were four poster sessions in which about 140 posters were presented. All of this was packed into four days, and there was some comment about the scarcity of time to ask questions of the speakers, discuss the posters and even to eat lunch. The compression of the program was a result of the continued desire of the organizing committees to avoid concurrent sessions. The congregation of participants into a single session increases interaction and generates a feeling of unity at these symposia. This is exemplified by the position of studies on isopentenoids within the framework of plant lipid biochemists. Only a few years ago research on isopentenoids was considered distinct from the work on fatty acids and glycerolipids. These areas become closer as we emphasize the interaction of lipid components with each other and with the proteins of biological membranes. The integration of these approaches is best exemplified by research with chloroplasts, where the biosynthesis and function of isopentenoid compounds is related to biosynthesis and function of glycerolipids.

A number of people who were attending the Symposium for the first time commented on the open and friendly atmosphere. There is a family feeling since the community of people who work on plant lipids is still fairly small. At this Symposium however, there were several friends of the family, people accustomed to reporting their research in other areas who have much to contribute to the field of plant lipids. We think in particular of the contribution of Leo Parks on the function of sterols in yeast, the presentation of Joe Kuć on the importance of lipids in host-pathogen interactions, and the paper of Rudy Demel on the dynamics of glycerolipid, sphingolipid and sterol interaction in membranes.

The organization of this Symposium was different from previous symposia in some aspects. The new sessions in interaction of government, industrial and university research, and on genetics and biotechnology are in some respects related. Clearly the field of plant lipids is entering a phase of research on the molecular biology of lipid biosynthetic enzymes. It is equally clear that industrial laboratories have a stake in the outcome of these investigations. Plant lipid biochemists recognize that part of their work interfaces with the large and profitable oleochemical industry. Progress has already been made toward tailoring oil seed crops for specific purposes. This will be an area for considerable research in the future. We may expect important results from both industrial and tax-supported institutions on genetic manipulations and genetic engineering of the lipid metabolism of higher plants.

Although we continue to understand the biosynthesis of lipids in an increasingly comprehensive and detailed manner, we still have few examples of specific functions of lipids in higher plants. Studies on degradation of specific lipids in thylakoids are making progress in this area, and the use of _Arabidopsis_ mutants holds further promise in this area. The proposal by Norio Murata that particular molecular species of phosphatidylglycerol (PG) are involved in chilling sensitivity has aroused a great deal of interest because (a) it suggests a specific role for a chloroplast lipid and (b) it suggests a simple biochemical reason for the difference between chilling sensitive and chilling resistant plants. Several papers in these Proceedings bear on this question.

This book presents the material discussed at the Symposium in Davis. It is a reflection of the state of research in plant lipids in a comprehensive and up-to-date fashion. It provides an opportunity for those who were not at the Symposium to review the status of the field, and for those who were at the Symposium it provides an opportunity to study details that may have been missed in the actual presentation.

Because of the _ad hoc_ nature of the Plant Lipid Conferences, financial support was essential in order to continue the practice of having an international meeting with speakers from all corners of the globe. We therefore wish to acknowledge with considerable gratitude travel funds donated by the following industrial sources: Arco Plant Cell Research Institute, DuPont, Calgene, Henkel, Proctor and Gamble, Monsanto and the Shell Agricultural Chemical Company as well as funds from state and federal sources: National Science Foundation, University of California Biotechnology Research and Education Program, Office of Naval Research (USA), Department of Energy (USA), and Department of Agriculture (USA).

In addition we wish to thank Mrs. Kathryn Gardner and her colleagues of the Campus Events and Information Office (UCD) for their great assistance in the organization of the conference and Mrs. Billie Gabriel for her usual excellent secretarial help. One of us is also indebted to Mrs. Ruth R. Stumpf for her yeoman service during the Conference in smoothing out ruffled feathers of some of the participants.

P.K. Stumpf
J.B. Mudd
W.D. Nes

CONTENTS

BIOSYNTHESIS AND FUNCTION OF STEROLS AND OTHER ISOPRENOIDS

Multiple Roles for Plant Sterols 3
  W.R. Nes

Stereochemistry of Monoterpene Cyclization 11
  R. Croteau

Diterpenes - The Gibberellin Biosynthetic Pathway in Zea mays . . 19
  B.O. Phinney and C.R. Spray

Synthesis of Plastoquinone-9, α-Tocopherol and Phylloquinone
    (Vitamin K_1) and Its Integration in Chloroplast Carbon
    Metabolism of Higher Plants 29
  D. Schulze-Siebert, U. Homeyer, J. Soll and G. Schultz

Carotenoid Biosynthesis and Carotenogenic Enzymes in Plastids . . 37
  H. Kleinig

Cell Walls and Secondary Products as Obstacles to Plant Enzyme
    Isolation: Problems and Solutions, Including a Simple
    Liquid Nitrogen Homogenizer for Bulk Tissue Extraction . . . 45
  R.P. Sandstrom and W.D. Loomis

Developmental Regulation of Sterol and Pentacyclic Triterpene
    Biosynthesis and Composition: A Correlation with Sorghum
    Floral Initiation 53
  R.C. Heupel, W.D. Nes and J.A. Verbeke

Effect of Intracellular Sterol Concentration on Sterol
    Esterification in Yeast 57
  L.W. Parks, T.A. Lewis, C. Low and K. Haeckler

Functional Organization of Carotenoids and Prenylquinones in
    the Photosynthetic Membrane 63
  H.K. Lichtenthaler

Arachidonic and Eicosapentaenoic Acids, Glucans and Calcium
    as Regulators of Resistance to a Plant Disease 75
  M. Zook and J. Kuć

Presence of Unusually High Levels of Cholesterol in the Shoot-
    Apices of Flowering Plants 83
  V.K. Garg, T.J. Douglas and L.G. Paleg

Changes in Sterol Biosynthesis From [2-^{14}C]Mevalonic Acid
  During Development of Cucurbita maxima Seedlings 87
    V.K. Garg and W.R. Nes

Inhibition of Growth of Celery Cells by Paclobutrazol
  and Its Reversal by Added Sterols 91
    P.A. Haughan, J.R. Lenton and L.J. Goad

The Steryl Esters of Higher Plants 95
    L.J. Goad, J. Zimowski, R.P. Evershed and V.L. Male

Synthesis and Biological Evaluation of Fungal Bioregulators
  of Sterol Biosynthesis 103
    E.J. Parish, P.K. Hanners and W.D. Nes

A Comparison of Sterol and Long Chain Fatty Alcohol
  Biosynthesis in Sorghum bicolor 107
    Y. Sauvaire, B. Tal, R.C. Heupel, R. England, P.K. Hanners,
    W.D. Nes and J.B. Mudd

Regulation of Terpenoid Biosynthesis in Tapped Latex 111
    G.J. Piazza, E.J. Saggese and M.P. Thompson

Triterpenoid Biosynthesis in Euphorbia lathyris 115
    C.L. Skrukrud, S.E. Taylor, D.R. Hawkins and M. Calvin

Phytosterol Structure and Composition in the Chemosystematics
  of the Caryophyllales 119
    J.H. Adler and T.A. Salt

Biosynthesis of Diterpene Phytoalexin Precursors in
  Cell-free Extracts of Rice 123
    K. Wickham and C.A. West

Endogenous Gibberellins in Wheat Shoots 127
    J.-T. Lin and A.E. Stafford

Molecular Associations in Lipid-carotenoid Monolayers 131
    M. Tomoaia-Cotişel, J. Zsakó, E. Chifu and P.J. Quinn

Separation and Identification of Carotenoid-esters in Red
  Pepper /Capsicum annum/ During Ripening 135
    P.A. Biacs, J. Bodnár, Á. Hoschke, A. Cs. Pavis, H. Daood,
    F. Hajdu and N. Kiss-Kutz

On the Biosynthesis of Fragrance Compounds in Ambrette
  (Hibiscus abelmoschus) Seeds 139
    M.R. Pollard, S. Jamil-Panah and T.Y. Nee

STRUCTURE AND FUNCTION OF LIPIDS

Structural and Dynamic Aspects of Membrane Lipids 145
    R.A. Demel

Lipid Saturation by Catalytic Hydrogenation and Its Effect on
  Membrane Structure and Function 153
    L. Vigh

Spatial Organization and Functional Roles of Acyl Lipids
  in Thylakoid Membranes 161
   P.A. Siegenthaler, A. Rawyler and C. Giroud

Ultrastructural Studies on Plant Membranes 169
   W.W. Thomson, K.A. Platt-Aloia and R.D. Bliss

Evidences for Different Acyl Lipid Domains in Spinach and Oat
  Thylakoid Membranes Supporting Various Photosynthetic
  Functions 177
   P.A. Siegenthaler, C. Giroud and J. Smutny

Phase Transition Behavior of Monogalactosyldiacylglycerol 181
   P.J. Quinn and L.J. Lis

Is Monogalactosyl Diacylglycerol Involved in the Packaging
  of Light-harvesting Chlorophyll Proteins in the
  Thylakoid Membrane? 185
   P.J. Dominy and W.P. Williams

Mechanisms of Lipid-protein Binding in Photosynthetic Membranes . . 189
   D.J. Murphy

Lipid Molecular Species Composition of Granal and Stromal
  Lamellae 193
   H.A. Norman, J.B. St. John, F.E. Callahan, A.K. Mattoo
   and W.P. Wergin

On the Function of Methyl-branched Chain Fatty Acids in
  Phospholipids of Cell Membranes of Higher Plants 197
   A. Radunz

Freezing Resistance and Lipid Changes in Choline-treated
  Wheat Seedlings 201
   W.P. Williams, I. Horvath, P.J. Quinn, P.G. Thomas
   and L. Vigh

Drought Stress Effects on Root Cell Membranes 205
   H. Svenningsson, M. Andersson and C. Liljenberg

Manipulating Membrane Fatty Acid Compositions of Soybean Plants . . 209
   W.B. Terzaghi

Plasma Membrane Lipid Alterations Following Cold Acclimation:
  Possible Relevance to Freeze Tolerance 213
   D.V. Lynch and P.L. Steponkus

Plasma Membrane and Tonoplast Fractions Isolated from Spinach
  Leaves by Preparative Free Flow Electrophoresis:
  Effect of Photoinduction 217
   C. Penel, G. Auderset, S. Kiefer, A. Sandelius,
   A. Brightman, H. Greppin and D.J. Morré

Study of the Intracellular Transfer of Lipids to the
  Plasmalemma 221
   P. Moreau, H. Juguelin, R. Lessire and C. Cassagne

Ca^{2+} and Inter-molecular Bridging of Membranal Phospholipids
  and Proteins 225
   Y.Y. Leshem

A Membrane-located, Calcium-/Calmodulin-activated Phospholipase
    Stimulated by Auxin 229
  D.J. Morré and B. Drobes

Regulation of Phospholipase Activity in Potato Leaves by
    Protein Phosphorylation-dephosphorylation
    and Proteolytic Activation 233
  R.A. Moreau

BIOSYNTHESIS OF COMPLEX LIPIDS

Synthesis and Degradation of Lipid Bodies in the Scutella
    of Maize 239
  A.H.C. Huang, R. Qu, S.-m. Wang, V.B. Vance, Y.-z. Cao
  and Y.-h. Lin

On the Control of Fatty Acid Compositions of Plant
    Glycerolipids 247
  G. Roughan

Lipid Distribution and Synthesis Within the Plant Cell 255
  R. Douce, C. Alban, R. Bligny, M.A. Block, J. Covès, A.-J.
  Dorne, E.-P. Journet, J. Joyard, M. Neuberger and
  F. Rebeillé

Regulation of Phospholipid Headgroup Composition in Castor
    Bean Endosperm 265
  T.S. Moore, Jr.

Molecular Species of Phosphatidylcholine In Plants: Biosynthesis
    and Role In Oleate Desaturation or Freezing Resistance 273
  C. Demandre, A.M. Justin, X.V. Nguyen, M. Gawer, A.
  Trémolières and P. Mazliak

Factors Affecting the Fatty Acid Composition of
    Phosphatidylglycerol as Related to Chilling
    Sensitivity in Higher Plants 283
  S.R. Thomas, J. Sánchez and J.B. Mudd

On the Synthesis of Digalactosyldiacylglycerol in
    Chloroplasts, and Its Relation to
    Monogalactolipid Synthesis 293
  J.F.G.M. Wintermans and J.W.M. Heemskerk

Characterization of Galactosyltransferases in Spinach
    Chloroplast Envelope Membranes — Applications of an
    Assay for UDPGal:Diacylglycerol Galactosyltransferase 301
  J.W.M. Heemskerk, M.A.M. Scheijen, F.H.H. Jacobs
  and J.F.G.M. Wintermans

Galactosyltransferase Activities in Intact Spinach
    Chloroplasts and Envelope Membranes 305
  J.W.M. Heemskerk, F.H.H. Jacobs, G. Bögemann and J.F.G.M.
  Wintermans

Biosynthesis of Sulfoquinovosyldiacylglycerol in
    Chloroplasts of Higher Plants 309
  K.F. Kleppinger-Sparace and J.B. Mudd

Sulfolipid Synthesis by Isolated Intact Spinach Chloroplasts 313
  J. Joyard, E. Blée and R. Douce

Calcium Chloride Effect on Glycerolipids Metabolism
    in Olive Tree Leaf 317
  B. Marzouk, M. Zarrouk, A. Cherif and P. Mazliak

Lipid Metabolism in Potato Leaf Discs: Effect of Calmodulin
    Antagonists 321
  G.J. Piazza and R.A. Moreau

Localization and Properties of Cholinephosphate
    Cytidylyltransferase Activity in Castor Bean Endosperm 325
  A.J. Kinney and T.S. Moore, Jr.

Choline Kinase Activity in Castor Bean Endosperm 329
  A.J. Kinney and T.S. Moore, Jr.

The Control of CTP:Cholinephosphate Cytidylyltransferase
    in Pea Stems 333
  M.J. Price-Jones and J.L. Harwood

Cocoa Butter Biosynthesis. Cocoa Seed Diacylglycerol
    Acyltransferase: Studies on the Microsomal
    Bound Enzyme 337
  L. McHenry and P.J. Fritz

Substrate Specificity of Plant Lipases 341
  Y.H. Lin, C. Yu, J. Olsen and A.H.C. Huang

Lipids from Rice Anthers 345
  S. Toriyama, K. Hinata, I. Nishida and N. Murata

Lipids of Soybean Inoculated with Microsymbionts 349
  R.S. Pacovsky and G. Fuller

Properties and in vitro Synthesis of Phospholipid
    Transfer Proteins 353
  F. Tchang, F. Guerbette, D. Douady, M. Grosbois,
    C. Vergnolle, A. Jolliot, J.P. Dubacq and J.C. Kader

Galactolipid Synthesis in Isolated Pea Chloroplasts 357
  J.P. Dubacq, R.O. Mackender and P. Mazliak

Triacylglycerol Biosynthesis in Developing Cotyledons
    of Safflower (Carthamus tinctorius) 361
  G. Griffiths, K. Stobart and S. Stymne

Long Chain Triacylglycerol Acyl Hydrolase (Lipase) Activity
    in Wheat Grain 365
  T. Galliard, M. Lond and D.M. Gallagher

Glycoprotein Nature of Lypolytic Acyl Hydrolases in
    Potato Tubers and Leaves 369
  R.A. Moreau and G. Nagahashi

Correlation of Metabolic Rate Changes and Membrane
    Transitions Determined by Microcalorimetric Methods 373
  R.W. Breidenbach, R.S. Criddle, E. Lewis
    and L. Hanson

Molecular Species Composition of Phosphatidylglycerol in
    Leaves of Camellia Species and Chilling Sensitivity 377
    J. Sekiya, H. Koiso, A. Morita and A. Hatanaka

OXYGEN REQUIRING SYSTEMS - OXYGENASES AND DESATURASES

The Lipoxygenase Pathway 383
    B.A. Vick and D.C. Zimmerman

Enzymic Oxygenative-cleavage Reaction of Linolenic Acid in
    Leaves - Chloroplastic Lipoxygenase and Fatty Acid
    Hydroperoxide Lyase in Tea Leaves 391
    A. Hatanaka, T. Kajiwara and J. Sekiya

Fatty Acid β-Oxidation in Higher Plants 399
    B. Gerhardt

Desaturation of Fatty Acids on Complex-lipid Substrates 405
    S. Stymne, G. Griffiths and K. Stobart

Rapid Enzymic Peroxidation of Polyunsaturated Fatty
    Acids on Hydration of Wheat Milling Products 413
    T. Galliard, S.P.C. Tait and D.M. Gallagher

Action of Boron on Ethylene Production and Lipoxygenase
    Activity in Microsomes From Sunflower Cotyledons 417
    A. Belver, P. Rodriguez, M. Roldán and J.P. Donaire

Preliminary Characterization of Lipoxygenase From the
    Entomopathogenic Fungus Lagenidium giganteum 421
    C.A. Simmons, J.L. Kerwin and R.K. Washino

Characterization of Tomato Lipoxygenase 425
    P.A. Biacs and H. Daood

Multiple Pathways of Linolenic Acid Synthesis Operate
    and Interact in Leaf Tissue 429
    H.A. Norman and J.B. St. John

The Effect of Temperature on Desaturation of Galactolipid
    Fatty Acids in Brassica napus 433
    J.P. Williams, K. Mitchell and M. Khan

Acyl Lipid Metabolism in Rhodotorula gracilis (CBS 3043) and
    the Effects of Methyl Sterculate on Fatty Acid
    Desaturation 437
    C.E. Rolph, R.S. Moreton, I.S. Small and J.L. Harwood

Metabolism of Eicosapolyenoic Acid Lipids in Race Specific
    Interactions Between Phytophthora infestans and Potato . . . 441
    R.M. Bostock

MEDIUM AND LONG CHAIN BIOSYNTHESIS

Modulation of Fatty Acid Synthesis in Plants by Thiolactomycin . . . 447
    M. Yamada, M. Kato, I. Nishida, K. Kawano, A. Kawaguchi
        and T. Ehara

Fatty Acid Synthesis in Developing Oilseeds 455
    M.R. Pollard and S.S. Singh

Medium and Long-chain Fatty Acid Synthesis 465
    J.L. Harwood

Cuticular Lipids in Plant-microbe Interactions 473
    P.E. Kolattukudy, W.F. Ettinger and J. Sebastian

Plant Elongases 481
    C. Cassagne, R. Lessire, J.-J. Bessoule and P. Moreau

Genes, Elongases and Associated Enzyme Systems in
    Epicuticular Wax Synthesis 489
    P. von Wettstein-Knowles

The Purification of Acetyl CoA:Acyl Carrier Protein
    Transacylase from Brassica campestris Leaves 499
    A.-M.A. Wolf and J.T. Perchorowicz

Regulation of Acetyl Coenzyme A Synthesis in Chloroplasts 505
    K.-P. Heise and H.-J. Treede

Lipid Precursors in Plant Cells: The Problem of Acetyl CoA
    Generation for Plastid Fatty Acid Synthesis 509
    B. Liedvogel

Partial Purification and Characterization of Acetyl-CoA
    Synthetase from Mature Spinach Leaves 513
    C.A. Zeiher and D.D. Randall

Acetyl-CoA Carboxylase and Biotin-containing Proteins
    in Carrot Somatic Embryogenesis 517
    B.J. Nikolau, J. Croxdale, T.H. Ulrich and E.S. Wurtele

Pyruvate Reversal of S-Ethyl Dipropylcarbamothioate (EPTC)
    Inhibition of Pyruvate Dehydrogenase Complex 521
    R.E. Wilkinson and T.H. Oswald

Acyl-CoA Elongation Systems in Allium porrum Microsomes 525
    R. Lessire, J.-J. Bessoule and C. Cassagne

Lipid Biosynthesis in Oil Palm Protoplasts 529
    R. Sambanthamurthi, K.-C. Oo and A.S.-H. Ong

Identification of Proteins Associated with Changes in
    the Linolenate Content of Soybean Cotyledons 533
    X. Wang, D.F. Hildebrand and G.B. Collins

Acyltransferases in Developing Seeds of Oilseed Rape 537
    D.J. Murphy

Properties of Acyl-(Acyl-Carrier Protein):Glycerol-3-Phosphate
    Acyltransferase from Greening Squash Cotyledons 541
    I. Nishida, M. Frentzen and N. Murata

Lipid Biosynthesis in Epidermal, Guard and Mesophyll Cell
    Protoplasts from Leaves of Vicia faba L. 545
    N. Sato

Ontogenetic Variations in the Chemical Composition
   of Maize Surface Lipids 549
   P. Avato, G. Bianchi and F. Salamini

Mechanism of Biosynthesis of β-Diketones and Alkan-2-OL
   Esters from Epicuticular Waxes 553
   G. Bianchi

Epicuticular Wax Formation on Needles of Picea abies
   and Pinus cembra 557
   M.S. Günthardt-Goerg

INTERACTION OF FEDERAL, INDUSTRIAL AND ACADEMIC RESEARCH

Inter-organizational Collaboration: A Key to Success in
   Technological Innovation 563
   R.L. Sampson

The Palm Oil Research Institute of Malaysia - A Unique
   Research Unit 571
   A. Ong

Biotechnology of Lipids in Industry 587
   R.D. Schmid

ALGAL LIPIDS

Little Known Facts of Plant Lipid Metabolism 599
   A.A. Benson

Unique Characteristics of Cyanobacterial Glycerolipids 603
   N. Murata

Lipids of Diatoms and of Halophilic Dunaliella Species 613
   M. Kates

Metabolism of Galactolipids in Dunaliella salina 623
   S.H. Cho and G.A. Thompson, Jr.

Sterol Synthesis and Distribution and Algal Phylogeny 631
   G.W. Patterson

Lipids of Acetabularia mediterranea. Composition,
   Cellular Localization and Biosynthesis 637
   W. Eichenberger and A. Gerber

The Effect of Environmental Conditions on Fatty Acid
   Composition of the Red Alga Porphyridium cruentum 641
   Z. Cohen, A. Vonshak and A. Richmond

Nile Red: A Fluorophore Useful in Assessing the Relative
   Lipid Content of Single Cells 645
   K.E. Cooksey, S.A. Williams and P.R. Callis

Lipid Accumulation in Silicon - Deficient Diatoms 649
   P.G. Roessler

Composition and Positional Distribution of Fatty Acids in
    Lipids from the Diatom <u>Phaeodactylum</u> <u>tricornutum</u> 653
  A. Kawaguchi, T. Arao and M. Yamada

Lipid Metabolism in Two Species of Red Marine Algae as
    Modified by Environmental Factors 657
  T.R. Pettitt and J.L. Harwood

Lipids of <u>Chattonella</u> <u>antiqua</u> (Raphidophyceae) 661
  N. Sato, Y. Nemoto and M. Furuya

THE FUTURE - GENETICS/BIOTECHNOLOGY

The Future - Genetics/Biotechnology 667
  P.K. Stumpf

Genetic Manipulation of Oilseed Quality 669
  R.K. Downey

Biotechnology in the Improvement of the Oil Palm 677
  L.H. Jones

Mutants of <u>Arabidopsis</u> Deficient in Fatty Acid Desaturation . . . 683
  C.R. Somerville, P. McCourt, L. Kunst and J. Browse

Acyl Carrier Protein as a Probe of the Molecular Biology
    of Plant Fatty Acid Synthesis 689
  J.B. Ohlrogge, P.D. Beremand, D.J. Hannapel, D.J. Guerra,
  D.E. Elmore and D.N. Kuhn

Oil Seed Rape Acyl Carrier Protein (ACP): Protein and
    Gene Structure 697
  A.R. Slabas, J. Harding, P. Roberts, A. Heliger, C. Sidebottom,
  C.G. Smith, R. Safford, J. deSilva, C. Lucas, J. Windust,
  C.M. James and S.G. Hughes

Complete Amino Acid Sequence of Non-specific Lipid Transfer
    Protein from Castor Bean Seeds 701
  M. Yamada, S. Watanabe, K. Takishima and G. Mamiya

A Possible Differential Role for Plant Acyl Carrier Protein
    Isoforms in Higher Plants 705
  D.J. Guerra, J.B. Ohlrogge and M. Frentzen

A Preliminary Characterization of Plant Holo-Acyl
    Carrier Protein Synthase 709
  S.A. Elhussein, J.A. Miernyk and J.B. Ohlrogge

Physiological and Transformational Analyses of Lipoxygenases . . . 715
  D.F. Hildebrand, M. Altschuler, G. Bookjans, G. Benzion,
  T.R. Hamilton-Kemp, R.A. Andersen, J.G. Rodriguez,
  J.C. Polacco, M.L. Dahmer, A.G. Hunt, X. Wang
  and G.B. Collins

INDEX 719

BIOSYNTHESIS AND FUNCTION OF STEROLS AND OTHER ISOPRENOIDS

MULTIPLE ROLES FOR PLANT STEROLS

William R. Nes

Department of Bioscience and Biotechnology
Drexel University
Philadelphia, PA 19104

INTRODUCTION

In certain ways all sterols are the same, yet in other respects they are quite different. To some extent we understand the reason for the similarities in that sterols must "fit" into the phospholipid leaflet which comprises the monolayer component of the common bilayer arrangement of natural membranes[1]. On the other hand, the reason or reasons for the differences in structure are still elusive. Although we are making a beginning, we still do not yet know why it is that, for instance, people have cholesterol as their major sterol[1], while in flowering poinsettia plants only about half the sterol is cholesterol[2], in cottonseed oil 93% of the sterol is 24α-ethylcholesterol (sitosterol)[1], and in the vine, Clerodendrum splendens, nearly all of the sterol is the 22,25(27)-bisdehydro derivative of the 24β-epimer of sitosterol, viz., 25(27)-dehydroporiferasterol[3].

There are two possible explanations for structural variation. One of them is that a "random" mutation occurred at some point in evolution leading to altered biosynthesis and a different end product. This new pathway has then just remained in the lineage to the present day. The other possibility is more complicated. It is that the detailed differences we see in structure, while also arising by mutation, are actively rather than passively retained, because they serve a purpose, i.e., a function. As a result they have been selected for and maintained through time presumably by Darwinian forces.

STRUCTURAL TYPES

End Products versus Intermediates

At the outset here I should like to distinguish between those compounds which are on the pathway only as intermediates and those which actually are the final products. There are in sterol biosynthesis a lot of sterols (as well as other compounds) which usually serve only as precursors to the end products[1,4]. In plants an example is 24-methylenecycloartanol. As in the case of most of the other intermediates, it usually does not accumulate and serves for the most part just as a passing way-station for carbon, hydrogen and oxygen on their journey to, say, sitosterol. If we exclude these intermediates, the number of sterols in Nature is much smaller than otherwise and lets us focus more meaningfully on the question at hand which has to do with function. The total number of possible sterols with known kinds of structural variants is in the thousands[5], but at the ends of the various pathways there are perhaps only about 100. Still, this is a very large number. It is, I think, much larger than many scientists realize. Although some plants have virtually a single sterol, it seems that most have a mixture as we and others have documented[1,4,6-10]. The mixture can range from fairly simple to quite complex. Thus, seeds in the family Cucurbitaceae can have as many as two dozen sterols[8-10] which for reasons of amount and structure are not likely to serve simply as a way-station for carbon.

Among the major sterols of eukaryotes there are several kinds of differences. They fall into the following principal categories: (a) differences in nuclear double bonds (principally Δ^5, $\Delta^{5,7}$, Δ^7, $\Delta^{5,22}$, $\Delta^{5,7,22}$, $\Delta^{7,22}$, Δ^5 or Δ^7 with $\Delta^{25(27)}$ with or without Δ^{22} and to a lesser extent arrangements including $\Delta^{24(28)}$, $\Delta^{7,9(11)}$ and $\Delta^{7,14}$) and (b) differences in alkylation of the side chain. Each of the carbons of the side chain except for C-21 are known to undergo alkylation, and for each of the carbons in the side chain except C-25 (when C-26 or C-27 is not alkylated) there is a chiral problem which doubles the number of possible isomers. If we consider just Δ^5-, $\Delta^{5,7}$-, and Δ^7-end products with 24α-CH$_3$, 24α-C$_2$H$_5$, 24β-CH$_3$, and 24β-C$_2$H$_5$-groups, and also having $\Delta^{25(27)}$ or $\Delta^{22,25(27)}$ for the 24β-alkyl cases[2] (an apparent mechanistic restriction), there are three dozen possibilities nearly all of which are known in plants as apparent end-products when the amounts and other factors are considered. The problem with mathematics, you see, is that by the simple expedient of considering the Δ^{22}-derivatives of these sterols (which are common in several structural types) we have doubled the possibilities to six dozen which is now rapidly approaching the figure of 100 without resort to the more exotic types (such as $\Delta^{7,14}$ or 22,23-cyclopropyl) which are well documented as major constituents of various sterol pools. We also have not considered such variants as those in which one or more methyl groups are found in the nucleus. These compounds also accumulate in a variety of cases as in the wool lipids of sheep which is perhaps the oldest example.

Common Features

Structural features which virtually all sterols have in common are not only the tetracyclic cyclopentanoperhydrophenanthrene ring system but also chirality such that the entire molecule has a flatness (a series of parallel planes containing all of the atoms) determined by trans-fusions at the ring junctions or the presence of double bonds at one or more of the ring junctions and by the existence of equatorial attachments of the polar head (3β-hydroxyl group) and nonpolar tail (17β-side chain). Natural sterols also have almost the same length to their side chains. There is never an alkyl group added to C-21, and, although the side chain can be shortened or lengthened on the other side (C-22) of C-20, the natural sterols all have either the carbon skeleton of squalene without shortening or lengthening (although it may be broadened by branches) or only one carbon has been added or subtracted from the length of the side chain. That is, the longest substituent on C-20 is 6 and the shortest is 4 carbon atoms. A 5-carbon length (as in cholesterol and sitosterol with alkyl branches at C-24 and C-25) is the norm, and in terrestrial organisms, especially higher organisms there is no departure from this norm, although one-carbon departures do occur in the more primitive organisms of the marine environment. Halosterol, for instance, the analog of cholesterol with one of the CH_2-groups in the side chain missing, is found frequently in salt water organisms as a minor sterol.

Differences in Structure

Structural variations are mainly in the position and extent of nuclear and side chain unsaturation and in the position and extent of alkylation in the side chain.

LIMITATIONS IMPOSED BY MEMBRANES

Length

Free sterols are almost always found mostly associated with membranous subcellular fractions and the majority of the free sterol is widely found in the plasma membrane as documented recently in animals, plants and fungi. The thickness of the common phospholipid monolayer is about 2.1 nm[11]. Since, as with the phospholipid itself, one would expect the polar end of the sterol to orient toward the aqueous interface and the nonpolar side chain to become imbedded in the alkyl groups of the fatty acyl moieties of the phospholipid which point toward the nonpolar side of the other monolayer, it is satisfying that the overall length of natural sterols centers on the long dimension of cholesterol which is virtually the same (ca. 2.1 nm) as that of the monolayer. Is this coincidence or cause and effect?

To test this question we have prepared a series of Δ^5-sterols bearing C-21 on one side of C-20 and an alkyl chain on the other side varying from C_4 to C_{12} without a branch, but two sterols, cholesterol and its C_{26}-analog, halosterol, were available with terminal branches [12]. Although we have not had a vascular plant system in which to examine this problem, we have been able to do so in protozoa[13] and yeast[14-16]. Some of the latter results are given in Table 1. Under conditions (lack of oxygen, mutation, chemical inhibition) in which biosynthesis of sterol is strongly inhibited, exogenous sterol is utilized and most appears in the plasma membrane[17]. We found[18] a pattern of growth support in very good agreement with the thickness of the membrane monolayer. Very short or very long side chains failed to support growth. The side chain which supported growth best had the same length (5 C-atoms added to C-20 ignoring C-21 and branches, giving an overall sterol length of about 2.1 nm) as that of the common sterols of vertebrates and terrestrial plants and fungi. Moreover, when the exogenous sterol had the bulk of its side chain held rigidly to the right by a double bond (E-17(20)-dehydrocholesterol), yeast grew well, but when the long dimension was shortened by placing the bulk on the left (Z-17(20)-dehydrocholesterol), no growth occurred. In the preferred conformation of natural sterols (20R-configuration) without a 17(20)-double bond the bulk of the side chain is to the right but to the left in the 20S-epimers[19]. None of the 20S-epimers of the Δ^5-sterols shown in Table 1 would support growth of yeast[18].

Other Dimensions

Since the flat 5α-sterols but not the bent 5β-analogs are (a) common and found in membranes, (b) support growth of sterol auxotrophs, and (c) induce appropriate properties in model membranes, it is generally thought that membranes impose a selection for flatness, although the detailed reasons for this remain obscure. Other aspects of structure and membranous function (thickness, role of the hydroxyl group, etc.) also are only partially understood.

Table 1[a]. Growth of Oxygen Deprived Yeast

Length of Sterol Side chain[b]	Cell Count at 72 Hours (Millions cells/ml)
0	0.1
4	63
5	78
6	50
7	0.5
9	0.1
12	0.4
4[c] (halosterol)	69
5[c] (cholesterol)	74
5[c,d] (ergosterol)	95

[a] From Nes et al.[18].

[b] Not counting C-20 and C-21. Thus, R is the variable in $CH_3-CH(C-17)(R)$. Unless otherwise noted the sterol is in the Δ^5-series and R is unbranched.

[c] Contains an additional C-atom as a terminal branch.

[d] Contains 24β-CH_3; in $\Delta^{5,7,22}$-series.

DISTRIBUTION AND EVOLUTION

Sterol Categories

To the extent that they have been investigated[1,4] most eukaryotes fall into three groups in so far as their sterols are concerned[7]. Category A contains mostly Δ^5-sterols; Category B has primarily Δ^7-sterols; and in Category C we have $\Delta^{5,7}$-sterols. Within each major category are subdivisions which depend on chirality introduced by alkylation of the side chain. If one or more sterols in the homologous series having H, CH_3, or C_2H_5 at C-24 are present such that the 24-alkylated sterols are mostly in the 24α-series, we have subdivision I. If they are mostly in the 24β-series, we have subdivision II.

The Main Line

Now, most eukaryotic cells are in Category A and most of the higher terrestrial plants are more specifically in Category A-I. Owing to this dominance, we have called subdivision I the "main line".

Sterols and Evolution

Main line plants seem to have been dominant from the beginning of vascularization, since ferns, gymnosperms including the primitive Ginkgo biloba, early angiosperms, e.g., Magnoliaceae, and highly evolved angiosperms are of either Categories A-1 or B-1 except in rare cases[1-4,6,7]. Thus, a typical, or preferably, a common sterol mixture of a tracheophyte without regard for its evolutionary position would be dominated by 24α-ethylcholesterol (sitosterol) often together with its 22E-dehydro derivative (stigmasterol). 24α-Methylcholesterol (campesterol) is usually next most abundant accompanied by perhaps half as much 24β-epimer (22-dihydrobrassicasterol). Lesser but significant amounts (1-10% of the sterol) of cholesterol, 24Z-ethylidenecholesterol, and such sterols as 24α-ethyllathosterol (22-dihydrospinasterol, the Δ^7-analog of sitosterol) are also frequently present. The overall similarity in the various plant sterol mixtures more or less precludes their use in chemotaxonomy although in some cases it is possible if done carefully.

The extensive similarities among tracheophytes especially in an evolutionary sense strongly suggest selective pressures have occurred through time which in turn points to a functional significance for the different sterols making up the mixture of end products.

Ontogenetic Changes

Functional significance is also indicated by changes which occur during development. Two families of higher plant have been the subject of substantial investigations. Although the seeds and mature tissues of many plants have much the same sterol mixture[1,4], in the families Cruciferae (especially the genus Brassica) and Cucurbitaceae developmental changes have been well documented[1,20]. In the former case 10-20% of the seed sterols are the 24β-$\Delta^{5,22E}$-sterol (brassicasterol) with little or no 22-dihydro derivative present. The monoenic 24-methylcholesterols, therefore, are composed almost entirely of the 24α-epimer (campesterol). However, in the mature leaf brassicasterol has all but disappeared and the 24-methylsterols have become the usual main line mixture of campesterol with smaller amounts of 22-dihydrobrassicasterol. It appears from this that in the synthesis of Brassica seed sterols a Δ^{22}-dehydrogenase is present which declines during ontogeny suggesting functionality for the sterols used and/or produced.

A more dramatic ontogenetic change occurs in the Cucurbitaceae[20]. The seeds of many species have been studied, and they all contain a complex mixture mostly of Δ^7-sterols but up to 18% of Δ^5-sterols[9]. The latter completely disappear during germination, at least in Cucurbita maxima which we have extensively studied[8,9,20], leading us to believe they play a role in the early part of the life cycle of the plant[20]. There is also a remarkable stereochemical change[20]. In C. maxima seed sterols virtually all of the components have a 24-ethyl group distributed in both the Δ^5 and Δ^7-series with a 24α/24β ratio of about 1/1. However, just after seed germination is complete and the seedling starts to develop, there is a surge of sterol biosynthesis which continues, though at a slower pace, during subsequent growth of the plant. This biosynthesis is of Δ^7-sterols with a 24α-ethyl group (spinasterol and 22-dihydrospinasterol) and to a much smaller extent of a sterol (24Z-ethylidenelathosterol, avenasterol) lacking chirality at C-24. The sterols then of the mature C. maxima plant unlike the seed have only traces of 24β-ethylsterols. This replacement of 24β- with 24α-alkylsterols during ontogeny reflects (recapitulates) the probable evolutionary transition of algae (mostly 24β) to tracheophytes (mostly 24α)[1].

Summarizing, we see that developmental changes are observable in terms both of double bonds and of configurations. In the latter case we also see parallel changes from algae to higher plants. It seems unlikely at least to me that such complicated alterations in sterol biosynthesis would occur during ontogeny and phylogeny unless there were meaningful selections for function in the course of it.

DIRECT EVIDENCE FOR MULTIPLE FUNCTIONS

The problem of multiple sterol function and its relation to sterol structure has been tackled most successfully with yeast. The work has clearly shown that a change in sterol structure can produce a qualitative not just a quantitative change in response. Thus, when sterol synthesis is prevented in Saccharomyces cerevisiae, the organism becomes auxotrophic for sterol. Ergosterol provided in the medium in sufficient quantity will permit growth, but in the presence of cholesterol growth can be induced by an amount of ergosterol which by itself is too small to permit growth[15-17,21]. Related results have been shown for enzymatic yeast assays[22] and with Gibberella fujikuroi[23]. Ergosterol (24β-methyl-7,22E-bisdehydrocholesterol) is thought to stimulate cell division as a regulatory function in addition to its incorporation in bulk as an architectural component of membranes. Sterols also stimulate growth of Phytophthora cactorum at the lag to log phase transition suggesting a change in the mitotic index[24]. While cholesterol and many other sterols such as sitosterol can play the architectural role[14,16] in yeast, ergosterol and other 24β-methylsterols seem to be the only or at least the best sterols to act in the role of a regulatory agent[15].

I should like to suggest that multiple roles for sterol may also exist in photosynthetic plants and that modulation of membrane properties may not be the only thing they do. Multiple roles both within membranes as well as extending into the cytosol would help to explain the diversity of sterol structures which we find in plants.

1. W. R. Nes and M. L. McKean, "Biochemistry of Steroids and Other Isopentenoids," University Park Press, Baltimore (1977).
2. B. C. Sekula and W. R. Nes, The identification of cholesterol and other steroids in E. pulcherrima, Phytochemistry 19:1509 (1980).
3. W. J. Pinto and W. R. Nes, 24β-Ethylsterols, n-alkanes and n-alkanols of C. splendens, Phytochemistry 24:1095 (1985).
4. W. R. Nes, Biochemistry of plant sterols, in: "Advances in Lipid Research, Volume 15," Academic Press, New York (1977).
5. T. H. Varkony, D. H. Smith, and C. Djerassi, Computer-assisted manipulation. Studies in the biosynthesis of natural products, Tetrahedron 34:841 (1978).
6. W. R. Nes, K. Krevitz, and S. Behzadan, Configuration at C-24 of 24-methyl- and 24-ethylcholesterol in tracheophytes, Lipids 11:118 (1976).
7. W. R. Nes, K. Krevitz, J. Joseph, W. D. Nes, B. Harris, G. F. Gibbons, and G. W. Patterson, The phylogenetic distribution of sterols in tracheophytes, Lipids 12:511 (1977).
8. V. K. Garg and W. R. Nes, Studies on the C-24 configurations of Δ^7-sterols in the seeds of C. maxima, Phytochemistry 23:2919 (1984).
9. V. K. Garg and W. R. Nes, Codisterol and other Δ^5-sterols in seeds of C. maxima, Phytochemistry 23:2925 (1984).
10. T. Akihisa, S. Thakur, F. U. Rosenstein, and T. Matsumoto, Sterols of cucurbitaceae: The configuration at C-24 of 24-alkyl-Δ^5-, Δ^7- and Δ^8-sterols, Lipids 21:39 (1986).

11. C. Huang and J. T. Mason, Geometric packing constraints in egg phosphatidylcholine vesicles, Proc. Natl. Acad. Sci. USA 75:308 (1978).
12. J. M. Joseph, Influence of side chain structure on metabolism of sterols in *Tetrahymena pyriformis*, Ph.D. Dissertation, Drexel University (1980).
13. W. R. Nes, J. M. Joseph, J. R. Landrey, and R. L. Conner, The effects of branching, oxygen, and chain length in the side chain of sterols on their metabolism by *T. pyriformis*, J. Biol. Chem. 255:11815 (1980).
14. W. R. Nes, B. C. Sekula, W. D. Nes, and J. H. Adler, The functional importance of structural features of ergosterol in yeast, J. Biol. Chem. 253:6218 (1978).
15. W. J. Pinto and W. R. Nes, Stereochemical specificity for sterols in *S. cerevisiae*, J. Biol. Chem. 258:4472 (1983).
16. W. J. Pinto, R. Lozano, B. C. Sekula, and W. R. Nes, Stereochemically distinct roles for sterol in *Saccharomyces cerevisiae*, Biochem. Biophys. Res. Commun. 112:47 (1983).
17. M. Ramgopal and K. Bloch, Sterol synergism in yeast, Proc. Natl. Acad. Sci. USA 80:712 (1983).
18. W. R. Nes, J. M. Joseph, and J. H. Adler, A comparison of the absorption of sterols with their ability to promote growth in *Saccharomyces cerevisiae*, Fed. Proc. 40:1561 (1981).
19. W. D. Nes, R. Y. Wong, M. Benson, J. R. Landrey, and W. R. Nes, Rotational isomerism about the 17(20)-bond of steroids and euphoids as shown by the crystal structure of euphol and tirucallol, Proc. Natl. Acad. Sci. USA 81:5896 (1984).
20. V. K. Garg and W. R. Nes, Changes in Δ^5- and Δ^7-sterols during germination and seedling development of *C. maxima*, Lipids 20:876 (1985).
21. R. J. Rodriguez, F. R. Taylor, and L. W. Parks, A requirement for ergosterol to permit growth of yeast sterol auxotrophs on cholestanol, Biochem. Biophys. Res. Commun. 106:435 (1982).
22. C. E. Dahl, H. P. Biemann, and J. S. Dahl, Stimulation of cell proliferation and polyphosphoinositide turnover in a yeast sterol auxotroph by ergosterol, Fed. Proc. 45:1886 (1986).
23. W. D. Nes, R. C. Heupel, and P. H. Le, A comparison of sterol biosynthesis in fungi and tracheophytes and its phylogenetic and functional implications *in*: "Structure, Function and Metabolism of Plant Lipids," P. A. Siegenthaler and W. Eichenberger, eds., Elsevier Science Publishers, Amsterdam (1984).
24. W. D. Nes, Biosynthesis and requirement for sterols in growth and reproduction of oomycetes, *in*: "Ecology of Lipids," G. Fuller and W. D. Nes, eds., ACS Monograph Series, in press (1986).

STEREOCHEMISTRY OF MONOTERPENE CYCLIZATION

Rodney Croteau

Institute of Biological Chemistry
Washington State University
Pullman, WA 99164-6340

INTRODUCTION

The vast majority of the several hundred naturally occurring monoterpenes are cyclic and represent a relatively small number of skeletal themes multiplied by a very large range of simple derivatives, positional isomers and stereochemical variants.[1] A unified conceptual framework for the origin of the various terpenoid types was put forward three decades ago by Ruzicka in formulating the "biogenetic isoprene rule."[2] As applied to the monoterpenes, this scheme (Fig. 1) posits intramolecular electrophilic attack of C1 of the neryl cation on the distal double bond to yield a monocyclic (α-terpinyl) intermediate, which by a series of subsequent internal additions, hydride shifts, and Wagner-Meerwein rearrangements give rise to the cationic equivalents of most known skeletal types. The latter species, by deprotonation to the corresponding olefin or capture by a nucleophile, could yield many of the common monoterpenes, and by subsequent, often oxidative, modification of these cyclic progenitors most other monoterpenes could be generated. This seminal contribution by Ruzicka set the foundation for nearly all subsequent biogenetic investigations, which have supported the broad outlines of this basic proposal. More recent investigations at the enzyme-level, however, have necessitated revision of some elements of the scheme, and, moreover, have allowed reformulation of the general model for monoterpene cyclization in explicit mechanistic and stereochemical terms. Detailed descriptions of the biosynthesis of cyclic monoterpenes are available;[3,4] the focus of this chapter is the stereochemical aspects of the cyclization process.

ISOMERIZATION-CYCLIZATION MODEL

Geranyl pyrophosphate (see Fig. 2) is the first C_{10} intermediate to arise in the general isoprenoid pathway, yet this compound cannot be converted directly to cyclohexanoid monoterpenes because of the trans-double bond geometry at C2. To circumvent this topological barrier to direct cyclization of geranyl pyrophosphate, a number of proposals were formulated for the origin of alternate cyclization precursors in which the cis-isomer, neryl pyrophosphate, or the tertiary allylic isomer, linalyl pyrophosphate, assumed a central role.[5,6] However, with the availability of partially purified cell-free enzyme systems in the late 1970s, it became apparent that geranyl pyrophosphate was efficiently converted to cyclic monoterpenes without preliminary isomerization to free neryl or

linalyl pyrophosphate, or the formation of any other detectable
intermediate.[6,7] Subsequent studies with a number of monoterpene
cyclases[8] have since indicated that geranyl pyrophosphate is, in most
cases, cyclized more efficiently than is neryl pyrophosphate and with
efficiencies comparable to that of (±)-linalyl pyrophosphate.[3,9,10]
It is therefore clear that monoterpene cyclases are capable of catalyzing
a multistep process whereby the enzyme carries out the isomerization of
geranyl pyrophosphate to a bound intermediate sterically capable of
cyclization, as well as the cyclization reaction itself. The recognition
of geranyl pyrophosphate as the universal precursor of cyclic monoterpenes
provided the critical insight which led to a new proposal for the origin
of cyclic monoterpenes based on the concept of a tightly coupled
isomerization-cyclization reaction.[3,4,8,11]

Such a scheme, in simplified form, is illustrated in Fig. 2.
According to this proposal, reaction of geranyl pyrophosphate is initiated
by ionization which is assisted by a divalent metal ion. The geranyl
cation:pyrophosphate anion pair then collapses to enzyme-bound linalyl
pyrophosphate, permitting transoid to cisoid rotation about the C2-C3
single bond. Subsequent ionization of the linalyl system and cyclization
affords the monocyclic α-terpinyl ion from which other monocyclic and
bicyclic skeletons are generated along the lines originally suggested by
Ruzicka.[2] The overall process is presumed to occur in stepwise fashion
via a series of carbocation:pyrophosphate anion paired intermediates where
topology is maintained between the initial ionization and the termination
steps.

This proposal is entirely consistent with all studies at the enzyme
level, and with chemical models of the cyclization process. Thus, Poulter

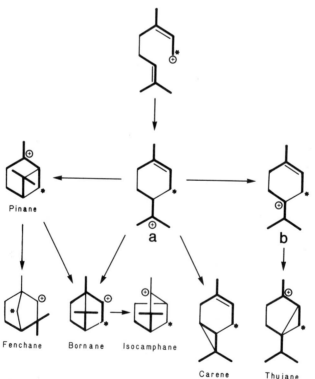

Fig. 1. Ionic scheme for the formation of several cyclic monoterpene
skeletons from the neryl cation via the α-terpinyl cation (a)
and the terpinen-4-yl cation (b).

and King [12,13] have shown that the direct cyclization of neryl derivatives to α-terpineol to be both stepwise and stereospecific, providing an elegant demonstration of the preference for ionic over concerted pathways for this reaction type. Geranyl derivatives also have been shown to give rise to monocyclic products, under solvolytic conditions, via preliminary conversion to the linalyl intermediate.[14,15] This mode of cyclization is favored where nucleophilic trapping is slow relative to re-ionization of the tertiary allylic system, and where stability of intermediate cationic species is favored by ion-pairing.

The two-step requirement of the enzymatic reaction sequence has been confirmed using a non-cyclizable substrate analog, 6,7-dihydrogeranyl pyrophosphate, to uncouple the normally cryptic isomerization step,[16] whereas employment of linalyl pyrophosphate as an alternate substrate has permitted focus on the cyclization component of the reaction scheme in the absence of the tightly coupled isomerization step normally required for the cyclization of geranyl pyrophosphate.[17,18] Evidence for the electrophilic nature of both the isomerization and cyclization steps of the enzymatic reaction sequence, and for ion-pairing of intermediates in the course of the reaction, has been obtained from a wide range of studies using analogs of both substrates and intermediates.[19-21]

ISOMERIZATION-CYCLIZATION STEREOCHEMISTRY

The labeling patterns of antipodal bornane and pinane monoterpenes derived from [1-^{3}H]geranyl pyrophosphate by partially purified cyclases have recently been determined [(+)- and (-)-bornyl pyrophosphate are labeled at C3, and (+)- and (-)-pinenes at C7].[22-24] These results indicate that the antipodes are derived via enantiomeric cyclizations involving antipodal linalyl and α-terpinyl intermediates, rather than by

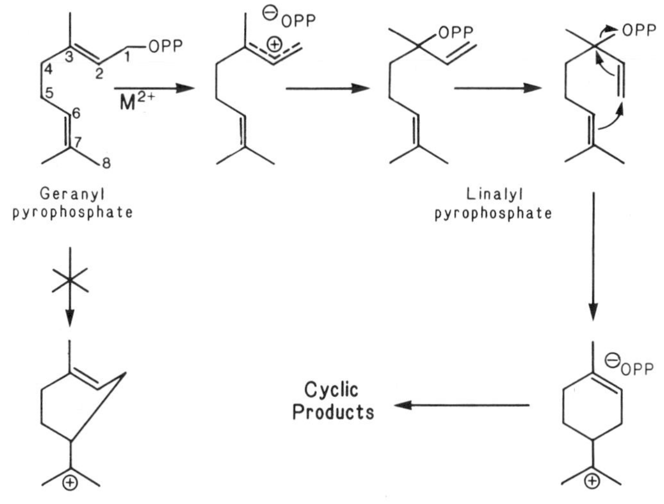

Fig. 2.   Scheme for the enzymatic isomerization-cyclization of geranyl pyrophosphate.

way of a hydride shift or other rearrangement from a common cyclic progenitor. The stereochemistry of a given monoterpene cyclization critically depends upon the precise folding of the acyclic geranyl substrate, which gives rise to one or the other antipodal linalyl intermediate and dictates the ultimate stereochemical outcome of the cyclization.[3-5,11] Although linalyl pyrophosphate is the first explicitly chiral intermediate in the cyclization scheme, the eventual configuration is pre-determined by the helical conformation of geranyl pyrophosphate achieved on binding. Conformational constraints imposed by the presence of two π-systems within the eight-membered carbon chain, along with the required alignment to effect 2p orbital overlap of the reacting double bonds, indicate that very few substrate (and intermediate) conformations can account for the majority of known monoterpene skeletons.[3,4,11] Incisive model studies by Arigoni and co-workers have, in fact, demonstrated that the anti-endo conformation of the linalyl system is preferred in the cyclization to α-terpineol.[25] This, along with the confirmation that allylic transposition reactions occur by a net suprafacial (syn) process,[26,27] allows deduction of the helical folding of geranyl pyrophosphate and thus, by summation, all of the basic stereochemical elements of the coupled isomerization-cyclization.

Based on these considerations, and on the assumption of least motion during the course of the enzymatic cyclization, there exists a direct one-to-one relationship between the absolute configuration of the cyclic product, the helical conformation of geranyl pyrophosphate, and the configuration and conformation of the cyclizing linalyl intermediate. For example, in the cyclization of geranyl pyrophosphate to (+)-bornyl pyrophosphate (the committed step in camphor biosynthesis[22]) by the corresponding cyclase from Salvia officinalis,[9] the prediction would be the left handed screw-sense isomer of geranyl pyrophosphate giving rise by syn-isomerization to (-)-3R-linalyl pyrophosphate folded in the anti-endo conformation (Fig. 3). For the antipodal cyclization to (-)-bornyl pyrophosphate by an enzyme from Tanacetum vulgare,[23] the correspondence would be the right handed screw-sense isomer of geranyl pyrophosphate and the (+)-3S-linalyl system, again folded anti-endo prior to cyclization. An assessment of the overall stereochemistry of a given cyclization requires a determination of the stereochemical alterations at C1 and C3 of the geranyl substrate in the course of the cyclization. Since most cyclases can utilize 3RS-linalyl pyrophosphate as an alternate substrate, the question of the configuration of the cyclizing intermediate can be addressed directly using the optically pure linalyl antipodes. Addressing

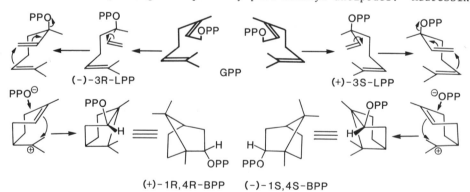

Fig. 3. Stereochemical scheme for the enzymatic cyclization of geranyl pyrophosphate (GPP) to bornyl pyrophosphate (BPP) via linalyl pyrophosphate (LPP).

conformational features of the general syn-isomerization, anti-cyclization model rests on the predicted retention of configuration at C1 of geranyl pyrophosphate in the cyclization (a consequence of the C2-C3 transoid-to-cisoid rotation of linalyl pyrophosphate which brings the face of C1 from which the pyrophosphate moiety has departed into juxtaposition with the 6,7-double bond).

The configuration of the tertiary linalyl intermediates in the bornyl cyclizations was initially examined using [3RS-1E-^{3}H;3R-8,9-^{14}C]linalyl pyrophosphate (^{3}H:^{14}C=5.2) as substrate, which was tested with the antipodal cyclases from each plant source (the cyclic product was converted to camphor and examined as the crystalline oxime).[18] This substrate yielded, with the T. vulgare-derived enzyme, (-)-bornyl pyrophosphate with ^{3}H:^{14}C>31, indicating specific utilization of (+)-3S-linalyl pyrophosphate as predicted (Fig. 3). With the (+)-bornyl pyrophosphate cyclase from S. officinalis the ^{3}H:^{14}C ratio of the product was about 4.2, indicating a preference for the predicted (-)-3R-enantiomer, but the ability also to utilize (+)-3S-linalyl pyrophosphate. Optically pure 3R- and 3S-[1Z-^{3}H]linalyl pyrophosphate were next prepared and each enantiomer separately compared to the achiral precursors [1-^{3}H]geranyl pyrophosphate and [1-^{3}H]neryl pyrophosphate as substrates for the cyclizations.[18] All functional precursors afforded optically pure (-)-1S,4S-bornyl pyrophosphate with the cyclase from T. vulgare (as determined by chromatographic separation of diastereomeric ketals of the derived ketone camphor). (+)-3S-Linalyl pyrophosphate was the preferred substrate, and (-)-3R-linalyl pyrophosphate was inactive. With the (+)-bornyl pyrophosphate cyclase from S. officinalis, geranyl, neryl and (-)-3R-linalyl pyrophosphates gave the expected (+)-1R,4R-stereoisomer as the sole product, and (-)-3R-linalyl pyrophosphate was the preferred substrate. However, 3S-linalyl pyrophosphate yielded (-)-1S,4S-bornyl pyrophosphate, albeit at much lower rates, indicating the ability of this enzyme to catalyze the anomalous enantiomeric cyclization. These results established the configurational preferences in the cyclization of the linalyl intermediates to the bornyl systems, and for both enantiomeric cyclases indicated that the cyclization of the preferred linalyl pyrophosphate enantiomer was faster than was the coupled isomerization-cyclization of geranyl pyrophosphate.

In examining conformational questions in the bornyl cyclizations, 1R-1-^{3}H;2-^{14}C- and 1S-1-^{3}H;2-^{14}C-labeled geranyl pyrophosphates were separately converted to (+)-bornyl pyrophosphate and (-)-bornyl pyrophosphate by partially purified preparations from S. officinalis and T. vulgare, respectively.[28] Each pyrophosphate ester was hydrolyzed, and the resulting borneol was oxidized to camphor (Fig. 4). The stereochemistry at C3 of the derived ketone (corresponding to C1 of the acyclic precursor)

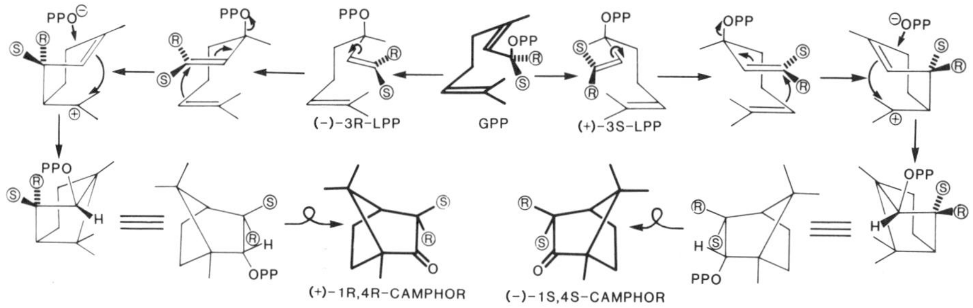

Fig. 4. Stereochemistry at C1 of geranyl pyrophosphate (GPP) in the cyclization, via linalyl pyrophosphate (LPP), to bornyl pyrophosphate (BPP).

was determined by taking advantage of the stereoselective base-catalyzed exchange of the exo-α-proton.[29,30] By comparison of such exchange rates to those of product generated from 1RS-1-^{3}H;2-^{14}C-labeled substrate it was demonstrated that geranyl pyrophosphate was cyclized to bornyl pyrophosphate with net retention of configuration at C1.[28] The observed stereochemistry therefore was entirely compatible with the proposed cyclization scheme and consistent with similar anti-cyclizations in the sesquiterpene[31] and diterpene[32] series.

The stereochemistry at C1 of the acyclic precursor in the cyclization to the (+)- and (-)-pinenes by enzymes from S. officinalis[10,24] was examined using 1R- and 1S-^{3}H-labeled geranyl pyrophosphate as before.[33] The (+)-α-pinene so obtained was converted to (+)-camphor (^{3}H:^{14}C ratio determined on the crystalline oxime) (Fig. 5). The (-)-α-pinene and (-)-β-pinene were similarly converted to (-)-camphor and the ^{3}H:^{14}C ratio determined. Tritium at C3 was then located, as before, by taking advantage of the selective exchange of the exo-α-proton and by comparing exchange curves to those of product generated from the racemic 1-^{3}H;^{14}C-labeled substrate. The pinane to bornane skeletal rearrangement proceeds with a degree of racemization (15-20%) placing the ^{3}H at C5 and thus inaccessible to exchange. This minor complication does not alter the results which indicated that cyclization of geranyl pyrophosphate to the (+)- and (-)-pinenes occurs with retention of configuration at C1.

[3RS-^{3}H;3R-^{14}C]Linalyl pyrophosphate was again employed as a substrate to investigate the configuration of the cyclizing tertiary intermediate in the pinane series.[33] In this case, the (+)-α-pinene produced, after conversion to the crystalline pinonic acid, bore a ^{3}H:^{14}C ratio of slightly greater than half of the starting material as expected for the enantioselective conversion of the 3R-enantiomer to the (+)-pinyl system. The (-)-β-pinene product, following conversion to nopinone oxime, was shown to contain primarily tritium, consistent with the enantioselective conversion of the 3S-enantiomer to the (-)-pinyl nucleus. The results with the pinenes indicated a somewhat higher degree of enantiomer discrimination by these cyclases than had been observed with the (+)-bornyl pyrophosphate cyclase; yet even here it was clear that the opposite enantiomer had participated in the cyclization to each bicyclic product. Conformational preferences in the construction of the antipodal pinenes were confirmed with optically pure 3R- and 3S-[1Z-^{3}H]linalyl pyrophosphates (3R→(+)-pinenes; 3S→(-)-pinenes) and, as before with the bornyl cyclases, the preferred linalyl enantiomer was a more efficient precursor than was geranyl pyrophosphate.

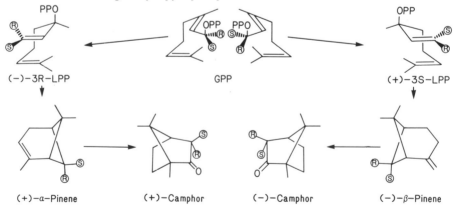

Fig. 5. Stereochemical scheme for the enzymatic cyclization of geranyl pyrophosphate (GPP), via linalyl pyrophosphate (LPP), to pinenes, and the conversion of these products to camphor.

The summation of the above studies has firmly established the overall stereochemistry of cyclization to bornane and pinane monoterpenes. Specifically, the results have permitted determination of the reacting helical conformation of geranyl pyrophosphate and the stereochemistry of the initial allylic rearrangement of geranyl to linalyl pyrophosphate (syn), the stereochemistry of the subsequent allylic displacement with C1-C6 cyclization (anti), and the configuration and conformation (anti-endo) of the cyclizing linalyl intermediate. Thus, a consistent three dimensional picture has emerged from this work which validates the stereochemical aspects of the proposed syn-isomerization-anti-endo-cyclization model and eliminates other stereochemical alternatives for this enzymatic reaction type.

REFERENCES

1. T.K. Devon and A. I. Scott, "Handbook of Naturally Occurring Compounds," Vol. II. Terpenes, Academic Press, New York (1972).
2. L. Ruzicka, A. Eschenmoser, and H. Heusser, The isoprene rule and the biogenesis of terpenic compounds, Experientia 9:357 (1953).
3. R. Croteau, Biochemistry of monoterpenes and sesquiterpenes of the essential oils, in: "Herbs, Spices and Medicinal Plants: Recent Advances in Botany, Horticulture and Pharmacology," Vol. 1, L.E. Craker and J.E. Simon, eds., Oryx Press, Phoenix (1985).
4. R. Croteau, Biosynthesis of cyclic monoterpenes, in: "Biogeneration of Aroma Compounds," T. Parliment and R. Croteau, eds., American Chemical Society, Washington, DC (1986).
5. D.E. Cane, The stereochemistry of allylic pyrophosphate metabolism, Tetrahedron 36:1109 (1980).
6. R. Croteau, Biosynthesis and catabolism of monoterpenes, in: "Isopentenoids in Plants," W.D. Nes, G. Fuller, and L.-S. Tsai, eds., Marcel Dekker, New York (1984).
7. R. Croteau, Biosynthesis of monoterpenes, in: "Biosynthesis of Isoprenoid Compounds," Vol. 1, J.W. Porter and S.L. Spurgeon, eds., Wiley-Interscience, New York (1981).
8. R. Croteau and D.E. Cane, Monoterpene and sesquiterpene cyclases, Methods Enzymol. 110:352 (1985).
9. R. Croteau and F. Karp, Biosynthesis of monoterpenes: Preliminary characterization of bornyl pyrophosphate synthetase from sage (Salvia officinalis) and demonstration that geranyl pyrophosphate is the preferred substrate for cyclization, Arch. Biochem. Biophys. 198:512 (1979).
10. H. Gambliel and R. Croteau, Pinene cyclases I and II. Two enzymes from sage (Salvia officinalis) which catalyze stereospecific cyclizations of geranyl pyrophosphate to monoterpene olefins of opposite configuration, J. Biol. Chem. 259:740 (1984).
11. D.E. Cane, Isoprenoid biosynthesis. Stereochemistry of the cyclization of allylic pyrophosphates, Accts. Chem. Res. 18:220 (1985).
12. C.D. Poulter and C.-H.R. King, Model studies of terpene biosynthesis. A stepwise mechanism for cyclization of nerol to α-terpineol, J. Am. Chem. Soc. 104:1420(1982).
13. C.D. Poulter and C.-H.R. King, Model studies of terpene biosynthesis. Stereospecific cyclization of N-methyl-(S)-4-([1'-^{2}H]neryloxy) pyridinium methyl sulfate to α-terpineol, J. Am. Chem. Soc. 104:1422 (1982).
14. R.L. Baxter, W.A. Laurie, and D. MacHale, Transformations of monoterpenoids in aqueous acid. The reaction of linalool, geraniol, nerol and their acetates in aqueous citric acid, Tetrahedron 34:2195 (1978).
15. R.C. Haley, J.A. Miller, and H.C.S. Wood, Phosphate esters. Part II. The formation of monoterpene hydrocarbons from geranyl and neryl diphenyl phosphates, J. Chem. Soc. 264 (1969).

16. C.J. Wheeler and R. Croteau, Monoterpene cyclases: Use of the noncylizable substrate analog 6,7-dihydrogeranyl pyrophosphate to uncouple the isomerization step of the coupled isomerization-cyclization reaction, Arch. Biochem. Biophys. 246:733 (1986).
17. D.M. Satterwhite, C.J. Wheeler, and R. Croteau, Biosynthesis of monoterpenes: Enantioselectivity in the enzymatic cyclization of linalyl pyrophosphate to (-)-endo-fenchol, J. Biol. Chem. 260:13901 (1985).
18. R. Croteau, D.M. Satterwhite, D.E. Cane, and C.C. Chang, Biosynthesis of monoterpenes: Enantioselectivity in the enzymatic cyclization of (+)- and (-)-linalyl pyrophosphate to (+)- and (-)-bornyl pyrophosphate, J. Biol. Chem. 261: in press (1986).
19. D.E. Cane, A. Saito, R. Croteau, J. Shaskus, and M. Felton, Enzymatic cyclization of geranyl pyrophosphate. Role of the pyrophosphate moiety, J. Am. Chem. Soc. 104:5831 (1982).
20. R.B. Croteau, J.J. Shaskus, B. Renstrøm, N.M. Felton, D.E. Cane, A. Saito, and C. Chang, Mechanism of the pyrophosphate migration in the enzymatic cyclization of geranyl and linalyl pyrophosphates to (+)- and (-)-bornyl pyrophosphates, Biochemistry 24:7077 (1985).
21. R. Croteau, C.J. Wheeler, R. Aksela, and A.C. Oehlschlager, Inhibition of monoterpene cyclases by sulfonium analogs of presumptive carbocationic intermediates of the cyclization reaction, J. Biol. Chem. 261:7257 (1986).
22. R. Croteau and F. Karp, Demonstration of a cyclic pyrophosphate intermediate in the enzymatic conversion of neryl pyrophosphate to borneol, Arch. Biochem. Biophys. 184:77 (1977).
23. R. Croteau and J. Shaskus, Biosynthesis of monoterpenes: (-)-Bornyl pyrophosphate cyclase in soluble enzyme preparations from tansy (Tanacetum vulgare), Arch. Biochem. Biophys. 236:535 (1985).
24. H. Gambliel and R. Croteau, Biosynthesis of (±)-α-pinene and (-)-β-pinene from geranyl pyrophosphate by a soluble enzyme system from sage (Salvia officinalis), J. Biol. Chem. 257:2335 (1982).
25. S. Gotfredsen, J.P. Obrecht, and D. Arigoni, The cyclization of linalool to α-terpineol. Stereochemical course of the reaction, Chimia 31:62 (1977).
26. D.E. Cane, R. Iyengar, and M.-S. Shiao, Cyclonerodiol biosynthesis and the enzymatic conversion of farnesyl to nerolidyl pyrophosphate, J. Am. Chem. Soc. 103:914 (1981).
27. S.E. Gotfredsen, The biosynthesis of lagopodine A, B, and C, coccinol and linalool, Ph.D. Thesis, ETH, Zurich, No. 6243.
28. R. Croteau, N.M. Felton, and C.J. Wheeler, Stereochemistry at C-1 of geranyl pyrophosphate and neryl pyrophosphate in the enzymatic cyclization to (+)- and (-)-bornyl pyrophosphate, J. Biol. Chem. 260:5956 (1985).
29. A.F. Thomas, R.A. Schneider, and J. Meinwald, Stereospecific α-hydrogen exchange in camphor, isofenchone, and carvonecamphor, J. Am. Chem. Soc. 89:68 (1967).
30. G.A. Abad, S.P. Jindal, and T.T. Tidwell, Steric effects in bicyclic systems. IV. Base-catalyzed enolization of bicyclo[2.2.2]octan-2-one, bicyclo[2.2.1]heptan-2-one, and bicyclo[2.1.1]hexan-2-one, J. Am. Chem. Soc. 95:6326 (1973).
31. D.E. Cane, Enzyme-level studies of the biosynthesis of natural products, in: "Enzyme Chemistry. Impact and Applications," C.J. Sucking, ed., Chapman and Hall, London (1984).
32. K.H. Overton, Concerning stereochemical choice in enzymic reactions, Chem. Soc. Rev. 8:447 (1979).
33. R. Croteau and D. M. Satterwhite, unpublished.

DITERPENES - THE GIBBERELLIN BIOSYNTHETIC PATHWAY IN ZEA MAYS

Bernard O. Phinney and Clive R. Spray

Department of Biology,
University of California,
Los Angeles, CA 90024.

INTRODUCTION

Gibberellins (GAs) are a class of tetracyclic diterpenes that are probably of universal occurrence in the plant kingdom. They are also present (although with unknown function) in the fungi, Gibberella fujikuroi and Sphaceloma manihoticola. As phytohormones, these gibberellins stimulate a wide variety of responses, including the mobilization of storage starch during germination, flowering (indirect), fruit set, fruit size and shoot elongation. There are numerous reviews that describe the effects of GAs on the growth and differentiation of higher plants (for example,[1-3]). There is also considerable literature on the biosynthesis of the gibberellins (for reviews and symposia reports see [4-10]). In contrast, information on the

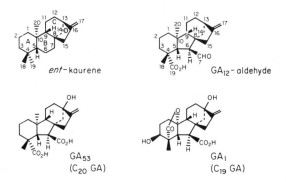

Fig. 1. Numbering system and nomenclature for the gibberellins and their precursors (Fig. from Phinney and Spray[11]). ent-Kaurene is the first tetracyclic hydrocarbon in the biosynthetic pathway. GA_{53} is a precursor to GA_1; the trivial names for the gibberellins are based on the order of identification, GA_1 to GA_n.

enzymology of gibberellin metabolism has proceeded at a slower pace, due in part to difficulties arising from the low levels of the cyclases and oxidases that control steps in the biosynthetic pathway, especially those subsequent to GA_{12}-aldehyde.

All gibberellins are built around the ent-gibberellane skeleton, for example, GA_{53} and GA_1 (Fig. 1); gibberellins (now numbering 72) have either 20 or 19 carbon atoms, the C_{20}-GAs being precursors to the C_{19}-GAs. Gibberellins also differ from each other in their oxidation state, and number and position of double bonds. They are numbered sequentially, based on order of publication. Interestingly, GA-conjugates have been found to be present amongst higher plants and absent from fungi; the role of conjugates in the control of plant growth has yet to be defined.

Gibberellins originate from the standard MVA biosynthetic pathway from which the diterpene GGPP is cyclized to ent-kaurene (Fig. 2). ent-Kaurene is then metabolized through a series of steps involving the oxidation of carbon-19 to the acid and rearrangement of ring B to GA_{12}-aldehyde. Several pathways then diverge from GA_{12}-aldehyde (Fig. 3); the pathways differ from each other in the patterns of hydroxylation, that is, no hydroxylation or initial hydroxylation at a particular position early in the pathway (C_{20}-GAs).

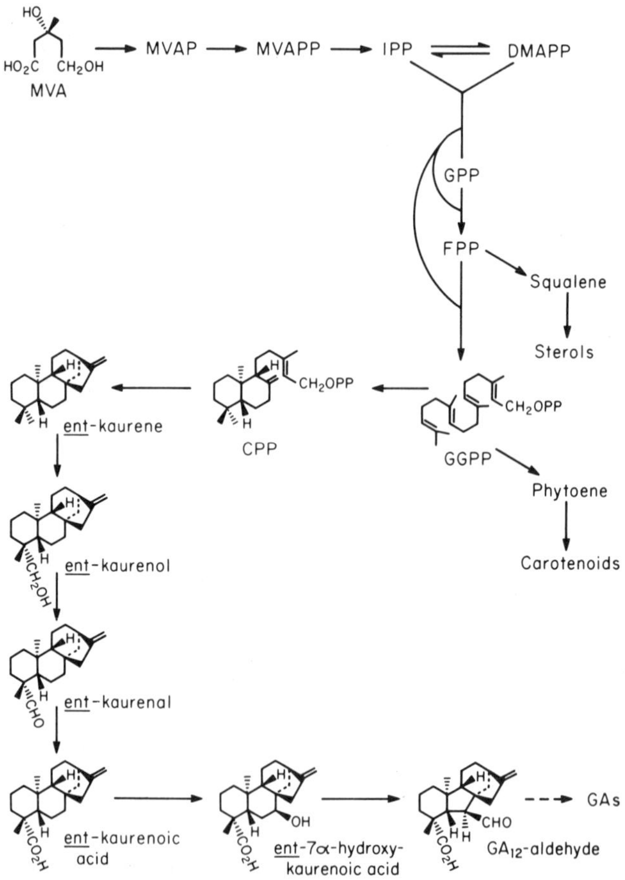

Fig. 2. The MVA biosynthetic pathway leading to the cyclic diterpene, ent-kaurene. GA_{12}-aldehyde is the common precursor for all known gibberellins (now numbering 72). (Figure from Phinney,[12]).

Gibberella fujikuroi was first used for studies on GA biosynthesis because of the copious quantities of gibberellins produced by the organism. Since these GAs diffuse into the culture media, a simple solvent partition of the culture filtrate provides sufficient purification for identification studies. The latter property was especially useful prior to the development of HPLC. The work with the fungus was also facilitated by the use of a mutant that blocked a step early in the biosynthetic pathway, thus virtually eliminating the occurrence of (endogenous) substrate dilution in feeding experiments. The extensive metabolic studies with the fungus have served as models for subsequent work with higher plants. Two pathways are present in the fungus, the early-3-hydroxylation pathway leading to GA_3, and the non-hydroxylation pathway leading to GA_9. Both pathways originate from GA_{12}-aldehyde.

The early investigations with plants led to the identification of a number of gibberellins with structures identical to the fungal GAs, thus implicating the presence of the two fungal pathways (Fig. 3). Metabolic studies then placed these GAs and their precursors into sequential steps in the pathways. At the same time, new, uniquely hydroxylated gibberellins were identified from plant material, suggesting the presence of additional pathways not present in the fungus. Metabolic studies have now documented the two fungal pathways in plants, also a third pathway unique to plants, the early-13-hydroxylation pathway and a possible fourth pathway initiated by 12α-hydroxylation. Additional pathways may also be present (Fig. 3). All pathways originate from the common gibberellin precursor, GA_{12}-aldehyde. It is also becoming clear that the number of pathways in plants (subsequent to GA_{12}-aldehyde) varies within taxonomic groups; also the presence of specific pathway(s) may be organ specific.

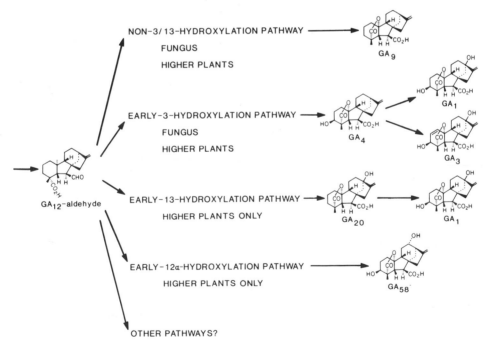

Fig. 3. GA biosynthetic pathways subsequent to GA_{12}-aldehyde. Classical feeds and refeeds have documented the details of the first three pathways. Evidence for the presence of additional pathways is based on the identification from plants of additional GAs that have unique hydroxylation patterns and for which there are no metabolic studies.

It is the purpose of this report to analyze the early-13-hydroxylation pathway in maize. (It is apparently the only pathway in vegetative shoots of maize). In addition, comments will be made on new approaches to the study of the enzymology of this most interesting class of plant hormones, the gibberellins.

THE EARLY-13-HYDROXYLATION PATHWAY IN MAIZE SHOOTS (Fig. 4)

Eight gibberellins (GA_1, GA_8, GA_{20}, GA_{29}, GA_{19}, GA_{17}, GA_{44} and GA_{53}) have been identified by full scan gas chromatography - mass spectrometry (GC-MS) from vegetative shoots of maize[13,14]. These GAs are members of the early-13-hydroxylation pathway. The pathway was originally defined in pea (Pisum sativum) [15,16]. In maize, seven steps in the gibberellin biosynthetic pathway have been defined by feeds and refeeds of labeled GAs and their precursors. The steps are: (1) cyclization of CPP to ent-kaurene[17]; (2) oxidation of ent-kaurenol to ent-kaurenal[18]; (3) oxidation of ent-kaurenal to ent-kaurenoic acid[18]; (4) oxidation of GA_{53} to GA_{44} (opened lactone)[13]; (5) hydroxylation of GA_{20} to GA_1[19]; (6) hydroxylation of GA_{20} to GA_{29}[13,19]; (7) hydroxylation of GA_1 to GA_8[14,20,21].

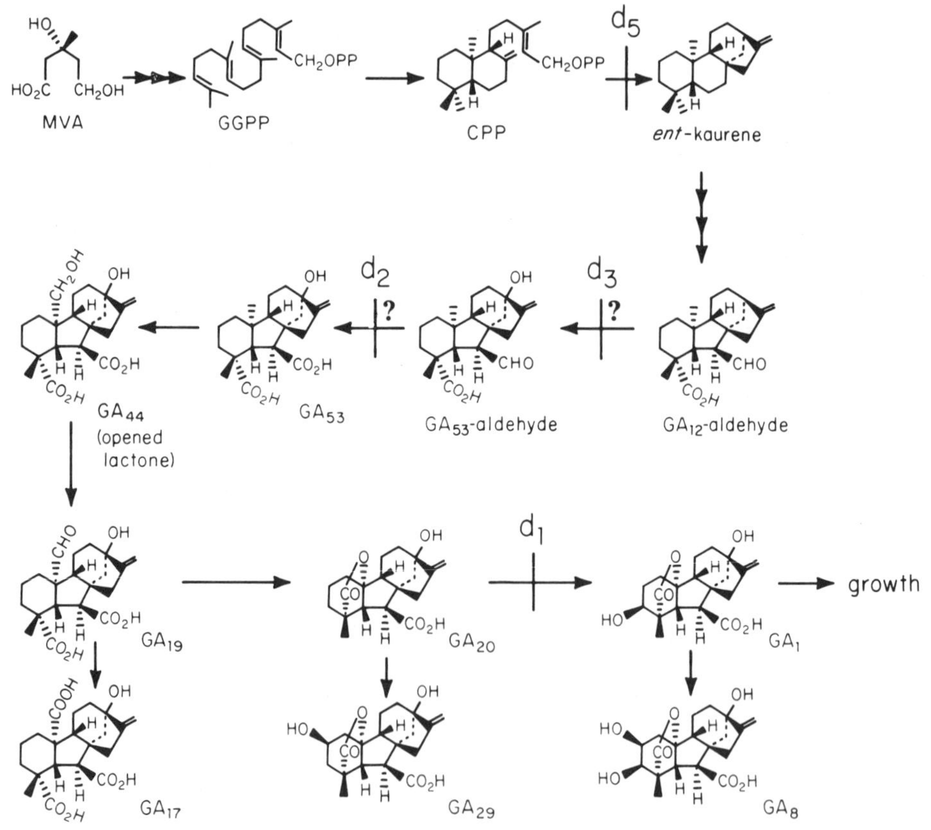

Fig. 4. Location of genetic blocks in the early-13-hydroxylation pathway. The anther ear-1 mutant (not shown) probably blocks a step prior to ent-kaurene. (Figure from Phinney and Spray,[22]).

GA_{17} and GA_{29} are shown as branches from the main pathway. GA_8 is a metabolite of GA_1. GA_{17}, GA_{29} and GA_8 are biologically inactive and not metabolized to active gibberellins (there is no evidence for reversibility of any of the steps shown in the pathway in Fig. 4). By contrast, ent-kaurene and all subsequent members of the main pathway are bioactive, the degree of activity depending on the organism and mutant used for bioassay. The bioactivity of individual members of the pathway raises the question of whether each member is active due to its presumptive direct binding to a substrate, i.e. per se activity, or whether each member is bioactive only as a consequence of its metabolism to the terminal active member, GA_1. These alternatives have been evaluated, at least for the maize system, through the use of five gibberellin mutants, mutants that block specific and different steps in the pathway.

The five mutants are dwarf-1 (d1), dwarf-2 (d2), dwarf-3 (d3), dwarf-5 (d5) and anther ear-1 (an1) (Fig. 5). Each mutant exhibits a dwarf habit of growth; internodes are shortened, leaves are small and relatively broad, and reproductive organs are reduced in size, but still viable. The dwarf phenotypes are expressed from the seedling stage to maturity. The mutants respond by normal growth to exogenous gibberellin (Fig. 6). Three kinds of data have been gathered to locate the positions of the genetic blocks in the GA biosynthetic pathway: (1) the use of each mutant as a bioassay in order to determine the relative activities of the native maize GAs and their precursors;

Fig. 5. The five non-allelic, simple recessive, gibberellin responding dwarf mutants of maize. (Figure from Phinney et al.[23]).

(2) radiofeeds of members of the pathway to normals and mutants; and (3) the presence or absence of endogenous gibberellins in the mutants as compared to normals. (The fourth and critical evidence on relative enzyme activities in normals and mutants has yet to be successfully studied). Data have been obtained from three approaches for the dwarf-1 and dwarf-5 mutants,[13,14,19,24] bioassay data only for the dwarf-2, dwarf-3 and anther ear-1 mutants[24]. The studies show that the dl gene blocks the 3β-hydroxylation step, GA_{20} to GA_1; the d2 and d3 genes control the C-20 oxidation and 13-hydroxylation steps; the d5 gene controls the cyclization step, CPP to ent-kaurene; and the anl gene controls a step prior to ent-kaurene.

Selected details documenting these generalizations are as follows: GA_{20} is less than 1% as active as GA_1 when assayed on dwarf-1 seedlings and essentially inactive when added continually throughout the growth of the mutant at levels that give a normal growth response for the dwarf-5 mutant (Fig. 6). Other members of the main pathway are inactive when assayed on dwarf-1 seedlings. In contrast, GA_{20} and GA_1 are equally active (and earlier members of the main pathway are active) when assayed on the dwarf-5 mutant[24]. These

Fig. 6. Response of dwarf-1 mutants to GA_1 (left set) and GA_{20} (right set). In each set the plants read from left to right, dwarf control, dwarf treated, normal treated, normal control. Each treated plant received a total of 250μg gibberellin, added at intervals and at dosages to give a normal-type growth response to GA_1. The total dosage per plant (GA_1 and GA_{20}) was the same. Gibberellins were added as microdrops (acetone:water; 1:1) to the uppermost unfolding leaves. (Figure from Phinney and Spray[22]).

response data support the position that the dwarf-1 lesion controls the 3β-hydroxylation step. Definitive evidence for this control comes from feeds of $[^{13}C,^3H]$-GA_{20} to dwarf-1, dwarf-5 and normal seedlings. The metabolite $[^{13}C,^3H]$-GA_1 can be recovered in feeds of $[^{13}C,^3H]$-GA_{20} to normal and dwarf-5 seedlings, but not from feeds to dwarf-1 seedlings[19]. Finally, the presence/absence of endogenous GAs in the mutants and normals supports the position of blockage by the dl gene; that is, GA_1, GA_{20}, GA_{19}, GA_{17} and GA_{44} are present in green shoots of normal seedlings and absent in dwarf-5 seedlings (albeit traces of GA_{20}). GA_{20}, GA_{19}, GA_{17} and GA_{44}, but not GA_1, are present in dwarf-1 seedlings[14].

The above kinds of studies have led to the conclusion that GA_1 is the only gibberellin in the maize shoot that is active per se in the control of stem elongation. Other native GAs are bioactive only through their metabolism to GA_1. If each member of the pathway were active per se, the different patterns of mutant bioactivities would not be observed. In other words, there would be no logical, simple explanation for the presence of a series of nonallelic dwarf mutants, each having a selective response to the maize endogenous GAs and their precursors.

While the data for shoot elongation in maize seem to clearly support the idea of one active gibberellin, GA_1, there are a number of unanswered questions that could alter this interpretation. It is not known what role conjugates play in the control of substrate levels in the maize biosynthetic pathway; it is not known if GA_1 is metabolized to a more oxidized bioactive gibberellin before presumptive binding to a substrate (for example, GA_3; however GA_3 has not been shown to be a metabolite in feeding studies to maize seedlings). Finally, the problems associated with the direct enzymatic approach remain an enigma in the evaluation of the mutant control of gibberellin biosynthesis in maize.

THE FUTURE

The enzymological studies of Graebe's group at Göttingen continue to make progress in this difficult field (for example,[9,15]). In addition, at least three laboratories are now developing and using new approaches to the study of the enzymology of GA-metabolism. The approaches can lead to exciting and significant advances by providing sufficient quantities of the low abundance proteins (GA enzymes) for biochemical studies. First, MacMillan's group (Department of Chemistry, University of Bristol, U.K.) are preparing monoclonal antibodies to the gibberellin 2β and 3β hydroxylases[25]. They are using the antibodies to isolate purified enzymes by affinity chromatography. Specific mRNAs will then be identified and cDNA clones constructed for studies on the gene expression for these enzymes. While the physiological role of the 2β-hydroxylases is not clear (they are inactivating enzymes, controlling the steps GA_{20} to GA_{29} and GA_1 to GA_8) the 3β-hydroxylase regulates shoot elongation by controlling the step, GA_{20} to GA_1. As mentioned earlier, GA_1 is the only endogenous gibberellin in maize that is active per se in stimulating shoot elongation. Second, West's group (Department of Chemistry and Biochemistry, U.C.L.A., U.S.A.) is using affinity chromatography to purify kaurene synthetase (and other similar diterpene cyclization enzymes)[26]. They plan to make antibodies to these enzymes and to select cDNA clones to study the regulation of synthesis of these enzymes. Kaurene synthetase may be a key regulatory enzyme since GGPP is a branch point metabolite in the biosynthetic pathway[4]. Third, our group (Department of Biology, U.C.L.A., U.S.A.) has initiated a program to clone the maize dwarfing genes and their normal alleles. We have isolated a series of GA-mutants (dwarf mutants) that originated as a result of the insertion of Robertson's mutator (Mu) into the genome[23]. Since the transposable element, Mu1, has been cloned and sequenced[27], it can be used as a probe in the cloning of the maize dwarfing genes[23].

25

In all three approaches the resulting clones can be used to make available large quantities of pure protein for further studies; they can also be used to investigate the regulation of the expression of the genes controlling each enzyme

REFERENCES

1. D. S. Letham, P. B. Goodwin and T. J. V. Higgins, eds. "Phytohormones and Related Compounds - A Comprehensive Treatise. Vol. II.," Elsevier/North Holland, Amsterdam (1978).
2. G. C. Martin, Commercial uses of gibberellins, in: "The Biochemistry and Physiology of Gibberellins, Vol II," A. Crozier, ed., Praeger, New York (1983).
3. R. P. Pharis and R. W. King, Gibberellins and reproductive development in seed plants, Ann. Rev. Plant Physiol. 36:517 (1985).
4. J. R. Bearder, Plant hormones and other growth substances - their background, structures and occurrence, in: "Encycopedia of Plant Physiology, New Series, Vol. 9, Hormonal Regulation of Development I Molecular Aspects," J. MacMillan, ed., Springer-Verlag, Berlin (1980).
5. R. C. Coolbaugh, Early stages of gibberellin biosynthesis, in: "The Biochemistry and Physiology of Gibberellins, Vol I," A. Crozier, ed., Praeger, New York (1983).
6. J. E. Graebe, Gibberellin biosynthesis from gibberellin A_{12}-aldehyde, in: "Plant Growth Substances 1985," M. Bopp, ed., Springer-Verlag, Berlin, Heidelberg (1986).
7. J. E. Graebe and H. J. Ropers, Gibberellins, in: "Phytohormones and Related Compounds - A Comprehensive Treatise. Vol. I," D. S. Letham, P. B. Goodwin and T. J. V. Higgins, eds., Elsevier/North Holland, Amsterdam (1978).
8. P. Hedden, J. MacMillan and B. O. Phinney, The metabolism of the gibberellins, Ann. Rev. Plant Physiol. 29:149 (1978).
9. C. A. West, Biosynthesis of diterpenes, in: "Biosynthesis of Isoprenoid Compounds. Vol. I," J. W. Porter and S. L. Spurgeon, eds., John Wiley and Sons, New York (1981).
10. C. A. West, Hydroxylases, monooxygenases and cytochrome P-450, in: "The Biochemistry of Plants - A Comprehensive Treatise. Vol. II," P. K. Stumpf and E. E. Conn, eds., Academic Press, New York (1980).
11. B. O. Phinney and C. R. Spray, Gibberellins (GAs), gibberellin mutants and their future in molecular biology, in: "Current Topics in Plant Biochemistry and Physiology. Vol. 4," D. D. Randall, D. G. Blevins and R. L. Larson, eds., University of Missouri, Columbia (1985).
12. B. O. Phinney, Gibberellin A_1, dwarfism and the control of shoot elongation in higher plants, in:" The Biosynthesis and Metabolism of Plant Hormones", A. Crozier and J. R. Hillman, eds., Society for Experimental Biology Seminar 23, Cambridge University Press, Cambridge (1984).
13. R. C. Heupel, B. O. Phinney, C. R. Spray, P. Gaskin, J. MacMillan, P. Hedden and J. E. Graebe, Native gibberellins and the metabolism of [^{14}C]gibberellin A_{53} and of [17-^{13}C,17-^{3}H]gibberellin A_{20} in tassels of Zea mays, Phytochemistry 24:47 (1985).
14. C. R. Spray, H. Yamane, S. Fujioka, B. O. Phinney, P. Gaskin and J. MacMillan, Unpublished data.
15. Y. Kamiya and J. E. Graebe, The biosynthesis of all major pea gibberellins in a cell-free system from Pisum sativum, Phytochemistry 22:681 (1983).
16. V. M. Sponsel, The localization, metabolism and biological activity of gibberellins in maturing and germinating seeds of Pisum sativum cv. Progress No. 9, Planta 159:454 (1983).

17. P. Hedden and B. O. Phinney, Comparison of ent-kaurene and ent-isokaurene synthesis in cell-free systems from etiolated shoots of normal and dwarf-5 maize seedlings, Phytochemistry 18:1475 (1979).
18. E. S. Wurtele, P. Hedden and B. O. Phinney, Metabolism of the gibberellin precursors ent-kaurene, ent-kaurenol, and ent-kaurenal in a cell-free system from seedling shoots of normal maize, J. Plant Growth Regul. 1:15 (1982).
19. C. Spray, B. O. Phinney, P. Gaskin, S. J. Gilmour and J. MacMillan, Internode length in Zea mays L. the dwarf-1 mutation controls the 3β-hydroxylation of gibberellin A_{20} to gibberellin A_1, Planta 160:464 (1984).
20. L. J. Davies and L. Rappaport, Metabolism of tritiated gibberellins in d-5 dwarf maize I, Plant Physiol. 55:620 (1975).
21. L. J. Davies and L. Rappaport, Metabolism of tritiated gibberellins in d-5 dwarf maize II, Plant Physiol. 56:60 (1975).
22. B. O. Phinney and C. Spray, Gibberellin biosynthesis in Zea mays: the 3-hydroxylation step GA_{20} to GA_1, in:" IUPAC Pesticide Chemistry, Human Welfare and the Environment", J. Miyamoto, ed., Pergamon, Oxford (1983).
23. B. O. Phinney, M. Freeling, D. S. Robertson, C. R. Spray and J. Silverthorne, Dwarf mutants in maize - the gibberellin biosynthetic pathway and its molecular future, in: "Plant Growth Substances 1985," M. Bopp, ed., Springer-Verlag, Berlin, Heidelburg (1986).
24. B. O. Phinney and C. Spray, Chemical genetics and the gibberellin pathway in Zea mays L., in:"Plant Growth Substances 1982", P. F. Wareing, ed., Academic Press, London (1982).
25. J. MacMillan, Personal communication.
26. C. A. West, Personal communication.
27. R. F. Barker, D. V. Thompson, D. R. Talbot, J. Swanson and J. L. Bennetzen, Nucleotide sequence of the maize transposable element Mul, Nucl. Acids Res. 12:5955 (1984).

SYNTHESIS OF PLASTOQUINONE-9, α-TOCOPHEROL AND PHYLLOQUINONE (VITAMIN K_1)
AND ITS INTEGRATION IN CHLOROPLAST CARBON METABOLISM OF HIGHER PLANTS[#]

D. Schulze-Siebert, U. Homeyer, J. Soll and G. Schultz

Botanisches Institut
Tierärztliche Hochschule Hannover
D 3000 Hannover 71, F.R.G.

INTRODUCTION

Plastoquinone-9, α-tocopherol and phylloquinone are known as plastidic prenylquinones fulfilling important functions: Plastoquinone-9 acts as mobile electron and proton carrier in photosynthetic electron transport and is involved in building up the electrochemical proton potential at the chloroplast cytochrom b_6/f complex /1,2/. α-Tocopherol is involved in inactivating energized oxygen species, formed in the light, by scavenging radicals and quenching singlett oxygen /3/. Phylloquinone is known as obligatory constituent of PS I (K_1/chlorophyll of PS I ratio about 1:100 /4/).

This report deals with the synthesis of α-tocopherol and plastoquinone-9 from homogentisate and of phylloquinone from 1,4-dihydroxy-2-naphthoate in chloroplasts. Furthermore, experimental data are presented to support earlier findings on the autonomic role of chloroplast carbon metabolism in forming plastidic isoprenoids. - The methods applied are described in /5,6,7,8,9/.

RESULTS AND DISCUSSION

Synthesis of α-Tocopherol and Plastoquinone-9 from Homogentisate at the Chloroplast Envelope Membrane

Homogentisate represents the aromatic intermediate in the formation of αT and PQ /10/ (Fig. 1). It is formed from Tyr via 4-hydroxyphenylpyruvate by

---

Abbreviations: DAHP, deoxy-D-arabinoheptulosonate-7-phoshate; DHAP, dihydroxyacetone phosphate; E´ase, 2-phosphoglycerate hydrolyase, enolase; rev. NADP GAPDH, reversible NADP glyceraldehyde 3-phosphate dehydrogenase; GG, geranylgeraniol; GGPP, geranylgeranyl-pyrophosphate; GK, glycerate kinase; HPP, 4-hydroxyphenylpyruvate; IPP, isopentenyl pyrophosphate; αKGA, 2-oxoglutarate; Me-6-PhQ(H_2) and isomers, 2-methyl-6-phytylquinone (quinol) and isomers; 2,3-Me_2-PhQ(H_2), 2,3-dimethyl-5-phytylquinone (quinol); MITO, mitochondrion; PDC, pyruvate dehydrogenase complex; PEP, phosphoenolpyruvate; PER, peroxysome; 3-PGA and 2-PGA, 3- and 2-phospho-D-glycerate; PK, pyruvate kinase; PGM, phosphoglycerate mutase; PQ(H_2), plastoquinone-9 (quinol-9); PS I, PS II, photosystem I and II; SAM, S-adenosylmethionine; SkA, shikimate; SORase, shikimate oxidoreductase; α, β, γ and δT, α-, β-, γ- and δ-tocopherol

---

[#] The author´s work was supported by the Deutsche Forschungsgemeinschaft.

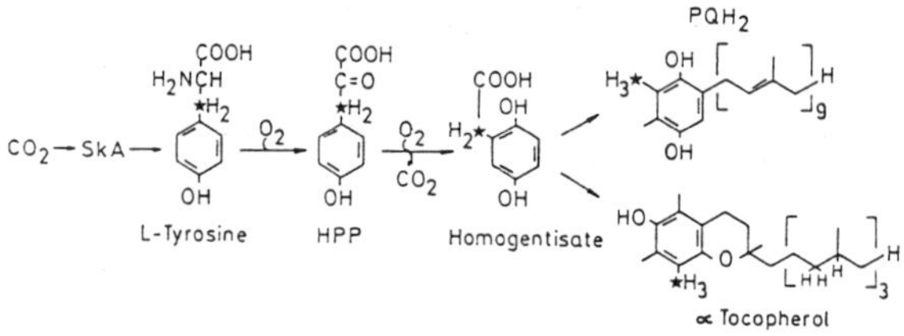

Fig. 1. Incorporation of /3-^{14}C/tyrosine into plastoquinol and α-tocopherol in Higher Plant /10/ chloroplasts /5,6,8/. Asterisks indicate ^{14}C label.

oxidases at the stromal face of chloroplast membranes /11/. All the following reactions of prenylquinone synthesis occur solely at the inner chloroplast envelope membrane /8/ (Fig. 4); chloroplast stroma or thylakoids are inactive /6,12/ (Fig. 4). In the prenyltransferase reaction, the carboxyl group of homogentisate is eliminated and a methylquinol is formed which is prenylated specifically at position 6 /6/. The prenyl side chain is always in trans configuration /13/. In this and the following methylation reaction only the quinol and not the quinone stage of the aromatic compound is active /5/. Phytyl-PP is the only prenyl-PP suited for αT synthesis in chloroplasts and nonaprenyl- (solanesyl-) PP in PQ synthesis /6/. Consequently, the introductory step of prenylquinone synthesis is strictly specific in respect to substrate and position of attack, and the following reactions are strongly directed to a homogenous pattern of products.

As can be seen from the methylation step by SAM in Fig. 2, 2-methyl-6-phytylquinol is strongly preferred to its isomers. Thus, the main product is 2,3-dimethyl-5-phytylquinol which undergoes ringclosure (only verified in intact chloroplasts /5/) to form γT which is methylated by SAM to form αT. The chromanol stage is prerequisite for the second methylation reaction; no trimethylphytylquinol was formed from dimethylphytylquinols /5,14/. The γ-tocopherol methyltransferase of Capsicum annuum was purified to homogeneity by d´Harlingue and Camara 1985 /15/. M_r has been determined to 33 kDa, K_m for γT to 13.7 uM and for SAM 2.5 uM. Thus the main sequence for αT synthesis in spinach is: Homogentisate → 2-Methyl-6-phytylquinol → 2,3-Dimethyl-5-phytylquinol → γT → αT (Fig. 3). Another sequence caused by changing methylation and cyclization reaction may occur additionally but at lower rates /16/: Homogentisate → 2-Methyl-6-phytylquinol → δT → γT → αT.

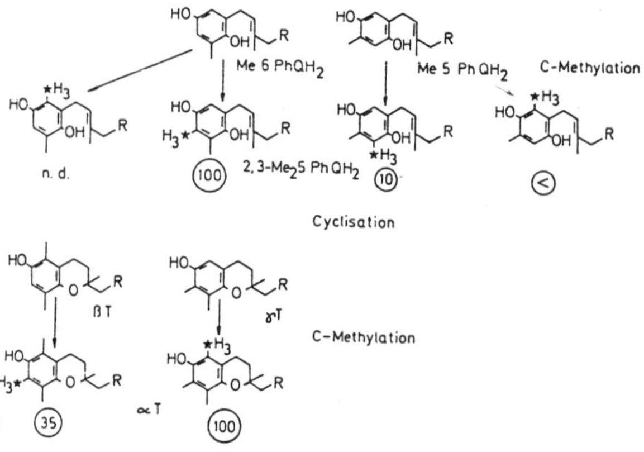

Fig. 2 Substrate specificity of the first (upper series) and second methylation reaction (lower series) of α-tocopherol synthesis in spinach chloroplasts /5/. /^{14}C/-Methyl from /Me-^{14}C/SAM was incorporated for labelling the substrates. The products were purified by co-chromatography adding reference substances and identified by radioscan. Asterisks indicate ^{14}C-label; the encircled figures indicate the ratio of products formed from different quinols (upper series) and tocopherols (lower series).

Phytyl-PP is formed from GGPP by hydrogenation at the envelope membrane /7/ (Fig. 3). Another pathway is the stepwise hydrogenation of GG-chlorophyllide a to form chlorophyll a at thylakoid membranes /17/ (Fig. 3). NADPH functions as electron donor in both reactions. A kinase which forms phytyl-PP from phytol plus ATP is localized in the stroma /6/. GGPP itself is formed from IPP by a recombinated system of envelope or thylakoid membranes plus stroma protein /18/.

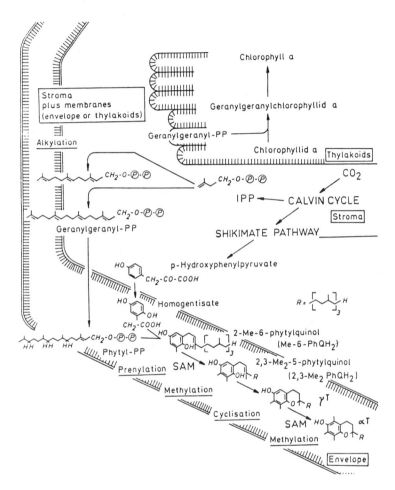

Fig. 3 Synthesis of α-tocopherol /5,6,12/ from homogentisate /10/ and phytyl-PP at the inner envelope membrane of chloroplasts /8/. Plastoquinone-9 (see text and Fig. 4) is synthesized from homogentisate and nonaprenyl-(solanesyl-)PP at the same membrane /6,8/. For synthesis of homogentisate /11/ and phytyl-PP /7/ in chloroplasts see text.

The reaction mechanism of PQ synthesis equals that of αT synthesis. The synthesis also occurs exclusively at the inner chloroplast envelope membrane /8/ (Fig. 4), however, it can be assumed that either prenylquinone is formed by its own enzyme garniture. 2-Methyl-6-nonaprenyl-(solanosyl-)quinol is formed from homogentisate plus nonaprenyl-(solanosyl-)PP. The quinol formed is then methylated by SAM to yield PQH_2 /6/ (Fig. 4). Even if the sequence in PQ synthesis is clarified: Homogentisate $\longrightarrow$ 2-Methyl-6--nonaprenylquinol $\longrightarrow$ PQH_2 no data are available for the synthesis of hydroxylated quinones Q_A and Q_B acting as primary electron acceptors in PS II /19/.

31

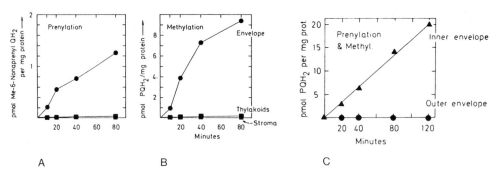

Fig. 4 Synthesis of plastoquinol-9 in spinach chloroplasts (calculated from /6,8/). (A) Prenylation reaction: Homogentisate + Nonaprenyl-(solanesyl-)PP ⟶ 2-Methyl-6-nonaprenylquinol; (B) Methylation reaction: The quinol of (A) + SAM ⟶ Plastoquinol-9. (C) Overall reaction: Homogentisate + Nonaprenyl-PP + SAM ⟶ Plastoquinol-9. The reactions occur at the inner membrane (C) /8/ of the chloroplast envelope (A, B) /6/; stroma and thylakoid membranes are inactive.

Synthesis of Phylloquinone (Vitamin K_1) in Higher Plants

Feeding experiments using total plants revealed that phylloquinone is formed in leaves from shikimate /20/ and 2-succinylbenzoate /21/. Just recently Leistner's group provided evidence from studies on E. coli /22/ that isochorismate and not chorismate reacts with 2-oxoglutarate to form 2-succinylbenzoate. The results from studies on cell cultures /23,24/ and chloroplasts /25,26/ are summarized in Fig. 5. The chloroplast envelope is the site of prenylation /25/ and the thylakoid membrane of methylation reaction /26/, however, compartmentation of the other reactions remains still unclear. The synthesis in plants resembles the microbial one /27/ though phytyl-PP is preferred as prenyl donor in plants /25/.

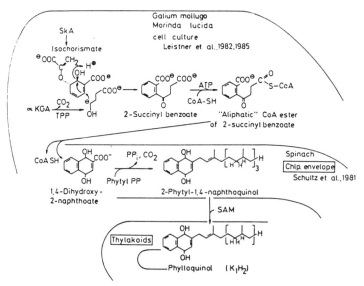

Fig. 5 Phylloquinone (K_1) synthesis in Higher Plants /23-26/. Probably isochorismate as in E. coli /22/ and 2-oxoglutarate (αKGA) forms 2-succinylbenzoate.

Role of Chloroplast Carbon Metabolism in Plastidic Prenylquinone Synthesis

The carbon flow from 3-phosphoglycerate, phosphoenolpyruvate, pyruvate and acetyl-CoA. Even if the synthesis of aromatic amino acids by shikimate pathway /28,29,30,31/ and also prenyl-PP synthesis via mevalonate /32,33,34/ has been established in chloroplasts by identification of respective plastidic enzymes, it is still a matter of discussion from where PEP origins to supply DAHP synthesis of the shikimate pathway and from where pyruvate is delivered to supply the plastidic pyruvate dehydrogenase complex (for isolation see Treede and Heise, this Conference). Because phosphoglycerate mutase (PGM) to form 2-PGA from 3-PGA could not be detected in chloroplasts /35/ and acetyl-CoA is preferably synthesized from added acetate by the actetyl-CoA synthetase /36/, particularly in spinach chloroplasts, it was argued that chloroplasts are dependent on import of these substrates from the external site. Evidence for PEP formation from 3-PGA within the chloroplast could be obtained by three different approaches (D. Schulze-Siebert, A. Heintze and G. Schultz, in preparation; D. Schulze-Siebert and G. Schultz, in preparation, for plastidic isoenzyme of PGM in Ricinus see /37/ and in Brassica /38/).

(i) The only enzyme of carbon metabolism hitherto questioned to be present in spinach chloroplasts, PGM, could be identified by the latency method. The enzyme thus identified exhibits an identical behaviour in comparison to reversible NADP D-glyceraldehyde 3-phosphate dehydrogenase and shikimate oxidoreductase as plastidic marker enzymes (Fig. 6).

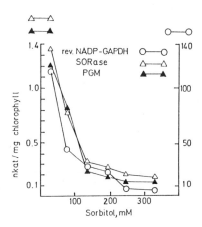

Fig. 6 Phosphoglycerate mutase (PGM) in spinach chloroplasts identified by the latency method. Purified, intact chloroplasts were subjected to stepwise osmotical shock by lowering the sorbitol conc. of the medium. The enzyme activity released from chloroplast was determined in the supernatant. In the latency studies, PGM exhibits an identical behaviour compared to reversible NADP-GAPDH and shikimate oxidoreductase (SORase) as marker enzymes for chloroplasts.

(ii) The carbon flow from 3-PGA to PEP and pyruvate was demonstrated within chloroplasts by adding labelled glycerate which is known to be phosphorylated by the glycerate kinase localized solely in the chloroplast stroma /39/. After spunning down the chloroplasts (Fig. 7), 3-PGA, 2-PGA, PEP and pyruvate were found at considerable amounts only in the chloroplast pellet and thus support above findings on plastidic PGM. Only 3-PGA was nearly equally distributed between chloroplast and suspension medium which can be attributed to action of the phosphate translocator. To reduce the activity of the translocator /40/ only 0.5 mM P_i was applied to the medium.

(iii) The increase of the synthesis of amino acids and prenylquinones in chloroplasts by omission of P_i and deminishing the exchange of triose-phosphates could be demonstrated earlier /41/.

Fig. 7 Experiment proving the carbon flow from 3-PGA to pyruvate via 2-PGA and PEP in spinach chloroplasts. Purified, intact chloroplasts were incubated with /1-^{14}C/glycerate which is phosphorylated by the glycerate kinase localized solely in the stroma /39/. 3-PGA thus formed served as substrate for C_3-metabolism within the chloroplast. Aliquots taken at indicated time were centrifuged for 1 min and 3-PGA, 2-PGA, PEP and pyruvate determined by modified enzymatic analysis (D. Schulze-Siebert, A. Heintze and G. Schultz, in preparation).

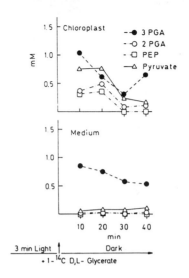

An external site of synthesis of PEP from photosynthetically formed 3-PGA can be detected by rising the P_i concentration up to 5 mM and adding exogenously excessive amounts of rabbit PGM and enolase to intact chloroplasts (10 and 2.5 units/50 ug chlorophyll) similar as in /42/. The activity of the shikimate pathway (measured as nmol Phe and Tyr formed) was considerably enhanced as a result of supply of high amounts of PEP by this enzyme reaction and re-import by the phosphate translocator. Consequently, this way might be considered as an additional site of supply optimized under in vitro conditions by adding enzymes in excess. Under in vivo conditions, the carbon flow from triosephosphates of photosynthetic carbon fixation is preponderantly directed to sucrose synthesis /43/ and less to PEP.

The origin of acetyl-CoA for isoprenoid synthesis. As shown in Table 1, highest amounts of PQ and ß-carotene were formed from added bicarbonate by spinach protoplasts in the light and only lower ones from added acetate. On the other hand, saturation of fatty acid formation is only achieved by adding acetate. Therefore, the hypothesis was raised that a more or less channelling of pyruvate dehydrogenase complex and isoprenoid synthesis in chloroplasts may exist. To prove this the dilution of /^{14}C/bicarbonate by increasing amounts of added acetate was studied. Inspite of increasing amounts of added acetate no dilution of /^{14}C/ activity in ß-carotene and PQ could be observed. Only fatty acids formed inside and sterols formed outside the chloroplasts were diluted as expected (Fig. 8). Also additional vice versa experiments (data not shown) pointed at the same direction.

Table 1 Formation of plastoquinone-9, ß-carotene, fatty acids (F.A.) and sterols from /^{14}C/bicarbonate, /2-^{14}C/acetate and /^{14}C/-mevalonate using spinach protoplasts at pH 7.6. Note the preference of bicarbonate in the isoprenoid and of acetate in the fatty acid synthesis. Plastoquinone-9 and ß-carotene are not formed if mevalonate is applied from the external site.

	nmol Acetate units formed per mg chlorophyll x h				
	Lipids	Sterols	F.A.	Carotene	PQ
NaH^{14}CO$_3$ 0.5 mM	5.1 ± 0.3	0.67	1.58	1.18	1.12
/2-^{14}C/Acetate 0.1 mM	33.3 ± 6.0	2.00	28.64	0.33	0.67
/2-^{14}C/Mevalonate	3.5 ± 0.6	1.80	0	0.03	0.03

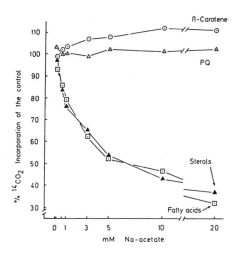

Fig. 8 Demonstration of the metabolic channelling of carbon metabolism and isoprenoid synthesis in chloroplasts by a competition experiment adding ^{14}C/bicarbonate and increasing amounts of acetate to spinach protoplasts. A competition effect from added acetate could only be observed in fatty acid and sterol synthesis. Plastoquinone-9 and ß-carotene synthesis remained unaffected.

The scheme in Fig. 9 summarizes the above results.

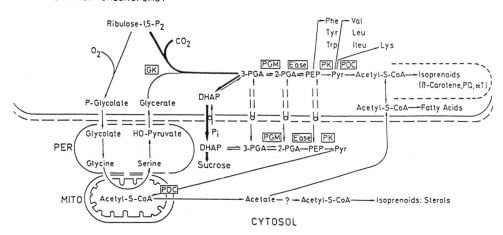

Fig. 9 Proposed scheme of carbon flow from primary photosynthetic products to isoprenoid compounds, fatty acids, amino acids and other compounds in spinach chloroplasts. Based on recent findings the scheme demonstrates the following points: (i) A carbon flow from 3-PGA to 2-PGA, PEP, and pyruvate within the chloroplast (and to some extent outside the chloroplast) exists to provide substrates for the synthesis of amino acids, isoprenoids and prenylquinones. (ii) Plastidic pyruvate dehydrogenase complex and plastidic isoprenoid synthesis are more or less channelled systems. (iii) The supply with acetate for fatty acid synthesis predominantly occurs from the external site.

REFERENCES

1. E.Hurt and G.Hauska, 1981, Eur.J.Biochem.,117:597.
2. I.Willms, R.Malkin and R.K.Chain, 1986, 7th Int.Congr.Photosynthesis, Providence, U.S.A., (Abstract).
3. C.S.Foote, 1976, in: Free Radicals in Biology, W.A.Pryor,ed., Academic Press, New York, Vol.II, p.85.
4. E.Interschick-Niebler and H.K.Lichtenthaler, 1981, Z.Naturforch.,36c:276.
5. J.Soll and G.Schultz, 1980, Phytochemistry,19:215.
6. J.Soll, M.Kemmerling and G.Schultz, 1980, Arch.Biochem.Biophys.,204:544.
7. J.Soll, G.Schultz, W.Rüdiger and J.Benz, 1983, Plant Physiol.,71:849.
8. J.Soll, G.Schultz, J.Joyard, R.Douce and M.A.Block, 1985, Arch. Biochem.Biophys.,238:290.
9. D.Schulze-Siebert, D.Heineke, H.Scharf and G.Schultz, 1984, Plant Physiol.,76:465.
10. G.R.Whistance and D.R.Threlfall, 1970, Biochem.J.,117:593.
11. E.Fiedler, J.Soll and G.Schultz, 1982, Planta,155:511.
12. J.Soll, R.Douce and G.Schultz, 1980, FEBS Lett.,112:243.
13. P.S.Marshall, S.R.Morris and D.R.Threlfall, 1985, Phytochemistry,24:1705.
14. J.Soll and G.Schultz, 1981, Biochem.Biophys.Res.Commun.,99:907.
15. A.d´Harlingue and B.Camara, 1985, J.Biol.Chem.,260:15200.
16. G.Schultz, J.Soll, E.Fiedler and D.Schulze-Siebert, 1985, Physiol. Plant.,64:123.
17. W.Rüdiger, P.Hedden, H.J.Köst and D.J.Chapman, 1977, Biochem.Biophys. Res.Commun.,74:1268.
18. M.A.Block, J.Joyard and R.Douce, 1980, Biochim.Biophys.Acta,631:210.
19. K.Tabata, S.Itoh, Y.Yamamoto, S.Okayama and M.Nishimura, 1985, Plant Cell Physiol.,26:855.
20. G.Thomas and D.R.Threlfall, 1971, Phytochemistry,13:807.
21. K.G.Hutson and D.R.Threlfall, 1980, Phytochemistry,19:535.
22. A.Weische and E.Leistner, 1985, Tetrahedron Lett.,26:1487.
23. L.Heide, R.Kolkmann, S.Arendt and E.Leistner, 1982, Plant Cell Reports, 1:180.
24. R.Kolkmann and E.Leistner, 1985, Tetrahedron Lett.,26:1703.
25. G.Schultz, B.H.Ellerbrock and J.Soll, 1981, Eur.J.Biochem.,117:329.
26. S.Kaiping, J.Soll and G.Schultz, 1984, Phytochemistry,23:89.
27. R.Bentley and R.Meganathan, 1982, Microbiol.Rev.,46:241.
28. E.Fiedler and G.Schultz,1985, Plant Physiol., 79:212.
29. C.L.Schmidt, C.Groß, H.Hennig, U.Homeyer, E.Fiedler and G.Schultz, 1986, Life Science Advances, Ser. B.,in press.
30. D.M.Mousdale and J.R.Coggins, 1985, Planta,163:241.
31. T.A.d´Amato, R.J.Ganson, C.G.Gaines and R.A.Jensen, 1984, Planta, 162:104.
32. R.J.Wong, D.K.McCormack and D.W.Russell, 1982, Arch.Biochem.Biophys., 216:631.
33. R.E.Arebalo and E.D.Mitchell, 1984, Phytochemistry,23:13.
34. D.Schulze-Siebert and G.Schultz, 1985, Plant Physiol.,77-S,Abstr.709.
35. M.Stitt and T.ap Rees, 1979, Phytochemistry, 18:1905.
36. D.N. Kuhn, M. Knauf and P.K. Stumpf, 1981, Arch.Biochem.Biophys.,209:441.
37. J.A.Miernyk and D.T.Dennis, 1982, Plant Physiol. 69,825.
38. E.P.Journet and R.Douce, 1985, Plant Physiol.,79:458.
39. M.R.Schmitt and G.E.Edwards, 1983, Arch.Biochem.Biophys.,224:332.
40. D.A.Walker, 1976, in: The Intact Chloroplast, J.Barber,ed., Elsevier, Amsterdam, p.235.
41. G.Schultz and H.Bickel, 1977, in: Proc.5th Hungarian Bioflavonoid Symp., Matrafüred,Hungary, p.271.
42. P.Baggè, K.Machera and C.Larsson, 1984, in: Advances in Photosynthesis Research, C.Sybesma,ed., Martinus Nijhoff/Dr.W.Junk Publishers, The Hague, Netherlands, Vol.III,p.897.
43. M.Stitt and H.W.Heldt, 1985, Planta, 164:179.

CAROTENOID BIOSYNTHESIS AND CAROTENOGENIC ENZYMES IN PLASTIDS

HANS KLEINIG

INSTITUT FÜR BIOLOGIE II UNIVERSITÄT FREIBURG
SCHÄNZLESTR. 1, D-7800 FREIBURG, FRG

INTRODUCTION

Within plant cells carotenoids are synthesized exclusively in three types of plastids (chloroplasts, chromoplasts, and etioplasts) from the general prenyllipid precursor isopentenyl pyrophosphate (IPP). In this brief review I shall tackle three points of view, the formation or cellular compartmentation of formation, respectively, of the precursor IPP, the carotenogenic enzymes in chromoplasts and the carotenogenic enzymes in other plastids.

CELLULAR COMPARTMENTATION OF IPP BIOSYNTHESIS

The formation of IPP proceeds according to the relatively well established scheme shown in Fig. 1, ultimately from acetyl-CoA. There are three sites or compartments in the plant cell where prenyllipids are synthesized: (i) the endoplasmic reticulum (ER) for sterol formation; (ii) the plastids for the synthesis of the side chains of chlorophylls, quinones and tocopherols, and of the carotenoids and fatty acid prenyl esters; (iii) the mitochondria for the synthesis of the ubiquinone side chains. Therefore, the question arises whether there are also three cellular sites for IPP formation, whereby the cytoplasm/ER is uncontested.

One finds in the literature two concepts for the intracellular compartmentation of the IPP pathway (Fig. 1). The first concept which has been called "enzyme segregation" by Goodwin's group (ref. 1-3) postulates that this pathway occurs in each compartment where IPP is used. More recent evidence in favour of this concept comes from experiments of Russel et al. who found two HMG-CoA reductase activities in pea seedlings which were characterized as a plastid and a ER enzyme, respectively (ref. 4). Results obtained by Lichtenthaler's group using radish seedlings and the HMG-CoA reductase inhibitor mevinolin are also interpreted in support of the existence of enzyme segregation (ref. 5). Results of Schultz et al. and others also point in the direction of this concept (e.g. ref. 6 and 7).

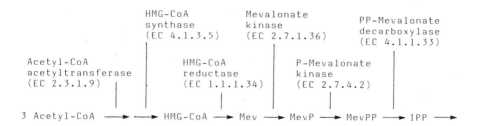

Fig. 1. Biosynthesis of isopentenyl pyrophosphate. HMG-CoA, hydroxymethylglutaryl-CoA; Mev, mevalonate.

The alternative concept (Fig. 1) claims that IPP is formed exclusively in the cytoplasm/ER and is distributed then to the various sites of prenyllipid formation. Gray and Kekwick e.g., failed to find isoenzymes of mevalonate kinase in Phaseolus (ref. 8). We have published several results in favour of this second concept using purified daffodil chromoplasts, spinach chloroplasts, and mitochondria from several plant tissues (ref. 9-11). These organelles are found to be not able to use HMG-CoA, mevalonate or phosphomevalonates for the synthesis of prenyllipids, whereas IPP is well accepted and incorporated. However, it may be argued that short-living chromoplasts from flowers and chloroplasts from more or less mature spinach leaves may have lost some enzymatic activities. We, therefore, extended our investigations to another plant system which is in an easily recognizable active state of growing, the seedlings of the white mustard Sinapis alba. These seedlings can be grown in continuous far-red light without significant protochlorophyllide/chlorophyllide conversion and, thus, without chlorophyll synthesis (ref. 12). The active form of phytochrome, P_{fr}, is present to some extent under these conditions and mediates an enhanced accumulation of carotenoids and an induction of several plastid enzymes. The plastids of these seedlings or their cotyledons, respectively, are true etioplasts with a prolamellar body. Upon illumination with white light, a burst of chlorophyll and carotenoid biosynthesis takes place in connection with the formation of the photosynthetic apparatus. These "preconditionned" etioplasts - as we call them - and the etiochloroplasts which are formed upon illumination are very active in prenyllipid synthesis in vivo and are expected to be so also in vitro upon isolation.

When etioplasts from the cotyledons of these seedlings are incubated in the presence of C-14 intermediates of the IPP pathway no prenyllipid synthesis can be observed, whereas IPP itself is well incorporated (Table 1). In the presence of cytoplasmic supernatant, however, which contains all the enzymes necessary for IPP formation the intermediates are also well incorporated. We conclude from such results that the pathway for IPP biosynthesis is exclusively localized within the cytoplasm of plant cells. It follows then that a mechanism has to exist which allows a regulated distribution of this central precusor, IPP, to the three sites of prenyllipid biosynthesis in the plant cell, the cytoplasm/ER, the plastids, and the mitochondria (Fig. 2). This concept means that the basic precursor pathway is used jointly and that a sort of enzyme segregation - to hold this term - takes place not until the level of prenyltransferases, when different chain lengthes are built up in the three compartments (Fig. 2). The presented results do not rule out, however, that another still unknown pathway may lead to the formation of IPP within the organelles. Some in vivo experiments may by interpreted to point into this direction.

Table 1. Sinapis cotyledon etioplasts. Prenyllipid synthesis from C-14 precursors in the absence and presence of cytoplasmic supernatant (CS). Chl, chlorophylls; car, carotenes. From Lütke-Brinkhaus and Kleinig, in press.

	Total incorp. (%) from 18.4 kBq	Distribution of radioactivity (%)			
		Polyprenol	Chl	Car	Others
+ [2-^{14}C]Mev + ATP	-	-	-	-	-
" " + CS	31.8	59.9	15.3	14.5	10.3
+ [2-^{14}C]MevPP + ATP	-	-	-	-	-
" " + CS	51.0	44.5	18.5	23.0	14.0
+ [1-^{14}C]IPP	41.1	40.5	12.5	29.8	17.2
" + CS	54.2	45.2	14.2	26.9	13.7

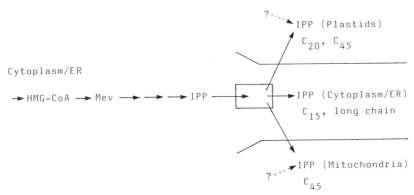

Fig. 2. Isopentenyl pyrophosphate biosynthesis and compartmentation in the plant cell. For explanation see text.

CAROTENOGENIC ENZYMES IN CHROMOPLASTS

The carotenogenic enzymes from the substrate IPP to ß-carotene are shown in Fig. 3. The further metabolism of carotenes to xanthophylls and other oxygenated carotenoids is scarcely understood on the enzyme level and will not be treated here.

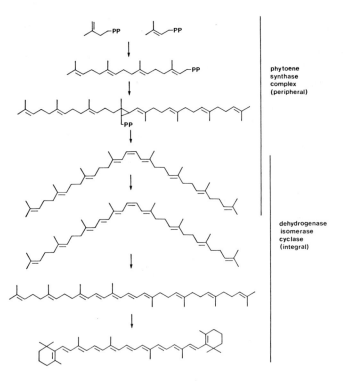

Fig. 3. Carotenogenic pathway from IPP to β-carotene. Not all carotene intermediates are shown.

The carotene pathway may be subdivided into two parts: (i) The so-called phytoene synthase complex which comprises the IPP isomerase leading to dimethylallylpyrophosphate, the prenyltransferase giving geranylgeranyl pyrophosphate, the prephytoene pyrophosphate synthase, and the phytoene synthase giving cis-phytoene. This complex is in chromoplasts a peripheral membrane protein as will be shown below. (ii) The subsequent enzymes which catalyze the dehydrogenation steps, the isomerization of cis-phytofluene to trans-phytofluene, and the cyclization reactions to β-carotene are integral membrane proteins.

Prephytoene pyrophosphate as an intermediate of the sequence was established by Gregornis and Rilling (ref. 13). Porter's group investigated the phytoene synthase complex which was obtained from a relatively crude chromoplast preparation from tomato fruits and from an acetone powder prepared thereof. The complex was partially purified and a molecular mass of about 166 000 Da was determined (ref. 14). The dissociation of an isomerase and prenyltransferase activity was achieved (molecular masses about 34 000 Da and 64 000 Da, respectively) (ref. 15). Furthermore, a strict dependency on Mn^{++} was established which is known for prenyltransferases from other sources. A severalfold stimulation of phytoene formation by ATP was observed, although the hydrolysis of ATP is not required within this enzymatic sequence (ref. 16).

More recently, new and more defined chromoplast systems for the in vitro carotenogenesis were introduced - more defined in the sense of cell biologists: the chromoplast from the daffodil flower (ref. 17 and 18) and the chromoplast from the red pepper fruit (ref. 19). A molecular mass of about 140 000 Da and an IEP of 5.3 for the phytoene synthase complex was determined. Especially, a strict lipid dependency of phytoene formation was observed. In the absence of lipids only pyrophosphate intermediates accumulate, particularly geranylgeranyl pyrophosphate. Lipids may be used in the form of chromoplast membranes or - more defined - in the form of pure liposomes prepared from chromoplast lipids. This lipid dependency classifies the phytoene synthase complex as a peripheral membrane protein which has lost his contact to the membrane upon desintegration of the organelle.

In contrast to the phytoene synthase complex the subsequent enzymes necessary for the formation of ß-carotene are truly integral membrane proteins (Fig. 3) and as such poorly characterized. The molecular masses, the electron acceptors of the dehydrogenase reactions, the possible oxygen requirement etc. are open questions. There are some inhibitors known for the dehydrogenase and cyclase reactions: norflurazon and other herbizides inhibit the dehydrogenation of phytoene and, e.g., azadihydrosqualene (ref. 20) or octyldiethylamine (ref. 21) inhibit the cyclase. Recently, however, we have achieved the solubilization of these integral enzymes using the zwitterionic detergent Chaps, and their functional reconstitution in liposomes prepared from chromoplast lipids (ref. 22). This opens new ways for the investigation of these enzymes.

CAROTENOGENIC ENZYMES IN OTHER PLASTID TYPES

Chromoplasts are ideal organelles for the study of carotene biosynthesis. The question arises what about chloroplasts in which in vivo apparently a carotenoid biosynthesis takes place. When chloroplasts or chloroplast subfractions are incubated in the presence of C-14 IPP a wide palette of prenyllipids is formed with the exception of carotenes (Table 2). Onyl in highly purified envelope fractions with very low contaminating chlorophyll synthase activities some phytoene synthase, phytoene dehydrogenase and cis-phytofluene isomerase activities have been described, when geranylgeranyl pyrophosphate, not IPP, was provided as the substrate (ref. 23).

Recently, we investigated etioplasts, etiochloroplasts, and chloroplasts from far-red light grown and white light grown Sinapis alba seedlings (see above) for their capacities of prenyllipid formation (Table 2). Etioplasts and etiochloroplasts actively synthesize, besides other prenyllipids, acrotenes and their stroma containes a phytoene synthase complex. Even in chloroplasts from cotyledons a small carotene forming activity is detectable and the pyhtoene synthase complex is still found in the stroma. Chloroplast from primary leaves, on the other hand, have totally lost these activities and their stroma is devoid of any phytoene synthase complex.

It is concluded that in chloroplasts the carotenogenic enzymes apparently have a drastically altered activity and topology when compared with yellow or greening plastid types. Although chlorophyll and carotenoid biosynthesis in vivo proceeds in a sort of concerted action in chloroplasts, carotenogenic enzymes can not be detected in vitro in these organelles, whereas chlorophyll formation is abundant. This situation parallels to some extent the findings of Apel et al. for the enzyme protochlorophyllide oxidoreductase which is abundant in etioplasts and nearly not detectable in chloroplasts (ref. 24).

Table 2. Sinapis alba etioplasts, etiochloroplasts, and chloroplasts. Formation of carotenes and other prenyllipids from C-14 IPP. C_{20}/C_{45}, geranylgeraniol/nonaprenol. From Lütke-Brinkhaus and Kleinig, in press.

	Total incorp. (%) from 37 kBq	Product pattern (total = 100%)				
		C_{20}/C_{45}	Chl	Phytoene	Desat. Car	Others
Etioplasts	67	20	10	6	32	32
Stroma	9	24	-	76	-	-
Etiochloroplasts	76	21	29	6	25	19
Stroma	7	37	-	63	-	-
Chloroplasts Cotyledons	68	24	66	1	2	7
Stroma	5	43	-	57	-	-
Chloroplasts Primary leaves	39	37	66	-	-	3
Stroma	2	100	-	-	-	-

REFERENCES

1. T.W.Goodwin, and E.J.Mercer, Symp.Biochem.Soc. 24: 37 (1963).
2. L.J.Rogers, S.P.J.Shah, and T.W.Goodwin, Intracellular localization of mevalonate-activating enzymes in plant cells. Biochem.J. 99: 381 (1966).
3. L.J.Rogers, S.P.J.Shah, and T.W.Goodwin, Mevalonate-kinase isoenzymes in plant cells. Biochem.J. 100: 14c (1967).
4. R.J.Wong, D.K.McCormack, and D.W.Russell, Plastid 3-hydroxy-3-methylglutaryl coenzyme A reductase has distinctive kinetic and regulatory features: properties of the enzyme and positive phytochrome control of activity in pea seedlings. Arch.Biochem.Biophys. 216: 631 (1982).
5. T.J.Bach, and H.K.Lichtenthaler, Inhibition by mevinolin of plant growth, sterol formation and pigment accumulation. Physiol.Plant. 59: 50 (1983).
6. H.Bickel, and G.Schultz, Biosynthesis of plastoquinone and ß-carotene in isolated chloroplasts. Phytochem. 15: 1253 (1976).
7. R.E.Arebalo, and E.D.Mitchell, Cellular distribution of 3-hydroxy-3-methylglutaryl coenzyme A reductase and mevalonate kinase in leaves of Nepeta cataria. Phytochem. 23. 13 (1984).
8. J.C.Gray, and R.G.O.Kekwick, Mevalonate kinase in green leaves and etiolated cotyledons of the french bean Phaseolus vulgaris. Biochem.J. 133: 335 (1973).
9. K.Kreuz, and H.Kleinig, On the compartmentation of isopentenyl diphosphate synthesis and utilization in plant cells. Planta 153: 578 (1981).

10. K.Kreuz, and H.Kleinig, Synthesis of prenyl lipids in cells of spinach leaf. Compartmentation of enzymes for formation of isopentenyl diphosphate. Eur.J.Biochem. 141: 531 (1984).
11. F.Lütke-Brinkhaus, B.Liedvogel, and H.Kleinig, On the biosynthesis of ubiquinones in plant mitochondria. Eur.J. Biochem. 141: 537 (1984).
12. H.Mohr, Control of chloroplast development by light - some recent aspects, in: "Photosynthesis V. Chloroplast Development", G.Akoyunoglou, ed., Balaban International Science Service, Philadelphia (1981).
13. D.E.Gregornis, and H.C.Rilling, The stereochemistry of trans-phytoene synthesis. Some observations on lycopersene as a carotene precursor and a mechanism for the synthesis of cis- and trans-phytoene. Biochemistry 13: 1538 (1974).
14. B.Maudinas, M.L.Bucholtz, C.Papastephanou, S.S.Katiyar, A. V.Briedis, and J.W.Porter, The partial purification and properties of a phytoene synthesizing enzyme system. Arch. Biochem.Biophys. 180: 354 (1977).
15. S.A.Spurgeon, N.Sathyamoorthy, and J.W.Porter, Isopentenyl pyrophosphate isomerase and prenyltransferase from tomato fruit plastids. Arch.Biochem.Biophys. 230: 446 (1984).
16. B.Maudinas, M.L.Bucholtz, C.Papastephanou, S.S.Katiyar, A. V.Briedis, J.W.Porter, Adenosine 5'-triphosphate stimulation of the activity of a partially purified phytoene synthase complex. Biochim.Biophys.Res.Commun. 66: 430 (1975).
17. P.Beyer, K.Kreuz, and H.Kleinig, ß-Carotene synthesis in isolated chromoplasts from Narcissus pseudonarcissus. Planta 150: 435 (1980).
18. K.Kreuz, P.Beyer, and H.Kleinig, The site of carotenogenic enzymes in chromoplasts from Narcissus pseudonarcissus L. Planta 154: 66 (1982).
19. B.Camara, F.Bardat, and R.Monéger, Sites of biosynthesis of carotenoids in Capsicum chromoplasts. Eur.J.Biochem. 127: 255 (1982).
20. B.Camara, O.Dogbo, A.d'Harlingue, and F.Bardat, Inhibition of lycopene cyclization from Capsicum chromoplast membranes by 2-aza-2,3-dihydrosqualene. Phytochem. 24: 2751 (1985).
21. B.Camara, O.Dogbo, A.d'Harlingue, H.Kleinig, and R.Monéger, Metabolism of plastid terpenoids: lycopene cyclization by Capsicum chromoplast membranes. Biochim.Biophys.Acta 836: 262 (1985).
22. P.Beyer, G.Weiss, and H.Kleinig, Solubilization and reconstitution of the membrane-bound carotenogenic enzymes from daffodil chromoplasts. Eur.J.Biochem. 153: 341 (1985).
23. F.Lütke-Brinkhaus, B.Liedvogel, K.Kreuz, and H.Kleinig, Phytoene synthase and phytoene dehydrogenase associated with envelope membranes from spinach chloroplasts. PLanta 156: 176 (1982).
24. K.Steinmüller, A.Batschauer, E.Mössinger, E.Schäfer, S.K. Rasmussen, and K.Apel, The light-induced greening of barley, in: "Molecular Form and Function of the Plant Genome", L.van Vloten-Doting, G.S.P.Groot, T.C.Hall, eds., Plenum Press, New York (1986).

CELL WALLS AND SECONDARY PRODUCTS AS OBSTACLES TO PLANT ENZYME ISOLATION: PROBLEMS AND SOLUTIONS, INCLUDING A SIMPLE LIQUID-NITROGEN HOMOGENIZER FOR BULK TISSUE EXTRACTION*

Richard P. Sandstrom[†] and W. David Loomis

Department of Biochemistry and Biophysics
Oregon State University
Corvallis, Oregon 97331 (U.S.A.)

INTRODUCTION

When we first undertook studies of monoterpene biosynthesis in peppermint (Mentha piperita), we and other plant biochemists were handicapped by inadequate methodology and by inadequate understanding of plants. Gas chromatography was not yet available. It was assumed that secondary metabolites were synthesized slowly and irreversibly, so tracer experiments were conducted over periods of days instead of minutes or hours. Finally, the unique problems of plant enzymology were not understood.

The problems caused by plant phenolic compounds during enzyme isolation are now generally recognized, though often ignored. Problems caused by other secondary products, including terpenes and organic isothiocyanates (mustard oils), are not generally appreciated. Further, it is not generally recognized that the plant cell wall is an ultrafilter which retains most proteins, and that homogenization procedures often are ineffectual in opening the cell wall envelope. The cell membrane is not the only barrier in plant tissues.

We intend here to briefly review these problems and their solutions. We will then describe the advantages of liquid-nitrogen homogenization as a valuable technique for combatting both secondary metabolites and cell walls, and a simple modification of the Waring Commercial Blendor to provide a convenient liquid-nitrogen homogenizer for large quantities of plant tissue. The examples cited will not all be from plant lipid studies, but they are relevant to any plant studies. Plants that produce interesting terpenoids (e.g. Mentha and Hevea) often contain high concentrations of phenolic compounds, while the terpenoid compounds themselves present additional problems. The use of rape as an oilseed crop [R. K. Downey, this symposium] was initially plagued by antinutritional factors in the oilseed meal, produced from organic isothiocyanates, and we have found that isothiocyanates bind to and modify horseradish peroxidase.

---

*Supported in part by grants from the National Science Foundation and the O.S.U. Environmental Health Sciences Center.

[†]Present address: Botany Department, University of Washington, Seattle, Washington 98195 (U.S.A.).

PROBLEMS CAUSED BY SECONDARY METABOLITES

In our early attempts to study monoterpene metabolism in peppermint we found that leaf extracts browned badly, and the only enzyme activity we could demonstrate was an uncharacterized phenol oxidase. After several years of frustration we found that the problems were due largely to plant phenolics. Many plant phenolics bind tightly to proteins by hydrogen bonding and are not removed by conventional procedures such as dialysis. If not removed, they may bind covalently and irreversibly by oxidative coupling reactions. We found that most of the peppermint phenolics could be removed by hydrogen-bonded adsorption on insoluble polyvinylpyrrolidone (PVPP, Polyclar AT)[1]. Peppermint extracts treated with Polyclar AT did not brown, and we were able to demonstrate several enzyme activities, including mevalonic kinase[1] and pulegone reductase[2]. However, the extracts were still contaminated by monoterpenes, bound to protein by hydrophobic interactions. Treatment with adsorbent polystyrene (XAD-4) removed the monoterpenes and made it possible to study monoterpene metabolism *in vitro* without the need for labeled substrates[3,4,5]. We also adopted metabisulfite as a superior antioxidant[6,7]. Rodney Croteau and his coworkers have continued to use these techniques in their recent studies of monoterpene metabolism. Adsorbent polystyrene also binds some plant phenolic compounds that are not removed by PVPP[4], as well as juglone from walnut hulls, isothiocyanates from horseradish roots, and detergents that may have been added during extraction procedures[4].

In addition to enzyme inactivation, plant secondary products often produce spurious protein analyses, and artifactual organelles and isozymes. Protein determination may be in error by orders of magnitude[7]. In particular, the popular method of Lowry et al. is an adaptation of the classical Folin method for measuring phenols, which is in fact a general method for reducing substances. The reducing substance in the Lowry-protein method consists primarily of complexes of Cu^{++} with peptide bonds (biuret complex)[8]. Interfering materials include plant phenolics, thiols, ascorbic acid, EDTA, and Tris buffer. The original paper of Lowry et al. reported that phenols interfere. Potty[9] described a modification of the Lowry method in which a Cu-free blank is used to correct for phenolic interference. Robinson[10] has reevaluated this problem and has recommended the Bradford dye-binding method using Coomassie Brilliant Blue G-250.

The phenol oxidase (catechol oxidase) of tea leaves provides a classic example of artifactual organelles, and probably of artifactual substrate specificity as well, both produced by protein-phenolic complexing. Li and Bonner[11] reported that the enzyme is insoluble, tightly bound to the chloroplast grana. Sanderson[12] demonstrated that conventional extracts from tea leaves contained no soluble protein, and that the "chloroplast" fraction consisted largely of protein-tannin complexes. Addition of hydrated nylon as a phenol adsorbent solubilized the enzyme completely. Coggon et al.[13] subsequently purified the tea catechol oxidase to homogeneity, using Polyclar AT as a phenol adsorbent, and found that the substrate specificity was narrower than previously reported by Gregory and Bendall. They quoted Gregory and Bendall as having reported that their enzyme fractions were "tanned", and suggested that contaminating phenolic substrates of the enzyme had mediated indirect and non-specific oxidation of other phenolic compounds.

Kaplan[14] in a chemotaxomic survey of 54 plant species, representing 37 genera, presented evidence that many of the apparent isozymes of aldolase were artifacts due to protein-phenolic interactions. The artifacts could be eliminated by procedures which included treatment with Polyclar AT and XAD-4. He suggested that there might be many such artifacts "masquerading as isozymes" in the literature. Smith and Montgomery[15] reported similar artifactual isozyme production with the polyphenoloxidase of d'Anjou pears. They were able to reduce the formation of artifacts by extracting at low pH (5.6-5.9) and removing phenolics with polymeric adsorbents. The most effective adsorbent in this case was the polystyrene-based anion exchange resin AG2-X8. We also found this to be the most effective adsorbent for removing quercetin or chlorogenic acid from solution in model systems[4]. Presumably its effectiveness is due to a combination of hydrophobic and polar binding by the ion exchange resin.

Experiences with horseradish peroxidase (HRP) provide an interesting case history. According to Maehly[16] this enzyme floats when precipitated by ammonium sulfate, suggesting the formation of a lipid-protein complex. Also according to Maehly it is essential to work outdoors or wear a gas mask when extracting the enzyme, and to protect the hands with rubber gloves. More recently, investigators have reported variable numbers of "isozymes" of HRP, from 7 to 42[17]. We have confirmed Maehly's report that ammonium-sulfate precipitates of horseradish extracts float. Untreated extracts had a sharp, pungent odor of isothiocyanates and a dark yellowish color. Treatment with XAD-4 plus Polyclar AT removed both the yellow color and the odor and produced extracts that behaved normally when subjected to ammonium sulfate fractionation[4]. We have also shown that either allylisothiocyanate (the principal component of horseradish oil) or ^{14}C-phenylisothiocyanate can cause interconversion of apparent "isozymes" of HRP at a pH of 8 or above. [Electrophoresis at pH 10.3 is a routine part of the traditional purification procedure[16].] The interconversions involve lowering of the enzyme pI, without loss of activity. In experiments with ^{14}C-phenylisothiocyanate[17] the dominant "C" band was largely converted to "B" bands within one day. At pH 10 the conversion of C to B "isozymes" was complete in one day. The augmented B bands were radioactive. Treatment of the horseradish homogenates with XAD-4 and Polyclar AT definitely reduced the formation of artifacts, but we are not yet certain that we have eliminated the artifacts [Darussamin[17], and in preparation]. The artifactual isozymes found by us, by Kaplan, and by Smith and Montgomery did not smear but formed sharp bands on electrophoresis or isoelectric focusing. Discrete stepwise modifications of protein R-groups are indicated.

Control of pH is especially important in plant enzyme isolation. As a generalization, low pH is desirable. Phenolic compounds, which are ubiquitous in plants, ionize at high pH, and this makes them subject to spontaneous air-oxidation (and subsequent covalent-addition reactions). Ionized phenolics will not form H-bonded complexes either with proteins or with adsorbent polymers. The former might be desirable, but the latter is not. A principal target of covalent addition reactions with proteins is the un-ionized -NH$_2$ group. This applies to 1,4-addition by quinones and sesquiterpene lactones and to the isothiocyanate addition reaction. Low pH converts amino groups to the unreactive ionized form. Obviously the general factors of protein chemistry also apply, such as possible denaturation or isoelectric precipitation.

THE PLANT CELL WALL AS AN OBSTACLE

It is universally recognized that isolation of plant organelles, especially plastids, is made difficult by the fact that techniques used to rupture the strong cell walls often disrupt the fragile organelles. Several authors have described tissue slicers that use razor blades to cut cells open, rather than grinding or hammering the tissues, thus providing high-quality organelles. The percent recovery of organelles with most of these slicers is low, but some of the motorized devices can process large amounts of tissue, thus producing large quantities of organelles in spite of low recoveries. They can also process the tissues quickly, thus further enhancing quality. Honda[18] described the use of 10 to 50 stacked stainless steel razor blades for hand-slicing of 10 g batches of leaf tissue. This cutting method is very slow, but reportedly cuts nearly every cell and yields very good chloroplasts. Kannangara and coworkers[19,20] described a modification of the Braun kitchen homogenizer, in which the standard cutting device is replaced by a holder for removable razor blades. This modified blender quickly chops batches of 25-250 g of leaf tissue and produced high quality chloroplasts from spinach or barley in 6-10% yield. For larger quantities of tissue, Leigh & Branton[21] designed a rotary tissue slicer, which they used in isolating plant vacuoles. The slicer holds eight replaceable razor blades and attaches to a drill press. Marty and Branton[22] subsequently described an improved, higher-yielding procedure for using the tissue slicer, and it is said to process 500 g of beet root tissue in less than 5 min[23]. The tissue slicer is effective with bulky plant tissues, and with large leaves (e.g. maize), but small leaves such as soy bean may slip through without being effectively sliced [D. Branton, personal communication].

Protoplasts, produced by enzymatic hydrolysis of the cell walls, have been used as a source of plant organelles[22] and could be used as a source of soluble enzymes, but the methods can only be applied to relatively soft tissues and require prolonged exposure of tissues to cell wall degrading enzymes[22,23]. In addition, most of the fungal enzyme preparations that are used contain such undesirable products as proteases, ribonucleases, peroxidases and phenolics[24].

It has not been generally recognized that the plant cell wall is an ultrafilter which retains not only organelles, but most soluble proteins as well, and that homogenization procedures often only separate cells, by parting the middle lamella, rather than opening the cells. Chibnall[25-27] described a method for selectively extracting small molecules from leaf tissues by disrupting cell membranes with ether. Small molecules from vacuole and protoplasm diffused out of the tissue rapidly (starting within seconds), while proteins were retained. A very small amount of protein was released from leaves of spinach or alfalfa[27], but in general all or nearly all of the leaf protein was retained by the cell wall ultrafilter, while small molecules diffused out. Other organic solvents were also tested. Chloroform had a similar action to ether, but the rest were slower acting or ineffective[25]. This phenomenon has been rediscovered recently[28] and interpreted in terms of "membrane pores" rather than cell wall pores, an interpretation which we consider doubtful. In the recent work, toluene, one of the slower-acting solvents in Chibnall's experiments[25], induced, after 10 min. exposure, leaching of small molecules, but not of proteins, from leaf tissues and *Chlorella* cells[28].

Kertesz[29] reported that blenders usually only part the middle lamella and may not effectively homogenize or extract plant tissues. We have observed that acetone powders of plant tissues, prepared by standard procedures using the Waring Blendor, contain many multicellular bits of tissue. In acetone powders of peppermint we have even observed intact oil glands, still filled with essential oil. Robin Hill stated in a 1961 lecture that spinach leaves are a superior tissue for enzyme isolation because they have strong middle lamellae, which promote cell breakage as opposed to cell-cell separation.

Carpita et al.[30] reported that "the limiting diameter of pores in the walls of living plants cells through which molecules can freely pass has been determined to be" from 35 to 52Å for various cells of several dicotyledenous species. They calculated that a pore diameter of 38-40Å would retain globular proteins of molecular weight about 17,000 or higher.

They concluded that "such a wall may represent a more significant barrier to cellular communication than has been previously assumed". Tepfer and Taylor[31], based on experiments in which cell wall preparations were used as gel filtration media, argued that "proteins of molecular weight up to 60,000 can permeate a substantial portion of the cell wall." Carpita, in rebuttal[32], pointed out that the two groups were measuring entirely different things. The 17,000 molecular weight figure of Carpita et al. represents the limiting size of molecules that can <u>pass through the limiting diameter</u> of the cell wall pores, while Tepfer and Taylor's 60,000 value represents the size of molecules that can <u>enter</u> the pores, but not necessarily pass through. Thus, it is valid to conclude that soluble proteins larger than about 17,000 molecular weight (i.e. most proteins) cannot be extracted from plant tissues without opening the cell wall envelope of the individual cells.

LIQUID-NITROGEN HOMOGENIZATION

Homogenization in liquid nitrogen is widely regarded as a convenient procedure in tissue extraction. However, the importance of opening the cell wall envelope, and the efficacy of liquid N_2 in accomplishing this, are not generally appreciated. Cryogenic homogenization and storage of tissues also help prevent protein modification, since chemical reactions are blocked at these low temperatures. Freezing in liquid N_2 is very rapid and does not produce ice crystals in the tissues.

Small quantities of non-bulky tissues such as leaves are readily ground to a very fine powder with liquid N_2 in a porcelain mortar. Two precautions should be observed. The mortar and pestle should be oven dried before use, to prevent their breaking due to freezing of water within the porous porcelain. Secondly, the frozen powder (after evaporation of remaining liquid N_2) should be poured from the mortar into the extracting buffer-adsorbent-antioxidant mixture. If the buffer is poured into the mortar it will freeze solid.

For bulky tissues the mortar and pestle are not satisfactory, and some kind of mill is needed. Janke and Kunkel in Germany make a small Analytical Mill with a cryogenic attachment for use of liquid N_2 (marketed in the U.S. by the Tekmar Company, Cincinnati, Ohio). Frozen samples are pulverized by a high-speed rotor turning in liquid N_2. The mill is convenient and effective for homogenizing small samples of plant tissue, but the chamber volume is only about 50 ml.

For studies of horseradish peroxidase, we wanted a mill that would perform like the Janke and Kunkel mill, but on a larger scale. The solution was found in modifying the cover of a Waring Commercial Blendor, Model CB-6, which has a 1 gallon (3.8 ℓ) stainless steel container. A round hole was cut in the center of the cover to allow insertion of a "chimney" vent similar to that on the Janke and Kunkel mill. The chimney was made of 2 3/8" O.D. x 0.050" stainless steel tubing (6.03 cm x 1.27 mm), 8 3/4" long overall (22.2 cm), with 6 3/4" (17.1 cm) projecting above the cover and 2" (5.1 cm) below, and attached with lead-free food-grade solder. The vertical dimensions of the chimney are more or less arbitrary. They were modelled after the Janke and Kunkel mill and found to be satisfactory. The diameter of the vent is not arbitrary; an earlier version used smaller tubing, which was found inadequate to release all of the gas pressure produced from boiling N_2 when the blender was in use. The 1 quart (950 ml) <u>stainless steel</u> Waring Blendor container could be modified in a similar fashion. We would not recommend using a glass blender container with liquid N_2.

We recommend that the homogenizer be used with a variable-voltage transformer, being sure that the transformer has adequate capacity. This allows a lower start-up speed than the Waring pushbutton controls. Prior to use, the blender blade assembly is dismantled, cleaned (but not oiled or greased), and oven dried. It is essential that there be no water in the bushing, or the whole assembly will freeze tight. The container is placed on the motor assembly, and the motor turned on at low speed. Liquid N_2 is added slowly to cool the container till the N_2 does not boil violently. The modified cover is then attached, and the motor speed increased to a medium range. Tissue, already frozen in liquid N_2, is added through the chimney. Bulky tissues such as root or fruit tissue should be cut into chunks of about 2 cm dimension before freezing, or broken up in an ice crusher after freezing. Additional liquid N_2 or more sample can be added through the chimney as desired. The chimney provides adequate venting of the boiling N_2 and prevents splashing. Final fine pulverization of the tissue is accomplished by turning the motor to high speed (ca. 20,000 rpm) for 2-3 min. When the blender has stopped, the liquid-N_2-containing vessel should be removed from the motor in order to minimize condensation of moisture in the motor bearings. Freezing of water in the blender blade bushing and the motor bearings is one of the greatest logistic problems, but it is easily avoided if one follows the precautions we have prescribed. As before, the liquid-N_2 powder should be poured into the buffer-adsorbent-antioxidant slurry, not <u>vice versa</u>.

This cryogenic blender reduces large quantities of frozen plant tissue quickly to an extremely fine powder, under super-cold conditions where no chemical reactions can occur. When the frozen liquid-N_2 powder is added to the buffer-adsorbent-antioxidant mixture, secondary products are rapidly removed, and the time available for reaction with proteins is reduced to a minimum. The blender is particularly useful for bulky tissues such as horseradish root, which cannot be ground satisfactorily in a mortar and pestle. However, it also satisfactorily homogenizes even such limp, non-resistant tissue as eelgrass (<u>Zostera marina</u>) leaves. The use of adsorbents as described above has reduced the "isozyme" variability of horseradish peroxidase. The use of this cryogenic blender has increased the efficiency of extraction. The acidic ("A") isozymes show a distinct increase in activity compared to extracts prepared with the blender at 4°. Liquid-N_2 homogenization also eliminates the seasonal variation in HRP yield reported by Jermyn and Thomas[33]. They reported that yields were maximal in winter and minimal in early summer. We find that the roots are very flexible in early summer, at the onset of renewed growth, and brittle at other seasons. The "flexible" period corresponds to Jermyn and Thomas' period of low HRP yields, and we assume that it coincides with rapid growth, and changes in the structure of the middle lamella and the primary cell wall. We find that at this time of year, conventional homogenization and extraction procedures yield negligible amounts of soluble peroxidase. They also yield a floating, sticky mass which contains substantial peroxidase activity. We suggest that this sticky mass contains cell debris with entrapped peroxidases, plus adsorbed isothiocyanates. When we homogenize horseradish roots in liquid nitrogen, extraction of HRP is efficient at any time of year, whether the roots are flexible or brittle. It appears that the differences are due to the influence of seasonal variations in cell wall structure on enzyme extractability rather than to seasonal variations in the actual enzyme content of the tissue.

CONCLUSIONS

We have not attempted here to describe all of the techniques for dealing with problems in plant enzymology, nor to provide "cookbook" methods. We emphasize that each plant tissue and enzyme is unique, and each investigator must understand the problems, and adapt the available techniques to his own needs. We refer the reader to several reviews for details[1,4,6,7,34,35]. The nitrogenase of soybean nodules can be cited as an example of an enzyme whose isolation required nearly all of the known "tricks of the trade": PVPP as a phenol adsorbent, ascorbate as an antioxidant, and an inert atmosphere[36,37]. In addition, Klucas et al.[37] found that the enzyme could be isolated from nodules stored at -70°, but the enzyme activity was destroyed if nodules were kept in ordinary frozen storage [H. J. Evans, personal communication]. Phenol oxidation and coupling reactions occur even in frozen tissues.

We have reviewed literature and experimental results which demonstrate that the plant cell wall is a greater obstacle to plant enzyme isolation than has been appreciated. We have also discussed the use of cryogenic techniques in combatting both secondary metabolites and cell walls, and described a convenient liquid-N_2 homogenizer for bulk extraction of plant tissues. In addition we have pointed out that plant phenolics are not the only secondary products that cause problems in plant enzymology.

BIBLIOGRAPHY

1. W. D. Loomis and J. Battaile, Plant phenolic compounds and the isolation of plant enzymes, Phytochemistry 5:423 (1966).
2. J. Battaile, A. J. Burbott, and W. D. Loomis, Monoterpene interconversions: Metabolism of pulegone by a cell-free system from Mentha piperita, Phytochemistry 7:1159 (1968).
3. R. Croteau, A. J. Burbott, and W. D. Loomis, Enzymatic cyclization of neryl pyrophosphate to α-terpineol by cell-free extracts from peppermint, Biochem. Biophys. Res. Comm. 50:1006 (1973).
4. W. D. Loomis, J. D. Lile, R. P. Sandstrom, and A. J. Burbott, Adsorbent polystyrene as an aid in plant enzyme isolation, Phytochemistry 18:1049 (1979).
5. A. J. Burbott and W. D. Loomis, Monoterpene interconversions by cell-free enzymes from peppermint (Mentha piperita), Plant Physiol. Suppl. 65:96 (1980).
6. J. W. Anderson, Extraction of enzymes and subcellular organelles from plant tissues, Phytochemistry 7:1973 (1968).
7. W. D. Loomis, Overcoming problems of phenolics and quinones in the isolation of plant enzymes and organelles, Methods in Enzymol. 31:528 (1974).
8. S.-C. Chou and A. Goldstein, Chromogenic groupings in the Lowry protein determination, Biochem. J. 75:109 (1960).
9. V. H. Potty, Determination of proteins in the presence of phenols and pectins, Anal. Biochem. 29:535 (1969).
10. T. Robinson, The determination of proteins in plant extracts that contain polyphenols, Plant Sci. Lett. 15:211 (1979).
11. L. P. Li and J. Bonner, Experiments on the localization and nature of tea oxidase, Biochem. J. 41:105 (1947).
12. G. W. Sanderson, Extraction of soluble catechol oxidase from tea shoot tips, Biochim. Biophys. Acta 92:622 (1964) and correction, 96:548 (1965).
13. P. Coggon, G. A. Moss, and G. W. Sanderson, Tea catechol oxidase: Isolation, purification and kinetic characterization, Phytochemistry 12:1947 (1973).
14. H. J. Kaplan, A. chemosystematic study of the phylogenetic position of Arabidopsis thaliana (L.) Heynh. (Brassicaceae) employing numerical methods, Ph.D. thesis, Oregon State University, Corvallis (1977).
15. D. M. Smith and M. W. Montgomery, Improved methods for the extraction of polyphenol oxidase from d'Anjou pears, Phytochemistry 24:901 (1985).

16. A. C. Maehly, Plant peroxidase, Methods in Enzymol. 2:801 (1955).
17. A. Darussamin, Reactions of horseradish peroxidase with the isothiocyanates of horseradish oil, Ph.D. thesis, Oregon State University, Corvallis (1985).
18. S. I. Honda, Fractionation of green tissue, Methods in Enzymol. 31:544 (1974).
19. C. G. Kannangara, S. P. Gough, B. Hansen, J. N. Rasmussen, and D. J. Simpson, A homogenizer with replaceable razor blades for bulk isolation of active barley plastids, Carlsberg Res. Commun. 42:431 (1977).
20. C. G. Kannangara, S. P. Gough, and D. von Wettstein, The biosynthesis of Δ-aminolevulinate and chlorophyll and its genetic regulation, in: "Developments in Plant Biology", vol. 2 (Chloroplast Development), G. Akoyunoglou and J. H. Argyroudi-Akoyunoglou, eds., Elsevier/North Holland, Amsterdam, New York, (1978), p. 147.
21. R. A. Leigh and D. Branton, Isolation of vacuoles from root storage tissue of Beta vulgaris L., Plant Physiol. 58:656 (1976).
22. F. Marty and D. Branton, Analytical characterization of beetroot vacuole membrane, J. Cell Biol. 87:72 (1980).
23. F. Marty, D. Branton, and R. A. Leigh, Plant vacuoles, in: "The Biochemistry of Plants", P. K. Stumpf and E. E. Conn, eds., vol. 1 ("The Plant Cell", N. E. Tolbert, ed.), Acad. Press, New York (1980), p. 625.
24. M. R. Davey and A. Kumar, Higher plant protoplasts--Retrospect and prospect, in: "Plant Protoplasts", K. L. Giles, ed., Acad. Press, New York (1983), p. 219.
25. A. C. Chibnall, A new method for the separate extraction of vacuole and protoplasmic material from leaf cells, J. Biol. Chem. 55:333 (1923).
26. A. C. Chibnall, Spinacin, a new protein from spinach leaves, J. Biol. Chem. 61:303 (1924).
27. A. C. Chibnall, "Protein Metabolism in the Plant", Yale Univ. Press, New Haven (1939).
28. H. R. Lerner, D. Ben-Bassat, L. Reinhold, and A. Poljakoff-Mayber, Induction of "pore" formation in plant cell membranes by toluene, Plant Physiol. 61:213 (1978).
29. Z. I. Kertesz, "The Pectic Substances", Interscience Publishers, New York, (1951) p. 100.
30. N. Carpita, D. Sabularse, D. Montezinos, and D. P. Delmer, Determination of the pore size of cell walls of living plant cells, Science 205:1144 (1979).
31. M. Tepfer and I. E. P. Taylor, The permeability of plant cell walls as measured by gel filtration chromatography, Science 213:761 (1981).
32. N. C. Carpita, Limiting diameters of pores and the surface structure of plant cell walls, Science 218:813 (1982).
33. M. A. Jermyn and R. Thomas, Multiple components in horse-radish peroxidase, Biochem. J. 56:631 (1954).
34. W. D. Loomis, Removal of phenolic compounds during the isolation of plant enzymes, Methods in Enzymol. 13:555 (1969).
35. M. J. C. Rhodes, The extraction and purification of enzymes from plant tissues, in: "Regulation of Enzyme Synthesis and Activity in Higher Plants", H. Smith, ed., Acad. Press, London (1977), p. 245.
36. B. Koch, H. J. Evans, and S. Russell, Reduction of acetylene and nitrogen gas by breis and cell-free extracts of soybean root nodules, Plant Physiol. 42:466 (1967).
37. R. V. Klucas, B. Koch, S. A. Russell, and H. J. Evans, Purification and some properties of the nitrogenase from soybean (Glycine max Merr.) nodules, Plant Physiol. 43:1906 (1968).

DEVELOPMENTAL REGULATION OF STEROL AND PENTACYCLIC TRITERPENE BIOSYNTHESIS AND COMPOSITION: A CORRELATION WITH SORGHUM FLORAL INITIATION

Rick C. Heupel[a], W. David Nes[a] and J. A. Verbeke[b]

[a] Plant and Fungal Lipid Group, Plant Development and Productivity Research Unit, WRRC, USDA/ARS, Albany, CA 94710. [b]Department of Biological Sciences, University of Illinois, Chicago, IL 60680.

Sterol and pentacyclic triterpene (PT) composition have been examined in vegetative[1-4] and reproductive tissues[3,5] of Sorghum bicolor (L.) Moench in various stages of development. Evidence for the developmental regulation of the amounts and biosynthesis of these polycyclic isopentenoids and its correlation with floral initiation have been obtained in this and our previous studies.[1-4] Sterol and PT composition of individual fully mature leaf blades was determined at various stages in the life cycle of S. bicolor (L.) Moench cv. G499 GBR (Figure 1). Free sterol content increased in the leaf blades logarithmically during the vegetative phase of the life cycle (day 7-34); this increase continued during the transition (Figure 2) of the vegetative to floral meristem (day 34-38), and then decreased during panicle development (day 48-104) (Figure 1). The relative ratio of the bulk sterols remained constant throughout the plant's development not only in the leaf blade but in other tissues, such as the spikelets (data not shown). Early in the plant's development the leaf blade free PT content was negligible. Following floral induction, free PT content increased to a level greater than the sterols as flowering progressed. The relative ratio for the individual free PT, unlike that of the sterols, did not remain constant.

Sterol biosynthesis from radiolabeled precursors and the possible interconversion of three of the five end product sterols by leaf blades from plants with inflorescences just emerging[1,3] (Table 1) and in germinating seed[3] have been examined.

The occurence data indicates that a critical mass of leaf free sterol is correlated with floral induction (Figure 1). The suggestion that a critical mass of free sterol plays a role in flowering in sorghum differs from the general view that if sterols regulate flowering they must be metabolized to a hormone.[6] We assume that sterols in sorghum, and other crops, may be utilized in multiple metabolic and non-metabolic roles, bulk and regulatory[1], each of which may have their functions expressed at different times during their life cycle.

Previous biosynthetic information[1,4] indicates that following floral induction, squalene-oxide (SO), the common precursor of sterols and PT, is shunted into PT pathways effectively decreasing the availability of SO for sterol production. These data, along with kinetic studies with [2-^{3}H]acetate (viz. 4, 6, 9, 24, 48 hr feeds to leaf blades of 66 day plants) indicated that 24-alkyl sterols were synthesized at a low rate (data not shown). The results of the [2-^{3}H] and [24-^{3}H]lanosterol and [2-^{3}H]cycloartenol feeds (Table 1) would indicate that

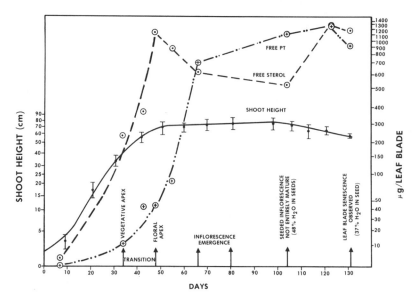

Figure 1. The occurence of free sterols and pentacyclic triterpenes in mature leaf blades of <u>Sorghum bicolor</u> cv G499 GBR with development. Plants were grown under greenhouse conditions as described previously.[1,2,4] At the indicated intervals fully mature leaf blades were harvested, extracted with acetone and the extracts analyzed. Day 7-leaf blade length 1.5 cm, 0.009 g. fr. wt., 0.001 gd. wt.; day 34-39.3 cm, 4.35 g. fr. wt., 0.65 g. d. wt.; day 43-49.0 cm, 4.38 g. fr. wt., 1.05 g. d. wt.; day 48-55.5 cm, 5.43 g. fr. wt., 0.88 g. d. wt.; day 55-55.0 cm, 4.75 g. fr. wt., 1.14 g. d. wt.; day 66-54.0 cm; 3.87 g. fr. wt., 0.92 g. d. wt.; day 104-68.0 cm, 8.82 g. fr. wt., 1.95 g. d. wt.; day 123-57.0 cm, 6.34 g. fr. wt., 1.51 g. d. wt.; day 131-58.0 cm, 4.42 g. fr. wt., 1.07 g. d. wt. Leaf material selected for days 7-55 were the top most mature blades; for days 66-131 blades from the second leaf from the flag were selected.

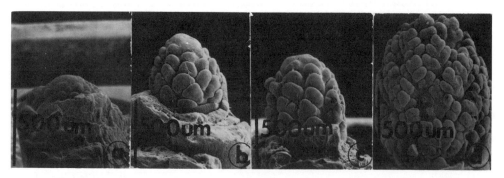

Figure 2. Scanning electron micrographs showing transition from vegetative to floral apical meristem in <u>S. bicolor</u> cv G499 GBR.
(a) 36 days, vegative meristem beginning transition to floral.
(b) 41 days and (c) 42 days. Intermediate stages of transition
(d) 46 days. Maturing floral apex.

Table 1. Distribution of label from feeds of radiolabeled precursors and sterols to mature leaves of S. bicolor cv G499 GBR with inflorescence just emerging.[a]

Substrate (Incubation Time)	Amount Fed[b]	Radioactivity in[c] 4,4-desmethylsterols	Radioactivity in HPLC Purified Products[d]		
			Cholesterol	24 ε-methylcholesterol[e]	Sitosterol	Stigmasterol

Substrate (Incubation Time)	Amount Fed[b]	Radioactivity in[c] 4,4-desmethylsterols	Cholesterol	24 ε-methylcholesterol[e]	Sitosterol	Stigmasterol
[2-³H] Acetate (48 hr)	1.31 x 10⁸ dpm 12 μg	80,562 dpm (0.06%)	3,309 dpm (0.003%)	3,324 dpm (0.003%)	10,551 dpm (0.008%)	315 dpm (0.0002%)
[2-¹⁴C] MVA (24 hr)	5.45 x 10⁷ dpm 71 μg	4.14 x 10⁵ dpm (0.8%)	8,055 dpm (0.015%)	14,256 dpm (0.026%)	21,162 dpm (0.038%)	1,584 dpm (0.003%)
[2-³H] Cycloartenol (72 hr)	0.92 x 10⁶ dpm 200 μg	2.71 x 10⁵ dpm (30.0%)	2,320 dpm[f] (0.25%)	1,872 dpm (0.20%)	3,071 dpm[f] (0.33%)	880 dpm (0.1%)
[2-³H] Lanosterol (72 hr)	3.16 x 10⁶ dpm 100 μg	2.09 x 10⁵ dpm (7.0%)	352 dpm (0.01%)	76 dpm (0.0002%)	151 dpm (0.005%)	26 dpm (-)
[24-³H] Lanosterol (72 hr)	2.98 x 10⁸ dpm 53 μg	7.74 x 10⁶ dpm (2.6%)	5,168 dpm[f] (0.002%)	678 dpm	187 dpm (0.00006%)	346 dpm (0.0001%)
[2-¹⁴C] Sitosterol (72 hr)	1.84 x 10⁷ dpm 55.7 μg	5.32 x 10⁶ dpm (29.0%)	0 dpm (0%)	0 dpm (0%)	3,113,111 dpm (17.0%)	29,622 dpm[f] (0.16%)
[4-¹⁴C] Cholesterol (72 hr)	2.78 x 10⁷ dpm 93 μg	1.31 x 10⁷ dpm (47.0%)	8,437,130 dpm (30%)	0 dpm (0%)	0 dpm (0%)	0 dpm (0%)

a. Radiolabeled substrates were applied to a spot 8 cm from the tip of the blade of the second leaf from the flag. Treated plants were incubated for 24, 48 or 72 hr. The leaf blades were ground and extracted with acetone to obtain a total lipid fraction (TLE).
b. Specific activities of the substrates were as follows: [2-³H] acetate 300 mCi/mMol; [2-¹⁴C] MVA 51 mCi/mMol; [2-³H] cycloartenol 0.9 mCi/mMol; [2-³H] lanosterol 6.0 mCi/mMol; [24-³H] lanosterol 1077 mCi/mMol; [2-¹⁴C] sitosterol 58 mCi/mMol; [4-¹⁴C] cholesterol 58.4 mCi/mMol.
c. 4,4-desmethylsterol fraction obtained from the TLE by TLC on SiO₃ gel with benzene:ether 9:1. The 4,4-desmethylsterol fractions were then purified by RP-TLC or AgNO₃-Anacil B cc prior to HPLC.
d. Individual sterols were obtained by HPLC of the purified and acetylated 4,4-desmethylsterol fractions. HPLC conditions: C18 column (Altex Ultrasphere ODS; 4.6 mm id x 25 cm); 100% MeOH as eluant (0.8 ml/min).
e. 24 ε-methylcholesterol ≡ 40:60 mixture of 24α-(campesterol) and 24 β-methylcholesterol (dihydrobrassicasterol). Because of the near background incorporation of [24-³H] lanosterol into the 24-alkyl sterols, it is likely that the label is associated with 24 α-methylcholesterol in each feed.
f. The indicated fractions were recrystallized to constant specific activity. In the [24-³H]lanosterol feed 15% of the radioactivity remained with cholesterol; in the [2-³H] cycloartenol feed 37% of the radioactivity remained with cholesterol, whereas only 2.5% remained with the sitosterol; in the [2-¹⁴C] sitosterol feed 43% of the radioactivity remained with stigmasterol.

dihydrobrassicasterol was not synthesized at this stage.[7] However, cholesterol (occuring at trace levels) was rapidly synthesized (same order of magnitude as PT[4]). PT may replace sterols in the leaf blade membranes (Figure 1) so that preformed sterol may be translocated to developing target organs and/or metabolized to other physiologically active compounds. Apparently, the shift of SO from sterol to PT production in leaves[1-4] and seeds[5] is periodic throughout the plant's development. The occurence and biosynthetic data[1,3] (Table 1) also indicates that there are separate post-cycloartenol (the first tetracyclic steroidal product of SO cyclization in sorghum) pathways to separate steroidal end products which are developmentally regulated. Additionally, the cyclization of SO to the individual PT may also be developmentally regulated.

References

1. R.C. Heupel and W.D. Nes, Evidence for differences in sterol biosynthesis and derivatization in sorghum, J. Nat. Prod. 47:292 (1984).
2. R.C. Heupel, Varietal similarities and differences in the polycyclic isopentenoid composition of sorghum, Phytochemistry 24:2929 (1985).
3. R.C. Heupel, Y. Sauvaire, P.H. Le, E.J. Parish and W.D. Nes, Sterol composition and biosynthesis in sorghum: Importance to developmental regulation, Lipids 21:69 (1986).
4. R.C. Heupel and W.D. Nes, The biosynthesis, metabolism and translocation of β-amyrin in Sorghum bicolor, Phytochemistry 24:2905 (1985).
5. M.A. Palmer and B.N. Bowden, Variations in sterol and triterpene contents of developing Sorghum bicolor grains, Phytochemistry 16:459-463 (1977).
6. E. Heftmann, Functions of steroids in plants, Phytochemistry 16:1925 (1977).
7. P. Benveniste, Sterol Biosynthesis, Ann. Rev. Plant Physiol. 37:275 (1986).

EFFECT OF INTRACELLULAR STEROL CONCENTRATION ON STEROL ESTERIFICATION IN YEAST

Leo W. Parks, Thomas A. Lewis, Christopher Low, and Kristen Haeckler

Department of Microbiology
North Carolina State University
Raleigh, NC 27695

Ergosterol (figure 1) was first isolated in 1889[1], and has been shown to be produced by 558 yeast cultures covering 60 species in 20 genera[2]. Yet, very little is known about the precise physiological role for this compound. Activities as a steroidal precursor, in the maintenance of optimal membrane fluidity, and as a modulator of enzymic activity have all been speculated upon as the principal function for sterols in yeast. Supporting experimental results have been obtained for each of these activities. It occurred to us that rather than a single function, sterols may in fact have multiple roles in the general economy of the cell. Such a conclusion was based on the results from many experiments which failed to provide a reasonable unifying model for sterol physiology. We elected to investigate the possibility of multiple roles for sterols in yeast, formulating our experiments on three assumptions. Firstly, if there were indeed multiple roles, it seemed highly unlikely that each of the functions would have the same quantitative requirement for sterol. Secondly, it would also be unlikely that the disparate roles would have identical structural affinities for different features of the sterol molecule. Lastly, it seemed most reasonable that ergosterol would be able to satisfy all of the sterolic qualitative and quantitative requirements for the various functions.

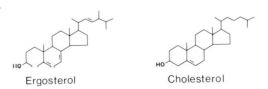

Figure 1: Sterols used in this study.

Table: Defined sterol functions in yeast

Sterol function	Required media concentration of sterol	Definition
Sparking	1 - 10 ng/ml	Ergosterol required for growth of cholestanol saturated cells.
Critical domain	100 ng/ml	Ergosterol required for growth of sterol-depleted cells cultured in the presence of lanosterol.
Domain	0.5 - 1.0 µg/ml	Miminal amount of sterol necessary for sustained growth of cells without other sterol sources. Cell conc.=<u>ca</u>.0.3 µg/mg d.w. cells
Bulk	15 µg/ml	Expandable pool of free sterol in absence of other sterol sources. Cell conc.=0.3 to 3.0 µg/mg d.w.

With these assumptions we prepared an auxotrophic organism with the appropriate genetic configuration for the experiments. Cholestanol was used as the simplest basic structure to feed to the organism as the starting point for ascertaining structural requirements for the sterols.

By systematically providing growing cultures of yeast with varying amounts of sterols with different structural features, we obtained evidence for at least four different functions for sterols in that organism[3,4]. These are described in the table. Each of these functions utilizes free sterol.

Ergosterol is known to occur in two forms in yeast, with the 3-OH group free and with it esterified to a long chain fatty acid, principally to C_{16} or C_{18} acids with the majority being unsaturated[5]. Our previous work indicated that a small, but virtually constant percentage of the total esterified sterol was present during active growth of the culture[6]. Under those conditions "minor" sterols or ergosterol precursors account for in excess of 90% of the ester fraction[7]. The levels of free sterols do vary with differing cultural conditions[8], but in all cases, a sharp increase in the rate of esterification occurs upon entry of the culture into the stationary growth phase[6]. Steryl ester formed during the late growth phases is principally ergosterol.

It has been speculated that steryl esters in yeast are simply superfluous and with no essential physiological function. Intuitively, this speculation is unsatisfying, since one would predict that cells lacking the esterification mechanism would have been positively selected in nature. Those cells without the capacity for synthesis of the esterifying enzyme and not needing the excess fatty acid requisite for ester synthesis would have been able to use that energy for essential functions and thereby gain a selective advantage. The conclusion that esterification is not essential in microbial systems was based on the observation that methyl ethers could replace ergosterol in the growth of a sterol auxotroph[9]. During the experiments there was degradation of the ether, however. Our calculations show that under aerobic conditions enough free sterol was produced from the ether to provide sufficient free sterol to satisfy the most essential functions of sterol in the cell. Additionally, we reasoned that if esterification was not essential to the cell, it should be relatively easy to obtain mutants that lacked the esterification process. Although we have made a substantial effort in that direction, to date no mutants of that type have been identified. This can not be considered conclusive, since it is possible our screening and selection procedures were not adequate to discern those specific mutants.

We investigated the ratio of free sterol to steryl ester in auxotrophic cells. Although the principal sterol of yeast is ergosterol, we have found that cholesterol can satisfy the sterol growth requirement of adapted yeast sterol auxotrophs[10] and is readily esterified. We chose to use cholesterol for these experiments because of its stability against oxidative degradation and availability as a high purity radiolabeled compound.

We had shown previously that the domain function of sterols in yeast represented, under standard growth conditions, a maintenance level of sterol[11]. Exogenously supplied sterol becomes growth limiting when the amount of proffered sterol is insufficient to allow approximately 0.3 µg cholesterol/mg dry weight of cells. In sterol limited cultures, the growth rate of the culture is constant, but the growth yield is directly proportional to the amount of sterol in the medium. Under these conditions there is virtually no steryl ester in the organisms[12].

Because it had been observed that auxotrophs could grow on only very small quantities of free sterol without apparent esterification, we reasoned that there might be a critical free sterol content of the membrane which the auxotroph must maintain. In those cells this could be accomplished at the expense of steryl esters or from exogenous free sterol. This would predict that a priority would exist for the maintenance of that free sterol pool. Under sterol limiting conditions, esterification would not occur. One means by which we could observe this prioritization of sterol distribution was to quantitate free sterol and steryl esters in cells grown under varying conditions with limiting to excess cholesterol concentrations. By plotting these specific quantities as a function of the total sterol content, we can determine if there is an observable change in the way sterol is incorporated into ester over the range of the cellular content. The experiment was done by adding sterol to sterol-depleted cells and either by following incorporation over a time period or by quantitating sterol in several batches of those cells after equal time exposure to different amounts of sterol. The results of both procedures are presented in figure 2, and show that the resulting patterns of sterol incorporation are identical. The inflection in the curves indicates a shift to enhanced esterification. The bulk function that we have described represents a quantitatively variable amount of free sterol and is found to range from 0.3 µg/mg cells to 3.0 µg/mg cells. The discontinuities occur at a free sterol content near 0.85 µg sterol per milligram dry weight cells. Before this critical sterol content is reached, sterol is incorporated almost entirely into the free fraction. Above that concentration, steryl ester accumulation exceeds free sterol accumulation.

When auxotrophic cells have been cultured to stationary phase in a medium containing an excess of sterols, a large pool of steryl ester is found in the cells. If those cells are used as an inoculum into a sterol-poor medium, the steryl ester is hydrolyzed to provide free sterol. Growth continues until the domain level of sterol in the cell is reached[12].

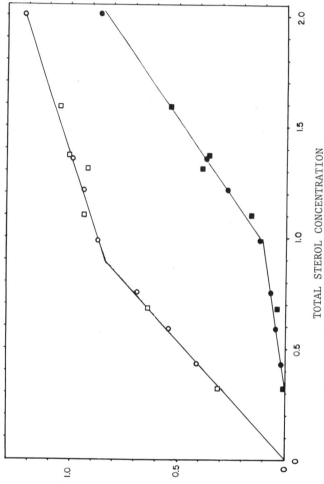

Figure 2: Relationship between free sterol concentration and steryl ester content of yeast cells cultured aerobically.

Yeast auxotrophic strain RD5R was grown to stationary phase on 1 μg/ml of ^{14}C-cholesterol^4. The cells were washed by centrifugation and varing amounts of additional cholesterol were added. Control cultures without additional sterol supplementation were collected and analyzed. Lipids were extracted, separated, and quantitated as previously described.[11] The open symbols represent free cholesterol and the closed symbols represent steryl ester. The circles represent samples that were obtained from cells given different amounts of cholesterol (0.25, 0.5, 1.0, 1.5, 2.0, 3.0, and 6.0 μg/ml) and collected after an equivalent amount of time. Squares represent data obtained from cells collected after different lengths of time (0.75 to 18 hours) after addition of 3.5 μg/ml cholesterol.

Under carefully specified conditions of sterol limitation with an auxotrophic strain of yeast it is possible to define cultural conditions wherein there is no steryl ester formation. These results could be interpreted to mean that steryl esters are indeed unimportant to the cell. We don't believe this is the case, however. The conditions that were required for these results are far from normal for a wild-type organism. However, it is important to note that endogenous amounts of free ergosterol in wild-type and sterol auxotrophic yeast under non-limiting growth conditions is ca. 2.5 µg/mg dry weight of cells. It should be recalled that this is very close to the saturating level of cholesterol in the bulk function that is described above.

Any model for the physiology of esterification must take into account: 1) esterification represents a sequestering of "excess" free sterol; 2) esterification increases dramatically during late exponential phase of culture growth; and 3) the limited esterification seen during exponential growth utilizes mostly non-alkylated ergosterol precursors.

ACKNOWLEDGMENTS

This work was supported in part by grants from the National Science Foundation (PCM 854394) and the National Institutes of Health of the U.S. Public Health Service (AM 37222). This is paper number 10611 of the Journal Series of the North Carolina Agricultural Research Service, Raleigh, NC 27695-7601

REFERENCES

1. C. Tauret, Sur un nuveau principe immediat de l'ergot de siegle, l'ergosterine, C. R. Seances Acad. Sci. 108:98 (1889).
2. E. L. Dulaney, E. O. Staply, and K. Simpf, Ergosterol production by yeasts, Appl. Microbiol. 2:371 (1954).
3. R. J. Rodriguez, F. R. Taylor, and L. W. Parks, A requirement for ergosterol to permit growth of yeast sterol auxotrophs on cholestanol. Biochem. Biophys. Res. Commun. 106:435-441 (1982).
4. R. J. Rodriguez, C. Low, C. D. K. Bottema, and L. W. Parks, Multiple functions for sterols in Saccharomyces cerevisiae. Biochim. Biophys. Acta 837:336 (1985).
5. L. W. Parks, Metabolism of sterols in yeast, CRC Crit. Revs. Microbiol. 6:301-341 (1978).
6. R. B. Bailey, Jr., and L. W. Parks, Yeast sterol esters and their relationship to the growth of yeast. J. Bacteriol. 124:606 (1975).
7. L. W. Parks, C. Anding, and G. Ourisson, Sterol transmethylation during aerobic adaptation of yeast. Europ. J. Biochem. 43:451 (1974).
8. P. B. Madyastha, and L. W. Parks, The effect of cultural conditions on the ergosterol ester components of yeast. Biochim. Biophys. Acta 176:858 (1969).
9. A. K. Lala, T. M. Buttke, and K. Bloch, On the role of the sterol hydroxyl group in membranes. J. Biol. Chem. 254:10582 (1979).
10. F. R. Taylor and L. W. Parks, Adaptation of Saccharomyces cerevisiae to grow on cholesterol: Selection for mutants defective in the formation of lanosterol. Biochem. Biophys. Res. Commun. 95:1437 (1980).
11. R. T. Lorenz, R. J. Rodriguez, T. A. Lewis, and L. W. Parks, Characteristics of sterol uptake in Saccharomyces cerevisiae. J. Bacteriol. 163:In press (1986).
12. T. A. Lewis, Sterol assimilation by Saccharomyces cerevisiae. Ph.D. Thesis, Oregon State University, 1986.

FUNCTIONAL ORGANIZATION OF CAROTENOIDS AND PRENYLQUINONES IN THE

PHOTOSYNTHETIC MEMBRANE

Hartmut K. Lichtenthaler

Botanisches Institut (Plant Physiology), Univ. of Karlsruhe
Kaiserstraße 12, D-7500 Karlsruhe, West-Germany

ABSTRACT
The three types of plant prenyllipids, the chlorophylls, carotenoids and prenylquinones, are integral components of a functional photosynthetic apparatus. The functional organization of these prenyllipids within the photosynthetic membrane is far from being well understood. Our present knowledge of the functional association and partition of individual carotenoids and prenylquinones with pigment proteins is reviewed. The importance of galacto- and phospholipids for a functional integration of pigments and prenylquinones in the photosynthetic membrane is underlined by results with the new grass-herbicide sethoxydim, which inhibits glycerolipid biosynthesis and blocks the accumulation - but not the biosynthesis - of pigments and prenylquinones.

INTRODUCTION

Chlorophylls and carotenoids, the photosynthetic prenylpigments and the prenylquinone derivatives plastoquinone-9, phylloquinone K1, α-tocoquinone and α-tocopherol form the group of functional chloroplast prenyllipids which are located in the thylakoids and are needed to perform the light reactions of photosynthesis (1,2). Many aspects of biosynthesis and functions of these plastidic prenyllipids had been reviewed at preceding plant lipid symposia. This report will therefore concentrate on some recent developments which might give new insights into function or metabolism of these thylakoid components. The three chapters deal with A) chlorophylls and carotenoids, B) prenylquinones, and C) the inhibition of prenyllipid accumulation and acyllipid biosynthesis in chloroplasts by the new grass herbicide sethoxydim.

A. CHLOROPHYLLS AND CAROTENOIDS

In algae and higher plants it is now well documented that chlorophylls and carotenoids are quantitatively bound within the thylakoids in various ways to different chlorophyll-carotenoid pigment-proteins. By polyacrylamide gel electrophoresis (PAGE) of digitonin-digested or sodium dodecylsulfate-treated thylakoid preparations one obtains 3 types of pigment proteins (3-7):

1. the P 700 - containing chlorophyll a/β-carotene proteins of photosystem I, CPI and CPIa; 2. the presumable reaction center of photosystem II CPa, which is a chlorophyll a/β-carotene protein, and 3. several forms of the light-harvesting chlorophyll a/b-lutein-neoxanthin proteins LHCPs (e.g. $LHCP_1$, $LHCP_2$, $LHCP_3$, and $LHCP_y$, see Fig. 1).

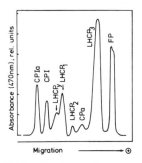

Fig. 1. Scan at 470 nm of the chlorophyll-carotenoid proteins of radish chloroplasts separated by the SDS-PAGE technique (after 6).

Table 1: Prenylpigment ratios (weight ratios) and percentage composition of carotenoids (weight %) in pigment-proteins isolated by SDS-PAGE from radish chloroplasts.

	Chloroplasts	CPIa	CPI	CPa	$LHCP_{1-3}$	FP
Pigment ratios:						
chlorophyll a/b	3.2	4-7	9-21	3-8	1.1 - 1.3	2.6
a+b/x+c	4	5-6	7- 9	3-5	3 - 5	3
a/c	12	12	9	4-8	60 - 180	14
% Carotenoid composition:						
β-carotene	30	56	70	75	1 - 2	24
lutein	45	24	15	13	56 - 65	37
neoxanthin	6	9	4	5	21 - 29	10
violaxanthin	11	5	6	4	4 - 5	20
others	8	6	5	3	6	9

β-Carotene is the main carotenoid component of the chlorophyll a-proteins of both photosystems of CPI + CPIa and of CPa. The xanthophylls lutein and neoxanthin as well as chlorophyll b are preferentially found together with the minor portion of chlorophyll a in the LHCPs. The pigment composition of these pigment proteins has been determined for Raphanus chloroplasts (Table 1). The carotenoid composition of the three LHCPs as lutein-neoxanthin-chlorophyll-protein varied little when the pigment proteins had been clearly separated. Occasionally the $LHCP_2$ band contained β-carotene up to 30% of the total carotenoid content. In these cases the ratio of chlorophyll a/β-carotene (a/c) exhibited low values of 8 to 9 instead of 60 to 180. Whether β-carotene is really bound to the $LHCP_2$ and sometimes remains there and sometimes not during electrophoresis or whether a second somewhat larger CPa aggregate sometimes moves with the $LHCP_2$ band remains open. A similar carotenoid distribution among the different pigment proteins has been found for other plants (see review 8), though there are minor differences which may be due to the different purity of a preparation and to varying LHCP content. The relative pigment content of a particular pigment protein may differ between normal plants and aurea mutants. In a dominant aurea mutant of tobacco (Su/su) we found a lower chlorophyll b content of the $LHCP_3$ than in normal green variety (su/su) (Fig. 2 and Table 2). The $LHCP_3$ of the Su/su mutant contains ß-carotene, which is present in the green form only in trace amounts (Table 2).

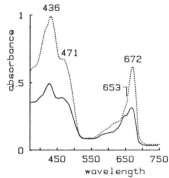

Fig. 2. Absorption spectra of the light-harvesting chlorophyll a/b-protein $LHCP_3$ isolated from a green tobacco (Su/su; ——). The $LHCP_3$ of the aurea mutant contains less chlorophyll b; the chlorophyll b shoulder at 653 nm is almost missing (·····).

Table 2: Pigment ratios and percentage composition of carotenoids (weight %) in the $LHCP_3$ pigment-proteins isolated by PAGE from the green tobacco form su/su and the dominant aurea mutant Su/su.

parameter	green form su/su	aurea form Su/su
Pigment ratios:		
chlorophyll a/b	1.4	4
a+b/x+c	9	5
a/c	41	11
% Carotenoids:		
ß-carotene	4	27
lutein	59	50
neoxanthin	33	15
violaxanthin	3	7
antheraxanthin and others	1	1

The xanthophyll violaxanthin is not enriched in one of the pigment proteins, it is mainly found in the free pigment fraction (Fig. 1 and Table 1). From this it appears that the apparently not protein-bound major portion of violaxanthin is identical with those 70 to 80% of violaxanthin, which can enzymatically be transformed into zeaxanthin at high light intensities (9).

Though chlorophyll a is found together with the larger proportion of chlorophyll b in the LHCPs (a/b-ratio 1.0 to 1.3), its main function seems to be in the chlorophyll a-proteins CPI + CPIa and CPa. The reaction center of photosystem II (RCII) contains chlorophyll a (the P 680) in addition to pheophytin a, which plays a role in the charge separation and electron transfer from P 680 (10,11). In earlier papers on chromatography of chlorophylls, various authors described a chlorophyll a'. The latter is a C 10-epimer of chlorophyll a and as a dimer it is identical with the reaction center chlorophyll P 700 of photosystem I, as was shown quite recently (12).

The pigment-protein band CPa is often seen as double or triplet band (13), this also applies to other pigment proteins. Thus the CPIa band has been separated into $CPIa_1$ and $CPIa_2$ in pea (14), and in spinach into 6 components (CPIa, b, c, d, e, f) (15) and in Chlamydomonas the $LHCP_1$ band into $LCHP_1$, 1a and 1b (16). In fact, much of the present research work on photosynthetic pigments concerns the isolation and characterization of more and more additional pigment proteins with particular chlorophyll a/b-ratios. Whether in all cases these are really new additional pigment proteins is questionable. It is known that mild digestion procedures allow the detection of a higher proportion of larger pigment protein aggregates (e.g. CPIa and $LHCP_1$), whereas at higher detergent concentrations and longer incubation times primarily the monomeric forms (e.g. CPI and $LHCP_3$) are obtained. By choosing intermediate conditions it is clear that one will detect transitional stages with a mobility between the larger aggregates and the monomeric forms. Furthermore, since the disaggregation of the larger pigment proteins proceeds continously during gel-electrophoresis, one may find small bands of monomeric forms (e.g. $LHCP_3$) which have not yet reached the position of the main $LHCP_3$ pool, and may wrongly indicate an additional pigment protein.

B. PRENYLQUINONES

In contrast to photosynthetic bacteria the photosynthetic apparatus of algae and higher plants comprises a uniform prenylquinone composition (Table 3) with plastoquinone-9 + its hydroquinone (PQ-9 + PQ-9.H_2) as the major component and smaller amounts of phylloquinone (vitamin K_1) and α-tocoquinone (α-TQ) (1,2). All three prenylquinones are potential photosynthetic electron carriers. The role of α-tocopherol, the cyclic form of the reduced α-tocoquinone, seems to be primarily that of a lipid antioxidant. In older, in senescent or chromoplast-bearing tissue other plastoquinone-forms (plastoquinone B and C) may show up, or in etiolated tissue there are biosynthetic precursors of K_1 and α-TQ (for literature see 2,17,18), but these minor quinone components do not play any role in photosynthesis.

According to present knowledge minor portions of plastoquinone-9 seem to function as the primary (Q_A) and secondary quencher (Q_B) at photosystem II. This would be analogous to the photosynthetic bacterium Rhodopseudomonas sphaeroides where a ubiquinone, the only prenylquinone present, takes over the function of Q_A and of Q_B (19,20). Other photosynthetic bacteria contain menaquinones (20) which may function as Q_A and as Q_B (Table 3). So far there is no experimental evidence that Q_B in higher plants is a plastoquinone-9. Therefore, the Q_B of higher plants could just be the phylloquinone K_1 or the α-tocoquinone as well. From a physiological point of view it would make sense if the electron (+ proton) flow into the large plastoquinone pool were controlled by another prenylquinone, which were not a plastoquinone (21).

Table 3: Prenylquinones in the photosynthetic membranes of bacteria and higher plants. UQ-10 = ubiquinone-10; MKn = menaquinone homologues (vitamin K_2); PQ-9 = plastoquinone-9; K_1 = phylloquinone (vitamin K_1); α-tocopherylquinone (= α-tocoquinone) = α-TQ.

Organism	Quinone	Function
Rhodospseudomonas sphaeroides (1 photosystem)	UQ-10	QA, QB
many other photosynthetic bacteria (1 photosystem)	MKn	Q_A, Q_B-type? cyclic electron flow?
green algae and higher plants (2 photosystems)	PQ-9	Q_A, Q_B ?, PQ-pool
	α-TQ	Q_C, Q_Z (Q-cycle) ?
	K_1	photosystem I, cyclic electron flow ?

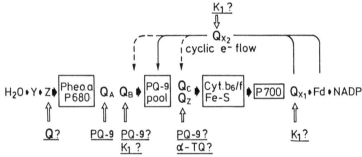

Fig. 3. Scheme of the photosynthetic electron transport chain of higher plants based on the Q-cycle concept with the quinones Q_A, Q_B, Q_C, Q_Z, Q_{x1} and Q_{x2}. Possibilities for the participation of the endogenous prenylquinones (PQ-9, α-TQ, K_1) in the electron transport chain are indicated.

The same applies to the emptying of the large plastoquinone pool (electron transfer and release of protons) via cytochrom b_6/f complex (including the Rieske center Fe-S). The introduction of the Q-cycles concept with the additional components Q_C and Q_Z (20,21) opens up new potential functional positions for the phylloquinone and α-tocoquinone, which are indicated in Figure 3. The possible participation of prenylquinones in photosystem I directly after P 700 is marked as position Q_{x1} and as Q_{x2}. Whether the cyclic electron flow reenters the electron transport chain directly at the PQ-pool or at the position of Q_C/Q_Z is not known. Our finding that the larger photosystem I aggregate isolated from Raphanus chloroplasts, the CPIa, contain 4 to 5 plastoquinone per 100 molecules chlorophyll a, points to a participation of plastoquinone in the cyclic electron transfer. The possibility that Z may be a prenylquinone (Q?) is also discussed (van Gorkom, personal communication). The many possibilities of quinone functions (Figure 3) have to be verified or excluded in further research with pigment proteins and subthylakoid preparations.

Table 4: Molar ratios of carotenoids (x/c) and prenylquinones in cotyledons of green, 12 d old radish seedlings, of etiolated plants and of greening radish cotyledons. c = ß-carotene; x = xanthophylls; K_1 = phylloquinone (vitamin K_1); PQ-9 = plastoquinone-9; α-TQ = α-tocoquinone.

Plants		x/c	PQ-9/K_1	PQ-9/ α-TQ
green plants	(12 d)	2.8	6-10*	5-9*
etiolated plants	(10 d)	16	25	17
greening plants		↓	↓	↓
10 d etiolated + 6 h light		8	8	7
10 d etiolated + 12 h light		5	5	6
24 h light		2.5	4	4
36 h light		2.6	6	5

* The ratios depend on the growth conditions and can vary within these values.

Excess amounts of plastoquinone-9 and of α-tocopherol are stored outside the thylakoids in the osmiophilic plastoglobuli (2,23). The latter contain K_1 and α-TQ only in trace amounts. In fact, the prenylquinones K_1 and α-TQ are minor but essential functional constituents of the photosynthetic membrane with a concentration similar to that of cytochroms and ferredoxin (24). In etiolated leaf tissue, they are present, similar to ß-carotene, only in trace amounts. During the light-induced thylakoid formation they are rapidly accumulated together with ß-carotene, as indicated by the decreasing values of the ratios of PQ-9 to K_1 and to α-TQ (Table 4); the functional ratios with values of 5 to 6 for plastoglobuli-free thylakoids are reached after 36 h of illumination.

It had long been shown that phylloquinone K_1 is enriched in photosystem I preparations obtained by digitonin treatment of chloroplasts (24,25), whereas PQ-9 is preferentially associated with the heavy particle fraction of photosystem II (24). α-TQ and α-tocopherol appear to be evenly distributed between the two photosystems (24). The larger portion of K_1 of chloroplasts is specifically bound to the CPI (+ CPIa) pigment proteins of photosystem I (25) and cannot be removed without extracting ß-carotene and chlorophyll a. However, in light harvesting pigment proteins we did not find plastoquinone, K_1 or α-TQ. A CPa preparation contained PQ-9 and low but distinct amounts of K_1 (Table 5). For PQ-9 similar results were found in a reaction center II preparation (26 and Table 5). The level of PQ-9 as well as K_1 and α-TQ in the CPa fraction or RCII preparation is much lower than in the photosystem II particle preparations or in whole thylakoids. This indicates that the major part of PQ-9, α-TQ or α-T is not specifically bound to particular pigment proteins of the thylakoids.

Table 5: Concentrations of chloroplast prenylquinones and β-carotene (moles per 100 of chlorophyll a) in plastoglobuli-free spinach thylakoid preparations, in photosystem II particles (= LHCPs + CPa), in a CPa preparation and in the reaction center II (RC II).

Compound	Thylakoids*	Photosystem II* particles (LHCP+CPa)	CPa*	RCII**
plastoquinone-9	5	6-8	3	3.7
phylloquinone K_1	0.8	0.2	0.3	< 0.1
α-tocoquinone	0.5	0.4	< 0.1	< 0.1
α-tocopherol	2.5	2	< 0.1	< 0.05
β-carotene	17	10	25-30	20(x+c)

* results of Lichtenthaler; ** of Omata et al. (26); (x+c = total carotenoids)

C. EFFECT OF SETHOXYDIM ON PRENYLLIPIDS AND ACYLLIPIDS OF CHLOROPLASTS.

The herbicide sethoxydim, a cyclohexane-1,3-dione derivative (Fig. 4) is applied for selective postemergence control of annual and perennial grass weeds in many broad leaf crops. Sethoxydim is absorbed by the leaves and

Fig. 4. Chemical structure of the herbicide sethoxydim.

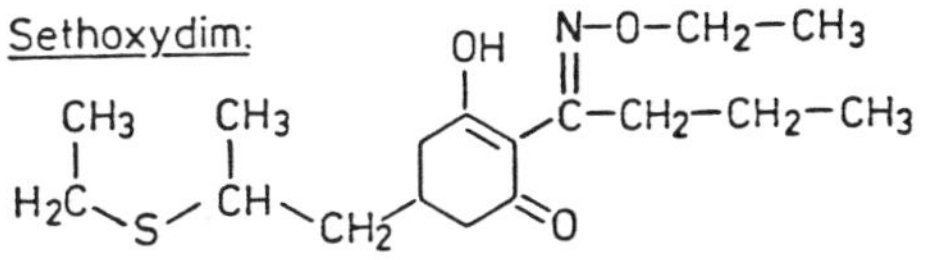

transported to the meristematic zones of the leaf, shoots and roots, where it blocks any further growth. Not only grass weeds but also maize, wheat and barley belong to the sethoxydim sensitive plants. Sethoxydim also inhibits chloroplast development (thylakoid formation and multiplication) at all developmental stages (27,28) and consequently blocks the accumulation of chlorophylls and carotenoids (Fig. 5). At sublethal dosis (spraying with ca. 16 to 32 g of a.i. per ha) leaf growth still proceeds, whereas pigment formation and chloroplast differentiation are blocked, which leads to the formation of white, pigment-free leaf parts. From these results we concluded that a major mode of action of sethoxydim might be the inhibition of chlorophyll and carotenoid biosynthesis. This is, however, not the case as was shown by a labelling experiment with ^{14}C-mevalonate.

Prenyllipid biosynthesis under the influence of sethoxydim: Most of the ^{14}C-label was found in the sterol fraction, whereby the highest mol-specific radioactivity was detected in the 4-dimethyl and 4-methyl sterols, which are known as biosynthetic precursors of the desmethyl sterols (Table 6). High label was also found in the mitochondrial ubiquinone, which in maize leaves consists of the two homologues Q-9 and Q-10 (29). Much less radioactivity is found in the chloroplast prenyllipids (chlorophylls, carotenoids, prenylquinones) as is expected and known, since exogeneously applied mevalonic

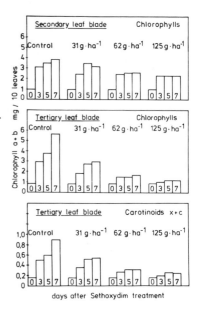

Fig. 5. Inhibition by sethoxydim of chlorophyll and carotenoid accumulation in secondary and tertiary leaves of maize seedlings (Zea mays L. var. Protador). The still growing tertiary leaves are more affected than the nearly fully developed secondary leaves. The plants (3 leaf stage) were sprayed with sethoxydim solutions corresponding to an outdoor treatment with 31, 62 and 125 g per ha. The numbers in the boxes indicate the days after treatment. Mean of 3 repetitions.

Table 6. Specific radioactivity (Bq/µMol) of prenyllipids after application of $2\text{-}^{14}C$-mevalonic acid to green excised maize shoots* from 10 day old plants with and without application of sethoxydim.

	Controls	Sethoxydim**
1. Chloroplast:		
Chlorophylls:		
chlorophyll a	8.3×10^3	32×10^3
chlorophyll b	1.7×10^3	5×10^3
Carotenoids:		
ß-carotene	13×10^3	27×10^3
lutein	3.3×10^3	8×10^3
neoxanthin	3.3×10^3	5×10^3
Prenylquinones:		
plastoquinone-9	10×10^3	13×10^3
plastohydroquinone	12×10^3	13×10^3
phylloquinone	7×10^3	10×10^3
α-tocopherol	0.3×10^3	0.7×10^3
Galactolipids:		
monogalactosyldiglyceride	< 10	< 5
digalactosyldiglyceride	< 10	< 5
2. Mitochondria:		
ubiquinone-9/10	6×10^4	11×10^4
3. Cytosol:		
4-dimethyl sterols	15×10^4	26×10^4
4-methyl sterols	24×10^4	23×10^4
desmethyl sterols	2.3×10^4	3.2×10^4

* the shoots were placed with the cut ends in a solution of 250 µCi $2\text{-}^{14}C$-mevalonate for 20 hours at white light of medium high intensity (15 klux, 40 W · m^{-2} PAR).
** The plants were sprayed with sethoxydim (equivalent to a field spraying of 32 g a.i. per ha) two days prior to the labelling experiment.

acid does not readily enter the chloroplast organelle. There are still some doubts whether chloroplasts synthesize their own isopentenylpyrophosphate (IPP) or import it from the cytoplasm (30,31), yet it is clear that ^{14}C-mevalonic acid can be applied to label the chloroplast prenyllipids. The higher ^{14}C-labelling of biosynthetic precursor compounds (e.g. 4-dimethyl sterols > desmethyl sterols; chlorophyll a > chlorophyll b; ß-carotene > neoxanthin) as well as the low proportion of the radioactivity incorporated into the galactolipids (MGDG, DGDG) indicate that the applied ^{14}C-mevalonate was mainly used for prenyllipid synthesis and only little metabolized.

In sethoxydim treated plants there are no differences in principle in the labelling pattern of the prenyllipids as compared to the controls (Table 6), though the specific labelling is generally higher. This indicates that the biosynthesis of the different prenyllipid classes is not affected by sethoxydim. It blocks the accumulation of chlorophylls and carotenoids, however.

Inhibition by sethoxydim of acyllipid formation: In our search for the mode of action of sethoxydim we investigated the acyllipid formation. Sethoxydim blocks the accumulation of the chloroplast glycolipids MGDG, DGDG and SQDG and of all phospholipids, including those, which are not present in chloroplasts (32). The inhibition of acyllipid accumulation is more drastic than that of the protein increase (33). This points to a possible inhibition by sethoxydim of accyllipid biosynthesis. In fact, lipid biosynthesis is blocked by sethoxydim as is shown by feeding ^{14}C-acetate to excised maize shoots (33,34).

Inhibition of fatty acid biosynthesis in chloroplasts: The de novo synthesis of fatty acids proceeds in the chloroplasts (plastids) (35). The biosynthesis of total fatty acids in isolated maize chloroplasts is inhibited by sethoxydim (Table 7). The I_{50}-value is lower than 10^{-5} molar. Thus in grass weeds the major biochemical target appears to be the fatty acid synthetase. Sethoxydim acts in a similar way to the structurally different herbicide diclofopmethyl (36). The block of chlorophyll and carotenoid

Table 7: Inhibition by sethoxydim of fatty acid biosynthesis from 2-^{14}C-acetate in isolated maize chloroplasts* (after 33).

condition	radioactivity (Bq per mg chlorophyll)	% inhibition
Control	3020	0
+ Sethoxydim:		
1 µM	606	79.9
10 µM	138	95.4
100 µM	26	99.2

* The incorporation of radio label was determined in freshly isolated intact chloroplasts during a 20 min incubation period in the light. ^{14}C-acetate and sethoxydim were simultaneously added to the incubation medium, which contained several cofactors e.g. ATP, NADPH (for details see 33).

accumulation and of chloroplast biogenesis by sethoxydim is therefore a consequence of the inhibition of fatty acid and acyllipid biosynthesis. Without newly formed acyllipids there is no differentiation and growth in the sethoxydim treated plants. Young leaves and meristematic tissue are more affected than older leaf tissue. This is to be expected, since meristematic cells are completely dependent on a de novo fatty acid biosynthesis, whereas older cells may utilise fatty acids from possible lipid stores in the cytoplasm. The results of the inhibition by sethoxydim of the pigment accumulation underline the importance of the chloroplast acyllipids for the functional integration of newly synthesized chloroplast pigments (and pigment proteins) in the photosynthetic membrane.

Acknowledgements

I wish to thank Dr. G. Retzlaff, BASF AG., D-6703 Limburgerhof, for providing sethoxydim, Mr. Guido Schmuck, Karlsruhe for the data on the tobacco mutant (Table 2, Fig. 2), Mrss. Ursula Prenzel and Inge Schürer, Karlsruhe, for excellent assistance and Mr. Philip Jackson, Karlsruhe, for checking the English text.

References

1. T.W. Goodwin, in "Lipids and Lipid Polymers in Higher Plants", M. Tevini and H.K. Lichtenthaler, eds., p. 29-47, Springer Verlag, Berlin 1977.
2. H.K. Lichtenthaler, in "Lipids and Lipid Polymers in Higher Plants", M. Tevini and H.K. Lichtenthaler, eds., p. 231-258, Springer Verlag, Berlin 1977.
3. J.P. Thornber, Annu. Rev. Plant Physiol. 26, 127-150 (1975)
4. J.M. Anderson, J.C. Waldron and S.W. Thorne, FEBS Letters 92, 227-233 (1978).
5. O. Machold, D.J. Simpson and B.L. Müller, Carlsberg Res. Commune. 44, 235-254 (1979).
6. H.K. Lichtenthaler, U. Prenzel and G. Kuhn, Z. Naturforsch. 37c, 10-12 (1982).
7. D. Siefermann-Harms and N. Ninnemann, FEBS Letters 104, 71-74 (1979).
8. D. Siefermann-Harms, Biochim. Biophys. Acta 811, 325-355 (1985).
9. D. Siefermann-Harms, in "Lipids and Lipid Polymers in Higher Plants", M. Tevini and H.K. Lichtenthaler eds., p. 218-230, Springer Verlag, Berlin 1977.
10. V.V. Klinov, A.V. Kleranik, U.A. Shuvalor and A.A. Krassnovsky, FEBS Letters 82, 183-186 (1977).
11. T. Omata, N. Murata and K. Satoh, Biochim. Biophys. Acta 765, 403 (1984).
12. T. Watanabe, M. Kobayashi, A. Hongu, M. Nakazato, T. Hiyama and N. Murata, FEBS Letters 191, 252-256 (1985).
13. K.J. Leto, Biochim. Biophys. Acta 766, 98-108 (1984).
14. T.-Y. Leong and J.M. Anderson, Photosynthesis Res. 5, 105-115 (1984).
15. K.S. Burton, W.S. Chow and B.R. Jordan, Physiol. Veg. 22, 793-800 (1984).

16. E. Lam, W. Ortiz, W. and R. Malkin, FEBS Letters 168, 10-14 (1984).
17. H.K. Lichtenthaler, In "Advances in the Biochemistry and Physiology of Plant Lipids", L.-A. Appelqvist and C. Liljenberg eds., pp. 57-58, Elsevier Biomedical Press, Amsterdam 1979.
18. H.K. Lichtenthaler, In "Handbook of Chromatography: Lipids and Technical Lipid Derivatives", Vol. II, H.K. Mangold ed., pp. 115-169, CRS Press, Boca Raton 1984.
19. B.A. Diner, C.C. Schenk and C. De Vitry, Biochim. Biophys. Acta 766, 9-20 (1984).
20. C.A. Wraight, In Photosynthesis Vol. 1 (Govindjee ed.), pp. 18-61, Academic Press, New York 1982.
21. H.K. Lichtenthaler, In Advances in Biogenesis and Function of Plant Lipids (P. Mazliak, P. Benveniste, C. Costes and R. Douce eds.) pp. 299-310, Elsevier, Amsterdam 1980.
22. A. Trebst, Rheinisch-Westfäl. Akademie der Wissenschaften Vortrag Nr. 343, 35-61 (1986).
23. H.K. Lichtenthaler and B. Sprey, Z. Naturforsch. 21b, 690-697 (1966).
24. H.K. Lichtenthaler, In "Progress of Photosynthesis Research", H. Metzner ed., Vol. 1, pp. 304-314 (1969).
25. E. Interschick-Niebler and H.K. Lichtenthaler, Z. Naturforsch. 36c, 276-283 (1981).
26. T. Omata, N. Murata and K. Satoh, Biochim. Biophys. Acta 765, 403-405 (1984).
27. H.K. Lichtenthaler, Z. Naturforsch. 39c, 492-499 (1984).
28. H.K. Lichtenthaler and D. Meier, Z. Naturforsch. 39c, 115-122 (1984).
29. S. Schindler and H.K. Lichtenthaler, In: Structure, Function and Metabolism of Plant Lipids, P.-A. Siegenthaler and W. Eichenberger eds., pp. 273-276, Elsevier, Amsterdam 1984.
30. K. Kreuz and H. Kleinig, Eur. J. Biochem. 141, 531-535 (1984).
31. T.J. Bach, Lipids 21, 82-88 (1986).
32. R.J. Burgstahler and H.K. Lichtenthaler, In "Structure, Function and Metabolism of Plant Lipids", P.-A. Siegenthaler and W. Eichenberger eds., pp. 619-622, Elsevier, Amsterdam 1984.
33. R.J. Burgstahler, Ph.D.-Thesis, University of Karlsruhe, in Karlsruhe Contributions to Plant Physiol. 13, 1-111 (1985), (ISSN 0173-3133).
34. R.J. Burgstahler, G. Retzlaff and H.K. Lichtenthaler, Pesticide Biochem. Physiol. (in preparation) 1986.
35. R. Lessire and P. Stumpf, Plant Physiol. 73, 614-618 (1982).
36. H.H. Hoppe, Pesticide Biochem. Physiol. 23, 297-308 (1985).

ARACHIDONIC AND EICOSAPENTAENOIC ACIDS, GLUCANS AND CALCIUM AS REGULATORS OF RESISTANCE TO A PLANT DISEASE

Michael Zook and Joseph Kuć

Department of Plant Pathology
University of Kentucky
Lexington, KY 40546

INTRODUCTION

A common response of plants, often associated with disease resistance, is the localized and rapid death of plant cells at the infection site. This reaction, referred to as a hypersensitive response (HR), includes the accumulation of phytoalexins (low molecular weight compounds which inhibit microbial development). In some plant-parasite interactions, different races of a pathogen can be distinguished by the type of response they elicit in different cultivars. Incompatible races of Phytophthora infestans elicit the accumulation of sesquiterpenoid phytoalexins, principally rishitin and lubimin, in tubers of resistant (R-gene) cultivars faster to higher levels following infection than do compatible races on susceptible cultivars.[1,2] Bostock et al.[3] reported that arachidonic acid (AA) and eicosapentaenoic acid (EPA) isolated from mycelium of P. infestans elicited the accumulation of rishitin and lubimin in potato discs. However, incompatible and compatible races of P. infestans do not differ in the quantity of these fatty acids. Thus, AA and EPA per se are not the determinants of race specificity in the potato-P. infestans interaction, but they constitute a major component of the "native" fungal elicitor; i.e., the part of the fungus which is "recognized" by the plant and is responsible for eliciting HR and phytoalexin accumulation. It is unclear how contact between an elicitor, such as AA, and potato tuber cells elicits these defense mechanisms.

Tomiyama[4] observed that potato cells died very rapidly after infection by incompatible races of P. infestans. In subsequent experiments,[5] ten percent of the cells of potato leaves and petioles were observed to die in an incompatible interaction within the first few minutes of penetration; most of the host cells at the infection site were dead after 40 to 60 min. In compatible interactions, a coexistence of pathogen and host was maintained for three or four days without either the appearance of injury to the host or an appreciable accumulation of phytoalexins.[6] Protoplasts prepared from tuber tissue or from callus tissues derived from tuber tissues reacted hypersensitively to hyphal wall components (HWC) from P. infestans.[7] Inhibition of this hypersensitive response by a short treatment of the protoplasts with a high molecular weight SH-binding reagent, dextran-bound p-chloromercuri benzoic acid, suggested that components with SH groups in the plasma membrane of potato tuber cells are associated with HR elicited by HWC.[8]

Table 1. Sesquiterpene Accumulation in Potato Discs Elicited by 20-Carbon Fatty Acids with Various Number and Position of Double Bonds

Double bond position	Concentration (mM)	Rishitin + lubimin		
		H_2O (μg/GFW)	Glucans (μg/GFW)	Specific activity[a]
5,8,11,14,17	3.3	190	---[c]	---
5,8,11,14	3.3	195	450	0.91
5,11,14,17	3.3	18	---	---
5,8,11	3.2	98	---	---
5,11,14	3.0	12	141	0.31
8,11,14	2.3	9	48	0.14
5,8,14	2.6	18	171	0.44
8,11	3.0	3	108	0.24
11,14	2.9	tr[b]	tr	---
5,11	3.9	3	54	0.09
8	2.8	tr	24	0.06
11	5.0	tr	tr	---
H_2O		tr	tr	---

Note. Discs (2.1 cm) were treated with glucans (50 μl, 1.0 mg-equivalent glucose/ml) 6 h after slicing, followed 18 h later by fatty acid (50 μl). Sesquiterpene determination was made 96 h after treatment with fatty acid.

[a] Specific activity, μg (rishitin + lubimin)/nmol fatty acid in the presence of glucans.
[b] tr, less than 0.1 μg (rishitin + lubimin) per gram fresh weight (GFW).
[c] Not determined.

Inoculation with a compatible race of P. infestans has been reported to inhibit the accumulation of sesquiterpenoid phytoalexins in response to a second inoculation with an incompatible race.[4,9] Water soluble glucans containing 17-23 glucose residues with predominately β 1-3 linkages and minor amounts of β 1-6 branches[10,11] have been obtained from mycelial extracts of P. infestans,[10,11] zoospores[10,13] and germination fluids of cystospores.[14] These glucan preparations suppressed HR and phytoalexin accumulation in response to inoculation with incompatible races in a race-specific manner. Doke and Tomiyama[7,15] published evidence, developed from work with nine resistance genotypes of potato and seven P. infestans races that supports the idea of race-specific suppression of the hypersensitive response.

The structural requirements in the fatty acid moiety of the native elicitor necessary for elicitation have been investigated.[16] Saturated and unsaturated 16- and 18-carbon fatty acids were inactive. A 20-carbon chain and free carboxyl group were necessary for activity. Of the five 20-carbon acids tested, those with a minimum double bond configuration of Δ 5,8,11 were by far the most active (Table 1).

Derivitives of AA generated by lipoxygenase are a common component of cellular activation in animal cells, and potatoes are rich in lipoxygenases which are specific for the double bond in the Δ5 position of AA.[17] These observations suggest that lipoxygenases may play a role in AA-elicited phytoalexin accumulation. However, unlike the situation in animal systems in which lipoxygenase-generated AA metabolites are produced in response to widely different external stimuli by many different cell types, the role of lipoxygenase in elicitation of HR and phytoalexin accumulation in potato is likely to be specific to AA or EPA elicitation since neither of these fatty acids occur naturally in potato. Arachidonic acid is rapidly metabolized upon application to potato tuber discs;[18] however, 64% of the AA was recoverable in phosphatidyl ethanol and phosphatidyl choline 5 h after treatment. Though the specific activity of AA-containing phosphatidyl choline is 2-3 fold that of AA, there is no conclusive evidence that the lipoxygenase products of AA possess greater elicitor activity than AA itself. Instead of the specificity of lipoxygenase enzymes accounting for the structural specificity of the fatty acid moiety of the elicitor from P. infestans (Table 1), it is possible that the potato possesses a mechanism for "recognizing" the elicitor; e.g., the elicitor may bind to a HR-triggering "receptor".

Glucans from P. infestans, inactive themselves as elicitors of sesquiterpenoid phytoalexin accumulation in potato tuber discs, enhanced the eliciting activity of unsaturated twenty-carbon fatty acids that had little or no elicitor activity in the absence of glucans[16] (Table 1). Application of these promoter glucans and 33 μmoles of AA per disc resulted in appreciable accumulation of rishitin and lubimin. The threshold of elicitor activity for AA alone was 3.3 pmoles per disc. The same fraction of partially-purified glucans from a compatible race of P. infestans suppressed rishitin and lubimin accumulation in response to infection by an incompatible race of P. infestans and enhanced the elicitor activity of AA.[19] Promoter and suppressor glucans from P. infestans may, however, be structurally different. Small differences in the carbohydrate structure of cell wall fragments from P. megasperma f. sp. glycinea have been demonstrated to strongly affect elicitor activity.[20]

A ROLE FOR CALCIUM IN THE ELICITATION OF PHYTOALEXIN ACCUMULATION IN POTATO[21]

It is now well established that the binding of specific external messengers to certain membrane proteins, known as "receptors", is linked to the production of "secondary messengers" which translate signals generated at the cell surface into a stimulation of the intracellular machinery which is responsible for cellular response. In the course of plant and animal evolution, it appears that certain aspects of cellular activation have been conserved; cells of both kingdoms have adapted activation mechanisms based on the concentration of calcium ions in in the cytosol. An integral part of calcium messenger systems are the intracellular calcium receptor proteins.[22] The most ubiquitous of these proteins is calmodulin, which has been characterized in both animals and plants.[23]

Application of Ca^{2+} (Fig. 1) or Sr^{2+} directly onto AA-treated potato tuber discs enhanced the accumulation of rishitin. Neither ion increased lubimin accumulation. The other ions tested, Mg^{2+}, Na^+, and K^+, were ineffective in enhancing the accumulation of rishitin. Poly-L-lysine, another elicitor of known structure, primarily elicited the accumulation of rishitin in potato tuber discs. Suprisingly, we discovered that Ca^{2+} and Sr^{2+} enhanced the accumulation of lubimin more than rishitin in potato tuber discs treated with poly-L-lysine. Sodium and potassium ions had no significant effect, whereas Mg^{2+} markedly inhibited the accumulation of rishitin.

In order to determine at what point in the elicitation process Ca^{2+} was active, potato tuber discs were treated with AA and then supplemented at later times with Ca^{2+}. Rishitin accumulation was enhanced only if Ca^{2+} was added within 24 h of the application of AA. Strontium had a similar effect on AA-elicited rishitin accumulation. Likewise, Ca^{2+} and Sr^{2+} had to be applied early after application of poly-L-lysine (within 6 h) to be effective in enhancing rishitin and lubimin accumulation.

Potato tuber discs were treated with AA in the presence of varying amounts of EGTA, a Ca^{2+}-specific ligand, in order to determine if Ca^{2+} is obligatory for the accumulation of rishitin and lubimin following elicitation. Inhibition of rishitin and lubimin accumulation by EGTA was concentration-dependent, and reached 90% at 5 µmoles EGTA/potato disc in the absence of added Ca^{2+}. The inhibition was overcome by the addition of excess Ca^{2+}.

In order to determine if the entry of Ca^{2+} into the potato cell is required for the accumulation of rishitin and lubimin, potato tuber discs were treated with La^{2+}, an inhibitor of calcium transport. The accumulation of rishitin was inhibited by La^{3+}, and it was reversed in a concentration-dependent manner by the addition of Ca^{2+}. Calcium was more effective at reversing the inhibition caused by 0.14 µmoles of La^{3+} than that caused by 0.20 µmoles of La^{3+}. Another observation which suggests

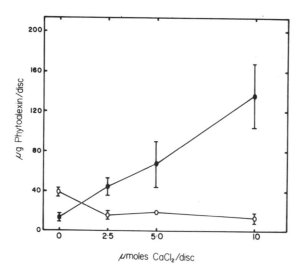

Fig. 1. Effect of $CaCl_2$ concentrations on arachidonic acid-elicited rishitin (-●-) and lubimin (-○-) accumulation. The salt was applied simultaneously with 10 µg of arachidonic acid in a total volume of 50 µl/potato disc. Discs were processed for determination of rishitin and lubimin accumulation 72 h after application of the treatments.

that La^{3+} is blocking the entry of Ca^{2+} into the potato cell and thus affecting rishitin accumulation is illustrated in Fig. 2. Potato tuber discs were first treated with AA in the presence of La^{3+}. Subsequently, different quantities of Ca^{2+} were added at either the time of elicitation or 12 h later. The Ca^{2+} was much more effective reversing La^{3+} inhibition of AA-elicited rishitin accumulation when added with La^{3+} than when added after 12 h. Lubimin accumulation elicited by AA was also inhibited by La^{3+} and the inhibition was reversed by Ca^{2+}. Lanthanum was only effective in inhibiting rishitin and lubimin accumulation when applied within 3 h of the application of AA. The greatest inhibition of phytoalexin accumulation occurred when La^{3+} was applied simultaneously with AA to potato tuber discs.

In light of the role of Ca^{2+} in transduction of external stimulus to cellular response in a wide variety of animal and plant cells,[24,25] it was reasonable to investigate whether Ca^{2+} is involved in the elicitation of phytoalexin accumulation. The stimulatory effects of Ca^{2+} and the inhibitory effects of Ca^{2+} antagonists on rishitin and lubimin accumulation suggest that Ca^{2+} mobilization is an essential part of the elicitation of phytoalexin accumulation in potato tuber tissue.

Strontium is the most effective element in substituting for calcium in calcium-dependent processes in biological systems.[26] Strontium can induce the conformational changes in calmodulin necessary for stimulation of bovine cyclic nucleotide phosphodiesterase and binding to the calmodulin antagonist trifluoperizine.

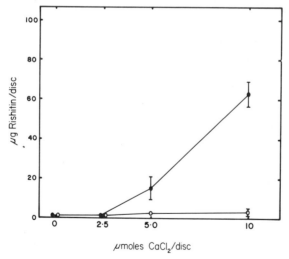

Fig. 2. A comparison of the effects of simultaneous application of $CaCl_2$ and 0.28 μmoles of $LaCl_3$/potato disc (-●-) with the application of $CaCl_2$ 12 h after addition of 0.28 μmoles of $LaCl_3$ (-O-) on the accumulation of rishitin elicited by 10 μg of arachidonic acid. Rishitin determined 72 h after application of arachidonic acid.

Rishitin and lubimin are highly reduced secondary metabolites. Their biosynthesis via the acetate-mevalonate pathway requires substrate in the form of acetyl CoA as well as a large expenditure of energy. The calcium ion regulates the activation of pyruvate dehydrogenase[27] and other enzymes in the TCA cycle[28] in animal cells. Thus, part of the calcium ion's function in regulating the activation of rishitin and lubimin biosynthesis may be to increase the amount of acetyl CoA that is available for the synthetic pathway. However, Ca^{2+} and Sr^{2+} are most effective in enhancing rishitin and lubimin accumulation when applied early after the application of AA or poly-L-lysine. These results suggest that Ca^{2+} mediates some early event in the elicitation of phytoalexin accumulation rather than directly affecting the activities of synthetic enzymes. Perhaps the elicitation process in potato tuber discs is hysteretic; i.e., the system responsible for phytoalexin accumulation remains activated at a lower $[Ca^{2+}]$ than that needed to activate it initially. Hysteretic behavior has been found for some calmodulin-regulated enzymes.[22]

The observation that both Ca^{2+} and Sr^{2+} enhance AA-elicited rishitin accumulation, but primarily promote lubimin accumulation in potato tuber discs treated with poly-L-lysine, remains unexplained. It is possible that both Ca^{2+} and Sr^{2+} have different effects in AA and poly-L-lysine elicitation on the levels and/or activities of the enzymes involved in the conversion of lubimin to rishitin.[29]

Experiments involving La^{3+} provide evidence that Ca^{2+} enters the potato cell after the application of AA and that the magnitude of the response (the level of rishitin accumulation) was dependent on the amount of Ca^{2+} that was able to gain access into potato tuber cells. Lanthanum and other lanthanides are thought to remain outside cells and block the entry of Ca^{2+} by occupying Ca^{2+}-binding sites on the plasma membrane.[30] The observation that Ca^{2+} was more effective in overcoming La^{3+} inhibition when applied simultaneously with La^{3+} than when applied 12 h after La^{3+} (Fig. 2) suggests that Ca^{2+} and La^{3+} are competing for the same binding site or channel.

There are probably a number of coordinated events which link the elicitation event; i.e., the initial host/parasite or plant/elicitor interaction, with the accumulation of phytoalexins in potato tuber discs. These include the mobilization of substrate and de novo synthesis and/or activation of the enzymes for the biosynthesis of the sesquiterpenoid phytoalexins. Calcium may regulate a number of these processes, and function as a secondary messenger. Elucidation of the mechanism whereby Ca^{2+} regulates the accumulation of phytoalexins may help us understand the expression of disease resistance in plants and suggest novel approaches for the control of disease.

REFERENCES

1. N. Sato, K. Kitazawa, and K. Tomiyama, The role of rishitin in localizing the invading hyphae of Phytophthora infestans in infection sites at the cut surface of potato tubers, Physiol. Plant Pathol. 1:289 (1971).
2. J. Kuć, Phytoalexins from the Solanaceae, in: "Phytoalexins", J. Bailey and J. Mansfield, eds., Blackie, Glasgow (1982).
3. R. M. Bostock, J. A. Kuć, and R. A. Laine, Eicosapentaenoic and arachidonic acids from Phytophthora infestans elicit fungitoxic sesquiterpenoids in potato, Science 212:67 (1981).

4. K. Tomiyama, Double infection by an incompatible race of Phytophthora infestans of potato cell which has previously been infected by a compatible race. Ann. Phytopathol. Soc. Japan 32:181 (1966).
5. K. Tomiyama, Further observation on the time requirement for hypersensitive cell death of potatoes infected by Phytophthora infestans and its relation to metabolic activity, Phytopathol. Z. 58:367 (1967).
6. K. Tomiyama, N. Doke, M. Nozue, and Y. Ishiguri, The hypersensitive response of resistant plants, in: "Recognition and Specificity in Host-Parasite Interactions", J. M. Dailey and I. Uritani, eds., University Park Press, Baltimore (1979).
7. N. Doke and K. Tomiyama, Effect of hyphal wall components from Phytophthora infestans on protoplasts of potato tuber tissue. Physiol. Plant Pathol. 16:169 (1980).
8. N. Doke, and N. Furuichi, Response of protoplasts to hyphal wall components in relation to resistance of potato to Phytophthora infestans, Physiol. Plant Pathol. 21:23 (1982).
9. J. Varns, and J. Kuć, Suppression of rishitin and phytuberin accumulation and hypersensitive response in potato by compatible races of Phytophthora infestans, Phytopathology 61:178 (1971).
10. N. Doke, N. A. Garas, and J. Kuć, Partial characterization and aspects of the mode of action of a hypersensitive-inhibiting factor (HIF) isolated from Phytophthora infestans, Physiol. Plant Pathol. 15:127 (1979).
11. G. Maniara, J. S. Rush, and J. Kuć, Partial structural characterization of hypersensitivity-modifying glucans from Phytophthora infestans, 1983 Annual Meetings of the Society for Complex Carbohydrates 53 (1983).
12. N. A. Garas, N. Doke, and J. Kuć, Suppression of the hypersensitive reaction in potato tubers by mycelial components from Phytophthora infestans, Physiol. Plant Pathol. 15:117 (1979).
13. N. Doke, Prevention of the hypersensitive reaction of potato cell to infection with an incompatible race of Phytophthora infestans, Physiol. Plant Pathol. 7:1 (1975).
14. N. Doke, N. A. Garas, and J. Kuć, Effect on host hypersensitivity of suppressors released during germination of Phytophthora infestans cystospores, Phytopathology 70:35 (1979).
15. N. Doke, and K. Tomiyama, Suppression of the hypersensitive response of potato tuber protoplasts to hyphal wall components by water soluble glucan isolated from Phytophthora infestans, Physiol. Plant Pathol. 16:177 (1980).
16. C. L. Preisig, and J. Kuć, Arachidonic acid-related elicitors of the hypersensitive response in potato and enhancement of their activities by glucans from Phytophthora infestans (Mont.) deBary, Arch. Biochem. Biophys. 236:379 (1985).
17. E. J. Corey, and P. T. Lansbury, Jr., Stereochemical course of 5-lipoxygenation of arachidonate by rat basophil leukemic cells (RBL-1) and potato enzymes. J. Amer. Chem. Soc. 105:4093 (1983).
18. C. L. Preisig and J. Kuć, Metabolism of arachidonic acid by potato tuber and activity of extracts as elicitors of the hypersensitive response, (submitted for publication).
19. G. Maniara, R. Laine, and J. Kuć, Oligosaccharides from Phytophthora infestans enhance the elicitation of sesquiterpenoid stress metabolite accumulation by arachidonic acid in potato, Physiol. Plant Pathol. 24:177 (1984).
20. J. K. Sharp, M. McNeil, and P. Albersheim, The primary structures of one elicitor-active and seven elicitor-inactive hexa (β-D-glucopyranosyl)-D-Glucitols isolated from the mycelial walls of Phytophthora megasperma f. sp. glycinea, J. Biol. Chem. 259:11321 (1984).

21. M. Zook, Elicitation and regulation of sesquiterpenoid phytoalexin accumulation in potato tuber tissue, Masters Thesis, University of Kentucky, Lexington 117 pp. (1986).
22. H. Rasmussen, and P. Q. Barrett, Calcium messenger system: An integrated view, Physiol. Rev. 64:938 (1984).
23. D. Marme, and P. Dieter, Role of Ca^{2+} and calmodulin in plants, in: "Calcium and Cell Function", Vol. 4, W. Y. Cheung, ed, Academic Press, New York (1983).
24. P. K. Hepler, and R. Wayne, Calcium and plant development, Annu. Rev. Plant Physiol. 36:397 (1984).
25. Y. Nishizuka, Studies and perspectives of protein kinase C, Science 233:305 (1986).
26. R. P. Rubin, "Calcium and Cellular Secretion", Plenum Press, New York (1982).
27. R. G. Hansford, and F. Castro, Role of calcium in pyruvate dehydrogenase interconversion in brain mitochondria and synaptosomes, Biochem. J. 227:129 (1985).
28. R. M. Denton, J. G. McCormack, and N. J. Edgell, Role of calcium ions in the regulation of intramitochondrial metabolism, Biochem. J. 190:107 (1980).
29. A. Murai, S. Sata, A. Osada, N. Katsui, and T. Masamune, Biosynthesis from solavetivone of the phytoalexin rishitin in potato. Implicit role of solavetivone as an activator, J. Chem. Soc. Chem. Commun. 1:32 (1982).
30. R. B. Mikkelsen, Lanthanides as calcium probes in biomembranes, in: "Biological Membranes", Vol. 3, D. Chapman and D. F. H. Wallach, eds, Academic Press, New York (1976).

PRESENCE OF UNUSUALLY HIGH LEVELS OF CHOLESTEROL IN THE SHOOT-APICES OF FLOWERING PLANTS

Vipin K. Garg*, Trevor J. Douglas* and Leslie G. Paleg

Department of Plant Physiology
Waite Agricultural Research Institute
University of Adelaide, SA 5064, Australia

INTRODUCTION

Although cholesterol, the principal animal sterol, has been isolated from many higher plant species, in most cases it was reported to occur only in trace amounts[1,2]. For this reason, cholesterol has generally been considered to be of relatively little significance in higher plants. In the course of our work on the sterol profiles of different plant tissues during floral development, we have identified very high levels of cholesterol in the shoot-apices of various plants. Sterol profiles of two monocot plants, Hordeum vulgare and Lolium temulentum, and one dicot plant, Xanthium strumarium, were studied. Our results suggest a specific association of cholesterol with meristematic and/or reproductive tissues.

MATERIALS AND METHODS

H. vulgare (cv. Clipper) and L. temulentum (strain Ceres, CSIRO, Australia) plants were grown in continuous light at 20°C in controlled environment growth chambers. X. strumarium plants were grown under 18h photoperiod at 25°C and induced to flower by the short-day treatment method of Salisbury[3]. In each species, total 4-desmethylsterols were analysed separately from leaves, stems and shoot-apices (terminal inflorescence carefully dissected-out under a microscope; Fig. 1), for several days during flowering. The flowering period studied ranged from floral induction to early floral differentiation (floret initials, etc.). Desmethylsterols were extracted and analyzed from the freeze-dried tissues as previously described[4]. The identity of the individual sterol components was established by GLC and mass spectroscopy.

RESULTS AND DISCUSSION

In the leaf and the stem tissues, as expected, sitosterol was the main sterol (50-80% of the total), followed by stigmasterol and campesterol (10-30% each), while cholesterol was mostly a minor component (<5% of the total). In contrast, cholesterol was highly abundant (15-45%) in the shoot-apices and matched the levels of the other three sterols throughout the sampling period. In all three species, on a per unit dry weight

---

*Present addresses: V.K.G. - ARCO Plant Cell Research Institute, 6560 Trinity Court, Dublin, CA, USA. T.J.D. - CSIRO, Division of Horticultural Research, Merbein, Victoria, Australia.

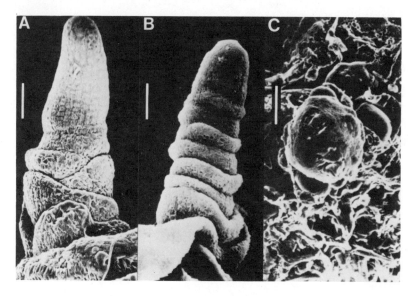

Fig. 1. Cholesterol rich shoot-apices (vegetative stage) of H. vulgare (A), L. temulentum (B) and X. strumarium (C). Bar = 50 µM (A,B); 200 µM (C).

basis, the cholesterol level of the apical tissue was several fold higher than that of the leaf and stem tissues (Table 1). The cholesterol level was highest in the young vegetative shoot-apices, and decreased during transition to flowering and subsequent floral development. The overall cholesterol level in the two monocot species, H. vulgare and L. temulentum, was much higher than the dicot species, X. strumarium, although, the general trend was the same in all three species (Table 1).

Table 1. Cholesterol Levels in Various Tissues of H. vulgare, L. temulentum and X. strumarium During Flowering. DW = Dry weight; ND = Not detectable.

Sampling Day	Apical Developmental Stage	ng Cholesterol (mg DW)$^{-1}$								
		Leaf	Stem	Shoot-apex	Leaf	Stem	Shoot-apex	Leaf	Stem	Shoot-apex
		(H. vulgare)			(L. temulentum)			(X. strumarium)		
1	Vegetative	48	452	9860	54	185	5680	ND	29	310
2	Transitional	38	267	8330	33	123	5150	ND	22	224
3	Floral	32	257	5560	30	93	3240	ND	17	232
4	Floral	22	131	3260	21	44	2950	ND	18	225
5	Floral	25	123	3750	17	31	1820	ND	12	213
6	Floral	29	83	1180	16	22	1500	ND	10	178
LSD (P=0.05)		4	17	33	2	7	33	-	2	16

It is clear from these results that in both the monocot and the dicot species studied, the young vegetative and early flowering apices were the richest source of cholesterol. In fact, to our knowledge, this work represents the first instance of the detection of such high level of cholesterol in a higher plant tissue. Since it has been suggested that differences in the sterol compositions of different tissues have functional significance[1,2], it is possible that the high levels of cholesterol are specifically associated with the functional differences in the membranes of meristematic and/or reproductive tissues. Previously, the data of Geuns[5], who examined the sterol composition in the different morphological regions of mung bean seedlings, also showed that in comparison with any other tissue, the meristematic stem tips contained the highest level of cholesterol. Similarly, in several members of the family Cruciferae[6], and in *Solanum andigena*[7], cholesterol levels were reported to rise in response to photomorphogenetic changes. These results further support a specific role of cholesterol in meristematic and morphogenetically changing tissues. When compared with other sterols, the flat (planar) molecular configuration of cholesterol has been shown to be most effective in controlling the permeability of plant membranes[2]. In model studies also, when steroids and terpenoids were added to lecithin-containing monolayers and bilayers, no other isopentenoid, synthetic or natural, exceeded the ability of cholesterol to modulate membrane fluidity[8,9]. Thus, it is tempting to suggest that the superior membrane stabilizing ability of cholesterol is specifically required by the membranes of actively growing and differentiating plant tissues.

Since the level of cholesterol was highest in the vegetative apices before their transition to flowering (Table 1), and cholesterol levels actually decreased during floral differentiation, it is also possible that cholesterol is in some way associated with the process of floral initiation. For instance, there is some evidence in the literature that the flower-inducing hormone may be an isopentenoid or steroid-like compound[10,11]. Various steroids occur in plants, and cholesterol has been shown to be a key intermediate in their biosynthesis[10,11]. Thus, it is possible that cholesterol was being used to make steroids, which may be involved in floral induction.

Finally, cholesterol could also be involved in plant tissues by exerting an effect on the nucleic acid and protein metabolism. It is now well established that the synthesis of cholesterol is an absolute requirement for cell proliferation and its associated nucleic acid and protein synthesis in animals[12,13]. Further, cholesterol has also been reported to bind to chromatin and to affect biosynthetic control mechanisms[14]. Similarly, a specific cholesterol-protein interaction[8,15] and a cholesterol effect on membrane bound enzymes is well documented[16]. Thus it is not difficult to visualize that a similar multi-faceted role of cholesterol is crucial for growth and differentiation at the shoot-apex of flowering plants.

REFERENCES

1. Bean, G.A. (1973) Adv. Lipid Res. 11:193-218.
2. Grunwald, C. (1980) in: "Encyclopedia of Plant Physiology", E.A. Bell and B.V. Charlwood, eds., New Ser. Vol. 8, pp. 221-256, Springer-Verlag, Berlin/New York.
3. Salisbury, F.B. (1965) Planta 66:1-26.
4. Garg, V.K. and L.G. Paleg (1986) Physiol. Plant. (In Press).
5. Geuns, J.M.C. (1973) Phytochemistry 12:103-106.
6. Ingram, D.S., B.A. Knights, I.J. McEvoy and P. McKay (1968) Phytochemistry 7:1241-1243.
7. Bae, M. and E.I. Mercer (1970) Phytochemistry 9:63-68.
8. Demel, R.A. and B. DeKruyff (1976) Biochim. Biophys. Acta 457:109-132.
9. Nes, W.D. and E. Heftmann (1981) J. Nat. Products 44:377-400.
10. Heftmann, E. (1975) Phytochemistry 14:891-901.
11. Geuns, J.M.C. (1978) Phytochemistry 17:1-14.
12. Kandutsch, A.A. and H.W. Chen (1977) J. Biol. Chem. 252:409-415.
13. Maltese, W.A., B.A. Reitz and J.J. Volpe (1981) Biochim. Biophys. Acta. 633:645-652.
14. Erickson, S.K., A.M. Davison and R.G. Gould (1975) Biochim. Biophys. Acta. 409:59-67
15. Quinn, P.J. (1981) Progr. Biophys. Mol. Biol. 38:1-104.
16. Sabine, J.R. (1977) Cholesterol, Marcel Dekker, New York.

CHANGES IN STEROL BIOSYNTHESIS FROM [2-^{14}C]MEVALONIC ACID DURING DEVELOPMENT OF CUCURBITA MAXIMA SEEDLINGS

Vipin K. Garg* and William R. Nes

Department of Biological Sciences
Drexel University
Philadelphia, PA 19104, U.S.A.

INTRODUCTION

We have recently characterized a large array of sixteen different sterols in the members of the family Cucurbitaceae[1-4]. These include the dominant Δ^7-sterols, some "main line" Δ^5-sterols, and sterols characteristic of organisms (algae and fungi) much lower on the evolutionary scale. This represents, by far, the most complex sterol profile found in a higher plant, and raises important questions regarding the sterol biosynthetic pathway in Cucurbitaceae. In the present investigation, we have examined the incorporation of [2-^{14}C]mevalonic acid (MVA) into various sterol fractions in seedlings of Cucurbita maxima.

MATERIALS AND METHODS

Cucurbita maxima var. True Hubbard seeds were obtained from W. Atlee Burpee Co. (Warminster, PA). Racemic [2-^{14}C]mevalonate as the dibenzylethylenediamine salt (0.25 mCi in 11.4 mg) was purchased from New England Nuclear Corporation. Sterol standards were either purchased from commercial sources or were isolated from C. maxima seeds as previously described[1,2].

Seeds were presented with [2-^{14}C]MVA in the water of imbibition. Sterile water (30 ml) containing 42.3 X 10^7 dpm [2-^{14}C]MVA was equally distributed between 10 petri dishes each containing 10 seeds and allowed to imbibe at 25°C. Within 12 hours the solution was completely absorbed. The swollen seeds were transferred to sterile petri dishes lined with moistened filter paper for the remainder of the germination period. The seedlings were further grown in small glass vials containing nutrient solution as previously described[3]. The total neutral lipids were extracted from the seedlings after saponification, as previously described[1]. Squalene, 4,4-dimethylsterols, 4 α-methylsterols and 4-desmethylsterols were separated from the neutral lipid mixture by TLC, and then analyzed by GLC and HPLC. The instrumentation and techniques for the TLC, GLC and HPLC have been described earlier[1,2,5]. Radioactive measurements were performed on a Beckman model LS 7500 Liquid Scintillation System.

---

*Present Address: ARCO Plant Cell Research Institute, 6560 Trinity Court, Dublin, CA 94568, U.S.A.

RESULTS AND DISCUSSION

As shown in Table 1, label from [2-^{14}C]MVA appeared primarily in squalene during the first day of germination (48 h after seeds were presented with water containing the radioactive substrate), while 4,4-dimethyl-, 4 α methyl- and 4-desmethylsterols together contained only about one third of the total label. This picture did not change much during the next 24 h (day 2), and it was only between days 2 and 4 that an increased proportion of the label was transferred to other sterol fractions (Table 1). On day 4, squalene represented only about 4% of the total label, while 4,4-dimethyl-, 4 α-methyl-and 4-desmethylsterols contained 36, 5 and 54% of the total label, respectively. As the seedlings grew, the incorporation into the desmethylsterols gradually increased, while that into the other three fractions decreased (Table 1). By day 12, desmethylsterols were labelled to an extent of 81%, while squalene was only to an extent of 1.5%. These results indicate that although squalene is synthesized very early in the germination of C. maxima, its further conversion into sterols does not occur until the onset of seedling elongation (between days 2 and 4). Similar delayed conversion of squalene to sterols has been observed previously during germination of seeds of Pisum sativum[6] and Pinus pinea[5]. While the biosynthesis of squalene is entirely nonoxidative, the first and many subsequent steps in its conversion to steroidal end products require molecular oxygen. Thus, it is tempting to suggest that the expression of the oxygen-requiring steps is developmentally regulated. It is possible that squalene performs some specific function during imbibition and seedling emergence and is subsequently converted to desmethyl-sterols that are required for new membrane synthesis during seedling elongation.

Table I. Incorporation of [2-^{14}C]Mevalonic Acid into Various Sterol Fractions in Seedlings of Cucurbita maxima.

Sterol fraction	Days After Germination					
	1	2	4	6	8	12
	(% of Total Radioactivity)					
Squalene	64.2	64.3	4.4	2.6	1.5	1.5
4,4-Dimethylsterols	19.8	24.1	36.1	28.8	21.3	16.9
4α-Methylsterols	5.3	3.0	5.2	2.7	1.0	0.9
4-Desmethylsterols:						
Δ^7-series	10.7	8.6	54.3	65.9	76.2	80.7
Δ^5-series	0.0	0.0	0.0	0.0	0.0	0.0

The two methylated sterols, 4,4-dimethly- and 4 α-methylsterols, are known to act as intermediates of desmethylsterols[7] and are probably acting as one in this case also. However, a relatively high level of incorporation into 4,4-dimethylsterols may indicate some functional significance for these sterols. When the 4-desmethyl fraction was further analysed into Δ^5- and Δ^7-sterols, the label was found to be associated only with the Δ^7- components, and failed to become incorporated into Δ^5-sterols (Table 1). This is consistent with our recent finding that while Δ^5-sterols are present in the seeds of C. maxima, they progressively disappear following germination[3]. Since Δ^5-sterols were also absent in mature plants, it does not appear that the Δ^5-sterols are synthesized at all in the green tissues of C. maxima. The only possibility that seems likely is that they are formed in the seed itself during its development prior to germination. While sterol synthesis in developing seeds of the family Cucurbitaceae has not been studied, it has been reported in the seeds of other tracheophyte families[8,9]. Among the Δ^7-sterols, the two types of C-24 epimers, 24α- and 24β-alkylsterols, have both been reported to occur in substantial amounts in the seeds of C. maxima[1], however, during seedling development and maturation, the relative distribution of the two epimers changes in favor of the 24α-stereochemistry[3]. Our further incorporation studies also support this ontogenetic shift in stereochemistry.

The results of the present investigation raise the possibility that the expression of sterol biosynthetic sequences in higher plants is temporally regulated. Clearly, different types of sterol structures are synthesized at different times during the life cycle of the plant. That these sterol structures have developmentally related functional significance is also very likely. For instance, in the Cucurbitaceae it seems that the Δ^5-sterols have a specialized function only during germination, while the Δ^7-sterols seem to perform other bulk sterol functions throughout the life cycle of the plant. Similarly, specialized functions for 24α- and 24β-alkyl structures are also possible. Developmentally related functions for isopentenoid compounds have become evident in many recent investigations[10-12]. It is suggested that there is a need to evaluate the overall expression of the isopentenoid pathway on molecular level in higher plants.

REFERENCES

1. Garg, V. K. and W. R. Nes (1984) Phytochemistry 23:2919-2923.
2. Garg, V. K. and W. R. Nes (1984) Phytochemistry 23:2925-2929.
3. Garg, V. K. and W. R. Nes (1985) Lipids 20:876-883.
4. Garg, V. K. and W. R. Nes (1986) Phytochemistry (In Press).
5. McKean, M. L. and W. R. Nes (1977) Lipids 12:382-385.
6. Nes, W. R., D. J. Baisted, E. Capstack, Jr., W. W. Newschwander and P. T. Russell (1967) In: "Biochemistry of Chloroplasts", T. W. Goodwin, ed., Vol. 2, pp. 273-282, Academic Press, New York.
7. Nes, W. R. and M. L. McKean (1977) "Biochemistry of Steroids and Other Isopentenoids" University Park Press, Baltimore, Maryland.
8. Grunwald, C. (1975) Annu. Rev. Plant Physiol. 26:209-236.
9. Goad, L. J. (1983) Biochem. Soc. Trans. 11:548-552.
10. Fenner, G. P., G. W. Patterson and P. M. Koines (1986) Lipids 21:48-51.
11. Heupel, R. C., Y. Sauvaire, P. H. Le, E. J. Parish and W. D. Nes (1986) Lipids 21:69-75.
12. Garg, V. K. and L. G. Paleg (1986) Physiol. Plant. (In Press).

INHIBITION OF GROWTH OF CELERY CELLS BY PACLOBUTRAZOL AND ITS REVERSAL BY ADDED STEROLS

Penny A. Haughan, John R. Lenton* and L. John Goad

Department of Biochemistry, University of Liverpool, P.O. Box 147, Liverpool L69 3BX, U.K. and *Long Ashton Research Station, Long Ashton, Bristol, BS18 9AF, U.K.

Paclobutrazol, (2RS, 3RS)-1-(4-chlorophenyl)-4,4-dimethyl-2-(1,2,4-triazol-1-yl) pentan-3-ol, is a synthetic plant growth retardant[1,2] which also has a high fungicidal activity.[3,4] The antifungal activity of paclobutrazol is related to its inhibition of sterol biosynthesis[4] by interaction with the cytochrome P-450 system required for 14-demethylation of sterol precursors.[3,5] This results in the accumulation of 14α-methylsterols in fungi.[4] Paclobutrazol also exerts plant growth regulatory effects by interfering with gibberellin biosynthesis at the step which is catalysed by the cytochrome P-450 requiring enzyme, kaurene oxidase.[2,3,6]

In addition to retarding growth of whole plants paclobutrazol has been reported to prevent cell division in a suspension culture of rose cells.[6] Other fungicides, plant growth retardants and hypocholesterolemic agents have been tested on plant tissue cultures and seedlings and shown to inhibit sterol synthesis with resultant changes in sterol composition and effects on growth.[7-10] We have now investigated the effects of paclobutrazol on the growth and sterol synthetic capacity of a suspension culture of celery (<u>Apium gravedens</u>).

Celery suspension cultures showed progressive inhibition of growth with increasing concentrations of paclobutrazol; no significant growth occurred at concentrations above 50µM paclobutrazol. GC-MS analysis of the sterols of control cultures harvested after 21 days revealed stigmasterol as the major sterol. In paclobutrazol-treated cells the amount of stigmasterol was lowered while obtusifoliol, cycloeucalenol and cycloartenol accumulated (Table 1) suggesting that paclobutrazol is probably interfering with removal of the 14α-methyl group as occurs in fungi.[4] This view was substantiated by incubation with [^{14}C]acetate or mevalonate; an accumulation labelled of $4\alpha,14\alpha$-dimethyl sterols was seen in inhibited cells when compared to the appropriate controls.

Addition of cholesterol to the cultures treated with paclobutrazol resulted in an uptake of cholesterol and partial restoration of growth (Tables 1 and 2) but in repeated experiments this has never exceeded 30-50% of the control value over a 21 day growth period. This contrasts with the report[10] that added cholesterol can completely reverse the block in cell division of rice, maize and soybean suspension cultures imposed by the growth retardant tetcyclacis.

Table 1. Effects of 50μM paclobutrazol and added sterols on the sterol content (μg/10⁴ cells) of celery tissue cultures harvested after 21 days.

	Sterol							
	Ch*	Ca	St	Si	Ob	CyE	CyA	TS
Control	6	57	157	96	-	-	-	316
+ PCB	11	102	74	79	128	31	18	443
+ PCB + Ch†	101	63	64	46	210	132	54	670
+ PCB + Ch†+St‡	94	54	226	48	90	17	9	538
+ PCB + St‡	8	49	163	39	20	-	1	280

* Ch = cholesterol; Ca = campesterol; St = stigmasterol; Si = sitosterol; Ob = obtusifoliol; CyE = cycloeucalenol; CyA = cycloartenol; TS = total sterol; PCB = paclobutrazol.
† 50μM Sterol added to medium at day 6
‡ 50μM Sterol added to medium at day 12.

Table 2. Growth of celery suspension cultures in the presence of paclobutrazol (50μM) and added sterol (50μM).

Experiment*	Days growth					
	6	9	12	15	18	21
A	7	61	100	98	96	84
B	0	0	0	4	4	2
C	0	0	0	8	17	22
D	0	5	44	85	102	98
E	4	6	5	32	58	80
F	0	0	1	26	72	100

* A = Control; B = plus paclobutrazol (PCB); C = PCB with cholesterol added at day 6; D = PCB with sitosterol added at day 6; E = PCB with cholesterol added at day 6 and stigmasterol at day 12; F = PCB with stigmasterol added at day 12. Growth was determined by increase in cell numbers and is expressed as a percentage of the control value at day 12.

When stigmasterol or sitosterol was added to paclobutrazol inhibited cells the cell growth was resumed and eventually reached the control level (Table 2). Moreover, when inhibited cells which had been treated with cholesterol at day 6 were further supplemented with stigmasterol in the growth medium at day 12, they increased their growth rate to almost reach the control values for both dry weight and cell number (Table 2). Also, cells inhibited with paclobutrazol for 12 days could still respond to added stigmasterol to reach the control values (Table 2). It is notable that the cells resuming growth after receiving sterol from the medium still contained significant amounts of 14α-methylsterols (Table 1).

Grossmann et al.[10] found that stigmasterol was very effective in reversing tetcyclacis inhibition of growth of plant cell cultures and application of stigmasterol has also restored stem growth to tobacco seedlings treated with growth retardants.[8] However, attempts to reverse the effects of fungicides on fungal growth by adding ergosterol to the medium have apparently been less successful.[11] There is mounting evidence that a C-24 methyl sterol is essential for fungal growth[12,13] and the present report, together with the results from other laboratories,[8-10] suggest that a C-24 alkyl sterol may also play an essential role in the events of the cell cycle in higher plants.

Acknowledgements: - We thank the Science and Engineering Research Council for financial support and ICI Plant Protection for the gift of paclobutrazol.

REFERENCES
1. B.G. Lever, S.J. Shearing and J.J. Batch, PP333-A new broad spectrum growth retardant. Proc. Br. Crop Protect. Conf. Pests Dis. 1: 3 (1982)
2. S.Y. Wang, J.K. Byun and G.L. Steffens, Controlling plant growth via the gibberellin biosynthesis system - II. Biochemical and physiological alterations in apple seedings. Physiol. Plant 63: 169 (1985)
3. B. Sugavanam, Diastereoisomers and enantiomers of paclobutrazol: their preparation and biological activity. Pestic. Sci. 15: 296 (1984)
4. B.C. Baldwin and T.E. Wiggins, Action of fungicidal triazoles of the diclobutrazol series on Ustilago maydis. Pestic. Sci. 15: 156 (1984)
5. T.E. Wiggins and B.C. Baldwin, Binding of azole fungicides related to diclobutrazol to cytochrome P-450. Pestic. Sci. 15: 206 (1984)
6. J. Dalziel and D.K. Lawrence, Biochemical and biological effects of kauarene oxidase inhibitors, such as paclobutrazol. Monograph of the British Plant Growth Regulator Group No 11, p.43 (1984)
7. H. Buchenauer and E. Röhner, Effect of triadimefon and triadimenol on growth of various plant species as well as gibberellin content and sterol metabolism in shoots of barley seedlings. Pest. Biochem. Physiol. 15: 58 (1981)
8. T.J. Douglas and L.G. Paleg, Inhibition of sterol biosynthesis and stem elongation of tobacco seedlings induced by some hypocholesterolemic agents. J. Exptl. Botany 32: 59 (1981).
9. P. Schmitt, A. Rahier and P. Benveniste, Inhibition of sterol biosynthesis in suspension cultures of bramble cells. Physiol. Veg. 20: 559 (1982)
10. a) K. Nitsche, K. Grossmann, E. Sauerbrey and J. Jung, Influence of the growth retardant tetcyclacis on cell division and cell elongation in plants and cell cultures of sunflower, soybean and maize. J Plant Physiol. 118: 209 (1985). b) K. Grossmann, W. Rademacher and J. Jung, Effects of NDA, a new plant growth retardant, on cell culture growth of Zea mays. J.Plant Growth Regul. 2: 19 (1983). c) K. Grossmann, E.W. Weiler and J. Jung, Effects of different sterols on the inhibition of cell culture growth by the growth retardant tetcyclacis. Planta 164: 370 (1985)
11. H.D. Sisler, N.N. Ragsdale and W.F. Waterfield, Biochemical aspects of the fungitoxic and growth regulatory action of fenarimol and other pyrimidin-5-ylmethanols. Pestic Sci. 15: 167 (1984)
12. R.J. Rodriguez, C. Low, C.D.K. Bottema and L.W. Parks, Multiple functions of sterols in Saccharomyces cerevisiae. Biochim. Biophys. Acta 837: 336 (1985)
13. W.D. Nes and R.C. Heupel, Physiological requirement for biosynthesis of multiple 24 β-methyl sterols in Gibberella fujikuroi. Arch. Biochem. Biophys. 244: 211 (1986)

THE STERYL ESTERS OF HIGHER PLANTS

L. John Goad, Jan Zimowski, Richard P. Evershed and Victoria L. Male

Department of Biochemistry, University of Liverpool, P.O. Box 147, Liverpool L69 3BX, U.K.

The fatty acid esters of cholesterol are widely distributed in animal tissues and they have been subjected to in depth biochemical studies because of their importance as constituents of plasma lipoproteins and their association with pathological conditions such as atherosclerosis. Steryl esters are also well established as constituents of various tissues from many plants. However, the biochemistry of the steryl esters in plants has been the subject of rather few studies when compared to the attention received by animal steryl esters. The information available on plant steryl ester biochemistry is thus somewhat fragementary[2] and although there has been speculation about the precise roles of these compounds in plants they cannot yet be defined. The steryl esters from many plant oils contain a mixture of the sterols typical of the plant and there is often a range of saturated and unsaturated fatty acids.[2-4] However, the proportions of different sterols may vary between the free and esterified forms indicating a degree of selectivity in the esterification process.

The analysis of plant steryl esters has typically[4] proceeded by isolation of the steryl ester fraction from a plant lipid extract followed by saponification to yield the sterol and fatty acid moieties. These are then identified and quantified by GC analysis, usually after preparation of the fatty acid methyl esters. However, this method of analysis has the disadvantage that the integrities of the steryl esters are lost. With the often multicomponent mixtures of sterols and fatty acid methyl esters resulting from a typical plant steryl ester mixture it is difficult to define with precision which sterols and fatty acids were originally linked. The analysis can be further complicated by the possibility of non-sterol fatty acyl ester contaminents accompanying the steryl esters during the initial isolation procedures and thus contributing a fatty acid complement upon saponification. Alternative methods of steryl ester analysis are now being explored which allow unambigious identification of intact steryl esters. A reverse phase HPLC method has been developed for the analysis of steryl esters[5,6] and both reverse phase and adsorption HPLC are being used in our laboratory for plant steryl ester preparative separations. Although good resolution of components can be achieved the sensitivity of the HPLC detector to steryl esters is relatively poor. Methods for GC and GC-mass spectrometry analysis of intact steryl esters are now being developed which employ fused silica capillary columns coated with thin films of bound apolar phases.[7-11] For the mass spectrometry of steryl esters electron impact ionisation is of little value as molecular ions are not observed. However, structural information can be obtained with chemical ionisation using preferably ammonia as the reactant gas[12] and operating the mass spectrometer in the positive[8,10] or negative[11] ion scanning mode. The GC/MS technique now provides a very sensitive method for the unambiguous identification and quantification of individual components in steryl ester mixtures and in our laboratory it has allowed the analysis of µg amounts of the plant steryl esters isolated from plant tissue subcellular fractions. This may permit the detection of subtle changes in steryl ester composition which would be important in defining roles for the steryl ester pools in plant cells.

In comparison to the extensive studies on the enzymology of animal steryl esters[1] the plant compounds have attracted rather scant attention so far. Garcia and Mudd[13-16] have studied plant steryl ester formation using acetone powder preparations from spinach leaves. They have provided evidence that the steryl ester synthase is membrane bound because activity was found in the 1000$\underline{g}$ (chloroplast), 3000$\underline{g}$ (mitochondrial) and particularly the 88000$\underline{g}$ (microsomal) pellets but not in the 88000$\underline{g}$ supernatant fraction.[14] The steryl esters are not produced from free sterol and fatty acid by a simple reversal of an esterase reaction[13] but require a 1,2-diacylglycerol as the acyl donor.[15] This contrasts with the production of steryl esters in animal tissues in which fatty acyl CoA or phosphatidylcholine are the acyl donor.[1] The specificity of the spinach leaf enzyme has been examined[16] and cholesterol, sitosterol and campesterol were equivalent substrates.

However, with different 1,2-diacylglycerols preference was observed for shorter chain fatty acids and the more unsaturated fatty acids.

In a study on the formation of steryl esters in roots of <u>Sinapis alba</u> Zimowski and Wojciechowski[17,18] found that the enzyme was again associated with the particulate fraction of the homogenate. However, in contrast to the spinach leaf preparation, a triacylglycerol was the preferred acyl donor; di- and monoacylglycerols were considerably less effective donors. The <u>Sinapis alba</u> enzyme has been solubilised and partially purified.[18] It has been shown to have a specificity towards sterols with a planar ring system but the side chain structure appears to be less important. Tripalmitoylglycerol is the best of the saturated fatty acid donors while among the C_{18} fatty acid donors the order of preference is for the desaturated fatty acids ($C_{18:1} > C_{18:2} > C_{18:3} > C_{18:0}$).

In our studies on plant steryl ester production we have examined steryl ester biosynthesis using an homogenate of <u>Zea mays</u> roots. The enzyme from this source was again membrane bound with the 105000<u>g</u> pellet showing the highest specific activity although considerable activity was also observed in the 20000<u>g</u> pellet and in the 105000<u>g</u> supernatent. The enzyme can utilise triacyl-, 1,2-diacyl- and 1,3-diacylglycerols as the acyl donor and it showed greater specificity for the unsaturated $C_{18:1}$, $C_{18:2}$ and $C_{18:3}$ fatty acids than for the saturated $C_{16:0}$ and $C_{18:0}$ fatty acids. Palmitoyl-CoA was a very poor acyl donor.

Steryl ester hydrolase activity has been reported in an homogenate of roots of <u>Sinapis alba</u>.[19] The enzyme is membrane bound but readily solubilised by treatment with Triton X-100. The specificity of the esterase has been tested towards a series of cholesteryl esters with saturated fatty acids; the highest rates were found with the myristate (C_{14}), palmitate (C_{16}) and stearate (C_{18}) esters but the shorter and longer chain fatty acid esters were very poor substrates.[19]

In the course of our investigations on sterol metabolism in germinating seeds we have measured steryl ester hydrolase activity in barley (<u>Hordeum vulgare</u>) and maize (<u>Zea mays</u>). Steryl ester hydrolase activity has been observed in homogenates of shoots, roots, scutella and endosperm plus aleurone of both plants. The roots showed the highest activity which rapidly increased during the first 10-days of seedling growth. The shoots contain somewhat lower steryl ester hydrolase activity while the endosperm, aleurone and scutellum have a relatively low activity which actually decreases markedly during the first 24-48 hours after imbibition of water. The maize root steryl ester hydrolase is membrane bound, principally in the 105000g pellet. It shows greatest activity towards the palmitate ester among the cholesterol esters of saturated fatty acids tested but it has somewhat greater specificity for the esters with the unsaturated fatty acids linoleate ($C_{18:2}$) and linolenate ($C_{18:3}$). The activity of the enzyme is little changed by the addition of Mg^{2+}-ATP but in most preparations it is lowered by upto 20% by addition of NaF. This does not appear to be an effect mediated through a plant phosphatase, which is inhibited by NaF, but to be a direct and reversible inhibition of the steryl ester hydrolyse.

The investigations outlined above reveal that plant tissues, particularly roots, contain enzymes for both the synthesis and hydrolysis of steryl esters. This leads to speculation that in some tissues the steryl esters may be in a dynamic state of rapid turnover. This is also indicated by the ready incorporation of radioactive sterol precursors into the steryl esters of plant tissues.[20-24] An interesting observation[3-4] is that several 4,4-dimethyl and 4α-methylsterol precursors of sitosterol, such as cycloartenol, cycloeucalenol, obtusifoliol and 24-ethylidinelophenol, appear in the steryl ester fraction of plant tissues and become labelled after incubation with [2-^{14}C]mevalonic acid.[21] This obervation has led to the suggestion that the steryl esters may be involved in some regulatory process in phytosterol production.[2,22] For example, oxidative loss of the C-4 methyl groups requires a free 3β-hydroxy group which can be converted to a 3-oxo intermediate during the C-4 demethylation process. Esterification will block this process and cycloartenol or other precursors produced in amounts in excess of those that can be accommodated by some rate limiting step operating at the later stages of the biosynthetic sequence might be diverted into a reserve steryl ester pool. This pool can then be subsequently remobilised by hydrolysis as the demands of the cell for more free sterol increases.

Steryl esters in seeds may also play a storage or reserve role to supply the sterols required to meet the demands of the developing seedling for membrane elaboration at the onset of germination. During cereal grain maturation free sterols are converted into steryl esters so that in some cases the dormant grain may contain a considerable proportion of the total sterol in the esterified form.[25,26] Changes in amounts and identities of the free sterols and steryl esters in germinating seeds have been the subject of several investigations.[1,27-31] Studies with germinating seeds of different Gramineae species have provided a variety of results. In Zea mays the free sterol content of the endosperm and scutellum decreased slightly with little change in the steryl ester content of the endosperm but there was a marked increase in the steryl esters of the scutellum as germination progressed.[27] We have recently confirmed that over a six day period there is an increase in steryl ester in the endosperm and scutella portion of the Zea mays seed. The steryl ester content of the developing axis (shoot and roots) falls significantly (70%) between days 2 and 3 of the growth period and remains at a low level up to day 6 while the free sterol content of these tissues increases.

An important observation by Paleg[32] is that the steryl ester and steryl glycoside content of wheat aleurone tissue decreases markedly and rapidly after imbibition of water by the dry seed. The fall in steryl ester and glycoside content is accompanied by a complementary increase in the free sterol content. These changes in steryl conjugate and free sterol concentrations in the aleurone cells are compatable with the suggestion that the former act as a store of the free sterol required for new membrane production at the onset of germination.[32] It is noteworthy that we have observed that the steryl ester hydrolase activity of barley endosperm and aleurone tissue is highest during the first few hours following water imbibition which is consistent with a mobilisation of reserve steryl esters.

With wheat, a phenotype with a high steryl ester content showed a decrease during germination in both free and esterified sterol whereas a phenotype with a low steryl ester content showed a negligible change in steryl ester concentration but there was a marked decrease in free sterol.[30] It was concluded[30] that the embryo contains sufficient sterol to provide the developing seedling with sterol during the initial stages of development and it was suggested that sterol is transferred from the scutellum to the roots and coleoptile. A decrease in steryl ester content and increase in free sterol was observed during the first 5 days of germination of barley[31] but subsequent investigations in our laboratory with other varieties of barley have yielded variable results with in some cases the steryl ester content staying constant. Thus, as already noted,[30] the seed phenotype may have an important bearing upon amounts and changes in steryl ester and free sterol during germination.

The suggestion that steryl esters may be the form required for transport between plant tissues has not gained support[2] but the possibility[27] of their involvement in the intracellular movement of sterols, perhaps in conjunction with a carrier protein, remains to be elucidated. The steryl ester mixtures present in various intracellular organelles in plant tissues have been examined[3,33,34] and differences in composition noted with cholesterol being a significant component of the nuclear[3] and mitochondrial fractions.[34]

There have been some studies on the effects of environmental factors on the steryl ester content and metabolism in plants[2] and light,[35-37] season[38] and water stress[39] have been found to affect the steryl ester composition of plants. Further studies of these factors can be anticipated and the application of the new HPLC and GC/MS methods of steryl ester analysis should help in the recognition of subtle changes in the identities and quantities of the individual steryl esters found in plant tissues.

Acknowledgements: We wish to thank the Agricultural and Food Research Council and the Science and Engineering Research Council for financial support for the work carried out in our laboratory.

REFERENCES

1. K.E. Suckling and E.F. Strange, Role of acyl-CoA: cholesterol acyl transferase in cellular cholesterol metabolism. J. Lipid Res. 26: 647 (1985)
2. C. Grunwald, Plant sterols. Ann. Rev. Plant Physiol. 26: 209 (1975)
3. a) R.J. Kemp and E.I. Mercer, The steryl esters of maize seedlings. Biochem. J. 110: 111 (1968). b) R.J. Kemp and E.I. Mercer, Studies on the sterols and steryl esters of the intracellular organelles of maize shoots. Biochem. J. 110: 119 (1968)
4. K. Staphylakis and D. Gegiou, Free, esterified and glucosidic sterols in cocoa butter. Lipids 20: 723 (1985)
5. R.M. Carroll and L.L. Rudel, Evaluation of a high performance liquid chromatography method for isolation and quantitation of cholesterol and cholesteryl esters. J. Lipids Res. 22: 359 (1981)
6. J.T. Billheimer, S. Avart and B. Milani, Separation of steryl esters by reversed-phase liquid chromatography. J. Lipid Res. 24: 1646 (1983)
7. N. B. Smith, Gas liquid chromatography of cholesteryl esters on non-polar capillary columns following on column injection. J. Chromatogr. 254: 195 (1983)
8. W.R. Lusby, M.J. Thompson and J. Kochansky, Analysis of steryl esters by capillary gas chromatography electron impact and chemical ionisation-mass spectrometry. Lipids 19: 888 (1984)
9. P.A. Cranwell, N. Robinson and G. Eglinton, Esterified lipids of the fresh water dinoflagellate Peridinium lomnickii. Lipids 20: 645 (1985)

10. S.G. Wakeham and N.M. Frew, Glass capillary gas chromatography - mass spectrometry of wax esters, steryl esters and triacylglycerols. Lipids 17: 831 (1982)
11. a) R.P. Evershed, M.C. Prescott, L.J. Goad and H.H. Rees, High temperature gas chromatography/mass spectrometry of steroid derivatives. Biochem. Soc. Trans. In press (1986). b) R.P. Evershed and L.J. Goad, Capillary gas Chromatography/Mass spectrometry of cholesteryl esters with negative ammonia chemical ionisation. Biomed. Mass Spec. Submitted (1986)
12. Y.Y. Lin, Identification of steroids by chemical ionization mass spectrometry. Lipids 15: 756 (1980)
13. R.E. Garcia and J.B. Mudd, Partial characterisation of steryl ester biosynthesis in spinach leaves. Plant Physiol. 61: 354 (1978)
14. R.E. Garcia and J.B. Mudd, Subcellular distribution of steryl ester biosynthesis in spinach leaves. Plant Physiol. 61: 357 (1978)
15. R.E. Garcia and J.B. Mudd, Identification of an acyl donor in steryl ester biosynthesis by enzyme preparations from spinach leaves. Plant Physiol. 62: 348 (1978)
16. R.E. Garcia and J.B. Mudd, Fatty acid and sterol specificity in the biosynthesis of steryl esters by enzyme preparations from spinach leaves. Arch. Biochem. Biophys. 190: 315 (1978)
17. J. Zimowski and Z.A. Wojciechowski, Acyl donors for sterol esterification by cell free preparations from Sinapis alba roots. Phytochemistry 20: 1795 (1981)
18. J. Zimowski and Z.A. Wojciechowski, Partial purification and specificity of triacylglycerol: sterol acyl transferase from Sinapis alba. Phytochemistry 20: 1799 (1981)
19. M. Kalinowska and Z.A. Wojciechowski, The occurrence of sterol ester hydrolase activity in roots of white mustard seedlings. Phytochemistry 22: 59 (1983)
20. K.J. Treharne, E.I. Mercer and T.W. Goodwin, Incorporation of [^{14}C]carbon-dioxide and [2-^{14}C]mevalonic acid into terpenoids of higher plants during chloroplast development Biochem. J. 99: 239 (1966)
21. R.J. Kemp, A.S.A. Hamman, L.J. Goad and T.W. Goodwin, Studies on phytosterol biosynthesis: observations on the esterified sterols of higher plants. Phytochemistry 7: 447 (1986)
22. L.J. Goad, How is sterol synthesis regulated in higher plants? Biochem. Soc. Trans. 11: 548 (1983)
23. M.A. Hartmann and P. Benveniste, Effects of ageing on sterol metabolism in potato tuber slices. Phytochemistry, 13: 2667 (1974)
24. A.M. Atallah, R.T. Axel, R.B. Ramsey and H.J. Nicholas, Biosynthesis of sterols and triterpenes in Pelargonium hortorum. Phytochemistry 14: 1529 (1975)
25. D.L. Davis and C.G. Ponleleit, Sterols in developing seed from low and high oil Zea mays strains. Phytochemistry 14: 1201 (1975)
26. J.V. Torres and F. Garcia-Olmedo, Genetic control of sterol esterification in developing wheat endosperm. Biochim. Biophys. Acta 409: 367 (1975)
27. R.J. Kemp, L.J. Goad and E.I. Mercer, Changes in levels and composition of the esterified and unesterified sterols of maize seedlings during germination. Phytochemistry 6: 1609 (1967)
28. P. Duperon, Nature et comportement des sterols libre et esterifies, un cours de la germination de divers types de semences. Physiol. Veg. 9: 373 (1971)
29. P.B. Bush and C. Grundwald, Sterol changes during germination of Nicotiana tobacum seedlings. Plant Physiol. 50: 69 (1972)

30. J.V. Torres, P. Carbonero and F. Garcia-Olmedo, Endosperm sterol phenotype and germination in wheat. Phytochemistry 15: 677 (1976)
31. M.A. Hughes and L.J. Goad, The hydrolysis of steryl esters during the germination of barley seed. Biochem. Soc. Trans. 11: 588 (1983)
32. S.P. Singh and L.G. Paleg, Low temperature-induced GA_3 sensitivity of wheat. V. sterol conversions in the wheat aleurone tissue during imbibition. Aust. J. Plant Physiol. 12: 549 (1985)
33. R. Duperon, M. Brillard and P. Duperon, Localisation intracellulaire des differents composes steroliques dans le tubercule de Pomme de terre, avant sa germination. C.R. Acad. Sci. Paris 274: 2321 (1972)
34. a) M. Hartman, M. Ferne, C. Gigot, R.D. Brandt and P. Benveniste, Isolement, caracterisation et composition en sterols de fractions subcellulaires de feuilles etiolees de Haricot. Physiol. Veg. 11: 209 (1973). b) R.D. Brandt and P. Benveniste, Isolation and identification of sterols from subcellular fractions of bean leaves (Phaseolus vulgaris). Biochim. Biophys. Acta 282: 85 (1972).
35. P.B. Bush, C. Grunwald and D.L. Davis, Changes in sterol composition during greening of etiolated barley shoots. Plant Physiol. 47: 745 (1971)
36. C. Grunwald, Shading influence on the sterol balance of Nicotiana tobacum. Plant Physiol. 61: 76 (1978)
37. P.B. Bush and C. Grunwald, Effect of light on mevalonic acid incorporation into the phytosterols of Nicotiana tabacum seedings. Plant Physiol. 51: 110 (1973)
38. L. Westerman and J.G. Roddick, Annual variation in sterol levels in leaves of Taraxacum officinale Weber. Plant Physiol. 68: 872 (1981)
39. C. Liljenberg, P. Karunen and R. Ekman, Changes in steryl lipids of oat root cells as a function of water-deficit stress. Physiol. Plant. 63: 253 (1985)

SYNTHESIS AND BIOLOGICAL EVALUATION OF FUNGAL BIOREGULATORS OF STEROL BIOSYNTHESIS

Edward J. Parish*, Patrick K. Hanners and W. David Nes**

Plant and Fungal Lipid Group
Plant Development and Productivity Research Unit
ARS/U.S. Department of Agriculture
Albany, CA 94710

Corey published a leading paper in 1967 demonstrating that 2,3-iminosqualene inhibited the cyclization of squalene-oxide to lanosterol in a cell-free system[1]. In the interim years between then and this seventh international plant lipid symposium, four groups-one each in the United States[2], Canada[3] another in France[4] and Italy[4], have firmly established the mechanistic and biological importance of using this and structurally related molecules, e.g. 25-azasteroids[5], to interfere with fungal sterol biosynthesis which results in diminished growth response. Industry and biotechnology firms have followed suit in recent years designing analogous inhibitors[6,7] which they believe may have potential promise in plant protection. On the premise that specific N-isopentenoids induce alterations in sterol biosynthesis which in turn may create a pathologic state in the structure and function of membranes of pathogenic fungi, we began a chemical synthesis program in 1984 by preparing blockers targeted at fungal lipid biosynthesis.

The synthetic routes (modified procedures-cf. ref. 3) and growth response to the four N-isopentenoids incubated with Phytophthora cactorum (a sterol-less fungus which requires exogenous sterol for sexual reproduction[7,8] and Gibberella fujikuroi (a sterol-synthesizing fungus which synthesizes multiple 24β-methylsterol end products[9] are shown in Scheme 1 and Table 1. While the research efforts are on-going, we have compiled sufficient preliminary data to conclude the following: (1) Both 2,3-iminosqualene (IS) and 2,3;22,23-diiminosqualene (DIS) inhibit G. fujikuroi growth and induce an accumulation of squalene-oxide indicating they may be (non-competitively) binding with the protosteroid cyclase. (2) [^{3}H]IS is metabolized to [^{3}H]24,25-iminolanosterol (IL) which indicates the epoxidase and cyclase are rather non-specific in terms of substrate recognition. (3) 24,25-iminolanosterol (IL) and 25-aminolanosterol (AL) inhibit C_1-transferase activity in G. fujikuroi and induce growth inhibition: IL is metabolized to AL and an unidentified 4-desmethyl N-containing sterol (m$^+$, 397). (4) IS (Fig. 1) but not IL inhibits growth of P. cactorum. The biochemical affect of IS is assumed to be by blocking the cyclization of squalene to a (as yet unidentified) pentacyclic triterpenoid.

---

*Visiting Scientist, Permanent address-Department of Chemistry, Auburn University, Auburn, Alabama 36830.
**The Plant and Fungal Lipid Group, Plants Physiology Research Unit, The Richard Russell Res. Lab., U.S. Department of Agriculture, College Station Road, P.O. Box 5677, Athens, Georgia, 30613

Table 1. Fungistatic properties of N-isopentenoids

Isopentenoid incubated[1]	Growth-Response in:		Comments
	G. fujikuroi	P. cactorum	
ergosterol	0	S	P.C. metabolizes $\Delta^{5,7,22}$ to $\Delta^{5,22}$
			G.F. metabolizes $\Delta^{5,7,22}$ to $\Delta^{5,7,9(11),22}$
squalene	0	0	G.F. metabolizes lanosterol to 24-methylenelanosterol; while P.C. fails to metabolize lanosterol IL and AL induces an accumulation of 24-methylene lanosterol in G.F.
squalene-oxide	0	0	
lanosterol	0	0	
24,25-iminolanosterol (IL)	I	0	
25-aminolanosterol (AL)	I	0	
2,3-iminosqualene (IS)	I	I	IS and DIS induces an accumulation of squalene-oxide in G.F.
2,3;22,23-diiminosqualene (DIS)	I	N.E.	

[1] Mycelia were grown on agar plates or cultured statically in Erlenmyer Flasks as previously described (8,9); compounds were incubated at 10 ppm; O-No effect, I - 50 per cent or greater inhibition relative to growth of control; N.E. - not examined; S- stimulatory by 20 per cent or greater relative to control.

```
                    Pentacyclic tritertenoids
                         ↑   IS
                         X   DIS
                         |
                    |---------|
                    |         |
Squalene ---→ squalene-oxide --X→ lanosterol --X→ 24(28)methylene24,25-dihydrolanosterol
                    ↑                     ↑
                    IS                    IL
                    DIS                   AL
```

Biosynthesis of 24-alkylated sterols and pentacyclic triterpenoids with the suggested interference by N-isopentenoids.

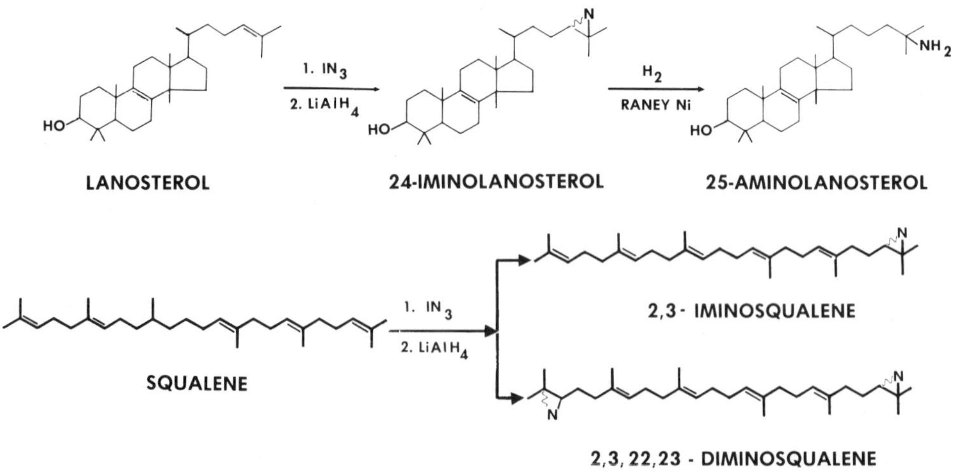

Scheme 1. Structures and Synthesis of N-containing Isopentenoids

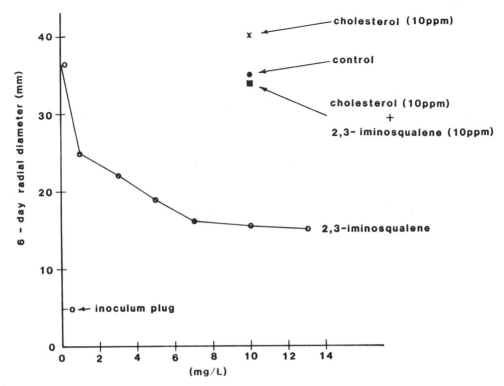

Figure 1. Growth-response of Phytophthora cactroum to Increasing Concentrations of 2,3-Iminosqualene Relative To Control and Cholesterol Supplementation.

References

1. E.J. Corey, P.R.O. de Montellano, K. Lin, and P.D. 2,3-Iminosqualene, a potent inhibitor of the enzymic cyclization of 2,3-oxidosqualene to sterols. J. Amer. Chem. Soc. 89:2797 (1967).
2. W.J. Pinto and W.R. Nes, Stereochemical specificity of sterols in Saccharomyces cerevisiae. J. Biol. Chem. 258:4472 (1983).
3. L. Avruch and A.C. Oehlschlager, A direct synthesis of 2,3-epiminosqualene synthesis 10:622 (1973).
4. A. Rahier, M. Taton, P. Bouvier-Nave, P. Schmitt, P. Benveniste, F. Schuber, A.S. Narula, L. Cattel, C. Anding and P. Place, Design of high energy intermediate analogues to study sterol biosynthesis in higher plants. Lipids 21:52 (1986).
5. A.C. Oehlschlager, R.H. Angus, A.M. Pierce, H.D. Pierce and R. Srinivasan, Azasterol inhibition of Δ^{24} sterol methyltransferase in Saccharomyces cerevisiae. Biochemistry 23:3582 (1984).
6. F. Paltauf, G. Daum, G. Zuder, G. Hogenauer, G. Schulz, and G. Seidl. Squalene and ergosterol biosynthesis in fungi treated with natifine, a new antimycotic agent. Biochim. Biophys. Acta 712:268 (1982).
7. W.D. Nes, P.H. Le, L.R. Berg, G.W. Patterson and J.L. Kerwin, A comparison of cycloartenol and lanosterol biosynthesis and metabolism by the Oomycetes. Experientia 42:556 (1986).
8. W.D. Nes and A.E. Stafford, Evidence for metabolic and functional discrimination of sterols by Phytophthora cactorum. Proc. Natl. Acad. Sci. 80:3227 (1983).
9. W.D. Nes and R.C. Heupel, Physiological requirement for biosynthesis of multiple 24β-methylsterols in Gibberella fujikuroi. Arch. Biochem. Biophys. 244:211 (1986).

A COMPARISON OF STEROL AND LONG CHAIN FATTY ALCOHOL BIOSYNTHESIS IN SORGHUM BICOLOR

Yves Sauvaire[1], Beni Tal[1], Rick C. Heupel[1], Roger England[1], Patrick K. Hanners[1], W. David Nes[1] and J. Brian Mudd[2]

Plant and Fungal Lipid Group
PDP-ARS/U.S. Department of Agriculture, Albany, CA 94710
and [2]Plant Cell Research Institute/ARCO, Dublin, CA 94568

There are two major lipid pathways which operate in crop plants viz., the isopentenoid and fatty acid pathways. To our knowledge, no previous investigation has attempted a comparative study of the composition, biosynthesis and function of the major end products of these pathways. However, in order to assess the physiological importance of having a select lipid profile to, for instance, proceed from one developmental stage into another, the lipid composition in toto-subcellular and surface wax, must be defined. As discussed in what follows, we have observed in this and our other work on sorghum[1-5] that the lipid profile changes dramatically during its development (Table 1). Thus, lipids essentially absent, i.e., pentacyclic triterpenoids, in the seedlings become quantitatively important bulk constituents of the blade's membrane and wax architectures as the plant enters into the flowering stage. Other membrane components have been identified by GLC-MS and 'HNMR eg., fatty acids (myristate, palmitate, oleate, linoleate and linolinate), long chain fatty alcohols (C_{22}-C_{32}-Table 2) and alkanes in the blades.

Table 1. A comparison of the lipid profile at two stages in sorghum's development.

	7-day Leaf[a]		66-day Leaf	
	µg/blade	µg/g dry wt.	µg/blade	µg/g dry wt.
Free sterol	3.7	3700	594.6	646.3
Esterified sterol	0.07	70	195.2	212.2
Free LCFA	10.8	10800	1391.5	1512.5
Esterified LCFA	1.9	1900	640.8	696.5
Free PT	0.5	520	710.3	788.9
Esterified PT	0.14	140	420.4	467.1
Fatty acid of Polar Lipids	37.8	37800	6500.1	7222.2
Fatty acid esters of sterols and other lipids	2.5	2500	225.0	250.0

[a]blades soaked for 20 h. with 12 µg of Na [2-^{14}C]acetate.

Unlike the free sterol content which increases then decreases in a uniform manner, these lipids appear then plateau quantitatively at flowering with their relative distribution e.g., chain length, fluctuating continually throughout the life cycle. Many of the lipids in the plasmalemma are shuttled into the wax as the plants ages.[6,7] However, the relative amount of the total lipid pool which is flip-flopped to the surface is small.

When isotopically labeled (^{14}C and ^{3}H) acetate and mevalonate were supplied to sorghum blades at developmentally different points in the life history of the plant, radioactivity was found to be associated with various sterols, pentacyclic triterpenoids, fatty acids and alcohols. As shown in Tables 3 and 4, most of the radioactivity from these feeds was associated with the isopentenoid end products-sterols and pentacyclic triterpenoids.

The ability for label from [2-^{14}C]mevalonate to be incorporated into docasonol (C_{22}) indicates that sorghum, like wheat, possesses the MVA-shunt pathway. Apparently, regulation of the shunt pathway is influenced by the biosynthetic events controlling fatty acid and alcohol production. The significantly higher incorporation of labeled acetate into sterols compared with docasonol (no radioactivity was associated with the higher homoloques, C_{28} to C_{32}, at any developmental stage examined) indicates: the exogenous uptake of acetate; and its in situ failure to be diluted by cytosolic acetyl CoA. However, it may be diluted in the chloroplast (as the ACP-derivative, thereby affecting fatty acid and alcohol production) which implies a compartmentalization of acetate pools in situ. This view is supported by additional work to be described elsewhere (note - for instance, differences in LCFA composition of 7-day leaves (Table 3) and those soaked in a solution of Na acetate (Table 1) and by the literature[8,9].

Table 2. Long chain fatty alcohol distribution (LCFA)* of sorghum blades during plant maturation.

LCFA	Day of Plant Development						
	6	7	35	48	66	121	131
C_{22}	53.0	20.0	33.0	21.0	17.3	17.7	35.5
C_{24}	0.5	6.0	1.6	2.0	3.3	1.5	0.8
C_{26}	1.5	8.0	2.9	3.2	3.8	2.0	3.1
C_{28}	21.0	37.0	38.6	57.2	60.3	65.0	55.2
C_{30}	4.5	8.0	8.5	14.7	14.9	12.5	5.3
C_{32}	19.5	21.0	16.0	2.0	0.4	1.3	0.1
Total µg/blade	0.73	6.36	487.4	1161.7	1391.5	1408.0	1263.2
Total µg/g dry wt.	734.0	4,837	1523.3	1320.0	1512.1	932.7	1181.2
Total µg/g fr. wt.	80.5	670	229.9	213.9	359.6	221.1	286.2
Length of blade in cm.	1.5	1.5	39.3	55.5	54.0	57.0	49.5

*As percent total primary LCFA: A mixture of unsaturated LCFA and LCFA with secondary alcohols were increasingly present in the older leaves, but were not quantified.

Table 3. Radioactivity associated with docasonol (C_{22}) and sterols of sorghum incubated with labeled acetate and/or mevalonate.

Treatment[1]	Period of incubation and amount of isotope added	Radioactivity in TLE (dpm)	Radioactivity on initial TLC plate:			Radioactivity (dpm) corresponding to chromatograhically purified docasonol[4]
			R_f[2] 0.30	R_f 0.46	R_f[3] 0.51	
[^{14}C] acetate (lightgrown) to - I	24h; 25μCi	16.8×10^6	1.63×10^4	3.52×10^4	5.19×10^4	5.0×10^3
[^{14}C] acetate (darkgrown) to - I	24h; 25μCi	1.37×10^6	5.5×10^4	2.6×10^4	4.0×10^4	4.8×10^3
[^{14}C]mevalonte (lightgrown) to - I	20h; 25μCi	3.12×10^6	2.66×10^5	1.52×10^5	4.14×10^5	3.1×10^3
[^{14}C]mevalonte (darkgrown) to - I	20h; 25μCi	4.39×10^6	3.51×10^5	1.24×10^5	6.32×10^5	3.1×10^3
[^{3}H]acetate and [^{14}C]mevalonte (lightgrown) to - I	20 h ^{3}H-25μCi ^{14}C-14.8μCi	^{3}H -9 $\times 10^5$ ^{14}C-3.8 $\times 10^6$	^{3}H-4.7 $\times 10^5$ ^{14}C-3.2 $\times 10^5$	^{3}H-4.1 $\times 10^4$ ^{14}C- 4.5 $\times 10^5$ [5]		^{3}H - b. g. ^{14}C-1.0 $\times 10^3$
[^{14}C]acetate to - II	20h; 25μCi	1.57×10^6	1.20×10^4	8.30×10^4	1.40×10^4	1.0×10^3
[^{14}C]-mevalonte to - II	72h; 24.5μCi	4.6×10^6	4.10×10^4	1.0×10^4	2.5×10^6	2.4×10^3
[^{3}H]acetate and [^{14}C]mevalonte to - II	20h ^{3}H-25μCi ^{14}C-6.3μCi	^{3}H -4.8 $\times 10^6$ ^{14}C-2.6 $\times 10^6$	^{3}H- 1.1 $\times 10^4$ ^{14}C- 2.19 $\times 10^5$	^{3}H -b.g. ^{14}C- 1.2 $\times 10^6$		^{3}H -b.g. ^{14}C - 7.3 $\times 10^3$

[1] Primary (7-day) leaves - I; 66-day leaf blades - II; blades were detached from the seedlings and incubated (floated) in a buffered solution containing the appropriate labeled substrates. 66-day blades were treated with the substrate dissolved in a silicon oil-locopherol solution.

Labeled substrates: [2-^{14}C]acetate, [2-^{3}H]acetate and [2-^{14}C]mevalonate.

[2] This band corresponds to 4-desmethyl sterols and contains NO LCFA.

[3] This band is a mixture of LCFA and pentacyclic triterpenoids. Most of the radioactivity was associated with the triterpenoids.

[4] Docasonol was purified by sequential chromatographies. Initial adsorption TLC (Benzene-diethylether, 9:1), bands at R_f 0.46 and at R_f 0.51 (in some cases 100 g of carrier C_{22} was added) corresponding to C_{22} and C_{24} were eluted from the plate and rechromatographed using RP-HPLC (4% aq. MeOH; Pump, 1.6 ml/min). The fraction at 19 min (C_{22} identify confirmed by GC) was acetylated and chromatographed using $AgNO_3$ TLC developed with Hexane:benzene (40:60).

[5] Scraped two bands as one.

Table 4. Specific activities (dpm/μg) of the major sterols isolated from [2-^{14}C]mevalonate and [2-^{3}H]acetate feeds to sorghum.

Sterol[1]	Isotope	7-day leaf blade	66-day leaf blade
Cholesterol	^{14}C	N.d.	923
	^{3}H	N.d.	144
Isofucosterol	^{14}C	1500	N.d
	^{3}H	300	N.d
Campesterol	^{14}C	3370	960
	^{3}H	356	259
Stigmasterol	^{14}C	379	24
	^{3}H	66	2
Sitosterol	^{14}C	2110	556
	^{3}H	297	97

[1]Sterols were sequentially purified by adsorption TLC, RP-HPLC, AgNO$_3$ TLC of acetylated HPLC fractions and recrystallization (in selected cases): carrier sterols were added before HPLC. See Table 3 for incubation procedures.

References

1. W.D. Nes, R.C. Heupel, M. Benson, A.E. Stafford and W.F. Haddon, The stereochemistry of migrated hopanes epimeric at C-21. J. Chem. Soc. Chem. Commun. 223 (1984.
2. R.C. Heupel, and W.D. Nes, Evidence for differences in sterol biosynthesis and derivatization in sorghum. J. Nat. Prod. 47:292 (1984).
3. R.C. Heupel and W.D. Nes, The biosynthesis, metabolism and translocation of ß-amyrin in Sorghum bicolor. Phytochemistry 24:2905 (1985).
4. R.C. Heupel, Varietal similarities and differences in the polycyclic isopentenoid composition of sorghum. Phytochemistry 24:2929 (1985).
5. R.C. Heupel, Y, Sauvaire, P.H. Le, E.J. Parish and W.D. Nes, Sterol composition and biosynthesis in sorghum: importance to developmental regulation. Lipids 21:69 (1986).
6. G. Bianchi, P. Avato, P. Bertorelli and G. Mariani, Epicuticular waxes of two sorghum varieties. Phytochemistry 17:999 (1978).
7. D.S. J Atkin and R.J. Hamilton, the changes with age in the epicuticular wax of Sorghum bicolor. Nat. Nat. Prod. 45:697 (1982).
8. W.D. Nes and T.J. Bach, Evidence for a mevalonate shunt in a tracheophyte. Proc. R. Soc. Lond. B. 225:425 (1985).
9. M.R. Pollard, T. McKeon, L.M. Gupta and P.K. Stumpf, Studies on biosynthesis of waxes of developing jojoba seed. II. The demonstration of wax biosynthesis by cell-free homogenates. Lipids 14:651 (1979).

REGULATION OF TERPENOID BIOSYNTHESIS IN TAPPED LATEX

George J. Piazza, Edward J. Saggese, and Marvin P. Thompson

Agricultural Research Service
U. S. Department of Agriculture
Philadelphia, PA 19118

INTRODUCTION

The tapped latex of Euphorbia lathyris can convert acetate to several structurally different triterpenols (TOH) and their fatty acid esters (TE) (1). Simple low speed centrifugation of the latex affords a pellet that utilizes mevalonic acid as a triterpenoid precursor (2). Biosynthetic activity in the latex pellet is absolutely dependent upon the presence of an osmoticum (0.4 M sorbitol), indicating that biosynthesis occurs in an osmotically sensitive organelle. Calcium constitutes about 7.5% of dry latex weight, and the addition of Ca^{2+} to an EGTA-treated latex pellet stimulates triterpene biosynthesis. This study was undertaken to investigate whether the calcium binding protein, calmodulin, mediates Ca^{2+} effects upon triterpenoid biosynthesis.

METHODS

Triterpene biosynthesis. Freshly tapped latex (100 µl per assay tube) was diluted three-fold with cold buffer containing 10 mM Na-phthalate (pH 5.5), 10 mM $MgCl_2$, 10 mM KCl, 30 mM $CaCl_2$ and 0.4 M sorbitol, and centrifuged for 5 min at 8,800 g. The pellet was resuspended in 220 µl buffer containing 1.0 nmol (0.01 mCi) R-(5-^{3}H)-mevalonic acid, 0.66 µmol dithiothreitol, and 22 nmol S-adenosylmethionine. Radiolabel in triterpenes was determined as described (1).

Phosphodiesterase Activity. Assays were performed at room temperature as described (3) except that $CaCl_2$ (18 µM) and calmodulin (0.1 µg or as indicated) were present.

Isolation of Calmodulin. Upper stems and leaves of E. lathyris were homogenized in the buffer (pH 6.5) described by Schreiber et. al. (4), also containing 20 mM mercaptoethanol. After centrifugation, the supernatant was heated to 85°C for 5 min, rapidly cooled, and recentrifuged. The supernatant was slurried with DE-52 cellulose, and the suspension was poured into a glass column. Elution of protein was achieved with a salt gradient of 0.1-0.5 M NaCl. The calmodulin fraction was purified to homogeneity by chromatography on Sephadex G-100 at pH 7.5 with a buffer containing EGTA.

RESULTS AND DISCUSSION

The addition of calmodulin antagonists (compounds I, II, III) to the latex pellet at a concentration of 200 µM severely inhibits TOH and TE biosynthesis. I_{50} values for compounds II and III are 150 and 55 µM, respectively, consistent with an action upon calmodulin, rather than a nonspecific mode of inhibition. Inhibition of biosynthesis by phenoxyalkylamines (compounds IV, V, VI) increases with increasing chlorine substitution. Assays using calmodulin-stimulated phosphodiesterase show that antagonism toward calmodulin increases in the same order.

A direct test of calmodulin involvement by an observed stimulation of biosynthetic activity by exogenously added calmodulin is not feasible due to the presence of a limiting membrane around the site of biosynthesis, and to date we have not observed biosynthesis in broken organelles.

Our goal was to demonstrate the presence of calmodulin in tapped latex. To obtain an appropriate standard, calmodulin was isolated and purified to homogeneity from E. lathyris. The isolated calmodulin is very acidic and has a molecular weight of about 17 KDa, similiar to calmodulins isolated from other plants (5). The amino acid analysis of E. lathyris calmodulin is shown following with the values in parentheses being those of spinach calmodulin: Asp 27(24), Thr 7(9), Ser 7(4), Glu 30(27), Pro 4(2), Gly 11(10), Ala 10 (11), Cys 1(1), Val 7(8), Met 5(8), Ile 6(7), Leu 9(11), Tyr 1-2(1), Phe 7(9), His 1(1), TML 1(1), Lys 9(9), Trp 0(0), Arg 4(5).

Fig. 1 shows that E. lathyris calmodulin is a good stimulator of bovine heart phosphodiesterase. The data were best fitted with two distinct dissociation constants, indicating an interaction by two different modes. In contrast, bovine brain calmodulin interacts with phosphodiesterase in a way that does not reveal any heterogeneity in binding sites.

When whole tapped latex is subjected to polyacrylamide gel electrophoresis in the presence of urea, a protein band corresponding to authentic calmodulin is observed (arrow, Fig. 2). Calmodulin is also detected when the latex is treated with EGTA. Calmodulin is not visible in the electophoresis of untreated latex, indicating that it remains bound to its target protein(s).

Table 1. Effect of Calmodulin Antagonists on Triterpene Biosynthesis

	COMPOUND		TRITERPENE BIOSYNTHESIS (control = 100%)
			%
I	Chlorpromazine	TOH	9 ± 2
		TE	7 ± 1
II	Fluphenazine	TOH	31 ± 3
		TE	26 ± 8
III	Trifluoperazine	TOH	3 ± 1
		TE	1 ± 1
IV	2-(2,6-dichlorophenoxy) ethyl N,N-diethylamine	TOH	87 ± 6
		TE	72 ± 8
V	2-(2,4,5-trichlorophenoxy) ethyl N,N-diethylamine	TOH	17 ± 6
		TE	20 ± 4
VI	2-(pentachlorophenoxy) ethyl N,N-diethylamine	TOH	1 ± 1
		TE	2 ± 1

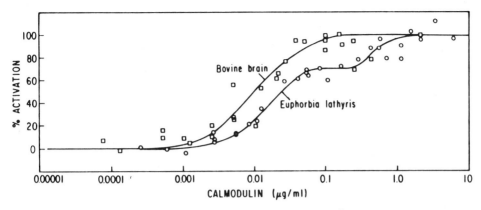

Figure 1. Activation of bovine heart phosphodiesterase-catalyzed hydrolysis of cAMP.

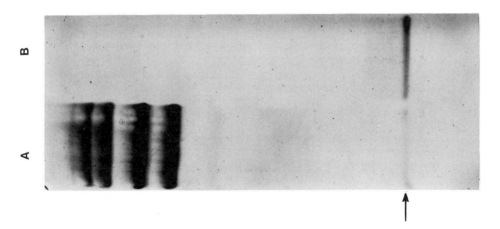

Figure 2. Polyacrylamide gel electrophoresis of whole latex (lane A) and purified E. lathyris calmodulin (lane B). Gel conditions: 4.5 M urea, 12% acrylamide. Running buffer: Tris-glycine, pH 8.3. Samples were dissolved in running buffer containing tracking dye, glycerol, 4.5 M urea, and 1.2 mM mercaptoethanol.

REFERENCES

1. E.K. Nemethy, C. Skrukrud, G.J. Piazza, and M. Calvin, Terpenoid biosynthesis in Euphorbia latex, Biochim. Biophys. Acta 760:343 (1983).
2. G. Ponsinet and G. Ourisson, Aspects particuliers de la biosynthese des triterpenes dans le latex D'Euphorbia, Phytochemistry 7:757 (1968).
3. R.W. Butcher, Cyclic 3',5'-nucleotide phosphodiesterase from bovine heart, Methods in Enzymology 38:218 (1974).
4. W.E. Schreiber, T. Sasagawa, K. Titani, R.D. Wade, D. Malencik, and E.H. Fisher, Biochemistry 20:5239 (1981).
5. G.J. Piazza, Calmodulin in plants, in: "Calcium Binding Proteins," M.P. Thompson, ed., CRC Press, Boca Raton, FL., in press.

TRITERPENOID BIOSYNTHESIS IN EUPHORBIA LATHYRIS

C. L. Skrukrud, S. E. Taylor, D. R. Hawkins, and M. Calvin

Lawrence Berkeley Laboratory and Department of Chemistry
University of California, Berkeley, CA

INTRODUCTION

Triterpenols and their esters make up 4-5% of the dry weight of Euphorbia lathyris.[1] A significant amount of these triterpenoids are synthesized and stored within laticifer cells where these compounds constitute up to 50% of the dry weight of the latex.[2] We are interested in the nature of these compounds and the organization and regulation of their biosynthesis. Here we report on recent work investigating the control of carbon allocation to these compounds, further elucidation of the identity of the triterpenols found in latex, evidence indicating that the conversion of HMG-CoA to mevalonate (MVA) is rate-limiting in the biosynthesis of triterpenoids in latex, and steps in the purification of HMG-CoA reductase (HMGR).

MATERIALS AND METHODS

Latex incubation, isolation and analysis of triterpenoids were performed as previously described.[2]

Authentic 24-methylenelanosterol was synthesized from lanosterol by the method of Barton.[3] ^{1}H-NMR spectra were determined at 250 MHz. Lanthanide isotope shifts (LIS) were measured in the presence of Eu(fod)$_3$.

Rates of acetate (Ac) and MVA incorporation into triterpenoids were determined by incubation of latex with saturating concentrations of ^{3}H-Ac (0.25 mM, 2 Ci/mmol) or ^{3}H-MVA (1 mM, 50 mCi/mmol) and 10 mM DTE. HMGR assay was done on the sonicated pellet of a 5000 x g, 15 min centrifugation of latex diluted 2:1 with 10 mM MES, 10 mM DTE, 30 mM EDTA, 0.4 M sorbitol, pH 6.5. HMGR incubation containing 0.3 mM ^{3}H-HMG-CoA (33 mCi/mmol) and 3.3 mM NADPH was quenched with KOH then acidified. Mevalonolactone (MVAL) was purified by silica gel TLC (1:1, acetone:benzene), further purified by HPLC using an organic acid column in 0.0025 N H$_2$SO$_4$, and analyzed by LSC.

For purification of HMGR, stem and leaf tissue were ground with buffer containing 10 mM K phosphate, 0.4 M sorbitol, 30 mM EDTA, 10 mM DTE, and 10% w/w insoluble PVP, pH 7.2. The filtered extract was centrifuged as described in Fig. 1. Pellets were resuspended in 0.1 M K phosphate, 30 mM EDTA, 10 mM DTE, and 0.04% Triton X-100, various pHs for pH curve. Solubilization of HMGR was tested by preparing an 18,000 x g pellet using buffer of pH 6.8,

Table 1. Evidence for HMGR as Rate-determining in Latex Triterpenoid Biosynthesis

Ac ⟶ Triterpenoids	0.02 nmol inc./100 μl latex/hr
MVA ⟶ Triterpenoids	0.55
HMG-CoA ⟶ MVA	0.03

then resuspending in buffer containing detergents listed in Table 2. Incubations sat 20 min on ice, were diluted 3-fold and centrifuged at 100,000 x g for 1 hr. For HMGR analysis, samples were incubated with either ^{3}H- or ^{14}C-HMG-CoA (0.6 mM, 5.6 mCi/mmol or 2.2 mCi/mmol, respectively) and 2.5 mM NADPH. Incubations were quenched with 6 N HCl and purified by silica gel TLC (2:1, CHCl$_3$:acetone). The MVAL band was eluted from the silica and analyzed by LSC. Protein was determined per Bradford.[4]

Glucose incorporation into latex was done by combining latex with ^{14}C-glucose (4 mM, 6 mCi/mmol), 2 mM MgCl$_2$, 1 mM NADPH, 1 mM ATP, and 0.33 M sorbitol. Two experimental conditions were used: unbuffered latex (pH 5.5); latex buffered to pH 7.0 (approx. cytoplasmic pH). Buffer (0.1 M MOPS) and fructose 2,6-bisphosphate (F26BP, 50 mM for pH 5.5, 2 mM for pH 7.0) were added as indicated in Table 3. Additional controls with Na pyrophosphate (PPi, 2 mM) and fructose 6-phosphate (F6P, 50 mM) were also performed. Incubations were run for 1 hr then analyzed for incorporation into triterpenoids.

RESULTS AND DISCUSSION

Nielsen et al.[5] identified the major components of latex as the tetracyclic triterpenols cycloartenol, 24-methylenecycloartenol, lanosterol, and an isomer of lanosterol. We have indentified this last compound as butyrospermol (5α-eupha-7,24-dien-3β-ol). We have also isolated 24-methylenelanosterol from latex. These structures were determined by comparison of HPLC elution, mass spectra, ^{1}H-NMR spectra, LIS, and optical rotations with either literature values or authentic samples.

The conversion of HMG-CoA to MVA catalyzed by HMGR is acknowledged as the major rate-limiting step in cholesterol biosynthesis in mammalian systems. As shown in Table 1, comparison of the conversion of Ac and MVA into triterpenoids establishes that the rate-limiting step in the pathway in E. lathyris latex occurs prior to MVA. As we have demonstrated previously[2], exogenously supplied HMG-CoA is not metabolized to triterpenoids in latex. However, HMGR activity measured in sonicated latex corresponds with the overall Ac conversion rate, suggesting that this step is rate-limiting in the formation of triterpenoids in latex.

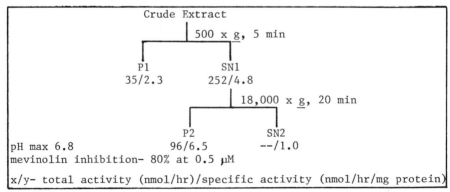

Fig. 1. Concentration of HMGR by differential centrifugation.

Table 2. Solubilization of HMGR from P2

Treatment of P2	HMGR Activity as a % of SN1	
	Supernatant	Pellet
none	5%	16%
1% Triton X-100	2	2
1% Triton X-100 + 2% SDS	2	1
0.25% Deoxycholate	25	30

In purification of HMGR from the whole plant, the first step can be achieved by obtaining an 18,000 x g pellet as in Fig. 1. As shown in Table 2, HMGR can be solubilized from the pellet. Deoxycholate treatment increased HMGR activity 2.6-fold over the untreated sample as well as increasing the ratio of solubilized HMGR to pelletable HMGR.

Most of the reduced carbon found in laticifer cells is in starch and triterpenoids, which are produced at near equal levels. It is possible that the rate of triterpenoid production is controlled at the initial conversion of sugar to the early precursors of triterpenoids or starch. One compound that has recently been found to control carbon utilization within a plant cell is F26BP which mediates the rate of conversion of F6P to fructose 1,6-bisphosphate.[6] To investigate the role of F26BP in latex we followed the incorporation of ^{14}C-glucose into triterpenoids with and without added F26BP. The results of the experiments (Table 3) indicate a role for F26BP in carbon allocation in latex. They also indicate that it is not a product of F26BP breakdown (F6P or PPi) that mediates the conversion.

Table 3. The Effect of Fructose 2,6-Bisphosphate on Glucose Incorporation into Triterpenoids

	Sterols (dpm above Background)	
	latex (pH 5.5)	latex + MOPS (pH 7.0)
+ ^{14}C-glucose	0	795
+ ^{14}C-glucose, fructose 2,6-dP	369	1443
+ ^{14}C-glucose, fructose 6-P	0	---
+ ^{14}C-glucose, Na pyrophosphate	---	772

Acknowledgement- This work was supported by the Assistant Secretary for Conservation & Renewable Energy, USDOE under Contract DE-AC03-76 SF00098.

REFERENCES

1. E. K. Nemethy, J. W. Otvos, and M. Calvin, Analysis of extractables from one Euphorbia, J. Am. Oil Chem Soc. 56:957 (1979).
2. E. K. Nemethy, C. Skrukrud, G. J. Piazza, and M. Calvin, Terpenoid biosynthesis in Euphorbia latex, Biochim. Biophys. Acta 760:343 (1983).
3. D. H. R. Barton, D. M. Harrison, G. P. Moss, and D. A. Widdowson, Investigations on the biosynthesis of steroids and triterpenoids, J. Chem. Soc. (C), 775 (1970).
4. M. M. Bradford, A rapid and sensitive method for the quantitation of microgram quantities of protein utilizing the principle of protein-dye binding, Anal. Biochem. 72:248 (1976).
5. P. E. Nielsen, H. Nishimura, J. W. Otvos, and M. Calvin, Plant crops as a source of fuel and hydrocarbon-like materials, Science 198:942 (1977).

6. C. C. Black, D. A. Smyth, and M-X Wu, Pyrophosphate-dependent glycolysis and regulation by fructose 2,6-bisphosphate in plants, in "Nitrogen Fixation and CO_2 Metabolism," P. W. Ludden and J. E. Burris, eds., Elsevier Sci. Pub. Co., New York (1985).

PHYTOSTEROL STRUCTURE AND COMPOSITION IN THE CHEMOSYSTEMATICS OF THE CARYOPHYLLALES

John H. Adler and Thomas A. Salt

Dept. of Biological Sciences
Michigan Tech. University
Houghton, MI   USA

Dept. of Botany
University of Maryland
College Park, MD   USA

INTRODUCTION

Angiosperms synthesize a diverse array of 4-desmethylsterols which vary primarily in position and number of double bonds as well as stereochemistry at C-24[1,2]. The high degree of structural chemistry[1,2] associated with sterols, their essentiality for proper eucaryotic cell function[3,4,5], and their ubiquitous distribution in the plant kingdom[1,2] make phytosterols potentially useful molecules for the characterization of taxonomically related organisms. Most angiosperms produce 24α-alkyl-Δ^5-sterols with relatively few species producing 24-alkyl-Δ^7-sterols as dominant sterols[1,2].

The order Caryophyllales is composed of 12 families[6] (Fig. 1) and was selected for a chemosystematic analysis of sterols based on early reports of 24-alkyl-Δ^7-sterols in three families i.e. Spinacia (Chenopodiaceae)[7], Lophocereus (Cactaceae)[8] and Phytolacca (Phytolaccaceae)[9]. Phytosterol structure and composition was analyzed with respect to each plant's taxonomic assignment. Isolation, identification methods, taxonomic classification, nomenclature systems and tribal arrangements are presented elsewhere [10,11,12]. Only mature photosynthetic tissue was analyzed to avoid variations in sterol content due to ontogeny.

RESULTS AND DISCUSSION

The sterol composition for 39 species in seven families of the order is reported in Table 1 and in detail elsewhere[10,11,12]. Twenty species from six families produce 24-alkyl-Δ^7-sterols as the exclusive or predominant sterols (>60% of total 4-desmethylsterols), while 13 species from three families produce Δ^5-sterols. Five species, three in Chenopodium and two in Salicornia, synthesize 1:1 ratios of Δ^7:Δ^5-sterols

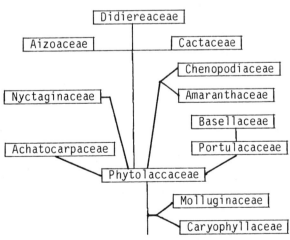

FIGURE 1. Proposed relationships of families in the order Caryophyllales[6].

STEROL COMPOSITION OF CARYOPHYLLALES
Percent of 4-Desmethylsterols

Nuclear Double Bond Configuration at C-24	Δ^7		Δ^5	Δ^0
	24α- a	24β-		
CARYOPHYLLACEAE				
Sclerantheae				
Scleranthus annuus	100	-	-	-
Paronychieae				
Paronychia virginica	100	-	-	-
Lychnideae				
Lychnis alba	100	-	-	-
Silene cucubalus	100	-	-	-
Diantheae				
Dianthus armeria	100	-	-	-
Gypsophilia paniculata	100	-	-	-
Saponaria officinales	100	-	-	-
Alsineae				
subtribe Sabulininae				
Spergula arvensis	20	80	-	-
Minuartia caroliniana	55	45	-	-
subtribe Stellariinae				
Myosoton aquaticum	100	-	-	-
Cerastium arvense	73	-	20	7
Cerastium vulgatum	53	-	23	24
Arenaria serpyllifolia	67	-	17	16
CHENOPODIACEAE				
Chenopodioideae				
Chenopodium ambrosioides	100	-	-	-
Chenopodium rubrum	100	-	-	-
Chenopodium album	53	-	45	2
Chenopodium urbicum	47	-	52	1
Chenopodium leptophyllum	43	-	57	-
Salicornia europaea	51	-	44	5
Salicornia bigelovii	41	-	51	8
Atriplex arenaria	19	-	81	-
Beta vulgaris	73	-	27	-
Bassia hirsute	5	-	95	-
Kochia scoparia	-	-	100	-
Salsoloideae				
Salsola kali	3	-	91	6
Suaeda linearis	-	-	76	24
CACTACEAE				
Echinopsis tubiflora	-	-	100	-
Epiphyllum sp.	8	-	88	4
Hylocereus undatus	-	-	100	-
Notocactus scopa	-	-	92	8
Opuntia comonduensis	-	-	100	-
Opuntia humifusa	-	-	100	-
Pereskia aculeata	-	-	100	-
Schlumbergera bridgesii	9	-	92	-
PHYTOLACCACEAE				
Phytolacca americana	100	-	-	-
AIZOACEAE				
Tetragonia expansa	100	-	-	-
BASELLACEAE				
Basella alba	>93	-	<7	-
PORTULACACEAE				
Portulaca grandiflora	>98	-	<2	-
Claytonia virginica	>98	-	<2	-

a) 24α- and planar C-24.

while _Cerastium vulgatum_ produces a 2:1:1 ratio of $\Delta^7:\Delta^5:\Delta^0$-sterols. The Amaranthaceae also synthesize Δ^7-sterols[13]. Most species in the Caryophyllaceae, Amaranthaceae and those few examined in the Phytolaccaceae, Aizoaceae, Basellaceae and Portulacaceae synthesize Δ^7-sterols. The 24β-alkyl-Δ^7-sterols appear to be limited to the subtribe Sabulininae in the Caryophyllaceae. The Cactaceae with one exception[8] synthesize Δ^5-sterols. Species in the Chenopodiaceae synthesize a diversity of sterols ranging from predominantly Δ^7-sterols → Δ^5-sterols and mixtures of both Δ^7- and Δ^5-sterols in relatively fixed percentage compositions[10]. Stanols (5αH) are only found in plants producing Δ^5-sterols[10,11,12] and are likely derived from Δ^5-sterols.

The sterol composition of the Caryophyllales may reflect differing functional requirements and regulatory controls, and/or may reflect the evolutionary origins of the plants[1]. If the putative familial relationships are correct[6] (Fig. 1), then the ancestors of the order are presumably Δ^7-sterol synthesizing plants. The Phytolaccaceae may therefore possess among its 125 species some plants which synthesize Δ^5-sterols. The ancestors of these Δ^5-sterol plants may have given rise to the Cactaceae and parts of the Chenopodiaceae. Likewise the Δ^7-sterol producing families may have arizen from Δ^7-sterol ancestors in the Phytolaccaceae. The 24β-alkyl-Δ^7-sterols in the Caryophyllaceae would support its more basal assignment since 24β-ethyl sterols occur in lower photosynthetic organisms[1,2]. A careful systematic analysis of additional species in this order will provide further insight into the significance of sterol structural and compositional data with respect to chemotaxonomy.

REFERENCES

1. W. R. Nes and W. D. Nes, "Lipids in Evolution", Plenum Press, New York (1980).
2. L. J. Goad, "The Biosynthesis of Plant Sterols", in "Lipids and Lipid Polymers in Higher Plants", Tevini, M., and Lichtenthaler, H.K. eds. pp. 146-168, Springer-Verlag, Berlin (1977).
3. R. J. Rodriguez, C. Low, C. D. K. Bottema, and L. W. Parks, "Multiple Functions for Sterols in _Saccharomyces cerevisiae_", Biochim. Biophys. Acta 837:336 (1985).
4. W. R. Nes, B. C. Sekula, W. D. Nes and J. H. Adler, "The Functional Importance of Structural Features of Ergosterol in Yeast", J. Biol. Chem. 253:6218 (1978).
5. C. H. Moeller and J. B. Mudd "Localization of Filipin - Sterol Complexes in the Membranes of _Beta vulgaris_ Roots and _Spinacia oleracea_ Chloroplasts", Plant Physiol. 70:1554 (1982).
6. A. Cronquist, "An Integrated System of Classification of Flowering Plants", Columbia University Press, New York (1981).
7. T. Itoh, T. Tamura, and T. Matsumoto, "Sterols, Methylsterols and Triterpene Alcohols in Three Theaceae and Some Other Vegetable Oils" Lipids 9:173 (1974).
8. C. Djerassi, G. W. Krakower, A. J. Lemin, L. H. Liu, J. S. Mills and R. Villotti, "The Neutral Constituents of the Cactus _Lophocereus schottii_,", J. Amer. Chem. Soc. 80:6284 (1958).
9. W. S. Woo and S. S. Kang, "Phytosterolins From _Phytolacca esculenta_" J. Pharmaceutical Soc. Korea 17:161 (1973).
10. T. A. Salt and J. H. Adler, "Diversity of Sterol Composition in the Family Chenopodiaceae" Lipids 20:594 (1985).
11. T. A. Salt and J. H. Adler, "Dominance of Δ^7-Sterols in the Family Caryophyllaceae", Lipids, In Press.
12. T. A. Salt and J. H. Adler, "Dominance of Δ^5-Sterols in Eight Species of the Cactaceae", Phytochemistry, In Press.
13. S. Xu, G. W. Patterson and K. Schmid, "Sterols in Amaranthaceae", Phytochemistry, In Press.

BIOSYNTHESIS OF DITERPENE PHYTOALEXIN PRECURSORS IN

CELL-FREE EXTRACTS OF RICE

Karen Wickham and Charles A. West

University of California, Los Angeles
Department of Chemistry and Biochemistry
Los Angeles, California 90024

INTRODUCTION

Rice leaves have been reported to accumulate two groups of diterpenoid phytoalexins, the momilactones and the oryzalexins, in response to infection with <u>Pyricularia oryzae</u> (Kono et al, 1985) or irradiation with UV-light (Cartwright et al, 1981). Scheme 1 illustrates proosed pathways and key intermediates in the biosynthesis of these phytoalexins based on known pathways of biosynthesis of polycyclic diterpenes such as ent-kaurene. Cyclization of geranylgeranyl-PP to two labadienyl pyrophosphates of appropriate stereochemistry followed by further cyclization of these intermediates could give rise to two tricyclic pimaradienes. A series of oxidative transformations of 9β H-pimara-7,15-diene and sandaracopimaradiene would be responsible for the formation of the momilactones and oryzalexins, respectively. This paper reports evidence for the formation of pimaradienes and ent-kaurene in cell-free extracts of UV-treated rice leaves.

MATERIALS AND METHODS

Cell free enzyme extracts were prepared from the foliar parts of rice plants, cultivar M101 (supplied by J.N. Rutger, USDA UC Davis). Leaves were exposed to 20 min of short wavelength UV radiation and the maintained in a growth chamber at $24°C$ in a 14 hr light - 10 hr dark cycle. Tissue was homogenized in liguid N_2 in the presence of half the tissue weight of insoluble polyvinylpyrollidone. A 50mM Hepes buffer, pH 6.8. (3ml/g tissue) containing 10mM $Na_2S_2O_5$, 10mM ascorbic acid, 5mM dithiothreitol, 150mM sucrose, 10mM glycerol by volume, 1μM leupeptin, 1mM EDTA, and 1mM phenyl-methanesulfonyl fluoride was added to the ground tissue. The suspension was filtered and centrifuged at 27,000g for 20 min to provide the soluble supernatant fraction that served as the source of enzymes. Emzyme extracts were frozen before use.

A typical assay contained 150μl of enzyme extract, 150μl of 10mM KH_2PO_4 byffer, pH 7.0, containing 5mM $MgCl_2$, and 100μl of 50mM (3H)-GGPP or (^{3}H)-CPP in a total volume of 0.5ml. Incubations were carried out at $30°C$ for 60 min and quenched with ethanol. Products were extracted with petroleum ether. The diterpene hydrocarbons were separated by argentation TLC on silica gel plates developed in hexane/benzene (13:7) and quantitated by measuring the radioactivity associated with appropriate regions of the plate.

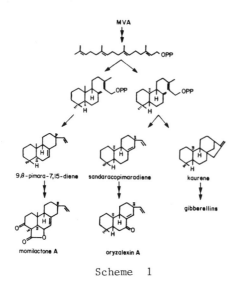

Scheme 1

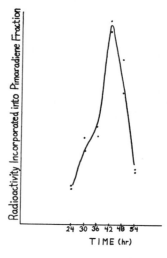

Figure 1. Time Course

RESULTS AND DISCUSSION

Authentic kaurene migrated 10cm from the origin in the TLC system used to analyze diterpene hydrocarbon products. Based on previous results we would expect the pimaradienes to be present in the region 8 to 10 cm from the origin. Extracts from UV-treated rice tissue incorporated substantial amounts of radioactivity into the pimaradiene region of the TLC plate, whereas extracts from untreated tissue yielded only about 3% as much radioactivity in this region (Table 1). Maximum cyclase activity occurs in extracts prepared 42 hr after UV-treatment (Fig. 1). Cyclase activity in freshly prepared extracts appears to be unstable with a half-life of about 19 hr at $4^\circ C$, but activity can be completely restored to partially inactivated preparations by freezing them.

Large scale incubations were undertaken with (^{3}H)-GGPP as substrate to obtain enough polycyclic diterpene hydrocarbons for identification. We predicted on the basis of Scheme 1 that 9 H-pimara-7,15-diene and sandaracopimaradiene should accumulate in extracts of UV-irradiated tissues along with kaurene. The products extracted from incubation mixtures with UV-treated tissue were resolved on a 15cm $AgNO_3$-silicic acid column run in hexane/benzene (30:1). Pooled fractions 3, 4 and 5 (Fig. 2) contained radiolabelled hydrocarbons that run in the pimaradiene region on TLC. GC-MS analysis indicates that fraction 3 contains two diterpenes of mass 272 (A and B), fraction 4 contains one diterpene of mass 272 (C), and fraction 5 contains three diterpenes of mass 272 (C, D and E). authentic samples of kaurene and sandaracopimaradiene had identical retention times with diterpenes B and D, respectively.

TABLE 1 : Effects of UV-treatment on Incorporation of Radioactivity into the Pimaradiene Region

Substrate	Diterpenes in Untreated Tissue (fmol)	Diterpenes in UV-treated Tissue (fmol)
GGPP	15	490
CPP	63	910

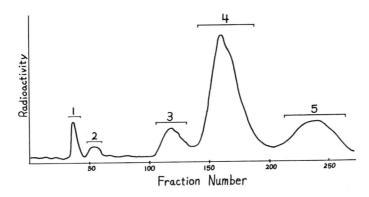

Figure 2. Silver nitrate-Silicic acid Column Chromatography of Hydrocarbons-Radioactivity Elution Profile

Comparison of the mass spectra for B and D with the mass spectra of authentic sanples of kaurene and sandaracopimaradiene confirms that B is kaurene and D is sandaracopimaradiene. The dominant fragment of mass 137 in the spectra of both A and D (sandaracopimaradiene) suggests they are structurally related with a site of unsaturation at 8(14). The mass spectra of C and E are similar to one another and consistent with the presence in each of a site of unsaturation at 7(8). C is most likely 9βH-pimara-7,15-diene, the precursor of the momilactones (Scheme 1).

GC analysis of the diterpenes produced from (^{3}H)-GGPP by extracts derived from untreated tissue shows that only kaurene is formed. The three major diterpenes were identified with CPP as a substrate.

SUMMARY

UV-treated rice leaves have a greatly increased capacity for biosynthesis of diterpenes in comparison with untreated controls. Among the diterpene hydrocarbon products, ent-kaurene, a gibberllin precursor, and sandaracopimaradiene, a proposed precursor of the oryzalexin group of phytoalexins, have been identified by GC-MS analysis. Three other metabolites tentatively identified as pimaradienes were also detected.

REFERENCES

Cartwright, D., Langcake, P., Pryce, R., Leworthy, D., and Ride, J., 1981, Isolation and Characterization of Two Phytoalexins from Rice as Momilactones A and B, Phytochem., 20:535.
Kono, Y., Takeuchi, S., Kodama, O., Sekido, H., and Akatsuka, T., 1985, Novel Phytoalexins (Oryzalexins A, B and C) Isolated from Rice Blast Leaves Infected with <u>Pyricularia oryzae</u>. Part II; Structural Studies of Oryzalexins, Agric. Biol. Chem., 49:1695.

ENDOGENOUS GIBBERELLINS IN WHEAT SHOOTS

Jiann-Tsyh Lin and Allan E. Stafford

Plant Development-Quality Research Unit
Western Regional Research Center, Agricultural Research
Service, U. S. Department of Agriculture
Albany, California 94710

INTRODUCTION

Identification of endogenous gibberellins (GAs) in developing wheat grains has been reported [1,2]. Both GA_1 and GA_3 have been identified in the vegetative tissues of mature wheat plants[3]. In this communication, we would like to report the identification of GAs in the shoots of wheat seedlings.

MATERIALS AND METHODS

Plant materials

Seedlings of Chinese Spring wheat (<u>Triticum aestivum</u> L.) were harvested 8 days after sowing (50 g or about 1400 seeds). They were grown at 20°C with a light intensity of approximately 400 µmol m^{-2} s^{-1} and 14 hr photoperiod in a growth chamber. The seedlings were cut at the surface of vermiculite and the shoots (133 g, about 14 cm tall) were obtained.

Extraction procedure

All glassware was cleaned with chromic acid before use. The shoots were homogenized with 1 liter of 80% methanol at 4°C. The extract was filtered through a sintered glass funnel (coarse). Fibers were stirred in methanol (1 liter) overnight at room temperature. The mixture was then filtered. The filtrates were combined and concentrated to an aqueous phase <u>in</u> <u>vacuo</u> below 40° C. The aqueous phase (about 200 ml) was adjusted to pH 8.0 and was then partitioned against hexane, 5 X 100 ml. The aqueous phase was then partitioned against water-saturated n-butanol, 5 X 100 ml. The aqueous phase was then adjusted to pH 3.0 and was partitioned against EtOAc, 5 X 100 ml. The acidic EtOAc fraction obtained was dried <u>in</u> <u>vacuo</u>.

PVPP slurry purification

The acidic EtOAc fraction (343 mg) was dissolved in about 250 ml of 0.1 M phosphate buffer, pH 8.0. Polyvinylpolypyrrolidone (PVPP) was added to the solution and the mixture was stirred for 30 min. The PVPP concentration was about 50 mg/ml of the buffer. The PVPP was removed by filtration through a sintered glass funnel (coarse). The PVPP slurry treatment was repeated for a total of 3 times. The final filtrate was adjusted to pH 2.5 and was extracted with EtOAC, 5 X 200 ml. The EtOAc extract was dried (192 mg).

Preparative C_{18} HPLC

The PVPP purified acidic EtOAc fraction was further purified by preparative HPLC, by using a reversed-phase C_{18} column (1 cm X 25 cm, Spherisorb S5 ODS2). The eluent used was a linear gradient from 35% MeOH (containing 0.05% of acetic acid) to 100% MeOH (containing 0.05% of acetic acid) in 40 min at a flow rate of 4 ml/min. Fractions were collected at 1 fraction/min. Fractions No. 6 to No. 30 were combined and dried (9.8 mg).

Analytical C_{18} HPLC

The purified fraction was then separated by HPLC on an analytical reversed-phase C_{18} column (0.46 cm X 25 cm, Ultrasphere ODS. The same linear gradient eluent described above in 40 min was used at flow rate of 1 ml/min. The fractions were collected at 1 fraction/min. Each fraction was transferred to a Reacti-vial (1 ml) and was then dried by a nitrogen evaporator. 20 µl of acetone:water (1:1) was added to each vial to dissolve the residue and 1 µl of the solution was tested with the Tan-ginbozu dwarf rice bioassay, using 5 rice plants for each fraction.

Derivatization and GC-SIM

HPLC fractions were dissolved in 100 µl of MeOH and methylated with excess ethereal diazomethane. The fractions were dried under nitrogen and then under high vacuum. The trimethylsilyl (TMSi) ethers of the methylesters were prepared using 25 µl of N-methyl-N-trimethylsilyl-trifluoroacetamide (MSTFA) (Pierce).

The derivatized HPLC fractions showing significant levels of the bioassay activity, and gibberellin standards were analyzed by GC-MS with a Hewlett-Packard 5970 quadrupole-based mass selective detector (MSD). The MSD was operated in the scan mode for the gibberellin standards and hydrocarbon series (C_{22} - C_{32}) to obtain full spectra and retention time data for Kovat's retention indices. The selected ion monitoring (SIM) mode was used for the analytical C_{18} HPLC fractions of the plant extract. Samples were introduced into the MSD using a HP 5790A gas chromatograph with a capillary direct interface operated at 280° C. A 12 m X 0.2 mm I.D. dimethyl silicone crosslinked column (0.33 µm film thickness) was used in these experiments. The column was programmed from 60°C to 180° C at 25° C/min, then 6° C/min to the final temperature of 280° C in the splitless mode. The injection temperature was 260° C. The ionization voltage is fixed at 70 eV, the ion source operates at fixed nominal temperature of 250° C and head pressure was 7.0 psi. SIM along with Kovat's Retention Indices (KRI) were used to detect and identify the gibberellins present in trace levels of the selected HPLC fractions from wheat seedling extracts.

RESULTS AND DISCUSSION

PVPP slurry purification, preparative and analytical C_{18} HPLC were used for the purification of the gibberellin extract from the shoots of Chinese Spring wheat seedlings. The fractions from analytical C_{18} HPLC were bioassayed by the Tan-ginbozu dwarf rice bioassay. Each biologically active fraction was derivatized and examined by GC-SIM. GA_1, GA_3, GA_{19}, GA_{20} and GA_{44} were identified as their Me ester TMSi ethers by comparison to GC retention times and SIM of GA standards. The GC-SIM for the endogenous GAs and GA standards used are shown in Table 1. GA_1, GA_{29} and GA_{34} show very similar GC-MS as their Me TMSI derivatives. However, GA_1, GA_{29} and GA_{34} can be separated by analytical C_{18} HPLC. The identification of GA_1 here is also based on G_{18} HPLC.

Table 1. GC-SIM for Endogenous GAs in the shoot of wheat seedlings and GA Standards

GA Standard and HPLC Fraction (Presumptive GA)	Time of Scan (Min)	Monitored Ions With Relative Abundances in Parenthesis				
GA_3	19.208	504(M^+,100)	489(10)	370(14)	347(20)	208(47)
10 (GA_3)	19.048	504(M^+,100)	489(9)	370(14)	347(20)	208(48)
GA_1	18.748	506(M^+,100)	491(11)	448(24)	377(23)	313(15)
11 (GA_1)	18.780	506(M^+,100)	491(10)	448(24)	377(30)	313(9)
GA_{20}	16.633	418(M^+,100)	403(16)	375(66)	359(16)	301(15)
19-20 (GA_{20})	16.486	418(M^+,100)	403(15)	375(68)	359(18)	301(16)
22 (GA_{44})	20.193	432(M^+,73)	417(12)	373(26)	238(43)	207(100)
GA_{19}	17.968	434(M^+,100)	402(40)	375(65)	374(76)	345(37)
23 (GA_{19})	17.910	434(M^+,100)	402(43)	375(69)	374(83)	345(44)

The conversion of [^{3}H] GA_1 to [^{3}H] GA_8 previously shown in wheat seedlings[4] and the identification of GA_1, GA_3, GA_{19}, GA_{20} and GA_{44} in wheat seedlings indicated the presence of early 13-hydroxylation pathway which occur in corn, peas, and rice.

The main pathway in wheat seedlings may be as follows: GA_{53} → GA_{19} → GA_{20} → GA_1 → GA_3. GA_3 may be converted from GA_1 by dehydrogenation or converted from GA_8 by dehyrdation. Since GA_8 is biologically inactive in various bioassay, the later is unlikely. GA_1 and/or GA_3 may be the native hormone(s) for stem elongation.

REFERENCES

1. P. Gaskin, P. S. Kirkwood, J. R. Lenton, J. MacMillan, and M. E. Radley, Identification of gibberellins in developing wheat grain. Agri Biol Chem 44:1589 (1980).
2. P. S. Kirkwood and J. MacMillan, Gibberellins A_{60}, A_{61}, and A_{62}: Partial syntheses and natural occurrence, JCS Perkin I 689 (1982).
3. H. Eckert, G. Schilling, W. Podlesak, and P. Franke, Extraction and L. and Secale cereale L. and changes in contents during ontogenesis, Biochem Physiol Pflanzen 172:475 (1978).
4. J. L. Stoddart, Growth and gibberellin-A_1 metabolism in normal and gibberellin-insensitive (Rht 3) wheat (Triticum aestivum L.) seedlings, Planta 161:432 (1984).

MOLECULAR ASSOCIATIONS IN LIPID-CAROTENOID MONOLAYERS

Maria Tomoaia-Cotişel, János Zsakó, Emil Chifu and *Peter J. Quinn

Department of Physical Chemistry, University of Cluj-Napoca 3400 Cluj-Napoca, Romania and *Department of Biochemistry King's College London, London W8 7AH, U.K.

INTRODUCTION

The properties of monomolecular films composed of binary mixtures of membrane components can provide useful information relevant to the interaction between molecules. We have used this system to examine interactions between 1,2-distearoyldigalactosylglycerol (DGDG) and 1,2-distearoylphosphatidylcholine (DSPC) and astaxanthin (AX; all-<u>trans</u>-3,3'-dihydroxy-4,4'-dioxo-β-carotene) and found that the collapse pressure of the mixed monolayers was greater than would be predicted on the basis of perfect behaviour of the mixture (Tomoaia-Cotişel et al., 1984). The increased stability of mixed monolayers of DGDG and AX could be explained on the basis of regular behaviour of the components allowing the derivation of an apparent interaction parameter, ξM, representing the attraction between molecules of DGDG and AX in the film. The numerical value of this parameter was -0.87, however, the standard deviation of the experimental data from the theoretical curve is rather high, viz. $\Delta = 1.106$ mN.m^{-1}. In the case of DSPC:AX monolayers the regular solution approximation does not fit at all. We have therefore undertaken an examination of the collapse pressure vs composition curves of AX with DGDG and DSPC to determine whether they conform to behaviour according to intermolecular associations.

RESULTS AND DISCUSSION

At the collapse of an insoluble monolayer it is generally assumed that a metastable equilibrium is established between the molecules that remain in a monolayer configuration (M) and those that are present in the newly created bulk phase (B). This being so, the chemical potentials of each of the components in the respective phases can be equated (Zsakó et al., 1984) and the collapse pressure will depend on the collapse pressure of the pure components, their respective mean molecular areas and the molar composition of the mixture. Data for monolayers of DGDG, DSPC and AX are presented in Table 1. The collapse pressure of mixed monolayers were invariably higher than could be predicted from theoretical curves derived on the basis of the particular composition of the monolayer (Tomoaia-Cotişel et al., 1984) and suggest that there is a strong cohesive interaction between the diacyl lipids and the xanthophyll.

The formation of $A_m B_n$ complexes in the monolayer will increase the collapse pressure and the composition of the monolayer at the point of

Table 1. Surface Characteristics, Stability Constants and Surface Pressure Increments of the Regular Associations and their Minimum S.D.

Component		Collapse Pressure ($mN.m^{-1}$)		Mean Mol. Area (nm^2)		K	$\Delta\pi$ $mN.m^{-1}$	Δ $mN.m^{-1}$
1	2	1	2	1	2			
DGDG	AX	68	45	0.39	0.25	1.17	0.05	0.172
DSPC	AX	58	45	0.38	0.25	1.45	1.00	0.213

maximum deviation from perfect behaviour will depend on both the stability of A_mB_n and on the ratio m:n (Prigogine et al., 1954). In the case of DSPC:AX films, the observed deviation is maximum when the AX component predominates in the mixture. If A represents DSPC and B the AX component and we assume an AB_n type molecular association where there is an equilibrium between the molecular species A, B, AB, AB_2,..., AB_n on the monolayer phase the reactions:

$$AB_{i-1} + B \rightleftharpoons AB_i$$

can be characterized by equilibrium constants of the form:

$$K_i = a_{AB_i} / a_{AB_{i-1}} \cdot a_B \qquad [1]$$

where a_{AB_i}, $a_{AB_{i-1}}$ and a_B represent the activity of the corresponding molecular species, which in the perfect solution approximation reduces to:

$$K_i = x^M_{AB_i} / x^M_{AB_{i-1}} \cdot x^M_B \qquad [2]$$

where x^M refers to the respective mole fraction in the monolayer phase. On the basis of the monolayer behaviour of phospholipids (Tomoaia-Cotişel et al., 1981) and xanthophylls (Chifu et al., 1983) a value of n = 6 appears to be the most reasonable for the AB_n complex.

The simplest approach in an analysis of this system is to consider the monolayer as a perfect solution of regular associations in which case:
1. All values of K_i are the same, i.e. K_i = K for $1 \leq i \leq 6$.
2. If B is associated with A or to AB_{i-1}, the same collapse pressure increment, $\Delta\pi$, as the pure monolayer will be obtained, i.e. $\pi_{AB_i} = \pi_A + i\Delta\pi$.
3. The molecular areas of the complex AB_i will be the sum of the molecular areas of the components, i.e. $A_{AB_i} = A_A + iA_B$.
Thus at the collapse pressure the following relationship pertains:

$$\sum_{i=0}^{6} x^M_{AB_i} \exp[(\pi-\pi_{AB_i})A_{AB_i}/kT] + x^M_B \exp[(\pi-\pi_B)A_B/kT] = 1 \qquad [3]$$

where the molar fractions $x^M_{AB_i}$ can be calculated by using the relation:

$$x^M_{AB_i} = (1-x^M_B)(x^M_B K)^i / \sum_{i=0}^{6} (x^M_B K)^i \qquad [4]$$

The equilibrium molar fraction of B, x^M_B, is obtained as the solution of a seventh order equation, resulting from the expression of K and from the material balance equations, i.e. it can be calculated as a function of K and of the apparent molar fraction x^M in the monolayer. Derivation of K and $\Delta\pi$ from equations [3] and [4] respectively from the experimental monolayer collapse curves were obtained using a double minimization procedure as follows. Values of x^M for which experimental data was available were calculated using equation [4] and designated values of K. Using equation [3] collapse pressures were then calculated for different values and the $\Delta\pi$ value giving the minimum standard deviation Δ from the experimental values obtained. Fig.1 shows plot of $\Delta\pi$ and Δ as a function

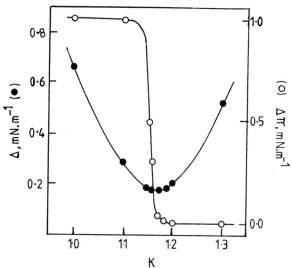

Fig. 1. Derivation of K and Δπ for regular associations in DGDG:AX films.

of K for DGDG:AX monolayers analysed according to the formation of AB_6 type regular associations. The stability constants and surface pressure increments obtained for DGDG:AX and DSPC:AX monolayers are presented in Table 1. It can be seen from the figure that Δ passes through a minimum at K = 1.17.

A curve calculated according to equation [3] using the data provided in Table 1 is in very good agreement with the experimental data published previously (Tomoaia-Cotişel et al., 1984) suggesting that the regular association model developed here provides an accurate description of the intermolecular interactions that take place between the diacyl lipids and astaxanthin. Accordingly, it appears that AB_6 type associations are slightly more stable with DSPC than with DGDG. This suggests that intermolecular interactions might involve formation of H-bonds between the OH group of AX and the carbonyl oxygens of the fatty acid ester bonds.

This conclusion can also be rationalized on the basis of a comparison of the monolayer properties of the two acyl lipids. The collapse pressure of DGDG is considerably greater than DSPC presumably because of hydrogen bonding between the sugar residues of the galactolipid (Tomoaia-Cotişel et al., 1983). Intermolecular interactions between DSPC molecules, on the other hand, are weaker and provide scope for AX molecules to increase cohesion via the formation of DSPC:AX associations.

REFERENCES

Tomoaia-Cotişel, M., Zsakó, J., Chifu, E. and Quinn, P.J., 1984, Mixed monolayers of 1,2-distearoyl digalactosyl glycerol and astaxanthin, Dev. Plant Biol., 9:421.

Zsakó, J., Tomoaia-Cotişel, M. and Chifu, E., 1984. Insoluble mixed monolayers. 1. Phase equilibria at the collapse of binary monolayers at gas/liquid interfaces, J. Colloid Interface Sci., 102: 186.

Prigogine, I., Defay, R. and Everett, D.H., 1954, "Chemical Thermodynamics", Longmans Green, London.

Tomoaia-Cotişel, M., Zsakó, J. and Chifu, E., 1981, Dipalmitoyl lecithin and egg lecithin monolayers at an air/water interface, Ann. Chim. Rome, 71:189.

Chifu, E., Zsakó, J. and Tomoaia-Cotişel, M., 1983, Xanthophyll films. 1. Single component monolayers at the air/water interface, J. Colloid Interface Sci., 95: 346.

Tomoaia-Cotişel, M., Sen, A. and Quinn, P.J., 1983, Surface-active properties of 1,2-distearoylgalactosylglycerols, J. Colloid Interface Sci., 94: 390.

SEPARATION AND IDENTIFICATION OF CAROTENOID-ESTERS IN RED PEPPER /CAPSICUM ANNUUM/ DURING RIPENING

Peter A.Biacs, János Bodnár, Ágoston Hoschke,
Anna Cs.Pavisa, Hussin Daood, Felix Hajdu and
Natalia Kiss-Kutz

Central Food Research Institute
Budapest /Hungary/

INTRODUCTION

Natural pigments of red pepper are mostly bound to fatty acids as carotenoid esters. In this paper the composition changes of red pepper lipids, carotenoids, mono- and disaccharides during ripening are discussed. The formation and the decomposition of more than 20 components were followed.

MATERIALS AND METHODS

The biosynthetic route of carotenoids is shown in Fig.1. The precursor of various xantophylls, β-carotene, responsible for the red colour of pepper undergoes oxidation and esterification yielding new pigments at different degrees of ripeness. The ripening process outlined in Fig.2. includes six main phases. The decrease of cellular respiration in the first four phases /1-4/ is accompanied by the appearance of red pigments and the gradual decomposition of chlorophyll. Phase 5 is characterized by ceasing cellular respiration in the course of which the synthesis of red pigments keeps going on. Post-ripening starts after harvesting the pod and continues during the open-air drying procedure.

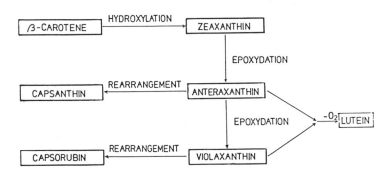

BIOSYNTHESIS OF CAROTENOIDS

Fig. 1.

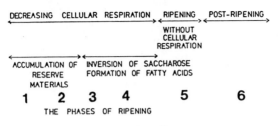

Fig. 2.

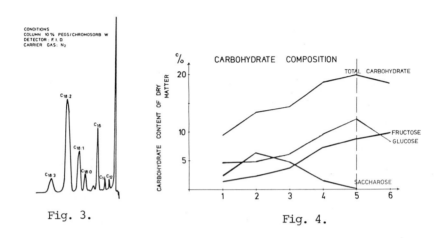

Fig. 3.

Fig. 4.

RESULTS AND DISCUSSION

The total lipid content of ground red pepper and those of separated parts of the fresh pod /peel, pulp and seeds/ were measured by GLC after extraction. /Fig. 3./

Red pepper lipids contain chiefly unsaturated fatty acids, the main component /40%/ of which is linoleic acid /$C_{18:2}$/. The level of fatty acids is increasing during the ripening process enabling the conversion of red carotenoid pigments into their biochemically more stable esters.

The determination of easily mobilizable individual carbohydrates also provides a characteristic picture of ripening. The overall amount of carbohydrates keeps increasing except that of saccharose, which is entirely consumed up to cover energy needs of biochemical processes by the time red pepper reaches complete ripeness. /Fig. 4./

Natural carotenoid pigments of red pepper are bound to long chain fatty acids. Pigment levels suffer remarkable changes during ripening /Fig. 5./ Carotenoids were separated by preparative TLC and the fatty acids of the individual red carotenoids were assayed by GLC. /Fig. 6./

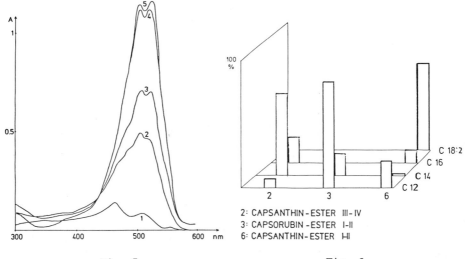

Fig. 5.   Fig. 6.

2: CAPSANTHIN-ESTER III-IV
3: CAPSORUBIN-ESTER I-II
6: CAPSANTHIN-ESTER I-II

Free and esterified components of red pepper carotenoids were separated and identified by HPLC. The amount of free capsanthin and capsorubin compound-group increases until the final stage of ripening. /Fig. 7. and Fig.8./ Lutein and saccharose are already not detectable in the last stage of ripening. During post-ripening the carotenoid synthesis increases as a result of using glucose as energy reserve. Oxidative processes of carotenoid decomposition were studied in vitro. The pigments of red pepper were oxidized by soya lipoxygenase and measured by UV spectroscopy. /Fig. 9./ Results of enzymatic oxidation indicate that capsanthin compound-group is more stable in the presence of lipoxygenase than capsorubin and its fatty acid esters. /Fig. 10./

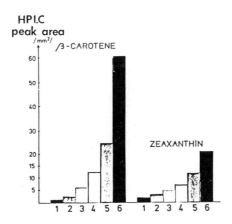

Fig. 7.

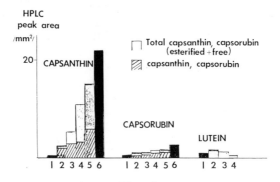

Fig. 8.

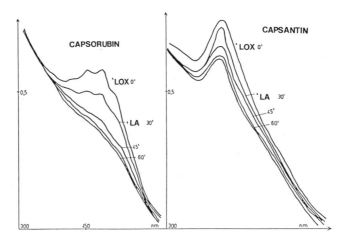

Fig. 9.

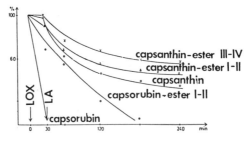

Fig. 10.

REFERENCES

Goodwin, T.W., 1980, Nature and distribution of carotenoids, Fd Chem., 5:3.
Linden, J.C. and Lewhead, C.L., 1975, LC of saccharides, J.Chromatogr., 105:125.

ON THE BIOSYNTHESIS OF FRAGRANCE COMPOUNDS IN AMBRETTE (HIBISCUS ABELMOSCHUS) SEEDS

Michael R. Pollard, Shahin Jamil-Panah and Thomas Y. Nee

ARCO Plant Cell Research Institute
6560 Trinity Court, Dublin, CA 94568, USA

INTRODUCTION

Most naturally occurring musk fragrances are macrocyclic lactones and ketones with a ring size between C_{14} and C_{18}.[1] These natural products, isolated from both plant and animal sources, are used in expensive perfumes, while synthetic musks, often of very different structures, are used to supply the bulk of the musk fragrance market. The first musk to be chemically characterised was ambrettolide (16-hexadec-7-cis-enolide), which occurs in ambrette (Hibiscus abelmoschus) seeds.[2] Further C_{10}-C_{14} fragrance compounds, including 14-tetradec-5-cis-enolide and alkyl acetates, were later identified in distillates from ambrette seed extracts by Maurer and Grieder.[3] With the purpose of investigating the biosynthesis of the macrocyclic lactone musks, we recently conducted an analysis of the fragrance compounds in ambrette seeds.[4] These are all monoesters, and are comprised of decyl (3%), dodecyl (3%) and tetradec-5-enyl (0.5%) acetates; 14-tetradec-5-enolide (0.5%), 16-hexadec-7-enolide (11%) and 18-octadec-9-enolide (7%); and 2-cis, 6-trans-(3%) and 2-trans, 6-trans-farnesyl acetate (72%). The numbers in parenthesis give a typical composition within the monoester fraction. The fragrance monoesters are present in the seed coat, at 0.3% of the dry weight of the whole seed. They are deposited during the period of embryo maturation (20-35 days after flowering) and not during the prior period of seed coat formation (5-20 days after flowering). This communication describes our first in vivo labeling experiments with developing ambrette seeds.

MATERIALS AND METHODS

Ambrette seeds were harvested at the appropriate time of development, which would correspond to the mid-point of monoester deposition or earlier. Assays with [1-^{14}C] substrates were run with 10 seeds per tube in 1 ml of 50 mM MES buffer, pH 6.0, at room temperature. At the end of the incubation, the seeds were rinsed with distilled water and dissected. The seed coats were ground in a mortar and pestle in hexane:isopropanol (3:2, v/v), and extraction of the lipids was performed as described by Hara and Radin.[5] Further details of the assays are described in "Results and Discussion". Labelled lipids were analysed by conventional GLC and TLC methods. The monoester fraction could be conveniently resolved by 2D-TLC. The [^{14}C] lipid extract was separated in the first dimension into neutral lipid classes by development with hexane:diethyl ether (9:1, v/v) over silica gel, and in the second dimension by development with toluene:ethyl acetate (95:5, v/v) over AgNO$_3$- silica gel. In this way, the monoesters were resolved into saturated alkyl acetates, hexadecenolide, octadecenolide, c, t-farnesyl acetate and t, t-farnesyl acetate.

RESULTS AND DISCUSSION

Table 1 shows two sets of assays with [1-^{14}C] acetate (5 μCi/assay; 56 Ci/mol) as substrate. In the upper experiment either the intact seed or the dissected seed coat was incubated for 5 hr. After the incubation the intact seed was dissected into its constituent parts. Both methods gave similar levels of acetate incorporation into total lipids and into monoesters by the seed coat. With the intact seed, [1-^{14}C] acetate could penetrate through to the embryo. However, the relatively low level of labelled lipids in the intermediate "perisperm plus endosperm" layer suggested that contamination of the seed coat lipids by other tissue layers would be minimal. Future incubations therefore used intact seeds, with post-incubation dissection. The lower experiment in Table 1 considers more carefully the distribution of label within the seed coat. The morphology of the seed coat has been described in detail by Singh.[6] The data show that the endogenous fragrance monoesters are located in the outer integument of the seed coat, and that biosynthetic activity is also largely restricted to the outer integument. The localisation of fragrance monoester biosynthesis and deposition in the outer integument indicates that monoester synthesis is a major metabolic activity for this tissue, as greater than 5% of the dry weight will be accounted for by these compounds. A more precise localisation of these compounds is now required, but it is known that they are not epicuticular.[4]

Table 1 [1-^{14}C] Acetate Incubations With Developing Ambrette Seeds

Incubation	c.p.m. Incorporated /hr/seed coat		Endogenous Lipids μg/seed coat	
	Total Lipids	Monoesters	Monoesters	Fatty Acids
Dissected Seed Coats	7530	1020		
Intact Seeds:-				
Complete Seed Coat	6630	1640		
Endosperm and Perisperm	625	N.D.		
Embryo	1970	N.D.		
Intact Seeds:-				
Complete Seed Coat	8020	1520	24.1	57.4
Outer Integument	6420	1300	20.5	13.2
Inner Integument	1405	170	2.2	44.3

N.D. Not Detected.

A variety of substrates were tested as precursors for fragrance monoesters. Table 2 shows that acetate is preferred. It labels all the monoester components, but at early time points farnesyl acetate has the greatest specific activity. Acetate was also incorporated into acyl lipids, including triacylglycerols, diacylglycerols and phosphatidyl choline. The fatty acids labelled were largely palmitate, oleate and linoleate. Decanoic, dodecanoic, tetradecenoic and hexadecenoic acids, which would correspond to the appropriate fragrance compounds, were not detected, either labelled or as endogenous constituents. [2-^{14}C] Mevalonic acid lactone was incorporated mainly into a triterpenoid alcohol fraction by the seed coats, and not into farnesyl acetate, indicating the inaccessibility of the latter pathway to exogenous mevalonic acid lactone. [1-^{14}C] Lauric and oleic acids showed some degradation, as label appeared in farnesyl acetate.

Acyl lipid labelling from [1-^{14}C] acetate was always several fold greater than monoester labelling in the seed coat, and both activities surged just prior to this period of endogenous monoester deposition, declining at the end of the period (Figure 1). A

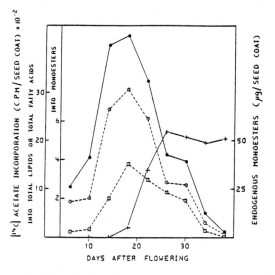
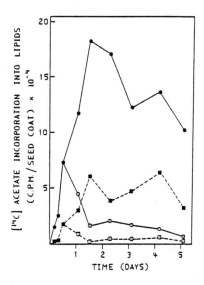

FIGURE 1. Biosynthetic activity during ambrette seed development, as measured by [14C] acetate incubation with intact seeds. Total labelled lipids (•———•), total labelled fatty acids (o----o), labelled monoester fragrance compounds (□----□), and the increase in endogenous monoester content (+———+) are shown.

FIGURE 2. Pulse chase experiment performed with [14C] acetate on intact seeds still attached to pods on the plant. After 12 hours of incubation the [14C] acetate droplets on the seeds were carefully removed to perform the chase. The regular time course shows total labelled polar lipids (•———•) and labelled monoester fragrance compounds (■----■), while the chase period shows total labelled polar lipids (o———o) and labelled monoester fragrance compounds (□----□).

Table 2: Comparison of the Effectiveness of Various [^{14}C] Substrates to Label Fragrance Monoesters in Developing Ambrette Seeds

Substrate	Assay Concentration (μM)	Percent Incorporation[2] Into	
		Total Lipids	Monoesters
[2-^{14}C] Acetate	172	3.66	0.84[1]
R-[2-^{14}C] Mevalonic Acid Lactone	98	15.8	0.06
[1-^{14}C] Lauric Acid	52	2.78	0.64[1]
[1-^{14}C] Oleic Acid	89	20.3	0.18[1]
[2-^{14}C] Acetate	172	1.05	0.22[1]
[2-^{14}C] Pyruvate	272	0.28	0.05
[U-^{14}C] Glucose	1098	0.13	0.01
[U-^{14}C] Sucrose	5.5	0.16	0.01

1. In all cases where there was sufficient label in the monoester fraction for 2D-TLC analysis farnesyl acetate comprised >90% of the label.
2. Assay time 12 hours.

pulse-chase experiment performed with [1-^{14}C] acetate on intact seeds still attached to pods on the plant showed a very rapid turnover and loss of label in both acyl lipid and monoester pools (Figure 2). This rather surprising turnover precludes satisfactory pulse chase experiments to search for an oleoyl lipid pool as a putative precursor for macrocyclic musk biosynthesis. However, the long term, in situ acetate labeling experiment does result in the specific activity of the individual monoesters reaching unity.

In conclusion, unequivocal proof of the pathway of biosynthesis of the macrocyclic lactone musks has yet to be obtained. However, given the positions of the double bonds in the various macrocyclic lactones ($C_{14:1}$, $C_{16:1}$ and $C_{18:1}$ musks all have an ω-9-cis double bond, suggesting a derivation by successive 2-carbon chain shortening of oleate), the temporal coincidence of both acyl lipid and monoester biosynthesis, and the high degree of catabolic activity, it is tempting to speculate that the musks are produced in the outer integument of the seed coat by release of oleate from acyl lipids, with subsequent partial chain shortening, omega-hydroxylation, and cyclisation.

REFERENCES

1. B. D. Mookherjee and R. A. Wilson, The chemistry and fragrance of natural musk compounds, in: "Fragrance Chemistry. The Science of the Sense of Smell," E. T. Theimer, ed., Academic Press, N.Y. (1982).
2. M. Kerschbaum, Ber. deutsch Chem. Ges. 60:902 (1927).
3. B. Maurer and A. Grieder, Z-5-Tetradecen-14-olide, a new macrocyclic lactone, and two unsaturated straight chain acetates from ambrette seed absolute, Helv. Chim. Acta 60:1155 (1977).
4. T. Y. Nee, S. Cartt and M. R. Pollard, Seed coat components of Hibiscus abelmoschus, Phytochem. (in press).
5. A. Hara and N. S. Radin, Lipid extraction of tissues with a low toxicity solvent, Anal. Biochem. 90:420 (1978).
6. B. Singh, The structure and development of Abelmoschus moschatus medic. seed, Phytomorphology 17:282 (1968).

STRUCTURE AND FUNCTION OF LIPIDS

STRUCTURAL AND DYNAMIC ASPECTS OF MEMBRANE LIPIDS

R.A. Demel

Laboratory of Biochemistry
State University of Utrecht
Padualaan 8, 3584 CH Utrecht
The Netherlands

INTRODUCTION

Current concepts on the structure of biological membranes depict the membrane as a liquid crystalline lipid bilayer with embedded protein structures. It is assumed that the proteins are responsible for specific properties as signal transduction and membrane transport, whereas the lipids provide a fluid matrix and impermeable barrier. Although one suitable type of lipid could accomplish the requirements of a fluid matrix biological membranes are built of complex mixtures of lipids, showing variations in polar headgroups, charge, length and unsaturation of paraffin chains and sterol content. Large differences in composition can be noted between membranes with different functions. This raises the question whether a complex lipid composition is a funcitonal requirement and if lipids contribute in a more specific way to membrane functions.

POLYMORPHISM OF MEMBRANE LIPIDS

Due to the amphipatic character, membrane lipids form association structures when dispersed in water, so that a continuum is formed by the polar and apolar part. The dispersions of the different lipids show a polymorphism in molecular organisation (1). Lipids like phosphatidylcholines, phosphatidylserines, sphingomyelines and diglycosyldiglycerides, spontaneously adopt the bilayer organization (Fig. 1). This structure can be formed by lipids of which the hydrated polar part balances the hydrophobic part.

Abbreviations:
MGDG	- monoglucosyldiglyceride
DGDG	- diglucosyldiglyceride
GalCer	- galactosylceramide
GalDG	- galactosyldiglyceride
PC	- phosphatidylcholine
DPPC	- 1,2-dipalmitoyl-L-phosphatidylcholine
DOPC	- 1,2-dioleoyl-L-phosphatidylcholine
MPPC	- 1-myristoyl-2-palmitoyl-L-phosphatidylcholine
PMPC	- 1-palmitoyl-2-myristoyl-L-phosphatidylcholine

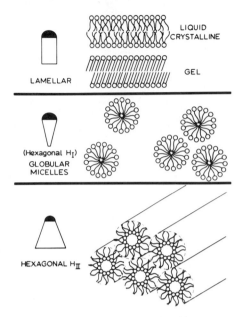
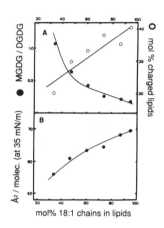

Fig. 1: Molecular association structures of membrane lipids in water, determined by the balance in size of the polar and apolar part of the molecules.

Fig. 2: Relationship between (A) the polar lipid composition of *A. laidlawii* membranes; (B) the mean molecular areas of monolayers of these lipids at 35 mN/m, and the acyl chain composition.

Membranes with a dominating size of the polar head groups as lyso lecithins and gangliosides preferentially form a hexagonal H_I phase, which is converted into a micellar structure in excess water.

On the other hand lipids with a relatively small polar moiety, like unsaturated phosphatidylethanolamines and monoglycosyldiglycerides form a hexagonal H_{II} phase also in excess water. This phase is composed of hexagonally organized tubes of phospholipids in which the lipids surround an aqueous channel with the polar head pointing to the inside of the cylinders.

The molecular shape is not a fixed parameter but a polar-apolar size balance, dependent on a number of conditions. Temperature and unsaturation largely affect the motion of the paraffin chains and accentuate the wedge shape properties and as a consequence several lipids adopt the bilayer organization at low temperatures, which is transferred into a hexagonal H_{II} phase at higher temperatures. Some negatively charged phospholipids form a lamellar organization, when dispersed as a sodium salt but are transferred into a hexagonal H_{II} phase by the addition of Ca^{2+} (2). Also lipid protein interactions are found to affect the lipid phase behaviour (3). It has been proposed that local regions of lipids forming non-lamellar structures may be advantageous to certain membrane functions (4).

The packing properties of *Acholeplasma laidlawii* membrane lipids can be regulated a) by varying the stoichiometric balance between MGDG and DGDG, and b) the balance between nonionic lipids and anionic lipids (5,6).

Glucolipids with equal amounts of palmitoyl and oleoyl chains form at the growth temperature of A. laidlawii a lamellar liquid crystalline phase. Large amounts of MGDG, high temperatures and a high degree of unsaturation favour a cubic or hexagonal liquid crystalline phase. The lipid composition of A. laidlawii secures always an optimal bilayer stability by alterations of the ratio of MGDG anmd DGDG and the amount of ionic lipids. A decrease in the MGDG/DGDG ratio will increase the average size of the head groups and counteract an increased acylchain unsaturation.

Increased acyl chain unsaturation in A. laidlawii increases the charged lipid fraction. This will increase the lipid lateral area due to increased electrostatic repulsion but the regulation of charged lipids could also be aimed to maintain a constant surface charge density (6). When the fatty acid chain unsaturation in the growth medium is increased so is the lipid acyl chain unsaturation. The MGDG/DGDG ratio decreases to maintain a stable lamellar phase (Fig. 2A). The amount of charged lipids (Fig. 2A) and the mean molecular area (Fig. 2B) increase with increasing acyl chain unsaturation. From the content of charged lipid and the molecular area an average lipid surface charge density of $5.7 \, 10^{-3}$ electronic charges/$Å^2$ can be calculated irrespective of the acyl chain composition of the membrane lipids.

The significance of a constant surface charge density is further underlined by the observation that increasing amounts of NaCl in the growth medium increase proportionally the molar fraction of charged membrane lipids. Na^+ exhibits a quenching effect on lipid surface charge with increasing concentration, thereby decreasing the polar head repulsion and thus the lateral packing areas. The cells restore the packing properties by an increased synthesis of charged lipids (6).

THE EFFECT OF STEROLS ON THE PHASE STRUCTURE OF MEMBRANE LIPIDS AND THEIR INFLUENCE ON THE BARRIER PROPERTIES

Sterols are present in most membranes of eucaryotic cells. Procaryotes, except for the mycoplasmas, are devoid of this lipid. The genus Acholeplasma lacks the growth requirement for cholesterol. It has nevertheless a capacity to incorporate cholesterol. Cholesterol is a flat rigid molecule compared to lipids containing flexible chains and does not form bilayers at physiological temperatures (4). In mixtures with phospholipids cholesterol by its condensing effect, increases the chain order in the liquid crystalline state and by its liquefying effect, decreases the chain order in the gel state. This has led to the concept that cholesterol containing membranes are in an intermediate state of fluidity (7). However, it has also been observed that cholesterol can destabilize the bilayer structure of dioleoyl phosphatidylethanolamine and of a mixture of soya phosphatidylethanolamine and egg yolk phosphatidylcholine (8,9).

The effects of different sterols on the bilayer → hexagonal H_{II} phase transition temperature of dioleoyl phosphatidylethanolamine show that cholesterol, 7-dehydrocholesterol and ergosterol have only a small effect on the transition temperature (10). Upon changing the 3ßOH into a 3αOH group (epicholesterol) the transition temperature is shifted downwards by 10°C. 3-Keto derivatives with one or two conjugated double bonds (cholest-4-en-3-one and cholest-4,6-dien-3-one) produce an additional temperature shift. As the polar part of the sterol molecule is much smaller than the hydrocarbon region, sterols are cone shaped. 3-Ketosteroids accupy a larger molecular area in monolayers compared to cholesterol. Due to fast axial rotation the angle of the dynamic cone is larger than of cholesterol. The correlation between the molecular shape and the phase preference of the lipid is clearly demonstrated (10).Yet in biological membranes large amounts of phosphatidylethanolamine and cholesterol are usually not found together. On the other hand, incorporation of cholesterol into membranes of *Acholeplasma laidlawii* having lipids with oleoyl chains, decreases the synthesis of MGDG and increase the synthesis of ionic lipids to counteract the bilayer destabilizing effect of cholesterol (11).

Incorporation of cholesterol into model membranes increase the order parameter of the hydrocarbon chains but leaves the lipid lateral diffusion almost unaffected (12). Therefore it can be concluded that the effect of cholesterol on the packing properties of the bilayers is more important than its influence on lipid bilayer dynamics.

Both model membrane experiments and studies on biological membranes have revealed that the sterol structure is of crytical importance in the sterol-lipid interaction. In general, a planar ring system, a 3ßOH group and a long flexible chain at C17 are prerequisite for both condensing and liquefying effect (13,14). These requirements are also found for sterols which promote growth of sterol-requiring mycoplasmas (15). It was found, however, that *mycoplasma capricolum* could grow nearly as well on media supplemented with cholesteryl methyl ether or cholesteryl acetate as on free cholesterol (16). It has to be noted, however, that microorganisms generally contain lower sterol concentrations than eucaryotic cells. It has been suggested also that in microorganisms sterols could fulfil additional functions to regulate membrane fluidity (17).

3ß-OH-blocked cholesterol ether derivatives show a decreasing film stability with increasing length of the atached moiety, namely, cholesteryl methyl ether, cholesteryl ethyl ether, cholesteryl-n-propyl ether, cholesteryl isopropyl ether and cholesteryl butyl ether in this order (18). Only cholesteryl methyl ether and cholesteryl ethyl ether show a condensing effect in mixtures with dioleoyl phosphatidylcholine. Cholesteryl acetate shows a low collaps pressure of 14.4 mN/m and no condensing effect. Cholesteryl methoxy methyl ether and cholesteryl (2'-hydroxy)-3- ethyl ether having an additional oxygen function as an ether or hydroxyl moiety, respectively, show an increased interfacial stability due to the increased polarity. Only cholesteryl (2'-hydroxy)-3-ethyl ether shows a condensing effect.

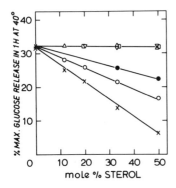 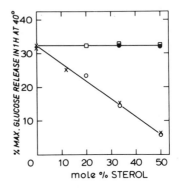

Fig. 3: Effect of cholesteryl ethers in liposomes prepared with DOPC on the relative amount of glucose released (A) (x) cholesterol; (o) cholesteryl methyl ether; (●) cholesteryl ethyl ether; (∇) cholesteryl-n-propyl ether; (□) cholesteryl isopropyl ether; (Δ) cholesteryl butyl ether; (B) (x) cholesterol; (o) cholesteryl (2'-hydroxy)-3-ethyl ether; (●) cholesteryl methoxy methyl ether; (□) cholesteryl acetate.

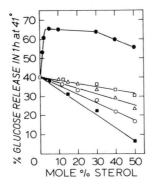

Fig. 4: Effect of oxysterols in liposomes prepared with egg PC on the relative amount of glucose released. (■) cholesterol; (o) 7-ketocholesterol; (Δ) 7ß-hydroxycholesterol; (□) 7α-hydroxycholesterol; (●) 25-hydroxycholesterol.

In agreement with the condensing effect measured in mixed monomolecular layers cholesteryl(2'-hydroxy)-3-ethyl ether, cholesteryl methyl ether and cholestery ethyl ether reduce the glucose permeability of dioleoylphosphatidylcholine liposomes (Fig. 3). These sterols also affected the heat content of the phase transition of DPPC. The other sterols had no effect on the glucose permeability nor on the phase transition of DPPC except for cholesteryl acetate and cholesteryl methoxy methyl ether. Although cholesteryl acetate and cholesteryl methoxy methyl ether have no ordering effect they do, however, influence the phospholipid phase transition. It can be concluded that a free 3ß-hydroxy group is not a prerequisite to observe a cholesterol-like effect in membranes. However, the interfacial stability and orientation of the oxygen moiety at the 3-position are important. Derivatives with an oxygen atom close to the oxygen at the 3-position have no chain ordering effect but do have a liquefying effect. Possibly these compounds act merely as spacer molecules.

The presence of an additional hydroxyl group added to the one at the 3ß position can have profound effects. Oxysterols as 7-ketocholesterol, 7ß-hydroxycholesterol, 7α-hydroxycholesterol and 25-hydroxycholesterol are readily formed when cholesterol is exposed to air, heat and radiation. Known physiological consequenses of oxysterols are decrease of cholesterol synthesis, effects on cell morphology and membrane associated functions and cytotoxicity towards smooth muscle cells and microorganisms.

The interfacial orientation of the oxysterols is illustrated by the pressure-area curves at the air-water interface (19). Compared to cholesterol the presence of a keto moiety at the 7-position, leads to a small increase in molecular area at low surface pressures. A remarkable difference is noted for the 7α- and 7ß-hydroxycholesterol isomers. 7ß-Hydroxycholesterol having an equatorial hydroxyl group is gradually forced into a perpendicular orientation at higher surface pressures. The axial orientation of the 7α-hydroxyl group leads to a stronger interaction with the aqueous phase and a tilted orientation is maintained to pressures of 18 mN/m. 25-Hydroxycholesterol with a hydroxyl group at both ends of the molecule retains a perpendicular orientation. Its interfacial stability is however reduced as can be concluded from the lower collaps pressure.

In mixed monolayers with phospholipids 7-ketocholesterol, 7ß-hydroxycholesterol and 7α-hydroxycholesterol, show a decreasing condensing effect in this order. 25-Hydroxycholesterol shows no condensing effect at all. In accordance with the monolayer data the incorporation of 7-ketocholesterol, 7ß-hydroxycholesterol or 7α-hydroxycholesterol into egg-PC liposomes, reduces the glucose permeability in that order. On the other hand a large increase in permeability of glucose is observed at low concentrations of 25-hydroxycholesterol. As a result of the perpendicular orientation of the oxysterol the 25-hydroxyl group is present in the hydrophobic core of the bilayer causing a local disorder of the membrane lipids, leading to the observed high permeability.

REDISTRIBUTION OF LIPIDS BETWEEN MEMBRANES

Due to the extremely low critical micelle concentration of natural phospholipids the spontaneous transfer of monomer phospholipid molecules between membrane surfaces through the aqueous phase is very slow. A number of proteins have been isolated which facilitate the transfer between membranes.

Recently a glycolipid-specific lipid transfer protein was purified which facilitates the transfer of various sphingoglycolipids and glyceroglycolipids. It could be measured that GalCer and GalDG are transferred from a monolayer to phosphatidylcholine vesciles in the subphase while there is no transfer of phosphatidylcholine (20). The protein functions as a carrier catalyzing a net mass transfer. The glycolipid transfer is influenced by the fluidity of the lipid interface and by the matrix lipid of the interface. GalCer transfer is stimulated in the presence of GalDG. The importance of GalCer distribution in a fluid lipid layer is supported when rat liver phosphatidylcholine is replaced by increasing amounts of GalDG. Although both GalCer and GalDG are transferred GalCer is transferred at higher rates from the monolayer containing GalDG.

The interfacial stability of membrane lipids is a delicate balance of the amphipatic properties. To measure how the oxydation of cholesterol affects its membrane stability radiolabelled oxysterols were incorporated in phospholipid monolayers and their rate of release from the interface was determined (19). In the absence of vesicles there is no release measurable. The addition of serum high density lipoprotein or small unilamellar vesicles to the subphase brings about a hardly measurable release of cholesterol ($\sim 0.5\%\ h^{-1}$). Much higher rates are found for 7-ketocholesterol, 7ß-hydroxycholesterol, 7α-hydroxycholesterol, and 25-hydroxycholesterol in this order. This order is similar to their interaction with DOPC, that is, the most cholesterol-like oxysterol 7-ketocholesterol shows the lowest transfer rate and the oxysterol with the greater distance between the hydroxyl groups, 25-hydroxycholesterol, the highest transfer rate. The transfer measured is consistent with the involvement of a water soluble intermediate.

PROPERTIES OF MIXED-CHAIN PHOSPHATIDYLCHOLINES

Lipids present in cell membranes are usually heterogenous with respect to fatty acyl composition. The structure and properties of 1,2-diacyl-L-phosphatidylcholine containing identical fatty acid chains have been studied in detail. The glycerol carbon atoms are orientated approximately perpendicular to layer plane. The initial part of the fatty acid at the 2-position extends layer-parallel but bends off at the second carbon atom. This gives rise to an apparent difference in chain length

(21). A reduction in the chain length of the *sn-2* fatty acid has a greater effect in this respect than does a comparable reduction at the *sn-1* position. The gel→liquid crystalline phase transition temperature of the two positional isomers of PC containing C-14 and C-16 fatty acyl chains compared to DPPC is: T_c (DPPC) 41°C > T_c(MPPC) 34°C > T_c (PMPC) 27°C (22).

The transfer of phosphatidylcholine from monolayer to vesicles catalyzed by phosphatidylcholine transfer protein shows remarkable differences for the positional isomers (23). The PC transfer protein acts as a specific carrier of PC between membrane interfaces. This protein has a recognition site for the phosphorylcholine headgroup and binding sites for the *sn-1* and *sn-2* fatty acyl chain. Lysophosphatidylcholine is not transferred. It was found that the protein transferred C10:0/C18:1 PC twice as fast as C18:1/C10:0 PC. Similar differences in rate were observed for C12:0/C18:1 PC and C18:1/C12:0 PC but not for isomers carrying myristic acid (Table I).

Table I: PC-TRANSFER PROTEIN MEDIATED TRANSFER OF PC ISOMERS FROM A MONOLAYER TO VESICLES

PC species	Initial rate of transfer (% monolayer PC/min)
C10:0 / C18:1-PC	0.9
C18:1 / C10:0-PC	0.4
C12:0 / C18:1-PC	2.6
C18:1 /C12:0-PC	1.1
C14:0 / C18:1-PC	8.8
C18:1 / C14:0-PC	8.8
C18:1 / C18:1-PC	4.8

To discriminate between isomers the acyl chains at the *sn-1* and *sn-2* position should differ sufficiently in length. The transfer rate is seen as a result between the hydrophobic forces at the interface, which decrease with acyl chain length and the binding by the protein such that a reduction in hydrophobicity of the PC molecule adversely affects the binding.

Taking together the results and extrapolating to the biological membrane, it can be concluded that such a membrane is a lipid-protein complex in which the actual organization of the lipids is a complicated equilibrium, not only dependent on the properties of the constituting lipids themselves but also on their interaction and the detailed structures of proteins. It can be speculated that fluctuations in local concentration, or protein conformation can trigger changes in the dynamic state and organization of the lipids, resulting in a specific control of membrane permeability.

REFERENCES

1. B. de Kruijff, P.R. Cullis, A.J. Verkleij, M.J. Hope, C.J.A. van Echteld and T.F. Taraschi, in Enzymes of Biological membranes, A.N. Martonosi (Ed.), Plenum Press, New York Vol. I, 43 (1984).
2. P.R. Cullis and B. de Kruijff, *Biochim. Biophys. Acta* 559, 399 (1979).
3. J.A. Killian and B. de Kruijff, *Biochemistry* 24, 7890 (1985).

4. J.N. Israelachvili, S. Marcela and R.G. Horn, *Q.Rev.Biophys.* 13, 121 (1980).
5. K. Khan, L. Rilfors, A. Wieslander and G. Lindblom, *Eur. J. Biochem.* 116, 215 (1981).
6. A. Christiansson, L.E.G. Eriksson, J. Westman, R.A. Demel, A. Wieslander, *J.Biol. Chem.* 260, 3984 (1985).
7. R.A. Demel and B. de Kruijff, *Biochim. Biophys. Acta* 457, 109 (1976).
8. P.R. Cullis and B. de Kruijff, *Biochim. Biophys. Acta* 507, 207 (1978).
9. P.C. Noordam, C.J.A. van Echteld, B. de Kruijff, A.J. Verkleij and J. de Gier, *Chem Phys. Lipids* 27, 221 (1980).
10. J. Gallay and B. de Kruijff, *FEBS Letters* 143, 133 (1982).
11. A. Wieslander, A. Christiansson, H. Walter and C. Weibull, *Biochim. Biophys. Acta* 550, 1 (1979).
12. G. Lindblom, L.B.A. Johansson and G. Arvidson, *Biochemistry* 20, 2204 (1981).
13. R.A. Demel, K.R. Bruckdorfer and L.L.M. van Deenen, *Biochim. Biophys. Acta* 255, 311 (1972).
14. J.C. Hsia, R.A. Long, F.E. Hruska and H.D. Gesser, *Biochim. Biophys. Acta* 290, 22 (1972).
15. S. Rothem, E.A. Pfend and L. Hayflick, *J. Bacteriol.* 105, 323 (1971).
16. A.K. Lala, T.M. Butke and K. Bloch, *J. Biol. Chem.* 254, 10582 (1979).
17. S. Razin, *Curr. Top. membranes Transp.* 17, 183 (1982).
18. R.A. Demel, A.K. Lala, S.N. Kumari and L.L.M. van Deenen, *Biochim. Biophys. Acta* 771, 142 (1984).
19. J.J.H. Theunissen, R.L. Jackson, H.J.M. Kempen and R.A. Demel, *Biochim. Biophys. Acta*, in press (1986).
20. T. Sasaki and R.A. Demel, *Biochemistry* 24, 1079 (1985).
21. H. Hauser, I. Pascher, R.H. Pearson and S. Sundell, *Biochim. Biophys. Acta* 650, 21 (1981).
22. E.N. Serrallach, G.H. de Haas and G.G. Shipley, *Biochemistry* 23, 713 (1981).
23. D. van Loon, R.A. Demel and K.W.A. Wirtz, *Biochim. Biophys. Acta* 856, 482 (1986).

LIPID SATURATION BY CATALYTIC HYDROGENATION AND

ITS EFFECT ON MEMBRANE STRUCTURE AND FUNCTION

Laszlo Vigh

Institute of Biochemistry
Biological Research Center of the
  Hungarian Academy of Sciences
H-Szeged, P.O. Box 521
6701 Hungary

INTRODUCTION

Our understanding of molecular organization of plant cell membranes and its relationship to various functions has undergone considerable advances in recent years. The progress in this subject was due to a great extent to refinement of modification techniques, applied to alter the lipid constituents of membrane systems.

Membrane lipids can be modified <u>in vivo</u> by using inhibitors of the biosynthetic pathways of membrane components, genetic selection of lipid auxotrophs or by adaptation of plants to various temperature and/or light conditions. A common drawback of all of these methods is, however, that they involve changes in the metabolic pathways associated with the synthesis of new lipids rather than the modification of existing lipids within preexisting membranes. Generally used <u>in vitro</u> techniques are to introduce sterols, fatty acids or complex lipids into the isolated membranes. Another way is lipase digestion and reconstitution of the treated membranes with various lipids. Methods listed above are unspecific and include the incorporation of foreign molecules into the membranes. Membrane modification by the technique of homogeneous catalytic hydrogenation which was pioneered by Chapman and Quinn (1) is a fundamentally different approach. It makes possible the rapid removal of unsaturated fatty acyl residues under conditions in which the essential structure of the organelle is preserved. The original catalysts, used to hydrogenate the lipids were, however, largely insoluble in water and had to be introduced into membrane systems in organic solvents. Furthermore, the catalyst could not be removed at the end of the reaction and its interference with biochemical functions could not be excluded (2). To avoid this problem water-soluble catalysts were synthetized by sulphonating the homogeneous Rh catalyst (3) and were used to hydrogenate model systems (4) and probe the disposition of lipids in lipoprotein structures (5). The water-soluble phosphine complexes prepared first by Joo and Beck (6) soon replaced their PPh$_3$-containing counterparts in biological hydrogenations (7,8). For an easier incorporation into the bilayers, amphiphilic, long chain, aliphatic tertiary phosphines were also synthetized to serve as ligands in a Rh-complex catalyst (9). Owing to its higher specific activity and air-stability a Pd-complex of sulphonated alizarine, Pd(QS)$_2$ (10) proved superior to the previously used catalysts. Its utility in hydrogenating unsaturated lipids of algal cell membranes (11,12), pea chloroplasts (12,14, 15,16) and tissue cultured plant cells (17) has already been established in a series of studies performed in our laboratory.

In this paper, I shall attempt to outline two aspects of these investigations. First, the relationship found between the chilling susceptiblity of the blue-green alga, Anacystis nidulans, and the cis-double bond content of their cytoplasmic membranes. The observations, suggesting that the saturation level of fatty acids rather than the level of microviscosity plays a crucial role in determining the structural organization and functional characteristics of photosynthetic membranes will then be described.

MEMBRANE SELECTIVE SATURATION AND ITS EFFECT ON CHILLING SUSCEPTIBILITY

Although the blue-green alga Anacystis nidulans is a prokaryotic organism, it can be regarded as a model system for studies which aim to elucidate the mechanism of thermo-adaptation and chilling susceptibility of higher plants. As reported by the Murata's group an adaptive mechanism exists in this alga that is reflected by a rapid shifting of the onset of the liquid-crystalline to gel phase transition of membrane lipis simultaneously with a shifting of the growth temperature. It has also been suggested, that it is particularly the occurrence of phase separation of the cytoplasmic membrane lipids which induces irreversible chilling injury in cyanobacteria (18,19). In fact, the determinative role of the cytoplasmic membrane in thermal acclimation of the algal cells has been confirmed recently (20), but hitherto no direct evidence has been presented indicating that chilling injury directly relates to lipid phase changes in the cytoplasmic membrane. Support for this view has come from hydrogenation experiments (12) presented below.

It was previously shown, that by using the sulphonated alizarine derivative of Pd(II) as a water-soluble hydrogenating catalyst, efficient hydrogenation can be achieved in the membranes of living A. nidulans cells, which preserved their integrity and physiological activity (11). According to a separate study (15), chlorophyll, carotenoids and plastoquinone were not hydrogenated by the complex. There were also some indications, that the extent of hydrogenation on intact cells is limited by the rate of penetration of the catalyst through the cell surface membranes. Therefore, we have separated the cytoplasmic and thylakoid membranes at different stages of the reaction expecting to establish experimental conditions under which saturation is confined to cytoplasmic membrane. In fact, under specific conditions, i.e., catalyst concentration 0.08 mM, pressur of hydrogen gas 0.3 MPa, cell density corresponding to 10 µg chlorophyll per ml and, provided the reaction time had a maximum of 60 min., satisfactory hydrogenation of fatty acyl residues located in the cytoplasmic membrane was obtained, while the double bond content in the thylakoids remained essentially constant. (For details, see refs. 11,12.)

It was suggested that the chilling induced absorption increae of zeaxanthin around 390 nm observed in A. nidulans is an indicator of the fluid to gel transition of lipids within the cytoplasmic membrane (21). Hence to test, whether lipid saturation brought about by hydrogenation raises the threshold temperature at which formation of gel-phase lipids initiates, absorbance changes were detected at 390 nm in both the control and hydrogenated populations and plotted against at different chilling temperatures. An upward shift could be observed for the midpoint values (4° and 11° C, respectively) of absorbance changes in accordance with the drop is cis-double bond content induced by the hydrogenation, within the cytoplasmic membrane. Also as a result of hydrogenation, specific for cytoplasmic membrane, cells became passively permeable to K^+ at much higher chilling temperatures (12°C vs 4°C). Finally, studies were carried out to investigate the relationship between the lipid phase change confined to cytoplasmic membrane and the photosynthetic activity (oxygen evolution) of the algal cells. In control and hydrogenated populations depression of photosynthetic oxygen evolution was observed at midpoint values of 4°C and 11°C, respectively. It was noted that the level of chilling susceptibility of the Anacystis cells in which saturation was confined to the cytoplasmic membrane resembled that of cells acclimated to high-temperature (Table 1 and refs. 18-20). As proposed by Murata and his co-workers, conversion of the lipids from their normal liquid-crystalline state to the gel state in the thylakoids reduces reversibly the photosynthetic activity and that of the cytoplasmic membrane induces irreversible damage. We have provided a unique experimental tool for supporting this hypothesis. Whereas neither compositional nor physical changes were detected in the lipid phase of thylakoids, partial, but irreversible inactivation of photosynthetic electron transport was found simultaneously.

HYDROGENATION OF LIPIDS IN THE PHOTOSYNTHETIC MEMBRANE: EFFECTS ON MEMBRANE STRUCTURE AND PHOTOSYNTHETIC ACTIVITY

The physical properties, lateral and transverse asymmetry and the possible functional role of acyl lipids in photosynthetic membranes have been the subject of considerable recent interest. It appears, that the mostly uncharged typically polyunsaturated acyl lipid complement of thylakoids not only fulfills a structural role in membrane organization, but may also play a functional role in the case of activity of several integral membrane protein complexes (22,23). Discovery of a pronounced lateral segregation of thylakoid macromolecular complexes, notably the light-harvesting chlorophyll a/b protein (LHCP), photosystem II (PS II), photosystem (PS I) and the cytochrome b_6/f complex has necessitated a revision of our ideas of the organization of photosynthesis at the molecular level. To account for a linear electron transport from water to $NADP^+$ the electron must be transferred rapidly between the two photosystems over a distance that ranges 100-500 nm or even more (24). The pool of plastoquinone located in the hydrophobic core of the thylakoid membrane is the most likely candidate for a mobile electron shuttle from PS II to cytochrome b_6/f, and the peripheral protein

Table 1. Midpoint values for the critical temperatures of lipid phase separation of cytoplasmic membrane, K^+-release, and rate of photosynthetic oxygen evolution tested in control and hydrogenated A. nidulans cells.

	Control	Hydrogenated
Sat/unsat ratio in the cyt. membranes	0.86	2.25
Sat/unsat ratio in the thylakoid membranes	0.96	0.96
Temp. of phase separation in cyt. membranes	4°C	11°C
K^+-release	4°C	12°C
Rate of O_2 evolution	4°C	11°C

plastocyanin from cyt b_6/f to PS I. In fact, long-range electron transfer by diffusive processes may be accepted in the case of a small, lipophilic molecule such as plastoquinon. It is more difficult to imagine the relatively bulky, hydrophilic plastocyanine as a mobile electron carrier (25). There are also arguments that the diffusion of plastoquinone is not a rate-limiting process (26,27), and in a complex phenomenon like photosynthesis no single process exerts a total control. It is suggested, that the bilayer core of thylakoid membrane, particularly in appressed regions, is appreciably less fluid than thought to be previously. According to some spin-label studies, in which the most appropriate probe molecules, i.e. thylakoid lipids were applied, a very high degree of motion restriction in appressed regions was suggested (28). In this paper I attempt to demonstrate that in situ modification of the level of unsaturated fatty acids by homogeneous catalytic hydrogenation of the photosynthetic membranes seems to be a promising technique for such studies. A relationship between fatty acid pattern, degree of saturation, and fluidity characteristics of the lipid matrix can be established. In addition, the consequences of this modification on determining the rate of electron-transport through the functional complexes as implicated above can also be directly investigated.

During hydrogenation, the basic structure of the pea chloroplasts remained unchanged. On the other hand, saturation of 10% of the double bonds induced a definite decrease in the dimension of both thylakoids and loculi. Microdensitometry showed that these structural changes arose from a thickening of the single membrane with a simultaneous decrease in the inter-membrane spacings (14). The orientational pattern of chlorophyll-a molecules was not altered by saturating up to 50% of fatty acyl double bonds in membrane lipids, as an indication that the energy-transfer processes amongst the chlorophyll molecules remained functional (14).

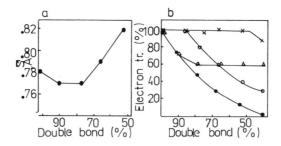

Fig. 1 (a) Changes in the order parameter (S_A^1) of fatty acid chains as a function of actual double bond content (%), measured with C-12 spin-labelled stearic acid.
(b) Effect of hydrogenation on the activity of the different parts of the photosynthetic electron-transport chain. (H_2O -> methyl viologen, ●——●; H_2O -> pBQ, o——o; DQH -> methyl viologen, ▲——▲; DCIP -> methyl viologen, ×——×.)

The time-course of hydrogenation was monitored by measurements of fatty acid residues of total lipid extract as well as the individual lipid classes. It could be seen (for details see refs. 13,14) that hydrogenation of the polyenoic fatty acyl residues took place relatively rapidly, while saturation of monoenic residues is much slower. The mono- and digalactosyldiacylglycerols were particularly susceptible to catalytic hydrogenation. This could be contrasted with the rates of reaction of the polyunsaturated acyl chains of the acidic lipids, i.e. sulphoquinovosyldiacylglycerol and particularly the phosphatidylglycerols (29). This may indicate a restricted access of the $Pd(QS)_2$ complex to these lipids which would support the model that certain lipids fulfill specific roles within chloroplast protein complexes. Rawyler and Siegenthaler showed (30) that SL and PG were less susceptible to lypolitic acid hydrolase attack than MGDG and DGDG. This is consistent with the relative accessibility of the different lipids to the catalyst applied in our case.

The physical-chemical state of lipids was characterized by incorporating spin-labelled stearic acid into the membrane and the orientational order parameter was calculated from the spectral anisotropy. The C-12 isomer seemed to be an optimal choice because this label is located in the depth of the double-bond region. The main point of interest in this part of our study was to reveal that the ordering state tested in C-12 region was not in the least a linear function of either the degree of unsaturation, or the actual level of linolenic acid. As seen on Fig. 1-a at moderate hydrogenation ($\geq$ 75% of the initial double bonds are present) S_A^1 values even slightly decreasing. However, a further decrease of unsaturation brings about an extensive increase of the orientational order. Coolbear and co-workers demonstrated (31) that it was the "first" double bond introducing to the alkyl chain which had a dramatic effect on the phase state, while the "second" double bond is less effective. The presence of a "third" double bond at C-15 resulted in even a slight increase of the temperature of the fluid to gel transition in the aqueous dispersions of phosphatidylcholine. In the base of this finding, it is understandable that eliminating predominantly the "third" double bond from 18:3 in the initial phase of hydrogenation explains why we could observe a slight decrease in the ordering state. According to Thomas and his co-workers (32) the formation of particle-free patches on the fracture faces on the membranes at room temperature only seen in samples subjected to extensive (>50%) hydrogenation. Since the gel-phase appeared even after relatively mild hydrogenation (cca. 30%) if tested in the isolated lipids, it was suggested, that the disordering effect of membrane proteins might interfere in the native chloroplast.

The effect of hydrogenation on the activity of the different parts of the photosynthetic electron-transport chain is presented in Fig. 1-b. The entire electron transport measured from H_2O to methyl viologen, gradually decreased as hydrogenation progressed. PS II electron transport measured from H_2O to p-benzoquinone was inhibited when more than 10% of double bonds were saturated and exhibited to gradual decrease upon further hydrogenation. PS I activity tested from duroquinol to methyl viologen showed an immediate drop upon the initial saturation, but did not change afterwards. If the PS I activity was measured from the cytochrome f-plastocyanin, the electron flow was only inhibited when more than 50% of double bonds were saturated. On the basis of these data, the following tentative conclusions are proposed. The detected loss of electron transport apparently can be traced back to the partial inhibition of the electron flow assayed from DQH to methyl viologen and to the loss of PS II. As a possible explanation, we assume that at the beginning of hydrogenation the polyunsaturated fatty acyl residues, required for the oxidoreductase activity of cytochrome b_6/f complex, are partially saturated, thus the reduced activity of this complex becomes rate limiting for electron transport. It is noted, that the functional importance of acyl lipids in the case of cyt b_6/f was evidenced recently (33,34). A decreased

PS II activity, resulting from saturation exceeding cca. 10% is in agreement with the results of Gounaris et al (35). They have found that stimulation of the O_2-evolution of the PS II preparation by the total lipid extract did not occur if they subjected lipids to hydrogenation prior to addition to the sample. It is also interesting to note, that in both PS II preparation and in the cyt. b6/f complex, DGDG and PC gave maximal stimulation of the electron transport.

Finally, we should point out two main consequences of our results. There seems to be no question about the functional importance of double bonds content of the acyl lipids. On the other hand, the contention that the fluidity per se is a key factor governing photosynthetic capacity does not have any experimental support. The rate of the electron flow between the photosystems was inhibited by saturation up to the level of cca. 20%, but the fluidity of the lipid matrix was even higher than at the beginning. Further increase in lipid saturation led to an effective enhancement of microviscosity in the deep hydrophobic core, but this has no apparent influence on the rate of steady-state electron flow from PS II to PS I. Instead a gradual decrease observed in the rate of electron transport confined to PS II. According to our most recent studies (16), catalytic hydrogenation affects the dissociation of reduced plastoquinone from the B protein and decreases electron transfer between the primary electron acceptor Q and the secondary acceptor, B.

CONCLUSION

In an effort to assess the role of adaptive changes in membrane lipid saturation within the cytoplasmic membrane of Anacystis nidulans in determining the chilling susceptibility of the cells, we applied the technique of homogeneous catalytical hydrogenation. We could find out conditions under which lipids were saturated selectively in the cytoplasmic membrane. After the hydrogenation the cytoplasmic membrane became leaky and phase seperation occurred at higher chilling temperatures. Irreversible inactivation of photosynthesis was found simultaneously. We have provided a proof for the hypothesis that the thermotropic properties of the lipids within the cytoplasmic membranes, and not those of thylakoids, control chilling susceptibility of the anacystis cells.

We have carried out a series of experiments in which the lipid composition of the photosynthetic membrane has been altered by the homogeneous catalytic hydrogenation of the unsaturated fatty acid residues of membrane lipids. Ordering state of lipids was not a linear function of the degree of unsaturation. Decrease of the double bonds beyond about 75% resulted only in a measurable increase of the orientational order. Progressive saturation of double bonds primarily inhibited electron transport between the photosystems, followed by the inhibition of electron flow around PS II. Our results underlined the functional importance of the presence of polyunsaturated fatty acyl chains, but the hypothesis, that the level of microviscosity in the hydrophobic core regulates the electron flow via controlling the rate of diffusion of mobile electron carriers has not been confirmed experimentally.

REFERENCES

1. Chapman, D. and Quinn, P. J., 1976, Proc. Natl. Acad. Sci. USA 73: 3971
2. Restall, C. J., Williams, W. P., Percival, M. P., Quinn, P. J. and Chapman, D., 1979, Biochim. Biophys. Acta 55: 119
3. Borowwski, A. F., Cole-Hamilton, D. J. and Wilkinson, G., 1978, Nouv. J. Chimie 2: 137

4. Madden, T. D., Peel, W. E., Quinn, P. J. and Chapman, D., 1980, J. Biochem. Biophys. Meth. 2: 19
5. Katagiri, C., Owen, J. S., Quinn, P. J. and Chapman, D., 1981, Eur. J. Biochem. 118: 335
6. Joo, F. and Beck, M. T., 1975, React. Kin. Catal. Let. 2: 257
7. Vigh, L., Joo, F., van Hasselt, P. R. and Kuiper, P. J. C., 1983, J. Molec. Catal. 22: 15
8. Vigh, L., Joo, F. and Cseplo, A., 1985, Eur. J. Biochem. 146: 241
9. Farin, F., van Gaal, H. L. M., Bonting, S. L. and Daemen, F. J. M., 1982, Biochim. Biophys. Acta 711: 336
10. Bulatov, A. V., Izakovich, E. N., Karklin, L. N. and Khidekel, M. L., 1981, Izv. Akad. Naukj. SSSR Ser. Khim. 9: 2032
11. Vigh, L. and Joo, F., 1983, FEBS Lett. 162: 423
12. Vigh, L., Gombos, Z. and Joo, F., 1985, FEBS Lett. 191: 200
13. Vigh, L., Joo, F., Droppa, M., Horvath, L. A., and Hovath, G., 1985, Eur. J. Biochem. 147: 477
14. Horvath, G., Droppa, M., Szito, T., Mustardy, L. A., Horvath, L. I. and Vigh, L., 1986, Biochim. Biophys. Acta 849: 325
15. Szalontai, B., Droppa, M., Vigh, L., Joo, F. and Horvath, G., 1986, Photobiochem. and Photobiophys. 10: 233
16. Hideg, E. Rozsa, Z. S., Vass, I., Vigh, L. and Horvath, G., 1986, Photobiochem. and Photobiophys., in press
17. Vigh, L., Horvath, I. and Joo, F., 1984, in: Siegenthaler, P.-A. and Eichenberger, W. (Eds.), "Structure, Function and Metabolism of Plant Lipids", Elsevier, p. 531
18. Ono, T. A. and Murata, N., 1981, Plant Physiol. 67: 182
19. Ono, T. A. and Murata, N., 1982, Plant Physiol. 69: 125
20. Gombos, Z. and Vigh, L., 1986, Plant Physiol. 80: 415
21. Ono, T. A. and Murata, N., 1981, Plant Physiol., 67: 176
22. Quinn, P. I. and Williams, W. P., 1983, Biochim. Biophys. Acta 737: 223
23. Murphy, D. J., 1986, Biochim. Biophys. Acta 864: 33
24. Millner, P. A. and Barber, J., 1984, FEBS Lett. 169: 1
25. Haehnel, W., 1982, Biochim. Biophys. Acta 682: 245
26. Mauro, S., Lannoye, R., Vandeloise, R. and Donckt, E. V., 1986, Photobiochem. and Photobiophys. 11: 83
27. Whitmarsh, J., 1985, in: Staehelin, A. and Arntzen, C. (Eds.), Encyclopedia of Plant Physiology, Photosynthetic Membranes, Springer, Berlin, in press
28. Murphy, D. J. and Knowles, P. F., 1984, in: Siegenthaler, P.-A. and Eichenberger, W. (Eds.), "Structure, Function and Metabolism of Plant Lipids", Elsevier, p. 425
29. Horvath, I., Mansourian, A. R., Vigh, L., Thomas, G. T., Joo, F. and Quinn, P. J., 1986, Chem. Phys. Lipids 39: 251
30. Rawyler, A. and Siegenthaler, P.-A., 1980, Eur. J. Biochem. 110: 179
31. Coolbear, K. P., Berde, C. B. and Keough, K. M. W., 1983, Biochemistry 22: 1466
32. Thomas, P. G., Dominy, P. J., Vigh, L., Mansourian, A. R., Quinn, P. J. and Williams, W. P., 1986, Biochim. Biophys. Acta 849: 131
33. Chain, R. K., 1985, FEBS Lett. 180: 321
34. Doyle, F. and Yu, C.-A., 1985, Biochim. Biophys. Res. Comm. 131: 700
35. Gournaris, K., Whitford, D. and Barber, J., 1984, in: Sybesma, C. (Ed.) Advances in Photosynthesis Research, Vol. III, Martinus Nijhoff, Dr. W. Junk, p. 107

SPATIAL ORGANIZATION AND FUNCTIONAL ROLES OF ACYL LIPIDS

IN THYLAKOID MEMBRANES

Paul-André Siegenthaler, André Rawyler and Christian Giroud

Laboratoire de Physiologie végétale
Université de Neuchâtel
Ch. de Chantemerle 20, 2000 Neuchâtel, Switzerland

The structural and functional asymmetry of protein complexes in the plane of and across the thylakoid membrane of higher plants[1] raises the question of whether a similar heterogeneity exists for acyl lipid distribution and if so, what should be its significance in terms of the overall structure and function of thylakoids. In spite of some recent progress concerning the spatial organization of acyl lipids, their role in photosynthetic functions is far from being elucidated.

Lateral asymmetry of acyl lipids in the thylakoid membrane

Several approaches have been adopted to assess the lateral heterogeneity of acyl lipids in the thylakoid membrane (for reports published until Sept. 1984, see ref. in Siegenthaler and Rawyler[2]) : (1) Fractionation of thylakoids into subchloroplast particles enriched in PSI or PSII activities; (2) Separation of appressed and non-appressed regions of thylakoids; (3) and the purification of one of the protein complexes found in the membrane[3,4,5]. The composition in acyl lipids (and proteins) of each of these fractions is determined, compared with the original composition of the thylakoid membrane and, eventually, related to its particular function. It is worth mentioning that almost all procedures for obtaining subchloroplast fractions rely on detergent solubilization of membrane and may therefore present drawbacks but also some advantages. On one hand, it may result in a differential displacement of certain acyl lipids, thereby influencing the final composition of lipids which are associated with the fractions or complexes. But, on the other hand, it may help to reflect the presence of tightly bound and specific associations of particular acyl lipids with membrane proteins.

As an example, a recent study by Murphy and Woodrow[6] illustrates this fractionation strategy. Compared to whole thylakoids, appressed regions

---

Abbreviations : BSA, bovine serum albumin; DGDG, digalactosyldiacylglycerol; EGTA, Na$_2$-ethylene glycol-bis (B-amino-ethyl ether)N,N,N',N' tetraacetate; LHCP, light harvesting chlorophyll a/b protein complex; MGDG, monogalactosyldiacylglycerol; PC, phosphatidylcholine; PG, phosphatidylglycerol; PL, phospholipid; PLA$_2$, phospholipase A$_2$; PSI, PSII, photosystem I,II; SQDG, sulphoquinovosyldiacylglycerol.

are characterized by a light-induced extrusion of protons, a depletion of acyl lipids, an equal amount of protein and an enrichment of chlorophyll, a lower (about half) lipid/protein ratio, a depletion of PSI complex and an enrichment of LHCP. In addition to this gross asymmetry, there are also considerable heterogeneities in acyl lipid class distribution between the two membrane fractions. On the basis of mol %, the most striking difference is an increase in PG of appressed regions whilst on the basis of protein there is a decrease in all lipid classes, except in PG. Appressed regions have a much higher MGDG/DGDG (2.4 vs 1.3), contain SQDG molecular species which are depleted in $C_{18:3}$ and enriched in $C_{16:0}$, and PG species which are slightly enriched in $3(t)-C_{16:1}$.

Recently, we made the proposal that each lipid class consists of several topologically distinct pools which are separated laterally and transversally in the membrane[2]. If this hypothesis is true, one should find an increased heterogeneity of acyl lipids in membrane particles of decreasing size and complexity. Four examples show that this proposal is likely to be correct : (1) The oligomeric form of LHCP is greatly enriched in MGDG and $(3t)-C_{16:1}$ PG species[7]; (2) A photochemically active reaction center complex II contain only two lipid classes (MGDG and SQDG) which are characterized by a low degree of unsaturation and by a complete absence of $C_{16:3}$ in MGDG species; (3) A purified CF_o-CF_1 preparation is highly enriched in SQDG and completely depleted in MGDG and phospholipids[4]; (4) A cytochrome b_6/f complex preparation contains only phospholipids[5]. A general feature of all these particles or complexes is that they are isolated as lipo-protein complexes.

In conclusion, the thylakoid membrane is characterized by considerable lateral heterogeneities in the distribution of all major membrane components, including acyl lipids.

Transversal asymmetry of acyl lipids in the thylakoid membrane

Several methods have been used to assess transversal localization of acyl lipids in the thylakoid membrane [for reports published until Sept. 1984, see ref. in (2)] : (1) Chemical labelling[8,9,10]; (2) Lipid exchange and (3) Immunological procedures; (4) Selective lipid extraction; (5) Enzymatic modification[8,10,12]. In our opinion, this last method has been the most rewarding since it allows one not only to assess the transmembrane distribution of acyl lipids but, simultaneously, to draw relationships between the hydrolysis (or depletion) of certain lipids and the impairment of particular photochemical functions in the thylakoid membrane. The precautions, criteria, special requirements, drawbacks and advantages of this approach have been discussed in detail recently[11,13]. Using this technique, the transversal distribution of galactolipids[10,11] and phospholipids[8,12,14,15] has been estimated, in several higher plant species.

As an example, the distribution in mol % of acyl lipids between the outer and inner monolayers of the spinach thylakoid membrane[10-12,14,15] is illustrated in Fig. 1A. It can be seen that the outer leaflet is mainly enriched in MGDG and PG whilst DGDG is confined essentially in the inner leaflet. When the distribution of each acyl lipid is expressed as % of total lipids, the results show the following striking features (Fig. 1B) : (1) Each class of lipids is asymmetrically distributed across the thylakoid membrane; (2) MGDG constitutes about 70% of the total lipids in the outer monolayer and only 38% in the inner leaflet; (3) Both

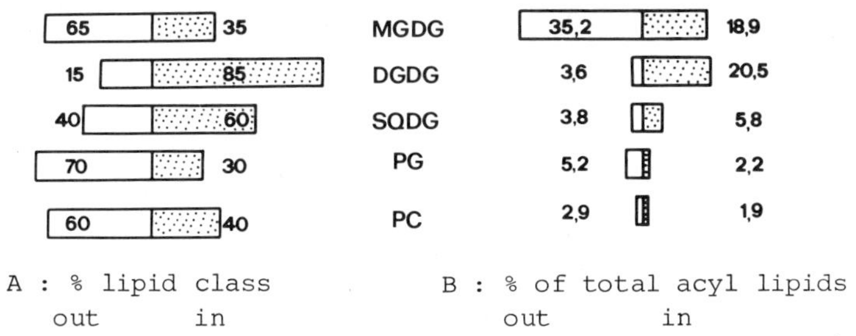

```
    65        35        MGDG      35,2        18,9
    15        85        DGDG       3,6        20,5
    40        60        SQDG       3,8         5,8
    70        30         PG        5,2         2,2
    60        40         PC        2,9         1,9
```

A : % lipid class   B : % of total acyl lipids
    out   in   out   in

Fig. 1. Distribution of acyl lipids in the two monolayers of the spinach thylakoid membrane

galactolipids (MGDG + DGDG) are equally distributed in both leaflets; (4) The ratio MGDG/DGDG is high (9.8) in the outer leaflet and very low (0.9) in the inner leaflet; (5) The ratio of bilayer/non-bilayer forming lipids is 0.44 and 1.61 in the outer and inner leaflets of the thylakoid membrane. Based on the results reported by Sprague and Staehelin[16] one can conclude that in the spinach thylakoid membrane, the acyl lipid distribution favors a lamellar configuration in the inner leaflet whilst in the outer one both hexagonal and lamellar configurations may coexist. Since the thylakoid membrane is a bilayer structure under physiological conditions, it is obvious that other components (e.g. proteins, ions, etc.) must contribute to the stabilization of lamellar configurations in the outer leaflet, as discussed recently[10].

Role of acyl lipids in photochemical activities

All the approaches proposed in the literature to scrutinize the relationships between acyl lipids and the functions of the thylakoid membrane are aimed at modifying the content, the structure or the composition of acyl lipids and measuring simultaneously one or several photochemical activities. The approaches are (up to Sept. 1984, see ref. in[2]) : (1) Aging of thylakoids in vitro and senescence[17]; (2) Extraction of membrane lipids by organic solvents; (3) Modulation of fatty acid composition by catalytic hydrogenation of acyl chains[18], seasonal variations and treatment by substituted pyridazinones[19], etc.; (4) Reconstitution studies[20,21]; (5) Lipid composition of thylakoids and subchloroplast fractions[3,4,5,22]; (6) Modification of the fluidity of the membrane; (7) Immunological studies; (8) Biogenesis studies; (9) Lipolytic treatments[23,24]. Although the importance of acyl lipids has been recognized to support the photosynthetic competence of leaves, chloroplasts and thylakoids (namely in maintaining the appropriate fluidity of the thylakoid membrane), most of these studies have failed to show a direct correlation between the presence of a certain lipid and a specific photochemical activity. However, very recently, we have succeeded to show that discrete and topologically distinct populations of phospholipids and MGDG are likely to be involved directly in electron transport activity.

Thylakoids were incubated in the presence of either the pancreatic phospholipase A_2 or the lipase from <u>Rhizopus arrhizus</u> under conditions (± bovine serum albumin, BSA, ± EGTA, appropriate pH, T and ionic

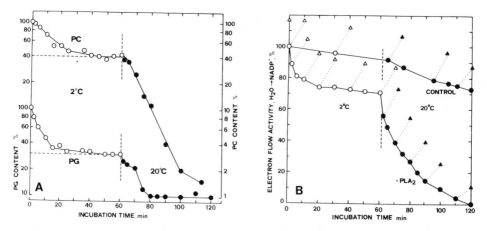

Fig. 2. Kinetics of the hydrolysis of PG and PC in phospholipase A_2-treated thylakoid membranes (A) and of the uncoupled non-cyclic electron flow activity $H_2O/NADP^+$ (B). $\triangle, \blacktriangle$: activity following BSA addition.

strength) allowing depletion of successive pools of phospholipids or MGDG both in the outer then in the inner monolayer of the thylakoid membrane. As an example, Fig. 2 shows that the hydrolysis of phospholipids which are localized in the outer monolayer of the spinach thylakoid membrane (60% of PC and 70% of PG) causes about 20-25% inhibition of the uncoupled non-cyclic electron flow activity. This inhibition is mainly due to hydrolysis products (free fatty acids and lyso-phospholipids) as shown by the restoration effect of BSA. The activity begins to be inhibited greatly only when the hydrolysis of these phospholipids located in the inner monolayer occurs. In the presence of BSA, the restoration of the activity is only partial indicating that phospholipid depletion is the cause of the irreversible BSA-insensitive inhibition. Fig. 3 shows the extent of uncoupled non-cyclic electron flow activity inhibition ($H_2O/NADP^+$) expressed as a function of phospholipids destroyed under conditions where, first, the outer pool of phospholipids and, then the inner pool is hydrolyzed. In the presence of BSA which removes completely free fatty acids and lyso-phospholipids, 4 populations of PG and PC molecules can be depicted : (1) Depletion of the 1st population (about 60% for PG and 40% for PC) which is exclusively located in the outer monolayer affects only slightly the activity ($<10\%$); (2) Depletion of the 2nd population (from 60 to 80% for PG and from 40 to 80% for PC) which is located in both monolayers, causes an inhibition of the activity up to about 30%; (3) Depletion of the 3rd population (from 80 to 90% for both PG and PC) inhibits the greatest part of the activity (30 to 100%); (4) The last population (5-15% remaining in the membrane) which is located in the inner monolayer does not support this type of activity.

In order to ascertain that the "barrier" properties of thylakoids are preserved during PLA_2 treatment, we have studied their osmotic response in sorbitol solutions (Fig. 4). It can be seen that under all experimental conditions, the mean packed volumes of thylakoids vary linearly with the reciprocal of sorbitol concentration. Phospholipase A_2-treated thylakoids have a larger volume after 60 min at 2°C (i.e., when only those phospholipids localized in the outer monolayer are hydrolyzed) and also after 120 min at 20°C (i.e., when all phospholipids are destroyed). These changes can be attributed to the hydrolysis products only since an

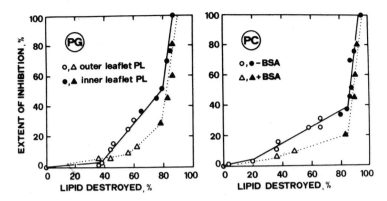

Fig. 3. Relationship between the extent of non-cyclic electron flow activity inhibition and successive phospholipid depletion in the outer and inner monolayers of the spinach thylakoid membrane.

addition of BSA to phospholipase A_2-treated thylakoids decreases the packed volumes close to the values obtained in the BSA-controls. Thus, the osmotic properties of the membrane (toward sorbitol) are essentially preserved in spite of extensive phospholipid breakdown. Therefore, the rapid hydrolysis of both PG and PC observed at 20°C under the conditions of Fig. 2 is not due to thylakoid disruption but to increased outward transbilayer movement of phospholipids.

CONCLUSIONS

(1) Acyl lipids are heterogenously distributed in the plane of the thylakoid membrane. Evidences come mainly from experiments carried out with subchloroplast fractions. A general feature is that all particles or complexes isolated from thylakoids are of lipoprotein nature. Depending on the degree of complexity of the particles, some of them are completely devoided or enriched in certain lipid classes which are generally highly saturated compared to those of the thylakoid membrane.

(2) Several approaches, but mainly the enzymatic one, have revealed that acyl lipids are asymmetrically distributed across the thylakoid membrane : 65 mol % MGDG, 15% DGDG, 70% PG, 60% PC and a maximum of 40% SQDG are localized in the outer monolayer. Glycolipids (MGDG + DGDG + SQDG) as well as phospholipids (PG + PC) are equally distributed in both monolayers but MGDG, a non-bilayer forming lipid, is by far the most predominant class in the outer leaflet. Its role remains to be elucidated. The ratio bilayer/non-bilayer forming lipids (in mol % of total acyl lipids) is 0.44 in the outer leaflet which favors hexagonal configurations. The ratio value which is 1.61 in the inner leaflet imposes lamellar configurations.

(3) The relationships between acyl lipids and photochemical activities are extremely complex. Several parameters are involved in sustaining maximal photosynthesis, the fluidity of the thylakoid membrane being only one of them. Based on "our" enzymatic approach and focusing our attention mainly on phospholipids, we can draw the following conclusions : (a) Phospholipids consist of at least 4 distinct populations

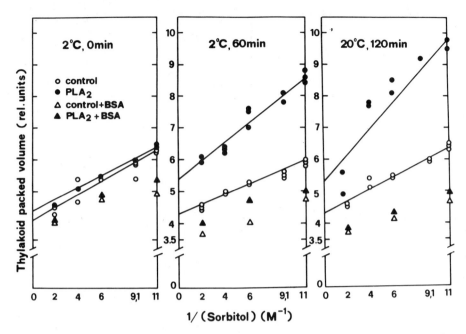

Fig. 4. Mean packed volumes of control and phospholipase A_2-treated thylakoids as a function of the reciprocal of sorbitol concentration. Temperature and time conditions are those of Fig. 2.

which are separated transversally (and laterally ?) in the membrane; (b) We can therefore visualize that each of these populations plays a specific role in the membrane; (c) This is illustrated by our finding that, in the case of uncoupled non-cyclic electron flow, the activity is directly dependent on the 3rd PG and PC populations which are entirely localized in the <u>inner</u> monolayer and which represent only 10% of the total phospholipid molecules. This may explain why in the past, it has been so difficult to establish a correlation between phospholipid depletion and a specific photochemical function; (d) It is noteworthy that in contrast to the above observation, the same activity depends on the MGDG molecule populations which are located in the <u>outer</u> monolayer[24]; (e) These observations do not exclude that other functions may depend on other phospholipid or galactolipid populations; (f) These results give further support to our proposal that acyl lipids are likely to be organized into distinct and discrete pools, each of them having a structural or/and a functional role in the thylakoid membrane[2].

ACKNOWLEDGEMENTS

The author thank Mrs Jana Smutny for her excellent technical assistance and Miss Christiane Bachmann for typing and arranging the manuscript. Part of this study was supported by the Swiss national science Foundation (Grant no. 3.417-0.83 to P.A.S.).

REFERENCES

1. D.J. Murphy, The molecular organisation of the photosynthetic membranes of higher plants, Biochim. Biophys. Acta 864:33 (1986).

2. P.A. Siegenthaler and A. Rawyler, Acyl lipids in thylakoid membranes: Distribution and involvement in photosynthetic functions, in: "Encyclopedia of Plant Physiology, New Series, vol. 19, Photosynthesis III", L.A. Staehelin and C.J. Arntzen, eds, Springer-Verlag, Berlin, Heidelberg, New York, Tokyo, pp. 693 (1986).
3. K. Gounaris and J. Barber, Isolation and characterization of a photosystem II reaction centre lipoprotein complex, FEBS Lett. 188:68 (1985).
4. U. Pick, K. Gounaris, A. Admon and J. Barber, Tightly bound sulpholipids in chloroplast CF_0-CF_1, Biochim. Biophys. Acta 808: 415 (1985).
5. M.F. Doyle and C.A. Yu, Preparation and reconstitution of a phospholipid deficient cytochrome b_6-f complex from spinach chloroplasts, Biochem. Biophys. Res. Commun. 131:700 (1985).
6. D.J. Murphy and I.E. Woodrow, Lateral heterogeneity in the distribution of thylakoid membrane lipid and protein components and its implications for the molecular organisation of photosynthetic membranes, Biochim. Biophys. Acta 725:104 (1983).
7. R. Rémy, A. Trémolières, J.C. Duval, F. Ambard-Bretteville and J.P. Dubacq, Study of the supramolecular organization of light-harvesting chlorophyll protein (LHCP), FEBS Lett. 137:271 (1982).
8. M.D. Unitt and J.L. Harwood, Sidedness studies of thylakoid phosphatidylglycerol in higher plants, Biochem. J. 228:707 (1985).
9. C. Sundby and C. Larsson, Transbilayer organisation of the thylakoid galactolipids, Biochim. Biophys. Acta 813:61 (1985).
10. A. Rawyler, M.D. Unitt, C. Giroud, H. Davies, J.P. Mayor, J.L. Harwood and P.A. Siegenthaler, The transmembrane distribution of galactolipids in chloroplast thylakoids is universal in a wide variety of temperate climate plants, Photosynthesis Res. (1986).
11. A. Rawyler and P.A. Siegenthaler, Transversal localization of monogalactosyldiacylglycerol and digalactosyldiacylglycerol in spinach thylakoid membranes, Biochim. Biophys. Acta 815:287 (1985).
12. P.A. Siegenthaler and C. Giroud, Transversal distribution of phospholipids in prothylakoid and thylakoid membranes from oat, FEBS Lett. 201:215 (1986).
13. P.A. Siegenthaler, Transmembrane distribution and function of lipids in spinach thylakoid membranes : rationale of the enzymatic modification method, in : "Biochemistry and Metabolism of Plant Lipids", J.F.G.M. Wintermans and P.J.C. Kuiper, eds, Elsevier Biomedical Press, Amsterdam, New York, Oxford, p. 351 (1982).
14. A. Rawyler and P.A. Siegenthaler, Transmembrane distribution of phospholipids and their involvement in electron transport, as revealed by phospholipase A_2 treatment of spinach thylakoids, Biochim. Biophys. Acta 635:348 (1981).
15. C. Giroud and P.A. Siegenthaler, Study of the distribution of phospholipids in prothylakoids and thylakoids from oat and spinach, in: "Structure, Function and Metabolism of Plant Lipids", P.A. Siegenthaler and W. Eichenberger, eds, Elsevier Science Publishers, Amsterdam, New York, Oxford, p. 413 (1984).
16. S.G. Sprague and L.A. Staehelin, Effects of reconstitution method on the structural organization of isolated chloroplast membrane lipids, Biochim. Biophys. Acta 777:306 (1984).
17. D.J. McRae, J.A. Chambers and J.E. Thompson, Senescence-related changes in photosynthetic electron transport are not due to alterations in thylakoid fluidity, Biochim. Biophys. Acta 810:200 (1985).

18. G. Horvath, M. Droppa, T. Szito, L.A. Mustardy, L.I. Horvath and L. Vigh, Homogeneous catalytic hydrogenation of lipids in the photosynthetic membrane : effects on membrane structure and photosynthetic activity, Biochim. Biophys. Acta 849:325 (1986).
19. G. Laskay and E. Lehoczki, Correlation between linolenic-acid deficiency in chloroplast membrane lipids and decreasing photosynthetic activity in barley, Biochim. Biophys. Acta 849:77 (1986).
20. I.J. Ryrie, Reconstitution of thylakoid components into artificial membranes, in: "Encyclopedia of Plant Physiology, New Series, vol. 19, Photosynthesis III", L.A. Staehelin and C.J. Arntzen, eds, Springer-Verlag, Berlin, Heidelberg, New York, Tokyo, p. 675 (1986).
21. R.K. Chain, Involvement of plastoquinone and lipids in electron transport reactions mediated by the cytochrome b_6-f complex isolated from spinach, FEBS Lett. 180:321 (1985).
22. J. Barber and K. Gounaris, What role does sulpholipid play within the thylakoid membrane ? Photosynthesis Res. 9:239 (1986).
23. P.A. Siegenthaler, J. Smutny and A. Rawyler, Involvement of hydrophilic and hydrophobic portions of phospholipid molecules in photosynthetic electron flow activities, in: "Structure, Function and Metabolism of Plant Lipids", P.A. Siegenthaler and W. Eichenberger, eds, Elsevier Science Publishers B.V., Amsterdam, New York, Oxford, p. 475 (1984).
24. P.A. Siegenthaler and C. Giroud, Evidences for different acyl lipid domains in spinach and oat thylakoid membranes supporting various photosynthetic functions, These proceedings (1986).

ULTRASTRUCTURAL STUDIES ON PLANT MEMBRANES

W. W. Thomson, K. A. Platt-Aloia and R. D. Bliss

Department of Botany and Plant Sciences
University of California
Riverside, CA 92521

The most accepted hypothesis for the general organization of membranes is the fluid-mosaic model of Singer and Nicolson.[1] In this model, membranes are viewed as a lipid bilayer with "integral" membrane proteins embedded in, as well as extending across, the lipid bilayer matrix. The model predicts that membranes can be asymmetrically organized with different lipids and proteins and amounts of each occurring in either half of the membrane. Also, Singer and Nicolson suggested that the lipid matrix was fluid with the lipid constituents as well as the proteins "floating" in this matrix, "freely" moving in the plane of the membrane. Thus, no long-range order or significant degrees of lateral heterogeneity as to the location of specific constituents would be expected in membranes unless constrained in place by extramembrane elements. In this paper, we examine the organization and lateral heterogeneity of plant membranes using electron microscopic methods.

OBSERVATIONS AND DISCUSSIONS

With the advent of transmission electron microscopy, thin sectioning, and heavy metal fixation and staining, various forms of membrane organization, including the bilayer construct, were deduced from density patterns in electron micrographs. In membrane systems such as the myelin sheath and myelin-like configurations, earlier X-ray diffraction studies and light polarization studies[2] indicated that these structures were repetitive bilayers in close surface association to each other. In micrographs of these structures, the repetitive dark-light-dark pattern was essentially supportive of the view that these were side-by-side aggregates of bilayered membranes (Fig. 1). The basic unit in these multi-layer systems was considered to be a single bilayer, and in thin sections, individual membranes such as the tonoplast and the plasmalemma (Fig. 2), characteristically appeared as two electron-dense lines separated by an electron-translucent region. However, not all plant membranes have this distinct trilaminate density pattern. For example, membranes of mitochondria and chloroplasts often appear to consist of globules or linear aggregates of some form of subunit (Fig. 3), and from such images it was proposed that these membranes have a subunit organization composed of a two-dimensional association of lipoprotein complexes.[3,4] The interpretative limitation of these observations is the lack of precise information as to the chemical interaction of fixatives and "stains" with the various

Fig. 1. Myelin-like configuration in a <u>Tamarix</u> salt gland. 60,000x

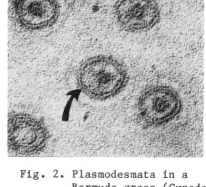

Fig. 2. Plasmodesmata in a <u>Bermuda</u> grass (Cynodon) leaf. Note the tripartite density pattern of the plasmalemma, arrow. 220,000x

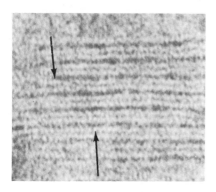

Fig. 3. Negative image of chloroplast granal membranes of <u>Phaseolus</u> (osmium vapor fixation, air dried). Note membrane "particles," <u>arrows</u>. 85,000x

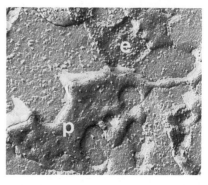

Fig. 4. E-face, <u>e</u>, and P-face, <u>p</u>, of ER in 4-day-old cowpea cotyledon. Note the P-face has more IMPs than the E-face. 32,000x

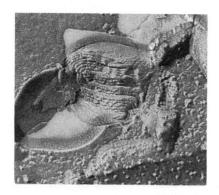

Fig. 5. Hexagonal phase lipid in a myelin-like configuration in a <u>Tamarix</u> salt gland. 100,000x

Fig. 6. E-face of the plasmalemma of a cell in a dry cowpea cotyledon. 42,000x

Fig. 7. E-face of the plasmlemma of a cowpea cotyledon cell after imbibition with 0.2 M NaCl. 49,000x

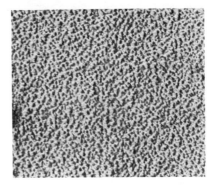

Fig. 8. P-face of the plasmalemma of a salt gland of Tamarix. Note the high density of IMPs. 120,000x

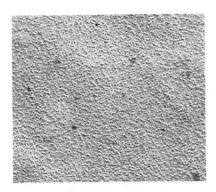

Fig. 9. P-face of the plasmalemma of a leaf mesophyll of Tamarix. Note density of IMPs as compared to Fig. 8. 25,000x

Fig. 10. E-face of the tonoplast in chilling injured avocado mesocarp cells. Note boundary regions (arrows). 30,000x

Fig. 11. P-face of the plasmalemma of a cell in a senescent cowpea cotyledon. Note the presence of particle-depleted areas, arrows. 43,000x

Fig. 12. Filipin deformations (arrows) near the nuclear pores in a barley aleurone layer. 40,000x

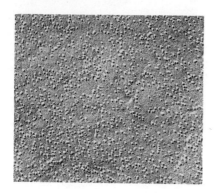 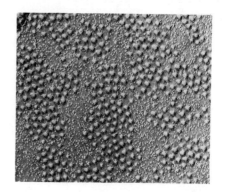

Fig. 13. E-face of the plasma-lemma of a mesocarp cell of a ripe avocado fruit treated with filipin. Note the induced deformations and the absence of IMPs in these areas. 40,000x

Fig. 14. E-face of the plasma-lemma of a mesocarp cell of a ripe avocado fruit without filipin treatment. Note the absence of particle-depleted area. 68,000x

membrane components. Thus, we cannot determine unequivocally the organiza- of membranes from density patterns; however, the variations in density patterns indicate that various membranes differ in organization.

Evidence that membranes are composed of lipid bilayers comes from freeze-fracture electron microscopy (FEM).[5] In this technique, the fracture plane moves through the center of membranes, and if membranes have a bilayer organization, planar images of the internal faces of the lipid layers are observed (Figs. 4-7, 9-14). A corollary to this is that other known configurations of membrane lipids, such as the hexagonal (H_{II}) phase, have recognizable images in freeze fracture (Fig. 5).[6]

Whether membranes in dry biological material are structurally composed of bilayers or have a different organization has been questioned. When dry seeds are placed in water, there is a leakage of solutes from the seeds to the ambient media. Simon[7] proposed that this could be explained if the membrane lipids in dry seeds were not organized in bilayers. This hypothesis was based on X-ray diffraction studies by Luzzati and Husson,[8] where they found that the membrane lipids undergo a transition, at low water content, from a bilayer configuration to a micellar configuration. From this, Simon reasoned that in dry seeds, membranes would have a micellar organization and, therefore, not show the property of semiperme- ability. However, in dry seed material,[9,10,11] as well as dry pollen,[12] using FEM, we found that all membranes had the typical internal E and P fracture faces characteristic of membranes with a bilayer organization (Fig. 6). The hexagonal H_{II} phase organization of lipids was never observed. Semipermeability is a characteristic of the membranes of all living cells. The only known organization of membrane lipids which confers this property is the bilayer configuration. We germinated barley seeds against concentrations of three osmotica---NaCl, mannitol, and betaine.[13] Our results showed that inhibition of germination during the early stages of imbibition was osmotic. These results and the FEM observations clearly establish that membranes in these dry, biological systems are bilayers.

FEM is also a method of determining whether membranes are asymmetric, at least as to the distribution of integral membrane proteins (IMPs). When membranes fracture along the central plane of the bilayer, two internal faces of the membranes are revealed. These are the P face, which remains attached to the cytoplasm and the E face which remains associated with spaces external to the cell.[14] In all membranes that we have examined, the number of IMPs per unit area of the membrane is always higher on the P face than the E face. This is true with the endoplasmic reticulum (Fig. 4) and the plasmalemma, tonoplast and chloroplasts (not shown).

It is well known that salinity deleteriously affects germination, growth, and development of many plants. One hypothesis advanced to explain these effects is that salt disrupts the integrity and organization of plant membranes.[15] In FEM studies of seeds imbibed with sodium chloride and within the limits of the procedure, we found no observable effect on the bilayer organization of the membrane.[10] With all treatments and material, smooth, planar images of the E- and P-faces of the membranes were observed (Fig. 7). However, we found that when radicles of dry cowpea seeds were hydrated, there was a large decrease of IMPs in the plasmalemma of the cells, but in the sodium chloride-treated material, the decrease was much less. Thus, NaCl has an effect on the IMPs, which are the functional units of the membrane.

One prediction of the fluid mosaic model is that no long-range pattern of differentiation would exist unless components of the membrane are anchored by exterior factors. However, photosynthetic membranes of higher plants are laterally heterogeneous in organization with the granal membranes and the interconnecting frets differentiated in both function and organization. Existing evidence suggests that this lateral segregation could well be a function of membrane surface interactions,[16] rather than of factors internal to the membrane. In other membranes, however, lateral differentiation in the organization of membranes also occurs. For example, in the P face of the plasmalemma of cells of the salt glands of Tamarix, the particle density is extremely high (Fig. 8). In fact, no lipid regions occur which suggests that the membranes may have a subunit organization. In the P face of the plasmalemma of the underlying mesophyll cells, the particle density is much less (compare Fig. 9 to Fig. 8), and smooth areas of the fracture faces indicate a bilayer organization (Fig. 9). Since most of the living cells of the plant are interconnected by plasmodesmata, bounded by a plasmalemma continuum (Fig. 2), this membrane system is clearly differentiated from cell to cell. This is not surprising since adjacent cells may have different functional capacities and specializations. Further, a high degree of differentiation might be expected in cells of salt glands where very high membrane fluxes have been reported.[17]

In regard to lateral heterogeneity in the organization of membranes, changes in membrane lipid fluidity may be directly or indirectly responsible for chilling injury in susceptible plant tissues.[18] Also, there are observations which show that with a change in fluidity a lateral segregation of lipids into gel-phase microdomains occurs.[19] Avocados are chilling-sensitive fruits, and we have found that particle-free, lipid microdomains form in the plasmalemma and tonoplast (Fig. 10) of the mesocarp cells of chilling injured fruit. We suggest that these microdomains represent chilling-induced, lateral phase separations. We assume that some portion of the membrane lipids undergo a phase transition at the chilling temperature and then sort out and "crystallize" laterally in the plane of the membranes. Along the same lines, it has been suggested that changes in membranes directly or indirectly influence senescent processes in plants.[20] We have found that as cotyledons senesce with germination, lipid-enriched microdomains appear in the plasmalemma of the parenchyma cells (Fig. 1).[21] These particle-free areas occur concomitant with the formation of gel-phase lipids.[22] The formation of microdomains in the plane of a cell membrane is an indication that lateral heterogeneity in organization can occur within membranes. These lateral phase separations can also be induced by chilling stress and are associated with developmental processes, in this case senescence. It is of interest that the formation of gel-phase membrane lipids is associated with the decrease in permeability of membranes[23] and that a loss of membrane permeability is consistently associated with both chilling injury[18] and senescence.[20]

Recently, a variety of compounds have been used to mark specific components of membranes. One such compound is filipin, which binds specifically to membrane sterols and brings about a defined deformation recognizable in freeze-fracture replicas.[24] Using filipin, it is possible to a certain extent, to mark the location as well as the distribution of sterols in membrane systems. We have noted an intriguing pattern with the nuclear envelope in cells of barley aleurone layers after imbibition for 3 hours. The filipin-sterol deformations primarily occurred around the orifices of the nuclear pores (Fig. 12). If this technique can be reliably used, this observation would be another example of the lateral heterogeneity in the composition of membranes. If sterols increase the rigidity of membranes, [25] it is interesting that here they are preferentially associated with nuclear pores.

When filipin is applied to the mesocarp tissue of ripe avocado fruit, highly ordered arrays of deformations occur in the E face of the plasmalemma (Fig. 13). Intramembrane protein particles tend to be excluded from these regions, which suggests that a lateral segregation has occurred. Such lipid-enriched, particle-depleted regions were not observed in controls (Fig. 14). As alluded to above, this raises questions as to how valid the use of filipin is for the localization of sterols, but does suggest that this chemical brings about an alteration and a change in the distribution of the constituents within the plane of the membrane.

SUMMARY

(1) In plants, the trilaminate density pattern observed in thin sections is commonly observed with the tonoplast and plasmalemma. Other membranes appear to have a more particulate organization based upon density patterns. These differences in density patterns suggest that some variation in membrane organization and composition exist between different membranes. (2) The planar, internal faces of membranes revealed by freeze fracture indicate that in dry seeds and pollen, and probably in all viable, dry biological material, membranes are organized as lipid bilayers. That membranes in dry seeds are bilayers is also supported by the observation that inhibition of germination during the early stages of imbibition is osmotic, a phenomenon dependent on the bilayer organization of membrane lipids. (3) All plant membranes that we have examined are asymmetric in regard to integral membrane proteins. In all cases, the P face of the membrane is enriched in proteins, compared to the E face. (4) At the freeze-fracture level, we were unable to detect primary changes in the lipid arrangement within the membranes of salt-treated seeds. However, the number of protein particles in the plasma membrane of the salt-treated material was considerably higher than in the water-treated controls. This may be of importance in regard to effects of salinity since proteins are the functional elements within membranes. (5) The plasmalemma is a continuous structure extending through plasmodesmata from cell to cell throughout most of the plant body. Within this continuum, the plasmalemma varies in organization from cell to cell. In some highly specialized cells, such as in salt glands, the particle density within the membrane is so high that these membranes may well have a subunit organization. (6) Membrane systems may be laterally-differentiated and heterogeneous in organization. With FEM, we have observed the formation of lipid-enriched microdomains in response to low temperature stress, and with the

developmental process of senescence. We conclude that these are phase separations in the plane of the membrane. (7) It can be shown in FEM that different membrane systems apparently have different amounts of sterol.[26] Further, in some instances sterols appear to be clustered in nuclear envelope membranes around the pores. However, some caution in interpretation is required since the marker compound, filipin, appears to induce the formation of lipid-enriched, particle-depleted areas in other membranes.

ACKNOWLEDGEMENT

This research was supported in part by a grant (PCM-8302145) from the National Science Foundation.

REFERENCES

1. S. J. Singer and G. L. Nicolson, The fluid mosaic model of the structure of cell membranes, Science 175:720 (1972).
2. D. Branton and R. B. Park, "Papers on Biological Membrane Structure," Little, Brown and Co., Boston (1968).
3. T. E. Weier and A. A. Benson, The molecular nature of chloroplast membranes, in: "Biochemistry of Chloroplasts," vol. I, T. W. Goodwin, ed., Academic Press, London and New York (1966).
4. A. A. Benson, The cell membrane: a lipoprotein monolayer, in: "Membrane Models and the Formation of Biological Membranes," L. Bolis and B. A. Pethica, eds., North Holland Publishing Co., Amsterdam (1968).
5. P. Pinto da Silva and D. Branton, Membrane splitting in freeze-etching, J. Cell Biol. 45:598 (1970).
6. D. W. Deamer, R. Leonard, A. Tardieu, and D. Branton, Lamellar and hexagonal lipid phases visualized by freeze-etching, Biochim. Biophys. Acta 219:47 (1960).
7. E. W. Simon, Phospholipids and plant membrane permeability, New Phytol. 73:377 (1974).
8. V. Luzzati and F. Husson, The structure of liquid-crystalline phases of lipid-water systems, J. Cell Biol. 12:207 (1962).
9. W. W. Thomson and K. A. Platt-Aloia, Ultrastructure and membrane permeability in cowpea seeds, Pl., Cell, Environ. 5:367 (1982).
10. R. D. Bliss, K. A. Platt-Aloia, and W. W. Thomson, Changes in plasmalemma organization in cowpea radicle during imbibition in water and NaCl solutions, Pl., Cell, Environ. 7:601 (1984).
11. E. L. Vigil, R. L. Steere, W. P. Wergin, and M. N. Christiansen, Structure of plasma membrane in radicles from cotton seed, Protoplasma, 129:168 (1985).
12. K. A. Platt-Aloia, E. M. Lord, D. A. DeMason, and W. W. Thomson, Freeze fracture observations on membranes of dry and hydrated pollen from Collomia, Phoenix and Zea, Planta (in press).
13. R. D. Bliss, K. A. Platt-Aloia, and W. W. Thomson, Osmotic sensitivity in germinating barley seeds, Pl., Cell, Environ., (in press).
14. D. Branton, S. Bullivant, N. B. Gilula, M. J. Karnovsky, H. Moor, K. Muhlethaler, D. H. Northcote, L. Packer, B. Satir, P. Satir, V. Speth, L. A. Staehelin, R. L. Steere, R. S. Weinstein, Freeze-etching nomenclature, Science 190:54 (1975).
15. E. V. Mass and R. H. Nieman, Physiology of plant tolerance to salinity, in: "Crop Tolerance to Suboptimal Land Conditions," G. A. Jung, ed., Amer. Soc. Agron. Specialist, Pub. 32.
16. L. A. Staehelin and C. J. Arntzen, Regulation of chloroplast membrane function: protein phosphorylation changes the spatial organization of membrane components, J. Cell Biol. 97:1327 (1983).

17. C. D. Faraday and W. W. Thomson, Morphometric analysis of *Limonium* salt glands in relation to ion efflux, *J. Exp. Bot.* 37:471 (1986).
18. J. M. Lyons, Chilling injury in plants, *Ann. Rev. Plant Physiol.* 24:445 (1973).
19. K. A. Platt-Aloia and W. W. Thomson, Freeze fracture evidence for lateral phase separations in membranes of chilling-injured avocado fruit, *Protoplasma* (in press).
20. R. D. Butler and E. W. Simon, Ultrastructural aspects of senescence in plants, *Adv. Gerontol. Res.* 3:73 (1971).
21. K. A. Platt-Aloia and W. W. Thomson, Freeze-fracture evidence of gel-phase lipid in membranes of senescing cowpea cotyledons, *Planta* 163:360 (1985).
22. R. F. Barber and J. E. Thompson, Senescence-dependent increase in the permeability of liposomes prepared from bean cotyledon membranes. *J. Exp. Bot.* 3:1305 (1980).
23. R. G. Miller, The use and abuse of filipin to localize cholesterol in membranes, *Cell, Biol. Int. Reps.* 8:519 (1984).
24. J. N. Israelachvili, S. Marcelja, and R. G. Horn, Physical principles of membrane organization, *Quart. Rev. Biophys.* 13:121 (1980).
25. E. M. Herman, K. A. Platt-Aloia, W. W. Thomson, and L. M. Shannon, Freeze fracture and filipin cytochemical observations of developing soybean cotyledon, *Eur. J. Cell Biol.* 35:1 (1984).

EVIDENCES FOR DIFFERENT ACYL LIPID DOMAINS IN SPINACH AND OAT THYLAKOID MEMBRANES SUPPORTING VARIOUS PHOTOSYNTHETIC FUNCTIONS

Paul-André Siegenthaler, Christian Giroud and Jana Smutny

Laboratoire de Physiologie végétale
Université de Neuchâtel
Ch. de Chantemerle 20, 2000 Neuchâtel, Switzerland

INTRODUCTION

One approach to studying the role of acyl lipids in thylakoid membranes has been to incubate intact thylakoids with lipolytic enzymes and then to determine the relationships between the changes occuring in their lipid content and composition and, simultaneously, in their photochemical activities[1-3]. Whilst phospholipases hydrolyze specifically the two phospholipids (PG and PC) encountered in the thylakoid membrane, lipolytic acyl hydrolases deacylate not only glycolipids (MGDG,DGDG,SQDG) but also phospholipids. The non-specificity of the enzymes used does not allow an unequivocal assessement of function to each of these lipids[4]. Moreover, in most studies, hydrolysis of acyl lipids have been carried out under conditions allowing the attack of all acyl lipids, i.e. of lipids which are localized indistinctly in both monolayers of the membrane. However, it is now well established that all acyl lipids of the thylakoid membrane present a transversal asymmetric distribution[1,3-8], suggesting that different acyl lipid domains may sustain different functions[1,3]. Thus, it would be of great interest to establish relationships between the hydrolysis of specific acyl lipids in the outer membrane monolayer only and the resulting changes occurring in various electron flow activities.

To this aim, we have designed conditions under which (1) only PG + PC or MGDG are hydrolyzed in the outer leaflet of the thylakoid membrane; (2) no transbilayer movement of acyl lipids occurs and (3) the deleterious hydrolysis products (e.g. free fatty acids, lysophospholipids or lyso-MGDG) are removed from the membrane by BSA. Simultaneously, we have measured several photochemical parameters (various electron transport activities and fluorescence characteristics, etc.). Results show that the selective hydrolysis of certain acyl lipids in the outer monolayer of the thylakoid membrane causes preferential effects on these activities.

Abbreviations : BSA, defatted bovine serum albumin; DAD, diaminodurene; DCPIP, 2,6-dichlorophenolindophenol; DGDG, digalactosyldiglyceride; MGDG, monogalactosyldiacylglyceride; PC, phosphatidylcholine; PG, phosphatidylglycerol; PLA_2, phospholipase A_2; SQDG, sulfoquinovosyldiacylglyceride; TMPD, N,N,N',N'-tetramethyl-p-phenylenediamine.

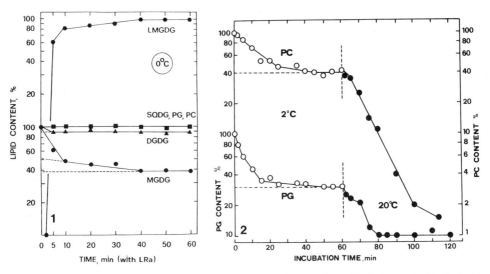

Fig. 1. (left side) Kinetics of MGDG hydrolysis (at 0°C) in oat thylakoid membranes, in the presence of the lipase from Rhizopus arrhizus (LRa) and BSA. The 100% value correspond to 1.11 $\mu mol/\mu mol$ Chl for MGDG.

Fig. 2. (right side) Kinetics of PG and PC hydrolysis (at 2° and 20°C) in spinach thylakoid membranes in the presence of porcine pancreas PLA_2. The 100% value correspond to 0.2 (PG) and 0.13 (PC) $\mu mol/\mu mol$ Chl.

RESULTS

Fig. 1 shows clearly that at 0°C, only MGDG is significantly digested by the lipase from Rhizopus arrhizus in the presence of BSA. After 40 min, 62% of the total MGDG are hydrolyzed. This amount corresponds to MGDG molecules localized in the outer monolayer of the oat thylakoid membrane. The addition of BSA removes all free fatty acids and about half of lyso-MGDG (LMGDG). Several changes in photochemical activities occur during MGDG depletion in the outer monolayer (not shown) : (1) If BSA is omitted, the two non-cyclic electron flow activities [$H_2O/NADP^+$ and H_2O/MV (methyl-viologen)-O_2] are inhibited more rapidly and at a greater extent than in the presence of BSA; this indicates that the BSA-insensitive inhibition of the activity is due to MGDG depletion; (2) PSII electron flow activities are also inhibited whatever the type of electron donors (H_2O, 1,5-diphenyl-carbazide, H_2O_2) or acceptors (2,5-dimethyl-p-benzoquinone, DCPIP, p-phenylene diamine, silicomolybdate) used. This shows that MGDG depletion in the outer monolayer does not affect the water splitting system; (3) PSI electron flow activities are inhibited rapidly whatever the type of electron donors (durohydroquinone, ascorbate-DCPIP, -DAD or -TMPD) used. The decreased activity can be fully restored by Triton X-100 (likely by allowing direct access of the donor to P_{700}) but not by an excess of plastocyanin; (4) Both rates and amplitudes of photooxidation and dark reduction of cytochrome f are decreased suggesting an inhibition site at the level of cyt b_6/f region; (5) The ratio of 77°K fluorescence F743/F689 increases, suggesting that energy is transferred to PSI at the expense of PSII, as indicated also by the lower quantum yield of the treated samples.

Fig. 2 shows that at 2°C, the hydrolysis of both phospholipids occurs in phospholipase A_2-treated spinach thylakoids. After 60 min 70% of PG and 60% of PC are digested. This amount corresponds to phospholipid molecules localized in the outer monolayer. This important phospholipid depletion causes only about 20-25% inhibition of the non-cyclic and PSII electron flow activities which can be restored partially in the presence of BSA. The activity begins to be inhibited greatly only when the hydrolysis of phospholipids located in the inner monolayer occurs (due to rapid transbilayer movement at 20°C, see Fig. 2). In the presence of BSA, the restoration of the activity is only partial suggesting that phospholipid depletion in the inner monolayer of the membrane is the cause of the BSA-insensitive inhibition.

In conclusion, our results show that optimal electron flow activity is dependent on the integrity of not only a population of MGDG molecules which is localized in the outer monolayer but also of a population of phospholipid molecules which is located in the inner monolayer. This finding is in agreement with our previous proposal[1,3].

ACKNOWLEDGEMENT

This research was supported by the Swiss national science Foundation (Grant No. 3.417-0.83 to P.A.S.).

REFERENCES

1. P.A. Siegenthaler and A. Rawyler, Acyl lipids in thylakoid membranes: Distribution and involvement in photosynthetic functions, in: "Encyclopedia of Plant Physiology, New Series, vol. 19, Photosynthesis III", L.A. Staehelin and C.J. Arntzen, eds, Springer-Verlag, Berlin, Heidelberg, New York, Tokyo, p. 693 (1986).
2. P.A. Siegenthaler, A. Rawyler and J. Smutny, Involvement of hydrophilic and hydrophobic portions of phospholipid molecules in photosynthetic electron flow activities, in: "Structure, Function and Metabolism of Plant Lipids", P.A. Siegenthaler and W. Eichenberger, eds, Elsevier Science Publishers B.V., Amsterdam, New York, Oxford, p. 475 (1984).
3. P.A. Siegenthaler, A. Rawyler and C. Giroud, Spatial organization and functional roles of acyl lipids in thylakoid membranes, These proceedings (1986).
4. A. Rawyler and P.A. Siegenthaler, Role of lipids in functions of photosynthetic membranes revealed by treatment with lipolytic acyl hydrolase, Eur. J. Biochem. 110:179 (1980).
5. M.D. Unitt, J.L. Harwood, Sidedness studies of thylakoid phosphatidylglycerol in higher plants, Biochem. J. 228:707 (1985).
6. C. Sundby and C. Larsson, Transbilayer organization of the thylakoid galactolipids, Biochim. Biophys. Acta 813:61 (1985).
7. A. Rawyler and P.A. Siegenthaler, Transversal localization of monogalactosyldiacylglycerol and digalactosyldiacylglycerol in spinach thylakoid membranes, Biochim. Biophys. Acta 815:287 (1985).
8. P.A. Siegenthaler and C. Giroud, Transversal distribution of phospholipids in prothylakoid and thylakoid membranes from oat, FEBS Lett. 201:215 (1986).

PHASE TRANSITION BEHAVIOUR OF MONOGALACTOSYLDIACYLGLYCEROL

Peter J. Quinn and *Leonard J. Lis

Department of Biochemistry, King's College London, Campden Hill, London W8 7AH, U.K. and *Physics Department, Kent State University, Kent, OH 44242, U.S.A.

INTRODUCTION

Monoglactosyldiacylglycerol is the major polar lipid component of the photosynthetic membrane and in higher plant chloroplasts it represents about half, by weight, of the total polar lipids. A particular feature of this lipid species is that, in certain plants such as bean and spinach, the fatty acyl composition is dominated by linolenic acid. This unusual polyunsaturated fatty acyl chain composition has been found to be a major factor in determining the phase behaviour of the lipid in aqueous systems (Gounaris et al., 1983). Studies of the fully saturated lipid species have been reported (Sen et al., 1983) which indicate the presence of metastable states. These states are characterized by decay rates that depend on the temperature and water content of the dispersion. This paper describes time-resolved X-ray diffraction experiments aimed to investigate phase transition behaviour in saturated monogalactosyldiacylglycerol.

MATERIALS AND METHODS

Saturated monogalactosyldiacylglycerol was prepared from spinach chloroplasts as described by Mansourian and Quinn (1986). Diffraction measurements were performed on fully hydrated specimens using the synchrotron radiation source at the Daresbury Laboratory, U.K. The camera configuration and data acquisition mode has been published elsewhere (Quinn and Lis, 1986a).

RESULTS AND DISCUSSION

The different phases formed by saturated monogalactosyldiacylglycerol in excess water and their interrelationships are illustrated in Fig.1. The structure of the α-phase is amorphous with no evidence of any long range order; the hydrocarbon chains are in a disordered configuration. The diffraction spacings in the low-angle and wide-angle region of the diffraction pattern of the α-phase are presented in Table 1 together with data for the other phases shown in Fig.1. Unlike other membrane lipids there is an appreciable scattering at intermediate angles which we have attributed to the packing of the galactose residue of the head group. The α-phase is regarded as the stable high-temperature phase. On cooling, the distearoyl derivative undergoes an exothermic

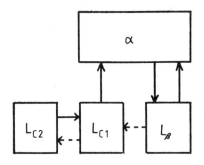

Fig. 1. Phases of saturated monogalactosyldiacylglycerol in excess water.

liquid-crystalline to gel phase transition at about 70°C to a metastable lamellar phase, L_β. Immediately reheating the L_β phase shows that the transition is reversible with a transition enthalpy of about 32 kJ.mole^{-1}.

TABLE 1. Diffraction spacings obtained from the different phases formed by aqueous dispersions of saturated monogalactosyldiacylglycerol.

Phase	d-spacing	Sugar Group	Acyl chain
α	2.8	0.67 0.59	0.48
L_β	6.25	0.69 0.61	0.41
L_{C1}	7.41	0.60	0.40 0.38
L_{C2}	6.00	0.83 0.75 0.60 0.56	0.43 0.40

The $L\beta$ phase is a metastable phase since on standing at 20°C for about 8 min. it undergoes a phase transition to a crystalline phase designated L_{C1}. Calorimetric studies do not reveal a sharp enthalpy change during the $L\beta \rightarrow L_{C1}$ transition although the structural changes are complete within about 30 s (Quinn and Lis, 1986b). The L_{C1} phase is converted directly to the α-phase on heating and has a transition temperature of about 82°C in the distearoyl derivative and an enthalpy of 67 kJ.mole^{-1} which is considerably greater than the $L\beta \longrightarrow \alpha$. transition. this reflects the greater cohesion between lipid molecules in the crystalline phase.

The mechanism and kinetics of the transition that occurs upon heating the L_{C1} phase has been studied in a temperature jump experiment the results of which are shown in Fig. 2. The L_{C1} phase is characterised by several orders of a lamellar repeat indexed by a reciprocal spacing of $s=0.135$ nm^{-1} and a sharp intense diffraction maximum at $s=1.64$ nm^{-1} corresponding to the spacing of the sugar groups. The acyl chain packing can be resolved into two peaks of reciprocal spacings of 2.63 and 2.50 nm^{-1} respectivly. When subjected to a temperature jump of about 1.6°.s^{-1} it can be seen that the L_{C1} phase is converted directly to the α-phase and there is no evidence for formation of L_β phase. The transit time for the conversion was of the order of 6s.

Prolonged storage of the L_{C1} phase at about 4°C brings about a progressive conversion to the L_{C2} phase which appears to be an equilibrium phase at this temperature. The pathway of transition between

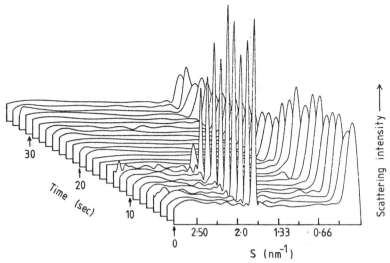

Fig. 2. Time-resolved X-ray diffraction of the $L_{C1} \rightarrow \alpha$-phase transition

the L_{C2} phase and the α-phase has been examined by calorimetry and time-resolved X-ray methods. An exothermic transition commences about $57^{\circ}C$ and precedes a main exothermic transition at about $82^{\circ}C$. The heat changes are consistent with diffraction data indicating a conversion mechanism in which the L_{C2} is transformed to the α-phase via an L_{C1} intermediate. Temperature jump experiments similar to that described in Fig.2. have shown that the transition is highly fluid with intermediate phases appearing with lifetimes of about 100 ms.

Comparison of the thermal data with the observed structural transitions indicate that the phase transitions in the saturated monogalactosyldiacylglycerol-water system are highly complex. Furthermore the data suggest that significant reorganizations of lipid and water may occur without large changes in enthalpy of the system.

ACKNOWLEDGEMENT

This work was aided by a grant from the Science and Engineering Research Council.

REFERENCES

Gounaris, K., Mannock, D.A., Sen, A., Brain, A.P.R., Williams, W.P. and Quinn, P.J., 1983, Polyunsaturated fatty acyl residues of galactolipids are involved in the control of bilayer/non-bilayer lipid transitions in higher plant chloroplasts, Biochim. Biophys. Acta, 732: 229.

Mansourian, A.R. and Quinn, P.J., 1986, Phase properties of binary mixtures of monogalactosyldiacylglycerols differing in hydrocarbon chain substituents dispersed in aqueous systems, Biochim. Biophys. Acta, 855: 169.

Quinn, P.J. and Lis, L.J., 1986a, Structural intermediates in phase transitions involving crystalline phase bilayers of monogalactosyldiacylglycerol in water, J. Colloid Interface Sci., in press.

Quinn, P.J. and Lis, L.J., 1986b, Time-resolved X-ray diffraction studies of phase changes in galactolipid-water systems, Biochem. Soc. Trans., 14: 650.

Sen, A., Mannock, D.A., Collins, D.J., Quinn, P.J. and Williams, W.P., 1983, Thermotropic phase properties and structure of 1,2-distearoylgalactosylglycerols in aqueous media, Proc. R. Soc. Lond. B, 218:349

IS MONOGALACTOSYL DIACYLGLYCEROL INVOLVED IN THE PACKAGING OF LIGHT-HARVESTING CHLOROPHYLL PROTEINS IN THE THYLAKOID MEMBRANE?

Peter J. Dominy and W. Patrick Williams,

Biochemistry Department, King's College London (KQC)

INTRODUCTION

The presence of high concentrations of the non-bilayer forming lipid MGDG in the thylakoid membrane of the chloroplast has led to two hypotheses regarding its membrane function. One suggests that the physical shape of the molecule aids membrane curvature and that MGDG is present in order to allow the high degree of membrane curvature associated with grana stack formation (1). The other suggests that MGDG is involved in the efficient packaging of the light-harvesting complexes of PS I and PS II in the membrane bilayer (2).

We have previously shown that inhibition of PS II leads to increased rates of photobleaching of a long-wavelength chlorophyll a component associated with PS I. Photobleaching of this type is normally accompanied by extensive lipid peroxidation (3). The light-harvesting units of PSI are known to be preferentially located in the non-appressed membranes of the end-membranes of the granal stacks and the stroma lamellae while those of PSII are located mainly in the appressed membranes of the grana stacks (4). This lateral separation of the two photosystems means that photo-oxidative damage preferentially associated with one of the two photosystems might be expected to be reflected in localised damage within the chloropast membrane system. With this in mind, we have examined the loss of the two main classes of galactolipid in order to see if there is any evidence to support the suggestion that MGDG is associated with the packaging of photosynthetic light-harvesting units within the thylakoid membrane (2).

MATERIALS AND METHODS

Plant Material. Plants of Pisum sativum (L. cv. Feltham First) and Phaseolus vulgaris (L. cv. Blue Lake) were grown and class D chloroplasts isolated according to procedures presented elsewhere (5).

Photobleaching. Ten ml aliquots of a 20 µg Chl. ml^{-1} suspension of thylakoids in assay media (0.33 M sorbitol, 3 mM MgCl$_2$, 10 mM NaCl, 30 mM phosphate buffer pH 6.5) were prepared just prior to irradiation. These samples were dispensed into 20 ml test tubes and placed in a thermostated water bath at 2°C. Irradiation was provided by a white light projector lamp delivering 10.2 mMole.m^{-2}.s^{-1} PAR.

Lipid Analysis. Aqueous suspensions of thylakoids were heated to 70°C for 5 minutes to de-activate endogenous lipases. Total lipid extracts were prepared by the method of Nichols (6) except that chloroform/methanol (2/1 v/v) was used instead of chloroform/propan-2-ol.

Special care was taken to minimise exposure to light. Samples were taken to dryness and resuspended in propan-2-ol/hexane/water/methanol (56/38.5/3.3/2.2 v/v) for HPLC separation. All solutions used for the extraction and storage of lipids contained 0.01% butylated hydroxytoluene and 0.2% α-tocopherol as antioxidants. Galactolipids were separated and analysed by a modification of the gradient HPLC method of Heemskerk et al. (7).

RESULTS AND DISCUSSION

The effects of high levels of illumination on the absorption spectrum of DCMU-poisoned pea thylakoids are illustrated in Fig.1. Calculations of the dark-minus-light difference spectra of samples irradiated for different times confirm our earlier reports (8,9) that there is a preferential bleaching of a long-wavelength chlorophyll a component. This component has a red absorption maximum at about 680 nm, close to that of P700-chlorophyll protein the main chlorophyll a complex of PSI (10). Prolonged irradiation leads to small shifts in the difference spectrum maximum to shorter wavelengths; possibly reflecting minor losses of chlorophyll-protein complexes associated with the core of PSII absorbing around 670 nm (11). Little bleaching is seen at 650 nm, corresponding to the red absorption maximum of chlorophyll b until most of the long-wavelength component is destroyed indicating that little or no damage is occurring to the main chlorophyll a/b light-harvesting complex in the early stages of photobleaching.

Addition of ascorbate (10 mM) or anaerobosis was found to effectively eliminate photobleaching indicating that the bleaching process is a reflection of photo-oxidative damage. Electronmicroscope studies performed on DCMU-poisoned chloroplasts (not shown) indicated progressive losses of stroma lamellae and disorientation of grana stacks with increased photobleaching but relatively little obvious damage to the grana stacks themselves. This is consistent with the occurrence of selective damage to PS I light-harvesting units located in the stromal membranes.

Attempts to determine quantitatively the levels of MGDG and DGDG in lipid extracts from photobleached thylakoids by conventional TCL and colourimetric analysis were unsatisfactory. Large variances were found in the total galactolipid recovered which appeared to be a result of

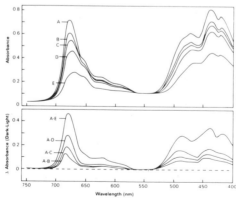

Fig.1 (a) Absorption spectra of samples of pea thylakoids (20μg.ml^{-1} chl) illuminated in strong white light in the presence of DCMU (40μM) at 25°C. Spectra A, B, C, D, E correspond to samples irradiated for 0, 20, 30, 40 and 60 min respectively, (b) corresponding dark-minus-light difference spectra.

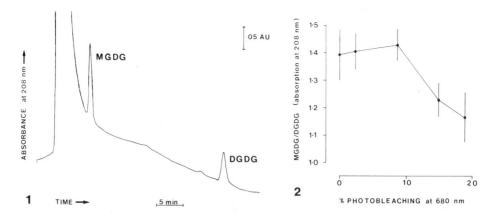

Fig.2 Typical HPLC separation of a total lipid extract of photobleached chloroplasts. See Materials and Methods for details.

Fig.3. The effect of photobleaching on the MGDG/DGDG ratio in bean thylakoids. See Materials and Methods for experimental details.

sample oxidation. Consequently a modification of the HPLC method of Heemskerk et al. (7) was adopted for the separation and analysis of MGDG and DGDG. A typical chromatogram of the total lipid extracts from bean thylakoids is presented in Fig. 2. All pigments, neutral lipids and antioxidants co-eluted with the solvent front but the MGDG and DGDG components were well separated. The MGDG to DGDG ratios of lipid extracts from photobleached bean thylakoids are shown in Fig. 3. No significant changes in this ratio was observed until chlorophyll photobleaching exceeded 10%; beyond this level there was a selective loss of MGDG. A decrease in the absorption of lipids at 208 nm is due to oxidation and/or peroxidation of the unsaturated acyl chains. Given that both lipids are predominantly di-linolenoyl forms, it is difficult to attribute a preferential photobleaching of MGDG to a higher inherent lability. Therefore, these results suggest that MGDG is more closely associated with the P700-chl a complex than DGDG. Further studies to investigate more fully the association of the less abundant thylakoid membrane lipids, SQDG and PG, as well as the individual molecular species of MGDG, with the light-harvesting chlorophyll protein complexes of PSI are currently underway.

REFERENCES

1 D. Murphy, FEBS Lett. 150:19 (1982).
2 P.J. Quinn and W.P. Williams, Biochim. Biophys. Acta 737:223 (1983).
3 R.L. Heath and L. Packer, Arch. Biochem. Biophys. 125:189 (1968).
4 B. Andersson and J.M. Anderson, Biochim. Biophys. Acta 593:427 (1980).
5 P.G. Thomas, P.J. Dominy, L. Vigh, A.R. Mansourian, P.J. Quinn and W.P. Williams, Biochim. Biophys. Acta 849:131 (1986)
6 B.W. Nichols, In "New Biochemical Separations" A.T. James and L.J. Morris, eds.) pp.321-337, Van Norstrand, New York (1964).
7 J.W.M. Heemskerk, G. Bogemann, M.A.M. Scheijen and J.F.G.M. Wintermans, Anal. Biochem. (in press).
8 W.P. Williams, A. Sen, and D.C. Fork, Photosynth. Res. in press.
9 D.C. Fork, A. Sen, and W.P. Williams, Photosynth. Res. in press
10 T.G. Dunahay and L.A. Staehelin, Plant Physiol. 78:606 (1985).
11 T.G. Dunahay and L.A. Staehelin, Plant Physiol. 80:429 (1986).

MECHANISMS OF LIPID-PROTEIN BINDING IN PHOTOSYNTHETIC MEMBRANES

Denis J. Murphy
Department of Botany, University of Durham
Durham DH1 3LE, United Kingdom

INTRODUCTION

The lipids of photosynthetic membranes can be divided into two groups on the basis of their protein-binding properties. Most of the non-acyl lipids, such as the chlorophylls, carotenoids and xanthophylls, are relatively tightly, albeit non-covalently, bound to protein ligands. On the other hand, the acyl lipids are not tightly bound to membrane proteins, although there may be specific interactions between them. Some lipids, such as the quinones fall into both categories. There are specific quinone-binding sites on several thylakoid proteins at which quinones or their derivitives are tightly bound, e.g., some proteins of the cytochrome b_6-f and the Q_A site of PSII (1). There are, however, other quinone-binding sites where the quinones are free to exchange with a mobile quinone population in the acyl lipid bilayer, e.g. the Q_B site of the 32kDa herbicide-binding protein of PSII (1).

Recent advances in the understanding of the molecular structure of thylakoid proteins have allowed for the description of possible mechanisms involved in lipid-protein binding in this important membrane system.

PIGMENT-PROTEIN BINDING

Pigments are non-acyl lipids and together account for 25-30% of the total thylakoid lipid. It is now known with some certainty that essentially all of the pigmented non-acyl lipid of thylakoid membranes is tightly but non-covalently bound to specific polypeptides. Some, but not all, of the polypeptides of the PSI, PSII and LHCII complexes contain bound pigments. For example, in the PSII complex, the 44kDa and 51kDa polypeptides apparently bind chlorophylls and carotenoids in a fixed stoichiometry, while the two 32kDa polypeptides and cytochome b_{559} do not (1). The detailed mechanism of chlorophyll-protein binding is not yet clear, but some clues have recently emerged. A hexapeptide region Asp-Pro-Thr-Thr-Arg-Arg has been found in the putative reaction-centre polypeptides of both PSI (2) and PSII (3). These regions contain H-bond donors and cationic amino acids, which may serve to bind the chlorophyll molecules which serve as the P700 and P680 reaction centres of PSI and PSII respectively. Since there is only one reaction centre chlorophyll per complex, one would expect a highly specific binding site. On the other hand, there are a further 20 or more antenna chlorophylls on each PSI pigment-binding polypeptide and here a less specific binding may be expected.

It is known that chlorophyll molecules are orientated at an average angle of 30-35° away from the bilayer in thylakoid membranes (4,5) and it has recently been proposed that this is due to a similar orientation of the transmembrane α-helical domains of the pigment-proteins (1,6). It is proposed that one tetrapyrrole ring fits on each side of the bilayer, aligned parallel to the principal axis of the protein and therefore at 30-35° to the bilayer normal - see Fig. 5, ref 1. Thermodynamic considerations would then favour the alignment of the phytyl chains of the chlorophylls parallel to the bilayer normal at the protein-acyl lipid interface where they would be free to interact with the acyl chains of the bilayer lipids (6).

Carotenoids, lutein and xanthophylls are also bound to thylakoid membrane proteins but the binding mechanism is less well understood. In the case of β-carotene, it is believed that it is orientated with its long axis parallel to the membrane normal in PSI and PSII. This implies a different binding mechanism to the chlorophylls.

QUINONE-PROTEIN BINDING

The major thylakoid quinone is plastoquinone A, which occurs in protein-bound and free forms. The free quinone pool accounts for 1-2% total bilayer lipid and outnumbers the bound quinone pool by about 3:1. Bound quinones are present in the PSII and cytochrome b_6-f complexes. The Q_A quinone binding site of PSII is probably shared between the 44kDa and 51kDa polypeptides. The 51kDa polypeptide contains a nine residue sequence His-Leu-Phe-Leu-Ser-Gly-Val-Ala-Cys, which may have a role in quinone binding, while the 44kDa polypeptide contains a similar well-conserved sequence at an identical location (3). Quinone-binding sites also exist on the two 32kDa polypeptides of PSII. It has been suggested that the His-Met dipeptide may be involved in reversible quinone binding and the sequence Ala-Met(Ile)-His-Gly is conserved at similar locations on both 32kDa polypeptides and also on quinone-binding polypeptides from bacteria (7).

It is expected that the Q_A site of PSII should differ from the Q_B site or the Rieske/cytochrome b_6 sites since the former site does not exchange with the free quinone pool, while the latter sites do exchange. In all cases, the putative quinone-binding sites are located near to the interface between the polar and non-polar regions of the thylakoid membranes. This environment would favour the quinone/quinol redox reactions, which will not readily occur in the aprotic thylakoid interior. Once released from the protein, however, the quinone species will tend to partition into the hydrophobic bilayer midplane and diffuse rapidly away in the lateral plane of the thylakoid membrane.

ACYL LIPID-PROTEIN BINDING

The general question of acyl lipid-protein binding in biological membranes is the subject of some controversy at present (8). The consensus view is that strong non-covalent binding interactions of the sort exemplified by thylakoid pigments are very rare in the case of acyl lipids (6,8). It is inevitable, however, in a protein-rich membrane like the thylakoid, that many acyl lipid molecules will of necessity interact with intrinsic proteins. This is especially true in the appressed thylakoid regions where the relative dearth of lipid means that, at any given instant, about 70% of the acyl lipid will be adjacent to a protein surface (1). The question of interest here is whether there is a tendency for some lipids to preferentially associate with certain proteins, giving rise to specific domains of particular acyl lipid classes around such proteins. Such interactions need not be very strong and may simply involve steric factors such as packaging (1,6) leading to a tendency for some lipid classes to remain adjacent to a particular protein for a longer time than others.

The heterogenous length of the protein hydrodrophobic domains (6) and their orientation away from the membrane normal (1) will induce considerable localised curvatures in the lipid bilayer and may favour the presence of certain lipid classes. There is some evidence from reconstitution studies that different thylakoid protein functions may be optimised in the presence of different acyl lipids (1,9). However, the signifigance of these data for thylakoid function in vivo remains to be proven. A more satisfactory and direct approach is to study lipid-protein interactions by techniques such as NMR and ESR spectroscopy. In a recent study, we reported on the extreme motion-restriction experienced by spin-labelled PG and PC molecules inserted into thylakoid membranes (10). This implies that the vast majority of the PG and PC pools are interacting directly with membrane proteins. We are currently studying the properties of galactolipid spin-labels in both native and artificial membrane systems in order to determine the extent of galactolipid-protein interactions in thylakoids.

In summary, we are now beginning to be able to address the molecular mechanisms of protein-lipid binding in photosynthetic membranes. It is apparent that there are several different binding mechanisms and given the right biophysical approaches it is likely that these will be considerably elucidated in the near future.

REFERENCES

1. Murphy, D.J. (1986) Biochim. Biophys Acta 864, 33-95
2. Fish, L.E., Kuk, U., Bogorad, L. (1985) J. Biol. Chem. 260, 1413-1421
3. Morris, J. and Herrmann, R.G. (1984) Nucleic Acids Res. 12, 2837-2850
4. Tapie, P., Choquet, Y., Breton, J., Delepelaire, P. and Willman, F.-A. (1984) Biochim. Biophys. Acta 767, 57-69
5. Nabredryk, E., Andrianambinintson, S. and Breton, J. (1984) Biochim. Biophys. Acta 765, 380-387
6. Murphy, D.J. (1986) In: From Structural Elucidation to Biological Function - Current Topics in Lipid Research, Klein, R. and Schmitz, B., eds, Royal Society of Chemistry, London, in press
7. Holschuh, K., Bottomley, W. and Whitfield, P.R. (1984) Nucleic Acids Res. 12, 8819-8834
8. Deveaux, P.F. and Seigneuret, M. (1985) Biochim. Biophys. Acta 822, 63-125
9. Murphy, D.J. (1986) Photosynthesis Res. 8, 219-233
10. Murphy, D.J. and Knowles, P.F. (1984) In: Structure, Function and Metabolism of Plant Lipids, pp 425-428, Siegenthaler, P.A. and Eichenberger, E., eds, Elsevier, Amsterdam.

LIPID MOLECULAR SPECIES COMPOSITION OF GRANAL AND STROMAL LAMELLAE

Helen A. Norman[1], Judith B. St. John[1],
Franklin E. Callahan[2], Autar K. Mattoo[2],
and William P. Wergin[3]

Weed Science Laboratory[1]
Plant Hormone Laboratory[2]
Plant Stress Laboratory[3]
USDA, ARS
Beltsville, MD 20705

INTRODUCTION

We are investigating the lipid biosynthetic pathways involved in both the development of chloroplast membranes and the lipid-protein interactions in a mature chloroplast. We report the molecular species compositions of monogalactosyldiacylglycerol (MGDG) and digalactosyldiacylglycerol (DGDG) from highly purified granal and stromal lamellae of Spirodela oligorrhiza. Characteristic differences in the lipids suggest that constituent lipid molecular species (formed via different pathways) are associated with the specialized structure and function of the chloroplast membrane systems.

MATERIALS AND METHODS

Spirodela oligorrhiza (Kurtz) Hegelm was cultured axenically for 15-20 days under steady state light (15 uE.m^{-2}.sec^{-1} PAR, 25° C) in half-strength Hunter's mineral medium[1] supplemented with a 5% sucrose. Whole thylakoids of Spirodela were fractionated into granal and stromal lamellae by a detergent solubilization and differential centrifugation method.[2] Homogeneity of fractions was verified by electron microscopic examination. HPLC methods were used to quantify lipid molecular species.[3]

RESULTS

Granal and stromal lamellae fractioned from whole thylakoids of Spirodela were $\geq$ 98% homogeneous based on electron microscopic examination (data not presented). Figure 1 reveals the distinct polypeptide compositions relative to the whole thylakoids. The granal lamellae are enriched in photosystem II (PS II) polypeptides (rc II, LHCP, PS II extrinsic 33KDa) while stromal lamellae are enriched in photosystem I (PS I) and ATPase (α,β) proteins.

Table 1. Lipid Molecular Species of Granal and Stromal Lamellae of Spirodela oligorrhiza. Values are given as nmol/mg Chl.

Molecular Species	Granal lamellae		Stromal lamellae	
	MGDG	DGDG	MGDG	DGDG
18:3/16:3	290	40	80	4
18:3/16:2	58		14	
18:2/16:2	103		48	
18:3/16:0	11	31	30	55
16:0/18:3	5	14	13	24
18:2/16:0	25	19	11	22
16:0/18:2	11	8	5	10
18:1/16:0	358	185	139	88
16:0/18:1	153	79	102	38
18:3/18:3	316	89	680	258
18:2/18:3	86	36	157	39
Totals	1416	501	1279	538

The ratio of MGDG/DGDG in granal and stromal lamellae were 2.83 and 2.37, respectively (Table 1). The molecular species composition of MGDG and DGDG from granal and stromal lamellae were the same qualitatively, however, considerable quantitative heterogeneity was observed. In particular, the total quantity of molecular species with 18/16 C pairings in the sn-1 and sn-2 positions versus the quantity of molecular species with 18/18 C pairings was strikingly different for the two membrane fractions. The predominant molecular species, in decreasing order, in granal MGDG were 18:1/16:0, 18:3/18:3, 18:3/16:3, and 18:2/16:2. The

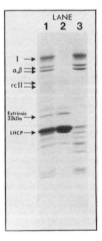

Fig. 1. Coomassie stained polypeptides of whole thylakoids (Lane 1), granal (Lane 2), and stromal (Lane 3) lamellae resolved by SDS-PAGE.[4] The positions of photosystem I reaction center polypeptide (I), α and β subunits of proton ATPase (α, β), photosystem II reaction center polypeptides (rc II), PS II extrinsic 33KDa protein of the water splitting complex, and light harvesting chlorophyll protein complex (LHCP) are indicated.

predominant molecular species in granal DGDG was 18:1/16:0 whereas 18:1/16:0 was decreased in stromal DGDG with a concomitant increase in 18:3/18:3 compared to granal DGDG. The ratio of the quantities of 18/16 C to 18/18 C species were: granal lamellae, 2.10 for MGDG and 2.20 for DGDG; stromal lamellae, 0.38 for MGDG and 0.57 for DGDG.

DISCUSSION

The biosynthetic pathways for production of saturated fatty acids occur in the chloroplasts of higher plants and are well understood. Subsequent desaturation reactions giving rise to trienoic fatty acids in galactolipids are not as well characterized. We have recently presented definitive evidence[3] (also see H. A. Norman and J. B. St. John, this volume) for different sites and substrates for linolenic acid (18:3) synthesis in Arabidopsis thaliana L. In the 'prokaryotic' pathway of MGDG synthesis, occurring totally in the chloroplast,[5] 18:1/16:0 MGDG was progressively desaturated to form 18:3/16:3 MGDG. A cytoplasmic, 'eukaryotic' pathway[5] utilized phosphatidylcholine as the substrate for formation of the 18:2/18:3 diglyceride precursor for galactosylation and further desaturation in the chloroplast to form 18:3/18:3 MGDG.

The presence of 18/16C and 18/18C molecular species of MGDG and DGDG in granal and stromal lamellae identify Spirodela as a plant with both pro- and eukaryotic pathways operative in the formation of polyunsaturated galactolipids. MGDG of granal membranes was enriched in lipid molecular species derived from the prokaryotic pathway; stromal membranes were enriched in eukaryotic lipids. Thus lipid molecular species formed via the different pathways are selectively distributed to granal and stromal lamellae. The enrichment of granal membranes in prokaryotic lipids/PS II polypeptides and stromal membranes in eukaryotic lipids/PS I polypeptides supports some specificity of lipid-protein interactions in the maintenance of functional photosystem complexes.

1. H. B. Posner, in: "Methods in Developmental Biology," F. A. Witt and J. N. Wessels, eds., Crowell, New York (1967).
2. K. J. Leto, E. Bell, and L. McIntosh, EMBO J. 4:1645 (1985).
3. H. A. Norman and J. B. St. John, Plant Physiol. 81:IN PRESS (1986).
4. A. K. Mattoo, U. Pick, H. Hoffman-Falk, and M. Edelman, Proc. Nat. Acad. Sci. USA 78:1572 (1981).
5. P. G. Roughan and C. R. Slack, Trends Biochem. Sci. 9:383 (1984).

ON THE FUNCTION OF METHYL-BRANCHED CHAIN FATTY ACIDS IN PHOSPHOLIPIDS OF CELL MEMBRANES OF HIGHER PLANTS

Alfons Radunz

Lehrstuhl Zellphysiologie, Fakultät für Biologie
Universität Bielefeld, D-4800 Bielefeld 1, FRG

SUMMARY

Mono-methyl branched fatty acids of the iso- and anteiso types with chain lengths of 12 to 22 carbon atoms were found in yellow-white leaves and petals of the plastome mutants *"Prasinizans"* of *Antirrhinum majus* and *"Xanthi"* of *Nicotiana tabacum*. These branched chain fatty acids are preponderantly esterified in the sn-2 position of the glycerol backbone of the membrane phospholipids namely phosphatidylcholine, phosphatidylethanolamine and phosphatidylinositol. The phospholipids appear to contain 16-24 per cent iso- and anteiso fatty acids, 19-47 per cent unsaturated and 38-63 per cent saturated fatty acids. The methyl-branched chain fatty acids seem to take over the function of the unsaturated fatty acids.

KEY WORDS

Iso- and anteiso fatty acids, phosphatidylcholine, phosphatidylethanolamine, phosphatidylinositol, cell membrane, plastome mutants of *Antirrhinum majus* *"Prasinizans"* and of *Nicotiana tabacum* *"Xanthi"*.

INTRODUCTION

In a previous publication we had reported that yellow-white leaves of the plastome mutant *"Prasinizans"* of *Antirrhinum majus* contained methyl-branched fatty acids (1). They make up for 6-8 per cent of the total fatty acids. The methyl-branched fatty acids were isolated via urea adducts and characterized by means of comparative gas chromatography with authentic fatty acids (2,3) by IR-spectroscopy, NMR-spectroscopy and by means of the measured optical activity as mono- and methyl-branched fatty acids of the iso- and anteiso-types (1).
Recent investigations have shown that these iso- and anteiso acids also occur in yellow-white leaves of the plastome mutant *"Xanthi"* of *Nicotiana tabacum*. The fatty acids were detected in the phosphatides of both plants and make up for 16-24 per cent of the fatty acids of the phospholipids. Phospholipids are known to be functional components of the lipid matrix of the cell membrane as well as of membranes of various cell organelles which allow biological activity in cooperation with other membrane components.

MATERIALS AND METHODS

1. <u>Plant material</u>: Plants of the plastome mutant "Prasinizans" of Antirrhinum majus strain 50 (4) and "Xanthi" of Nicotiana tabacum (5) were grown in a greenhouse under normal sunlight conditions. The plastome mutant "Prasinizans" characterized by the fact that leaves green in the early developmental states with retardation. The mutant "Xanthi" of Nicotiana tabacum we succeeded, by cutting back the primary sprouts, in producing either exclusively whitish yellow, yellow-green, variegated or fully green side sprouts.

2. <u>Preparation of lipids</u>: The whitish yellow and green leaves as well as the petals were lyophilized and extracted with methanol and dietylether. From the crude extract the lipids were extracted with diethylether. The isolation of the phospholipids and glycolipids was obtained by means of column and thin layer chromatographic methods on silica gel g (6,7).

3. <u>Fatty acid analyses</u>: The lipids were saponified with methanolic 0.5 NaOH. Thereafter, fatty acids were transformed with 5% methanolic HCl or with N,N-Dimethyl foramide-dimethylacetyl (Macherey und Nagel, Düren) into the methyl esters. The analysis by gas chromatography was carried out with a Hewlett-Packard Type 5750 on columns of Reoplex and ethylenglycol succinate (6). For the identification we used authentic fatty acids, as a 16-methyl-heptadecanoic acid and a 14-methyl-hexadecanoic acid isoplated by Hansen et al. (2,3). For the isolation of the methyl branched components the unsaturated components were separated from the fatty acids mixture via their adducts with mercury II-acetate. Subsequently we cristallized the branched chain acids via their urea adducts from the mixture of the n-fatty acids (1).

4. <u>Positional Analysis of Fatty Acids in Phospholipids</u>: Phospholipid samples (15 mg) were suspended in 2 ml 50 mM Tris-HCl-buffer pH 7.4 and were sonicated for 20 sec. To this suspension 6 mg Phospholipase A_2 (from pig pankreas, Boehringer, Mannheim, FRG) dissolved in the same buffer were added. After a 15 min hydrolysis 20 ml isopropanol were added in order to stop the enzyme reaction. After removal of the solvent the products of the enzymic cleavage were dissolved in $CHCl_3$/MeOH (1/1 v/v) and filtered. The fatty acids were separated from the lyso-phospholipid by thinlayer chromatography on silicic acid in the solvent system $CHCl_3$/MeOH/ACOH/H_2O (63/35/4/4 v/v). Thereafter, the fatty acid composition was analyzed by gas chromatography.

RESULTS

The yellow-white leaf areas are in comparison to green leaves not only characterized by a lower chlorophyll and lipid content but also by the fact that the lipid and fatty acid composition shows appreciable differences (table I). The lipids of the yellow-white leaf areas are characterized by a higher content of phospholipids and a lower content of glycolipids. This difference in comparison to green leaves is due to the lower chloroplast number in the yellow-white leaves. Earlier investigations have shown that approximately half of the leaf lipids are located in chloroplasts (7). Among the fatty acids of the phospholipids 16-24 per cent methyl-branched acids are observed, which are completely lacking in the phospholipids of the green leaves (table II). Whereas the phospholipids of green leaves contain in general 60-80 per cent unsaturated acids, the yellow-white leaves contain only 20-50 per cent unsaturated fatty acids. Thus, the fatty acids of the phospholipids of the yellow-white leaves are characterized by a higher degree of saturation.

The occuring methyl branched fatty acids are characterized, due to their behaviour in the comparatitive GLC with the authentic 16-methyl-heptadecanoic and the 14-methyl-hexadecanoic acid (2,3) and due to IR, NMR spectroscopy and the measurement of optical activity as being mono-methyl-branched fatty acids of the iso- and anteiso types with chain lengths of 12 to 22 carbon atoms. Degradation experiments of the lipids with phospholipase A_2 have shown that the iso- and anteiso acids are essentially esterified at the sn-2 of the glycerol of phosphatidylcholine, phosphatidylethanolamine and phosphatidylinositol just as unsaturated fatty acids (table II).

The listed phospholipids are localized in the lipid matrix of the cell membranes of the plasmalemma, of the tonoplast, of the endoplasmatic reticulum as well as in membranes of different cell organelles such as the nucleus, mitochondria, peroxisomes, glyoxysomes and oleosomes. In chloroplasts only relatively small amounts of phospholipids are present whereas the main membrane lipids of chloroplasts are monogalactolipids, digalactolipids, sulfolipids. These glycolipids do not contain methyl branched fatty acids.

Table I: Composition of Lipids of Yellow-White Leaves and Petals of the Plastome Mutants "Prasinizans" of *Antirrhinum majus* and "Xanthi" of *Nicotiana tabacum*.

	Nicotiana tabacum-"Xanthi"		Antirrhinum majus-"Prasinizans"		
	yellow-white leaves	green leaves	yellow-white leaves	green leaves	Petals
Lipids (in per cent of freeze-dried leaves)	9.7	14.0	5.0	11.5	2.0
Chlorophyll[1]	0.3	13.0	1.4	14.0	*
Glycolipids[2]	15.9	27.0	23.4	43.1	*
Phospholipids[3]	19.0	6.8	24.7	17.6	*
saturated n-fatty acids[4]	25.0	12.0	25.0	16.0	50.5
iso- and anteiso acids[4]	4.2	--	5.7	--	9.0

1. in per cent of lipids; 2. Glycolipids: Monogalactolipid, Digalactolipid, Sulpholipid (in per cent of lipids); 3. Phospholipids: Phosphatic acid, Phosphatidylethanolamine, Phosphatidylglycerol, Phosphatidylcholine, Phosphatidylinositol, Cardiolipin (in per cent of lipids); 4. in per cent of fatty acids.
*not tested

Table II: Positional Distribution of Fatty Acids between the sn-1 and sn-2 Position at the Glycerol Backbone in Phospholipids of Petals and Yellow-White Leaves of the Plastome Mutant "Prasinizans" of *Antirrhinum majus*.

	Phosphatidylcholine		Phosphatidylinositol		Phosphatidylethanolamine	
	sn-2	sn-1	sn-2	sn-1	sn-2	sn-1
iso- and anteiso fatty acids	14.4	1.3	17.5	0.8	18.2	5.3
saturated n-fatty acids	18.4	19.3	18.2	44.5	16.5	27.5
unsaturated n-fatty acids	17.2	29.4	14.3	4.7	15.1	17.2

The fatty acid content is given as per cent of the total fatty acid content of the respective phospholipids. The iso-acids were identified by their retentions times in GLC to have the following chain lengths: 10-methyl-undecanoic acid (iso-lauric); 12-methyl-tridecanoic acid (iso-myristic); 14-methyl-pentadecanoic acid (iso-palmitic); 16-methyl-heptadecanoic acid (iso-stearic); 20-methyl-heneicosanoic acid (iso-behenic). The anteiso-acids were: 12-methyl-tetradecanoic acid; 14-methyl-hexadecanoic-acid; 16-methyl-octadecanoic acid; 19-methyl-heneicosanoic acid.

DISCUSSION

Our studies have shown that the iso- and anteiso acids are esterified in the sn-2 position of the glycerol of the phospholipids. In membrane phospholipids they take over the position of unsaturated fatty acids in two respects. Firstly the methyl-branching of the carbon chain influences the molecular packing of the lipids in a lipid double layer. A dense stretched packing of the fatty acid residues is disturbed which leads to a change in the hexagonal arrangement of the molecules which is necessary for the ordered gel phase. A similarly loose arrangement in layers of the hydropholic paraffine chains is achieved by increasing double bonds in multiple unsaturated fatty acids. Furthermore, it was shown, that iso- and anteiso acids have a considerably lower melting point in comparison to straight chained fatty acids with the same carbon number. Whereas the melting point of an octadecanoic acid ($C_{18}H_{36}O_2$) which is a stearic acid, is 69.6ºC, the melting point of the 16-methyl-heptadecanoic acid, which is the isostearic acid, is found to be only 43.5-44.7ºC (8). On the other hand, the introduction of several double bonds into long chained fatty acids also leads to a considarable lowering of the melting point, e.g. from the octadecanoic acid ($C_{18}H_{36}O_2$) via the octadecenoic ($C_{18}H_{34}O_2$) to the octadienoic acid ($C_{18}H_{32}O_2$) from 69.6ºC via 13ºC to -11ºC.

A higher content of multiple unsaturated fatty acids in the lipids increases the fluidity of biomembranes. A higher fluidity of membranes allows inproved ion transport between cell compartments and also changes the activity of phospholipid dependent membrane bound enzyme systems. Thus is due to the fact that enzyme reactions depend on the chemical composition of the lipid-matrix. Since the investigated membrane phospholipids in the yellow-white leaves of the mutants contain considerably less unsaturated fatty acids than the phospholipids of normally greened leaves it is obvious to conclude from the presented data, that iso- and anteiso fatty acids increase just as unsaturated fatty acids the fluidity of cell membranes.

REFERENCES

1. A. Radunz: Über das Vorkommen von verzweigtkettigen Fettsäuren bei *Antirrhinum majus*. Hoppe-Seyler's Z. für Physiol. Chemie 341:192(1965)
2. R.P. Hansen, F.B. Shorland and N.J. Cooke: The Branched Chain Fatty Acids of Mutton Fat. 1. The Isolation of (+)-14-methylhexadecanoic Acid.Biochem. J. 52:203 (1952)
3. R.P. Hansen, F.B. Shorland and N.J. Cooke: The Branched Chain Fatty Acids on Mutton Fat. 3. The Isolation of 16-Methylheptadecanoic Acid (iso Stearic Acid).Biochem. J. 64:214 (1956)
4. A. Wild: Untersuchungen zweier albomakulater Linien von *Antirrhinum majus* auf ihr Verhalten in Teilreaktionen der Photosynthese. Beitr. Biol. Pflanzen 35:137 (1959)
5. L.G. Burk, R.N. Stewart and H. Dermen: Histogenesis and Genetics of a Plastid-Controlled Chlorophyll Variegation in Tobacco. Amer. Journ. Bot. 51:713 (1964)
6. A. Radunz: Localization of the Tri- and Digalactosyl Diglyceride in the Thylakoid Membranewith Serological Methods. Z. Naturforsch. 31c:589 (1976)
7. A. Radunz: Chlorophyll- and Lipidgehalt der Blätter und Chloroplasten von *Antirrhinum majus* in Abhängigkeit von der Entwicklung. Pflanzenphysiol. 54:395 (1966)
8. S. Abrahamson, S. Ställberg-Stenhagen und E. Stenhagen: The Higher Saturated Branched Chain Fatty Acids, in: Progress in the Chemistry of Fats and other Lipids. Vol. 7, R.T. Holman, ed., Pergamon Press, Oxford, London, New York, Paris (1963)

FREEZING RESISTANCE AND LIPID CHANGES IN CHOLINE-TREATED WHEAT SEEDLINGS

W.P. Williams[a], I. Horvath[b], P.J. Quinn[a], P.G. Thomas[a] and L. Vigh[b]

[a]King's College London (KQC), [b]Hungarian Academy of Sciences Szeged

INRODUCTION

A number of studies have appeared over the years suggesting that the ability of seedlings to resist low temperatures can be enhanced by exposure to amino alcohols such as choline and ethanolamine (1-3). The idea behind these studies was that treatment with such compounds leads to an increase in phosphatidylcholine synthesis at the expense of phosphatidylethanolamine synthesis.
In this study, we report the effects of such treatment on seedlings of two winter-wheat cultivars, Miranovskaja 808 and Brigand, of differring cold-hardiness

MATERIALS AND METHODS

Seeds of the winter wheat cultivars Brigand and Miranovskaja 808 were surface sterilised and germinated in the dark on wet blotting-paper. They were then transferred to water culture and grown for 8 to 10 days in the presence of 0, 20 or 40 mM choline chloride as described elsewhere (2). Plasma membrane preparationsof leaf tissue were prepared according to the procedure of Gronewald et al. (4). Lipid analyses and the determination of LT_{50} values were performed as described by Horvath et al (2). Protoplasts were isolated from the leaves of 14 to 18 day old seedlings by a modified version of the procedure of Sethi and Maeda (5). Protoplasts from seedlings grown in 0, 20 and 40 mM choline chloride were maintained in solutions of sorbitolin Marashinge and Skoog salts of osmolarities 600, 790 1030 mOsm. kg^{-1} respectively. The cryomicrocope system used to determine the freezing resistance of the protoplasts has been detailed elsewhere (6).

RESULTS AND DISCUSSION

Lipid Analysis: The total phospholipid content (nmoles/g leaf tissue) of Miranovskaja 808 seedlings increases markedly on growth in the presence of choline chloride (Fig. 1). This increase is largely due to higher levels of PC. Small increases in PG and PI also occur but PA and DPG levels remain effectively constant. In the case of Brigand, there is a decrease in total phospholipid content. This appears to reflect a drop in PA and

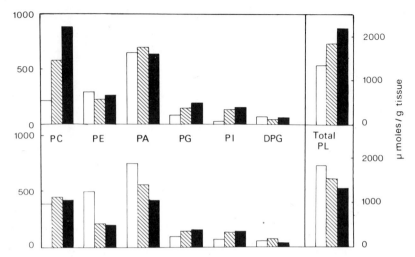

Fig. 1 Measurements of the lipid content of leaf tissue from Brigand and Miranovskaja 808 seedlings grown hydroponically in media containing no choline chloride (unhatched), 20 mM (hatched) and 40 mM (solid) choline chloride.

PE levels. Small increases in PG and PI levels occur but PC and DPG levels remain unchanged.

Measurements performed on a plasma membrane fraction obtained from the leaf tissue of Brigand seedlings (Table 1) shows a 35% increase in total phospholipid/mg protein following choline treatment. The PC/PE ratio remains effectively constant. A small decrease occurs in the double bond index (d.b.i.) of the phospholipid fraction and there is a 13% decrease in sterol content.

LT_{50} values: The LT_{50} values of Brigand and Miranovskaja 808 were calculated from ion-leakage studies. The LT_{50} of Miranovskaja treated with 15 mM choline chloride for 8 days at 25°C drops from a control value of -2.2°C to -12.3°C while that of Brigand treated with 40 mM choline chloride falls only from -3.6°C to -7.8°C.

Cryomicroscopy Study: Protoplasts isolated from the leaf tissue of Brigand seedlings grown in the presence of 0, 20 mM or 40 mM choline chloride were cooled on the stage of the cryomicroscope at a rate of -10°C.min^{-1}. The stability of the plasma membranes of the protoplasts was assessed by following intra-cellular ice formation. The cytoplasm of undamaged cells initially remains clear reflecting super-cooling of the cell contents. Damage to the plasma membrane results in a nucleation of

TABLE 1. Effects of choline chloride treatment* on the lipids of the plasma membranes of laef tissue of Brigand.

Choline chloride (mM)	PC	PE	sterols	PC/PE	Sterols/Total PL	d.b.i.
	(umoles/g, protein)					
0	187	83	172	2.25	0.47	1.83
40	268	115	150	2.32	0.30	1.64

*Treatment for 10 days at 22°C.

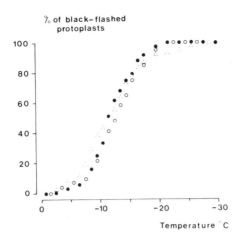

Fig.2 Plot showing the total percentage of protoplasts that have undergone "black-flashing" at decreasing temperatures. Protoplasts were isolated from Brigand seedlings grown in the presence of 0 mM ($\triangle$), 20 mM ($\bullet$) and 40 mM ($\circ$) choline chloride.

ice-formation in the cytoplasm. Ice rapidly spreads across the cell which becomes opaque. This process, often referred to as black-flashing, was followed as a function of temperature. Small variations were seen in the freezing patterns of treated and untreated cells but little difference was seen in the mean temperatures for the different cell populations (Fig. 2).

Addition of choline chloride to the growth medium leads to increased cold-hardiness in winter wheat. The extent of this increase varies between cultivars; probably reflecting their different genetic capacities for hardening. In the case of Miranovskaja, which shows a high frost-tolerance, a large increase in phospholipid content and a marked stimulation of PC synthesis occurs. This is in agreement with earlier findings for this cultivar (2). A quite different pattern of changes is seen, however, for the less frost-tolerant cultivar Brigand.

There is increasing evidence in the literature to suggest that cold-hardiness is affected by changes in plant hormonal levels (7-9). Choline is known to act as a weak growth retardant and is probably acting by influencing hormonal levels rather than as a specific perturbant of lipid synthesis. This would explain the diversity of response seen in the different cultivars. The very different LT_{50} values for isolated protoplasts and intact tissue suggests that the critical step in membrane damage is probably associated with recovery from low temperatures rather than exposure to low temperature per se.

REFERENCES

(1) R. Ilker, A.J. Waring, J.M. Lyons and R.W. Briedenbach, Protoplasma, 90:229 (1976).
(2) I. Horvath, L. Vigh, and T. Farkas, Planta 151:103 (1981).
(3) I. Horvath and P.R. Van Hasselt, Planta 164:83 (1985).
(4) J.W. Gronewald, W. Abou-Khalil, E.J. Weber and J.B. Hanson, Phytochem. 21:859 (1982)
(5) M. Sethi and E. Maeda, Japan. J. Crop Sci. 52:187 (1983).
(6) G.E Coulson, G.J. Morris and d. Smith, J. Gen. Microbiol. 132:183 (1986).
(7) A. Rikin and A.E. Richmond, Plant Physiol. 38:95 (1976).
(8) H.H. Chen and L.V. Gustav, Plant Physiol. 73:71 (1983).
(9) A.H. Markhart, Plant Physiol. 74:81 (1984).

DROUGHT STRESS EFFECTS ON ROOT CELL MEMBRANES

Helén Svenningsson, Maria Andersson and Conny Liljenberg

Department of Plant Physiology, Botanical Institute
University of Göteborg, Carl Skottsbergs Gata 22
S-413 19 Göteborg, Sweden

INTRODUCTION

Drought stress or consecutive periods of drought stress implies drastic changes in the root cell membranes of higher plants. We have found, in previous investigations on lipid analysis of whole roots, changes in the total amount of acyl lipids, the unsaturation degree and the dry weight ratio root/shoot, the molar ratio free sterols/phospholipids and the level of abscisic acid. These investigations were performed on oats and rape (Liljenberg and Kates, 1982; Liljenberg and Kates, 1985; Svenningsson and Liljenberg, 1986).

MATERIAL AND METHODS

Rape seeds (Brassica napus L.) were sown in gravel in plastic pots with holes in the bottom. The pots filled with gravel to a height of 12 cm were placed in plastic bins containing nutrient solution to a depth of 5 cm. The nutrient solution was changed every 3 days for 19 days after which the stress program was started.

The water stress was applied by placing the pots on a coarse stainless steel net whereby excess nutrient solution was drained through the bottom holes of the pots permitting the gravel to air-dry effectively. The water-deficit stress program consisted of two cycles of 24 h stress and rewatering for 24 h, followed by a final 24 h stress. During the stress program samples were taken every 24 h for water potential determination (measured by the pressure bomb technique) and for measurement of the ratio fresh weight to dry weight. A detailed examination of the first 24 h stress period was done by measuring water potential and fresh weight/dry weight ratio of shoot and root every 3 h.

The plasma membrane enriched fraction was prepared by discontinuous sucrose density gradient centrifugation of a microsomal fraction prepared from total root homogenate freed from mitochondria, plastids and nuclei by differential centrifugation. Except for minor modifications the isolation procedure was as described by Hodges et al. (1972). The plasma membrane fraction was identified by enzyme reactions and by electron microscopy with PACP staining. Total lipids were extracted from the suspended membranes

with boiling 2-propanol and afterwards further extracted with chloroform-
-methanol (2:1, v/v). The lipids were analysed as described by Svenningsson
and Liljenberg (1986). Proteins were determined according to Lowry (1951).

RESULTS AND DISCUSSION

The oscillations in water potential of the shoot and the changes in the
fresh weight/dry weight of the roots during the stress program were very
similar and in phase with each other. The stress treatment did not imply any
detectable injuries of the root system and after rewatering and growth for
another 7 days the plants appeared similar to the controls but were somewhat
shorter. After three stress periods the ratio dry weight of roots/dry weight
of shoot had increased 69 %.

The plasma membrane enriched fraction (PMEF) from the control root cells
contained 1.33 nmol phospholipids/µg protein. This value decreased after
1 x 24 h stress to 1.11 nmol and after 3 x 24 h stress to 0.67 nmol.

The lipid analysis was concentrated to the major groups, the phospholipids and the free sterols. The determinations were performed on individual
membrane fractions and means and standard deviations were thus calculated
from isolated membranes of different plant samples. That might explain the
sometimes high standard deviation.

Qualitative thin layer chromatography of neutral lipids revealed one
dominant spot of free sterols and more faint spots of triacylglycerols, free
fatty acids and steryl esters (not shown here). Total mol phospholipids decreased after one stress period to 83 % of the control and decreased after
three stress periods to 50 %. The phospholipid composition and how this
changes with stress is shown in Tab. 1. Very little change after 24 h stress.
After three stress periods however, the PC/PE mol ratio changed from 1.8
(control) to 1.2. In addition, the percentage of PI increased significantly.
Tab. 2 shows the fatty acid coposition of individual phospholipids from the
PMEF after different amounts of water-deficit stress.

The free sterols consisted of campesterol (24 %), stigmasterol (4 %)
and β-sitosterol (72 %). This composition changed very little with one or
three stress periods. PMEF from plant cells are usually characterized by a
high ratio free sterols/total phospholipids. In our system the molar ratio
free sterols/total phospholipids of the PMEF was high in the control membranes, 0.58, decreased somewhat after one stress treatment, 0.51, but increased after three stress periods to 0.70.

Table 1. Phospholipid composition (mol %) of a plasmamembrane enriched
fraction from rape root cells

	Control	1 x 24 h stress	3 x 24 h stess
Phospatidylcholine	45.3 ± 0.5	45.9 ± 0.5	33.4 ± 1.4
Phosphatidylethanolamine	25.5 ± 0.1	23.7 ± 0.2	28.5 ± 1.5
Phosphatidylinositol	7.8 ± 1.9	6.3 ± 0.3	12.1 ± 1.2
Phosphatidic acid	21.5 ± 2.4	24.4 ± 0.3	26.2 ± 1.7

Table 2. Fatty acid composition of individual phospholipids from a plasmamembrane enriched fraction of rape root cells

	Fatty acid, mol %						
	14:0	16:0	16:3	18:1	18:2	18:3	DBI
Phosphatidylcholine							
Control	tr	35±7.6	5±1.2	12±1.2	17±1.4	31±8.7	4.40
Stress 1 x 24 h	tr	30±0.5	8±0.5	10±1.0	20±0.5	31±1.5	5.47
Stress 3 x 24 h	2±0	49±2.0	6±0	4±2.0	17±1.0	31±8.7	2.63
Phosphatidylethanolamine							
Control	1±0	48±10.7	5±1.6	8±3.7	14±2.9	23±6.6	2.45
Stress 1 x 24 h	2±0	54±5.0	8±0.5	5±0	16±2.5	16±1.0	1.95
Stress 3 x 24 h	4±0	58±11.5	7±0	6±0	14±3.5	20±0.5	1.85
Phosphatidylinositol							
Control	4±1.5	78±7.3	14±4.2	4±0.9	tr	tr	0.56
Stress 1 x 24 h	3±0	77±5.5	10±0.5	3±0	4±2.5	5±0	0.70
Stress 3 x 24 h	14±4.5	66±15.0	16±5.9	5±1.4	2±0.5	3±1.0	0.83
Phosphatidic acid							
Control	6±1.5	70±3.5	10±1.0	6±0.5	5±1.0	8±0	0.92
Stress 1 x 24 h	2±0	52±7.5	11±1.0	6±0.5	14±2.5	16±3.0	2.13
Stress 3 x 24 h	4±3.5	55±3.0	9±1.0	4±0	12±1.5	16±3.5	1.88

The lipid/protein ratio of the PMEF from control rape root cells was in the same range as previously published values of root cell plasma membranes (Travis and Booz, 1979; Gronewald et al., 1982). This ratio decreased dramatically after repeated stress. The phospholipid composition changed with stress too. The unsaturation degree of the major phospholipids decreased after stress. All the mentioned changes together with the changed sterol/phospholipid ratio and with the dehydration effects per se on the complex membrane lipids, might imply alterations in the properties and function of the plasma membrane.

References

Gronewald, J. W., Abou-Khalil, W., Weber, E. J. and Hanson, J. B., 1982, Lipid composition of a plasma membrane enriched fraction of maize roots, Phytochem., 21:859.

Hodges, T. K., Leonard, R. T., Bracker, C. E. and Keenan, T. W., 1972, Purification of an ion-stimulated adenosine triphosphatase from plant roots: association with plasma membranes, Proc. Nat. Acad. Sci., USA, 69:3307-3311.

Liljenberg, C. and Kates, M., 1982, Effect of water stress on lipid composition of oat seedling root cell membranes, in: "Biochemistry and Metabolism of Plant Lipids", J. F. G. M. Wintermans and P. J. C. Kuiper, eds., Elsevier Biomedical Press B. V., Amsterdam.

Liljenberg, C. and Kates, M., 1985, Changes in lipid composition of oat root membranes as a function of water-deficit stress, Can. J. Biochem. Cell Biol., 63:77.

Lowry, O. H., Rosebrough, N. J., Farr, A. L. and Ramdall, R. J., 1951, Protein measurement with the Folin phenol reagent, J. Biol. Chem., 193:265.

Svenningsson, H. and Liljenberg, C., 1986. Membrane lipid changes in root cells of rape (Brassica napus) as a function of water-deficit stress, Physiol. Plant., 47: .

Travis, R. L. and Booz, M. L., 1979, Partial characterization of a potassium stimulated adenosine triphosphatase from plasma membrane of meristematic and mature soybean root tissue, Plant Physiol., 63: 573.

MANIPULATING MEMBRANE FATTY ACID COMPOSITIONS OF SOYBEAN PLANTS[1]

William B. Terzaghi[2]

Department of Biology
University of Utah
Salt Lake City, UT 84112

INTRODUCTION

A system for altering the membrane fatty acid composition of cultured soybean cells by adding Tween-fatty acid esters to their growth medium has been recently described (4). Cells transferred large amounts of fatty acids from Tweens to all acylated lipids detected, resulting in major changes in membrane fatty acid composition. However, although this system seemed useful for a variety of studies related to membrane and lipid metabolism, its usefulness was limited by the requirement to work with cultured cells. This paper reports that Tween-fatty acid esters may also be used to modify the membrane fatty acid compositions of whole plants. Exogenous fatty acids supplied as Tweens were incorporated into membranes of both the tissues to which they were applied, and tissues elsewhere in the plant.

MATERIALS AND METHODS

Plant growth. Surface-sterilized soybean seedlings (cv Noir 1) were grown axenically in quart glass jars (covered with a 100 mm glass petri dish lid) containing 50 mls of Leggett and Frere hydroponic medium (2), modified by the addition of Fe (as Sequestrene) to 90 µM. Plants were maintained at 25° C under continuous illumination from fluorescent bulbs (Sylvania cool-white).

Chemicals. Tweens carrying 17:0 as sole fatty acid were synthesized as described (4). All solvents were reagent grade.

Lipid extraction and analysis. Lipids were extracted and analysed as described (4). Briefly, lipids extracted with

---

[1] Supported by NIES grant #01498 to Dr. Karl G. Lark.
[2] Supported by NSF Predoctoral fellowship SPE 835 0132 and by a University of Utah graduate research fellowship. Present address: Carnegie Institution of Washington, Department of Plant Biology, Stanford, CA 94305

Table I. Incorporation of 17:0 into membrane lipids.

Tween-17:0 was applied to the primary leaves or added to the nutrient solution (to 10 mM 17:0) of 10 day old seedlings (cv Noirl) growing axenically in glass jars. After 8 days lipids were extracted and analyzed as described in "Materials and Methods". Roots and shoots of untreated plants and those with Tween added to their medium were extracted separately. Roots, primary leaves and trifoliolates of those with Tweens applied to their primary leaves were extracted separately.

Table I

TREATMENT	SAMPLE	16:0	16:1	17:0	18:0	18:1	18:2	18:3
				MOLE %				
UNTREATED	ROOTS	25	ND	ND	6	tr	32	36
	LEAVES	16	1	ND	4	tr	17	62
ADDED TO MEDIUM	ROOTS	16	ND	12	5	1	31	34
	LEAVES	11	tr	2	5	tr	27	53
APPLIED TO 1° LEAVES	ROOTS	16	ND	36	3	1	22	23
	1° LEAVES	10	1	10	2	tr	10	67
	TRIFOLIOLATES	11	tr	10	2	tr	19	58

hexanes:isopropanol (3:2, v/v) were fractionated by TLC and those comigrating with authentic Tweens were discarded (to prevent unmetabolized Tweens from contributing fatty acids to the analsis). Fatty acid methyl esters were prepared from the remaining lipids and analyzed by GLC. Neutral lipids comigrate with Tweens in the TLC systems used, consequently fatty acids were prepared solely from phospho- and galactolipids (i.e., membrane lipids).

RESULTS

Tweens carrying 17:0 as sole fatty acid synthesized and applied directly to leaves or added to the nutrient solution of soybean plants growing hydroponically under axenic conditions in glass jars (to avoid confusion due to microbial Tween metabolism). Table I shows that 17:0 was incorporated into both roots and shoots of plants treated with Tween-17:0 applied to either roots or leaves. Greater amounts of 17:0 were incorporated when Tween 17:0 was applied to the leaves, and larger amounts of 17:0 were observed in roots and in leaves younger than those to which Tween was applied. Incorporation of 17:0 proceeded slowly, and maximum amounts of 17:0 in membrane lipids were not observed until up to 8 days after Tween was applied to the plants (data not shown).

DISCUSSION

This paper described a system for modifying the membrane fatty acid composition of whole plants in a controlled manner. Plants treated with Tween carrying 17:0 transferred this fatty acid from the Tween to membrane lipids, both at the site of application and elsewhere in the plant (Table I). Incorporation of 17:0 into individual lipids was not investigated, but by analogy with the results from cultured cells it is anticipated that exogenous fatty acids will be incorporated into all "eucaryotic" lipids (see 3 for review of prokaryotic and eucaryotic plant lipids), and that other types of fatty acids may be introduced into membrane lipids by this means (4).

This system should be useful for a variety of studies related to plant membranes ranging from influences of fatty acid composition on growth temperature limits to effects on specific membrane proteins. It also allows results obtained with tissue cultures (which are inherently easier to work with) to be tested in whole plants. The ability to modify tissues which were not directly treated was especially useful, as these tissues would presumably be relatively unaffected by secondary effects of Tween application.

Incorporation of 17:0 throughout plants treated with Tween-17:0 was a surprising result, as lipids are not thought to be transported in plants (1), and it will be of interest to determine how this occurs. It is suggested that transport occurs in the phloem, as 17:0 appeared in tissues both above and below the point of application (Table I). It is also suggested that fatty acids are transported to the most active site of lipid synthesis, accounting for the greater labeling of the expanding tissues above and below the site of application than in the primary leaves themselves, which expanded very little during the experiment (Table I).

<u>Acknowledgements</u>. I wish to thank Dr. Karl G. Lark, my thesis advisor, for his support and encouragement. I am also endebted to Dr. Pete D. Gardner for many helpful discussions and for the generous use of his lab for Tween synthesis and fatty acid analysis, and to Dr. Reid Palmer for his gift of seeds of cultivars Noir 1 and Minsoy.

REFERENCES

1. Harwood JL, NJ Russell (1984) Lipids in Plant and Microbes George Allen & Unwin, London

2. Leggett JE, MH Frere (1971) Growth and nutrient uptake by soybean plants in nutrient solutions of graded concentrations. Plant Physiol 48:457-460

3. Roughan PG, CR Slack (1984) Glycerolipid synthesis in leaves. TIBS 9:383-386

4. Terzaghi WB (1986) A system for manipulating the membrane fatty acid composition of soybean cell cultures by adding Tween-fatty acid esters to their growth medium: basic parameters, and effects on cell growth. Plant Physiol. In press

PLASMA MEMBRANE LIPID ALTERATIONS FOLLOWING COLD ACCLIMATION:

POSSIBLE RELEVANCE TO FREEZE TOLERANCE

Daniel V. Lynch and Peter L. Steponkus

Department of Agronomy
Cornell University
Ithaca, New York 14853

INTRODUCTION

The plasma membrane plays a central role in cellular behavior during a freeze/thaw cycle and lysis or alterations in its semipermeable characteristics is a primary cause of freezing injury[1]. In protoplasts isolated from nonacclimated (NA) rye leaves, injury over the range of 0° to −5°C is a consequence of freeze-induced osmotic contraction resulting in endocytotic vesiculation of the plasma membrane – with sufficiently large area contractions being irreversible. As a result, lysis of the protoplasts occurs during osmotic expansion following thawing of the suspending medium. Alternatively, cooling to temperatures below −5°C results in destabilization of the plasma membrane so that the protoplasts are osmotically unresponsive during thawing of the suspending medium. This form of injury is associated with several changes in the ultrastructure of the plasma membrane, including the formation of lateral phase separations, aparticulate lamellae, and hexagonal$_{II}$ configurations[2]. Cold acclimation dramatically alters the behavior of the plasma membrane during freeze-induced osmotic contraction and dehydration. In protoplasts from acclimated (ACC) rye leaves, osmotic contraction results in the formation of exocytotic extrusions of the plasma membrane; and at severe levels of dehydration, lateral phase separations, aparticulate lamellae, or H_{II} configurations are not observed.

We believe there is a possible commonality between the propensity for endocytotic vesiculation and H_{II} formation in the plasma membrane of NA protoplasts that is related to its lipid composition, and further suggest that the difference in behavior between NA and ACC protoplasts (endocytotic vesiculation vs. exocytotic extrusion and the differential propensity for H_{II} formation) is, in part, a consequence of changes in the lipid composition of the plasma membrane following cold acclimation. Preliminary studies contrasting the cryobehavior of liposomes prepared from the plasma membrane lipids of NA and ACC rye demonstrate that the differential behavior of the plasma membrane observed in NA and ACC protoplasts is also apparent in the respective liposomes and provide strong evidence for a causal relationship between alterations in the lipid composition of the plasma membrane and cold acclimation. Our current objectives are to determine the influence of cold acclimation on the lipid composition of the plasma membrane and to provide a mechanistic explanation for the increased cryostability of the plasma membrane.

EXPERIMENTAL AND DISCUSSION

Plasma membrane fractions were isolated from leaves of nonacclimated and acclimated rye seedlings (Secale cereale L. cv Puma) using a polyethylene glycol-dextran partition system[3]. Lipids were extracted and separated into classes by column chromatography and TLC. Lipids were analyzed using TLC and gas chromatography[4].

The results of the lipid analyses reveal that a) the lipid composition of the plasma membrane is unique relative to other plant membranes, and b) cold acclimation results in numerous changes in the lipid composition of the plasma membrane. Free sterols, steryl glucosides, and acylated steryl glucosides comprised >50 mol% of the total lipids in both NA and ACC fractions. There were, however, substantial changes in the proportions of the three classes following cold acclimation. Free sterols increased from 34 to 45 mol% of the total lipid, while steryl glucosides and acylated steryl glucosides decreased from 16 to 6 mol% and 4 to 1 mol%, respectively. The predominant free sterols were β-sitosterol, stigmasterol, and campesterol, with trace amounts of cholesterol. Following cold acclimation, β-sitosterol increased from 21 to 32 mol% of total lipid, campesterol increased slightly from 9.1 to 10.2 mol%, and stigmasterol decreased from 2.4 to 1.2 mol%. Following cold acclimation, the mol% of steryl glucosides and acylated steryl glucosides containing β-sitosterol and campesterol decreased in amounts equivalent to the increases in the corresponding free sterols, suggesting interconversion among the lipid classes during cold acclimation.

A distinctive feature of the plasma membrane is the relatively high content of glucocerebrosides. With acclimation, the content of glucocerebrosides decreased from 17 to 7 mol%. The associated acyl chain moieties were hydroxy fatty acids of 16, 20 and 22 carbons (tentative identification based on chromatographic properties). Following acclimation there were substantial differences in the proportions of these constituents, with increases in the longer chain fatty acids.

The phospholipid content of the plasma membrane increased from 30 to 42 mol% following cold acclimation, although the relative proportions of the individual phospholipids did not change appreciably. The predominant phospholipids were phosphatidylcholine (PC) and phosphatidylethanolamine (PE), which increased from 14 to 19.3 mol% and from 10.3 to 15.5 mol% total lipid, respectively. Phosphatidylglycerol (PG), phosphatidyl serine (PS), and phosphatidylinositol (PI) were present in lesser amounts, each constituting <2 mol% total lipid.

Numerous changes in the molecular species of PC and PE occurred after cold acclimation. The relative proportions of 34-carbon species (16:0/18:2 and 16:0/18:3) of PC and PE decreased with proportional increases in 36-carbon species (18:1/18:2, 18:2/18:2, 18:2/18:3, and 18:3/18:3). When considered on the basis of mol% total lipids, the content of C_{34}-species remained relatively constant, whereas the content of C_{36}-species more than doubled. PE, and to a lesser extent PC, also contained 22:0, typically paired with 18:2 and 18:3 as 40-carbon molecular species. For PG, the principal 34-carbon species (16:0/18:1, 16:0/18:2 and 16:0/18:3) increased with corresponding decreases in 32-carbon species.

The complexity of the lipid composition of the plasma membrane and the numerous changes following cold acclimation preclude the possibility that any simple correlative analysis of the changes will establish their role in the cold acclimation process. Instead, a mechanistic approach must be taken to determine what changes in the lipid composition are responsible for the differential behavior and stability of the plasma membrane during a freeze/thaw cycle. Given the observed changes in the ultrastructure of the plasma membrane following severe freeze-induced dehydration, it is obvious that changes in lipid composition cannot be viewed solely from the perspective of membrane "fluidity" - as is often the case. Instead, more contemporary views of lipid mesomorphism[5] and dehydration-induced structural alterations[6] must be considered - especially lyotropic L_α - L_β and L - H_{II} phase transitions. It is well established that dehydration increases the L_α - L_β phase transition temperature and decreases the L - H_{II} phase transition temperature.

Recently, Crowe and Crowe[6] have proposed that dehydration-induced destabilization of biological membranes occurs as a result of a sequence of events: lyotropic L_α - L_β phase transitions in phospholipids such as PC result in demixing of the lipid and the localized enrichment of species such as PE which, upon further dehydration, undergo an L - H_{II} phase transition. Although the scheme is rather simplistic and remains to be verified experimentally, it provides one way in which to consider the dehydration-induced ultrastructural changes observed in NA protoplasts. The major species of PC (16:0/18:2, 16:0/18:3, and 16:0/18:1) have T_m's below -10°C. Freeze-induced dehydration would raise the T_m's of these species sufficiently that they form regions of gel phase lipid, facilitating demixing. Although the phase properties of glucocerebroside species of the plasma membrane remain to be determined, similar species from myelin have high T_m's. Thus, they also may undergo phase transitions as a result of dehydration. Demixing of PC, and, possibly GC, results in the enrichment of lipid species which have a propensity to form H_{II} structures. PE species account for a major proportion of the membrane phospholipids and are potent H_{II}-forming lipids, especially those containing two unsaturated acyl chains.

Freeze-induced dehydration of acclimated protoplasts does not lead to lateral phase separations and formation of H_{II} structures, in spite of the fact that the same C_{34} species of PC are present in similar amounts (although glucocerebrosides decrease by ~60%) and PE species increased by ~50%. This suggests that alterations in the lipid composition of the plasma membrane following cold acclimation increase membrane cryostability by modulating the phase behavior of certain constituents, maintaining miscibility during dehydration, and preventing demixing. It is conceivable that the increase in free sterols following acclimation may serve to prevent demixing during dehydration. Cholesterol is known to interact in a stoichiometric fashion with PC to diminish thermotropic transitions, and removal of cholesterol from myelin lipid extracts results in pronounced thermotropic transitions in the remaining lipids. Physical studies of the phase properties and miscibilities of plasma membrane constituents are in progress to determine these possibilities.

(This material is, in part, based on work supported by the United States Department of Energy under Grant DE-FG02-84ER13214 and USDA Competitive Research Grant No. 85-CRCR-1-1651.)

REFERENCES

1. P.L. Steponkus, Ann. Rev. Plant Physiol. 35:543 (1984)
2. W.J. Gordon-Kamm and P.L. Steponkus, Proc. Natl. Acad. Sci. (USA) 81:6373 (1984)
3. M. Uemura and S. Yoshida, Plant Physiol. 75:818 (1984)
4. D.V. Lynch and G.A. Thompson, Jr., Plant Physiol. 74:193 (1984)
5. S.M. Gruner, P.R. Cullis, M.J. Hope, C.P.S. Tilcock, Ann. Rev. Biophys. Biophys. Chem. 14:211 (1985)
6. J.H. Crowe and L.M. Crowe, in: "Membranes, Metabolism and Dry Organisms", A.C. Leopold ed. p. 188, Comstock Pub. Assoc., Ithaca (1986)

PLASMA MEMBRANE AND TONOPLAST FRACTIONS ISOLATED FROM SPINACH LEAVES

BY PREPARATIVE FREE FLOW ELECTROPHORESIS: EFFECT OF PHOTOINDUCTION

C. Penel,* G. Auderset,* S. Kiefer,* A. Sandelius,**
A. Brightman,*** H. Greppin* and D. James Morré***

*University of Geneva, Switzerland, **University of
Göteborg, Sweden, and ***Purdue University, West Lafayette
Indiana, 47907, U.S.A.

INTRODUCTION

A change in membrane structure as a result of photoinduction of the flowering response in plants was predicted from physiological studies (Penel and Greppin, 1974; Lenk et al., 1981; Karege et al., 1982). With the availability of purified fractions of plasma membrane and tonoplast from green leaves (Auderset et al., in press), a study was initiated to examine these membranes for structural and compositional changes associated with photoinduction. Spinach (Spinach oleracea) was investigated as an example of a species induced to flower by continuous light (so-called long-day plant) and Japanese morning glory (Pharbitis nil) as an example of a plant photoinduced by a long night (so-called short-day plant).

METHODS

Plants were grown in a growth chamber with artificial illumination (Karege et al., 1982). Plants were photoinduced as described in results. Control plants remained under non-inducing illumination.

For isolation of membranes, leaves were homogenized (blender) in a medium containing 0.3 M sucrose, 10 mM KCl, 1 mM $MgCl_2$, 50 mM Hepes, pH 7.5 (NaOH), filtered, centrifuged at 10 min at 6,000 g_{max} to remove plastids and centrifuged 30 min at 60,000 g_{max}. The final 60,000 g_{max} pellet was resuspended in electrophoresis chamber buffer for free-flow electrophoretic separation of membranes as described (Sandelius et al., 1986).

Membrane fractions or tissues were fixed for electron microscopy in a mixture of 2% (w/v) buffered glutaraldehyde + 0.2% (w/v) osmium tetroxide in 0.1 M sodium cacodylate, pH 7.3 for 2 h at 4°C (Franke et al., 1969). Measurements were on negatives photographed at a primary magnification of 50,000 and enlarged 20 times to give a final magnification of 1,000,000 times. Measurements were between the outer edges of the stained portions of the membrane profiles to the nearest 0.5 mm. A minimum of 300 measurements was averaged for each determination. Other assays were as described (Sandelius, et al., 1986).

RESULTS

The 60,000 g_{max} (30 min) fraction for the electrophoretic separation consisted of 30% plastid-derived material (thylakoids and recognizable chloroplast-derived fragments), 15% plasma membranes (based on thick, 9-11 nm membranes staining with phosphotungstic acid at low pH, a characteristic of the plant plasma membrane), 15% tonoplast (7-9 nm thick membranes unstained with phosphotungstic acid at low pH), 15% mitochondria (morphology, 6 nm membranes), 13% endoplasmic reticulum (morphology, 6 nm membranes) 4% peroxisomes (morphology) and 8% unidentified membranes with thin ($\leq$ 6 nm) membranes. When separated by free-flow electrophoresis (Fig. 1), the tonoplast membranes were the most electronegative and collected in fractions A and B, whereas plasma membranes were least electronegative and were found in fractions D and E. Contaminating membranes including 95% of the chlorophyll were located in the middle of the separations (fractions B-D). With induction, the separations were similar but usually slightly broadened both with the long-day spinach (Auderset et al., in press) (Fig. 1) and with the short-day Pharbitis. The basis for the broadening is not known.

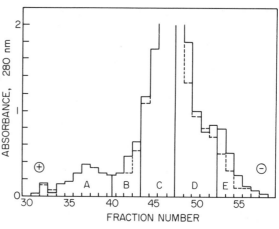

Fig. 1. The absorbance at 280 nm of fractions from free-flow electrophoretic separations comparing photoinduced (24 h continuous illumination) (solid lines) and non-induced (dashed lines) spinach plants fractionated in parallel. The designation A-E along the fraction number axis indicate pooled fractions.

Evidence for a direct response of photoinduction on membranes came from measurements of membrane thickness where the membranes of plasma membrane vesicles of fraction E (and elsewhere on the separation) were about 10% thicker from the photoinduced leaves than from leaves not photoinduced (Table 1). The change observed first with the isolated fractions was seen

Table 1. Membrane dimensions of plasma membrane vesicles before and after photoinduction with 24 h fluorescent light (10 mmol $m^{-2}s^{-1}$). Values are averages in nm ± SD of 300 measurements each for 3 experiments for tissue (in situ) and 5 experiments for fractions.

Photoperiod	Plasma membrane	
	in situ	isolated vesicles
Not photoinduced (short day)	8.7 ± 0.3	8.9 ± 0.2
Photoinduced (long day)	10.5 ± 0.2	10.3 ± 0.2

also in situ. Tonoplast membranes, either in situ or isolated by electrophoretic separation (Fraction A) were much less affected, if at all. When analyzed by densitometry, the change in dimensions was primarily in the dark portion of the dark-light-dark region of the membrane due to an increased density of a shoulder layer on the vesicle interior (for fraction E) that gives asymmetry to the distribution of density across the plant plasma membrane. The increase in membrane dimensions was seen with lead citrate-stained membranes or with membranes stained with phosphotungstic acid at low pH. A similar increase in membrane thickness was observed in the short-day plant, Pharbitis nil, upon induction with a single long night.

DISCUSSION

The change in thickness of the plasma membranes of spinach and Pharbitis provides direct evidence of an early membrane change associated with photoinduction of flowering. Common to both a long- and a short-day species, the thickening in situ and in isolated fractions was nearly identical. The change, whatever its biochemical basis, was stable to isolation and did not depend for its maintenance on energy charge or on a transient association with cytoplasmic or cytoskeletal elements.

Work is ongoing to determine the biochemical basis for the membrane change. Comparing photoinduced and non-induced plants, patterns of protein bands on SDS-PAGE gels show no major alterations nor are protein phosphorylation or amino acid incorporation rates significantly altered for the membranes upon photoinduction. Rates of choline incorporation are similar as are overall glycolipid patterns determined by thin layer chromatography. Thus, a substantial increase in the rate of membrane biogenesis or turnover of major membrane constituents does not appear to be involved in the dimensional alteration. Thus, other bases for the conformational change in terms of chemical alterations or rearrangement of existing constituents are being sought as the basis for the observed differences in membrane thickness.

Supported by a grant from the National Science Foundation PCM 8260222.

REFERENCES

Auderset, G., Sandelius, A. S., Penel, C., Brightman, A., Greppin, H., and Morré, D. J., Isolation of plasma membrane and tonoplast fractions from spinach leaves by preparative free-flow electrophoresis and effect of photoinduction, Physiol. Plant. (in press).

Franke, W. W., Krien, S., and Brown, R. M., Jr., 1969, Simultaneous glutaraldehyde-osmium tetroxide fixation with postosmication. An improved fixation procedure for electron microscopy of plant and animal cells, Histochemie, 19:162-164.

Karege, E., Penel, C., and Greppin, H., 1982, Floral induction in spinach leaves by light, temperature and gibberellic acid: Use of the photocontrol of the basic peroxidase activity as a biochemical marker, Z. Pflanzenphysiol., 107:357-365.

Lenk, R., Bonzon, M., and Greppin, H., 1981, Irreversible thermodynamic and biophysical evolution in spinach leaves as studied by NMR, Z. Pflanzenphysiol., 101:108-118.

Penel, C., and Greppin, H., 1974, Variation de la photostimulation de l'activité des peroxydases basiques chez l'epinard, Plant Sci. Lett., 3:75-80.

Sandelius, A. S., Penel, C., Auderset, G., Brightman, A., Millard, M., and Morré, D. J., 1986, Isolation of highly purified fractions of plasma membrane and tonoplast from the same homogenate of soybean hypocotyls by free-flow electrophoresis, Plant Physiol., 81:177-185.

STUDY OF THE INTRACELLULAR TRANSFER OF LIPIDS TO THE PLASMALEMMA

P. Moreau, H. Juguelin, R. Lessire, and C. Cassagne

I.B.C.N. (C.N.R.S.)
1 rue Camille St-Saëns
33077 Bordeaux-Cedex
France

INTRODUCTION

It has been shown in vitro that in leek (Allium porrum L.) epidermis cells very long chain fatty acids (VLCFA) are synthesized in the endoplasmic reticulum.[1] Moreover, VLCFA are more abundant in the plasma membrane (up to 20% of the total fatty acids) than in the endoplasmic reticulum. It was suggested that an eventual transfer of VLCFA from their site of synthesis (ER) to the plasma membrane could occur.[2]

A new system (seven-day-old etiolated leek seedlings) has been employed to study the intermembrane transfer of lipids and fatty acids in vivo.[3] By in vivo pulse-chase experiments, followed by membrane fractionation on linear sucrose gradients, intermembrane transfer events of VLCFA between the light fraction and the heavier ones was demonstrated.

However, these results did not establish that VLCFA were transferred to the plasmalemma. The characterization of the membranes obtained after linear sucrose gradients using marker enzymes showed that the heavy membrane fraction contains almost all of the membrane vesicles originating from the plasma membrane, but also that it is relatively heterogeneous.

The purification of the plasmalemma by phase partition in an aqueous two-polymer phase system was undertaken, and the study of the intermembrane transfer of lipids and fatty acids to the plasma membrane has been studied.

RESULTS AND DISCUSSION

Purification of the Plasmalemma by the Phase Partition Method

Plasma membrane vesicles are isolated according to their surface charge properties.[4,5] After phase partition of a homogeneous solution of polyethylene-glycol (5.6%, w/w) and dextran (5.6%, w/w), the plasma membrane has more affinity for the PEG-enriched upper phase whereas other membranes are accumulated at the interface and in the dextran-enriched lower phase.

The purification of the plasmalemma was undertaken by this method, using either the heavy membrane fraction obtained on linear sucrose gradients, or the microsomal pellet.

Table 1. Purification of the Plasmalemma by the Phase Partition Method

	GS II Activity	CDP-Choline Diglyceride Transferase Activity	Sterols	VLCFA	Saturated Fatty Acids
Dextran-enriched lower phase + interface	1	1	1	1	1
PEG-enriched upper phase	3.8	0.14	2.1	3	2.6

Marker enzyme activities (per mg of proteins) and lipid or fatty acid contents are expressed with respect to those of the lower phase, which are taken as the reference and equal to 1.

The efficiency of plasma membrane isolation was identical in both cases, so the results given in Table 1 refer to a compilation of all the data so obtained. Table 1 presents biochemical data showing the purification of the plasmalemma in the PEG-enriched upper phase. All data concerning the upper phase are given with respect to those of the lower phase, which are taken as the reference and equal to 1. Glucane-synthetase II activity is nearly 4 times higher in the upper phase than in the lower phase, whereas CDP-choline diglyceride transferase activity is at least 7 times lower. Table 1 also shows some results from the lipid and fatty acid analyses of the membranes recovered in the upper and lower phases. The percentage of sterols, VLCFA, and of saturated fatty acids are higher in the upper phase than in the lower phase.

The higher levels of GS II activity[6] and the higher proportions of sterols,[7] VLCFA,[8] and saturated fatty acids[9] in the upper phase are in agreement with the purification of the plasma membrane. The low CDP-choline diglyceride transferase activity in the upper phase indicates the very low contamination by membrane vesicles originating from the endoplasmic reticulum. Lastly, other analyses have suggested that this purified plasma membrane fraction is only slightly contaminated by plastidial, mitochondrial, and tonoplastic membranes.[10]

In Vivo Transfer of Lipids to the Plasmalemma

After pulse-chase experiments, realized as described earlier,[3] the purified plasma membrane fraction was prepared directly from the microsomal pellets and the radioactivity of the lipids was determined. Table 2 shows the radioactivity of the lipids of the plasmalemma as a function of the chase time, following a 2-hour labeling period. The radioactivities of the lipids at the end of the labeling period are taken as the reference and are given the value of 100.

An increase of the total radioactivity of the plasma membrane lipids is observed during the chase, whereas the total radioactivity of the lipids of the microsomal pellets remains unchanged over the same period. These results demonstrate that a transfer of lipids to the plasmalemma has occurred in vivo.

Table 2. Radioactivity of the Lipids of the Plasma Membrane as a Function of the Chase Time Following a 2-Hour Labeling Period

	Total Lipids	Neutral Lipids	PC	PE
Labeling period (2 hours)	100	100	100	100
Chase time (min)				
30	116	123	104	109
60	178	166	162	190
120	244	228	283	254

The radioactivities found at the end of the labeling period were taken as references equal to 100.

Table 2 also shows that the lipids that were transferred were essentially the neutral lipids (including the sterols) PC and PE. Since the latter lipids are constitutive of the plasma membrane, these results demonstrate an in vivo transfer of lipids to the plasmalemma.

Finally, the analysis of the fatty acid label showed that C16, C18 fatty acids and VLCFA were transferred to the plasmalemma. During the chase period, an enrichment of labeled VLCFA in the plasma membrane, compared to labeled C16 and C18 fatty acids, was also observed.

Since plasma membranes isolated from etiolated leek seedlings are incapable of synthesizing VLCFA,[10] as is the case of the plasma membrane isolated from leek epidermis,[1] it follows that the VLCFA were transferred to the plasmalemma in vivo.

The results presented here demonstrate, for the first time, the existence of in vivo transfers of lipids and fatty acids to the plasma membrane of a higher plant. Hence, etiolated leek seedlings appear to be a good model for the study of the intracellular pathways of lipid and fatty acid transport to the plasmalemma in vivo and consequently should provide information that is complementary to what is known about the intracellular transport of proteins.

REFERENCES

1. C. Cassagne and R. Lessire, Arch. Biochem. Biophys. 191: 146 (1978).
2. R. Lessire, T. Abdul-Karim, and C. Cassagne, in: "The plant cuticle," D.F. Cutler, K.L. Alvin, C.E. Price, eds., Academic Press, New York (1982).
3. P. Moreau, H. Juguelin, R. Lessire, and C. Cassagne, Phytochemistry 25: 387 (1986).
4. S. Yoshida, M. Uemura, T. Niki, A. Sakai, and L.V. Gusta, Plant Physiol. 72: 105 (1983).
5. L.E. Korner, P. Kjellbom, C. Larsson, and I.M. Moller, Plant Physiol. 79: 72 (1985).
6. P.H. Quail, Ann. Rev. Plant Physiol. 30: 425 (1979).
7. M.A. Hartmann, G. Normand, and P. Benveniste, Plant Sci. Lett. 5: 287 (1975).
8. R. Lessire, M.A. Hartmann-Bouillon, and C. Cassagne, Phytochemistry 21: 55 (1982).

9. P. Mazliak, in: "lipids and lipid polymers in higher plants," M. Tevini and H.K. Lichtenthaler, eds., Springer-Verlag, Berlin (1977).
10. P. Moreau, University Thesis, University Bordeaux II (1986).

Ca^{2+} AND INTER-MOLECULAR BRIDGING OF MEMBRANAL PHOSPHOLIPIDS AND PROTEINS

Y.Y. Leshem

Department of Life Sciences, Bar-Ilan University
Ramat-Gan, Israel

Membranal stability, microviscosity ($\bar{\eta}$) and bulk lipid phase state may be controlled by Ca^{2+}. All negatively charged phospholipids (PL's) such as phosphatidylinositol or phosphatidylserine, by virtue of electrostatic crosslinking with the $^-O-P-OH$ section of the headgroup induce physiologically meaningful structural rearrangements in membranal architecture. Three types of molecular bridging may occur:

1) A multiple PL-PL bonding between PL's themselves [$^-O-P-O-Ca-O-P-O^-$] inducing partial separation of charged PL's into discrete domains. At high Ca^{2+} concentrations this causes membrane rigidification.

2) PL-protein binding between the membranal PL's and the carboxyl tails of membrane proteins [$^-O-P-O-Ca-OOC-protein$] which restricts protein motility.

3) A protein-cytoskeleton bridging of protein chemical nature as in '2', or an indirect effect promoting protein-cytoskeleton binding via the γ-glutamyl- ε -lysine bridges which is Ca^{2+}-activated transglutaminase mediated. Such 'anchoring' besides limiting the protein's motility may also cause its deeper insertion within the bilayer and thus less exposure for interaction with effectors on the membrane's exterior.

The acidic nature of several hormones (IAA, GA and ABA) may induce displacement of bridging Ca^{2+}. Low hormone concentrations would decrease membrane rigidity and promote liquid crystalline configurations. At higher ones the Ca^{2+} stabilizing effect would be lost, anchored protein hormone receptors could be ejected from the bilayers and released Ca^{2+} triggers the "phospholipase-lipoxygenase cascade" resulting in ethylene production and senescence. The following schemes represent the various types of bridging:

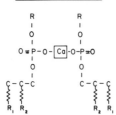
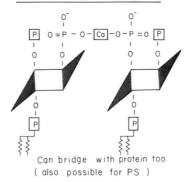
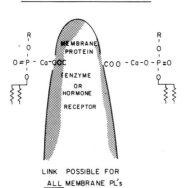
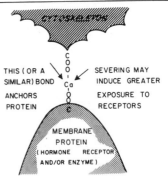
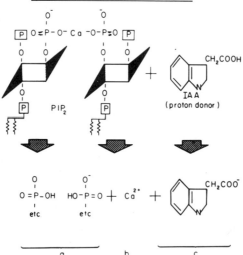

a. $\bar{\eta}$ (microviscosity) decrease, membrane relaxation, increase of embedded receptor exposure.

b. Activation of calmodulin, phospholipases A_2, C, and protein kinases. Triggering of "phosphatidyl" linoleyl(-enyl) cascade."

c. Recycles and/or may bind to other sites.

EXPERIMENTS AND RESULTS

Employing the compressibility principle whereby hydrostatic pressure selectively ejects integral membrane protein[1] (including the putative ethylene-forming enzyme and/or ethylene receptor) from fluid phospholipid domains, isolated liquid crystalline pea foliage microsomal membranes[2] were subjected to 1500 bar pressurization in a French Press for 20 min in the presence or absence of 10 ppm indoleacetic acid (IAA). Ethylene production of the protein-depleted membranes was compared to that of non-pressurized membranes. Ca^{2+} release to the supernatant after pressurization was determined with tetramurexide.

Results (Fig. 1-left) indicate that without IAA, ethylene production was equal in both pressurized and non-pressurized membranes. However, application of IAA significantly decreased ethylene in both non-pressurized and pressurized treatment, the latter being particularly effective and more than halving ethylene formation.

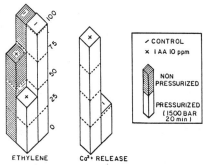

Fig.1: ETHYLENE & Ca^{2+} RELEASE FROM MEMBRANES AS EFFECTED BY AUXIN & HYDROSTATIC PRESSURE

Results are interpreted in the light of the suggested auxin effect (see introductory scheme) whereby IAA acting as a proton donor releases ethylene-forming protein or receptor by severing of electrostatic bridging. This hypothesis is furthermore supported by the observation (Fig. 1-right) that pressurization in the presence of auxin increased amount of Ca^{2+} release. Clearcut IAA-Ca interdependence has, moreover, been shown in rapid auxin action in maize roots.[3]

Experimental procedure and results will be presented in greater detail elsewhere.[4]

ACKNOWLEDGEMENT: The author wishes to thank Prof. Meir Shinitzky, of the Membrane Research Department, Weizmann Institute, Rehovoth, Israel, for his help and advice in pressurization techniques.

REFERENCES

1. M. Deckman, R. Haimovitz and M. Shinitzky, Selective release of integral proteins from human erythrocyte membranes by hydrostatic pressure, Biochim. Biophys. Acta, 821:334 (1985).
2. Y.Y. Leshem, S. Sridhara and J.E. Thompson, Involvement of calcium and calmodulin in membrane senescence, Plant Phys., 75:329 (1984).
3. K.H. Hasenstein and M.L. Evans, Calcium dependence of rapid auxin action in maize roots, Plant Phys., 81:439 (1986).
4. Y.Y. Leshem and G. Bar-Nes, Hormone receptor manipulation by hydrostatic pressure: Interaction between Ca^{2+} membrane components and phosphatidyl inositol in pea foliage membranes, in: "Plant Hormone Receptors," D. Klämbt, ed., NATO Advanced Research Workshop, Springer Verlag, Berlin (1987).

A MEMBRANE-LOCATED, CALCIUM-/CALMODULIN-ACTIVATED PHOSPHOLIPASE STIMULATED BY AUXIN

D. James Morré and Barbara Drobes

Department of Medicinal Chemistry, Purdue University
West Lafayette, IN 47907, U.S.A.

INTRODUCTION

Lipase-catalyzed hydrolyses of membrane phospholipids yield substances implicated as second messengers including calcium ions,[1] diglycerides[2] and inositol phosphates[3]. These substances are capable, under appropriate circumstances, of amplifying and translating received hormonal stimuli into a response cascade potentially important to growth control. Our work has focused on an examination, both *in vivo* and *in vitro*, of phospholipid breakdown[1] and calcium release[4] associated with the action of auxin in stimulation of elongation growth in plant stem segments.

MATERIALS AND METHODS

Results presented here compare auxin-stimulated hydrolysis of endogenous choline-labeled phospholipids and the external water-soluble synthetic phospholipase substrate, p-nitrophenylphosphorylcholine. Experiments were with membrane vesicles freshly and rapidly prepared from etiolated hypocotyls of 4 day old soybean (<u>Glycine max</u>) seedlings. Homogenization was with mortar and pestle in 0.3 M sucrose containing 1 mM $MgCl_2$ and 25 mM Tris, pH 6.5, ± 10 mM KCl, ± 100 mM LiCl. Homogenates were filtered, precentrifuged 10 min, 6,000 g_{max} and centrifuged 20 min, 45,000 g_{max}.

For estimation of phosphatidylcholine breakdown, 1 g tissue sections was pre-incubated in 1 ml distilled water containing 2 μCi [^{14}C]choline for 4 h. Isolated membranes were resuspended in distilled water and incubated in Eppendorf cups ± various additions for 15 min at 25°C. Reactions were stopped with 10% TCA, the membranes pelleted 5 min at 10,000 g_{max}, 4°C and the pellets washed twice with 1 ml portions of cold distilled water.

The phospholipase C assay contained in a final volume of 1 ml, 4 mM p-nitrophenylphosphorylcholine (Sigma), 80 mM Tris, pH 7, 0.4 M sucrose and 300-800 μg protein. Incubations were for 20 min at 25°C and the reactions were stopped by addition of 0.25 N sodium hydroxide. Absorbance was measured at 420 nm and corrected for minus membrane and minus substrate controls.

Studies in vitro. The in vitro hydrolysis of in vivo-labeled choline-containing phospholipids of isolated soybean membrane vesicles was rapid and essentially complete within a 15 min incubation (Fig. 1) and accelerated either by the auxin, 2,4-D, Ca^{2+}-calmodulin (CM) or both (Table 1). The label lost in the 15 min incubation was 20 to 40% of the total incorporated. Radioactivity lost from the membrane pellet was recovered in the supernatant.

With the exogenous synthetic substrate, p-nitrophenylphosphorylcholine (pNPP-choline), an auxin response also was seen but only a 10 to 20% increase in hydrolysis. A similar stimulation was given by calcium + CM (Table 2). The auxin-induced component of pNPP-choline hydrolysis exhibited a pH optimum at 7 (Fig. 2) and an optimum stimulation by auxin at 1 µM (Fig. 3).

Studies in vivo. With excised but intact hypocotyl segments incubated with or without 1 µM 2,4-D, incorporation of [^{14}C]-choline or [^{3}H]inositol was accelerated 25 to 30% over 4 h of incubation with 1 µM auxin. Steady-state levels of diglyceride were increased 10-50%.

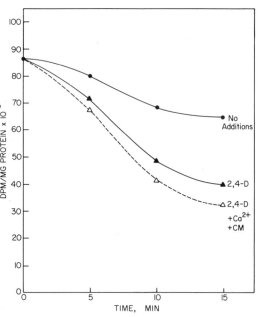

Fig. 1. Kinetics of loss of radioactivity from [^{14}C]choline-prelabeled soybean membranes as a function of time of incubation with or without 1 µM 2,4-D alone or in the presence of 1 µM Ca^{2+} + 1 µM calmodulin.

DISCUSSION

While the net effect may be the same in terms of diglyceride formation and calcium or inositol release, the present results do not support a specific effect of auxins on a closed cycle of inositol phosphatide turnover as is currently under investigation in animal cells. Rather, the phospholipid changes appear to represent a general response of phospholipid degrading enzymes to auxin. Choline-containing lipids are affected to an extent equal to or greater than those containing inositol,[1] presumably via a D-type lipase. C-type activities, assayed either by pNPP-choline hydrolysis or with exogenous PI (4% stimulation in vitro; from a study with H. Pfaffmann and E. Hartmann) were influenced much less by auxin.

The precise mechanism by which the hydrolysis of endogenous choline phospholipids takes place is unknown. The ultimate products appear to be choline, diglyceride and inorganic phosphorous. Since soybean membranes

Table 1. Loss of [^{14}C]choline from prelabeled membranes.

Additions	cpm/mg protein		
	membrane	lost	supernatant
Initial	27,025		
Control incubation	24,597	2,428	3,881
+ 1 µM 2,4-D	21,893	5,132	9,627
+ 1 µM Ca^{2+}	20,768	6,257	7,728
+ 1 µM CM	19,219	7,806	5,828
+ 2,4-D + Ca^{2+}	21,754	5,271	5,305
+ 2,4-D + CM	18,581	8,445	7,728
+ CM + Ca^{2+}	19,208	7,817	9,812
+2,4-D + Ca^{2+} + CM	17,076	9,949	12,715

Table 2. Hydrolysis of p-nitrophenyl phosphorylcholine by soybean vesicles.

Additions	μmoles/h/mg protein
Freshly isolated:	
None	1.63 ± 0.06
+ 10 μM Ca^{2+}	+ 0.20 ± 0.03
+ 3 μM CM	+ 0.27 ± 0.02
+ 1 μM 2,4-D	+ 0.29 ± 0.04
+ Ca^{2+} + CM	+ 0.30 ± 0.14
+ Ca^{2+} + 2,4-D	+ 0.28 ± 0.01
+ CM + 2,4-D	+ 0.25 ± 0.01
+ Ca^{2+} + CM + 2,4-D	+ 0.32 ± 0.01
EDTA-washed	
None	1.64 ± 0.12
+ 10 μM Ca^{2+}	+ 0.03 ± 0.01
+ 3 μM CM	+ 0.01 ± 0.01
+ 1 μM 2,4-D	+ 0.00 ± 0.0
+ Ca^{2+} + CM	+ 0.07 ± 0.03
+ Ca^{2+} + 2,4-D	+ 0.20 ± 0.05
+ CM + 2,4-D	+ 0.22 ± 0.14
+ Ca^{2+} + CM + 2,4-D	+ 0.25 ± 0.20

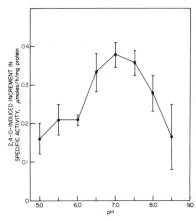

Fig. 2. pH dependency of the 2,4-D-induced increment of p-nitrophenylphosohorylcholine hydrolysis by soybean membranes.

Table 3. Phosphatase activities of soybean membranes (μmoles/h/mg protein).

Phosphatidic acid	0.05
myo-Inositol-2-monophosphate	0.35
Choline phosphate	0.15

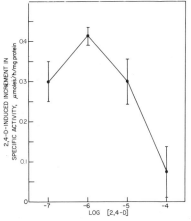

Fig. 3. Dose dependency of the 2,4-D-induced increment of p-nitrophenylphosphorylcholine hydrolysis by soybean membranes.

contain both active cholinephosphate and inositolphosphate phosphatases as well as phosphatidic acid phosphatase (Table 3), these products could arise from either a D- or a C-type primary cleavage.

REFERENCES

1. D. J. Morré, B. Gripshover, A. Monroe, and J. T. Morré, Phosphatidylinositol turnover in isolated soybean membranes stimulated by the synthetic growth hormone 2,4-dichlorophenoxyacetic acid, J. Biol. Chem., 259:15364-15368 (1984).
2. D. J. Morré, J. T. Morré, and R. L. Varnold, Phosphorylation of membrane-located proteins of soybean in vitro and response to auxin, Plant Physiol., 75:265-268 (1984).
3. W. F. Boss, and M. O. Massel, Polyphosphoinosotides are present in plant tissue culture cells, Biochem. Biophys. Res. Commun. 132: 1018-1023 (1985).
4. D. J. Morré, Calcium modulation of auxin-membrane interactions in plant cell elongation, in: "Molecular and Cellular Aspects of Calcium in Plant Development," A. J. Trewavas, ed., Plenum, New York (1986).

Work supported by the National Science Foundation PCM 8206222.

REGULATION OF PHOSPHOLIPASE ACTIVITY IN POTATO LEAVES BY PROTEIN PHOSPHORYLATION-DEPHOSPHORYLATION AND PROTEOLYTIC ACTIVATION

Robert A. Moreau

Eastern Regional Research Service
Agriculture Research Service
U.S. Department of Agriculture
Philadelphia, PA 19118

INTRODUCTION

We have recently reported that calcium and calmodulin stimulate the rate of autolysis of phospholipids in potato leaf homogenates (1). Further studies (2) revealed that a soluble phospholipase activity in potato leaves could be stimulated to the same degree (30-50%) by either calmodulin or protein kinase (+ATP). Two other plant enzymes, quinate:NAD+ oxidoreductase (3) and isofloridoside-phosphate synthase (4), have also been shown to be similarly stimulated by both calmodulin and protein phosphorylation. However, the latter enzyme was also shown to be stimulated to an even greater degree by proteolysis with trypsin or chymotrypsin (4). This study was undertaken to investigate whether the phospholipase activity in potato leaves may also respond to proteolytic activation.

MATERIALS AND METHODS

Potato (<u>Solanum</u> <u>tuberosum</u> c.v. Kennebec) plants were grown as previously described (1,2). Young (1-2 cm) leaves were ground with a mortar and pestle in chilled buffer containing 0.3 M sucrose, 0.1 M HEPES pH 7.5, and 5 mM each of DTT and β-mercaptoethanol. The homogenate was centrifuged at 100,000 g for 50 minutes and the supernatant fraction was used as a source of enzyme. Phospholipase activity was assayed by a new technique (5) which utilizes the fluorescent phospholipid analogue, C_6-NBD-Phosphatilycholine. The reaction mixture contained 5 µM C_6-NBD-PC, 50 mM HEPES pH 7.0, and 10-30 µl enzyme. The other conditions of the assay were as previously described (2).

RESULTS

Time-course studies were conducted in order to study the effects of various treatments on the levels of phospholipase activity in 100,000 g supernatant fractions from potato leaves (Fig. 1). The phospholipase activity in the 100,000 g supernatant remained constant for 60 minutes at 25° C (data not shown). In other control experiments the addition of 3 mM $MgCl_2$ and 0.3 mM ATP to the 100,000 g supernatant fraction had no immediate or delayed effect on phospholipase activity (Fig. 1). When NaF, a phosphatase inhibitor, was added to the mixture of 100,000 g supernatant, $MgCl_2$, and ATP, phospholipase activity increased gradually during the 60 minute study as previously reported (2). The subsequent addition of either protein kinase or calmodulin caused a rapid increase (about 50% in phospholipase activity which peaked at 10 and 20 minutes, respectively, as previously described (2). In order to investigate the possible proteolytic activation of pholipase activity, 25 µgrams of trypsin was added to either the 100,000 g supernatant or the 100,000 g supernatant plus $MgCl_2$ and ATP. With both treatments the addition of trypsin caused a rapid increase in phospholipase activity (about 3-fold increase in the first 20 minutes) (Fig. 1). Similar results were obtained with chymotrypsin, but the degree of stimulation was less (about 50%) than that obtained with comparable quantities of trypsin.

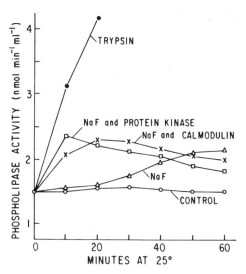

Fig. 1. Time-course study of the stimulation of phospholipase activity in the 100,000 g supernatant fraction from potato leaves. Control (0) = 1 ml 100,000 g supernatant fraction containing 3 mM Mg^{2+} and 0.3 mM ATP, control plus 50 mM NaF (Δ), control plus 50 mM NaF and the catalytic subunit of cyclic AMP-dependent protein kinase (500 u) (□), control plus 50 mM NaF and calmodulin (10,000 u) (X), control plus 25 µgrams trypsin (●). All reported concentrations are those which occur in the final volume of 1.1 ml.

DISCUSSION

This study served to verify our previous report (2) that phospholipase activity in potato leaves is stimulated by protein phosphorylation and by calmodulin. More importantly, it demonstrated that the degree of stimulation of the same enzyme activity by proteolytic activation was even greater than the degree of stimulation with protein kinase or calmodulin. Further work is required in order to elucidate the mechanisms of these three types of stimulation. We previously suggested that the calmodulin and protein kinase stimulations could be explained by phosphorylation of a phospholipase enzyme by an endogenous calmodulin-stimulated protein kinase (2). However, in light of the current evidence of proteolytic activation, a more complex mechanism of activation is necessary to best explain the three types of activation. The only other plant enzyme which has been reported to be stimulated by the same three treatments is isofloridoside-phosphate synthase (4) as previously described. Those authors presented evidence for the presence of an endogenous acid protease which could stimulate the enzyme. They also proposed (4) that calmodulin and protein kinase somehow stimulated this protease which in turn proteolytically-activated the isofloridoside-phosphate synthase. Such a mechanism is also conceivable for the potato leaf phospholipase. The only other plant enzyme which has been reported to be stimulated by proteolytic activation is a glucan synthase in soybean cells (6). It is also interesting to note that an animal phospholipase, pancreatic phospholipase A_2, is activated by the proteolytic removal of a heptapeptide from the amino terminus (7).

REFERENCES

1. R. A. Moreau and T. S. Isett, Autolysis of membrane lipids in potato leaf homogenates: effects of calmodulin and calmodulin antagonists, Plant Sci., 40:95 (1985).
2. R. A. Moreau, Regulation of phospholipase activity in potato leaves by calmodulin and protein phosphorylation-dephosphorylation, Plant Sci., (in press).
3. R. Ranjeva, G. Refino, A. M. Boudet, and D. Marme, Activation of plant quinate NAD+ 3-oxidoreductase by Ca^{2+} and calmodulin, Proc. Natl. Acad. Sci., 80:5222 (1983).
4. H. Kauss, Volume regulation in Poterioochromonas: involvement of calmodulin in the Ca^{2+}-stimulated activation of isofloridoside-phosphate synthase. Plant Physiol., 71:169 (1983).
5. L. A. Wittenauer, K. Shira, R. L. Jackson, and J. D. Johnson, Hydrolysis of a fluorescent phospholipid substrate by phospolipase A_2 and lipoprotein lipase, Biochem. Biophys. Res. Comm., 118:894 (1984).
6. H. Kauss, H. Kohle, and W. Jeblick, Proteolytic activation and stimulation by Ca^{2+} of glucan synthase from soybean cells, FEBS Letters 158:84 (1983).
7. G. H. DeHaas, N. M. Postema and W. Nieuwenhuizen, and L. L. M. VanDeenen, Purification and properties of an anionic zymogen of phospholipase A_2 from porcine pancreas, Biochem. Biophys. Acta., 159:118 (1968).

BIOSYNTHESIS OF COMPLEX LIPIDS

SYNTHESIS AND DEGRADATION OF LIPID BODIES IN THE SCUTELLA OF MAIZE

Anthony H. C. Huang, Rongda Qu, Shue-mei Wang, Vicki B. Vance, Yi-zhi Cao, and Yon-hui Lin

Biology Department
University of South Carolina
Columbia, SC 29208,

INTRODUCTION

Most seeds contain storage lipids in the form of triacylglycerols, which usually comprise 20-50% of the total seed dry weight (1,2,3,4). This lipid reserve is synthesized during seed maturation, and is rapidly mobilized to provide energy and carbon skeleton for the growth of the embryo during germination. The triacylglycerols are densely packed in subcellular organelles called lipid bodies (oleosomes, spherosomes, oil bodies). The spherical lipid body is about 0.5-1 µm in diameter, and is surrounded by a "half-unit" membrane of one monolayer of phospholipids about 3 nm thickness (5). The fatty acyl moieties of the membrane phospholipids are believed to orient themselves toward the matrix so that they can form hydrophobic interaction with the internal triacylglycerols.

The ontogeny of the lipid bodies is still unclear. There is little documentation to indicate that the lipid bodies in maturing seeds contain enzymes for triacylglycerol biosynthesis. Instead, it is generally believed that the fatty acids are synthesized in the plastids (1,2,3,4). The subsequent formation of mono-, di-, and tri-acylglycerols from activated fatty acids occurs in the microsomes, which presumably represent the endoplasmic reticulum (ER). The mechanism of transport of fatty acid from the plastids to the ER, and of triacylglycerol from the ER to the lipid bodies is unknown. The origin of the lipid body membrane is also unclear until recently. The membrane phospholipids are generally assumed to be synthesized on the ER. It has been suggested that the newly synthesized triacylglycerols in the ER are sequestered between the two phospholipid layers of the membrane at a particular region, such that a budding vesicle of triacylglycerols surrounded by a layer of phospholipid is formed (6,7). The vesicle is then detached off to become a lipid body. An alternative postulation states that the lipid bodies arise directly in the cytoplasm by condensation of triacylglycerol molecules followed by formation of the surrounding membrane (4,8). The fate of the lipid body membrane after lipolysis in post-germination is also unclear until recently.

Using the scutella of maize, our laboratory has been studying the synthesis and degradation of the lipid bodies during seed maturation and germination. In order to delineate the mechanism of deposition and utilization of triacylglycerols in the lipid bodies, we study the last enzyme of triacylglycerol biosynthesis in seed maturation (diacylglycerol acyltransferase), and the first enzyme of triacylglycerol degradation during germination (lipase). In order to investigate the biosynthesis of the membrane of the lipid bodies, we characterize the major proteins of the lipid body membrane.

DIACYLGLYCEROL ACYLTRANSFERASE

Diacylglycerol acyltransferase (EC 2.3.1.20) catalyzes the final step in the synthesis of triacylglycerols in oil seeds (1,2,3,4). It is also the only known enzyme unique to the long biosynthetic pathway of triacylglycerols, since the diacylglycerol produced could also be used to produce phospholipids or galactolipids. The enzyme, alone or in conjunction with the enzymes in preceding metabolic steps, has been detected in the microsomes from several seed species (see review, 4). Generally, the in vitro activity of the enzyme alone was either too low to account for the activity required to catalyze the sequence of reaction from glycerol phosphate to triacylglycerol (4), or un-detectable (9). An important but unknown (until now) aspect of the enzyme is its subcellular location. In general, the microsome fractions were used to study the enzyme activity, and they presumably contained vesicles of the ER as well as membranes of other subcellular particles, including broken plastids. In spinach leaves, the enzyme was shown to be present in the outer membranes of the plastids (10).

An assay has been developed to detect the activities of diacylglycerol acyltransferase in the scutella of maize, the cotyledons of peanut and soybean, and the endosperm of castor bean during seed maturation (11). The detected activities are high enough to account for the rate of in vivo triacylglycerol synthesis. Subcellular fractionation of the total maize scutellum extract and castor bean endosperm extract in sucrose density gradients has been performed. By comparing the migration of the enzyme between rate and equilibrium centrifugation, and between equilibrium centrifugation in the presence and absence of magnesium ions in the preparative media, the enzyme has been shown to be associated with the rough ER. As to be described, this subcellular location of the enzyme has some bearing on the biosynthesis of the lipid bodies.

MEMBRANE PROTEINS OF THE LIPID BODIES

Characterization

The monolayer of phospholipids of the lipid body membrane contains no unusual phospholipid components. In castor bean, the major components are phosphatidylcholine, phosphatidylethanolamine, and phosphatidylinositol (12). Besides the monolayer of phospholipids, the membrane also contains proteins. In several seed species examined, the protein components of the lipid bodies are different from those in other subcellular organelles such as the microsomes, mitochondria, and glyoxysomes, as revealed by SDS polyacrylamide gel electrophoresis (8,12, 13). Also, as shown in identical electrophoretic analyses, lipid bodies isolated from 11 taxonomically diverse species contain distinctly different protein patterns (14).

The membrane proteins of maize lipid bodies have been subjected to intensive studies (14). By SDS polyacrylamide gel electrophoresis (Figure 1), the proteins are resolved into several major protein bands, three of low M_r's (19,500, 18,000, and 16,500), and one of higher M_r (40,000). The low M_r proteins have alkaline pI values, and behave as hydrophobic integral proteins, as shown by their resistance to solubilization after repeated washing, amino acid composition, and partitioning in a Triton X-114 system.

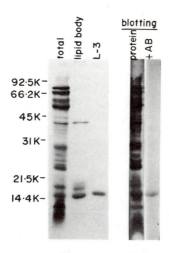

Fig. 1. Left panel: SDS-polyacrylamide gel of total homogenate, isolated lipid bodies, and isolated L-3 protein (Mr 16,500) of maize scutella. The positions of Mr markers are shown on the left. Right panel: Blotting of SDS polyacrylamide gel of total scutellum homogenate. After blotting, the strip was stained for protein or allowed to react with antibodies (+AB). (14)

Biosynthesis

It is assumed that most of the proteins of the lipid body membrane are synthesized during seed maturation and not germination. Some proteins, of important but of minor quantities (e.g., lipase, to be described) or of unknown functions (15) are apparently synthesized during germination.

The maize protein of Mr 16,500 represents about 40% of the total membrane proteins (Figure 1). It was chosen for detailed studies (14). The protein has been purified to apparent homogeneity, and rabbit monospecific antibodies have been raised against it. Labeling in vivo with 35-S methionine and translation in vitro using extracted RNA in a wheat-germ system reveal that the protein is synthesized during seed maturation and not germination. The proteins synthesized in vivo and in vitro have no appreciable difference in their mobilities by two-dimensional gel electrophoresis (isoelectric focusing and molecular sieving). The protein is synthesized predominantly, if not exclusively, by RNA derived from bound polyribosomes and not from free polyribosomes.

Molecular Biology

Using molecular cloning techniques, a c-DNA clone tentatively identified to be that of the maize lipid body protein of Mr about 16,500 has been obtained (16). The clone has a length of 780 nucleotides, and apparently includes the corresponding 3'-terminal poly-A region of the original mRNA. Northern blotting analysis using the above clone shows that the original m-RNA has a length of about 950 nucleotides. The mRNA is present in maturing seeds, persists in dormant seeds, and disappears 2 days after germination.

The nucleotides of the above c-DNA clone have been sequenced. Of the three possible amino acid sequences deduced from the nucleotide sequence, only one comes close to the Mr of the authentic protein and its amino acid composition. From the analyses of the hydrophobicity of the amino acid residues and the secondary structure of the presumed polypeptide, three structural segments could be deduced. The first segment contains an α helix of about 25-30 amino acids; the initial 2/3 consisting of hydrophilic amino acids and the remaining 1/3 hydrophobic amino acids. The second segment of the polypeptide has about 70 amino acids which are all hydrophobic. The third segment of the C-terminus has about 35 amino acids with a sequence having alternate hydrophobic and hydrophilic amino acids. A three dimensional analysis of this third segment shows that the amino acids arrange themselves into an α helix which has one side being hydrophobic and the other side being hydrophilic. If this presumed polypeptide is indeed the authentic protein, the hydrophobic second segment would penetrate into the lipid body; the third segment of amphipathic helix would locate on the surface of the lipid body, interacting directly with the triacylglycerols inside. The exact identity of this deduced polypeptide is being pursued by analyses of the amino acid sequence of the isolated protein and the entire 5' terminus of the m-RNA.

LIPASES

General

During germination, lipase (EC 3.1.1.3) hydrolyzes the storage triacylglycerols (17,18,19). Most of the products fatty acid and glycerol are converted to carbohydrates which support the growth of the embryonic axis. In most seeds, lipase activities are absent in the ungerminated seeds and increase during the early stage of germination. The only well-documented exception to this developmental pattern is castor bean which has active lipase in the ungerminated seeds (20). The lipases from several representative seed species have been shown to be relatively specific on triacylglycerols containing the major fatty acid components of the storage triacylglycerols in the same species.

In subcellular fractionation of the storage tissues of many germinated seeds, the lipase activity is present either in the soluble fraction or associated with the membranes of the lipid bodies (see reviews, 18,19). In seeds where the lipase is found in the soluble fractions, the enzymes in vivo will still have to come in contact with the membrane of the lipid bodies during catalysis. In those seeds where the lipase is associated with the membrane of the lipid bodies, the enzyme may be loosely (e.g., rapeseed, mustard seed) or tightly (e.g., castor bean and maize) associated with the organelles.

Characterization of lipase

The classical studies of castor bean lipase have been well-documented (20). The only plant lipase that has been purified to homogeneity and its biosynthesis studied is the enzyme from maize. In the scutella of maize, lipase activity is absent in ungerminated seeds and increases during germination (21). At the peak stage of lipolysis, about 60% of the lipase activity can be recovered in the lipid body fraction after flotation centrifugation. The lipase is tightly bound to the lipid bodies, and resists solubilization by repeated washing with buffers or NaCl solution. The lipase has been purified 272 fold to apparent homogeneity (22). The enzyme in sodium deoxycholate has an approximate

Mr of 270,000 by sucrose gradient centrifugation and Mr of 65,000 by SDS-polyacrylamide gel electrophoresis. The amino acid composition as well as a biphasic partition using Triton-X 114 reveals the enzyme to be a hydrophobic protein.

Biosynthesis of lipase

The biosynthesis of the maize lipase has been pursued using monospecific rabbit antibodies raised against the purified maize lipase (23). Using an in vitro protein synthesis system, the mRNA for the lipase has been detected in germinated but not maturing seeds (Figure 2). The in vitro and in vivo synthesized lipase exhibit the same Mr (65,000) by SDS-polyacrylamide gel electrophoresis, and thus there is no apparent co- or post- translational processing of the lipase. The enzyme is synthesized by mRNA extracted from free and not bound polyribosomes. Apparently, after its synthesis, the lipase will attach itself specifically to the membrane of the lipid bodies and not other cell organelles.

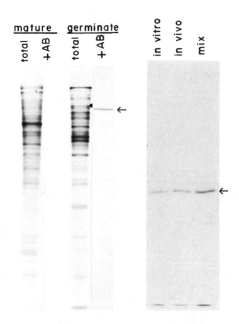

Fig. 2. Left panel: Fluorography of SDS-polyacrylamide gel (12.5%) of radioactive polypeptides synthesized in vitro using RNA extracted from scutella of maturing or germinating maize. The total reaction product or immunoprecipitated lipase was loaded onto the gel. Right panel: Fluorography of SDS-polyacrylamide gel (8%) of immunoprecipitated lipase synthesized in vitro using extracted RNA or in vivo by labeling the tissue with radioactive amino acid. A mixture of the in vitro and in vivo synthesized lipase was also used. The positions of lipase on the gels are indicated by arrows.(23)

The maize lines, Illinois High Oil, Illinois Low Oil, and their F-1 generation, contain about 18%, 0.5%, and 10%, respectively, of kernel lipids. Lipase activity which appears during germination is proportional to the lipid content in each maize line (24). This proportionality does not hold for catalase and isocitrate lyase, which are the same in the three maize lines. Thus, there is a co-selection for high lipid content and high lipase activity through breeding. Since the lipase is synthesized <u>de novo</u> during germination, it is possible that the enzyme is synthesized or degraded in proportion to the available substrate or anchoring sites (i.e., lipid body membrane).

Fate of lipase and lipid body membrane

The maize lipase activity starts to appear 2 days after imbibition, concomitant with the decrease in total lipids (23). The activity reaches a maximum at about day 5-6. The rise and peaking of lipase activity at the initial stage of seedling growth parallels those of catalase and isocitrate lyase, two enzymes known to be involved in gluconeogenesis from lipids. However, after reaching the maximum at day 5-6, lipase activity remains almost unaltered from day 5 to 10, whereas catalase and isocitrate lyase activities drop off rapidly. Apparently, either lipase is continuously being synthesized or more likely the old lipase is not being degraded.

At the peak stage of lipolysis (day 5-6), about 60% of the lipase activity can be recovered in the isolated membrane of the lipid bodies, and the remaining activity (presumably representing nascent enzyme and enzyme derived from consumed lipid bodies) is present in the 10,000g pellet, 120,000g pellet, and 120,000g supernatant (23). As seedling grows beyond day 5, the proportion of lipase in the lipid bodies decreases (concomitant with the decrease in triacylglycerol), whereas it increases in the 10,000g pellet, 120,000g pellet and 120,000g supernatant. Since there is no change in the total lipase activity from day 5-10, it is likely that there is a transfer of the membrane of the consumed lipid bodies together with

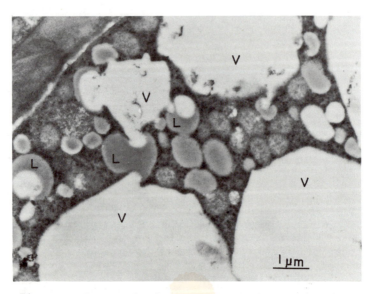

Fig. 3. Electron micrograph of a portion of a scutellum cell of 6-day-old maize seedling, showing the fusion of degrading lipid bodies and the cell vacuoles. L, lipid body; V, vacuole. (23)

the lipase to a fragile compartment in the cell, presumably the vacuoles. Electron microscopic observation does show a physical connection between the membrane of degrading lipid bodies and the membrane of enlarging cell vacuoles (Figure 3). During the fusion, the lipid body membrane should re-arrange itself to form a double phospholipid layer.

OVERALL ASSESSMENT OF THE SYNTHESIS AND DEGRADATION OF LIPID BODIES IN MAIZE KERNEL DURING MATURATION AND GERMINATION

Our findings on maize lipid bodies are consistent with a model of the synthesis and degradation of the organelle during seed maturation and germination (Figure 4). In maturation, the membrane protein of the lipid bodies are synthesized, without co- or post- translational modification, on the polyribosomes on the rough ER. The rough ER also synthesizes triacylglycerols and membrane phospholipids of the lipid bodies. These lipid body components are sequestered at a localized region in the ER, such that a lipid body is formed in which the triacylglycerol matrix is surrounded by a membrane of one phospholipid layer with embedded proteins. During germination, lipase is synthesized on free polyribosomes. Without apparent co- or post- translational processing, the lipase moves to attach itself specifically to the membrane of the lipid body. During or after lipolysis of the lipid body, the organelle membrane together with the lipase fuses with the membrane of an enlarging cell vacuole. During this fusion, re-arrangement of the lipid body membrane occurs, so that the monolayer of phospholipids is converted to a double layer of phospholipids of the vacuolar membrane. While the findings can best fit to this model, they do not totally eliminate but render very unlikely the alternate proposal that the lipid body is synthesized as a naked triacylglycerol droplet to be subsequently encircled by a membrane. In the first place, the original electron microscopic observation of such a proposed biosynthetic model (8) is disputed subsequently (7). Secondly, the observed in vitro formation of naked oil droplets in microsomes supplied with acyl CoA and glycerol (4) can be explained by the lack of a simultaneous formation of membrane phospholipids and proteins.

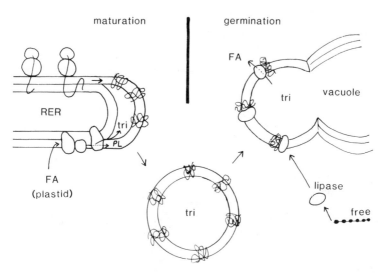

Fig. 4. Schematic illustration of the synthesis and degradation of a lipid body in the maize scutellum during seed maturation and germination. See text for explanation. RER, rough endoplasmic reticulum; FA, fatty acid; tri, triacylglycerol, PL, phospholipid; free, free polyribosomes.

An intriguing question is the function of the proteins (other than lipase which has an obvious function) on the lipid body membrane. Seeds store triacylglycerols in many small lipid bodies instead of one single large lipid droplet per cell as in mammalian white adipose tissues. By doing so, the small lipid bodies provide ample surface area for the binding of lipase, so that a rapid mobilization of the storage lipid can occur. The partition of the lipids into small droplets can be accomplished by a phospholipid layer without specific proteins. These proteins therefore may have other purposes. During germination, lipase synthesized on free polyribosomes attaches itself specifically to the membrane of the lipid bodies, and not the membranes of other subcellular structures. The specific recognition signals should be present on both the lipase and the lipid bodies. It is unlikely that the recognition signal on the lipid bodies resides on the phospholipids, since the membrane do not have uncommon phospholipids and it is generally agreed upon that the common phospholipids do not have sufficient signal for specific binding. Thus, it is logical to speculate that some of the membrane proteins of lipid bodies serve the role of providing specific binding signal for the lipase during germination. If so, the hydrophilic portions of the membrane proteins exposed to the cytosol are likely candidates for binding sites for the lipase. Further test of this binding hypothesis is warranted.

ACKNOWLEDGMENT — Our work has been supported by grants from the National Science Foundation, the US Department of Agriculture, and Pioneer Hi-Bred International. Inc.

REFERENCES

1. Appelqvist LA. 1975. In Recent Advances in the Chemistry and Biochemistry of Plant Lipids. Galliard T, Mercer EI, eds. Academic Press, London. pp 247-283.
2. Gurr MI. 1980. In The Biochemistry of Plants. Stumpf PK, Conn EE, eds. Academic Press, New York. Vol 4. pp 205-248.
3. Roughan PG, Slack CR. 1982. Annu. Rev. Plant Physiol. 33:97-132.
4. Stymme S. 1987. In The Biochemistry of Plants. Stumpf PK, Conn EE, eds. Academic Press, New York. Vol. 9. (in press).
5. Yatsu LY, Jack TJ. 1972. Plant Physiol. 49:937-943.
6. Schwarzenbach AM. 1971. Cytobiologie 4:145-147.
7. Wanner G, Formanek H, Theimer RR. 1981. Planta 151:109-123.
8. Bergfeld R, Hong TN, Kuhnl T, Schopfer P. 1978. Planta 143:297-307.
9. Martin BA, Wilson RF. 1983. Lipids 18:1-6.
10. Martin BA, Wilson RF. 1984. Lipids 19:117-121.
11. Cao YZ, Huang AHC. 1986. Plant Physiol. (in press)
12. Moreau RA, Liu KDF, Huang AHC. 1980. Plant Physiol. 65:1176-1180.
13. Slack CR, Bertaud WS, Shaw BD, Holland R, Browse J, Wright R. 1980. Biochem. J. 190:551-561.
14. Qu R, Wang SM, Lin YH, Vance VB, Huang AHC. 1986. Biochem. J. 235:57-65.
15. Sturm A, Schwennesen, Kindl H. 1985. Eur.J. Biochem. 150:461-468.
16. Vance VB, Huang AHC. (in preparation)
17. Galliard T. 1980. In The Biochemistry of Plants. Stumpf PK, Conn EE, eds. Academic Press, New York, Vol 4, pp. 85-116.
18. Huang AHC. 1983. In Lipases. Brockman HL, Borgstrom B., eds. Elsevier Press, Amsterdam. pp. 419-442.
19. Huang AHC. 1987. In the Biochemistry of Plants. Stumpf PK, Conn EE, eds. Academic Press, New York, Vol 9 (in press).
20. Ory RL. 1969. Lipids 4:177-185.
21. Lin YH, Wimer LT, Huang AHC. 1983. Plant Physiol. 73:460-463.
22. Lin YH, Huang AHC. 1984. Plant Physiol. 76:719-722.
23. Wang SM, Huang AHC. (in preparation)
24. Wang SM, Lin YH, Huang AHC. 1984. Plant Physiol. 76:837-839.

ON THE CONTROL OF FATTY ACID COMPOSITIONS OF PLANT GLYCEROLIPIDS

Grattan Roughan

Division of Horticulture and Processing
Mt Albert Research Centre
DSIR, Private Bag
Auckland, New Zealand

INTRODUCTION

Just a few years ago, a rational explanation of the sometimes confusing fatty acyl compositions of plant glycerolipids seemd a long way off. The perversity of these lipids was legendary[1]; different phospholipids within the same plant had their acyl groups arranged differently; both the digalactosyl diacylglycerol (DGD) and the sulfolipid (SL) had different dispositions of fatty acids depending upon their source; the hexadeca-trans-3-enoate (t16:1), which is unique to plant phosphatidylglycerol (PG), was strictly confined to the sn-2 position; the monogalactosyl diacylglycerol (MGD) of some plants contained up to 40% of its constituent fatty acids as hexadecatrienoate (16:3) which however, was completely missing from the MGD of other plants. Even so, there were sufficient clues to suggest that the problem of fatty acid distribution within plant glycerolipids might be soluble.

Whereas a mixture of plant glycerolipids will yield 6 or 7 different major fatty acids, a particular glycerolipid may contain a spectrum of fatty acids that immediately enables an experienced analyst to hazard a reasonable identification of the glycerolipid. Whilst no-one today would win a prize for identifying PG on the basis of its t16:1 content or of MGD on the basis of its 16:3 content, both MGD and DGD may normally be identified, even in 18:3 plants, by their very high content (>80% of total fatty acids) of trienoic fatty acids. In SL and phosphatidylinositol (PI), palmitate (16:0) frequently approaches 50% of total fatty acids, but SL will normally contain the higher proportion of α-linolenate (18:3). Phosphatidylcholine (PC) and phosphatidylethanolamine (PE) are identified, almost by default, as containing about 20% 16:0 and having either linoleate (18:2) or 18:3 as the major fatty acid. Currently, given a positional analysis of fatty acid composition, most plant lipid researchers should be able to identify the particular glycerolipid with reasonable confidence. Therefore, there are likely to be some constraints governing these fatty acid compositions, just waiting to be discovered.

Prior to 1980 it was widely believed that oleate (18:1) and 16:0 were desaturated in plants as their CoA esters. It was acknowledged however, that plant glycerolipids seemed to be first assembled from radioactive precursors as relatively saturated molecular species and only approached a degree of unsaturation matching that revealed by chemical analysis after a considerable lag. A convenient explanation of that apparent dilemma invoked a series of deacylation/reacylation reactions, involving the glycerolipids and CoA, which would both permit the desaturation of those fatty acids already incorporated into glycerolipids and reshape those anomalous molecular species initially synthesised into what we know they should be. The alternative explanation that fatty acids of glycerolipids were desaturated in situ was apparently less palatable. However, there is now compelling evidence to suggest that desaturation of glycerolipid-bound fatty acids does occur. By accepting that 18:1 can be desaturated without first being removed from the glycerolipid, 18:1, 18:2 and 18:3 may reasonably be considered as equivalent; which species will predominate at a particular position of a glycerolipid is simply a function of time and temperature. Similarly, 16:0 and t16:1 at sn-2 on PG may be considered as equivalent, as may be the series from 16:0 to 16:3 at sn-2 in MGD of 16:3 plants. This now greatly reduces the number of molecular species that must be considered by any theory professing to explain the fatty acid compositions of plant glycerolipids. The two-pathway hypothesis for glycerolipid metabolism in plants attempts such an explanation.

THE EVOLUTION OF THE TWO-PATHWAY MODEL

Between 1975 and 1978 it became increasingly more evident that net synthesis of PC, probably within the endoplasmic reticulum, played an intermediary role in the synthesis of MGD in leaves[2] and of triacylglycerol in some oil seeds[3]. It was also realized[4] by 1978 that long-chain fatty acids synthesised from labeled acetate by isolated spinach chloroplasts were directed either into unesterified fatty acids (UFA) or into glycerides, primarily 1,2-diacylglycerol (DAG) and phosphatidate (PA). Which of these lipid classes predominated depended upon the presence of sn-glycerol 3-phosphate (G3P) in incubation media. Increasing concentrations of G3P (50-500 µM) had two linked effects on chloroplast lipid metabolism; a stimulation of glyceride synthesis at the expense of UFA synthesis, and a stimulation of the accumulation of labeled 16:0 at the expense of labeled 18:1. The latter was a consequence of the glycerides containing equal amounts of labeled 16:0 and 18:1 whereas the UFA invariably contained 75-90% labeled 18:1. The distribution of label amongst the UFA synthesised by isolated spinach chloroplasts bore a striking resemblence to that occurring within the fatty acids of PC synthesised in vivo when leaves of a variety of species were provided with labeled acetate[5-7]. It therefore seemed reasonable to assume that chloroplast UFA could be the precursors of that PC which rapidly became labeled in vivo. This assumption received some support from the discovery that those chloroplast UFA were converted to acyl-CoAs at the cytoplasmic side of the chloroplast envelope and were then incorporated in vitro into PC of added microsomes[8].

The glycerides synthesised by isolated chloroplasts, on the other hand contained 18:1 at sn-1 and 16:0 at sn-2[8,9], a distribution by chain length which was analogous to that known to occur in PG of all chloroplasts and in about 50% of the MGD of spinach chloroplasts. Indeed, it was soon shown that the acyl-labeled DAG synthesised by isolated spinach chloroplasts was almost quantitatively converted to MGD in the presence of UDPgalactose[8,9]. Further, it was subsequently found that the 18:1/16:0 MGD thus synthesised was readily desaturated in vitro to 18:3/16:3 MGD[10], a major constituent of the chloroplast membranes.

At this stage then, it was possible to speculate that long-chain fatty acids synthesised from labeled acetate de novo as acyl carrier protein (ACP) esters within spinach chloroplasts, were destined for one of two fates (Figure 1). Either they would be used directly for glycerolipid synthesis within the chloroplasts or they would be released as the free acids which would then be exported to support glycerolipid synthesis in other cellular compartments. This branch point in chloroplast lipid metabolsim appeared to be controlled by the intracellular concentration of G3P. Just such a process now provided a much better explanation for an observation made some years previously[6]. This was that mature leaves on actively growing and fruiting pumpkin plants directed about 6-fold as much exogenous, labeled acetate into PC as into the next most highly labeled glycerolipid, PG. However, in large leaves induced on small, pot-grown plants by pinching off all the new shoots, exogenous acetate was directed into PC and PG about equally. It could now be argued that removing the "sinks" for photosynthate from the latter plants had resulted in their leaves becoming replete with metabolites, including G3P. The higher intracellular concentration of G3P had in turn, diverted chloroplast lipid metabolism in the direction of glycerides, and hence PG, at the expense of UFA, and hence PC, synthesis. The idea of two independent pathways of glycerolipid synthesis in plant cells first surfaced in 1980[9]. Evidence supporting the idea had come from a number of very careful, in vivo labeling studies[2,7] and from results obtained using isolated spinach chloroplasts possessing very high biosynthetic activities[4,9]. The chloroplast pathway was termed "procaryotic" because the probable acyl donors were ACP esters, because the resulting glycerolipids had a fatty acid disposition reminiscent of that in some cyanobacteria, and out of deference to the putative origin of higher plant chloroplasts. In addition, the relatively uncomplicated fatty acid compositions of cyanobacteria had earlier[11], and independently, been defined as procaryotic. Consequently, the extra-chloroplast pathway, probably confined largely to the edoplasmic reticulum, was termed "eucaryotic." However, the two-pathway concept was not initially greeted with much enthusiasm. Firstly there was a reluctance to concede a requirement for a pathway additional to that already discovered in spinach chloroplast envelopes by Joyard and Douce[12], and secondly, others working mainly with 16:3 plants[13], had been unable to confirm an intermediary role for PC in MGD synthesis, such as had been inferred from in vivo labeling studies using leaves of 18:3 plants[2,5-7]. Instead, they had found a more direct labeling of MGD independent of PC. Also, there appeared to be a problem with PC being an intermediate in MGD synthesis in spinach and Anthriscus since the localization of radioactive 16:0 at sn-1 of PC did not conform with the appearance of labeled 16:0 at sn-2 of MGD[13,14].

By 1982 however, it was realized that chloroplast lipid metabolism was significantly different in 16:3 and 18:3 plants. In 16:3 plants, such as spinach, a variable proportion of the total MGD was synthesised wholly within chloroplasts and incorporated labeled acetate quite rapidly in vivo[13]. The remainder of the MGD, which has 18:3 almost exclusively at both sn-1 and sn-2 positions, would be synthesised from DAG derived from extra-chloroplast PC as is the case in 18:3 plants[6,7]. Consequently, that latter fraction would incorporate labeled acetate relatively slowly. In spite of that, a movement in vivo of acyl label out of PC of microsomes and into MGD of chloroplast fractions isolated from expanding spinach leaves had been demonstrated as early as 1975[7]. It now proved possible to manipulate in a predictable manner the metabolism of exogenous labeled acetate by leaves. Treatments which

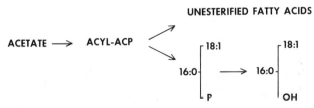

Figure 1. Showing where newly synthesised fatty acids diverge for the synthesis of different glycerolipids.

raised the intracellular concentrations of G3P resulted in a greater proportion of the supplied label entering glycerolipids synthesised by the procaryotic pathway[15]. In the meantime, other workers had also begun to appreciate the differences in lipid metabolism between 16:3 and 18:3 plants[16].

Something of a breakthrough occurred in 1982. During a visit to New Zealand, Ernst Heinz isolated highly-active chloroplasts from a selection of 16:3 and 18:3 plants and discovered a basic different in their metabolism of glycerolipids synthesised from labeled acetate. When incubated in a glyceride-synthesising mode, chloroplasts from 16:3 plants accumulated predominantly DAG whereas chloroplasts from 18:3 plants accumulated PA[17]. The 18:3 plants had a greatly reduced activity of the chloroplast phosphatidate phosphatase! Here then was a logical genetic mutation separating 16:3 and 18:3 plants; the absence of the phosphatase from chloroplast envelopes of 18:3 plants meant that a procaryotic DAG could not be generated to allow the synthesis of a procaryotic MGD, e.g. 18:3/16:3 MGD, or SL. However, procaryotic PA was available to permit the synthesis of PG in the chloroplasts[18]. The recent confirmation[19] that MGD, DGD and SL in 18:3 plants have virtually no procaryotic molecular species argues for the complete absence of chloroplast phosphatidate phosphatase _in vivo_ in those plants. It may not be coincidental either, that the DAG:UDPgalactose galactosyl transferase appears also to have been deleted from the inner membrane of chloroplast envelopes of the 18:3 plants, pea[20].

THE TWO-PATHWAY MODEL

It seemed possible by 1982 that the complicated fatty acid compositions of some plant glycerolipids might be explained by making some simple

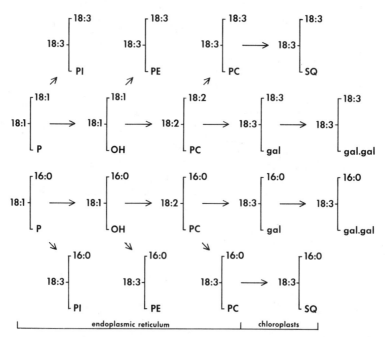

Figure 2. Abbreviated pathways for the synthesis of eucaryotic glycerolipids in leaves.

assumptions[21]. These were (i) that fatty acids are not promiscuous but once in O-ester linkage, remain there indefinitely; (ii) that C16 fatty acids are strictly excluded from the sn-2 position in the eucaryotic pathway whereas C18 fatty acids are strictly excluded from the sn-2 position in the procaryotic pathway; (iii) that 18:1 may be desaturated when it occurs at either position on most glycerolipids, and 16:0 may be desaturated when it occurs at the sn-2 position of MGD and PG; (iv) that MGD, DGD and SL may be synthesised from DAG arising either from the eucaryotic or procaryotic pathways (see Figures 2 and 3).

The operation of the PC:acyl-CoA acyl transerase reaction in developing oil seeds[22] appears to be incompatible with the long-term stability of ester linkages in glycerolipids, and yet the specificity of the reaction is such that assumption (ii) remains valid. Assumption (i) is consistent with in vivo labeling data for leaves[13], and has been confirmed for acyl groups of E. coli phospholipids[23]. Assumptions (i) and (ii) taken together, imply that fatty acyl configurations are impressed on plant lipids at the earliest stage in their synthesis, i.e. the acylation of G3P. This is supported by the specific transfer of 16:0 to 1-acyl G3P in chloroplast envelopes[24], and by experiments on G3P acylation by microsomal preparations[25,26]. Assumption (ii) is supported by positional analyses of glycerolipids newly synthesised in vivo from labeled acetate[13] and by positional analyses of glycerolipids synthesied by isolated chloroplasts[8,9,17]. The exclusion of C16 fatty acids from sn-2 of eucaryotic glycerolipids, however, is most evident in careful positional analyses of fatty acids in PC, PE and PI. Those results suggest that in vivo, both 16:0 and 18:1 are initially transferred to the sn-1 position but that C18 unsaturated fatty acids alone are transferred to the sn-2 position of 1-acyl G3P. Assumption (iv) seemed to be the most likely explanation for the existance of two types of MGD molecular species in 16:3

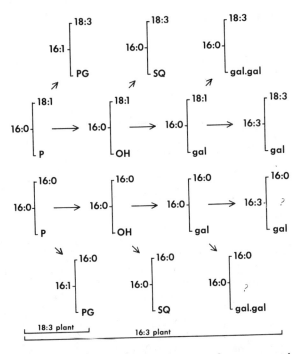

Figure 3. Abbreviated pathways for synthesis of procaryotic glycerolipids in chloroplasts.

plants[1]. It would also explain the occurence of 16:0 at both sn-1 and sn-2 of DGD and SL in 16:3 plants and its localization to sn-1 in 18:3 plants[1].

What predictions may be made from the model? One obvious prediction is that the contributions of the procaryotic and eucaryotic pathways to the make-up of a particular glycerolipid is given by the proportions of C16 and C18 fatty acids at the sn-2 position. Chloroplast PG must be synthesised entirely by the procaryotic pathway, and MGD will be synthesised either entirely by the eucaryotic pathway (maize, pea etc.) or predominatly (>90%) by the eucaryotic pathway (fat hen) or about equally by both pathways (spinach, tobacco) or predominantly (>75%) by the procaryotic pathway (chervil, parsley, Arabadopsis). It has now been shown that whereas MGD is synthesised largely (>75%) by the procaryotic pathway in Arabadopsis, DGD is derived largely (>75%) from the eucaryotic pathway[27]! This seems to be a general rule[1,19] and any proposed mechanism of DGD synthesis therefore, must take this observation into account. Since there must be a selection, probably at the outer membrane of the chloroplast envelope, of eucaryotic MGD molecular species for DGD synthesis, the actual contribution of the eucaryotic pathway to MGD synthesis is underestimated by a positional analysis of MGD alone; both galactolipids must be considered in estimates of the flow of DAG moieties into MGD. The model would seem best served by separate DAG:UDPgalactose galactosyl transferases in the inner and outer membranes of the envelope in 16:3 plants. This would then spatially separate procaryotic and eucaryotic MDGs thereby facilitating the selection of molecular species for DGD synthesis. It would also make sense for the phosphatidate phosphatase and DAG:UDPgalactose galactosyl transferase to be eliminated concomitantly from the inner membrane of chloroplast envelopes of 18:3 plants.

Other predications are (i) that SL in 16:3 plants must be synthesised from a sugar nucleotide and a mixture of procaryotic and eucaryotic DAGs, and (ii) that disaturated molecular species of plant glycerolipids will be synthesised by the procaryotic pathway alone. Whereas there is no convincing evidence for disaturated PC or PE in plants, it is now known that disaturated molecular species of PG may account for up to 50% of the total chloroplast PG in some plants[28].

GLYCEROLIPID SYNTHESIS BY AMARANTHUS CHLOROPLASTS

A. lividus is a C4 chilling-sensitive plant that synthesises a highly saturated chloroplast PG; the sum of saturated fatty acids (16:0+t16:1+18:0) amounts to 75% of total fatty acids, indicating that 50% of the total PG exists as disaturated molecular species. By comparison, the PGs synthesised by chloroplasts within spinach and pea leaves have only 50-55% saturated fatty acids and have almost exclusively C18, unsaturated fatty acids at the sn-1 position. The model predicts that disaturated PG will by synthesised by the procaryotic pathway within chloroplasts and probably arises because the PA from which the PG is synthesised contains a high proportion of disaturated molecular species. That PA would result either because the G3P:acyl-ACP acyl transferase has different specificities in pea and Amaranthus chloroplasts or because 16:0 is abnormally high in the acyl-ACP pool of Amaranthus chloroplasts.

Attached, expanding leaves of A. lividus incorporated 25% of the labeled acetate supplied into total lipids within 60 minutes. The labeled glycerolipids were mainly PC, PG and MGD; 60, 12 and 12% respectively. Saturated fatty acids accounted for more than 70% of the label in PG, but for only 14 and 8% of that in PC and MGD. Chloroplasts isolated from expanding leaves of A. lividus incorporated labeled acetate primarily into UFA but also into glycerolipids. Whereas the UFA typically contained 80% 18:1, the glycerides contained more than 80% 16:0 plus 18:0. In the presence of both G3P and CTP, up to 12% of the incorporated label was recovered as PG; an identical proportion to that recovered in vivo. This PG contained 85% saturated fatty acids and was synthesised from a PA which also contained 85% saturated fatty acids (Figure 4). A comparison of the acyl-ACP and lyso-PA pools during fatty acid synthesis (Figure 4) indicates that the G3P:acyl-ACP acyl transferase of A. lividus chloroplasts discriminates against steroyl-ACP but transfers palmitoyl- and oleoyl-ACPs with about equal facility. Although the G3P acyl transferase activity of membrane-free extracts of A. lividus chloroplasts was unstable, it was possible to demonstrate that the enzyme discriminated against 18:0 alone when presented with a mixture of 16:0-, 18:0- and 18:1-thioesters.

A high content of disaturated PG may confer chilling sensitivity upon a plant. Chilling sensitivity is normally an undesirable characteristic in those plants that are so afflicted since it may limit growth and development of the whole plant as well as reducing the shelf life of the fruit of the plant. The chloroplast G3P acyl transferase of chilling-sensitive plants therefore, now becomes a target for the genetic engineers; by conferring upon it a similar fastidiousness to that already present in the enzymes from spinach and pea it may just be possible to reduce the chilling sensitivity of a number of important horticultural crops.

CONCLUSION

The two-pathway hypothesis, although still in its infancy, provides a useful description of plant lipid metabolism and an explanation for the control of fatty acid compositions in different plant glycerolipids. Its simplicity is appealing. A number of previous observations that were difficult to interpret now seem obvious. The advances that have been made on the genetics of lipid metabolism in Arabadopsis[29] would not have been possible without the aid of this model.

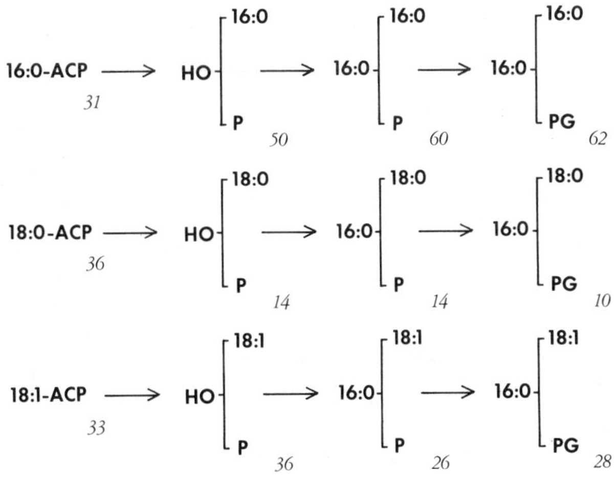

Figure 4. Synthesis of glycerolipids by Amaranthus chloroplasts. The figures in parentheses show the percentage distribution of products measured at each step in the synthesis.

REFERENCES

1. Heinz, E., in "Lipids and Lipid Polymers in Higher Plants." M. Tevini and H.K. Lictenthaler Eds. pp. 102-120. Springer Verlag, Berlin (1977).
2. Slack, C. R., P. G. Roughan, and N. Balasingham, Biochem. J., 162:289-296 (1977).
3. Slack, C. R., P. G. Roughan, and N. Balasingham, Biochem. J., 170:421-433 (1978).
4. Roughan, P. G., R. Holland, and C. R. Slack, Biochem. J., 184:193-202 (1979).
5. Williams, J. P., G. R. Watson, and S. P. K. Leung, Plant Physiol., 57:179-184 (1976).
6. Roughan, P. G., Lipids, 10:609-614 (1975).
7. Slack, C. R., and P. G. Roughan, Biochem. J., 152:217-228 (1975).
8. McKee, J. A. W., and J. C. Hawke, Arch. Biochem. Biophys., 197:322-332 (1979).
9. Roughan, P. G., R. Holland, and C. R. Slack, Biochem. J., 188:17-24 (1980).
10. Roughan, P. G., J. B. Mudd, T. T. McManus, and C. R. Slack, Biochem. J., 184:571-574 (1979).
11. Zepke, H.D., E. Heinz, A. Radnuz, M. Linscheid, and R. Pesch, Arch. Microbiol., 119:157-162 (1978).
12. Joyard, J., and R. Douce, Arch. Biochem. Biophys., 486:273-285 (1977).
13. Seibertz, H. P., and E. Heinz, Z. Naturforsch., 32c:193-205 (1977).
14. Seibertz, H. P., E. Heinz, M. Linscheid, J. Joyard, and R. Douce, Eur. J. Biochem., 101:429-438 (1979).
15. Gardiner, S. E., P. G. Roughan, and C. R. Slack, Plant Physiol., 70:1316-1320 (1982).
16. Williams, J. P., and M. U. Khan, Biochem. Biophys. Acta, 713:177-184 (1982).
17. Heinz, E., and P. G. Roughan, Plant Physiol., 72:273-279 (1983).
18. Mudd, J. B., and R. DeZaks, Arch. Biochem. Biophys., 240:584-591 (1981).
19. Bishop, D. G., S. A. Sparace, and J. B. Mudd, Arch. Biochem. Biophys., 240:851-858 (1985).
20. Cline, K., and K. Keegstra, Plant Physiol., 71:366-372 (1983).
21. Roughan, P. G., and C. R. Slack, Ann. Rev. Plant Physiol., 33:97-132 (1982).
22. Stymne, S., and G. Glad, Lipids, 16:298-305 (1981).
23. Cronan Jr., J. E., and J. H. Prestegard, Biochemistry, 16:4738-4742 (1977).
24. Frentzen, M., E. Heinz, T. A. McKeon, and P. K. Stumpf, Eur. J. Biochem., 129:629-636 (1983).
25. Frentzen, M., W. Hares, and A. Schiburr, in "Structure, Function and Metabolism of Plant Lipids". P-A. Siegenthaler and W. Eichenberger Eds. pp 105-110. Elsevier, (1982).
26. Griffiths, G., A. K. Stobart, and S. Stymne, Biochem. J., 129:379-388 (1985).
27. Browse, J., N. Warwick, C. R. Somerville, and C. R. Slack, Biochem. J., 234:25-31 (1986).
28. Murata, N., Plant Cell Physiol., 24:81-86 (1983).
29. Browse, J., P. McCourt, and C. Somerville, Plant Physiol., (in press) (1986).

LIPID DISTRIBUTION AND SYNTHESIS WITHIN THE PLANT CELL

Roland Douce, Claude Alban, Richard Bligny, Maryse A. Block, Jacques Covès, Albert-Jean Dorne, Etienne-Pascal Journet, Jacques Joyard, Michel Neuburger and Fabrice Rebeillé

Laboratoire de Physiologie Cellulaire Végétale, UA CNRS 576 DRF/CENG and USTMG, 85 X, 38041 Grenoble-cedex, France

INTRODUCTION

The higher plant cell contains numerous distinct organelles or membranes[1], but only some of these have been properly purified and characterized. Determination of the in vivo glycerolipid composition of these plant cell organelles or membranes, and their role in lipid metabolism is not simple, in contrast to what is often believed.

The purpose of this article is to summarize results obtained in our laboratory on plant cell organelles or membranes that we are able to prepare in highly purified and metabolically competent forms. The methods we have developed allowed a better understanding of the specific properties and composition of each membrane system within the plant cell.

GENERAL CONSIDERATIONS FOR THE ISOLATION OF PLANT CELL ORGANELLES

A common feature of all plant cells is the presence of a rigid wall which must be disrupted to liberate cytoplasmic organelles into the grinding medium. The methods used to break the cell wall lead inevitably to the rupture of the vacuole which is known to contain harmful products such as hydrolytic enzymes, phenolic compounds, tannins, alkaloids and terpenes. These products, released in the grinding medium inevitably interact with mitochondrial or plastid membranes, and care must be taken to avoid or minimize such interactions by protective agents able to strongly chelate inhibiting substances or to block the functionning of hydrolytic enzymes. For instance, polyvinylpyrrolidone (PVP), which acts as a strong scavenger of phenols and tannins, has been shown to improve the quality of mitochondria extracted from tissues with high phenolic content. In addition, the vacuole is very acidic, mainly due to organic acids of the TCA cycle or to inorganic acids (such as phosphate). Following rupture of the vacuoles during the grinding step, these compounds would lead to a dramatic shift of the pH towards acidic values in the absence of a strong additional buffer in the grinding medium. This problem is a real one since some enzymes of either the vacuole (such as proteinases, various lipolytic acylhydrolases) or organelles (such as the galactolipid:galactolipid galactosyltransferases or acyltransferases) are active at acidic pH values and produce rapid transformation or destruction of the membrane glycerolipids of mitochondria or plastids.

Numerous details, which are often sufficient to minimize the effects of such enzymes on the glycerolipid content or the functioning of isolated organelles or membranes, have been published[2-4]. We believe that the following conditions should be taken into account for the preparation of large amounts of intact, highly purified, and physiologically competent organelles.

1. The grinding step should be reduced to a minimum : prolonged grinding procedures increases the yield of recovered cell organelles but considerably increases the percentage of envelope-free mitochondria and of plastids that have resealed following rupture and the loss of stromal content.
2. Since short grinding times lead to a low yield of organelles extracted, large quantities of tissues are needed to isolate sufficient amounts of highly intact organelles.
3. Assessment of organelle integrity should be carefully done.
4. Control of purity of the fraction is essential and purification of the different organelles by Percoll gradient procedures is highly recommended.

A good example of the importance of organelle purification is given by the preparation of mitochondria from potato tuber[3]. When prepared from non-green tissues, washed mitochondria are often yellow, due to the presence of carotenoids. These pigments have often been claimed to be genuine constituents of mitochondrial membranes. Purification, and especially percoll purification, strongly reduces the amount of carotenoids present in the preparation (Figure 1). In addition, Figure 1 also demonstrates the higher efficiency of percoll to remove most of the contaminants.

The same observation can be made with the lipolytic acyl hydrolases which also contaminate washed or sucrose-purified mitochondria (Figure 2). Such hydrolases produce rapid hydrolysis of membrane phospholipids and dramatically impair the mitochondrial functionning, and release large amounts of free fatty acids[5] which are then oxidized to fatty acid hydroperoxides[6] in the presence of molecular oxygen. Figure 2 clearly demonstrates the ability of percoll to remove such vacuolar enzymes which have harmful effects on the glycerolipid composition of plant organelles.

Purification is most important when the presence of organelles having the same size and density leads to the preparation of crude pellets containing different types of intact organelles. This is often the case in non-green tissues, such as potato tubers or cauliflower buds. In these examples, the so-called mitochondrial pellet consist in fact of intact mitochondria and peroxisomes (potato tubers) or of intact

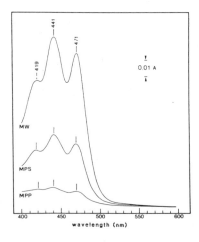

Figure 1: *Absorption spectra of ethanol extracts of potato tuber mitochondria.* M.W, washed mitochondria (3.2 mg protein, 0.32 µg carotenoid) ; MPS, mitochondria purified on sucrose (3.2 mg protein, 0.14 µg carotenoid) ; MPP, mitochondria purified on percoll (3.2 mg protein, 0.03 µg carotenoid). A second percoll purification would lead the carotenoid content to non-detectable levels. The carotenoids present in these preparation actually derive from membranes of non-green plastids, such as *amyloplasts*. Reproduced from Neuburger et al (1982).

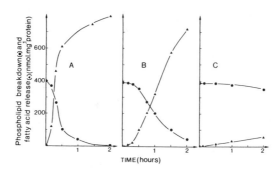

Figure 2 : *Rate of endogenous phospholipid breakdown during ageing of potato tuber mitochondria induced by 5 mM Ca^{2+}. A, washed mitochondria ; B, sucrose purified mitochondria ; C, percoll purified mitochondria.* Reproduced from Neuburger et al (1982).

mitochondria and plastids (cauliflower buds) which have to be separated from each other[3].

Modification of the osmoticum in the percoll purification gradients leads to a modification of the sedimentation behavior of the different organelles which allows a clear separation of peroxisomes from mitochondria[3] or of plastids from mitochondria[7].

To conclude, we believe that the passage of plant cell organelles preparations through percoll density gradients is always an advangeous step to separate quickly and efficiently the organelle to be purified from the other contaminating membranes, soluble compartments or other intact organelles. We have used such procedures with success for mitochondria from non-green tissues[3] and leaves[8], for peroxisomes[3], for chloroplasts[9] and for plastids from non-green tissues[7]. All these organelles have been used for estimation of polar lipid content, studies on lipid metabolism or for preparation of the different constituting compartments. In addition, combination of percoll purification procedures with the use of protoplasts to produce highly intact and almost pure organelles is a very efficient tool to study lipid metabolism within plant cells.

GLYCEROLIPID COMPOSITION OF PURIFIED ORGANELLES FROM PLANT CELLS

We have analyzed the polar lipid composition of mitochondria (from both green and non-green tissues), peroxisomes, chloroplasts and non-green plastids. Some experiments have also been performed with isolated cells from sycamore, cultivated in liquid suspension. In this last case, a specific parameter for the culture, i.e. the oxygen content of the medium, has to be taken into account as far as the fatty acid content of the membrane glycerolipids is concerned[10].

A more detailed analysis have been done for mitochondria from non-green tissues (sycamore cells, mung bean) and for spinach chloroplasts since in these organelles the different membranes have been purified and the membrane fractions properly characterized. The plant cell contains several other membranes such as golgi membranes, tonoplast, endoplasmic reticulum and plasmalemma. However none of these membranes have been as thoroughly characterized and purified as mitochondrial and plastid membranes, thus leading to contradictory results which are mostly due to the presence of contaminating membranes. Therefore, such membranes will not be considered in this short discussion.

A specific treatment of plastids is necessary prior to lipid analyses. We have demonstrated that a galactolipid:galactolipid galactosyl transferase, which manipulates galactolipid molecules and leads to the formation of diacylglycerol and unnatural galactolipids[11], is localized on the cytosolic side of the outer envelope membrane[12]. To prevent the functionning of the enzyme during the course of plastid fractionation, we have used a mild proteolytic treatment of the intact organelles with thermolysin[12]. After this treatment, no further modification of the glycerolipid content of the organelle is possible, thus leading to a composition as close as possible to the in vivo situation. Interestingly, in collaboration with J. Soll, we have demonstrated that the same enzyme is present on the cytosolic side of the outer envelope membrane of etioplasts. Thus it is recommended to incubate almost all kind of plastids with thermolysin if the membrane glycerolipids have to be quantified seriously.

Being aware of such limitations, we have analyzed the glycerolipid content of membrane fractions from different cell organelles (Table I). A striking feature of plant mitochondria is that their glycerolipid pattern in membranes from both green or non-green tissues are almost identical, and also virtually identical to that from human heart mitochondria. Presumably, these basic features of mitochondrial membranes are essential for their functionning in energy transduction. The major glycerolipids in mitochondria are PE and PC. PE is present in almost all membrane systems, including peroxisomes, except in membranes from chloroplasts or from non-green plastids. The salient feature of plant mitochondria is the high concentration of cardiolipin (DPG), that is not found in any other cellular organelle[2]. DPG is actually localized on the inner mitochondrial membrane (Table I). This is not surprizing since the number of cardiolipin molecules per molecule of cytochrome oxidase (a specific marker of the inner membrane) is the same in the whole Sycamore cell, in the intact mitochondria isolated from these cells, and in the inner membrane (about 80). Phosphatidic acid is sometimes reported, but its presence is probably due to the functionning of a phospholipase D during the preparation procedure. Galactolipids and sulfolipid are missing from pure plant mitochondria; indeed, since these glycolipids are only present in plastid membranes, their presence in mitochondrial preparation (for instance in those from cauliflower buds) reflects contamination by small amounts of plastid membranes (and often from envelope membranes).

Plastid membranes are characterized by a high glycolipid content and a low phospholipid content. All the membranes from chloroplasts, etioplasts or other non-green plastids are devoid of PE. Its presence, for instance in cauliflower bud plastids, reflects contamination by extraplastidial membranes. Interestingly, except for PC, the glycerolipid composition of chloroplasts is almost identical to that of cyanobacteria, or of Prochloron[13,14], which are supposed to be related to the ancestors or chloroplasts. In non-green plastids, the high level of DGDG, compared to the situation in chloroplasts, is due to the envelope membrane (and more precisely to the outer envelope membrane, since the inner one is almost identical to thylakoids with respect to their glycerolipid content).

There is a similar situation in all the membranes in contact with the cytosol: PC is a major constituent in the outer membrane from plastids and mitochondria and in the peroxisomal membrane (table I). One can question whether this situation has any physiological significance or results from the functionning of the phospholipid transfer proteins that are present in the cytosol. There is almost no redistribution of this PC whithin the chloroplast since phospholipase C treatment destroy almost all the envelope PC without rupture or modification of the outer envelope membrane structure[15].

Table I

Glycerolipid composition of plant cell organelles from various plants

ORGANELLE	MGDG	DGDG	SL	PC	PG	PI	PE	DPG
MITOCHONDRIA								
- human heart[a]	0	0	0	43	tr	5	34	18
- pea leaves	0	0	0	40	2	6	46	14
- cauliflower buds	tr	tr	0	37	2	8	38	13
- mung bean hypocotyls								
total	0	0	0	36	1	2.5	46	14
inner membrane	0	0	0	29	1	2	50	17
outer membrane	0	0	0	68	2	5	24	0
- Sycamore cells								
total	0	0	0	43	3	6	35	13
inner membrane	0	0	0	41	2.5	5	37	14.5
outer membrane	0	0	0	54	4.5	11	30	0
PEROXISOMES								
potato tubers	0	0	0	52	0	0	48	0
PLASTIDS								
- spinach chloroplasts								
thylakoids	52	26	6.5	4.5	9.5	1.5	0	0
inner envelope membrane	49	30	5	6	8	1	0	0
outer envelope membrane	17	29	6	32	10	5	0	0
total envelope	32	30	6	20	9	4	0	0
- pea etioplasts								
prothylakoids	42	35	6	9	5	2	0	0
envelope	34	31	6	17	5	4	0	0
- cauliflower bud non-green plastids	30	29	6	19	8	5	2	0.6
PROKARYOTES	MgluDG							
- *Prochloron sp.*[b]	3.2	55.3	10.4	25.7	0	5.4	0	0
- *Anabaena variabilis*[c]	1	54	17	11	0	17	0	0
- *Anacystis nidulans*[d]	0	50.8	19	15.9	0	14.3	0	0

Ref. a) Rouser et al[16] ; b) Murata and Sato[13] ; c) Murata and Sato[14] ; Allen et al.[17]

The picture which emerges for the data discussed above is that each membrane contains characteristic glycerolipids in fixed molar ratios that are probably determined genetically : the almost identical glycerolipid composition of all mitochondria, the high similarity between envelope membranes from spinach chloroplasts, pea etioplasts and non-green plastids (which contain almost only envelope membranes as membrane constituents) strongly support this hypothesis. On the contrary, these observations differ strikingly from "the overwhelming impression of a great uniformity among plant cell membranes" which was commonly accepted a few years ago. Purification of plant cell organelles was really a breakthrough for an understanding of the glycerolipid content of plant membranes.

However, if the total glycerolipid composition of a given membrane is rather stable, its fatty acid content is more susceptible to environmental variations or modifications. Plant membrane glycerolipids have been subjected to analyses with respect to proportions, distribution and pairing of fatty acids. The uniqueness of chloroplast membranes, with specific fatty acids such as 16:3 in some molecular species of MGDG and $16:1_t$ in PG, and their high level of unsaturation, has focused almost all the attention of scientists towards these glycerolipids, which have been used to classify structures in "prokaryotic" and "eukaryotic" structures[18]. One can question, however, whether or not mitochondria contain "prokaryotic" structures for some of their glycerolipids. So far, no detailed analyses have been done for these organelles. In addition, pronounced changes in plant cell membrane fatty acid composition have been observed under very special circumstances, such as when they are grown in tissue culture. A good example is the modification of the fatty acids of all glycerolipids from Sycamore cells cultivated in suspension in the presence of either saturating or limiting O_2 concentrations. A complete modification of the pattern was observed in PC as well as in MGDG[10] (Table II).

Table II

Modification of fatty acid pattern of Sycamore glycerolipids under different O_2 concentration in the culture medium.

The values are expressed as weight per cent of total fatty acids in a given phospholipid.

LIPID	16:0	18:0	18:1	18:2	18:3	Oxygen concentration
Total	24	3	44	15	12	25 µM
	27	1.5	3.5	45	23	250 µM
PC	26	5	33	23	13	25 µM
	26	1.5	8	44	20	250 µM
MGDG	7	5	46	19	12	25 µM
	3	tr	1	10	85	250 µM

We believe that the modification and specifity of the fatty acid pattern of plant glycerolipids reflects the regulation of the biosynthetic pathways, especially at the level of desaturases involved in the mechanisms of fatty acid desaturation. Insertion of double bonds into an acyl chain is achieved by a family of fatty acid desaturases. In fact, all the double bonds (except the initial desaturation of stearic acid which occurs at the level of acyl-ACP)[19], each of which is inserted by a different enzyme, are added after the fatty acid becomes esterified to a glycerolipid[20]. We believe that the desaturation process in plant cells is catalyzed by membrane bound enzymes which use polar lipids for substrates ; these same lipids are structural components of the membrane.

BIOSYNTHESIS OF PLANT GLYCEROLIPIDS

Biosynthesis of glycerolipids requires the tight functioning of numerous enzymes involved in the formation of fatty acids, sn-glycerol 3-phosphate and polar head groups precursors (usually containing nucleotides). The first steps lead to the formation of phosphatidic acid (PA) and diacylglycerol (DG) via the Kornberg-Pricer pathway[21]. Originally, this pathway was thought to be restricted to "microsomal" membranes, a membrane fraction often considered to be endoplasmic reticulum, although numerous other membrane fractions are also present. Achievement of purification procedures for plastids and mitochondria clearly demonstrated that the Kornberg-Pricer pathway was also present in these organelles. Furthermore, purified Golgi membranes are also able to acylate sn-glycerol 3-phosphate to almost the same extent as endoplasmic reticulum-rich fractions[22]. The acylation of sn-glycerol 3-phosphate to phosphatidic acid which is the initial step for the synthesis of glycerolipids de novo, proceeds in a non-random manner. For instance the microsomal and the mitochondrial acylation system has a high preference for palmitic acid at the 1 position and for oleic acid at the 2-position whereas the plastidial acylation system has a very high preference for palmitic acid at the 2-position. In addition, the synthesis of PA was found to occur equally well in both the inner and outer mitochondrial (or plastid) membranes. Therefore one can conclude that each plant cell contains numerous sites of phosphatidic acid synthesis. In this short survey, we will focus on glycerolipid biosynthesis in young plastids from non-green material (i.e. cauliflower buds).

Glycerolipid biosynthesis by cauliflower bud plastids

The ability of plastids to synthesize fatty acids is not restricted to chloroplasts since non-green plastids, and those from cauliflower as an example, are able to incorporate 14C-acetate into 16:0 and 18:1[19]. We have shown[7] that all the enzymes of the Kornberg-Pricer pathway are present in these plastids: phosphatidic acid and diacylglycerol are the main glycerolipids to be formed by intact plastids from cauliflower buds (Figure 3). Fractionnation of these plastids has demonstrated that the first acyltransferase, responsible for lyso-PA formation is, as in chloroplasts, released in the swelling mixture and recovered as a soluble enzyme. The second acyltransferase, which forms PA, is tightly membrane bound and is recovered with envelope membranes (identified by the presence of the polypeptide E30 involved in phosphate transport, and by the presence of the galactosylation activity). However, it is possible that the greenish membranes, deriving from and connected to the inner envelope membrane also contain a small level of PA synthesis, suggesting that the young membrane fractions deriving from the inner envelope membrane have the same biosynthetic capabilities as the inner membrane. A more detailed study is therefore necessary

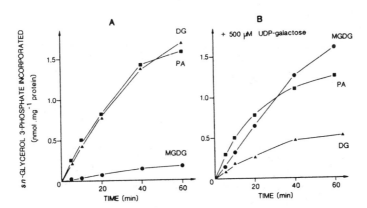

Figure 3 : *Synthesis of glycerolipids from sn-glycerol 3-phosphate by cauliflower bud plastids*. *In experiment B, UDP-gal was added to the incubation mixture. The cauliflower bud plastids are non-green and are mostly devoid of internal membranes. However, they behave almost like chloroplasts with respect to the Kornberg-Pricer pathway enzymes and MGDG formation.*

before this hypothesis can be rigorously tested. As in mature spinach chloroplasts, addition of UDP-gal to cauliflower bud plastids also results in the formation of MGDG. Again, this property is restricted to the inner envelope membrane.

Cauliflower bud plastids have not yet been investigated for their ability to form PG, the major plastid phospholipid, and SL, another unique plastid constituent. It is likely that results similar to those obtained with mature spinach chloroplasts would be obtained. Indeed, plastids are able to synthesize their major constituants (galactolipids, sulfolipid, PG) as far as they contain the so-called prokaryotic structure.

CONCLUSION

Studies of glycerolipid biosynthetic capabilities of plant cell membranes have led to a rather complex picture : there are several sites of biosynthesis within the plant cell, each of which is involved in the formation of various glycerolipids having a specific fatty acid composition (the best example is given by chloroplasts, which are known to contain the specific prokaryotic structure of glycerolipids). It is still an open question how lipid biosynthesis in different compartments is coordinated and how the parallel systems are regulated. Detailed molecular and genetic investigations will be necessary to elucidate these problems. Finally, an exhaustive study of all the membrane specific desaturases involved in the desaturation of polar lipids is required to understand further the assembly of cell membranes. As a matter of fact, all the various molecular species of polar lipids must be regarded as the structural partners of proteins during membrane biogenesis.

REFERENCES

1. Gunning, B.E.S. and Steer, M.W. (1975) "Ultrastructure and the Biology of Plant Cells". Arnold, London.
2. Douce, R. (1985) "Mitochondria in Higher Plants. Structure, Function and Biogenesis". Academic Press, New York.
3. Neuburger, M., Journet, E.P., Bligny, R., Carde, J.P. and Douce, R. (1982) Arch. Biochem. Biophys. 217, 312-323.
4. Douce, R. and Joyard, J. (1982) in "Methods in Chloroplast Molecular Biology" (M. Edelman, R. Hallick, and N.-H Chua, eds.), pp. 239-256. Elsevier/North-Holland, Amsterdam.
5. Bligny, R., and Douce, R. (1978) Biochim. Biophys. Acta 529, 419-428.
6. Dupont, J., Rustin, P., and Lance, C. (1982) Plant Physiol. 69, 1308-1314.
7. Journet, E.P., and Douce, R. (1985) Plant Physiol. 79, 458-467.
8. Day, D.A., Neuburger, M., and Douce, R. (1985) Aust. J. Plant Physiol. 12, 219-228.
9. Mourioux, G., and Douce, R. (1981) Plant Physiol. 67, 470-473.
10. Bligny, R., Rébeillé, F., and Douce, R. (1985) J. Biol. Chem. 260, 9166-9170.
11. Van Besouw, A., and Wintermans, J.F.G.M. (1978) Biochim. Biophys. Acta 529, 44-53.
12. Dorne, A.J., Block, M.A., Joyard, J., and Douce, R. (1982) FEBS Lett. 145, 30-34.
13. Murata, N., and Sato, N. (1982) in "Biochemistry and Metabolism of Plant Lipids" (J.F.G.M. Wintermans and P.J.C. Kuiper, eds), pp. 165-168. Elsevier Biomedical Press, Amsterdam.
14. Murata, N., and Sato, N. (1983). Plant Cell Physiol. 24, 133-138.
15. Dorne, A.J., Joyard, J., Block, M.A., and Douce, R. (1985) J. Cell Biol. 100, 1690-1697.

16. Rouser, G., Nelson, G.J., Fleisher, S., Simon, G. (1968) in "**Biological membranes**" (D. Chapman, ed.), pp. 5-69. Academic Press, London and New York.
17. Allen, C.F., Hirayama, O., and Good, P. (1966) in "**Biochemistry of Chloroplasts**" (T.W. Goodwin, ed.), pp. 195-200, Academic Press, London.
18. Heinz, E., and Roughan, P.G. (1983) Plant Physiol. 72, 273-279.
19. Stumpf, P.K. (1980) in "**The Biochemistry of Plants**, Vol. 4. **Lipids : Structure and Function**" (P.K. Stumpf, ed.), pp. 177-204, Academic Press, New York.
20. Heinz, E. (1977) in "**Lipids and Lipid Polymers in Higher Plants**", pp. 102-120, Springer Verlag, Berlin.
21. Kornberg, A., and Pricer, W.E., Jr. (1953) J. Biol. Chem. 204, 345-357.
22. Sauer, A., and Robinson, D.G. (1985) J. Exp. Bot. 36, 1257-1266.

REGULATION OF PHOSPHOLIPID HEADGROUP COMPOSITION

IN CASTOR BEAN ENDOSPERM*

        Thomas S. Moore, Jr.

        Department of Botany
        Louisiana State University
        Baton Rouge, LA 70803

INTRODUCTION

    As a general rule, the phospholipid composition of cellular membranes, as defined by headgroups, is highly conserved. Repeated isolates of a membrane fraction at a given developmental stage will, within experimental error, provide the same phospholipid composition. On the other hand, there are reports that in some instances specific changes in phospholipid headgroup composition can take place as a response to environmental or hormonal influences. Indeed, in mammalian systems phosphatidylinositol metabolism responds dramatically to hormonal influences and has been demonstrated to be an intermediary in signal transmission[2]. These considerations lead to the conclusion that the phospholipid composition of cells must be highly regulated.

    The primary phospholipids of plants are PC**, PE, PI, PG, BPG, and PS. While the general composition of these phospholipids remains the same for a specific plant tissue, there have been reports of modifications as a response to certain conditions. For example water stress, temperature stress, hormones, etc. have been reported to lead to modifications in phospholipid composition or phospholipid turnover rates[14]. These studies have referred to changes in both the headgroup and the fatty acid composition of the phospholipids. These two portions of the molecule may be altered independently, but it also is possible to modify the fatty acid content of a particular phospholipid type by modifying the headgroup. For example:

    PE + 3SAM --> PC + 3SAH

is a reaction which modifies the headgroup of phosphatidylethanolamine by the addition of three methyl groups and results in the formation of phosphatidylcholine. If such a reaction were selective for the fatty acid composition of the substrate PE, and this were different from that of the PC pool, then it could provide a specific fatty acid composition to that pool.

---

*Supported by NSF grant DMB-8402001
**Abbreviations: Eth, ethanolamine; Ins, inositol; Ser, serine; PC, phosphatidylcholine; PE, phosphatidylethanolamine; PG, phosphatidylglycerol; PI, phosphatidylinositol; BPG, <u>bis</u>-phosphatidylglycerol (cardiolipin); PS, phosphatidylserine.

Therefore, headgroup additions or modifications could lead to changes in both the concentration of a phospholipid type and also to specific species within that type.

Regulation of phospholipid synthesis could occur in many forms. These include compartmentalization, regulation of substrate levels, control of the ionic environment, control of the phospholipids associated with key enzymes, and others. Several aspects of regulation of phospholipid synthesis, along with some recent advances in these areas, are discussed below.

COMPARTMENTALIZATION

Relatively little recent information has been added to our knowledge of the intracellular compartmentalization of the final headgroup additions in phospholipid synthesis[14]. The primary exception to this statement is the recent publication by Sauer and Robinson[21], in which they demonstrate the occurrence of cholinephosphotransferase activity in a density gradient fractionation which they identify as having been derived from the Golgi apparatus. These data would appear to support the demonstration of similar compartmentalization by Montague and Ray[11] some years ago. A summary of the information available concerning compartmentalization of headgroup additions is contained in Figure 1. This diagram identifies three primary compartments as being active in the final phospholipid steps, these being the plastids, the mitochondria, and the endomembrane system (which includes the endoplasmic reticulum and the Golgi apparatus). The single phospholipid which has been demonstrated to be synthesized by all three of these compartments is phosphatidylglycerol[14]. The relative activities of the compartments is such that the endomembrane system generally is much more active than the other oraganelles in phospholipid synthesis, except where a quantity of a given phospholipid is required for a special purpose within a particular organelle. For example, the synthesis of PG, presumably as a precursor of cardiolipin (BPG), in the inner mitochondrial membrane is by a pathway which is as active as the PG synthetic pathway of the ER[14].

Another point of interest is the recent comfirmation by Morre et al.[16] that the inositol exchange enzyme does occur in plants. This activity, previously reported only from castor bean endosperm[22], catalyzes the unusual exchange of free inositol for the inositol of PI. Morre et al.[16] have cautioned that such activities could provide changes in turnover for PI that might be attributed to the inositol phosphate secondary messenger role and so adequate testing must be performed.

An area which is receiving increased attention is the compartmentalization of the enzymes leading to synthesis of CDPcholine for production of PC. It has been argued for mammalian systems that the final step in this pathway, catalyzed by CTP:cholinephosphate cytidylyltransferase, regulates the synthesis of phosphatidylcholine through the nucleotide pathway[23]. For these tissues it has been proposed that an inactive form of the enzyme exists which is soluble in the cytoplasm, and that it is activated through promotion of its attachment to microsomal membranes. Tests to elucidate the intracellular location of this enzyme in plant cells have led to mixed results (Table I). Data have been presented in favor of this enzyme being soluble[18], attached to the endoplasmic reticulum[9], in an unidentified light membrane fraction[10], associated with the Golgi apparatus[21], or in a variety of fractions[17]. A. J. Kinney, in my laboratory, recently investigated the cytidylyltransferase from castor bean endosperm, a tissue which is actively engaged in phosphatidylcholine synthesis[9]. Contrary to previous results with this tissue[10], greater than 95% of the activity was found associated with membrane fractions with 70-80% being in the endoplasmic reticulum fraction and the remainder located in an unidentified light ($1.10 g/cm^3$) membrane fraction which has been called "Band A"[10]. When several isolations

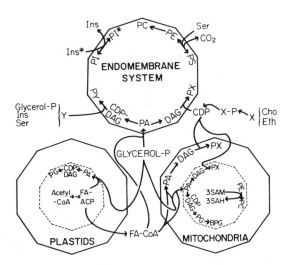

Figure 1. Compartmentalization of phospholipid biosynthesis in plant cells. The endomembrane system, which includes the endoplasmic reticulum and the Golgi apparatus, is the most active site of synthesis and presumably supplies phospholipids to a number of membranes. The mitochondria appear to produce primarily phospholipids for use within that organelle, particularly the inner membrane. The plastids have a restricted capacity for direct phospholipid synthesis, but supply fatty acids for the lipid produced by other membranes. The abbreviations are those given footnote 2 on the first page of this paper.

Table I. Compartmentalization of cytidylyltransferase

TISSUE	FRACTION	REFERENCE
Onion roots	Various	17
Castor bean endosperm	Band "A"	10
Rye roots	Microsomes	9
Pea stems	Cytosol	18
Maize roots	Golgi/ER	21

were compared, we found that the appearance or occurrence of activity in Band A increased in parallel with a loss of activity in the ER fraction. From this result we feel that the most likely in situ location for the enzyme in this tissue is the ER. Explanations of what is happening upon isolation include the disruption of the endoplasmic reticulum, thereby breaking it into high density and low density fractions, with the cholinephosphotransferase being located in the low density band; or the disruption of the normal enzyme attachment to the ER, thereby allowing it to associate with some other low density membrane fraction. In either case, the occurrence of the activity in the low density fraction appears to be an artifact and care must be taken in interpretation of such results from this and other tissues as being a part of another organelle.

SUBSTRATE AVAILABILITY AND COMPOSITION

Recently, in an attempt to understand the pathways leading to the production of the substrates for headgroup addition, we performed some pulse-chase experiments using potential precursors for specific headgroups[8]. We particularly were interested in synthesis of phosphatidylcholine and the origin of the CDPcholine. We anticipated that feeding radioactive serine to the endosperm would result in labeling of choline, cholinephosphate, and CDPcholine and then ultimately from these derivatives PC would be produced. We found, however, that although serine did become incorporated into PC, the pathway did not involve CDPcholine, but appeared to go through PE and then presumably transmethylation to form PC[8]. The proposal of operation of this pathway was supported by the use of inhibitors, in particular hemicholinium-3, in order to block the nucleotide pathway of synthesis by inhibiting production of CDPcholine. The addition of this inhibitor had little effect on the incorporation of radioactivity from serine into PC, but did strongly inhibit the incorporation of choline. In the absence of the inhibitor, choline readily entered the phospholipid product. An interesting sidelight of these investigations was the fact that when ethanolamine was added to the tissue it was incorporated into PC, but this incorporation was not affected by the inhibitor despite the fact that CDPethanolamine synthesis would also be inhibited. This result suggests that there is an active exchange enzyme present which can incorporate ethanolamine into a phospholipid fraction without going through the nucleotide pathway. Recently we have tested for that enzyme and found it to be present (data not presented). It is worth noting that both the serine and ethanolamine incorporations leading to PC synthesis are exchange reactions. It is possible that the same enzyme catalyzes both reactions, since we know that ethanolamine will compete with serine in PS synthesis[12].

These pulse-chase experiment results confirm and extend the previous data of Vandor and Richardson[24] and relate to those of Hansen's group dealing with water stress[4], but do not lend themselves to an easy interpretation. From our results we conclude that the precursors for phospholipid synthesis occur in different pools within the cell and that these pools do not always have ready access to each other. For example, exogenously added serine can enter into phosphatidylserine and thereby arrive in PC, but it does not appear to arrive at PC through the nucleotide pathway of synthesis[8]. Therefore, it appears not to be contributing immediately to the choline pool used for PC synthesis. This does not mean, however, that endogenously produced serine would not be available for choline production. We have yet to sort out exactly where or how these pools are compartmented, but the results raise some intriguing questions for the future.

Another mechanism for regulating the production of the phospholipids might be through regulating the availability of the lipid substrate. For example, if the species of diacylglycerol (DAG) were to be controlled and the cholinephosphotransferase, which adds the choline headgroup to form PC, were to recognize a specific species of DAG, then the presence or absence of that particular species of DAG might regulate the formation of PC. Available information in this area leads to mixed conclusions. For example, Justin et al.[7] have demonstrated that potato tuber microsomes *in vitro* produce PC molecular species in a mixture that is constant throughout a 30-minute period and the ratios of these species appear to be quite similar to the ratios of the diacylglycerols available in those membranes to act as substrate. Thus they conclude that the abundance of a species of diacylglycerol determines the abundance of PC synthesized by the enzyme[7]. We have taken a different approach by attempting to strip the enzyme of its endogenous substrate, utilizing detergents to accomplish this, and then adding back known diacylglycerols to determine whether or not the enzyme will utilize them. We found (Table II) that the enzyme demonstrated a distinct preference for the 1-16:0, 2-18:1 species of diacylglycerol. Other species of DAG also were utilized as long as some unsaturated bonds were present in the fatty acids, but if both acyl groups were saturated the enzyme appeared not to incorporate them. These results are of particular interest with respect to the proposed role of phosphatidylcholine in fatty acid desaturation[19] because the 1-16:0, 2-18:1 species of DAG would be the one utilized for production of PC so that the acyl unit in the 2 position could be further desaturated to produce the 18:2 species of fatty acid.

Unfortunately, it is difficult to make a direct comparison of our data and those of Justin, et al.[7] since there is little overlap between the molecular species in their system and those we have used so far. It is possible, however, that the most active substrate DAG's in these membranes had been depleted and were not replenished in the isolated system. We know that DAG is not at its optimal substrate level in castor bean[13]. On the other hand, such differences in cholinephosphotransferase found in studies of the mammalian enzymes have led to the suggestion of tissue differences[1]. The possibility of artifacts also must be considered.

ENVIRONMENT

The environment for the phospholipid synthesizing enzymes plays a major role in their activity. By environment I refer to pH, ion, and membrane lipid requirements. I previously have discussed the pH and ion requirements, in particular the dichotomy of requirements for magnesium and manganese by the DAG and CDP-DAG pathways of phospholipid synthesis [15]. Calcium plays a role in exchange reactions[14,15] and has been described as

Table. II. Stimulation of cholinephosphotransferase activity with various diacylglycerols

DAG	% of control
1-16:0, 2-18:1	100
1-16:0, 2-16:0	3
1-18:1, 2-18:1	66
1-18:1, 2-16:0	30
1-18:0, 2-18:0	2
1-18:0, 2-18:1	30
1-18:0, 2-18:2	41

inhibiting cholinephosphotransferase[14]. Recently, Choinski investigated this latter phenomenon more closely and found inhibition at concentrations as low as 0.5uM Ca^{2+}, which becomes physiologically significant (J. Choinski, personal communication).

There appear to be specific non-substrate lipid requirements. A requirement for phospholipids in order to obtain cholinephosphotransferase activity has been demonstrated by two different approaches. Mazliak's group[6] has treated microsomal membranes with phospholipase C in order to remove the phospholipid headgroups and in so doing demonstrated a loss of activity of the enzyme. They were able to recover activity by the addition of phospholipids which previously had been extracted from similar membranes of the same tissue. We have taken a different approach by treating ER from castor bean endosperm with detergent and then attempting to recover the lost activity by the addition of specific phospholipids. In so doing we have found that our best recovery of activity occurs with cardiolipin, PS, and PG. We get no increase of activity by the addition of PC or PE and often obtain inhibition. Unfortunately, we have been unable to maintain the activity of the enzyme for extended periods of time by these treatments or, as yet, to recover activity from enzyme which has been left in the detergent for long periods of time.

HORMONES

As mentioned in the introduction, phosphatidylinositol and phosphorylated derivitives of that lipid have been found to play a role in hormone signal transmission in mammalian cells. Despite the fact that phospholipid responses to hormones have been known for plant cells for some years, comparable information to that in mammalian systems is lacking. Gibberellins, auxins, and cytokinins all have been found to affect phospholipids in some manner[14], but the precise effects or significance of the results have, as yet, not been elucidated. Attempts to demonstrate the role of inositol phosphates in signal transmission in plants has only recently begun to be explored. Indeed it was only in 1985 that the intermediates for the pathways, the inositol phosphates, were first demonstrated by Boss[3] and additional publications in that area have only recently appeared[5,20]. Convincing evidence of a role for these intermediates in plants, however, has not as yet been presented. This is a largely untapped field and one in which considerable promise remains; on the other hand, it could be as disappointing as has been the cyclic AMP story.

CONCLUSIONS

In general it may be concluded that our understanding of the regulation of phospholipid synthesis, including both the addition of the headgroups and acyl modifications, is in its infancy. We are beginning to gain information on the physical requirements of the enzymes, their substrate specificities, and have hints of some effects of plant hormones on their metabolism. It is possible that within the very near future our knowledge in these areas will be greatly expanded and hopefully we will achieve an understanding of this important area of metabolism comparable to what is being achieved with mammalian and microorganism systems.

REFERENCES

1. G. Arthur and P. C. Choy, 1984, Acyl specificity of hamster heart CDPcholine:1,2-diacylglycerol phosphotransferase in phosphatidylcholine biosynthesis, Biochim. Biophys. Acta, 782:221-229.
2. M. J. Berridge and R. F. Irvine, 1984, Inositol triphosphate, a novel second messenger in cellular transduction, Nature, 312:315-321.
3. W. F. Boss and M. O. Massel, 1985, Polyphosphoinositides are present in plant tissue culture cells, Biochem. Biophys. Res. Comm., 132:1018-1023.
4. A. D. Hansen and W. D. Hitz, 1982, Metabolic responses of mesophytes to plant water deficits, Ann. Rev. Plant Physiol., 33:163-203.
5. S. Heim and K. G. Wagner, 1986, Evidence of phosphorylated phosphatidylinositols in the growth cycle of suspension-cultured plant cells, Biochem. Biophys. Res. Comm., 134:1175-1181.
6. A. Jolliot, A.-M. Justin, E. Bimont and P. Mazliak, 1982, Regulation by lipids of plant microsomal enzymes III. Phospholipid dependence of the cytidine-diphosphocholine phosphotransferase of potato microsomes, Plant Physiol., 70: 206-210.
7. A. M. Justin, C. Demandre, A. Tremolieres and P. Mazliak, 1985, No discrimination by choline- and ethanolamine phosphotransferases from potato tuber microsomes in molecular species of endogenous diacylglycerols, Biochim. Biophys. Acta, 836:1-7.
8. A. J. Kinney and T. S. Moore, 1986, Phosphatidylcholine synthesis in castor bean endosperm I. Metabolism of L-serine, Plant Physiol., in press.
9. A. J. Kinney and T. S. Moore, 1986, Localization and properties of cholinephosphate cytidylyltransferase activity in castor bean endosperm, This volume.
10. J. M. Lord, T. Kagawa, and H. Beevers, 1972, Intracellular distribution of enzymes of the cytidine diphosphate choline pathway in castor bean endosperm, Proc. Natl. Acad. Sci. USA, 69:2429-2432.
11. M. J. Montague and P. Ray, 1977, Phospholipid-synthesizing enzymes associated with Golgi dictyosomes from pea tissue, Plant Physiol., 59:225-230.
12. T. S. Moore, 1975, Phosphatidylserine synthesis in castor bean endosperm, Plant Physiol., 56:177-180.
13. T. S. Moore, 1976, Phosphatidylcholine synthesis in castor bean endosperm, Plant Physiol., 57:383-386.
14. T. S. Moore, 1982, Phospholipid biosynthesis, Ann. Rev. Plant Physiol., 33:235-259.

15. T. S. Moore, 1984, Biochemistry and biosynthesis of plant acyl lipids, in: Structure, Function and Metabolism of Plant Lipids, P. A. Siegenthaler and W. Eichenberger, eds., Elsevier Science Publ., New York, pp. 83-91.
16. D. J. Morre, B. Gripshover, A. Monroe and J. T. Morre, 1984, Phosphatidylinositol turnover in isolated soybean membranes stimulated by the synthetic growth hormone 2,4-dichlorophenoxyocetic acid, J. Biol. Chem., 259:15364-15368.
17. D. J. Morre, S. Nyquist and E. Rivera, 1970, Lecithin biosynthetic enzymes of onion stem and the distribution of phosphoryl-choline-cytidyl transferases among cell fractions, Plant Physiol., 45:800-804.
18. M. J. Price-Jones and J. L. Harwood, 1985, Purification of CTP:cholinephosphate from pea stems, Phytochem., 24:2523-2527.
19. P. G. Roughan and C. R. Slack, 1982, Cellular organization of glycerolipid metabolism, Ann. Rev. Plant Physiol., 33:97-132.
20. A. S. Sandelius and M. Sommarin, 1986, Phosphorylation of phosphatidylinositols in isolated plant membranes, FEBS Lett., 201:282-286.
21. A. Sauer and D. G. Robinson, 1985, Subcellular localization of enzymes involved in lecithin biosynthesis in maize roots, J. Exptl. Boty., 36:1257-1266.
22. J. C. Sexton and T. S. Moore, 1981, Phosphatidylinositol synthesis by a Mn^{++}-dependent exchange enzyme in castor bean endosperm, Plant Physiol, 68:18-22.
23. D. E. Vance and S. L. Pelech, 1984, Enzyme translocation and the regulation of phosphatidylcholine biosynthesis, Trends Biochem. Sci., 9:17-20.
24. S. L. Vandor and K. E. Richardson, 1968, Incorporation of ethanolamine 1,2-^{14}C into plant microsomal phospholipids, Can. J. Physiol., 46:1309-1315.

MOLECULAR SPECIES OF PHOSPHATIDYLCHOLINE IN PLANTS :

BIOSYNTHESIS AND ROLE IN OLEATE DESATURATION OR FREEZING RESISTANCE

C. Demandre, A.M. Justin, X.V. Nguyen, M. Gawer,
A. Trémolières and P. Mazliak

Laboratoire de Physiologie Cellulaire (UA 1180), Université
Paris 6, 4 place Jussieu, T 53/3, 75252 Paris Cedex, France

INTRODUCTION

High performance liquid chromatography has recently offered the possibility to separate all the molecular species (differing by their component fatty acids) forming a peculiar lipid class in a plant tissue. Thus a real molecular biology of lipids is now on progress in several laboratories[1,3]. The purpose of this paper is to present some recent data obtained in our laboratory, concerning the biosynthesis of various phosphatidylcholine molecular species on one side and the role played by these molecules in oleate desaturation or freezing-resistance, on the other side.

MOLECULAR SPECIES OF PHOSPHATIDYLCHOLINE IN PLANT TISSUES

Table I shows that within the same plant tissue, one can find as many as 11 molecular species of phosphatidylcholine. The most unsaturated species contains 6 double bonds (dilinolenoylphosphatidylcholine, 18:3/18:3 PC) ; fully saturated species are very rare.

Considering the molecular species of phosphatidylcholine from different plant tissues, cells or membranes analyzed in our laboratory, some generalizations can be proposed, tentatively :

1.- In non-green tissues (soya seeds, potato tubers, apple embryos) molecular species of PC containing linoleic acid (18:2/16:0, 18:2/18:2 and 18:2/18:1) are predominant.

2.- In entire green tissues (leaves), molecular species of PC containing linolenic acid (18:3/16:0, 18:3/18:2 and 18:3/18:3) are predominant.

Interestingly, in tobacco cell suspensions harvested at stationary phase, *i.e.* when isolated cells are greening, the predominating molecular species of phosphatidylcholine are the same as in entire leaves.

3.- When comparing PC molecular species distribution in green leaves and in microsomes prepared from these leaves, one can see that microsomal PC from green tissues has a molecular species composition approaching that

Table I

MOLECULAR SPECIES OF PHOSPHATIDYLCHOLINE FROM VARIOUS PLANT TISSUES, CELLS OR MEMBRANES

mol %

Molecular species	Soya beans	Potato tuber tissue	Potato tuber microsomes	Tobacco leaves	Tobacco microsomes	Tobacco cells	Pea leaf	Pea microsomes	Apple embryos
18:3/18:3	tr	tr	tr	14.5	5.5	14.8	11.1	6.2	--
18:3/18:2	14.9	13.7	13.5	17.5	8.7	24.3	24.4	17.8	--
18:2/18:2	29.5	32.5	31.8	8.5	6.8	5.4	18.9	21.2	35.4
18:3/16:0	tr	10.4	10.2	42.2	37.9	49.1	26.7	16.0	--
18:3/18:1	4.4	tr	tr	--	--	--	--	2.2	--
18:2/16:0	33.7	36.3	35.5	17.3	41.1	6.4	7.0	28.0	15.6
18:2/18:1	6.1	tr	tr	--	--	--	2.5	3.2	27.9
18:3/18:0	--	--	2.1	--	--	--	6.5	2.6	--
18:2/18:0	11.4	7.1	6.9	--	--	--	2.9	tr	6.5
18:1/16:0	--	--	tr	--	--	--	--	1.3	6.7
18:1/18:1	--	--	--	--	--	--	--	--	7.8
18:1/18:0	--	--	--	--	--	--	--	1.5	--

of microsomal PC from non green tissues (e.g. potato tuber) : 18:2/16:0 PC is the predominating species in the microsomes of both types of tissues.

Since there are obvious differences in molecular species distributions between phosphatidylcholines from different tissues or organelles, we have looked for an eventual selectivity displayed by one of the enzymes active in PC biosynthesis.

DISCRIMINATION BY MICROSOMAL CHOLINE-PHOSPHOTRANSFERASE BETWEEN MOLECULAR SPECIES OF ENDOGENOUS DIACYLGLYCEROLS

The final step in the biosynthesis of phosphatidylcholine is catalyzed by CDPcholine : 1,2-diacylglycerol-choline phosphotransferase (EC 2.7.8.2). This enzyme is mostly microsomal and catalyzes the following reaction :
$$\text{CDP-choline} + \text{diacylglycerol} \longrightarrow \text{PC} + \text{CMP}$$

We have asked ourselves whether there were any selectivity of choline phosphotransferase for some molecular species of diacylglycerols when PC was synthesized ? To answer this question we decided to investigate the biosynthesis of the different molecular species of PC within potato tuber microsomes[4].

We first compared the molecular species distributions in microsomal PC and endogenous diacylglycerols. Fig. 1 shows that these distributions are identical, which would indicate that choline-phosphotransferase is synthesizing PC from the endogenous diacylglycerol pool without displaying any selectivity towards any type of diacylglycerol molecular species.

To confirm this lack of selectivity, the biosynthesis of the various molecular species of PC was followed in potato tuber microsomes, starting from CDP (^{14}C) choline, under such conditions where only the endogenous pool of diacylglycerols could be utilized (no exogenous diacylglycerol was added to the incubation media[4]). Results shown in Fig. 2 demonstrate that all seven molecular species are synthesized in parallel, in the same rythm. From the shorter incubation periods (1 mn), the percentages of total PC radioactivity found in the various molecular species of PC are the same as the mass percentages of endogenous diacylglycerols within microsomes. Those percentages of radioactivity do not vary with time. It can thus be concluded that CDP-choline phosphotransferase from potato microsomes brings, within the membranes, phosphocholine groups onto all diacylglycerol molecules with the same velocity. Thus, the more abundant the species of diacylglycerol the more abundant that of PC synthesized by the enzyme.

Different results were obtained with a CDP-choline transferase from pea leaf microsomes. First we noticed a discrepancy between diacylglycerol molecular species distribution and PC molecular species distribution in pea leaf microsomes : molecular species containing linolenic acid were more abundant in diacylglycerols than in PC ; on the contrary, linoleic acid containing species are more abundant in PC than in diacylglycerols. In good agreement with these mass analyses, we have found that the CDP-choline phosphotransferase from pea leaf microsomes <u>preferred 16:0/18:2 diacylglycerol as a substrate over 18:3/18:2 diacylglycerol.</u>

These data indicate that differences between phosphatidylcholine molecular species distributions in various plant tissues or organelles could result both from different diacylglycerol pools existing in those tissues or organelles and also, in certain tissues, from the selectivity towards diacylglycerols displayed by choline-phosphotransferase.

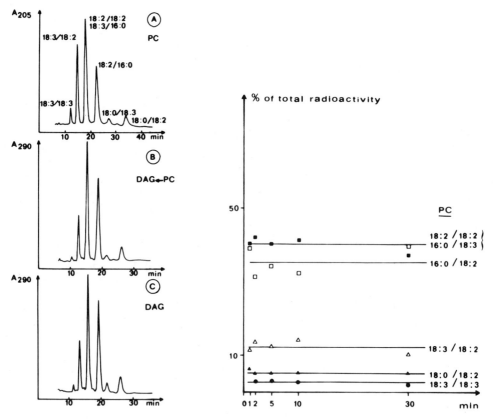

Fig. 1 (left) - HPLC separations of the molecular species of PC (A), diacylglycerols (DAG) prepared by phospholipase C treatment of PC (B) and endogenous diacylglycerols (C) from potato tuber membranes.

Fig. 2 (right) - Biosynthesis of the different molecular species of phosphatidylcholine in potato tuber microsomes. Evolution with time of the percentages of total radioactivity found in each molecular species. Incubation media (final volume : 1.6 ml) contained 0.05 M Tris-HCl (pH8), 5 uMol $MgCl_2$, 20 umol DTT, 2 umol CDP-^{14}C (choline 200,000 cpm) and 1 mg of microsomal protein.

Table II

PATTERN OF LABELLED MOLECULAR SPECIES OF PHOSPHATIDYLCHOLINE APPEARING IN POTATO MICROSOMES AFTER 10 MIN INCUBATION WITH ^{14}C OLEOYL CoA, IN ABSENCE OF NADH.
Results are expressed as percents of total radioctivity.

Molecular species of phosphatidylcholine (PC)		%
18:3/18:1	1 ^{14}C oleoyl-2,linolenoyl-PC 1,linolenoyl-2, ^{14}C oleoyl-PC	1.6 2.4
18:2/18:1	1 ^{14}C oleoyl-2,linoleoyl-PC 1,linoleoyl-2, ^{14}C oleoyl-PC	10.0 31.0
16:0/18:1	1,palmitoyl-2 ^{14}C oleoyl-PC 1 ^{14}C oleoyl-2,palmitoyl-PC	50.1 4.9

Incubation conditions : Microsomal suspensions (1.5-2mg of proteins) were incubated with 10 mg of BSA, 2 mg $MgCl_2$ and 2 nmol of ^{14}C -oleoyl CoA (57 mCi/nmole) in 0.1 M phosphate buffer pH 7.2 (final volume 0.5 ml). After incubation, each molecular species separated by HPLC was hydrolyzed with purified lipase A_1 from Rhizopus arrhizus and radioactivity was determined by scintillation counting.

Table III

PATTERN OF MOLECULAR SPECIES OF PHOSPHATIDYLCHOLINE, CONTAINING ^{14}C LINOLEOYL-RESIDUES, WHICH APPEARED IN POTATO MICROSOMES AFTER 10 MIN INCUBATION WITH ^{14}C OLEOYL CoA, IN PRESENCE OF NADH.
Results are expressed as percents of total radioactivity of ^{14}C linoleoyl-molecular species.

Molecular species of phosphatidylcholine (PC)		%
18:3/18:2	1 ^{14}C linoleoyl-2,linolenoyl-PC 1,linolenoyl-2 ^{14}C linoleoyl-PC	1.8 3.5
18:2/18:2	1 ^{14}C linoleoyl-2-,linoleoyl-PC 1,linoleoyl-2 ^{14}C linoleoyl-PC	12.3 36.8
16:0/18:2	1,palmitoyl-2 ^{14}C linoleoyl-PC 1 ^{14}C linoleoyl-2,palmitoyl-PC	38.8 6.8

Incubation conditions as in table II except that 1 uM NADH was added.

ROLE OF PHOSPHATIDYLCHOLINE MOLECULAR SPECIES IN OLEATE DESATURATION

Plant oleate desaturase is a membrane-bound enzyme system present in the reticulum of developing seeds[5], aged storage tissues[6] or growing leaves[7,8]. Most evidence suggests that oleoyl-phosphatidylcholine is the true substrate of oleate-desaturase[9]. Before subsequent desaturation oleoyl residues are inserted into phosphatidylcholine molecules either by an acyl-exchange mechanism between oleoyl-CoA and phospholipids[10] or by acylation to lysophosphatidylcholine[11]. HPLC allowed us to follow the time-course of oleate acylation and desaturation in the various molecular species of phosphatidylcholine from potato microsomes[12].

Since molecular species containing oleoyl residues do not accumulate normally in potato membranes (Table I), it can be concluded that these species are rapidly desaturated to linoleoyl-PC. To prevent this rapid desaturation, ^{14}C oleoyl-CoA was furnished to potato microsomes in absence of NADH in the incubation media. Table II shows that six labelled molecular species were formed in these conditions after 10 min incubation ; 83.5 % of ^{14}C -oleoyl residues were esterified in position 2 of <u>sn</u> glycerol.

When NADH was added to the incubation medium, oleate desaturation occurred and several new molecular species appeared which contained labelled linoleate (Table III). Labelled linoleoyl residues were found mainly (but not exclusively) in position 2 of <u>s.n.</u> glycerol.

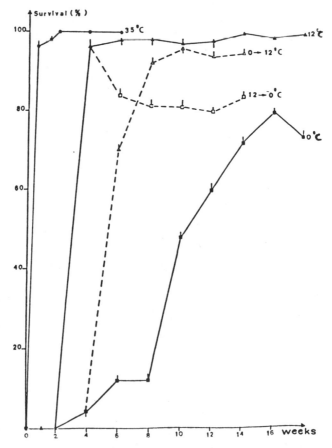

Fig. 3 - Development of freezing-resistance in apple embryos stored at different temperatures.

These data show clearly that six molecular species of PC containing oleoyl residues either in position 2 or in position 1 of s.n. glycerol can be utilized as substrates by potato oleate-desaturase. In presence of NADH, the molecular species of PC containing linoleoyl residues are formed by direct desaturation of the previous six molecular species. The main molecular species formed after desaturation are the following : 1,palmitoyl-2,^{14}C linoleoyl PC (38.8 %), 1,linoleoyl-2 ^{14}C linoleoyl PC (36.8 %) and 1, ^{14}C linoleoyl-2,linoleoyl PC (12.3 %).

Working on pea leaf oleate desaturase, Murphy et al.[13] pointed to 1,palmitoyl(stearoyl)-2,oleoyl-PC as the main substrate for oleate desaturase. In good agreement with these authors, we have found that 16:0/18:1 PC, in potato tuber, did represent 55 % of all molecular species used as substrates.

ROLE OF PHOSPHATIDYLCHOLINE MOLECULAR SPECIES IN FREEZING-RESISTANCE

Apples (cv. Golden Delicious) were harvested in October in a french orchard, immediatly transferred to rooms with controlled temperature (0,12 or 35 °C) and stored for several weeks in the dark. In some experiments, fruits were transferred after some period of storage at 0°C to 12°C or the reverse. Every 15 days, seeds were taken out from fruits and their survival was determined after freezing at -7°C for 30 min.[14].

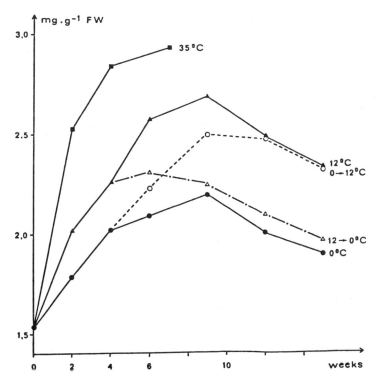

Fig. 4 – Evolution of total phospholipid contents of apple embryos stored at different temperatures.

The percentages of survival of embryos treated at -7°C and taken at different dates along storage is given in fig. 3. At harvest time all embryos were killed by the freezing treatment. During the storage of fruits, the frost sensitivity of the embryos gradually disappeared, the rate of disappearance depending on storage temperature. In fruits stored at 0°C, survival of a 80 % embryos did not occur until after 14 weeks of storage. At 12°C the same level of frost resistance was obtained after 2 or 3 weeks and at 35°C, after 3 days only. Change of storage temperature after 4 weeks from 0 to 12°C caused a pronounced rise in frost resistance of embryos. Transfer of fruits to a lower temperature of storage (from 12 to 0°C) resulted in some lowering of the survival percentage of embryos.

It was interesting to search whether the different effects of storage temperature on freezing resistance could be correlated with changes in the levels and compositions of some classes of lipids. Fig. 4 shows the changes in total phospholipid content of apple seeds during storage of the fruits at various temperatures. It can be seen that an increase of storage temperature, while decreasing the levels of non-polar lipids as shown elsewhere[15], promoted the accumulation of total phospholipids. When apples were transferred from 0 to 12°C, a clear accumulation of phospholipids could be observed ; on the contrary a transfer from 12 to 0°C led to a pronounced decrease of phospholipids. Among apple seed phospholipids, phosphatidylcholine, the major lipid class, followed exactly the evolution of freezing resistance of embryos during storage. Thus we decided to look for which molecular species of phosphatidylcholine do accumulate in embryos when freezing resistance is developing at 35°C.

Fig. 5 shows clearly that the least unsaturated species (18:1/18:1 and 16:0/18:1) do accumulate at 35°C whereas the more unsaturated species (18:2/18:2) progressively decreased. Just reverse results were observed at 0°C (data not shown).

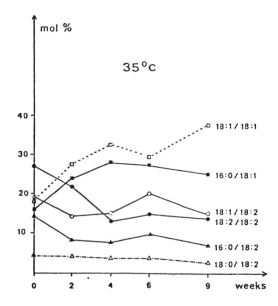

Fig. 5 - Evolution of phosphatidylcholine molecular species percentages in apple embryos stored at 35°C.

In conclusion, we propose that the relatively saturated molecular species of phosphatidylcholine (18:1/18:1 PC and 16:0/18:1 PC) which accumulate in apple embryos when freezing resistance is being acquired, play some anti-freeze role in cell-membranes. Surely PC accumulation stabilize the bilayer organization of phospholipids in membrane. On the other side, PC is well known to induce a more fluid state when it is added to a phospholipid mixture ; in the peculiar case of apple seeds the fluidizing effect of PC would be somehow counterbalanced by the preferential accumulation of relatively saturated molecular species, which could restrict membrane permeability. All together these biophysical changes would favour freezing resistance.

REFERENCES

(1) D. V. Lynch, R. E. Gundersen, and G. A. Thompson, Separation of galactolipid molecular species by high-performance liquid chromatography, Plant Physiol., 72 : 903 (1983)

(2) J. Kesselmeier, and E. Heinz, Separation and quantitation of molecular species from plant lipids by high-performance liquid chromatography, Analytical Biochemistry, 144 : 319 (1985)

(3) C. Demandre, A. Tremolieres, A. M. Justin, and P. Mazliak, Analysis of molecular species of plant polar lipids by high-performance and gas liquid chromatography, Phytochemistry, 24 : 481 (1985)

(4) A. M. Justin, C. Demandre, A. Tremolieres, and P. Mazliak, No discrimination by phosphocholine and phosphoethanolamine transferases from potato tuber microsomes in molecular species of endogenous diacylglycerols, Biochim. Biophys. Acta, 836 : 1 (1985)

(5) J. K. Vijay, and P. K. Stumpf, Nature of the substrate and the product of oleoylCoA desaturase from Carthamus tinctorius, J. Biol. chem., 246 : 2910 (1971)

(6) A. B. Abdelkader, A. Cherif, C. Demandre, and P. Mazliak, The oleoyl-CoA desaturase of potato tubers, Europ. J. Biochem., 32 : 155 (1973)

(7) A. Tremolieres, D. Drapier, J. P. Dubacq, and P. Mazliak, Oleoyl-CoA metabolization by subcellular fractions from growing pea leaves, Plant Sci. Letters, 18 : 257 (1980)

(8) D. J. Murphy, K. D. Mukherjee, and I. E. Woodrow, Functional association of a monoacylglycerophosphocholine acyltransferase and the oleoylglycerophosphocholine desaturase in microsomes from developing leaves, Europ. J. Biochem., 139 : 373 (1984)

(9) P. G. Roughan, and C. R. Slack, Cellular organization of glycerolipid metabolism, Ann. Rev. Plant Physiol., 33 : 97 (1982)

(10) A. K. Stobart, and S. Stymne, Regulation of the fatty acid composition of the triacylglycerols in microsomal preparations from avocado mesocarp and the developing cotyledons of safflower, Planta, 163 : 119 (1985)

(11) C. P. Rochester, and D. G. Bishop, The role of lysophosphatidylcholine in lipid synthesis by developing sunflower (Helianthus annuus L.) seed microsomes, Arch. Biochem. Biophys., 232 : 249 (1985)

(12) C. Demandre, A. Tremolieres, A. M. Justin and P. Mazliak, Oleate

desaturation in six phosphatidylcholine molecular species from potato tuber microsomes, Biochim. Biophys. Acta, (in the press) (1986)

(13) D. J. Murphy, I. E. Woodrow, and K. D. Murkherjee, Substrate specificities of the enzymes of the oleate desaturase system from photosynthetic tissue, Biochem. J., 225 : 267 (1985)

(14) X. V. Nguyen, and D. Côme, Opposite effects of temperature on breaking of dormancy and induction of frost resistance in apple embryos, Physiol. Plant., 62 : 79 (1984)

(15) X. V. Nguyen, D. Côme, S. Lewak, and P. Mazliak, Dormancy breaking and frost resistance induction in apple embryos as related to changes in reserve and polar lipids, Physiol. Plant., 62 : 566 (1984)

FACTORS AFFECTING THE FATTY ACID COMPOSITION OF PHOSPHATIDYL-GLYCEROL AS RELATED TO CHILLING SENSITIVITY IN HIGHER PLANTS

Steven R. Thomas, Juan Sánchez and J. Brian Mudd

ARCO Plant Cell Research Institute
6560 Trinity Court
Dublin, California 94568

INTRODUCTION

Plant biochemists and physiologists have recently been presented with a working model which may explain the mechanism for chilling sensitivity and resistance in higher plants. The model is based on the observation that membranes from chilling sensitive plants contain a significantly higher proportion of di-saturated molecular species of PG than do those from chilling resistant species (Murata, 1983). [For the purposes of this study "saturated" fatty acids are defined as 16:0, 18:0 and $16:1^{3t}$.] Exceptions to this general rule occur among the *Solanaceae* and the C4 grasses (Roughan, 1985).

Molecular species of PG, which are entirely synthesized in the plastid, always contain a "saturated" (16:0 or $16:1^{3t}$) fatty acid esterified at the sn-2 position of glycerol-3-phosphate. Therefore, the only variability in molecular species of PG of any significance to the chilling phenomenon occurs at the sn-1 position. These observations have focussed attention on the role of sn-glycerol-3-phosphate : acyl-ACP acyltransferase (E.C. 2.3.1.15; henceforth, "AT1") in determining the relative proportions of saturated and unsaturated fatty acids which are esterified at the sn-1 position of molecular species of PG.

Murata's hypothesis predicts that the AT1 from a chilling resistant species nearly always esterifies an unsaturated fatty acid (18:1) at the sn-1 position, and that the AT1 from a chilling sensitive species does so at a lower frequency, tending to neutrality in terms of acyl preference. Thus, if molecular species of PG are involved in chilling sensitivity and resistance, then AT1 is potentially capable of regulating the relative abundances of di-saturated and unsaturated molecular species of PG, and therefore of determining whether a plant is chilling sensitive or chilling resistant.

Other workers (Frentzen *et al.*, 1983) have previously shown that purified stromal AT1 activities from spinach and pea, two chilling resistant species, clearly prefer to use 18:1-ACP in preference to either 16:0- or 18:0-ACP in the synthesis of lysophosphatidic acid (LPA). We set out to see if partially purified stromal AT1 from chilling sensitive species exhibit properties which are consistent with Murata's hypothesis. We have compared the results from these experiments with the fatty acid composition of PG synthesized from ^{14}C-acetate by isolated chloroplasts.

MATERIALS AND METHODS

Chemicals. Palmitic acid, oleic acid, sn-glycerol-3-phosphate (G3P), BSA (fatty acid free), DTT, Trizma base, MOPS, PIPES and palmitoyl-CoA, were obtained from Sigma. [2-^{14}C] sodium acetate (56.8 mCi/mmol), [U-^{14}C]-sn-glycerol-3-phosphate (170 mCi/mmol), [1-^{14}C]palmitic acid (57.4 mCi/mmol) and [1-^{14}C]oleic acid (52.7 mCi/mmol) were obtained from Amersham. ACP-SH from was obtained from Calbiochem. *E. coli* B cell paste (late log phase) was purchased from Grain Processing, Inc. (Muscatine, IA) and aqueous counting scintillant (Aquamix) from Westchem (San Diego, CA). DEAE-Sepharose (fast flow), prepacked PD-10 desalting columns and octyl-Sepharose were purchased from Pharmacia. DE-52 anion exchange resin and precoated silica gel TLC plates (type K6) were purchased from Whatman.

Plants. For AT1 preparations plants were grown from seed in flats in a glasshouse. Tobacco (*N. tabacum* cv. Burley 21), cucumber (*C. sativus* cv. Marketmore), cantaloupe (*C. maxima* cv. Rocky Ford Green Flesh), spinach (*S. oleracea* L.cv. Hybrid 424), and pea (*P. sativum* cv. Laxton's Progress no. 9) leaves were harvested at 5, 3, 3 1/2, 6 and 2 weeks after sowing, respectively.

Preparation of partially purified stromal AT1. Flats of plants were placed in darkness for 1-2 days prior to harvesting leaves in order to deplete starch grains and therefore maximize the yield of intact chloroplasts. Enzymes were purified from 0.4 - 2 kg of expanding leaves from young plants, essentially as described previously by Bertrams and Heinz (1981), except that chloroplasts were isolated from leaves using the procedure of Nakatani and Barber (1977).

General assay for AT1 activity. The standard assay for acyl transferase activity was identical to that of Bertrams and Heinz (1981). Reactions were incubated for 30 minutes at 25°C. After removal of the upper phase from the extraction 0.8 ml of the organic phase was blown dry with air, redissolved in liquid scintillant and counted.

Preparation of labelled acyl-ACPs. [1-^{14}C] labelled and unlabelled acyl-ACPs were prepared by the procedure of Rock, *et al*. (1981), using acyl ACP synthetase freshly prepared from *E. coli* B cell paste as described by Rock and Cronan (1981).

Determination of kinetic constants. K_m and V_{max} determinations for the acyl-ACP substrates were performed in reactions which contained 0.25 M MOPS, pH 7.4, sn-G3P at 8 mM (3.05 mCi/mmol) and unlabelled acyl-ACP at various concentrations. No BSA was present in these reactions, which were carried out for 15 minutes at 25°C with appropriately diluted enzyme preparations. The enzyme diluent was 50% glycerol in 10 mM Tris (pH 7.8), 2.5 mM DTT. Reactions were terminated and processed as for the general AT1 assay.

Acyl substrate selectivity analysis. Acyl substrate competition assays employing various mixtures [1-^{14}C]acyl-ACPs contained 5 µM total ACPs, 0.25 M MOPS, pH 7.4, 2 mM sn-G3P, and no BSA. BSA has been omitted from the reaction mixes because the partially purified enzymes all behaved well in its absence and because BSA is not physiological in this system. These assays were carried out for 5 minutes at 25°C. After extraction into chloroform-methanol, the reaction products were dried under N_2, redissolved in chloroform-methanol (2:1) and separated by TLC as described by Frentzen *et al.* (1983). Radioactivity co-migrating with an LPA standard was detected using a Berthold radioactivity scanner, scraped from the plate, saponified, methylated and analyzed as described in the following section.

De novo synthesis of total fatty acids and fatty acids in PG from [1-^{14}C]acetate in isolated chloroplasts. Intact chloroplasts were isolated from expanding leaves and purified using a Percoll step gradient as described previously (Mudd and Dezacks, 1979). Incubations were carried out in the presence of 175 µM [2-^{14}C] sodium acetate, 33 mM Tricine-KOH, pH 7.9, 330 mM sorbitol, 2 mM $MgCl_2$, 0.2 mM K_2HPO_4, 0.5 mM DTT, 5 mM $NaHCO_3$, 2 mM ATP, 0.5 mM CoA, 2mM G3P, and chloroplasts equivalent to 80-300 µg of chlorophyll (chl), in a volume of 250 or 500 µl. Incubations were carried out for 60 minutes at 25°C, under illumination (200 µEs m^{-2} s^{-1}) with constant shaking. At the end of the incubation, lipids were separated, saponified, methylated and analyzed by radio - gas liquid chromatography.

RESULTS AND DISCUSSION

Estimation of AT1 Kinetic Constants

Estimations of enzyme kinetic constants have been used previously to explain the substrate selectivity exhibited by pea and spinach stromal AT1 (Frentzen, *et al.*, 1983). We have determined the acyl substrate affinity constants and maximal velocities for the enzymes isolated from chloroplasts of three chilling sensitive species and will rely upon previously published values for spinach and pea for our discussion. We have not determined K_m values for G3P in the presence of acyl-ACPs for any of these enzymes.

We have used 8 mM G3P in the determinations for the chilling sensitive species, which may not represent saturating levels for this substrate, but are certainly well above the K_m values for G3P for spinach and pea (Frentzen, *et al.*, 1983). Table 1 summarizes the kinetic parameters derived for the stromal acyltransferases of these plants as well as for spinach, and pea (Frentzen, *et al.*, 1983). K_m and V_{max} values were derived using a computer program (Williams, 1985), which employs the direct linear method of Eisenthal and Cornish-Bowden (1974). The values listed in Table 1 are high in comparison to those published previously for spinach and pea (Frentzen, *et al.*, 1983) and for squash (Nishida, *et al.*, this volume). For the purposes of this discussion the ratios between pairs of K_m values and V_{max} values for any one enzyme preparation are more important than are the absolute values of those parameters. Under different assay conditions these ratios may be more repeatable than the absolute values for the kinetic constants (Frentzen *et al.*, 1983).

As might have been expected, the pattern for K_ms is reversed for both cucumber and cantaloupe AT1, in comparison with spinach and pea (Table 1). These extremely chilling sensitive species contain an AT1 whose K_m for 16:0-ACP is lower than that for 18:1-ACP, indicating a higher affinity for the former. Interestingly enough, however, the pattern is also reversed for the V_{max} parameter. The V_{max} in the presence of 18:1-ACP is greater, in the case of both enzymes, than that in the presence of 16:0-ACP. Thus, while the substrate affinity of any of the AT1s, except tobacco, is higher for the appropriate acyl substrate, the V_{max} is significantly less in the presence of that substrate. This can be visualized most easily as the ratio between the two K_m parameters or the two V_{max} parameters (Table 1). This apparent paradox is resolved in the case of spinach by the fact that the K_m for G3P is 100 fold lower in the presence of 18:1-ACP (compared with 16:0-ACP, Frentzen, *et al.*, 1983). This fact virtually guarantees that the spinach AT1 will preferentially utilize 18:1-ACP in the synthesis of LPA under physiological conditions. In contrast, the situation in the case of squash is further confounded by the fact that the K_m for G3P is again 75 fold less in the presence of 18:1-ACP (compared with 16:0-ACP; Nishida, *et al.*, this volume).

Table 1

Summary of Kinetic Parameters of Stromal AT1 from Various Plant Species

Plant	Acyl-ACP				Ratio 16:0/18:1	
	16:0-ACP		18:1-ACP			
	K_m (μM)	V_{max}**	K_m (μM)	V_{max}**	K_m	V_{max}
Spinach*	3.2	5.3	0.3	1.5	10.7	3.53
Pea*	5.6	6.5	0.7	0.6	8.0	10.8
Tobacco	7.2	36.8	7.2	33.6	1.0	1.09
Cucumber	6.3	28.0	20.2	60.2	0.31	0.46
Cantaloupe	16.7	58.3	25.4	78.5	0.65	0.74

TABLE 1. The values for V_{max} and K_m are comparable for any one enzyme, but are not comparable between any two enzymes, due to differences in specific activities between enzyme preparations.
* Values cited are those of Frentzen, et al., 1983.
** Units are nkat/mg for data of Frentzen, et al., (1983), pmol/min for data reported here.

From the values in Table 1, values from the literature regarding G3P concentrations in chloroplasts (0.1-0.2 mM; Sauer and Heise, 1983) and acyl-ACP concentrations (Table 5; Soll and Roughan, 1982) one can easily see that the spinach AT1 should preferentially incorporate 18:1 from a mixture. In fact, it does so, when faced with the choice in selectivity assays (Frentzen et al., 1983). Using the appropriate kinetic parameters and the same values for concentrations of G3P and acyl-ACPs as approximations in the cucumber and cantaloupe systems, it is not so easy to predict what these enzymes will do when faced with a mixture of acyl-ACP substrates. The tobacco enzyme, on the other hand, presents virtually indistinguishable values for K_m and V_{max}. On the basis of the information at hand, this enzyme provides no evidence for potential substrate selectivity. This would suggest that any substrate selectivity exhibited by the tobacco AT1 might be governed entirely by its K_m for G3P and the concentration of G3P. The selectivity assays described in the following section decide the issue.

We can now group the AT1 from these five species into three categories based on their kinetic parameters:
1. spinach and pea, two chilling resistant species, which display an 8-10 fold lower K_m for 18:1 and 3-10 fold higher V_{max} in the presence of 16:0-ACP (Frentzen, et al., 1983);
2. cucumber and cantaloupe, two extremely chilling sensitive species, which exhibiting a 2-3-fold lower K_m for 16:0 and an approximately two-fold higher V_{max} in the presence of 18:1-ACP; and
3. tobacco, a borderline chilling sensitive species, which displays indistinguishable K_m and V_{max} values, indicative of neutrality in substrate preference.

The pattern which emerges here on inspection of the kinetic parameters is complex. There is a good correlation between the acyl substrate K_m values and chilling sensitivity; the sensitive plants clearly show a greater affinity for 16:0-ACP than for 18:1-ACP, and vice-versa for the chilling sensitive species, with tobacco displaying neutral characteristics. In contrast to the pattern provided by the K_m data, the V_{max} values comprise a pattern of the opposite polarity. Under conditions of saturating substrate concentrations the resistant plants apparently operate much faster with 16:0-ACP and the sensitive plants apparently operate somewhat faster using 18:1-ACP.

Estimation of AT1 acyl-ACP selectivity using competition assays

A logical extension which emerges from a consideration of Murata's hypothesis is that AT1s from chilling sensitive chloroplasts should be less selective in their preference for 18:1-ACP, but not to the extent that they actually display a selectivity for 16:0-ACP. This conclusion is reached from the realization that while di-saturated species of PG are present in higher proportions in chilling sensitive plants than in chilling resistant plants, no plant species much exceeds the value of 75%. This maximum in di-saturated molecular species describes an AT1 which is neutral in its acyl substrate selectivity. These expectations are fulfilled.

In these experiments we chose to use a concentration of G3P of 2 mM, representing higher than physiological concentrations which should tend to favor incorporation of 16:0-ACP into LPA (Frentzen, et al., 1983). Figure 1 illustrates the results of competition experiments which were carried out in order to determine the relative acyl substrate preference of AT1 from chilling sensitive and chilling resistant species when confronted with various mixtures of 16:0- and 18:1-ACP. As previously reported by other workers (Frentzen, et al., 1983), we find that the spinach enzyme quite efficiently selects 18:1-ACP in preference to 16:0-ACP from a mixture containing any of several combinations of the two. In agreement with Frentzen, et al. (1983), we also find that the pea enzyme is less selective for the unsaturated acyl substrate than is the spinach AT1.

Under the conditions of these assays, the AT1 from both cucumber and cantaloupe are less selective than the enzyme from spinach, but not neutral. Surprisingly, they behave very much like the pea enzyme in this assay. This fact seems to argue against Murata's hypothesis that AT1 is involved in determining relative proportions of PG molecular species. To confuse the issue further, the tobacco AT1 is only slightly less selective than the spinach enzyme. However, since several solanaceous species have recently been recognized as exceptions to the correlation between molecular species of PG and chilling sensitivity (Roughan, 1985), it is not clear whether data relative to tobacco AT1 can be considered as pertinent to the problem of chilling sensitivity.

Inspection of the data for all of the enzymes in Figure 1 shows that the distribution of unsaturated fatty acid in the LPA product is altered by the relative pool sizes of acyl-ACPs. Therefore, at constant concentrations of total acyl-ACP and G3P, any changes acyl-ACP pool sizes in chloroplasts could alter the proportions of LPA molecular species synthesized. Consequently, we must assume that the chloroplast is capable of extremely tight regulation of the acyl-ACP pool sizes if AT1 is to control the molecular species distribution of PG.

We conclude from the selectivity experiments that:
1) all of the AT1s, to a greater or lesser extent, preferentially select 18:1-ACP from a mixture of 18:1 and 16:0-ACP;
2) the behavior of the tobacco and pea AT1 show that there is a poor correlation between degree of substrate selectivity and chilling resistance;
3) because of the above, and the possible effect which acyl-ACP pool sizes could have on the degree of AT1 selectivity, AT1 cannot by itself provide the entire explanation for chilling sensitivity and resistance in higher plants.

In vitro labelling of PG from newly synthesized fatty acids

Data from other laboratories indicate that leaf PG from different plant species is composed of characteristic proportions of molecular species (Murata, et al., 1982; Murata, 1983; Kenrick and Bishop, 1986; Roughan, 1985). We asked the question whether or not isolated chloroplasts are capable of synthesizing molecular species of PG in proportions similar to those found in vivo.

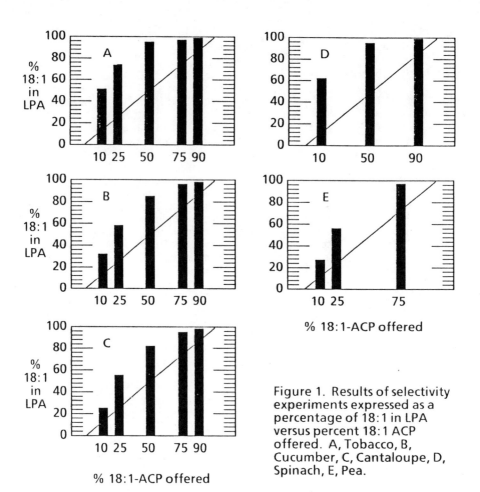

Figure 1. Results of selectivity experiments expressed as a percentage of 18:1 in LPA versus percent 18:1 ACP offered. A, Tobacco, B, Cucumber, C, Cantaloupe, D, Spinach, E, Pea.

Fatty acid analysis of total fatty acids synthesized *in vitro* by chloroplasts (Table 2) reveals the fact that, in all cases, more label is incorporated into unsaturated fatty acids (16:1 9c, 18:1, 18:2, 18:3 and 20:1) than into saturated fatty acids (16:0 and 18:0). The ratio of saturated to unsaturated fatty acids in all lipid classes varies with the plant (Table 2) and with incubation conditions (unpublished data).

The pattern of total synthesis is closely reflected by the acyl-CoA fraction (data not shown). However, when compared with total fatty acid synthesis, PA, DG and PG are always relatively enriched in saturated fatty acids (Table 2). Although we have not actually performed molecular species analysis on the products of the chloroplast incubations, most, but not all, of this enrichment for saturated fatty acids can be explained by the apparent absolute specificity of the plastid 1-acyl-sn-glycerol-3-phosphate:acyl-ACP acyltransferase (henceforth, "AT2") for palmitoyl-ACP. Given that chloroplast AT2 always incorporates palmitate at the sn-2 position of PA, then the first 50% of the 16:0 observed in PG, and in other lipid classes, in these experiments is probably esterified at the sn-2 position. When 16:0 is present at less than 50% we have supplemented the other fatty acids at sn-2, first with 14:0, then 18:0 and lastly with 18:1 to make up the difference. Any remaining 16:0 must be esterified at the sn-1 position. By this reasoning, the percentage of di-saturated molecular species of PG can be estimated from the data in Table 2. These manipulations produce numbers for molecular species of PG which compare favorably with data taken from the literature (Table 3). Therefore, isolated, intact chloroplasts are able to synthesize molecular species of PG in proportions approximating those found *in vivo*.

Assuming a lack of selectivity for molecular species by enzymes downstream from AT1 in the pathway to PG, for which there is no direct evidence in plants, and next that the acyl-ACP pools in these chloroplasts are nearly equally represented as they occur in spinach chloroplasts (see Table 3), then we can inspect the distributions of saturated and unsaturated molecular species of PA, DG and PG derived from data in Table 2 as described above, for evidence as to which chloroplastic enzymes may actually control the proportions of molecular species of PG *in vivo*. These values will be compared with the results from the competition assays using partially purified stromal AT1. In particular, we can get an idea of the importance of substrate selectivity of AT1 in this process.

Spinach. An efficient mechanism exists for ensuring that the PA pool is predominantly populated by molecular species containing an unsaturated fatty acid at the sn-1 position. This is most likely accomplished through the substrate selectively exhibited by AT1 (see above; Frentzen, et al., 1983). Considering the low levels of di-saturated species of PA, DG and PG present (Table 3), it is not possible to state conclusively that any mechanism exists for the enrichment of di-saturated species in PG and DG pools relative to those present in PA.

Pea. These chloroplasts also possess a mechanism for producing unsaturated molecular species of PA, but do so to the exclusion of di-saturated species. However, AT1 from pea, while selecting 18:1-ACP from a mixture in favor of 16:0 ACP, does not display nearly the selectivity of the spinach and tobacco enzymes (Figure 1). These facts hint that AT2 may possibly select for 18:1-LPA in the synthesis of PA. From the molecular species composition of DG and PG (Table 3), it is obvious that a small amount of di-saturated species of PA must exist in these chloroplasts. There is a marked enrichment for di-saturated molecular species at the level of the headgroup enzymes in these chloroplasts.

Tobacco. In selectivity assays tobacco AT1 discriminates aginast 16:0 in the synthesis of LPA (see Figure 1), and this bias is reflected here in the PA pool of chloroplasts (Table 3). In direct contrast to this obvious selectivity by AT1 (and, again, possibly AT2) is the observation that the newly synthesized PG and DG pools from isolated tobacco chloroplasts are relatively enriched in di-saturated species in comparison with the PA pool. This indicates to us that the bias of AT1 for 18:1 at sn-1 is partially overridden by a preference for di-saturated species by the headgroup enzymes.

Cucumber. The situation in cucumber chloroplasts is similar to that in tobacco chloroplasts (Table 3). The upstream selectivity for 18:1 at the sn-1 position of PA is again partially overcome by a selectivity for di-saturated species by the headgroup enzymes. The molecular species composition of the PA pool would indicate a prior selection for either 18:1-ACP at AT1, and/or for 18:1-LPA by AT2. We cannot discriminate between these possibilities from the chloroplast experiments, but the competition experiments again show that cucumber AT1 selects for 18:1-ACP in mixtures of acyl substrates.

In support of the assumption that the data presented here are not biased in favor of incorporation of 16:0-ACP by the action of AT2 utilizing pre-existing LPA is the evidence which shows that LPA pool sizes are vanishingly small in chloroplasts (Gardiner, et al., 1984). Our data show that isolated intact chloroplasts are able to mimic the *in vivo* pattern of PG biosynthesis fairly faithfully, and therefore that there exists a set of self-contained mechanisms for the maintenance of steady state levels of molecular species of PG in the chloroplast. These mechanisms may perhaps vary from plant to plant, but it is clear solely from the chloroplast data that they exceed the influence of AT1 alone and are probably localized after the point of PA synthesis in chloroplasts. The headgroup enzymes

Table 2

Fatty Acid Composition of ^{14}C-Acetate Labelled Lipid Classes

Plant species	Lipid class	% total label incorp.	Fatty Acid Distribution (%)					% Sat. FA	Ratio Sat./Unsat.
			14:0	16:0	16:1*	18:0	Unsat. C_{18} & C_{20}		
Spinach	Total FAs	100	-	27	-	4	69	31	0.45
	PA	1	-	43	-	8	49	51	1.04
	DG	38	-	51	-	3	35	54	1.17
	PG	4	-	53	-	4	43	57	1.32
Pea	Total FAs	100	t	22	-	6	72	28	0.39
	PA	2	-	38	-	8	54	46	0.85
	DG	22	4	41	-	14	41	59	1.44
	PG	7	t	44	-	8	48	52	1.08
Tobacco	Total FAs	100	-	46	-	1	52	47	0.88
	PA	t	-	63	-	-	37	63	1.70
	DG	28	-	75	-	-	25	75	3.0
	PG	2	-	75	-	-	25	75	3.0
Cucumber	Total FAs	100	2	39	3	5	50	46	0.85
	PA	2	-	58	4	5	33	63	1.70
	DG	16	9	62	3	2	24	73	2.70
	PG	9	-	75	8	6	11	81	4.26

-, not detected;   t, trace = <1%;   *, $16:1^{9t}$

Table 3

Distribution of Molecular Species, Frequency of Occurence at sn-1

Plant Species	Fatty acid (at sn-1)	Lipid Class			
		ACYL-ACPs*	PA	DG	PG
Spinach	14:0	30	0	0	0
	16:0	32	0	2	6
	18:0	11	2	6	8
	18:1	28	98	90	86
Pea	14:0	30	0	0	0
	16:0	21	0	0	0
	18:0	11	0	18	4
	18:1	28	100	82	96
Tobacco	14:0	30	0	0	0
	16:0	32	26	50	50
	18:0	11	0	0	0
	18:1	28	74	50	50
Cucumber	14:0	30	0	18	0
	16:0	32	16	24	50
	18:0	11	10	4	12
	18:1	28	74	54	38

*data for spinach acyl-ACP pool sizes from Soll and Roughan (1982) are substituted for pea, tobacco and cucumber.

are positioned at a branch point in the biosynthetic pathway where they could potentially regulate the molecular species composition of membrane lipids. The evidence presented here argues that these latter enzymes do exercise a large degree of control over the distribution of molecular species into PG, even to the extent that the bias introduced into the pathway by AT1 (in the case of cucumber and tobacco and pea) is actually partially reversed by them.

REFERENCES

Bertrams, M., and Heinz, E., 1981, Positional specificity and fatty acid selectivity of purified sn-glycerol 3-phosphate acyltransferases from chloroplasts, Pl. Physiol. 68:653-7.

Eisenthal, R., and Cornish-Bowden, A., 1974, The direct linear plot. A new graphical procedure for estimating enzyme kinetic parameters, Biochem. J. 139:715-20.

Frentzen, M., Heinz, E., McKeon, T.A., Stumpf, P.K., 1983, Specificities and selectivities of glycerol-3-phosphate acyltransferase and monoacylglycerol-3-phosphate acyltransferase from pea and spinach chloroplasts, Eur. J. Biochem. 129:629-36.

Gardiner, S.E., Roughan, P.G., Browse, J., 1984, Glycerolipid labelling kinetics in isolated intact chloroplasts, Biochem. J. 224:637-43

Kenrick, J.R., and Bishop, D.G., 1986, The fatty acid composition of phosphatidylglycerol and sulfoquinovosyldiacylglycerol of higher plants in relation to chilling sensitivity, Pl. Physiol. 81:946-9.

Mudd, J.B., and Dezacks, R., 1981, Synthesis of phosphatidylglycerol by chloroplasts from leaves of *Spinacea oleracea* L. (spinach), Arch. Biochem. Biophys. 209:584-91.

Murata, N., 1983, Molecular species composition of phosphatidylglycerols from chilling-sensitive and chilling-resistant plants, Pl. Cell Physiol. 24:81-6.

Murata, N., Sato, N., Takahashi, N., Hamazaki, Y., 1982, Compositions and positional distributions of fatty acids in phospholipids from leaves of chilling-sensitive and chilling-resistant plants, Pl. Cell Physiol. 23:1071-9.

Nakatani, H.Y., and Barber, J., 1977, An improved method for isolating chloroplasts retaining their outer membranes, Biochim. Biophys. Acta 461:510-2.

Rock, C.O., Garwin, J.L., and Cronan, J.E., Jr., 1981, Preparative enzymatic synthesis of acyl-acyl carrier protein, Meth. Enzymol. 72:397-403

Rock, C.O., and Cronan, J.E., Jr., 1981, Acyl-Acyl carrier protein synthetase from *Escherichia coli*, Meth. Enzymol. 71:163-8

Roughan, P.G., 1985, Phosphatidylglycerol and Chilling Sensitivity in Plants, Pl. Physiol. 77:740-6.

Roughan, P.G., in press, Acyl lipid synthesis by chloroplasts isolated from the chilling-sensitive plant *Amaranthus lividus* L.

Sauer, A, and Heise, K.-P., 1983, The influence of the glycerol-3-phosphate level in the stroma space on lipid synthesis of intact chloroplasts, Z. Naturforsch. 38c:399-404.

Soll, J., and Roughan, G. 1982, Acyl-acyl carrier protein pool sizes during steady-state fatty acid synthesis by isolated spinach chloroplasts, FEBS Lett. 146:189-92.

Williams, P.A., 1985, ENZPACK (computer software and manual), Elsevier Science Publishers, BV, Amsterdam.

ON THE SYNTHESIS OF DIGALACTOSYLDIACYLGLYCEROL IN CHLOROPLASTS, AND ITS RELATION TO MONOGALACTOLIPID SYNTHESIS

Jef F.G.M. Wintermans and Johan W.M. Heemskerk

Botanisch Laboratorium, University of Nijmegen, Toernooiveld 6525 ED Nijmegen (The Netherlands)

About a quarter century after the first observation that galactolipids are specific membrane lipids of chloroplasts (1), it seems appropriate to make a short survey of the knowledge gained about their biosynthesis. In contrast to other papers (2-5) we shall emphasize on the synthesis of digalactosyldiacylglycerol (DGDG*), which has been a subject of much confusion.

SYNTHESIS OF GALACTOLIPIDS IN CHLOROPLAST ENVELOPE MEMBRANES

A major breakthrough in elucidating the role of the chloroplast envelope membranes in galactolipid synthesis was made by Douce et al. (6,7), who first isolated these membranes. Now it is well established that the chloroplast envelope is the site of *de novo* synthesis of diacylglycerol, and also of the monogalactosyldiacylglycerol (MGDG) forming enzyme UDPGal : diacylglycerol galactosyltransferase (UDGT) (Table 1, reaction *(i)*) (8,9). This diacylglycerol made in the chloroplasts is characterized by C16 fatty acids at sn-2 (10). Such a configuration is typically found in *Cyanobacteria*, and therefore is called prokaryotic.

DGDG arises by galactosylation of MGDG. The stereochemical specificity of the inner β-D-galactopyranosyl group and the outer α-D-galactopyranosyl group points to a different enzyme for the second galactosylation step. This was already denoted by Ongun and Mudd (11), who found the DGDG synthetase activity of spinach chloroplasts to be more soluble than the MGDG forming activity. Since most chloroplast preparations make both MGDG and DGDG when supplied with galactose-labeled UDPGal, it was attractive to assume an analogous reaction for both enzymes. Such an UDPGal : MGDG galactosyltransferase activity *(ii)* has been reported by Siebertz and Heinz (12) in soluble preparations of pea seedlings, although rates were quite modest compared to rates of galactolipid synthesis in envelope membranes (9,13).

On the other hand it has been established that envelope membranes can transform MGDG into DGDG in the absence of UDPGal by action of a galactolipid : galactolipid galactosyltransferase (GGGT) (13,14). During the galactosyl-transfer reaction diacylglycerol is liberated simultaneously with DGDG production *(iii)*. Tri- (TGDG) and

---

* Abbreviations are defined in Table 1.

Table 1. Enzymatic reactions of the galactolipid metabolism of spinach chloroplast envelope membranes.

No.	Reaction	Enzyme
(i)	UDPGal + diacylglycerol → MGDG[a] + UDP	UDGT
(ii)	UDPGal + MGDG → DGDG + UDP	?
(iii)	MGDG + MGDG ↔ DGDG + diacylglycerol	GGGT
(iv)	DGDG + MGDG ↔ TGDG + diacylglycerol	GGGT
(v)	TGDG + MGDG ↔ TeGDG + diacylglycerol	GGGT
(vi)	MGDG + MGDG → 6-O-acyl-MGDG + lyso-MGDG	GGAT
(vii)	DGDG + MGDG → 6-O-acyl-DGDG + lyso-MGDG	GGAT

[a]Abbreviations: GGAT, galactolipid : galactolipid acyltransferase; GGGT, galactolipid : galactolipid galactosyltransferase; MGDG, DGDG, TGDG and TeGDG, mono-, di-, tri-, and tetragalactosyldiacylglycerol; PC and PG, phosphatidylcholine and -glycerol; UDGT, UDPGal : diacylglycerol galactosyltransferase.

tetragalactosyldiacylglycerol (TeGDG) can be formed in a similar reaction *(iv-v)*. GGGT is strongly activated when chloroplasts are broken and the envelope membranes are isolated. Hence, early reports of low MGDG content in envelope membranes, and high contents of diacylglycerol, TGDG and TeGDG (15,16) can be ascribed to this activation. By treating chloroplasts with the proteinase thermolysin, which inactivates GGGT, before isolation of the envelope membranes Dorne et al. (17) established that endogenous contents of diacylglycerol, TGDG and TeGDG are very low. The effect of thermolysin-treatment on the lipid composition of envelope membranes is illustrated in Fig. 1.

Some properties and characteristics of UDGT and GGGT have been determined in our laboratory using separate assay methods for both galactosyltransferases (14,19,20). Such specific assays were desirable, since - in isolated envelopes - MGDG, the product of UDGT, is a substrate for GGGT and inversely the diacylglycerol produced by GGGT can be a substrate for UDGT (13,14). Among the most conspicuous properties of the spinach enzymes were the ready galactosylation of especially saturated and oligoene diacylglycerol species by UDGT, and the high galactosylation rate of polyene MGDG, compared to oligoene/satured MGDG, in the GGGT assay. GGGT is activated by a number of mono- and divalent cations, whereas UDGT is only moderately stimulated by Mg^{2+} and Mn^{2+} (20). The remarkable sensitivity of GGGT to ionic conditions is understandable by the influence of cations on the physical state of the envelope lipids. Especially the observed cation-induced phase separations of MGDG (41) may be relevant. In the UDGT assay the presence of UDPGal did not influence the rate of transformation of MGDG into DGDG, which means that we were unable to measure the UDPGal : MGDG galactosyltransferase reaction *(ii)* (14).

Following the fate of di[^{14}C]acylglycerol incorporated into envelope membranes, we always observed that some label appeared in MGDG in the absence of UDPGal (20,28). This agrees with other experiments in which chloroplasts were labeled with acetate (29). The UDPGal-independent MGDG formation could be prevented by previous treatment of the chloroplasts with thermolysin. The MGDG labeling increased in proportion to added diacylglycerol up to very high concentrations (>3000 nmol/mg envelope protein), and was not affected by the presence of UDP. Since UDGT is affected by UDP, is not affected by thermolysin, and has a high affinity for diacylglycerol, our earlier explanation of this phenomenon involving UDGT activity (28) may not be correct. The most parsimonious explanation is a reverse action of GGGT *(iii)*, although a separate, specific enzyme cannot be excluded yet. There are no thermodymanic nor stereochimical objections to such reversibility, and we will point to a possible physiological relevance lateron. Other experimental indications for a reversibility of GGGT are the conversion reactions in envelope membranes of incorporated DGDG into MGDG (14), and of incorporated TeGDG into TGDG, DGDG and MGDG (Fig. 2).

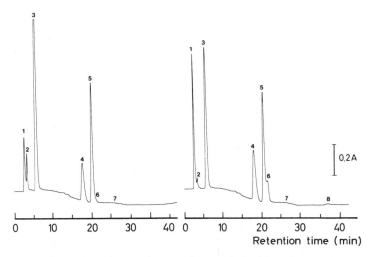

Fig. 1. Lipid pattern of spinach chloroplast envelope membranes, obtained from thermolysin-treated (left) and untreated (right) chloroplasts. Chromatograms are shown obtained by HPLC analysis; only non-acidic lipids were eluted from the aminopropyl-modified silica column (18). Peak labels: 1, carotenoids and diacylglycerol; 2, unknown; 3, MGDG; 4, PC; 5-6, DGDG; 7, PG and sulfolipid (overload); 8, TGDG. Note the small peak 1, the high MGDG peak, and the absence of TGDG in the thermolysin-treated membranes.

Envelope membranes contain another enzyme catalyzing transfer reactions between galactolipids, galactolipid : galactolipid acyltransferase (GGAT) (21). The enzyme, first seen by Heinz (22), produces from two galactolipid molecules a lyso-galactolipid (galactosyl monoacylglycerol) and an acylated galactolipid (6-O-acyl galactosyl diacylglycerol) *(vi-vii)*, which probably are no natural constituents of the chloroplast membranes (15). Activity at pH 6 has been reported so far, but we recently noticed high activities at pH 7.2 in the presence of Fe^{2+} ions (19). Up to the present only these three enzymatic activities with a galactolipid as a substrate or a product have been defined clearly in envelope membranes: UDGT, GGGT and GGAT.

Table 2. Localization of enzymes involved in galactolipid metabolism in the chloroplast envelope membranes.

Enzyme	Plant species	Outer envelope membrane	Inner envelope membrane	Ref(s).
UDPGal : diacylglycerol galactosyltransferase *(i)*	spinach		+[a]	23,24
	pea	+		25
Galactolipid : galactolipid galactosyltransferase *(iii-v)*	spinach	+[b]		17,26
	pea	+		25
Galactolipid : galactolipid acyltransferase *(vi)*	spinach	+[a]		26

[a]Not inactivated by thermolysin; [b]inactivated by thermolysin, after treatment of the intact chloroplasts.

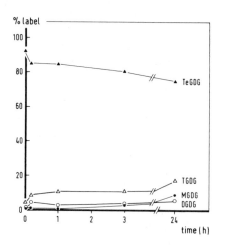

Fig. 2. Conversion of TeGDG incorporated into spinach envelope membranes. [Galactose-^{14}C]TeGDG was solubilized and incorporated into envelope membranes as described elsewhere (14). Label conversion was followed after the addition of 10 mM $MgCl_2$.

Recently, various studies have been performed to localize the galactolipid synthetizing enzymes in the outer or inner envelope membrane of the chloroplast. A survey is given in Table 2. Unfortunately, localization studies have been limited to only two plant species, spinach (a 16:3-plant) and pea (a 18:3-plant), so that the reported different location of UDGT in the outer membrane of pea and in the inner membrane of spinach chloroplasts may be a consequence of the specific physiology of both types of plants. Some remarks should be made about the localization of the spinach GGGT, which encountered difficulties. Although the enzyme was fully inactivated by treating chloroplasts with the non-penetrating proteinase thermolysin (17), in our first studies we measured highest conversion rate of MGDG in fractions rich in inner envelope membranes (24). The discrepancy could be resolved when the unavoidable cross-contamination of the isolated inner and outer membrane fractions, and the high dependence of GGGT on its substrate concentration were taken into account. MGDG concentration was found very low in fractions containing mostly outer membranes, and hence these fractions were low in GGGT activity. However, when the enriched inner and outer membranes were tested at high, comparable MGDG concentrations highest GGGT activity was measured in the outer membranes, proving that GGGT is localized there (26).

The idea that UDGT and GGGT could function *in vivo* in mutual regulation, and control the MGDG/DGDG ratio in the chloroplast, seems not *a priori* supported by their localization in distinguishable - though proximate - membranes. Therefore, co-operativity of UDGT and GGGT in galactolipid synthesis was studied with intact chloroplasts.

GALACTOLIPID SYNTHESIS IN CHLOROPLASTS

Intact spinach chloroplasts incorporate galactose from UDP[^{14}C]Gal into MGDG, DGDG, TGDG and TeGDG. Similarly as in isolated envelope membranes, UDGT and GGGT co-operate to optimalize galactolipid synthesis. As is described elsewhere (27), specific effectors of GGGT can regulate the UDPGal incorporation. Table 3 indicates that a number of cations stimulate both total galactolipid synthesis and GGGT activity in the chloroplasts. It may be recalled that UDGT itself is moderately stimulated only by Mg^{2+} and Mn^{2+}. Apparently, in the sequential operation of UDGT and GGGT, the rate limiting factor is diacylglycerol, which is being continuously regenerated by GGGT (27). Indeed, when the chloroplasts where first treated with thermolysin, galactolipid synthesis was strongly inhibited (27), although UDGT was not inactivated (unpublished results and Ref. (17)). As with isolated envelope membranes, we could not observe UDP-Gal : MGDG galactosyltransferase activity in such experiments (27). So far, we can

Table 3. Influence of added cations on galactolipid synthesis by spinach chloroplasts. Intact spinach chloroplasts were incubated with UDP[^{14}C]Gal and indicated cations. GGGT, incorporation of label into DGDG+TGDG+TeGDG.

Cation added	UDPGal incorporation	GGGT
(mM)	(nmol/mg chlorophyll)	
$MgCl_2$ (10)	69	22
$BaCl_2$ (10)	70	30
$CaCl_2$ (10)	68	22
$FeCl_2$ (10)	20[a]	5
$MnCl_2$ (10)	11	5
No cation added	3	0.5
$CoCl_2$ (10)	2[a]	0.4
$ZnCl_2$ (10)	0.2	0
NaCl (50)	10	2
NH_4Cl (50)	7	1.5
KCl (50)	4	0.4

[a]Addition of $FeCl_2$ and $CoCl_2$ resulted in 6-O-acyl-MGDG synthesis of, respectively, 5 and 0.1 nmol/mg chlorophyll.

resume that in chloroplasts, in the absence of an alternative source of diacylglycerol, cooperative action of UDGT and GGGT results in an accumulation and redistribution of galactosyl moieties, taken from UDPGal via MGDG into DGDG and higher homologues.

An interesting, but still unresolved, question arises now, how the diacylglycerol content is kept low in intact chloroplasts, *i.e.*, by what means activation of GGGT is prevented in the absence of UDPGal. Apparently, the inactivation mechanism is distorted and GGGT becomes active upon rupturing of the chloroplasts, so that diacylglycerol, TGDG and TeGDG may accumulate in the isolated envelope membranes, independent on UDGT activity. It is conceivable that the low affinity of GGGT for MGDG and its vectorial orientation inside the intact chloroplast contribute to prevent GGGT activation (26).

Chloroplasts synthetize diacylglycerol *de novo* from the precursors acetate and *sn*-glycerol-3-phosphate, in the prokaryotic 18:1/16:0 (*sn-1*/*sn-2*) fatty acid composition (10,29,30). Heemskerk et al. (24) labeled spinach chloroplasts with UDP[^{3}H]Gal and [^{14}C]acetate, under conditions permitting *de novo* diacylglycerol synthesis. Whereas the galactose-label was found in the galactolipids MGDG, DGDG and T(e)GDG, the fatty acid label was mostly restricted to MGDG. This agrees with earlier results of Renkonen and Bloch (31) with *Euglena*. The unequal galactosylation of the two MGDG labels can be understood now from a preference of GGGT for the hexaene [^{3}H]MGDG species (produced by the UDGT-GGGT couple) above the *de novo* oligoene [^{14}C]MGDG.

GALACTOLIPID : GALACTOLIPID GALACTOSYLTRANSFERASE IN THE GALACTOLIPID METABOLISM *IN VIVO*

So far discriminating research towards the biochemical mechanism of DGDG synthesis, either by dismutation of galactolipids *(iii)* or by galactosyl transfer from UDP-Gal to MGDG *(ii)*, has been performed mainly with spinach chloroplast membranes. As described above, GGGT is an active enzyme in spinach, where the alternative mode was below the level of detection. Activity of GGGT now has also been reported in pea envelope membranes (25), and in chromoplast membranes of *Narcissus pseudonarcissus* (32). Unfortunately, extensive studies with solubilized enzymes with high activities are not available yet. It is important to extend further research to DGDG synthesis in other plant species and to more purified enzyme preparations. Here we shall discuss data from literature in the light of our own experiments, since the functioning of GGGT in synthesis of

Table 4. Synthesis of TGDG and TeGDG in spinach chloroplasts in dependence of ionic conditions. Intact chloroplasts were incubated with UDP[^{14}C]Gal and various concentrations of MgCl$_2$ and pyrophosphate.

Mg^{2+}	PP_i	UDPGal incorp.	MGDG	DGDG	TGDG+TeGDG	(TGDG+TeGDG)/DGDG
(mM)		(nmol/mg chl/h)	(% of label)			(ratio)
0	0	11.2	89	10	1	0.08
	5	18.2	83	15	2	0.12
	10	21.6	83	15	2	0.14
5	0	180.9	66	18	16	0.91
	5	42.4	76	18	6	0.35
	10	25.8	82	17	2	0.15
10	0	173.8	74	11	15	1.32
	5	78.0	73	17	10	0.56
	10	36.8	77	18	5	0.26

DGDG *in vivo* raises some questions.

Douce and co-workers (5,17) have pointed already to the high synthesis of TGDG and TeGDG in chloroplast envelope membranes. Although virtually absent in membranes in their natural lipid composition, these lipids may even become major products after longer incubations with UDP[^{14}C]Gal/MgCl$_2$ (13,14). Taking into account also the rather low pH optimum of GGGT and its peripheral localization in the chloroplast (GGGT is accessible to thermolysin), the authors suggest involvement in catabolism of galactolipids, rather than in anabolism. We suppose that synthesis of TGDG and TeGDG is mainly due to an artificial hyperactivation of GGGT by high levels of unchelated divalent cations. For instance, Table 4 illustrates that the addition of a chelator to chloroplasts synthetizing galactolipids lowers the synthesis of TGDG and TeGDG more than that of other galactolipids. As to the peripheral localization of GGGT, it can be remarked that thermolysin-accessibility does not necessarily imply that it is the active site of the enzyme which is oriented towards the cytosol.

Chloroplast envelope membranes from the unicellular algae *Euglena gracilis* (33) and *Chlamydomonas reinhardii* (34) incorporated UDP[^{14}C]Gal, although at relatively low rates, mostly into DGDG, with little incorporation into MGDG and no TGDG synthesis. This suggests direct reactions of UDPGal both with diacylglycerol and MGDG. However, in the case of *Euglena*, pulse label experiments did not indicate a direct exchange of galactose from UDPGal to MGDG. Taken together also the K_m values reported for MGDG and DGDG synthesis, the pH dependence of galactolipid synthesis, and the low diacylglycerol content in the *Euglena* envelope membranes, it is not unlikely that the UDGT-GGGT couple also is active in *Euglena*. High DGDG synthesis then is explained by a high turnover of the synthetized MGDG, possibly induced by the scarcity of diacylglycerol. A similar reasoning can be valid for explaining a relatively high ratio of labeled DGDG/MGDG, produced from UDP[^{14}C]Gal, in membranes isolated from chromoplasts of *Narcissus pseudonarcissus* (32,35).

Synthesis of DGDG from UDPGal and MGDG has been reported with soluble cell preparations of spinach (11) and pea (12), although rates for pea did not exceed 8 nmol/mg protein/h. It is conceivable that those preparations are similarly low in diacylglycerol, but also relatively enriched in outer envelope membranes, and therefore enriched in GGGT. Newly formed MGDG may be readily acted upon by the excess GGGT and donate galactose to another MGDG molecule. An argument in favor of GGGT in the spinach enzyme preparation (11) is that inclusion of MGDG resulted in the production of DGDG and TGDG, and that inclusion of DGDG resulted in formation of TGDG (11). It may be noted that GGGT was not known at the time of the authors' experiments.

In vivo labeling experiments in which leaves of various plants were fed with $^{14}CO_2$ and radioactive lipids were analyzed, support the view that DGDG is formed by galactosylation of MGDG (36-38). Results with the 18:3-plant *Vicia faba* (36) and with the 16:3-plants spinach (37) and *Brassica napus* (38) indicate that DGDG synthesis occurs in two phases: rapid galactosylation of a pool of newly formed (oligoene) MGDG, and a more random galactosylation of pre-existing, highly unsaturated MGDG. Although such studies could not discriminate between GGGT activity and alternative modes of DGDG synthesis, our experiments suggest that at least one of the galactosylation reactions is catalyzed by GGGT.

The labeling of intact cells of the *Cyanobacterium, Anabaena variabilis* with $^{14}CO_2$ led to the conclusion that DGDG is made from MGDG by transfer of a newly synthetized galactose unit from an unknown donor (39). It is interesting to note that glycolipid synthesis in *Anabaena* starts with monoglucosyldiacylglycerol, which is converted to MGDG by epimerization of glucose to galactose (39,40). In agreement with Sato and Murata (40), we were unable to measure incorporation of UDP[^{14}C]Gal either into MGDG or DGDG with membrane fractions of *Anabaena*, such in contrast to the ready incorporation of UDP[^{14}C]Glc into glycolipids (G. Caerteling, unpublished). However, Sato and Murata (40) report a low rate of DGDG synthesis from UDPGal with soluble fractions of *Anabaena* cells.

The apparent reversibility of the GGGT reaction (see above), *i.e.* the MGDG formation from diacylglycerol in the absence of UDPGal, invites to some speculation about a possible function. From research on 18:3-plants much information has been gained regarding the synthesis of galactolipids with an eukaryotic fatty acid pattern. Eukaryotic diacylglycerol moieties are made in the *endoplasmic reticulum* (4), but it is still unclear whether they are transported to the chloroplast as PC, PA, diacylglycerol or otherwise (5). Unknown is also the site of deposition, although the outer envelope membrane seems to be a logical site when the transport is mediated by a phospholipid transfer protein (41). Since the apparent reverse GGGT reaction is an outer envelope membrane reaction, a reaction with eukaryotic diacylglycerol would result in eukaryotic MGDG, made in the absence of UDPGal. This would imply an alternative forward and reverse GGGT reaction in the formation of eukaryotic galactolipids.

Chloroplasts have, during the aeons of their symbiotic dwelling within the plant cells more or less retained their bacterial membrane system. Plastid development, however, is tightly regulated by multiple factors from the cytosol. The idea arises that important regulating systems and sensors are localized at the interface of host and symbiont, *i.e.* in the outer plastid envelope membrane. One of these systems may be GGGT which seems to determine the overall rate of galactolipid synthesis.

REFERENCES

1 Wintermans, J.F.G.M. (1960) Biochim. Biophys. Acta 44, 49-54.
2 Heinz, E. (1977) *In* Lipids and Lipid Polymers in Higher Plants (Tevini, M. and Lichtenthaler, H.K., eds.) pp. 102-120, Springer Verlag Berlin.
3 Douce, R. and Joyard, J. (1980) *In* The Biochemistry of Plants, Vol. 4, Lipids: Structure and Function (Stumpf, P.K., ed.) pp. 321-362, Academic Press, New York.
4 Roughan, P.G. and Slack, C.R. (1982) Annu. Rev. Plant Physiol. 33, 97-132.
5 Douce, R., Block, M.A., Dorne, A.-J. and Joyard, J. (1984) *In* Subcellular Biochemistry, Vol. 10 (Roodyn, D.B., ed.) pp. 1-84, Plenum Press, New York.
6 Douce, R., Holtz, R.B. and Benson, A.A. (1973) J. Biol. Chem. 248, 7215-7222.
7 Douce, R. (1974) Science 183, 852-853.
8 Joyard, J. and Douce, R. (1977) Biochim. Biophys. Acta 486, 273-285.

9   Joyard, J. and Douce, R. (1976) Biochim. Biophys. Acta 424, 125-131.
10 Frentzen, M., Heinz, E., McKeon, T.A. and Stumpf, P.K. (1983) Eur. J. Biochem. 129, 629-636.
11 Ongun, A. and Mudd, J.B. (1968) J. Biol. Chem. 243, 1558-1566.
12 Siebertz, M. and Heinz, E. (1977) Hoppe Seyler's Z. Physiol. Chem. 358, 27-34.
13 Van Besouw, A. and Wintermans, J.F.G.M. (1978) Biochim. Biophys. Acta 529, 44-53.
14 Heemskerk, J.W.M., Bögemann, G. and Wintermans, J.F.G.M. (1983) Biochim .Biophys. Acta 754, 181-189.
15 Siebertz, H.P., Heinz, E., Linscheid, M., Joyard, J. and Douce, R. (1979) Eur. J. Biochem. 101, 429-438.
16 Wintermans, J.F.G.M., Van Besouw, A. and Bögemann, G. (1981) Biochim. Biophys. Acta 663, 99-107.
17 Dorne, A.-J., Block, M.A., Joyard, J. and Douce, R. (1982) FEBS Lett. 145, 30-34.
18 Heemskerk, J.W.M., Bögemann, G., Scheijen, M.A.M. and Wintermans, J.F.G.M. (1986) Anal. Biochem. 154, 88-92.
19 Heemskerk, J.W.M. (1986) Ph. D. Thesis, University of Nijmegen, Nijmegen (The Netherlands).
20 Heemskerk, J.W.M., Scheijen, M.A.M., Jacobs, F.H.H. and Wintermans, J.F.G.M. (This volume).
21 Heinz. E., Bertrams, M., Joyard, J. and Douce, R. (1978) Z. Pflanzenphysiol. 87, 325-331.
22 Heinz, E. (1973) Z. Pflanzenphysiol. 69, 359-376.
23 Block, M.A., Dorne, A.-J., Joyard, J. and Douce, R. (1983) J. Biol. Chem. 258, 13281-13286.
24 Heemskerk, J.W.M., Bögemann, G. and Wintermans, J.F.G.M. (1985) Biochim. Biophys. Acta 835, 212-220.
25 Cline, K. and Keegstra, K. (1983) Plant Physiol. 71, 366-372.
26 Heemskerk, J.W.M., Wintermans, J.F.G.M., Joyard, J., Block, M.A., Dorne, A.-J. and Douce, R. (1986) Biochim. Biophys. Acta 877, 281-289.
27 Heemskerk, J.W.M., Jacobs, F.H.H., Bögemann, G. and Wintermans, J.F.G.M. (This volume).
28 Van Besouw, A. and Wintermans, J.F.G.M. (1979) FEBS Lett. 102, 33-37.
29 Douce, R. and Guillot-Salomon, T. (1970) FEBS Lett. 11, 121-124.
30 McKee, J.W.A. and Hawke, J.C. (1979) Arch. Biochim. Biophys. 197, 322-332.
31 Renkonen, O. and Bloch, K. (1969) J. Biol. Chem. 244, 4899-4903.
32 Wintermans, J.F.G.M., Van Besouw, A., Bögemann, G. and Aerts, J. (1980) In Biogenesis and Function of Plant Lipids (Mazliak, P., Benveniste, P., Costes, C. and Douce, R., eds.) pp. 49-56, Elsevier, Amsterdam.
33 Blee, E. and Schantz, R. (1978) Plant Sci. Lett. 13, 247-255.
34 Mendiola-Morgenthaler, L., Eichenberger, W. and Boschetti, A. (1985) Plant Sci. 41, 97-104.
35 Liedvogel, B. and Kleinig, H. (1977) Planta 133, 249-253.
36 Williams, J.P., Khan, M. and Leung, S. (1975) J. Lipid. Res. 16, 61-65.
37 Siebertz, H.P., Heinz, E., Joyard, J. and Douce, R. (1980) Eur. J. Biochem. 108, 177-185.
38 Williams, J.P. and Khan, M.U. (1982) Biochim. Biophys. Acta 713, 177-184.
39 Sato, N. and Murata, N. (1982) Biochim. Biophys. Acta 710, 271-278.
40 Sato, N. and Murata, N. (1982) Plant Cell Physiol. 23, 1115-1120.
41 Gounaris, K., Sen, A., Brain, A.P.R., Quinn, P.J. and Williams, W.P. (1983) Biochim. Biophys. Acta 728, 129-139.

ACKNOWLEDGEMENT

J.H. acknowledges a grant of the Stichting Hugo de Vriesfonds and of the Dutch Society for Cell Biology.

CHARACTERIZATION OF GALACTOSYLTRANSFERASES IN SPINACH CHLOROPLAST ENVELOPE MEMBRANES
APPLICATIONS OF AN ASSAY FOR UDPGal : DIACYLGLYCEROL GALACTOSYLTRANSFERASE

Johan W.M. Heemskerk, Martin A.M. Scheijen, Frans H.H. Jacobs and Jef F.G.M. Wintermans

Botanisch Laboratorium, University of Nijmegen, Toernooiveld 6525 ED Nijmegen (The Netherlands)

INTRODUCTION

Two enzymes involved in the galactolipid metabolism of spinach chloroplast envelope membranes were studied. UDPGal : diacylglycerol galactosyltransferase (UDGT) synthetizes monogalactosyldiacylglycerol (MGDG) from UDPGal plus diacylglycerol (1,2):

UDPGal + diacylglycerol → UDP + MGDG.

The enzyme has been localized on the inner envelope membrane of spinach chloroplasts (3,4). Galactolipid : galactolipid galactosyltransferase (GGGT) forms di-, tri- and tetragalactosyldiacylglycerol (DGDG, TGDG and TeGDG) by transgalactosylation of galactosyldiacylglycerols (5,6):

MGDG + MGDG → DGDG + diacylglycerol
DGDG + MGDG → TGDG + diacylglycerol.

Recently, GGGT could be localized definitively on the outer envelope membrane of spinach chloroplasts (7). In most studies to galactolipid synthesis, the incorporation of UDP-Gal into total galactolipids has been followed. However, then UDGT and GGGT activities are measured together: GGGT produces diacylglycerol which is used by UDGT for MGDG synthesis, and MGDG again is a substrate for GGGT (5,8). The recent possibility to measure UDGT and GGGT separately cleared up the way for a further characterization of the enzymes. Some of the results obtained (8) are presented.

MATERIAL AND METHODS

Intact spinach chloroplasts and unseparated envelope membranes were isolated as before (6). GLC and HPLC analysis of lipids were as described elsewhere (9). UDGT was assayed by pre-incubation of a mixture of phosphatidylcholine (0.7 mg soybean PC) liposomes and envelope membranes (0.2 mg protein) with phospholipase C (from *Bacillus cereus*), and incubation with 0.12 mM UDPGal (pH 7.2, 30 °C) (8,9). Radioactive label was in [*glycerol*-^{3}H]PC, di[1-^{14}C]oleoyl-PC or UDP[U-^{14}C]Gal. GGGT was assayed with galactose labeled [^{14}C]MGDG as before (6). Results were quantified by taking into account the conversion of [^{14}C]MGDG and of endogenous MGDG (7).

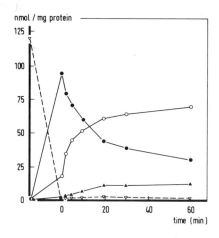

Fig. 1. Time course of galactolipid synthesis from diacyl[^{3}H]glycerol. Envelope membranes (50 μg protein) were sonicated with liposomes of [*glycerol*-^{3}H]PC (4.8 μg), and phospholipase C was added. At t=0, UDPGal (0.12 mM) was added. Radioactivity is given for PC (∇--∇), diacylglycerol (●——●), MGDG (○——○), and DGDG (▲——▲).

RESULTS AND DISCUSSION

Since we had only an assay for GGGT (3), it was desirable to develop a separate method for the measurement of UDGT activity. For such an assay, a first requirement is incorporation of a controlled amount of substrate diacylglycerol into the envelope membranes. enzyme. This was achieved by sonication of a mixture of envelope membranes with PC liposomes, and treatment of the mixture with phospholipase C. As shown in Fig. 1, the resulting diacylglycerol was a good substrate for UDGT after the addition of UDPGal. A second requirement is that the assay measures real UDGT activity, instead of the co-operative activities of UDGT and GGGT. Under standard conditions with soybean PC and UDP[^{14}C]Gal, 90% of the incorporated label was in MGDG (the remainder was in DGDG and TGDG). When 10 mM MgCl$_2$ was added, which strongly increases GGGT activity in unmodified membranes, label incorporation into MGDG remained at 89% (8). So the procedure is rather specific for measurement of MGDG synthesis.

Table 1. 50% Inhibition concentrations of inhibitors for galactosyltransferases

Inhibitor	UDGT	GGGT
UDP	30 μM	8 mM[a]
UMP	2.5 mM	10 mM[a]
UDPGlc	1.5 mM	-
4-(Hydroxymercuri)-benzoic acid	7 μM	5 μM
N-Ethyl maleimide	0.3 mM	5 mM
ZnCl$_2$	6 mM	0.6 mM
CdCl$_2$	8 mM	2 mM
Oleic acid	0.3 mM	2 mM
Ethanol (v/v)	7 %	25 %

[a]Inhibition can be attributed to chelating properties at this concentration

Using the assays for UDGT and GGGT, some kinetic properties of the enzymes could be determined (8). UDGT appeared saturated at a low concentration of diacylglycerol. Activity of GGGT was strongly dependent on the MGDG concentration. Saturation of the enzyme was above 3300 nmol MGDG/mg envelope protein, indicating a rather low affinity for MGDG (7). Velocitties were high compared to that of UDGT.

UDGT activity was slightly stimulated by Mg^{2+} and $Mn2+$. Other tested cations were ineffective or inhibitory (8). GGGT was strongly stimulated by a number of mono- and divalent cations in the order:

$$Ca^{2+}, Ba^{2+}, Mn^{2+}, Mg^{2+} > Fe^{2+} > K^+, Na^+, NH_4^+ > \text{control}.$$

Within a certain concentration range UDP, N-ethyl maleimide and oleic acid were most effectively as specific inhibitors for UDGT (Table 1). Zn^{2+} and chelating anions (6) can act as specific inhibitors for GGGT activity. Uridine-derivatives - including UDPGal (6,8) - had no or little effect on GGGT. SH reagentia, especially 4-(hydroxymercuri)-benzoic acid were inhibitory to both enzymes (Table 1).

The specificity of both galactosyltransferases for particular molecular species of substrate was tested. UDGT activity was highest with distearoylglycerol, although differences with other species were small. Increasing the desaturation degree, and shortening of the acyl chain length lowered the activity. GGGT preferently used hexaene species of MGDG as galactosyl acceptor. Saturated/monoene species were converted slower (8). So application of the assays for UDGT and GGGT have revealed a number of characteristics of the enzymes, which can be compared easily with properties of the galactolipid synthesis in unfractionated spinach chloroplasts (10).

When a mixture of [*glycerol*-^{3}H]PC and envelope membranes was treated with phospholipase C; a small amount of MGDG was formed even in the absence of UDPGal (Fig. 1, $t=0$). Heat-treatment of the membranes (15 min at 90°C) abolished this UDPGal-independent MGDG formation. Further, the activity increased with higher amounts of incorporated diacylglycerol, and it was not inhibited by UDP (8). Although the source of the galactosyl-group for the UDPGal-independent MGDG synthesis is not determined yet, the above properties may argue for a reversed GGGT activity.

REFERENCES

1 Joyard, J. and Douce, R. (1976) Biochim. Biophys. Acta 424, 125-131.
2 Frentzen, M., Heinz, E., McKeon, T. and Stumpf, P.K. (1983) Eur. J. Biochem. 129, 629-636.
3 Block, M.A., Dorne, A.-J., Joyard, J. and Douce, R. (1983) J. Biol. Chem. 258, 13281-13286.
4 Heemskerk, J.W.M., Bögemann, G. and Wintermans, J.F.G.M. (1985) Biochim. Biophys. Acta 835, 212-220.
5 Van Besouw, A. and Wintermans, J.F.G.M. (1978) Biochim. Biophys. Acta 529, 44-53.
6 Heemskerk, J.W.M., Bögemann, G. and Wintermans, J.F.G.M. (1983) Biochim. Biophys. Acta 754, 181-189.
7 Heemskerk, J.W.M., Wintermans, J.F.G.M., Joyard, J., Block, M.A., Dorne, A.-J. and Douce, R. (1986) Biochim. Biophys. Acta 877, 281-289.
8 Heemskerk, J.W.M. (1986) Ph. D. Thesis (Nijmegen, The Netherlands).
9 Heemskerk, J.W.M., Bögemann, G., Scheijen, M.A.M. and Wintermans, J.F.G.M. (1986) Anal. Biochem. 154, 85-91.
10 Heemskerk, J.W.M., Jacobs, F.H.H., Bögemann, G. and Wintermans, J.F.G.M. (This volume).

GALACTOSYLTRANSFERASE ACTIVITIES IN INTACT SPINACH CHLOROPLASTS AND ENVELOPE MEMBRANES

Johan W.M. Heemskerk, Frans H.H. Jacobs, Gerard Bögemann and Jef F.G.M. Wintermans

Botanisch Laboratorium, University of Nijmegen, Toernooiveld 6525 ED Nijmegen (The Netherlands)

INTRODUCTION

Two galactosyltransferases producing galactolipids have been studied extensively in chloroplast envelope membranes (1). UDPGal : diacylglycerol galactosyltransferase (UDGT) produces monogalactosyldiacylglycerol (MGDG) (2); and galactolipid : galactolipid galactosyltransferase (GGGT) producess di-, tri- and tetragalactosyldiacylglycerol (DGDG, TGDG and TeGDG) (3-6). In the isolated envelope membranes, UDGT and GGGT co-operate in the production of MGDG, DGDG, TGDG and TeGDG (4,5).

Since GGGT produces TGDG, TeGDG and diacylglycerol, which are no major envelope lipids *in vivo*, and since the enzyme is accessible from the outside of the chloroplast (5,6), its physiological significance for DGDG synthesis has raised questions. We studied the galactolipid synthesis in unfractionated spinach chloroplasts and compared the results with studies to galactosyltransferase activities in isolated envelope membranes. Here we present evidence for an active role of GGGT in the UDPGal-dependent galactolipid formation in the isolated chloroplast.

MATERIAL AND METHODS

Intact spinach chloroplasts were purified by Percoll-gradient centrifugation (7). Chloroplasts were broken by resuspension in 5 mM Tricine (pH 7.2)/ 4 mM $MgCl_2$ and incubation on ice for 30 min. Thermolysin treatment of the intact chloroplasts was in 330 mM sorbitol, further as described by Dorne et al. (5). Galactolipid synthesis was measured with chloroplasts (40 µg chlorophyll) in 50 mM Tricine/NaOH (pH 7.2), 10 mM $MgCl_2$ and 0.12 mM UDP[U-^{14}C]Gal for 20 min at 30°C.

RESULTS AND DISCUSSION

Intact spinach chloroplasts synthetize MGDG, DGDG, TGDG and TeGDG from UDPGal (8). After treatment with the non-permeating protease thermolysin, the chloroplasts have lost their capacity of DGDG and T(e)GDG synthesis and produce only small amounts of MGDG (Fig. 1). Since GGGT activity of isolated envelope membranes is strongly affected by thermolysin and UDGT activity is not (5,6), these results suggest a deficient diacylglycerol supply by GGGT after thermolysin treatment. The diacylglycerol content of chloroplasts during galactolipid synthesis was determined. It was low, 15-22

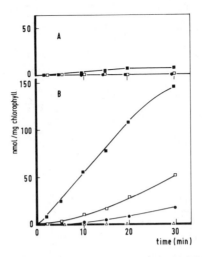

Fig. 1. Galactolipid synthesis by thermolysin-treated (A) and untreated (B) intact chloroplasts. Indicated are: incorporation of UDPGal into total galactolipids (■——■); synthesis of DGDG+TGDG+TeGDG (□——□); synthesis of TGDG+TeGDG (●——●); and synthesis of 6-O-acyl-MGDG (△——△).

nmol/mg chlorophyll (<0.3% of chloroplast acyllipids). Thus is seems that in the untreated chloroplasts, as in isolated envelopes, diacylglycerol produced by GGGT is used effectively by UDGT again. This points to a close co-operation of UDGT and GGGT in the plastid system.

In chloroplasts the rates of both total galactolipid synthesis and of DGDG+TGDG+TeGDG synthesis were dependent on the UDPGal concentration (Fig. 2). Apparent K_m values for UDPGal incorporation and for DGDG+TGDG+TeGDG synthesis were calculated at 16 μM. The equal apparent K_m values for both processes indicate that there is only one affinity site for UDPGal, and thus we could not measure activity of UDPGal : MGDG galactosyltransferase in the chloroplast.

Table 1. Characteristics of galactolipid synthesis by intact chloroplasts and envelope membranes

	Chloroplasts	Envelope membranes	
	UDPGal incorporation	UDGT	GGGT
Optimal pH	6.5	7-9	6-7
Mg^{2+}	++[a]	+	++
Ca^{2+}, Ba^{2+}	++	0	++
Mn^{2+}	+	+	++
Na^+, K^+, NH_4^+	+	0	+
Zn^{2+}	−	−	−−
EDTA	−	0	−
UDP	−−	−−	0
N-Ethyl maleimide	−−	−−	−
4-(Hydroxymercuri)-benzoic acid	−−	−−	−−
Influence of thermolysin treatment	−−	0	−−

[a] (+) stimulation, (0) no effect, (−) inhibition

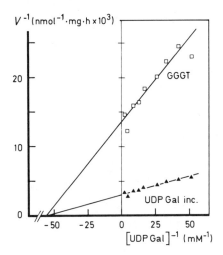

Fig. 2. Lineweaver-Burk plot of galactolipid synthesis by intact chloroplasts. Values are given for: incorporation of UDPGal into total galactolipids (▲), and for synthesis of DGDG+TGDG+TeGDG (GGGT) (□).

If properties of the UDPGal incorporation by intact chloroplasts are compared with those of UDGT and GGGT in isolated envelope membranes (Table 1), it is clear that most properties of GGGT are also found for the plastidial galactolipid synthesis. However, specific effectors of UDGT - like UDP (1) - also influence the plastidial system. As a conclusion, these and other (Heemskerk et al., submitted) experiments show an active role of GGGT in the UDPGal-derived galactolipid synthesis in the spinach chloroplast.

REFERENCES

1 Heemskerk, J.W.M., Scheijen, M.A.M., Jacobs, F.H.H. and Wintermans, J.F.G.M. (This volume).
2 Joyard, J. and Douce, R. (1976) Biochim. Biophys. Acta 424, 125-131.
3 Van Besouw, A. and Wintermans, J.F.G.M. (1978) Biochim. Biophys. Acta 529, 44-53.
4 Heemskerk, J.W.M., Bögemann, G. and Wintermans, J.F.G.M. (1983) Biochim. Biophys. Acta 754, 181-189.
5 Dorne, A.-J., Block, M.A., Joyard, J. and Douce, R. (1982) FEBS Lett. 145, 30-34.
6 Heemskerk, J.W.M., Wintermans, J.F.G.M., Joyard, J., Block, M.A., Dorne, A.-J., and Douce, R. (1986) Biochim. Biophys. Acta BBA 52279 (in press).
7 Heemskerk, J.W.M., Bögemann, G. and Wintermans, J.F.G.M. (1985) Biochim. Biophys. Acta 835, 212-220.
8 Wintermans, J.F.G.M., Bögemann, G. and Heemskerk, J.W.M. (1984) *In* Advances in Photosynthesis Research (Sybesma, C., ed.) pp. III.2.147-150, Nijhoff/Junk Publ., The Hague (The Netherlands).
1 Heemskerk, J.W.M., Scheijen, M.A.M., Jacobs, F.H.H. and Wintermans, J.F.G.M. (This volume).
2 Joyard, J. and Douce, R. (1976) Biochim. Biophys. Acta 424, 125-131.
3 Van Besouw, A. and Wintermans, J.F.G.M. (1978) Biochim. Biophys. Acta 529, 44-53.
4 Heemskerk, J.W.M., Bögemann, G. and Wintermans, J.F.G.M. (1983) Biochim. Biophys. Acta 754, 181-189.
5 Dorne, A.-J., Block, M.A., Joyard, J. and Douce, R. (1982) FEBS Lett. 145, 30-34.
6 Heemskerk, J.W.M., Wintermans, J.F.G.M., Joyard, J., Block, M.A., Dorne, A.-J., and Douce, R. (1986) Biochim. Biophys. Acta 877, 281-289.
7 Heemskerk, J.W.M., Bögemann, G. and Wintermans, J.F.G.M. (1985) Biochim. Biophys. Acta 835, 212-220.
8 Wintermans, J.F.G.M., Bögemann, G. and Heemskerk, J.W.M. (1984) *In* Advances in Photosynthesis Research (Sybesma, C., ed.) pp. III.2.147-150, Nijhoff/Junk Publ., The Hague (The Netherlands).

BIOSYNTHESIS OF SULFOQUINOVOSYLDIACYLGLYCEROL IN CHLOROPLASTS OF HIGHER PLANTS

Kathryn F. Kleppinger-Sparace and J. Brian Mudd

ARCO Plant Cell Research Institute
6560 Trinity Court
Dublin, CA 94568, USA

INTRODUCTION

Sulfoquinovosyldiacylglycerol (SQDG) was first isolated and characterized 27 years ago (Benson et al. 1959). But the pathway of biosynthesis and the function have not yet been satisfactorily demonstrated.

Biosynthesis of SQDG in Euglena appears to depend on the use of cysteic acid as a precursor (Davies et al. 1966). This activity led to the postulation of a "sulfoglycolytic" pathway to explain the incorporation of the three carbon moiety into sulfoquinovose by reactions analogous to those of glycolysis. The evidence from studies with higher plants does not suggest that cysteic acid is a satisfactory precursor of SQDG and argues against the existence of a sulfoglycolytic pathway (Mudd et al. 1980). It has been demonstrated that isolated chloroplasts can synthesize SQDG from radioactive sulfate (Haas et al. 1980, Kleppinger-Sparace et al. 1985). In these studies, the need for illumination was shown. The question remained whether the light was required for the generation of ATP or reductant or both. We have therefore attempted to synthesize SQDG using chloroplasts incubated with radioactive sulfate in darkness.

METHODS

Chloroplasts were isolated as described previously (Kleppinger-Sparace et al. 1985). The incubation mixtures were as described earlier except for the additions noted in the Tables (Kleppinger-Sparace et al. 1985). Assay of SQDG synthesis was as described before (Kleppinger-Sparace et al. 1985), essentially using organic solvent to remove the lipid from the aqueous reaction mixture (leaving the precursor sulfate in the aqueous fraction). The SQDG was purified by thin layer chromatography before assay of the incorporation of radioactivity.

RESULTS

Table I demonstrates that when isolated spinach chloroplasts are incubated with $^{35}SO_4^{2-}$ in darkness there is low incorporation into SQDG. The incorporation can be increased by addition of ATP or by illumination. Neither ADP nor AMP significantly improves incorporation in darkness. While ATP stimulates incorporation in illuminated chloroplasts, ADP and AMP inhibit the incorporation. One might expect that the first step of assimilation of $^{35}SO_4^{2-}$ into SQDG would be the formation of adenosylphosphatosulfate (APS) and so the stimulation by added ATP, which is reasonably well taken up by the chloroplast (Heldt, 1969) is to be expected.

Table I. Effect of Adenine Nucleotides on Incorporation of $^{35}SO_4^{2-}$ into SQDG.

Additions	Dark	Light
	nmol/mg chl/h	
None	0.36	1.11
ATP	1.34	1.83
ADP	0.55	0.70
AMP	0.23	0.48

Reaction mixture contained 0.3M sorbitol, 2mM $MgCL_2$, 0.5 mM DTT, 33 mM tricine-HCL pH 7.9, 100 μM $^{35}SO_4^{2-}$ (0.25 μCi/nmol), 2mM nucleotide, and chloroplasts equivalent to 66 μg chlorophyll in a reaction volume of 1.0 ml. Incubation was for 30 min at 25° in a photosynthetic reaction bath. Light intensity was 300 μE/m^2/sec.

The requirement for ATP in the assimilation of $^{35}SO_4^{2-}$ into SQDG suggested that an ATP generating system "the dihydroxyacetonephosphate (DHAP) shuttle" (Werdan et al. 1975), would be effective. The results show that chloroplasts incubated in darkness will synthesize SQDG from sulfate using either ATP or the ATP generating system. When the chloroplasts are incubated in darkness, the addition of ATP or the DHAP shuttle stimulates SQDG synthesis tenfold higher than the dark control. All components of the shuttle (DHAP, OAA, and Pi) are necessary to facilitate incorporation of $^{35}SO_4^{2-}$ into SQDG in chloroplasts incubated in the dark. However, the illuminated control is also stimulated both by ATP and even more by the DHAP shuttle. It is possible that the shuttle components may have some stimulatory effect in addition to the generation of ATP. Indeed the use of APS and PAPS as precursors for SQDG synthesis is more effective in the presence of DHAP, but not in the presence of other shuttle components (OAA and Pi). The original design of the DHAP shuttle (Werdan et al. 1975) was for the generation of ATP and involved the cycling of NADP(H) using OAA and the malate dehydrogenase in the chloroplast. Thus NADPH generated by glyceraldehyde-3-phosphate dehydrogenase does not accumulate. It might appear that in the presence of the DHAP shuttle (or in the presence of ATP) no reductant is generated in the chloroplast and the incorporation of $^{35}SO_4^{2-}$ into SQDG does not require reductant. However the levels of incorporation are low and a small amount of reductant may suffice. Further research is necessary to clarify this point.

Table II. Incorporation of $^{35}SO_4^{2-}$ into SQDG Stimulated by ATP or the DHAP Shuttle.

Additions	Dark	Light
	nmol/mg chl/h	
None	0.3	2.3
2mM ATP	2.7	4.1
DHAP Shuttle	3.3	7.8

Incubation conditions were the same as for Table I. Components of the shuttle were 2 mM DHAP, 1mM OAA, and 2mM Pi. Chloroplasts were equivalent to 102μg chlorophyll.

Since we expect that APS will be a necessary intermediate for the synthesis of SQDG, we tried the use of ^{35}S-APS and ^{35}S-PAPS as precursors of SQDG. The results show that ^{35}S-APS was the best precursor of SQDG in this experiment (Table III) and the incorporation was not diluted by the addition of cold sulfate or cold PAPS. The incorporation from ^{35}S-PAPS was not greater than from ^{35}SO$_4^{2-}$ in this experiment.

Table III. Incorporation of Sulfate, APS, or PAPs into SQDG.

Additions	pmol/mg chl/h
^{35}SO$_4^{2-}$	10
AP^{35}S	115
+10µM sulfate	97
+10µM PAPS	107
PAP^{35}S	9.3
+10µM sulfate	6.8*
+10µM APS	2.1

Reaction conditions were similar to those described for Table I. The labelled precursors were present at 10µM. The reactions included chloroplasts equivalent to 193µg chlorophyll and were incubated for 30 min at 25° in darkness.

The use of more sophisticated precursors will be needed to delineate the steps of incorporation of sulfate into SQDG. Our best guess is that a nucleotide-sulfoquinovose is the donor to diacylglycerol (DG), but it is an open question at what point the sulfonic acid moiety is transferred to the carbohydrate. Broken chloroplast systems would facilitate an understanding of the site(s) of SQDG biosynthetic reactions, but so far broken chloroplasts have given only poor incorporation.

REFERENCES

Benson, A.A., Daniel, M., and Wiser, R., 1959, A sulfolipid in plants, Proc. Natl. Acad. Sci. U.S. 45:1582.

Davies, W.H., Mercer, E.I., and Goodwin, T.W., 1966, Some observations on the biosynthesis of the plant sulfolipid by Euglena gracilis, Biochem. J. 98:369.

Haas, R., Siebertz, M.P., Wrage, K., and Heinz, E., 1980, Localization of sulfolipid labeling within cells and chloroplasts, Planta 148:238.

Heldt, H.W., Adenine nucleotide translocation in spinach chloroplasts, 1969, Biochim. Biophys. Acta. 5:11.

Kleppinger-Sparace, K.F., Mudd, J.B., and Bishop, D.G., 1985, Biosynthesis of sulfoquinovosyl diacylglycerol in higher plants: The incorporation of ^{35}SO$_4^{2-}$ by intact chloroplasts, Arch. Biochmem. Biophys. 240:859.

Mudd, J.B., deZacks, R., and Smith, J., 1980, Studies on the biosynthesis of sulfoquinovosyl diacylglycerol in higher plants, in: "Biogenesis and Function of Plant Lipids," P. Mazliak, P. Benveniste, C. Costes, and R. Douce, eds., Elsevier/North Holland Biomedical Press, Amsterdam.

Werdan, K., Heldt, H.W., and Milovancer, M., 1975, The role of pH in the regulation of carbon dioxide fixation in the chloroplast stroma. Studies on CO_2 fixation in the light and dark, Biochem. Acta, 396:276.

SULFOLIPID SYNTHESIS BY ISOLATED INTACT SPINACH CHLOROPLASTS

Jacques Joyard[a], Elisabeth Blée[b] and Roland Douce[a]

a Laboratoire de Physiologie Cellulaire Végétale, UA CNRS 576, DRF/CENG and USTMG, 85 X, 38041 Grenoble-cedex, France

b Laboratoire d'Enzymologie Moléculaire et Cellulaire, UA CNRS 1182, Institut de Botanique, 28 rue Goethe, 67083 Strasbourg-cedex, France

INTRODUCTION

Isolated intact chloroplasts are able to incorporate, under illumination, $^{35}SO_4^{2-}$ into sulfoquinovosyldiacylglycerol (SQDG)[1,2]. However, up to now the origin of the diacylglycerol (DG) used to form SQDG was not clearly established. The purpose of this article is to determine whether DG molecules actively synthesized through the plastid Kornberg-Pricer pathway[3,4] can be incorporated into SQDG.

MATERIAL AND METHODS

<u>Chloroplasts</u> - They were prepared from young spinach grown in a greenhouse by methods, including Percoll purification, already described[5].

<u>Reaction mixture</u> - Intact purified chloroplasts (0.1-0.2 mg chlorophyll) were incubated, under illumination, at 25°C for 15 min, in 1 ml of the basic incubation medium[2]: 0.33 M mannitol; 30 mM tricine-NaOH (pH 7.9); 2 mM $MgCl_2$; 10 mM $NaHCO_3$; 0.5 mM DTT; 0.1 mM Na_2SO_4; 0.2 mM Na_2HPO_4 and 2 mM ATP. To this basic medium, 1 mM <u>sn</u>-glycerol 3-phosphate, 0.2 mM Na-acetate, 0.13 mM triton X-100 and 0.1 mM UDP-gal were added as specified. Two sets of experiments were run : 1) with $^{35}SO_4^{2-}$, 2) with [1-^{14}C]-acetate as labeled substrates. The details for lipid extraction and analyses are described by Joyard et al[6].

RESULTS

<u>^{35}S labeling of SQDG in isolated intact spinach chloroplasts</u> - In agreement with Kleppinger-Sparace et al[2], two chloroform-soluble sulfur-containing compounds were formed by intact chloroplasts incubated under illumination in the basic medium containing $^{35}SO_4^{2-}$. The major one behave as authentic SQDG in all the chromatographic systems used[6] and represents about 80% of the label incorporated into the chloroform phase (Table I). The minor one run at the front of the solvent used and behave as elementary sulfur (S_0), which is very hydrophobic. Addition of acetate and/or <u>sn</u>-glycerol 3-phosphate to the incubation mixture induced a stimu-

Table I

Synthesis of chloroform-soluble sulfur-containing compounds from $^{35}SO_4^{2-}$ by isolated intact spinach chloroplasts

ADDITIONS TO THE BASIC REACTION MEDIUM	$^{35}SO_4^{2-}$ INCORPORATED INTO SQDG (PMOL/MG CHLOROPHYLL/H)	X
NONE	1006	194
+ ACETATE	1145	99
+ ACETATE, SN-GLYCEROL 3-PHOSPHATE	1285	83
+ ACETATE, TRITON X-100	816	69
+ ACETATE, TRITON X-100, SN-GLYCEROL 3-PHOSPHATE	1445	169
+ ACETATE, TRITON X-100, SN-GLYCEROL 3-PHOSPHATE, UDP-GAL	545	215

lation of the incorporation of $^{35}SO_4^{2-}$ into chloroform-soluble molecules (Table I). With the complete incubation mixture, about 90% of the radioactivity was incorporated into the compound running as SQDG. Further identification of radioactive SQDG was provided by analyses of the water soluble deacylates obtained after mild alkaline hydrolysis (performed according to Maruo and Benson[7]) of the chloroform-soluble sulfur-containing compounds: the major compound formed was identical to glyceryl-quinovoside-6sulfonate[6]. Finally, addition of UDP-gal to the reaction mixture induced a marked inhibition of SQDG formation, thus demonstrating a competition between SQDG and galactolipid biosynthesis at the level of their common substrate, i.e. DG.

^{14}C-acetate labeling of SQDG by isolated intact spinach chloroplasts - In presence of $[1-^{14}C]$-acetate, SQDG formation was promoted in experimental conditions which allow the functioning of all the enzymes of the envelope Kornberg-Pricer pathway: between 4 and 5% of the total radioactivity incorporated into chloroplast lipids was recovered as SQDG (Table II). Again, addition of UDP-gal strikingly reduced the amount of SQDG formed (Table II). These experiments suggest a tight coupling of SQDG formation with the enzymes responsible of DG synthesis via the Kornberg-Pricer pathway. Further evidences were provided by the analyses of the newly synthesized fatty acids: SQDG synthesized under our experimental conditions has almost the same fatty acids as phosphatidic acid

Table II

Glycerolipid synthesis from $[1-^{14}C]$-acetate by intact spinach chloroplasts

ADDITIONS TO THE BASIC REACTION MIXTURE CONTAINING ACETATE	ACETATE INCORPORATED INTO								FATTY ACIDS	
	LIPIDS								SATURATED (MOSTLY 16:0)	18:1
	LYSO-PA	PA	DG	SQDG	MGDG	DGDG	UFA	TOTAL (NMOL ACETATE/ MG CHLOROPHYLL/H)		
	%								%	
NONE	<1	<1	27.6	2.8	<1	<1	67.1	1139	32.5	67.5
+ SN-GLYCEROL 3-PHOSPHATE	4.4	5.6	47.9	4.1	6.2	<1	31.7	1392	51.5	48.5
+ TRITON X-100	1.8	2.2	40.8	2.6	<1	<1	50.5	1325	35.6	64.4
+ SN-GLYCEROL 3-PHOSPHATE, TRITON X-100	8.4	5.3	60.7	4.8	<1	<1	20.5	1395	48.6	51.4
+ SN-GLYCEROL 3-PHOSPHATE, TRITON X-100, UDP-GAL	2.8	1	14.1	1.1	55.6	2.5	22.8	1551	45.1	54.9

Table III

Fatty acid composition and positional distribution in glycerolipids synthesized from [1-^{14}C]-acetate by intact spinach chloroplasts

Lipid analyzed	RADIOACTIVITY INCORPORATED WITHIN THE DIFFERENT FATTY ACIDS (%)											
	TOTAL				sn-1 POSITION				sn-2 POSITION			
	<16:0	16:0	18:1	18:2	<16:0	16:0	18:1	18:2	<16:0	16:0	18:1	18:2
LYSO-PA	0.5	4.6	94.9	-	0.5	4.6	94.9	-	-	-	-	-
PA	0.8	52.9	46.3	-	<0.5	12.0	87.9	-	1.5	93.0	5.5	-
DG	1.9	51.4	46.7	-	0.9	11.7	87.4	-	2.8	91.1	6.0	-
SQDG	-	57.4	41.6	1	-	21.3	76.7	2	<0.5	93.5	6.4	-

(PA) and DG (Table III). This relationship is even more obvious when the positional distribution of the newly synthesized molecules is analyzed: after *Rhizopus arrhizus* lipase hydrolysis (performed as in ref. 8) of PA, DG and SQDG, the remaining sn-2 lyso derivatives contain almost exclusively 16:0, thus demonstrating their common origin and their formation through the envelope Kornberg-Pricer pathway (which is responsible for the formation of the "prokaryotic" structure of glycerolipids[9]). Finally, Table III also show that 18:2 is present in limited, but consistent, amounts after 15 min incubation in SQDG, but not in PA nor DG, thus suggesting that desaturation of 18:1 at sn-1 position is possible.

CONCLUSION

We have demonstrated the ability of isolated intact chloroplasts from spinach to incorporate DG formed through the envelope Kornberg-Pricer pathway into SQDG: experimental conditions which promote DG formation induce a stimulation of SQDG formation whatever the labeled substrate was. In addition, newly synthesized SQDG contains 18:1 and 16:0 respectively at sn-1 and sn-2 position of the glycerol backbone, corresponding to the structure also found in PA and DG formed through the envelope Kornberg-Pricer pathway.

REFERENCES

1. Haas, R., Siebertz, H.P., Wrage, K. and Heinz, E. (1980) Planta 148, 238-244.
2. Kleppinger-Sparace, K.F., Mudd, J.B. and Bishop, D.G. (1985) Arch. Biochem. Biophys. 240, 859-865.
3. Douce, R. and Guillot-Salomon, T. (1970) FEBS Lett. 11, 121-126.
4. Joyard, J. and Douce, R. (1977) Biochim. Biophys. Acta 486, 273-285.
5. Douce, R. and Joyard, J. (1980) in **Methods in Chloroplast Molecular Biology** (Edelman, M., Hallick, R.B. and Chua, N.-H, eds.) pp. 239-256, Elsevier/North-Holland Biomedical Press, Amsterdam.
6. Joyard, J., Blée, E. and Douce, R. (1986) submitted for publication.
7. Benson, A.A. and Maruo, B. (1958) Biochim. Biophys. Acta 41, 328-333.
8. Tulloch, A.P., Heinz, E. and Fisher, W. (1973) Hoppe-Seyler's Z. Physiol. Chem. 354, 1115-1123.
9. Heinz, E. and Roughan, P.G. (1983) Plant Physiol. 72, 273-279.

CALCIUM CHLORIDE EFFECT ON GLYCEROLIPIDS METABOLISM IN OLIVE TREE LEAF

Brahim Marzouk[*], Mokhtar Zarrouk[*], Abdelkader Cherif[*] and Paul Mazliak[**]

[*] C.B.R.G. INRST 1, avenue de France 1000- Tunis Tunisie
[**] Université P. et M. Curie 4, Place Jussieu 75230 Paris

INTRODUCTION

Many works about the effect of saline environment on higher plants shown that salt stress affects several aspects of their physiological and biochemical functions (1,2,3). So, for the purpose of a better understanding of the olive tree physiology under saline conditions, we have studied the effects of calcium chloride on some glycerolipids metabolism aspects in olive tree leaf.

MATERIAL AND METHODS

Young olive tree plants two years aged were cultivated on sand and irrigated with nutrient solutions of variable concentrations : 0,25,50,75 and 100 mM. Young leaves were incubated during 48 hours with Na $[1-^{14}C]$ acetate (36,4 mCi/mM) placed as microdroplets on the surface of the leaf. Lipid extraction and separation, radioactivity and methyl esters analysis have been described in a previous paper (4).

RESULTS AND DISCUSSION

Effect of $CaCl_2$ on the fatty acids composition

The principal changes in the fatty acids composition consist essentially in dropped linolenic acid percentage in favour of oleic and linoleic ones (Table I). This leads to a decreasing fatty acids unsaturation degree when $CaCl_2$ concentration is increasing.

Effect of $CaCl_2$ on the fatty acids contents

The total fatty acids (TFA) content is not practically modified up to 50 mM of $CaCl_2$ concentration. Above 50 mM, the TFA content shows a considerable decrease (figure 1). The main change in fatty acids contents concerns the continuous decreasing of linolenic acid.
These results are in agreement with those mentioned by other authors.

Table 1. Calcium Chloride effect on fatty acids composition of Olive tree leaf (in % of TFA)

$CaCl_2$ (mM)	$C_{16:0}$	$C_{18:0}$	$C_{18:1}$	$C_{18:2}$	$C_{18:3}$
0	23,3	2,0	16,7	10,5	47,5
25	24,4	1,9	22,3	11,8	39,6
50	21,4	1,7	28,0	14,7	34,2
75	24,0	2,0	30,5	14,7	28,8
100	29,1	2,9	29,4	15,8	22,8

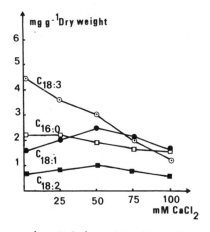

Figure 1. Calcium chloride effect on fatty acid contents in olive tree leaf.

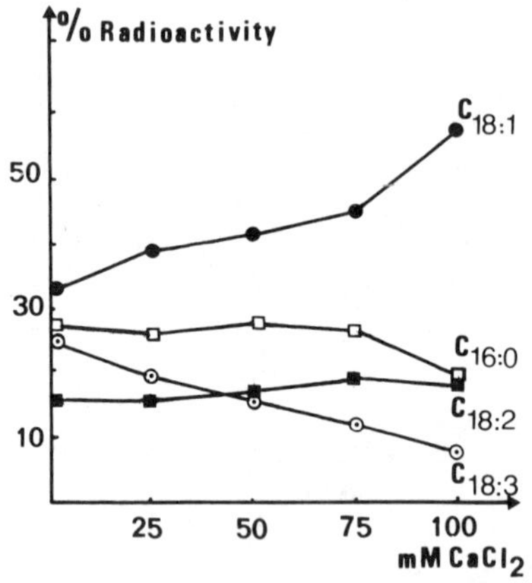

Figure 2. Calcium chloride effect on the incorporation of $[1-^{14}C]$ acetate into the main fatty acids of olive tree leaf.

Effet of $CaCl_2$ on the fatty acids biosynthesis

The $CaCl_2$ increasing concentrations provoke an important $[1-^{14}C]$ acetate incorporation into the oleic acid which incorporates more than 50% of the TFA radioactivity (Figure 2).

Effect of $CaCl_2$ on glycerolipids biosynthesis

The $[1-^{14}C]$ acetate incorporation is decreasing in phospholipids and glycolipids (MGDG et DGDG) and increasing in neutral lipids (Table II). PC and PA incorporate 50% to 60% of $|1-^{14}C|$ acetate radioactivity of polar glycerolipids for high $CaCl_2$ concentrations. On the other hand, there is slowing down of the biosynthesis of PG, PE, MGDG and DGDG when $CaCl_2$ levels increase.

Table 2 : Calcium chloride effect on the incorporation of [1-^{14}C] acetate into the glycerolipids classes in olive tree leaf.

CaCl$_2$ (mM)	Percent of total					Glycerolipids radioactivity		
	PI	PC	PG	PE	PA	MGDG	DGDG	NL
0	3	22	16	5	0	16	10	28
25	2	19	6	4	2	16	6	45
50	2	17	5	5	2	17	5	47
75	3	19	7	3	9	13	4	42
100	3	17	6	2	10	5	2	55

CONCLUSION

Glycerolipidic metabolism in olive tree is affected by the calcium chloride action. It exerces an inhibitory effect on the phospholipids biosynthesis enzymes and particularly on CDP-choline and CDP-ethanolamine phosphotransferases (5,6). Phospholipids acylation reactions are widely dropped in favour of neutral lipids at increasing CaCl$_2$ levels (6).

REFERENCES

1. M. Zarrouk et A. Cherif, Action du chlorure de sodium sur la teneur en lipides de plants d'Olivier (Olaea europea L.) Z. Pflanzenphysiol. 105 : 85 (1980)
2. C.E.E. Stuiver, P.J.C Kuiper, H. Merschner and A. Kylin, Effects of salinity and replacement of K$^+$ by Na$^+$ on lipid composition in two sugar beet inbred lines, Physiol.Plant. 52: 77 (1981).
3. M. Gharsalli et A. Cherif, Action du chlorure de sodium sur la croissance et la teneur en lipides de plants de tournesol (Helianthus annuus L.), Physiol. Veg. 17 : 215 (1979)
4. B. Marzouk et A. Cherif , La lipogenèse dans l'Olive, Rev.Fr.Corps Gras, 11 : 487 (1980).
5. A. Oursel, A. Tremolieres et P. Mazliak, Inhibition par le calcium des CDP-choline et CDP-ethanolaminephosphotransferases de microsomes isoles de racines de feverole (Vicia faba L. var. minor), Physiol. Veg. 15: 377 (1977).
6. A. Oursel-Thibaudin, "Effets du calcium sur le metabolisme des lipides dans les racines de feverole ou de lupin", These Doct. Etat, Paris 6 (1979).

LIPID METABOLISM IN POTATO LEAF DISKS:
EFFECT OF CALMODULIN ANTAGONISTS

George J. Piazza and Robert A. Moreau

Agricultural Research Service
U. S. Department of Agriculture
Philadelphia, PA 19118

INTRODUCTION

Dibucaine (nupercaine), a local anesthetic, was recently shown to alter the lipid composition of barley-root membranes (1). When excised roots were placed in dibucaine there was an increase in the proportions of palmitic, stearic, and oleic acids and a decrease in the proportions of linoleic and linolenic acids. We have recently shown that the rate of autolytic degradation of phospholipids in potato leaf homogenates is inhibited by calmodulin antagonists (which include dibucaine) and stimulated by calmodulin (2). This study was undertaken in order to investigate the effect of dibucaine and other calmodulin antagonists on the polar lipid composition of potato leaves.

MATERIALS AND METHODS

Potato (Solanum tuberosum c.v. Kennebec) plants were grown in a greenhouse for 30-40 days. Twelve mature leaves (4-6 cm) were removed and 8 disks (9 mm in diameter) were cut from each leaf with a brass cork borer (#6). One disk from each leaf was floated (top-side up) in a petri dish containing 20 ml of various test solutions (pH 6.0). The petri dishes were placed in a vacuum chamber (640 mm Hg) for 2 minutes and then removed and incubated in the dark at 25°C. After the desired times disks were removed, placed in 1 ml of hot isopropanol and heated to 70°C for 10 min to inactivate the lipolytic enzymes. The lipids were extracted and separated by thin layer chromatography as previously described (1). Phosphatidylcholine and digalactosyl diacylglycerol were identified by comparison with known standards, removed from the TLC plates and analyzed quantitatively (3,4). Trifluoperazine sulfoxide was a generous gift from Smith, Kline, and French Laboratories. All other reagents were obtained from Sigma Chemical Co.

RESULTS

In the first experiment (Table 1) leaf disks were incubated in various test solutions for 18 hours. The rates of hydrolysis of phosphatidylcholine (PC) and digalactosyl diacylglycerol (DGDG) in dark controls were very low (about 4%). Each of the eight treatments increased the hydrolytic rates, however some caused much more pronounced increases. For each treatment the

Table 1. Hydrolysis of phosphatidylcholine (PC) and digalactosyl diacylglycerol (DGDG) in potato leaf disks which were floated in various test solutions for 18 hours. Number in parentheses is the relative % hydrolysis with the control set equal to zero.

Treatment	% hydrolyzed per 18 hours[a]	
	PC	DGDG
Control (H$_2$O)	3.9 (0)[b]	4.2 (0)[c]
2 mM dibucaine	75.6 (71.7)	62.9 (58.7)
50 µM chlorpromazine	32.4 (28.5)	30.5 (26.3)
50 µM w$_7$	21.5 (17.6)	33.4 (29.2)
50 µM trifluoperazine	43.7 (39.8)	32.8 (28.6)
50 µM trifluoperazine sulfoxide	7.1 (3.2)	11.5 (7.3)
10 mM CaCl$_2$	11.7 (7.8)	10.5 (6.3)
50 µM calcium ionophore, A 23187	10.9 (7.0)	15.3 (11.1)
1 mM indomethacin	42.2 (38.3)	40.7 (36.5)

[a] The values reported are the means of three separate experiments
[b] The level of PC at 18 hours was 229 nmol/12 disks
[c] The level of DGDG at 18 hours was 288 nmol/12 disks

rates of hydrolysis of PC and DGDG were comparable, which probably indicates that they are both degraded by the same enzyme. The rates of hydrolysis of PC and DGDG were stimulated the most by the dibucaine treatment. The other three calmodulin antagonists (chlorpromazine, w$_7$, and trifluoperazine) also caused significant rate increases (21.5-43.7%). The hydrolytic rates were much lower with trifluoperazine sulfoxide than trifluoperazine, which is suggestive of an authentic calmodulin interaction rather than a nonspecific effect of the drugs. The calcium ionophore and CaCl$_2$ each caused a small increase in the hydrolytic rates. However, indomethacin, a calcium antagonist, caused a large stimulation in the rates of hydrolysis, comparable to those observed with the calmodulin antagonists.

Because dibucaine caused the largest stimulation of hydrolysis in Table 1, a time-course study of its effects on potato leaf disks was conducted (Fig. 1). This study revealed that the rate of hydrolysis of PC was linear from 0 to 6 hours (about 13% of the PC was hydrolyzed in 6 hours) and suggested that the rates of hydrolysis observed in Table 1 were probably also linear for 0 to 18 hours.

DISCUSSION

In this study we demonstrate that four calmodulin antagonists and one calcium antagonist stimulate the rate of breakdown of PC and DGDG in potato leaf disks. These results are very different from our in vitro studies with potato leaf homogenates (1) where calmodulin antagonists inhibited the rate of autolytic PC breakdown. In another related study fluphenazine (also a calmodulin antagonist) when applied to senescing pea leaves was shown to delay membrane deterioration as measured by several criteria (5). In the study of barley roots (1), dibucaine was shown to increase the proportions of saturated fatty acids. It is conceivable that such effects were caused by the selective stimulation of an acyl hydrolase which is specific for the esters of polyunsaturated fatty acids. Very little is known about the properties of the lipolytic enzymes in pea leaves and barley roots, but it is likely that they are very different than those that we have studied in potato leaves (2). The different effects caused by the drugs used in these four studies could easily by explained by different sensitivities of the lipolytic enzymes of each type of tissue to the drugs.

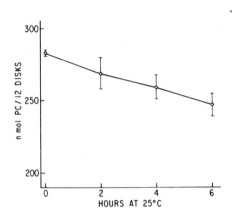

Figure 1. Time-course study of the rate of hydrolysis of phosphatidylcholine (PC) in potato leaf disks floated in 2 mM dibucaine. Data points are the means of three experiments ± S.D.

REFERENCES

1. P.C. Jackson and J.B. St. John, Anesthetics alter the lipid composition of barley-root membrane, Planta 162:415 (1984).
2. R.A. Moreau and T.S. Isett, Autolysis of membrane lipids in potato leaf homogenates: effects of calmodulin and calmodulin antagonists, Plant Sci. 40:95 (1985).
3. J.C. Dittmer and A.W. Wells, Quantitative and qualitative analysis of lipids and lipid components, in "Methods in Enzymology" S.P. Colowick and N.O. Kaplan, eds., Academic Press, New York (1969).
4. P.G. Roughan and R.D. Batt, Quantitative analysis of sulfolipid (sulfoquinovosyl diglyceride) and galactolipid (monogalactosyl and digalactosyl diglycerides) in plant tissues, Anal. Biochem., 22:74 (1968).
5. Y.Y. Leshem, S. Sridhara, and J.E. Thompson, Involvement of calcium and calmodulin in membrane deterioration during senescence of pea foliage, Plant Physiol. 75:329 (1984).

LOCALIZATION AND PROPERTIES OF CHOLINEPHOSPHATE CYTIDYLYLTRANSFERASE
ACTIVITY IN CASTOR BEAN ENDOSPERM*

Anthony J. Kinney and Thomas S. Moore, Jr.

Department of Botany
Louisiana State University
Baton Rouge LA 70803 USA

INTRODUCTION

The reaction catalyzed by CTP:cholinephosphate cytidylyltransferase (E.C. 2.7.7.15) has been postulated to be a control reaction in the synthesis of phosphatidylcholine (PtdCho) in animals and some plant tissues (1,2). A number of studies utilizing animal tissues have indicated that the enzyme is regulated by translocation of a soluble enzyme to the cytosol, where it becomes activated (2). We report here a study of this enzyme in postgermination castor bean endosperm.

MATERIALS AND METHODS

Seeds of castor bean endosperm (Ricinus communis L. var. Hale) were removed from their seedcoats and germinated under aseptic conditions for 3-days at 30°C. Thirty endosperm halves were placed in a petri dish on ice and chopped with a razor blade for 10 min in 10ml of 0.15M Tris buffer (pH 7.5) containing 0.5M sucrose, 3mM EDTA and 10mM KCl. The homogenate was filtered through 2 layers of Dacron cloth. For differential centrifugation the filtered homogenate was successively centrifuged at 250gx10min, 800gx10min, 12,000gx10min and 100,000gx60min. The pellets from each step were resuspended in 1ml of homogenization fluid. For sucrose density gradient centrifugation, 5ml of the filtered homgenate were layered onto 10ml of 0.6M sucrose which topped a 1.0-2.25M linear sucrose gradient (20ml) and a 2.0ml, 2.25M sucrose cushion. The sucrose contained 3mM EDTA, pH 7.5. The gradient was centrifuged at 20,000rpm for 3 hours. Cytidylyltranferase was assayed according to the method of Sleight and Kent (3) and protein by the method of Bradford (4).

For determination of the cholinephosphate and CDPcholine levels in vivo, three-day old endosperms were frozen and ground to a powder in liquid nitrogen. The powder was extracted using 15% TCA/ethyl ether (5). The final extract was evaporated to dryness in vacuo, redissolved in 20mM HEPES buffer (pH 7.0) and separated by ion-exchange and thin-layer chromatography (6). Cholinephosphate and CDPcholine were assayed by periodide precipitation of choline (5) after incubation of the compounds with alkaline phosphatase (Sigma type I-S, 15U) and phosphodiesterase (Sigma type IV, 0.08U).

---

*Supported by NSF grant DMB-8402001

RESULTS AND DISCUSSION

Following differential centrifugation, 70% of the cytidylyltransferase activity was in the 12,000gx10min pellet, 10% was in the 100,000gx60min pellet and no activity was detected in the 100,000gx60min supernatant. Recovery of activity after centrifugation was >90%. After density gradient centrifugation most of the cytidylytransferase (60-75%) was localized in the endoplasmic reticula (ER) band and about 20% in a light membrane band known as "band A" (7). No soluble activity was observed at the top of the gradient. The absence of soluble enzyme suggests that the cytoplasm-microsome model of regulating cytidylytransferase activity does not apply to this tissue. This cannot be proven at this time, however, since the enzyme may be switched fully on by being completely membrane bound or a "soluble" form of the enzyme may still be loosely associated with (rather than bound to) a membrane fraction such as the ER or band A.

The activity in band A was variable (12-24%). An increase in band A activity could be correlated with an increase in the visible size of that band as well as a reduction in the activity in the ER fraction, and vice versa. We propose that at least part of band A is a subfraction of the ER which contains the cytidylyltransferase but not the cholinephosphotransferase (we have never observed cholinephosphotransferase activity in band A). Minor changes in homogenization medium and technique could have resulted in the discrepancies between this study and that of Lord et al (7), which reported most of the cytidylyltransferase activity in the band A fraction. The possibilty of easily obtained compartmentalization artifacts must therefore be considered when intracellular localization studies of this enzyme are performed.

The ER enzyme was Mg^{2+}-dependent (optimal stimulation at concentrations above 2mM). The pH optimum was 7.0 and the apparent Km's for cholinephosphate and CTP were 2.5mM and 0.22mM respectively (calculated from double reciprocal plots). The final cytoplasmic concentration of cholinephosphate was estimated from the amount of cholinephosphate in the tissue (174nmol/g fresh wt), the water content of the tissue (80%) and the amount of tissue which was cytoplasm (from electron micrographs, 50%). The resulting estimated value of 0.21mM is 10x lower than the Km for this substrate. However, the concentration of cholinephosphate available to the cytidylyltransferase might be higher if the cholinephosphate were concentrated in a cell compartment, such as the ER lumen. We do not know if this situation exists in this tissue, although we do have some evidence for multiple pools of cholinephosphate and CDPcholine (6).

It is possible that the activity of cytidylyltransferase is regulated by the CTP supply, since the Km for CTP (0.22mM) is 10x lower than the Km for cholinephosphate and is near the expected concentration of CTP in this tissue (determination of the CTP concentration in this tissue is in progress).

The level of CDPcholine in this tissue was 180nmol/g fresh wt, which is more than twice the estimated level of PtdCho (70nmol/g fresh wt). This suggests that conversion of cholinephosphate to CDPcholine is not "rate limiting" for PtdCho biosynthesis in this tissue. We seek to confirm this with choline pulse-chase experiments.

REFERENCES

1. T. S. Moore, 1982, Phospholipid biosynthesis, Ann. Rev. Plant Physiol., 33:235-259.
2. D. E. Vance and S. L. Pelech, 1984, Enzyme translocation in the

regulation of phosphatidylcholine biosynthesis, Trends Biochem. Sci., 9:17-20.
3. R. Sleight and C. Kent, 1983, Regulation of phosphatidylcholine biosynthesis in mammalian cells. II. Effects of phospholipase C treatment on the activity and subcellular distribution of CTP:phosphocholine cytidylyltransferase in Chinese hamster ovary and LM cell lines, J. Biol. Chem., 258:831-835.
4. M. M. Bradford, 1976, A. rapid and sensitive method for the quantitation of microgram quantities of protein utilizing the principle of protein-dye binding, Anal. Biochem., 72:248-254.
5. A. J. Barek and D. J. Tuma, 1981, Determination of choline, phosphorylcholine, and betaine, Meth. Enzymol., 72:287-292.
6. A. J. Kinney and T. S. Moore, 1986, Phosphatidylcholine synthesis in castor bean endosperm, Plant Physiol., in press.
7. J. M. Lord, T. Kagawa and H. Beevers, 1972, Intracellular distribution of enzymes of the cytidine diphosphate choline pathway in castor bean endosperm, Proc. Natl. Acad. Sci. USA, 69:2429-2432.

CHOLINE KINASE ACTIVITY IN CASTOR BEAN ENDOSPERM*

Anthony J. Kinney and Thomas S. Moore, Jr.

Department of Botany
Louisiana State University
Baton Rouge LA 70803 USA

INTRODUCTION

The reaction catalyzed by choline kinase (E.C. 2.7.1.32: ATP:cholinephosphotransferase) is the first committed reaction of the nucleotide pathway for phosphatidylcholine (PtdCho) biosynthesis. Various theoretical considerations have led to the suggestion that this reaction is far from equilibrium and, along with the cytidylyltransferase reaction, may be rate-limiting for PtdCho biosynthesis in some animal tissues (1). As part of our studies on the regulation of PtdCho biosynthesis in plants we have investigated the properties of this enzyme and the <u>in vivo</u> concentrations of metabolites involved in the reaction in postgermination castor bean endosperm.

TISSUE PREPARATION

Castor bean endosperms were homogenized in 150mM Tris-HCl buffer, pH 7.5, which contained 0.5M sucrose, 3mM EDTA and 10mM KCl. The homogenate was filtered through 2 layers of cheesecloth and then centrifuged at 3°C and 150,000xg for 60 minutes in a Sorval OTD65B ultracentrifuge. The supernatant was concentrated by ultrafiltration (Amicon B15 filter) at 2°C for 8 hours (approx. 100x concentration). The concentrate was diluted with 100mM glycylglycine buffer (pH 10.0) to a final protein concentration of 10mg/ml.

CHOLINE KINASE REACTION

Choline kinase was assayed in a final volume of 20ul which contained 10mM [^{14}C]choline-Cl (0.25 Ci/mol), 10mM Tris-ATP, 5mM $MgCl_2$, 50ug of protein and 100mM glycylglycine, pH 10.0. The reaction was incubated at 30°C for 0-30 min and was stopped by adding 15ul of ice-cold water followed by 150ul of allyl cyanide-tetraphenylboron. Labeled choline phosphate formed during the reaction was extracted and measured by the method of Burt and Brody (2). The forward reaction was Mg^{2+}-dependent with maximal stimulation observed at 3mM and above. Manganese and a variety of other divalent cations, including calcium, could not substitute for the Mg^{2+}. The enzyme had a narrow pH optimum, with maximal activity observed at pH 10.0, and

---

*Supported by NSF grant DMB-8402001

apparent Km's (calculated from double reciprocal plots) for choline and ATP of 1.0mM and 0.29mM respectively. The maximal velocity of the forward reaction, at saturating substrate concentrations, was 41.7pkat.mg^{-1} protein. The enzyme was inhibited <u>in vitro</u> 70% by 1mM hemicholinium-3 (a choline analog), but not by low (0-4mM) concentrations of ADP. We have observed that 50mM hemicholinium-3 applied directly to the abaxial surface of endosperm tissue caused a 66% inhibition of [^{14}C]choline incorporation into PtdCho whereas incorporation of cholinephosphate was inhibited only 40%. Incorporation of other precursors into PtdCho was inhibited 15-50% by hemicholinium-3 (3).

CHOLINE KINASE REVERSE REACTION

The choline kinase also catalysed a reverse reaction. The incubation medium contained 10mM [^{14}C]cholinephosphate, 10mM ADP, 5mM MgCl$_2$, 50ug protein and 100mM glycylglycine buffer, pH 10.0, in a final volume of 20ul. The reaction was incubated at 30°C for 0-15 min and stopped with 15ul of ice cold water followed by 150ul allyl cyanide-tetraphenylboron. Labeled choline formed during the reaction was measured in the organic phase after microcentrifugation. At pH 10.0 phosphatase activity accounted for less than 5% of the choline formed (determined by zero-ADP controls), the final rates were corrected for this. The reaction was also observed using unlabeled cholinephosphate and measuring the ATP formed by the luciferase technique (4). The enzyme had apparent Km's for cholinephosphate and ADP of 4.0mM and 1.0mM respectively. Approximately 1 mole of ATP was formed for every mole of choline when cholinephosphate and ADP were present in saturating amounts. The maximal velocity of the reverse reaction was 111.0pkat.mg^{-1} protein.

SUBSTRATE POOL SIZES AND THE REGULATION OF CHOLINE KINASE ACTIVITY

Endosperms were frozen with liquid N$_2$ and extracted as previously described (5). The in vivo concentrations of ATP, ADP and AMP were measured using the luciferase reaction (4); choline and cholinephosphate by periodide precipitation (6). Concentrations were estimated from tissue water content and approximate cytoplasmic volumes calculated from electron micrographs of the tissue (5). The concentrations of choline and cholinephosphate were 0.04mM and 0.21mM respectively. The concentrations of ADP and AMP were 0.06mM and 0.04mM respectively. The [ATP] was 0.26mM, which was close to the apparent Km of the kinase for ATP (0.29mM). This suggests that, under some conditions, the activity of this enzyme, and hence PtdCho biosynthesis, may be regulated by the availability of ATP. Choline kinase activity may also be regulated by free-Mg^{2+} since, over the physiological range of [Mg^{2+}] (0.5-2mM, ref.1), we have observed the enzyme to be very sensitive to changes in the [Mg^{2+}] of the incubation medium.

In 3-day old endosperm the choline kinase reaction is probably near equilibrium (the disequilibrium ratio estimated from the above values is 0.2; although this does not take into account any possible intracellular compartmentation of substrates). It seems likely therefore that this reaction does not determine the rate of PtdCho biosythesis in this tissue.

REFERENCES

1. J. P. Infante, 1977, Rate-limiting steps in the cytidine pathway for the synthesis of phosphatidylcholine and phosphatidylethanolamine, <u>Biochem. J.</u>, 167:847-849.
2. A. M. Burt and S. A. Brody, 1975, The measurement of choline kinase

activity in rat brain: The problem of alternate pathways of ATP metabolism, Anal. Biochem., 65:215-224.
3. A. J. Kinney and T. S. Moore, 1986, Phosphatidylcholine synthesis in castor bean endosperm. I. Metabolism of L-serine, Plant Physiol., in press.
4. H. Spielman, U. Jacob-Miller and P. Schulz, 1981, Simple assay of 0.1-1.0 pmol of ATP, ADP, and AMP in single somatic cells using purified luciferin luciferase, Anal. Biochem., 113:172-178
5. A. J. Kinney and T. S. Moore, Localization and properties of cholinephosphate cytidylyltransferase activity in castor bean endosperm, This volume.
6. A. J. Barek and D. J. Tuma, 1981, Determination of choline, phosphorylcholine, and betaine, Meth. Enzymol., 72:287-292.

THE CONTROL OF CTP: CHOLINEPHOSPHATE CYTIDYLYLTRANSFERASE IN PEA STEMS

Molly J. Price-Jones and John L. Harwood

Department of Biochemistry,
University College,
Cardiff, CF1 1XL, U.K.

INTRODUCTION

Phosphatidylcholine is the major phospholipid of the non-photosynthetic membranes of higher plants[1], and its main route of synthesis is via the CDP-base pathway[2]. Studies with the plant growth-promoting compound, indol-3-yl acetic acid (IAA), using the third internode region of pea (<u>Pisum sativum</u> L.) stems, have shown that in the presence of IAA the incorporation of [Me-^{14}C]choline into phosphatidylcholine was reduced within one hour of treatment[3]. This was shown to be due to a change in the activity of cytidylyltransferase.

This enzyme has been more widely studied in mammalian tissues where it has also been found to be rate-limiting for phosphatidylcholine biosynthesis. Several methods for control of the enzyme in mammalian tissues have been proposed; these include aggregation of the enzyme to higher molecular weight forms which have greater activity, activation by translocation from the cytosol to the endoplasmic reticulum and a phosphorylation-dephosphorylation cycle[5].

We have examined these putative control mechanisms in the pea stem tissue and also the possibility of regulation by alteration of cytidylyltransferase enzyme protein levels and allosteric regulation by nucleotides.

MATERIALS AND METHODS

Tissue preparation and assay.

Seeds were germinated, explants excised and treated, and subcellular fractions prepared as previously[4]. Cytidylyltransferase was assayed by the method of Infante and Kinsella[6] modified as previously described[7].

Preparation of exogenous lipids.

Potassium oleate was prepared from oleic acid by the method of Pelech et al.[8] and resuspended in 0.32M sucrose/2mM Tris/HCl (pH 7.4). Suspensions of monoacylphosphatidylethanolamine and phosphatidylglycerol from soybean were prepared in 100mM Tris/HCl (pH 7.4) by the method of Feldman et al.[9].

Measurement of enzyme protein.

Soluble cytidylyltransferase protein levels were measured using a competitive inhibition enzyme-linked immunosorbent (ELISA) method[10]. A polyclonal antibody was raised in rabbit to the purified cytidylyltransferase and this was used in the ELISA with a goat anti-rabbit IgG-horseradish peroxidase conjugate. The colorimetric reaction was obtained with a substrate containing hydrogen peroxide and O-phenylenediamine[11].

Protein phosphorylation.

This was carried out[12] using [γ-^{32}P]ATP prepared by the method of Glynn and Chappell[13]. Gels were fluorographed using Amplify.

RESULTS AND DISCUSSION

Cytidylyltransferase is mainly soluble in pea stems[7]. IAA-treatment caused a decrease in overall cytidylyltransferase activity, accompanied by a significant increase (from 2% to 5.2%) in the activity of the microsomes. However, there was no significant decrease in the soluble activity of treated material, which represented 89.5% of the total cytidylyltransferase activity. This was the reverse of the situation in mammals[5].

The presence of oleate was found to stimulate the activity of cytidylyltransferase in the post-mitochondrial supernatant from the pea stems. When the post-mitochondrial supernatant was incubated with 2mM oleate and then fractionated further, the soluble cytidylyltransferase activity was significantly increased from 0.43 to 0.96 nmoles CDP-choline formed/min/mg protein. Again this contrasted with the situation in mammals where stimulation by oleate was due to translocation of the enzyme

Table 1. Summary of the effects of possible methods of control on cytidylyltransferase in pea stems

Method tested	Effect on cytidylyltransferase activity
Alteration of subcellular distribution	Microsomal activity up from 2% to 5% of total in IAA-treated material
Oleate-mediated	223% stimulation of soluble activity at 2mM. 67% loss of activity from microsomes
Phospholipid-mediated	No effect
Phosphorylation-dephosphorylation cycle	No observed phosphorylation of cytidylyltransferase protein
Alteration in enzyme protein level	Small (5%) decrease in soluble enzyme protein in IAA-treated material
Regulation by nucleotides	CMP and ATP inhibitory

Table 2. The effect of 0.2mM CMP or 1.5mM ATP on the kinetic parameters of cytidylyltransferase from pea stems (Means ± S.D.)

	Phosphorylcholine substrate		
	Control	+ CMP	+ATP
K_m(mM)	2.97±0.58	0.81±0.12 (p 0.05)	0.42±0.10 (p 0.02)
V_{max} (μmoles/min/mg protein)	0.30±0.07	0.03±0.06 (p<0.05)	0.02±0.01 (p<0.05)
	CTP substrate		
K_m(mM)	0.36±0.09	2.92±1.00 (p 0.10)	1.22±0.60
V_{max} (μmoles/min mg protein)	0.20±0.03	0.17±0.05	0.05±0.01 (p<0.05)

from the cytosol to the endoplasmic reticulum[8]. The addition of 0.25mM monoacyl PE or 0.25mM PG to assays using the post-mitochondrial supernatant from the pea stems had no effect on cytidylyltransferase activity.

The subunit molecular weight of pea cytidylyltransferase is 56,000[7]. No proteins of this molecular weight were phosphorylated under the conditions used.

There was only a 4.5% reduction in soluble cytidylyltransferase protein within 1h of IAA-treatment of pea stems, despite a 49% reduction in soluble activity.

The effects of ATP, ADP, AMP, CDP, CMP and GTP at concentrations of 0-2mM were investigated. 0.2mM CMP or 1.5mM ATP gave maximal inhibition but no effect was obtained with any of the other nucleotides tried. CMP was found to act as a competitive inhibitor towards the CTP substrate, but as an uncompetitive inhibitor towards the phosphorylcholine substrate. ATP was also an uncompetitive inhibitor towards the phosphorylcholine substrate but appeared to be a mixed inhibitor with CTP as substrate (Table 2).

The results show that cytidylyltransferase in pea stems is regulated differently from that in mammals. Activity was not increased by intracellular translocation nor by exogenous lipids. In fact, the cytosolic enzyme was more important in pea stems than in mammalian tissues.

IAA was responsible for a decrease in cytidylyltransferase protein levels but this could not account for the large decrease in activity and nucleotides may also be involved.

No direct phosphorylation of the enzyme was demonstrated, but ATP inhibited the purified enzyme. However, the physiological significance of control by ATP is doubtful as the maximal effect in vitro was obtained at 1.5mM and the endogenous level of ATP was found to be 2.15μM. CMP also reduced activity and may act as a negative feedback inhibitor.

ACKNOWLEDGEMENT. We are grateful to the A.F.R.C. for financial support.

REFERENCES

1. J.L. Harwood, Plant acyl lipids:structure, distribution and analysis, in: "Biochemistry of Plants", Vol. 4, P.K. Stumpf and E.E. Conn, eds., Academic Press, New York (1980).

2. J.B. Mudd, Phospholipid biosynthesis, in: "Biochemistry of Plants," Vol. 4, P.K. Stumpf and E.E. Conn, eds., Academic Press, New York (1980).

3. T.S. Moore, M.J. Price-Jones, and J.L. Harwood, The effect of indole acetic acid on phospholipid metabolism in pea stems, Phytochem., 22:2421 (1983).

4. M.J. Price-Jones and J.L. Harwood, Hormonal regulation of phosphatidylcholine synthesis in plants. The inhibition of cytidylyltransferase by indol-3-yl acetic acid, Biochem.J., 216:627 (1983).

5. S.L. Pelech and D.E. Vance, Regulation of phosphatidylcholine biosynthesis, Biochim.Biophys.Acta, 779:217 (1984).

6. J.P. Infante and J.E. Kinsella, A novel method for determining equilibrium constants CTP:phosphorylcholine cytidylyltransferase Biochim.Biophys.Acta, 526:440 (1978).

7. M.J. Price-Jones and J.L. Harwood, Purification of CTP:cholinephosphate cytidylyltransferase from pea stems, Phytochem., 24:2523 (1985).

8. S.L. Pelech, H.W. Cook, H.B. Paddon, and D.E. Vance, Membrane-bound CTP:phosphocholine cytidylyltransferase regulates the rate of phosphatidylcholine synthesis in HeLa cells treated with unsaturated fatty acids, Biochim.Biophys.Acta, 795:433 (1984).

9. D.A. Feldman, C.R. Kovac, P.L. Dranginis, and P.A. Weinhold, The role of phosphatidylglycerol in the activation of CTP:phosphocholine cytidylyltransferase from rat lung, J.Biol.Chem., 253:4980 (1978).

10. A. Voller, D.A. Bidwell, and A. Bartlett, Enzyme immunoassays in diagnostic medicine:theory and practice, Bull.Wld.Hlth.Org., 53:55 (1976).

11. M.J. Price-Jones and J.L. Harwood, Changes in CTP:cholinephosphate cytidylyltransferase protein levels in pea stems treated with indole-3-acetic acid, Phytochem., (in press).

12. K. Veluthambi and B.W. Poovaiah, Calcium- and calmodulin-regulated phosphorylation of soluble and membrane proteins from corn coleoptiles, Plant Physiol., 76:359 (1984).

13. I.M. Glynn and J.B. Chappell, A simple method for the preparation of ^{32}P-labelled adenosine triphosphate of higher specific activity, Biochem.J., 90:147 (1964).

COCOA BUTTER BIOSYNTHESIS
COCOA SEED DIACYLGLYCEROL ACYLTRANSFERASE: STUDIES ON THE MICROSOMAL BOUND ENZYME

Lauren McHenry and Paul J. Fritz

ACRI Cocoa Research Laboratory
Department of Food Science
The Pennsylvania State University
University Park, Pennsylvania 16802

INTRODUCTION

Triacylglycerols make up about 96% of mature cocoa (Theobroma cacao) seed lipid, more than half the seed dry weight, and thus are the major storage products of the seed. These storage products are a mixture of glycerol esters of oleic, palmitic, stearic, and much smaller amounts of linoleic acid. The collection of esters is called cocoa butter, a solid fat, and, at more than $4 a pound, is the most expensive edible fat in the world. Annual world production of cocoa butter is more than 2 billion pounds.

Our laboratory has undertaken a comprehensive study of the enzymes, and their genes, of cocoa butter biosynthesis and in this paper we present some observations on diacylglycerol acyl transferase (DAGAT), the final enzyme in the pathway.

MATERIALS AND METHODS

Cocoa seeds were obtained from pods harvested at various stages of development and supplied to us by the Hershey Foods Corporation Agribusiness Division from their research farm in Belize, Central America. Seed maturity was estimated by a series of visual clues including color and size of pods, color and consistency of seeds, and consistency of pulp surrounding seeds. From these clues, an approximate days after pollination (DAP) designation was assigned.

Microsomes were prepared and protein determined as previously described (Fritz et al., 1985). DAGAT activity was assayed by measuring the rate of incorporation of [C-14] labelled fatty acids into triacylglycerol (TAG) in the presence of acyl-CoA's, diacylglycerol (DAG), and microsomes. Radio-labelled products were co-chromatographed with authentic standards on silica gel TLC plates (Analtech) in chloroform/acetone (96:4 V/V). Triacylglycerols were scraped from the plates and counted in a Beckman LS-100 with 5 ml ACS II.

TABLE 1 Cocoa Seed Maturity and DAGAT Activity

Visual Clues	Estimated DAP	DAGAT Activity
Pods green Pulp hard Seeds liquid & clear	100	.396*
Pods green Pulp hard Seeds semi solid, beginning to show color	110-120	1.82*
Pods beginning to yellow Pulp hard Seeds solid, not fully colored	120-150	2.48**
Pods yellow Pulp soft Seeds solid, fully colored	150-180	.585**

DAGAT activity is nmol oleate (*) or palmitate (**) incorporated into triacylglycerol per minute per mg microsomal protein. Figures are averages of 2 values obtained in assays using 50 and 100 micrograms of microsomal protein per reaction.

RESULTS AND DISCUSSION

Four stages of seed development were identified and an approximate DAP assigned (Table 1). Microsomes prepared from seeds at all four stages of development were capable of catalyzing TAG formation from DAG. This DAGAT mediated activity increased during seed development, peaking between 120 and 150 DAP a period during which rapid accumulation of TAG in cocoa seeds is known to occur (Lehrian and Keeney, 1980).

Microsomal preparations from seeds between 110 and 120 DAP could catalyze formation of TAG in the absence of added DAG. Addition of increasing amounts of 1-palmitoyl-2-oleoyl DAG to the reaction mixture resulted in typical substrate saturation kinetics with a 5 mM plateau. The data suggest a non-saturating pool of endogenous DAG available for DAGAT mediated acylation (Figure 1).

When glycerol-3-phosphate (G-3-P) was substituted for DAG in the standard assay, microsomes from seeds between 110 and 150 DAP catalyzed acyl incorporation into both lysophosphatidic acid (LPA) and TAG, in contrast to fully mature seeds (greater than 180 DAP) which could support only LPA synthesis. The inability of mature seeds to synthesize TAG from G-3-P, as well as the waning ability of seeds 150-180 DAP to catalyze the final acylation reaction in the Kennedy pathway, might be predicted since at these late developmental stages the seed has made and stored amounts of TAG sufficient for its needs. We previously reported large amounts of the first enzyme in the pathway, glycerol-3-phosphate acyltransferase (G-3-PAT), in the post-microsomal supernatant fraction of mature cocoa seeds (Fritz et al., 1986). Whether the microsomal enzyme reported here is a different enzyme is presently unknown but the occurrence of G-3-PAT activity over a broader range of seed development then found for DAGAT is consistent with

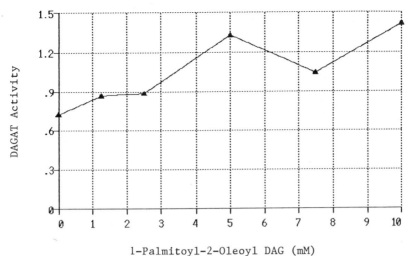

Figure 1. DAG Concentration Curve.

the dual role of G-3-PAT in neutral lipid and phospholipid biosynthesis.

We have also studied the effect of temperature on the acyl specificities and selectivities of cocoa microsomal G-3-PAT and DAGAT. The data from these experiments indicate that both enzymes show broad specifications for the three most common fatty acids present in cocoa butter: palmitic, stearic, and oleic. A trend of increasing selectivity for the unsaturated fatty acid at lower assay temperatures is evident and may be related to the observance of lower melting point cocoa butter obtained from seeds grown in cooler climates (Alvim and Berbert, 1972).

REFERENCES

Alvim, P., and Berbert, P. R. F., 1972, Factores que Afetam o Indice de Iodo da Manteiga de Cacau do Brazil, Revista Theobroma, 2:3-10.

Fritz, P. J., Fritz, K. A., Kauffman, J. M., Patterson, G. R., Robertson, C. A., Stoesz, D. A., and Wilson, M. R., 1985, Cocoa Seeds: Changes in Protein and Polysomal RNA During Development, J. Food Science, 50:946-950.

Lehrian, D. W., and Keeney, P. G., 1980, Changes in Lipid Components of Seeds during Growth and Ripening of Cacao Fruit, J. Am. Oil Chem. Soc., 57:61-65.

SUBSTRATE SPECIFICITY OF PLANT LIPASES

Y.Y. Lin, Charles Yi, Julie Olsen, and Anthony H.C. Huang

Biology Department, University of South Carolina
Columbin, SC 29208

INTRODUCTION

In plants, lipases (EC 3.1.1.3) are active in lipid-storing tissues in seeds during germination. These lipases catalyze the hydrolysis of storage triacylglycerols to fatty acids, which are converted to sugars to support the growth of the embryo. Seed lipases have not been studied intensively, and our knowledge of them lags behind that of the mammalian and microbial enzymes (1,2,3). A unique feature of seed lipases may be that the enzyme from a certain plant species is relatively specific on the characteristic storage triacylglycerols of the same species. Our experimental results show that this specificity is indeed species-dependent, and is not due solely to the physical characteristics of the specific triacylglycerol in the substrate micelles.

MATERIALS AND METHODS

Seeds of inbred maize (Zea mays) MO-17, rapeseed (Brassica napus L. cv. Dwarf Essex), mustard (Brassica juncea), castor bean (Ricinus communis L. cv. Hale), and elm seeds (Ulmus americana) were used. The enzymes were prepared as described (4, 5). Lipase activity was assayed using a colorimetric method (5, 6). The product fatty acids in the reactions catalyzed by maize, castor bean, rapeseed, elm, and pancreas lipases were confirmed by thin layer chromatography and gas chromatography (5). In the hydrolysis of trierucin catalyzed by rapeseed lipase, erucic acid was the dominant product. However, at the very early stage of hydrolysis when only a few percent of the trierucin had been hydrolyzed, a fatty acid tentatively identified as oleic acid by gas chromatography was dominant (5). This oleic acid may be the hydrolytic product of the remaining internal triacylglyerols in the enzyme preparation, or secondary fatty acid generated from the initial product erucic acid by an unknown enzymatic reaction.

RESULTS

Selection of lipases for the current studies

The seed lipases from five plant species were selected because of their differences in the fatty acid composition of the storage triacylglycerols. Castor bean contains about 80% ricinoleic acid; maize seed possesses about

Table I. Hydrolysis of various triacylglycerols by lipases

Source of lipase

Substrate	Porcine pancreas	Castor bean	Maize	Rapeseed	Mustard	Elm
			relative activity			
tricaprin	207	43	127	89	–	100
trilaurin	92	60	0	31	–	4
trimyristin	50	26	0	92	–	3
tripalmitin	5	46	0	27	39	0
tristearin	0	62	0	36	40	0
triolein	97	55	38	44	96	4
trilinolein	100	57	100	89	89	6
triricinolein	53	100	0	83	–	0
tribehenin	–	–	0	16	–	0
trierucin	92	36	45	100	100	0

For each lipase, the activity on one particular triacylglycerol was chosen to represent a 100% relative activity. Relative activities of 100 represent 172 (pancreas), 96 (castor bean), 145 (maize), 32 (rapeseed), 28 (mustard), and 143 (elm) nmol/min.

50% linoleic acid and 30% oleic acid; rapeseed has about 20% each of erucic acid, oleic acid, and linoleic acid; elm seed houses about 60% capric acid (decanoic acid). In addition, commercially prepared porcine pancreatic lipase of well-known substrate specificity (7,8) was used. The hydrolysis of different triacylglycerols, present individually and in mixtures, by various lipases under an identical assay condition in the same experiment was studied.

Hydrolysis of different triacylglycerols by various lipases

As shown in Table I, in general each of the five seed lipases exhibits high activities with triacylglycerols containing the major storage fatty acids found within each respective plant species. Additional evidence of species-dependent substrate specificity of lipases comes from experimental results of the hydrolysis of mixtures of triacylglycerols (5). Such a specificity is of physiolgical significance. Our findings from a direct comparison of triacylglycerols as lipase substrate negate the possibility that the lack of high lipase activity on triacylglycerols such as tristearin and tripalmitin is due only to the low solubility of these substrates. Furthermore, castor bean lipase and rapeseed lipase actually hydrolyze tristearin and tripalmitin at rates comparable to that of the more soluble triolein. The relatively high activities of castor bean, rapeseed, and maize lipases on tricaprin presumably are due in part to the physical nature of tricaprin, as in the case of the pancreatic lipase. However, this factor of physical properties is negligible in the case of elm lipase which is active almost exclusively on tricaprin; such a specificity should be due to inherited structural properties of the enzyme.

Hydrolysis of diacylglycerols, monoacylglycerols, fatty acyl 4-methylumbelliferone and lecithin

The above pattern of fatty acyl specificity was also observed on diacylglycerols, monoacylglycerols, and fatty acyl 4-methylumbelliferone, although the pattern is less distinctive (5). The seed lipases were inactive on lecithins. The gradual loss in the fatty acyl specificity of the seed lipases from tri- to di- to monoacylglycerols may be of physiological significance. Each storage triacylglycerol molecule generally is not composed of only the major fatty acid, but rather of the major faty acid and other fatty acids in the same molecule. Thus, after the lipase has hydrolyzed the major fatty acid from a triacylglycerol molecule, it should still have high capacity to hydrolyze the remaining diacylglycerol and monoacylglycerol containing the other fatty acids.

Positional specificity in the hydrolysis of trilinolein, dilinolein, and monolinolein

Both maize and rapeseed lipases were more active on tri-, than on di- or monolinolein. Castor bean lipase was equally active on tri-, di-, and monolinolein, as has been reported realier (9). These relative activities are different from those of pancreatic lipase which is more active on tri- than dilinolein, and is inactive on monolinolein.

In general, the seed lipases release linoleic acid from both primary and secondary positions from trilinolein. As judged from the kinetics of hydrolysis of rac-glyceryl-2,3-stearate-1-oleate and rac-glyceryl-1,3-stearate-2-oleate, and of trilinolein and dilinoleins, maize lipase exerts some degree of preference in releasing fatty acid from the primary position of a triacylglycerol. At the primary position, maize lipase is more active on oleyl ester than stearyl ester.

ACKNOWLEDGMENT

The work was supported by a grant from the National Science Foundation.

REFERENCES

1. Galliard T. (1980) in The Biochemistry of Plants (Stumpf PK and Conn EE, eds), Vol 4, Academic Press, New York. pp 85-116.
2. Huang AHC. (1984) in Lipases (Brockman HL and Borgstrom B, eds). Elsevier, Amsterdam. pp. 419-442.
3. Huang AHC. (1987) in The Biochemistry of Plants (Stumpf PK and Conn EE, eds), Vol 9, Academic Press, New York. (in press)
4. Lin YH, Huang AHC. (1983) Arch. Biochem. Biophys. 225:360-369.
5. Lin YH, Yu C, Huang AHC. (1986) Arch. Biochem. Biophys. 244:346-356.
6. Dixon M, Chen SHP. (1979) Anal. Biochem. 97:403-409.
7. Entressangles B, Pasero L, Savary P, Sarda L, Desnuelle P. (1961) Bull. Soc. Chim. Biol. 43:581-591.
8. Brockerhoff H., Jensen RG. (1974) Lipolytic Enzymes, Academic Press, New York.
9. Ory RL, St. Angelo AJ, Altschul AM. (1962). J. Lipid Res. 3:99-105.

LIPIDS FROM RICE ANTHERS

S. Toriyama[1], K. Hinata[1], I. Nishida[2], and N. Murata[2]

[1]Laboratory of Plant Breeding, Faculty of Agriculture, Tohoku University, Sendai 980, and [2]National Institute for Basic Biology, Okazaki 444, Japan.

INTRODUCTION

The phosphatidylglycerols (PG), but no other lipid classes, from the chilling-sensitive plants, contain a large proportion of saturated molecular species, 1,2-dipalmitoyl-PG and sn-1-palmitoyl-2-(trans-3)hexadecenoyl-PG, which undergo a thermotrophic phase transition at the room temperature range or above (Murata et al. 1982; Murata 1983; Murata and Yamaya 1984). The phase transition of cellular membranes induced by these saturated lipid molecules is proposed to be the primary event of the chilling injury.

Rice is one of the most chilling-sensitive crops, and the chilling damage in cool summer limits its productivity in the temperate regions. The chilling sensitivity of rice plant varies with organs, developmental stages and cultivars. Among the life cycle of rice plants, the most sensitive stage is the phase from tetrad to early microspore of pollen mother cells in anthers (Satake and Hayase 1970; Satake 1974).

In the present study we compared fatty acid compositions of glycerolipids from rice anthers and leaves, and found that the lipid molecules having high phase transition temperatures are more abundant in the anthers than the leaves. This is consistent with the finding that the anthers are more sensitive to chilling than the leaves.

MATERIALS AND METHODS

Oryza sativa L. cv. Hayayuki (Japonica) was grown in the experimental field of Tohoku University. Anthers and anther walls were collected from the spikelets by the sucking method (Toriyama and Hinata 1984) one day before and 1-2 hours after anthesis, respectively. The collected anthers and anther walls were stored at $-25^{\circ}C$ until lipid extraction. For collection of leaves, the seedlings were grown in plastic nursery pans in a greenhouse under natural light conditions. The 3rd, 4th and 5th leaves of 4.5-leaf stage seedlings were collected.

Anthers (1 g fresh weight) and anther walls (0.13 g fresh weight) were homogenized with a glass homogenizer in 10 ml chloroform/methanol (1:2, v/v).

The homogenate was centrifuged at 2000 xg for 10 min. The supernatant was collected, while the pellet was re-extracted as above, and centrifuged. The supernatants were combined. Leaves (100 g fresh weight) were cut into small pieces and homogenized in 1200 ml chloroform/methanol (1:2, v/v). The homogenate was filtered, and the filtrate was collected. Chloroform and water were added to the supernatant or filtrate to give chloroform:methanol: water = 1:1:0.9 (v/v). The chloroform layer was collected, evaporated, dried under vacuum, and weighed. The lipid extract from the anthers (15 mg) and that from the anther walls (6.5 mg) were dissolved in 1 ml chloroform. The lipid extract from leaves (1.8 g) was dissolved in 10 ml chloroform. The lipids in chloroform were stored at $-80^{\circ}C$ or $-25^{\circ}C$ until use. The total lipid extracts were separated into lipid classes according to the method of Murata et al. (1982), except that DEAE-Toyopearl 650M was used instead of DEAE-Sepharose CL-6B for column chromatography.

Aliquots of the total lipids and separated lipid classes were subjected to methanolysis at $80^{\circ}C$ for 2 hours in 5% hydrochloric acid in methanol. The resultant fatty acid methyl esters were extracted with hexane and analyzed by gas liquid chromatography and gas chromatography-mass spectrometry. Positional distribution of fatty acids was analyzed by selective hydrolysis of the ester linkage at the C-1 position with <u>Rhizopus delemar</u> lipase (Fischer et al. 1973; Murata et al. 1982).

RESULTS AND DISCUSSION

The fatty acid compositions of the total lipids as well as the separated lipids are presented in Table I, II and III. Palmitic acid (16:0), trans-3-hexadecenoic acid (16:1t, only in leaf PG), stearic acid (18:0), oleic acid (18:1), linoleic acid (18:2) and α-linolenic acid (18:3) were major components. A characteristic feature of the fatty acid composition was the high content of 18:0 in all lipid classes of the anthers and anther walls. In the total lipids from the anther walls, 18:0 amounted to 28% of the total fatty acids, whereas in the total lipids from leaves it did not exceed 2%. 18:0 distributed in all lipid classes. The positional distribution of fatty acids examined by the selective enzymatic hydrolysis indicated that 18:0 existed in both C-1 and C-2 positions of the glycerol backbone of monogalactosyl diacylglycerol (MGDG), digalactosyl diacylglycerol (DGDG), phosphatidylcholine (PC) and phosphatidylethanolamine (PE) from the anthers (data not shown).

Another characteristics in the fatty acid composition was the difference in the degree of fatty acid saturation among the lipids from the leaves, anthers and anther walls. The proportion of saturated fatty acids was higher in anthers and anther walls than leaves in each of sulfoquinovosyl diacylglycerol (SQDG), MGDG, DGDG, PC and PE. On the contrary, no great difference among the tissues was observed in the fatty acid saturation of phosphatidylinositol (PI).

The proportion of saturated fatty acids in PG was the same in the anthers and leaves, if 16:1t was regarded as a saturated fatty acid (Murata et al. 1982). However, a part of 16:0 and 16:1t in leaf PG was substituted by 18:0 in the anther PG. Since elongation of a fatty acid increases the phase transition temperature of lipids, the high content of 18:0 in anther PG may account for the high sensitivity of anthers to chilling. The higher proportion of saturated fatty acids in SQDG, MGDG and DGDG from the anthers than in those from the leaves may contribute to the higher chilling sensitivity of anthers than leaves. The very high contents of 18:0 in anther walls may suggest that the anther walls, rather than the pollen grains, are the chilling-sensitive tissues at the meiotic stage of pollen mother cells.

TABLE I. The Fatty Acid Composition (Mole Percent) of Lipids from Leaves

	14:0	16:0	16:1t	17:0	18:0	18:1	18:2	18:3	18:4	Saturated
Total	0	17	3	0	2	2	13	62	1	22*
P G	0	36	28	1	2	3	7	18	5	67*
SQDG	0	34	0	1	2	2	8	53	0	37
MGDG	0	1	0	0	0	1	4	94	0	1
DGDG	0	8	0	0	3	1	5	83	0	11
P I	0	47	0	1	2	3	24	23	0	50
P C	0	30	0	0	2	5	37	26	0	32
P E	0	25	0	0	2	5	48	20	0	27

*16:1t is included

TABLE II. The Fatty Acid Composition (Mole Percent) of Lipids from Anthers

	14:0	16:0	16:1t	17:0	18:0	18:1	18:2	18:3	18:4	Saturated
Total	0	22	0	2	9	6	18	41	2	33
P G	1	58	0	1	7	8	10	15	0	67
SQDG	1	43	0	1	8	9	12	26	0	53
MGDG	0	18	0	0	3	4	20	54	1	21
DGDG	1	36	0	1	5	12	20	25	0	43
P I	0	42	0	1	5	4	13	35	0	48
P C	0	33	0	1	9	9	18	29	1	43
P E	0	32	0	0	7	4	21	35	1	39

TABLE III. The Fatty Acid Compositions (Mole Percent) of Lipids from Anther Walls

	14:0	16:0	16:1t	17:0	18:0	18:1	18:2	18:3	18:4	Saturated
Total	0	18	0	1	28	0	9	42	2	47
P G	0	40	0	2	19	4	6	29	0	61
SQDG	0	29	0	2	21	7	6	35	0	52
MGDG	0	7	0	1	11	5	7	69	0	19
DGDG	0	11	0	1	27	3	6	52	0	39
P I	0	23	0	3	27	2	6	39	0	53
P C	0	15	0	1	23	6	9	45	1	39
P E	0	20	0	1	15	7	12	44	1	36

REFERENCES

Fischer, W., Heinz, E. and Zeus, M., 1973, Hoppe-Seyler's Z. Physiol. Chem., 354:1115-1123.
Murata, N., Sato, N., Takahashi, N. and Hamazaki, Y., 1982, Plant Cell Physiol., 23:1071-1079.
Murata, N., 1983, Plant Cell Physiol., 24:81-86.
Murata, N. and Yamaya, J., 1984, Plant Physiol., 74:1016-1024
Satake, T. and Hayase, H., 1970, Proc. Crop Sci. Japan, 39:468-473.
Satake, T., 1974, Proc. Crop Sci. Soc. Japan, 43:31-35.
Toriyama, K. and Hinata, K., 1984, Plant Cell Physiol., 25:1215-1221.

LIPIDS OF SOYBEAN INOCULATED WITH MICROSYMBIONTS

Raymond S. Pacovsky and Glenn Fuller

Western Regional Research Center
U. S. Department of Agriculture, Agricultural Research Service
800 Buchanan Street
Albany, CA 94710

INTRODUCTION

In many soils, especially tropical lateritic soils, crop productivity is limited by low N and P availability. The growth of legumes can be improved substantially by the presence of both symbiotic N_2-fixing bacteria and endomycorrhizal fungi. In most experiments with these microsymbionts, nodulated or mycorrhizal hosts are compared to N- or P-deficient controls. Many of the physiological differences between these plants are derived from nutritional differences. In this study non-symbiotic controls were given N and/or P fertilizer at rates designed to produce equal plant growth in all treatments.

The purpose of this study was to examine the individual and interactive effects of infection by a N_2-fixing bacterium or an endomycorrhizal fungus on the quality and quantity of extractable-lipid fatty acids (ELFA) from soybeans. The intent was to determine if there are organ-specific shifts in host-plant lipid metabolism following infection by these microsymbionts.

MATERIALS AND METHODS

Soybean (Glycine max L. Merr. cv. Amsoy 71) seeds were planted in a P-fixing soil that had been inoculated with either Bradyrhizobium japonicum (10^9 cells/pot) or Glomus fasciculatum (200 spores/pot). Other plants received both of the microsymbionts or were left non-inoculated (control). All plants were furnished with a basal (N- and P-free) nutrient solution. Soybeans not inoculated with Bradyrhizobium received 1.5 mM NH_4NO_3 (Pacovsky et al 1986a), while plants not inoculated with Glomus received 0.2 mM KH_2PO_4 (Pacovsky et al. 1986b), such that plants of all treatments were of similar size and developmental stage. Soybeans were grown in a walk-in style growth chamber (650 µmol m^{-2} s^{-1}, 16 h photoperiod) and were harvested after 15 wks. There were seven replicates per treatment for a total of 28 soybeans. Plant parts were harvested separately and quickly frozen in liquid N_2. Either 200 mg (leaf, pod, seed) or 400 mg (root) of dried tissue was wetted with 0.8 ml H_2O (4°C) and 200 µg of margaric acid (17:0) was added (internal standard). The sample was extracted with 5 ml of hexane:isopropanol (3:2, v/v) for 2 min at 4°C. Following a second extraction of the residue, the organic phases were combined, washed (6% Na_2SO_4,w/v) and the solvent was vacuum distilled.

ELFA were transesterified using methanol and BF_3 in benzene. FA methyl esters were analyzed by GLC using a Hewlett-Packard 18835B gas chromatograph equipped with a flame ionization detector (270°C). FA methyl esters were separated on a 12 m X 0.2 mm glass column packed with 12QC2/BP20 (0.25 μm film thickness). A fused silica dimethylsilicone column (25m x 0.2 mm) was used to verify peak assignments. The identity of the major FA were confirmed using a Hewlett-Packard 5970 quadrupole-based mass selective detector (70 eV ionization voltage) employing a selective ion monitoring mode. Samples were introduced using a Hewlett-Packard 5790A gas chromatograph with a capillary direct interface (280°C). An analysis of variance (ANOVA) and a Student's t test based on four random replicate samples from the seven replications in each treatment were used to process the data.

RESULTS

The objective of growing all plants to equal weight was nearly achieved (Table 1). Nodulated soybeans had the highest N concentration of any treatment suggesting the N provided by N_2 fixation was sufficient for the requirements of the host. Concentrations of P for Glomus-inoculated plants were the lowest of any treatment suggesting the P input by the mycorrhizae was immediately used for new growth (Table 1).

Table 1. ELFA FROM SOYBEANS INOCULATED WITH MICROSYMBIONTS.

	PHOSPHORUS + NITROGEN	GLOMUS + NITROGEN	PHOSPHORUS + BRADYRHIZOBIUM	GLOMUS + BRADYRHIZOBIUM
Total dry wt (g)	69.4 a	65.7 ab	67.3 ab	65.5 b
Leaf N (%)	1.72 c	1.76 c	2.30 a	2.05 b
Leaf P (%)	0.16 a	0.10 b	0.18 a	0.09 b
	($mg\ g^{-1}$)			
Leaf 16:0	2.53 b	2.33 b	3.05 a	2.80 a
Leaf 18:2(9,12)	3.75 c	4.62 b	3.59 c	5.51 a
Root 16:1(11c)	0.0 c	3.55 a	0.0 c	1.48 b
Root 18:3(6,9,12)	0.0 b	0.35 a	0.0 b	0.28 a
Pod 16:0	35.0 d	40.2 c	45.0 b	49.7 a
Seed 16:0	8.60 b	11.04 ab	13.2 a	13.06 a
Seed 18:2(9,12)	25.5 c	35.4 b	46.6 a	49.6 a
Seed 18:3(9,12,15)	10.8 b	11.5 b	14.3 a	13.1 a
Total leaf FA	23.9 b	26.0 a	25.1 ab	28.0 a
Total root FA	4.2 c	10.1 a	3.9 c	6.8 b
Total seed FA	57.9 c	79.1 b	98.6 a	105.9 a

Soybeans inoculated with Bradyrhizobium had a higher content of 16:0 in the leaves than did non-nodulated plants (Table 1). Mycorrhizal soybean leaves contained higher levels of 18:2 (9c, 12c) (Table 1) and 18:1 (9c) (data not shown) than did non-mycorrhizal plants.

Five specific FA were present in soybean roots infected with Glomus [16:1 (11c); 18:3 (6c,9c,12c); 20:3 (8c,11c,14c); 20:4 (5c,8c,11c,14c); and 20:5 (5c,8c,11c,14c,17c)] that were absent in non-infected roots (Table 1). FA 16:1 (11c) comprised 29% of all root FA in Glomus-infected roots. These FA were not found in leaves, pods, or seeds of either Glomus-infected or non-infected soybeans. Similar to the results with citrus roots (Nordby et al. 1981), P-amended plants did not produce these unusual FA. It is not certain if these FA are derived from fungal lipids or if the FA were produced by the plant in response to fungal infection. However, the fact that no organs other than infected roots contain these unusual FA suggests that these FA originate from fungal lipids.

N_2-fixing soybeans had higher amounts of 16:0 in the pods than non-nodulated plants (Table 1). Mycorrhizal soybeans contained less 18:3 (9c,12c,15c) and less 20:0 in the pod, but comparatively more 22:0 than did P-fertilized plants. Soybeans inoculated with Bradyrhizobium had a higher content of 16:0 and 18:0 in the seed than did non-nodulated plants (Table 1). N_2-fixing plants contained more 18:1 (11c), 18:2 (9c,12c), and 18:3 (9c,12c, 15c) in the seed than did N-fertilized plants.

In general, Glomus-inoculated plants contained more lipids per unit weight in the leaf and roots than the P-fertilized soybeans (Table 1). Nodulated plants contained more total FA in the pod and seed than the N-fertilized plants. Except for the total ELFA content of the roots, dually-inoculated plants had the highest total lipid content in all plant parts (Table 1). The relatively high production of lipids in these dually-inoculated soybeans suggests an altered utilization of plant lipids in these symbiotic associations. These lipids may play a potential role in the infection process or maintenance of the microsymbionts.

DISCUSSION

A number of physiological and biochemical effects of nodulation by Rhizobium or infection by mycorrhizae may have been overlooked in previous studies examining N and P nutritional effects in these beneficial symbioses. More attention should be paid to increases in plant hormone levels, chlorophyll content, and lipid composition in two- and three-way symbioses.

Soybean roots infected with Glomus fasciculatum were shown to contain five ELFA that were unusual to higher plant material and which constituted over 35% of the total root FA. The presence of γ-linolenic acid instead of α-linolenic acid and other C_{20} and C_{22} FA (Nordby et al. 1981) indicate the presence of a phycomycete (Weete, 1974) in the mycorrhizal roots. The response of the host plant to the bacterial or fungal endophyte must be further understood if optimum performance from Glomus- or Bradyrhizobium-inoculated soybeans is to be obtained under field conditions.

CONCLUSIONS

1. The data on mineral and fatty acid composition and turnover in symbiotic plants confirms that the host response to either endophyte cannot be accounted for solely as a P or N effect.
2. Nutrient requirements, amount of lipid storage material and membrane composition may have changed in symbiotically-grown plants such that different modes of metabolism are favored in these hosts.
3. The presence of specific ELFA is indicative of an infection by a phycomycete. Quantifying these fatty acids may be one way of determining fungal infection and effectiveness with time.
4. Legumes that are mycorrhizal and nodulated must be included in future studies of nutrient interactions or lipid composition because of the important and unique role that mycorrhizae and N_2-fixing bacteria play in these systems.

Acknowledgements - The authors would like to thank S. Finlayson, E. Grey, G. Secor and A. Stafford for their support and technical assistance.

REFERENCES

Nordby H.E., Nemec S. & Nagy S. 1981. J. Agric. Food Chem. 29: 396-401.
Pacovsky R.S., Paul E.A. & Bethlenfalvay G.J. 1986a. Crop Sci. 26: 145-150.
Pacovsky R.S., Bethlenfalvay G.J. & Paul E.A. 1986b. Crop Sci. 26: 151-156.
Weete J.D. In: Fungal Lipid Biochemistry, J.D. Weete, Ed. New York: Plenum. Press, pp. 74-79, 158-161.

PROPERTIES AND IN VITRO SYNTHESIS OF PHOSPHOLIPID TRANSFER PROTEINS

F.Tchang, F.Guerbette, D.Douady, M.Grosbois, C.Vergnolle,
A.Jolliot, J.P.Dubacq and J.C.Kader

Laboratoire de Physiologie Cellulaire (U.A.CNRS 1180)
Université P. et M. Curie, 4 pl.Jussieu, 75005 Paris, France

INTRODUCTION

Phospholipid transfer proteins (PLTP) have been isolated to homogeneity from various plant sources : spinach leaves[1], maize seedlings[2,3] and castor bean endosperm[4]. These proteins are able to facilitate an in vitro transfer of phospholipids between membranes. This led to the hypothesis that these proteins participate in vivo in the turnover and biogenesis of membranes[5]. In order to study the function of these proteins, it is essential to know their biochemical properties as well as their biogenesis. In this paper, we compare the properties of the transfer proteins isolated from the three tissues and we indicate new informations concerning the in vitro synthesis of the protein isolated from maize seedlings.

MATERIALS AND METHODS

Phospholipid transfer proteins were purified to homogeneity from spinach leaves and maize seedlings by chromatography on carboxymethyl-Sepharose as previously described [1,3]. The isolation of RNA, the in vitro synthesis and high performance liquid chromatography were performed according to Tchang et al [6].

RESULTS

Properties

A remarkable homogeneity was noted for maize and spinach proteins, studied in this work and for castor bean protein, isolated by Yamada's group[4] (Table) ; a value close to 9 kDa was found for the three proteins. Another common feature is the isoelectric point which is high varying from 8.8 to 10.5; basic PLTP are also present in animal cells[7]. However, acidic PLTP have been also found in one plant tissue (castor bean)[5] and in various animal sources[7]. All the basic PLTP from plants are non specific towards phospholipids[1,3,4]. The three proteins are able to transfer phosphatidylcholine, phosphatidylinositol and phosphatidylethanolamine ; in addition, spinach and castor bean proteins transfer respectively phosphatidylglycerol and phosphatidic acid.

Table. Properties of phospholipid transfer proteins from plants

	Maize[a]	Spinach[b]	Castor bean[c]
Molecular mass (kDa)	9-10	9	9
pI	8.8	9	10.5
Transferred lipids	PC,PI,PE	PC,PI,PE,PG	PC,PI,PE,PA galactolipids

[a]: ref.2 ; [b]: ref.1 ; [c]: ref.4.

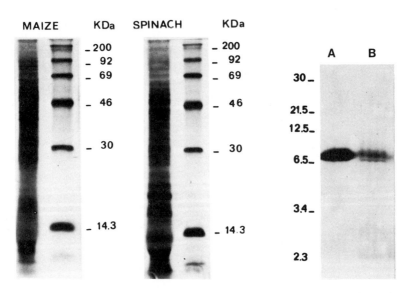

Fig.1.(Left). Fluorogram of SDS-electrophoresis showing in vitro translation of maize and spinach poly(A)+RNAs.
 0.5 µg of poly(A)+ RNAs were translated in 9 µl of reticulocyte lysate in the presence of ^{35}S-methionine (20 µCi). The fluorogram was obtained after a 2-day exposure on a Kodak X-Omat R film.

Fig.2.(Right). Analysis of in vitro synthesized proteins by high-performance liquid chromatography followed by SDS-electrophoresis and fluorography.
A : ^{14}C-methylated maize PLTP. B : pooled fractions containing PLTP.

In vitro Synthesis

Total RNAs have been extracted from shoots of 3-day-old maize seedlings and spinach leaves. After oligo (dT)cellulose chromatography, poly(A)$^+$ RNAs were translated in a reticulocyte lysate system. The in vitro products were analyzed by SDS-electrophoresis. As shown in Fig.1, several proteins were synthesized, among which high-molecular mass polypeptides were observed. This result indicates that the preparation of poly(A)$^+$ RNAs had a good quality.

In vitro products, to which 0.1 mg of maize PLTP were added, were analyzed by high performance liquid chromatography using a reverse phase column. The fractions corresponding to the elution time of the PLTP were pooled and analyzed by SDS electrophoresis, followed by fluorography. A radioactive band, comigrating with a ^{14}C-methylated maize PLTP, was seen (Fig.2).

CONCLUSION

The recent progress on the knowledge of plant PLTP has led to the conclusion that, at least in three different plants, PLTP have very similar properties. This reinforces our ideas concerning their basic role in the lipid metabolism of plant cells. The detection of RNAs coding for these proteins opens new pespectives in the study of their biogenesis. It will be possible in future studies to try to establish a correlation betwen their synthesis and the biogenesis of intracellular membranes.

REFERENCES

1. J.C.Kader, M.Julienne and C.Vergnolle, Purification of a spinach-leaf protein capable of transferring phospholipids from liposomes to mitochondria or chloroplasts, Eur.J.Biochem. 139 : 411(1984).
2. D.Douady, M.Grosbois, F.Guerbette and J.C.Kader, Purification of a basic phospholipid transfer protein from maize seedlings, Biochim.Biophys.Acta 710 :143 (1982).
3. D.Douady, F.Guerbette, M.Grosbois and J.C.Kader, Purification of phospholipid transfer protein from maize seeds using a two step chromatographic procedure, Physiol.Vég. 23 : 373 (1985).
4. S.Watanabe and M.Yamada, Purification and characterization of a non specific lipid transfer protein from germinated castor bean endosperms which transfers phospholipids and galactolipids,Biochim. Biophys.Acta , 876 : 116 (1986).
5. J.C.Kader, Lipid-binding proteins in plants, Chem.Phys.Lipids, 38 : 51 (1985).
6. F.Tchang, M.Laroche-Raynal, C.Vergnolle, C.Demandre, D.Douady, M.Grosbois ,F.Guerbette,M.Delseny and J.C.Kader, In vitro synthesis of a plant phospholipid transfer protein : a study by high performance liquid chromatography Biochem.Biophys.Res.Commun. 133:75 (1985).
7. K.W.A.Wirtz, Phospholipid transfer proteins ,in : "Lipid-protein interactions" P.Jost and O.H.Griffiths,ed., Wiley-Interscience,New York (1982).

GALACTOLIPID SYNTHESIS IN ISOLATED PEA CHLOROPLASTS

J.P Dubacq, R.O. Mackender and P. Mazliak

Physiologie Cellulaire, Université P; et M. Curie, Paris
France - Botany Department, Queen's University, Belfast
U.K.

INTRODUCTION

UDPgal transferase (UDPGT) activity has been localized in the outer chloroplast envelope membrane (OEM) of pea (18:3 plant) (1) and in the inner chloroplast envelope membrane (IEM) of spinach (16:3 plant) (2). This seems logical in the context of current ideas concerning the sources of diglyceride for the galactosylation reaction (3). However, analysis of the molecular species of MGDG extracted from thylakoid membranes isolated from sequential segments of young oat seedling shoots (4) shows that although the 18:3/18:3 molecular specie predominates at all developmental stages, there are significant proportions of 18:1/16:0 and 18:3/16:0 molecular species in thylakoids from the younger plastids. This suggested to us that UDPGT may be present in both membranes each galactosylating the diglycerides available to it. The following experiments have been carried out with plastids isolated from expanding leaves of pea with the aim of corroborating previous research with pea chloroplasts.

METHODS

Chloroplasts from Pisum sativum L. var Kelvedon Wonder were isolated and purified on Percoll layer. They were incubated (4) at 1 mg chlorophyll per ml with 10^6 dpm of UDP-^{14}Cgalactose. Then envelope membranes were separated and purified as described previously (5) except that the washings of the chloroplasts following incubation with radioactive label were centrifuged at 100.000 g for 1 hr and the pellets added to the washed chloroplasts after the freeze thaw treatment and prior to density gradient centrifugation. To avoid any further metabolism of galactolipids during isolation of membranes the transferase activity was inhibited at each time point by adding the sample to isotonic medium containing 10 mM N-ethylmaleimide (NEM) which at 7.5 mM inhibits galactosylation by more than 98%. NEM was also added to all media (including gradient solutions) used post incubation. In some experiments, HPLC allied to a radiochemical detector system, was used to ascertain (i) which molecular species were labelled and (ii) whether this labelling was selective.

RESULTS AND DISCUSSION

Table : Ratio ^{14}C-MGDG/^{14}C-DGDG in different chloroplast subfractions in continuous labelling experiment

Time (min)	Washed thylakoids	Crude envelopes	Purified envelopes
0.5	4.98	1.97	---
1	5.35	2.68	1.14
2	5.93	2.80	1.15
4	5.49	2.26	1.38
8	5.93	2.41	1.60
16	5.49	3.27	1.24

Three types of experiments have been carried out: (a) isolated chloroplasts have been continuously labelled for upto 20 min by incubating in UDP-^{14}Cgal. At various times the chloroplasts were removed and fractionated into crude envelopes, purified envelopes or OEM, IEM and washed thylakoids membranes (WT) and the labelling in the MGDG and DGDG determined in each. (b) As in (a) except that after 10 secs exposure to UDP-^{14}Cgal, the label was removed by the addition of nucleotide phosphatase. (c) Determination of Km and Vm using isolated OEM, IEM and WT.

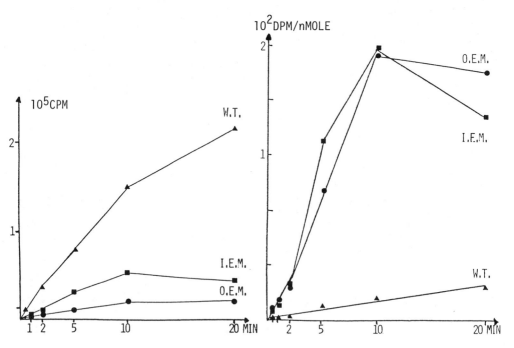

Fig.1 (left): The time course of total ^{14}C-galactose incorporation into OEM, IEM and WT in continuous labelling experiment

Fig.2 (right): The specific activity of MGDG in OEM, IEM and WT in continuous labelling experiment

1-Continuous labelling experiments show: (a) that all membranes were labelled within 30 secs and that even in this time the highest proportion of label (>50%) was in the thylakoid membranes (Fig.1); (b) there was more label in MGDG and DGDG of IEM than OEM; (c) that the specific activity of the galactose in MGDG was similar in both OEM and IEM (Fig.2) and increased (x20) in a similar way during the incubation; (d) the specific activity of the MGDG in WT increased during the incubation but was ten times less than that of the OEM and IEM. Moreover, the ratio of activity MGDG/DGDG largely differs in each membrane (Table). HPLC analysis (not shown) indicates that the molecular species of MGDG or DGDG labelled in each membrane are identical.

The most interesting point to note is that even with the shortest sampling times used the highest proportion of total label is in the thylakoids. If it is accepted that there is not UDPGT activity in the WT, this must infer that there is a very rapid transfer of MGDG from envelopes to thylakoids (6,7).

2-Envelope membranes and WT (twice purified on glycerol sucrose gradients) were assayed for UDPGT and LFACoA synthetase activities. Both enzyme activities were detected in both envelope membrane fractions but neither in WT. On the assumption that LFACoA snthetase activity is restricted to the OEM, the IEM fraction was only contaminated with OEM to about 10%. In addition Mg^{2+}-ATPase was absent from OEM which proved that this fraction was very pure. Using UDPgal concentrations up to 500 μM the Km for UDPGT was about 70 μM in OEM and about 105 μM in IEM.(data not shown).

Hence, on the basis of these experiments, we would conclude that UDPGT activity is present in both envelope membrane fractions. Until we carried out experiments with added diglycerides, we cannot come to the same firm conclusion with regard to thylakoids. The Km values do not in fact mean very much in themselves except perhaps to give an indication of the relative pool sizes of endogenous diglycerides. True Km and Vm for UDPGT activity in these membrane fractions can only be achieved by adding non limiting concentrations of diglycerides.

3-Pulse chase experiments show that label continues to appear in MGDG and DGDG of all membranes throughout the chase period, but is much reduced compared with continuously labelled chloroplasts. The specific activity of the galactose in MGDG of OEM and IEM are about the same, but greater than the specific activity in thylakoids (data not shown). These results seem rather inconclusive because the label continues to increase in all membranes, despite attempts to remove UDPgal with nucleotide pyrophosphatase and dilution. A part of the UDPgal seems unaccessible to the destructive enzyme. As in continuous labelling experiments thylakoids contained the highest activity.

CONCLUSION

Therefore we conclude from these experiments that there were two locations of UDPGT in pea chloroplasts, one in each envelope membrane and that, in the absence of clear evidence for the presence of UDPGT activity in thylakoids, that the transfer of label between envelope and thylakoid membranes is extremely rapid.

REFERENCES

1- Cline K. et al., Plant Physiol., 71, 366-372 (1983)
4- Gillanders B. et al., in Biochem. and Metab. of Plant Lipids (Wintermans and Kuiper, eds) Elsevier Biomedical Press, Amsterdam, pp. 191-197 (1982)
3- Roughan P.G. et al., in Ann. Rev. Plant Physiol., 33, 97-132 (1982)
2- Block M.A. et al., J. Biol. Chem., 258, 13273-13286 (1983)
5- Cline K. ct al., PNAS, 78, 3595-3599 (1981)
6- Bertrams M. et al., Z. Naturforsch., 36, 62-70, (1981)
7- Joyard J. et al., Eur. J. Biochem., 108, 171-176, (1980)

TRIACYLGLYCEROL BIOSYNTHESIS IN DEVELOPING COTYLEDONS OF SAFFLOWER
(Carthamus tinctorius)

Gareth Griffiths, Keith Stobart and Sten Stymne*

Department of Botany, University of Bristol, U.K.
*Department of Food Hygiene, Swedish University of
Agricultural Sciences, Uppsala, Sweden

INTRODUCTION

Triacylglycerol (TAG) deposition in the maturating seeds of safflower occurs over a narrow but well defined period. This report presents evidence to establish (a) the optimum seed age at which to conduct studies on active TAG biosynthesis (b) the use of such material in in-vivo experiments to examine the current in-vitro based models proposed by Stymne and Stobart (1) for linoleate-TAG formation.

MATERIALS AND METHODS

Safflower plants var. Gila were grown under greenhouse conditions with an average day and night temperature of 28 and 22°C and an 18h photoperiod. Developing cotyledons were harvested at intervals after flowering (DAF) and about 0.2 g fresh weight of sliced cotyledons were incubated in the dark in 1 ml of P-buffer, pH 7.2, containing the radioactive substrate at 30°C. Lipids were extracted in an acidified chloroform/methanol mixture and analysed by TLC and radioassayed by scintillation counting in PCS/xylene.

RESULTS

To establish the optimum seed age at which to conduct studies on TAG biosynthesis, the cotyledons were harvested at intervals from seeds from 14-28 DAF and incubated with [^{14}C]glycerol. The results (Fig. 1) clearly show a marked change in the pattern of radioactive glycerol incorporation into the lipid fraction with cotyledon age. In the early stages of development (14-16 DAF), high rates of TAG synthesis were observed (Fig. 1) and both PC and DAG were labelled similarly indicating a close metabolic relationship between these two lipids. No phosphatidate was detected in these incubations. At later stages of maturation however the incorporation of glycerol into TAG was substantially reduced. At 26-28 DAF, high amounts of radioactivity accumulated in PA and DAG and in addition

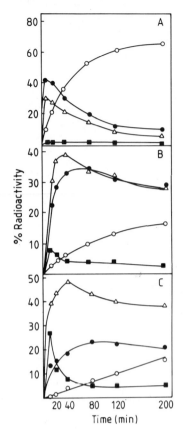
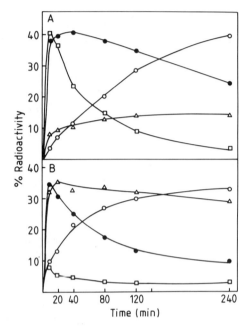

Fig.1.(Left) Utilisation of [^{14}C]glycerol by developing safflower cotyledons, (A) 14-16 DAF, (B) 20-22 DAF, (C) 26-28 DAF, ■, PA; △, DAG; ○ TAG; ●, PC.

Fig.2.(Right) Metabolism of (A) [^{14}C]oleate, (B) [^{14}C]linoleate by slices of developing safflower cotyledons. □, FFA; _, DAG; ○, TAG, ●, PC.

the labelling of DAG and PC was now dissimilar. These results suggest a decrease in the activity of phosphatidase and DAG-acyltransferase during seed maturation. The difference in the labelling of the DAG and PC however could also be explained on the basis of a decrease in the activity of a choline phosphotransferase, an enzyme which has been demonstrated <u>in-vitro</u> (2) and which under conditions of active TAG synthesis appears to catalyse a freely reversible exchange between these two lipids (3).

It is implicit from these results that cotyledons should be selected from young developing seeds in which oil-deposition has only just commenced and which gives good rates of TAG synthesis. The practice of freezing whole seeds for storage prior to experimental use is to be avoided since poor and artifactual incorporation of labelled substrate is obtained (4). This applies equally well to the use of whole tissue for <u>in-vivo</u> work and also for the preparation of microsomal membranes with adequate biosynthetic activity.

Metabolism of [^{14}C]oleate and [^{14}C]linoleate

The metabolism of radioactive oleate and linoleate was examined in tissue-slices from cotyledons with active TAG biosynthesis (14-16 DAF). Typical results from time-course studies are presented in Fig. 2. With oleate it is evident that PC is the most heavily labelled lipid over the initial time period and that at later times the radioactivity rapidly accumulated in TAG. Interestingly, over 90% of the radioactivity in the PC was located at the position sn-2 (Table 1). The rapid turnover of acyl

Table 1. Positional distribution of ^{14}C-fatty acids in phosphatidylcholine from safflower cotyledons.

Time	^{14}C-Fatty acid	Position (%) sn-1	sn-2
10	18:1	7	93
	16:0	63	37
80	18:1	10	90
	16:0	76	24
240	18:1	17	83
	16:0	81	19

groups at position sn-2 of sn-PC concomitant with the biosynthesis of TAG is consistent with the proposals for the entry of oleate into PC for desaturation and the availability of the linoleate, via acyl exchange, for TAG synthesis. Using [^{14}C]linoleate it can be seen that both PC and DAG were labelled to an almost equal extent over the initial time periods with the linoleate ultimately accumulating in TAG. The kinetics of oleate and linoleate incorporation by the cotyledon tissue reflect the rapid transfer of oleate to PC for desaturation (over 70% after 4h) prior to linoleate utilisation in the acylation of glycerol 3-phosphate (G3P) and the formation of TAG. Linoleate was however rapidly utilised for DAG formation via the reactions of the Kennedy pathway as indicated by the similarity in label in DAG and PC over the initial time periods. These data are in agreement with the preferential utilisation of linoleate in the acylation of G3P that has been observed in microsomal membranes from safflower(5-6). It should be pointed out that over 90% of the [^{14}C]linoleate was located at position sn-2 of PC.

Recently it was demonstrated (5) that palmitate was exclusively acylated to position sn-1 of G3P in microsomes supplied with a mixed acyl-CoA substrate. Since DAG equilibrates with the PC pool during active TAG synthesis (3) then the palmitate that initially esterifies position sn-1 of G3P should also enter PC at this position. Positional analysis of the PC obtained from safflower cotyledons that had been incubated with [^{14}C]palmitate (Table 1) shows that substantial amounts of this fatty acid had entered PC at the sn-1 position. Small amounts of palmitate were also present at position 2 and this probably had occurred through acyl exchange in the absence of competing unsaturated fatty acids as reported previously (5).

CONCLUSIONS

The results obtained in this study with tissue slices of safflower are in strong agreement with the proposals for the synthesis of TAG rich in linoleate that were based previously on in-vitro work (1). It should also be pointed out that an acritical approach in the growing of experimental material at precise and synchronised stages of development will lead to conclusions which may have little relevance to the situation which occurs during active TAG synthesis. It is important therefore that workers are fully aware of such problems before embarking on experimentation designed to elucidate the regulatory mechanisms and enzymology of oil biosynthesis.

ACKNOWLEDGEMENT

G. G. acknowledges the receipt of an SERC studentship. The authors are grateful to the Swedish Natural Science Council, Swedish Council for Forestry and Agricultural Research and The National Swedish Board for Technical Development.

REFERENCES

1. S. Stymne, and A. K. Stobart, Triacylglycerol Biosynthesis, in: "The Biochemistry of Plants: a Comprehensive Treatise, Vol. 10," P. K. Stumpf, ed., Academic Press, N. Y. (1986).
2. C. R. Slack, P. G. Roughan, J. A. Browse, and C. R. Gardiner, Some properties of cholinephosphotransferase from developing safflower cotyledons, Biochim. Biophys. Acta 833:438 (1985).
3. A. K. Stobart, and S. Stymne, The interconversion of diacylglycerol and phosphatidylcholine during triacylglcerol production in microsomal preparations of developing cotyledons of safflower, Biochem. J. 232:217 (1985).
4. G. Griffiths, Ph. D. Thesis "Aspects of Triacylglycerol Biosynthesis in Developing Oil-Seeds," University of Bristol, U.K. (1986).
5. G. Griffiths, A. K. Stobart, and S. Stymne, The acylation of sn-glycerol 3-phosphate and the metabolism of phosphatidate in microsomal preparations from the developing cotyledons of safflower, Biochem. J. 230:379 (1985).
6. A. Stobart, S. Stymne, and G. Glad, The synthesis of linoleate and phosphatidic acid and its relationship to oil production in the microsomes of developing seeds of safflower, Biochim. Biophys. Acta 754:292 (1983).

LONG-CHAIN TRIACYLGLYCEROL ACYL HYDROLASE (LIPASE) ACTIVITY

IN WHEAT GRAIN

T. Galliard, M. Lond and D.M. Gallagher

R.H.M. Research Ltd., Lord Rank Research Centre
High Wycombe, HP12 3QR, England, U.K.

INTRODUCTION

In response to nutritional recommendations for increased amounts of dietary fibre, particularly of cereal origin, the consumption of wheat wholemeal and bran-enriched products has increased dramatically in many countries. However, wheat milling products containing bran and germ are unstable and deteriorate on storage, giving inferior baking performance and undesirable off-flavours.

Recent studies in our laboratory [1-4], have shown that a major factor that determines rates of deterioration of whole grain products is the level of activity of a lipase enzyme, located specifically in the bran component. This paper describes some of the characteristics of wheat bran lipase.

MATERIALS AND METHODS

Materials. Wheat grain, wholemeal flour, bran and germ were obtained from commercial flour mills. Milling products were stored at $-20^{\circ}C$ within 12h of milling, or were stored at $+20^{\circ}C$, 65% relative humidity.

Methods. Lipolytic activity measurements using tri-^{14}C-oleoylglycerol and methylumbelliferyl esters in enzyme assays, or by measurement of total lipid and free fatty acid (FFA) levels were as described recently [2,3]. Oxygen uptake by aqueous suspensions of wholemeal, bran, etc. correlates very closely ($r \geqslant 0.96$; $P < 0.001$) with FFA content and has been used as a rapid and simple routine method for measuring deterioration of such materials during storage [2,3].

RESULTS

Lipolysis in stored wholemeal and bran

Lipolysis rates vary between samples of wheat grain after milling, depending upon the lipase activity in the bran component. Over 50% of the endogenous triacylglycerol (TC) is converted, quantitatively, to FFA during storage of wholemeal at $+20^{\circ}C$ (12-14% moisture) over 8-20 weeks, depending on the lipase activity. In bran, lipolysis is more rapid; for example in

4 weeks at 20°C the TG content fell from 61% to 13% of the total lipids, whereas FFA increased from 10% to 67% [2]. FFA accumulate in material stored at normal moisture levels with little further conversion until hydrated (see below).

Location of long-chain triacylglycerol (LCTG) lipase in bran

When ^{14}C-LCTG was incubated with different, purified milling fractions, only the bran component contained hydrolase activity towards LCTG (see Table 1). Although the germ component of ungerminated wheat contains hydrolase activity towards a range of fatty acid esters and short-chain TG, it lacks lipolytic activity with endogenous LCTG.

This is supported by FFA and O_2-uptake (see Methods) measurements on bran, germ and mixtures during storage at 20°C. For example, O_2-uptake (μmol/10 min/g) of aqueous suspensions increased from 2 in freshly milled bran to 18 after storage for 4 weeks; corresponding figures for pure germ were 2 and 5 [1]. (N.B. commercially obtained 'germ' contains substantial amounts of bran).

However, there is a synergistic effect between bran and germ in FFA formation (and the corresponding O_2-uptake figures). A 50:50 (w/w) mixture of the bran and germ (above) stored for 4 weeks gave an O_2-uptake value of 30 (cf 18 and 5 for bran and germ, respectively). The synergism is due to oil-rich germ providing additional LCTG substrate for the bran lipase.

Comparison of enzyme assays for 'lipase' activity

Chromogenic or fluorogenic acyl esters are commonly used to measure 'lipase' activity. Table 1 shows that when LCTG (the natural substrates) are used to assay the lipase in wheat fractions the ratio of activity in bran to germ is 14:1. With MUH as substrate, the ratio was 0.3:1. These data illustrate the importance of using an assay that correlates with the process occurring naturally.

Table 1. Comparison of tri-^{14}C-oleoylglycerol (TG) and 4-methyl-umbelliferyl heptanoate (MUH) as substrate for 'lipase' in wheat.

Substrate	Hydrolase activity		Bran/Germ Ratio
	Bran	Germ	
TG (μmol FFA/24h/g)	1.4 (± 0.4)	0.10 (± 0.03)	14
MUH (ΔF^a/min/mg × 10^{-3})	1.0 (± 0.2)	3.3 (± 0.7)	0.3

a ΔF = change in fluorescence units.

Action of bran lipase at low moisture levels

Unlike most enzymes, some lipases are active at very low water activities. Deterioration of wholemeal increases linearly from a level of moisture (5%), well below the range (12-14%) found in normal storage, or necessary for microbial growth ($\geq$ 16%) [3]. Previous work [6] has shown that wheat lipase has maximal activity at a_w=0.8 (approx. 17% H_2O).

Heat stability of bran lipase

At natural moisture levels, the FFA content of wholemeal increases with storage temperature, and no inactivation of lipolysis is evident, even when heated at 80°C for 7 days [4]. However, at higher moisture contents, the enzyme is inactivated relatively rapidly. For example, at 80°C and 50% moisture, the lipase activity is inactivated within 15 min. Autoclaving (110°C, 10 min) gives adequate stabilisation [4].

DISCUSSION

Although present at relatively low levels, the lipase from wheat bran is active in milled grain from the time of milling and during subsequent storage under ambient temperature and moisture conditions.

The activity at low moisture levels is due, presumably, to the fact that the substrate (TG) of wheat is present as an oil which can diffuse through the milled materials without the requirement of free water as solvent/carrier. Previous work with model systems [7] consisting of pancreatic lipase on powdered cellulose has shown that liquid, but not solid phase TG are hydrolysed at low moisture levels. The fact that the rate of hydrolysis is inversely related to the particle size distribution of the bran component of wholemeal [4], provides further evidence for a diffusion-limited process.

The rate at which milling products deteriorate under given conditions of storage is determined by level of lipase activity [2,3] and particle size distribution [4] of bran. In practical terms, storage of wholemeal at very low moisture and temperature is not feasible for several reasons. However, bran and germ may be stabilised by heat-moisture treatment and this is practised commercially. The relatively high heat stability of the lipase, precludes heat-treatment of wholemeal flour, because the essential functionality of wheat proteins is destroyed by less severe conditions than those required to inactivate the lipase.

The major products of lipase action are free (unesterified) fatty acids of which polyunsaturated fatty acids (PUFA) represent 60% of the total. When water is added to the milled products, a rapid oxidation of PUFA occurs, the extent of which is determined by the concentration of PUFA and, hence, by the storage history of the material. The oxidation process is described elsewhere in these proceedings.

ACKNOWLEDGEMENTS

Most of the work reported here was sponsored by the U.K. Ministry of Agriculture, Fisheries and Food and is Crown Copyright © 1986.

References

1. Galliard,T. in "Chemistry and Physics of Baking" (J.M.V.Blanshard, P.J.Frazier and T.Galliard, eds.) Royal Society of Chemistry, London, 1986, 199-215.
2. Galliard,T. J.Cereal Sci. 4 (1986) 33-50.
3. Galliard,T. J.Cereal Sci. 4 (1986) 179-192.
4. Galliard,T. Proc. 1st Congress of European Federation of Food Science & Technology (in press).
5. Saunders,R.M. and Helvted,F. J.Cereal Sci. 3 (1985) 79-86.
6. Drapron,R. Ann.Tech.Agric. 21 (1972) 487-499.
7. Acker,L. and Wise,R. Z.Lebensm.-Unters Forsch. 150 (1972) 205-211.

GLYCOPROTEIN NATURE OF LIPOLYTIC ACYL HYDROLASES IN POTATO
TUBERS AND LEAVES

Robert A. Moreau and Gerald Nagahashi

Eastern Regional Research Center
Agriculture Research Service
U.S. Department of Agriculture
Philadelphia, PA 19118

INTRODUCTION

The leaves and tubers of potatoes contain high levels of lipolytic acyl hydrolase activities (1,2,3). These enzymes are capable of hydrolyzing all endogenous phospholipids and galactolipids. It was recently reported that all of the lipolytic acyl hydrolase activity in potato tubers is associated with a glycoprotein fraction called "patatin" which comprises about 30% of the soluble protein in tubers (4). This study was undertaken in order to verify whether this finding was valid for other varieties of potatoes (in ref 4 Kennebec was the only variety studied). In addition, experiments were also conducted to determine whether the lipolytic enzymes in potato leaves were also associated with patatin or other glycoproteins.

MATERIALS AND METHODS

Potato (<u>Solanum tuberosum</u> c.v. Russet Burbank) plants were grown from seed tubers as previously described (3). Leaves (5 g) were homogenized in a chilled mortar with a buffer (20 ml) containing 25 mM potassium phosphate (pH 7.0) and 2mM sodium metabisulphite. Tubers (40 g) were homogenized in a Waring Blendor with the above buffer (80 ml). Homogenates from leaves or tubers were filtered through 2 layers of cheesecloth and centrifuged at 100,000 g for 50 min. An aliquot (5 ml) of the 100,000 g supernatant fraction was desalted on a Sephadex G-50 column (1.5x30 cm) eluted with 25 mM potassium phosphate (pH 7.0). This desalted fraction was applied to a DEAE cellulose (DE-52) column (1.5x4 cm) rinsed with 3 bed volumes of the above buffer, 2 bed volumes of 0.25 M NaCl, and finally, 2 bed volumes of 0.5 M NaCl. The fraction eluted with 0.25 M NaCl was applied to a Con A Sepharose column (1x6 cm) which was eluted with 3 bed volumes of the above buffer, 2 bed volumes of 20 mM α-methyl glucose (to yield patatin fraction as described in ref 4), and finally 2 bed volumes of 300 mM α-methyl glucose. Small samples of each of the above fractions were retained for accurate book-keeping of enzyme yields and recoveries. The rates of hydrolysis of PNP-laurate, 4-MU-laurate, and C_6-NDB-PC were assayed as described (3,4,5).

RESULTS AND DISCUSSION

Patatin was prepared as previously described (4), and total protein and three different liploytic enzyme assays were measured in each of the steps of its preparation (Table 1). Two additional steps were added which did not appear in the original report of lipolytic acyl hydrolase in patatin (4). The 0.5 M NaCl eluate and the 300 mM α-methyl glucose eluate were included to remove protein which may have still been retained on the columns after elution with 0.25 M CaCl and 20 mM α-methyl glucose (Meglc), respectively. Most of the PNP-laurate hydrolase activity was retained on the Con A column as previously reported, but only about half of the activity was eluted in the patatin (20 mM Meglc eluate) fraction and the rest was eluted with 300 mM Meglc. In the previous report essentially all of the lipolytic acyl hydrolase (PNP laurate hydrolase) activity was associated with patatin (4). A possible reason for the differences in this study and the previous one may be due to the fact that different cultivars of potatoes were used (Russet Burbank here versus Kennebec). We are currently investigating this possibility, especially in light of our previous report of substantial varietal differences in the total levels of lipolytic enzymes (6). In this experiment and in the next an appreciable loss of enzyme activity occurred with each chromatographic step, perhaps due to removal of a cofactor, proteolytic degradation, or other problems caused by instability of the enzymes. The distribution of C_6-NDB-PC hydrolase in the various fractions was very similar to that of PNP-laur hydrolase, except that more activity was found in the 0.5 M NaCl DEAE eluate. When 4-methyl umbelliferyl laurate hydrolase activity was assayed in the various fractions 38% of the activity was apparently nonglycosylated (Con A Sepharose effluent). It therefore appears that the three different substrates were hydrolyzed at different rates by lipolytic enzymes which differed according to charge and degree of glycosylation. This is consistent with the previous report (5, also with Russet tubers) that three different lipolytic enzymes with different substrate specificities could be separated on DEAE cellulose (peaks at 0, 0.15, and 0.25 M CaCl). The conclusion of this experiment is that most of the lipolytic enzymes in potato tubers are glycoproteins as recently reported (4), but not all of them appear to be associated with patatin (at least in Russet Burbank).

In the second experiment (Table 2) patatin was purified from potato

Table 1. Phospholipase activity in the various stages of purification of the glycoprotein "patatin" from potato tubers (c.v. Russet Burbank).

Fraction	protein (mg/fraction)	esterase activity		phospholiase activity
		PNP-Laur units[a]	4-MU-Laur units[a]	C_6-NBD-PC units[a]
100,000 g supernatant	20.4	17.8	7.69	8.84
Sephadex G-50 void	5.89	12.8	5.29	7.35
DEAE cellulose				
Effluent	1.35	0	0	0
Eluate (0.25 M NaCl)	3.67	10.6	4.13	6.00
(put on Con A Seph)				
Eluate (0.50 M NaCl)	0.16	0.49	0	1.47
Con A Sepharose				
Effluent	1.26	0.30	1.10	0
Eluate (20 mM Meglc)	1.08	3.41	0.82	2.00
(= patatin)				
Eluate (300 mM Meglc)	0.87	3.79	1.01	2.42

[a] enzyme activity = μmol min^{-1} fraction^{-1}

Table 2. Phospholipase activity in the various stages of purification of the glycoprotein "patatin" from potato leaves (c.v. Russet Burbank).

Fraction	protein (mg/fraction)	esterase activity 4-MU-Laur units[a]	phospholipase activity C_6-NBD-PC units $\times 10^{-3}$
100,000 g super	30.5	87.7	19.8
Sephadex G-50 void	14.6	42.6	16.5
DEAE cellulose			
Effluent	1.72	0	0
Eluate (0.25 M NaCl) (put on Con A Seph)	10.7	29.9	5.59
Eluate (0.50 M NaCl)	1.03	0	1.24
Con A Sepharose			
Effluent	10.19	30.1	2.01
eluate (20 mM Meglc) (= patatin)	0.23	0	0.65
Eluate (300 mM Meglc)	0.03	0	0.13

[a] enzyme activity = μmol min^{-1} fraction^{-1}

leaves and two different lipolytic enzymes were measured in the various steps. Since PNP-laurate hydrolase was barely detectable it was not reported. All of the 4-MU-laurate hydrolase activity occurred in the 0.25 M NaCl DEAE eluate and the Con A Sepharose effluent, indicating that it was not associated with patatin. Much lower levels of C_6-NBD-PC hydrolase activity were detected. Although most of this activity also was found in the Con A Sepharose effluent, significant levels were also found in the 0.5 M NaCl DEAE eluate, and the patatin fractions. This experiment indicates that unlike potato tubers, very little of the lipolytic activity in potato leaves is associated with patatin or other glycoproteins which bind to Con A.

In a final experiment we investigated the possible occurence of other common (nonlipolytic) hydrolases in the patatin fraction from potato tubers. High levels of acid phosphatase and N-acetyl glucosaminidase activities were detected (data not shown). The occurence of at least 3 different enzyme activities in the patatin fraction causes one to question whether the 6-10 isoforms (ionic forms) which comprise the patatin fraction actually have very much in common (4).

REFERENCES

1. T. Galliard, The enzymatic deacylation of phospholipids and galactolipids in plants. Purification and properties of a lipolytic acyl-hydrolase from potato tubers, Biochem. J. 121:379 (1971).
2. H. Matsuda and O. Hirayama, Purification and properties of a lipolytic acyl hydrolase form potato leaves, Biochem. Biophys. Acta. 573:155 (1979).
3. R. A. Moreau, Regulation of phospholipase activity in potato leaves by protein phosphorylation-dephosphorylation, Plant Science (in press).
4. D. Racusen, Lipid acyl hydrolase of patatin, Can. J. Bot. 62:1640 (1984).
5. E. P. Hasson amd G. G. Laties, Separation and characterization of potato lipid acylhydrolases, Plant Physiol. 57:142 (1976).
6. R. A. Moreau, Membrane-degrading enzymes in the tubers of various cultivars of Solanum tuberosum, J. Ag. Food. Chem. 33:36 (1985).

CORRELATION OF METABOLIC RATE CHANGES AND MEMBRANE

TRANSITIONS DETERMINED BY MICROCALORIMETRIC METHODS

R.W. Breidenbach, R.S. Criddle, E. Lewis and L. Hanson

Plant Growth Laboratory and Department of Biochemistry
University of California, Davis, CA, and Department of
Chemistry, Brigham Young University, Provo, UT

Metabolic Rate Changes

Heat evolution from cells is interpretable as a measurement of metabolic rate[1]. Recent advancements in microcalorimeter design allow rapid nondestructive measurements of heat produced by as little as 4 mg dry wt plant tissues under steady state isothermal conditions. This provides a novel means for sensitive and precise evaluations of metabolic responses to varied environmental factors, chemical treatments and genetic alterations[2].

Fig. 1 shows a typical measurement of heat release from cultured tomato cells at 23.1°. The system attains thermal equilibrium in approximately 40 min and a steady state metabolic rate is observed. Repeated measurements of metabolic heat release by these same cells can subsequently be made using different conditions to determine effects on cell metabolism.

Measurements of the effects of temperature on metabolic rates of tomato and carrot cells are shown in Fig. 2 with data plotted in Arrhenius form. Differences in responses to temperature for the two

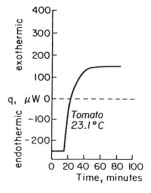

Fig. 1. Calorimeter response while measuring metabolic heat evolution from tomato cells maintained at 23.1°C. The value of the steady state rate of heat evolution was determined after 40 min.

373

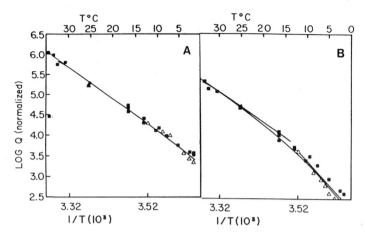

Fig. 2. Effect of temperature on metabolic rates of (A) cultured carrot cells (■) and root sections (△) and (B) cultured tomato cells (■) and leaf sections (△). Tomato cells and leaf tissue were from leaf explants of a L. esculentum/L. pyruvaniam hybrid[3]. Carrot cells were a diploid line of wild carrot, Dacus carrota[4]. Carrot root tissue was purchased at a local market. The cell data were in each case collected from three separate cell cultures.

species are clear. A sharp departure from linearity is noted near 12° for the tomato cells that is not observed for the carrots. Fig. 2 also includes data obtained using tomato leaf tissue and carrot root sections. Data for the different tissue section and cell preparations data were normalized to adjust them to the same scale. This is possible since the use of the logarithmic transformation in the Arrhenius calculation yields values of an intensive property with slopes independent of sample size.

Changes in metabolic heat rates over the range from 0° to 30° were completely reversible. Experimental points collected for Fig. 2 were obtained in random order. Repeated points at selected temperatures were reproducible within ±5% whether approached from higher or lower temperatures. Measured effects of temperature on cell metabolism correlate well with generally observed long-term temperature responses of the intact plants.

While the data shown here represents metabolic responses to changes in temperature, the method has general applicability to measurement of responses to many experimental variables, such as inhibitors, herbicides, and genetic alterations. It can be used to make rapid nondestructive measurements, repeatedly measure changes in a single sample, and examine reversibility of test parameters. Little or no sample preparation is required.

Membrane Transitions

Heat capacity changes in test samples can be determined by differential scanning microcalorimetry. Newer designs in calorimetry equipment now allow sensitive measurement of thermal transitions with small amounts

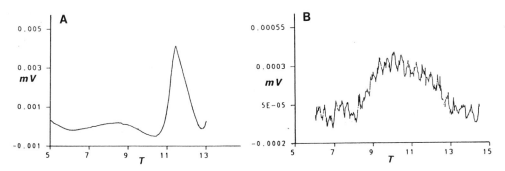

Fig. 3. Differential scanning calorimetry scans (increasing T) of (A) tomato microsomes (2°/h, 700 μg protein), and (B) azide treated tomato cells (2°/h, 1 ml packed cells).

of biological materials. A wide range of sizes (100 μg to one gram) of solid and liquid samples can be scanned with increasing or decreasing temperatures, using very slow scan speeds (as low as 1° per hour) when required.

Previous studies of plant membrane transitions have been largely limited to purified lipids, lipid mixtures or lipid extracts. Fewer measurements have been made using intact membrane preparations. Results of such studies are often extrapolated to interpret behavior of lipids in cells. We know of no previous reports of measurements of transitions in situ to correlate with in vitro studies.

Fig. 3 presents differential scanning experiments on isolated microsomes, and azide-inhibited, cultured cells from tomatoes. The temperature induced transitions for membranes are sharper than those for cells. The thermal transitions observed in cells and membranes probably represent lipid phase transitions. The transition temperatures agree quite well with the break in the Arrhenius plot of Fig. 2B.

Discussion

Calorimetry has been used to examine short term rates of cell metabolism and properties of plant cells. Isothermal studies illustrated the ability to correleate changes in cell metabolism with temperatures commonly observed to alter growth properties of carrots and tomatoes. Scanning calorimetry was able to relate the temperature effects to thermal transitions and correlate them with transitions in situ.

References

1. J.D. Loike, S.C. Silverstein and J.M. Sturtevant, Proc. Natl. Acad. Sci. 78:5958 (1981).
2. R.S. Criddle, R.W. Breidenbach, E. Lewis, D. Etaugh and L. Hansen, submitted for publication (1986).
3. N.G. Hogenboom, Euphytica 21:397 (1972).
4. A gift from H. Bonnet, University of Oregon.

MOLECULAR SPECIES COMPOSITION OF PHOSPHATIDYLGLYCEROL IN LEAVES OF

CAMELLIA SPECIES AND CHILLING SENSITIVITY

Jiro Sekiya, Hiroaki Koiso[*], Akihiko Morita[*]
and Akikazu Hatanaka[*]

Department of Agricultural Science
Okayama University
Tsushima, Okayama 700
Japan

[*]Department of Agricultural Chemistry
Yamaguchi University
Yamaguchi 753
Japan

MGDG isolated from chloroplast thylakoid of camellia species was 18:3 and this fatty acid accounted for 90% of fatty acid constituents of MGDG. PG showed characteristic fatty acid composition; 16:0, (3E)-16:1 and 18:1 were major fatty acids. (3E)-16:1 attached to C-1 position of PG and 18:1 to C-1. Major molecular species of PG was 1-18:1-2-(3E)-16:1-PG. This molecular species accounted for 40-70% in tea leaves and showed seasonal dependent change.

The hypothesis has been postulated that molecular species of PG correlates with chilling sensitive of higher plant leaves (1-3). Leaves of evergreen camellia species such as tea and camellia survive in cold winter season and life time of these leaves is usually more than one year. The functions of the wintered-over tea leaves are recovered in spring; for example, photosynthetic light reaction which reduced in winter were recovered in spring (4). However, relationship between molecular species of PG and chilling-resistance in evergreen tree leaves has not been reported.

This report describes molecular species of PG in leaves of camellia species, particularly tea leaves.

Abbreviations. 16:0, palmitic acid; (9Z)-16:1, (9Z)-hexadecenoic acid; (3E)-16:1, (3E)-hexadecenoic acid; 18:0, stearic acid; 18:1, oleic acid; 18:2, linoleic acid; 18:3, linolenic acid; PG, phosphatidylglycerol; MGDG, monogalactosyldiacylglyceride; DNB, 3,5-dinitrobenzoyl.

MATERIALS AND METHODS

Leaves of tea (Camellia sinensis var. sinensis), camellia (C. japonica) and sasanqua (C. sasanqua) were plucked in the Yamaguchi University Agricultural Experiment Station in the seasons indicated. Chloroplast thylakoid was prepared as 4,000 g pellet as described previously (5).

Total lipids were extracted essentially by the method of Allen and Good (6) and separated into individual glycolipid class by silica gel TLC (6) or silica gel column chromatography with the elution solvent of $CHCl_3$/MeOH.

Fatty acid composition of lipid was determined by GLC analysis, after transesterification with 5% H_2SO_4 in MeOH. A glass column (3 mm x 3 m) was packed with 15% DEGS or 5% BDS on Chromosorb W AW.

Positional distribution of fatty acids was determined by GLC analysis, after position-specific hydrolysis with snake venom phospholipase A_2 (7).

Molecular species of PG was analyzed by the method of Kito et al. (8,9). Each DNB derivative separated by HPLC was pooled and subjected to fatty acid determination. Molecular species were determined by fatty acid analysis and comparison of retention times on HPLC with those of authentic compounds.

RESULTS AND DISCUSSION

MGDG isolated from chloroplast thylakoids of camellia species did not contain 16:3 and 18:3 accounted for 90% of fatty acid constituents (Table I). Therefore, MGDG from these leaves was a 18:3 type.

Table 1. Fatty acid compositions of MGDG and PG isolated from chloroplast thylakoids of camellia species.

Plant*	Fatty acid composition, %						
	16:0	(9$\underline{Z}$)-16:1	(3$\underline{E}$)-16:1	18:0	18:1	18:2	18:3
MGDG							
Tea	2	1	0	0	2	7	89
Camellia	4	1	0	1	1	4	89
Sasanqua	4	1	1	0	1	2	91
Spinach	2	1	0	0	23**	2	72
PG							
Tea	21	0	29	1	21	13	15
Camellia	26	0	17	0	28	13	16
Sasanqua	25	1	26	1	29	11	7
Spinach	15	0	47	1	3	3	31

*Leaves were harvested in winter.

**The value means the sum of 16:3 and 18:3.

Digalactosyldiacylglyceride, sulfoquinovosyldiacylglyceride, phosphatidylcholine and phosphatidylethanolamine contained 16:0 and 18:3 as major fatty acids. These lipid classes seem to contain enough amounts of polyunsaturated fatty acids to maintain the membrane fluidity in the cold season.

PG which is postulated to be a major lipid class responsible for chilling sensitivity composed of 16:0, (3E)-16:1 and 18:1 as major fatty acid constituents (Table 1). Annual plants such as spinach did not contain 18:1 in a significant amount. Thus, PG in leaves of camellia species has a characteristic fatty acid composition. PG isolated from leaves of sakaki (Cleyera japonica), camphor tree (Cinnamomum comphora) and mirica (Mirica rubra) showed fatty acid composition similar to that of camellia species. In PG from tea leaves, 18:1 exclusively attached to C-1 position of PG and (3E)-16:1 to C-2 position (Table 2).

Table 2. Positional distribution of fatty acids in PG isolated from tea chloroplast thylakoid.

Position	Fatty acid composition, %					
	16:0	(3E)-16:1	18:0	18:1	18:2	18:3
Whole	25	31	2	30	6	6
C-1	33	0	4	61	1	1
C-2	30	60	0	5	4	1

Molecular species analysis of tea PG revealed that a major molecular species was 1-18:1-2-(3E)-16:1-PG. This molecular species accounted for 70% of total molecular species of PG in summer leaves and 40% in winter leaves. More than 80% of PG molecules contained unsaturated fatty acids with (9Z)-double bond. The molecular species composition of tea PG is different from that of annual plants (1). Molecular species compositions of PG from camellia and sasanqua leaves may be similar to that of tea PG based on fatty acid composition. According to Murata et al. (1), membrane lipids containing this type of PG could be chilling resistant. In addition molecular species of PG with polyunsaturated fatty acids increased in winter tea leaves with the decrease in PG containing 18:1. This change consisted with acquisition of higher chilling resistance in winter leaves (10).

Thus, characteristic molecular species of PG may correlate with chilling resistance of leaves of camellia species. Further molecular species analyses are now in progress.

REFERENCES

1. N. Murata, Plant and Physiol., 24, 81-86 (1983).
2. N. Murata and J. Yamaya, Plant Physiol., 74, 1016-1024 (1984).
3. P.G. Roughan, Plant Physiol., 77, 740-746 (1985).
4. S. Aoki, Japanese J. Crop Sci., 53, 396-402 (1984).
5. J. Sekiya, S. Numa, T. Kajiwara and A. Hatanaka, Agric. Biol. Chem., 40, 185-190 (1976).
6. C.F. Allen and P.G. Good, Methods in Enzymology Vol. XXIII, ed. by A. San Pietro, Academic Press, New York, pp. 523-547, 1971.
7. S.A. Sparace and J.B. Mudd, Plant Physiol., 70, 1260-1264, (1982).
8. M. Kito, H. Takamura, H. Narita and R. Urade, J. Biochem. 98, 327-331 (1985).

9. H. Takamura, H. Narita, R. Urade and M. Kito, <u>Lipids</u>, <u>21</u>, 356-361 (1986).
10. T. Shimura and N. Sugiyama, <u>Japanese J. Breeding</u>, <u>15</u>, 232-240, (1965).

OXYGEN REQUIRING SYSTEMS - OXYGENASES AND DESATURASES

THE LIPOXYGENASE PATHWAY

Brady A. Vick and Don C. Zimmerman

United States Department of Agriculture, Agricultural Research Service, Metabolism and Radiation Research Laboratory, State University Station, Fargo, ND 58105 (USA)

INTRODUCTION

After more than fifty years since its discovery, the study of the lipoxygenase pathway remains one of the most engaging, yet elusive areas of plant lipid research. Since 1932 when Andre and Hou (1) first reported on the existence of oxidizing enzymes which alter soybean oil, Chemical Abstracts has cited more than 1000 research articles under the subject of plant lipoxygenase. Despite the abundance of research effort on this topic, reviews (2-6) over the decades have been forced to report that the physiological role of lipoxygenase in plants is unknown.

There is reason for optimism, however. During the past ten years advances have been made in identifying metabolic routes through which the products of lipoxygenase are transformed into biologically active compounds. From the variety of oxygenated fatty acids characterized, it is clear that lipoxygenase participates in many physiological responses. The ultimate product produced as a result of the initial lipoxygenase reaction probably depends on the nature of the cellular stimulus, the available substrates, and the intracellular site of the reactions. The purpose of this short review will be to summarize current knowledge of the reactions by which the fatty acid hydroperoxide products of lipoxygenase are further metabolized, and to speculate on the physiological roles of these metabolites in plant metabolism. Finally, some comparisons will be drawn between oxygenated fatty acid metabolism in plants and mammals.

THE LIPOXYGENASE REACTION

The substrates for lipoxygenase are fatty acids containing a cis,cis-1,4-pentadiene structure. In plants the two most common fatty acids with this structure are linoleic and linolenic acid. Lipoxygenase catalyzes the incorporation of molecular oxygen into the substrate to form an unsaturated fatty acid hydroperoxide (Fig. 1). During the reaction, the cis double bond attacked by oxygen moves into conjugation with the neighboring cis double bond and assumes a trans configuration. Depending upon the plant source, the isozymes present, and the reaction conditions, oxygen incorporation can occur at either carbon 9 (e.g., tomato, maize, barley) or at carbon 13 (e.g., soybean, flaxseed, watermelon). In either case the hydroperoxide group has the S stereoconfiguration. In some plants there are isozymes of

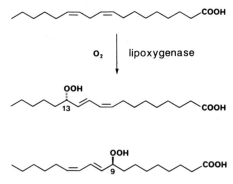

Figure 1. The reaction of linoleic acid and oxygen catalyzed by lipoxygenase.

lipoxygenase that can act on linoleic or linolenic acid even when they are esterified to glycerol as triacylglycerols or phospholipids, but generally the substrates for plant lipoxygenases are thought to exist in the free fatty acid form. An important component of lipoxygenase is an atom of iron that alternates between the Fe(II) and Fe(III) states. Another feature of the lipoxygenase reaction is an anaerobic component. Under anaerobic conditions a number of free radical intermediates are produced that lead to the formation of oxodienoic acids and dimers. The free radical intermediates are also thought to be responsible for another characteristic reaction of lipoxygenase: the cooxidation of pigments.

Most plants contain lipoxygenase, and experiments in which activity could not be found probably reflect insensitive or unreliable detection methods (3). Lipoxygenase is not confined to a particular plant organ, for it has been demonstrated in roots, cotyledons, hypocotyls, and leaves. The precise location of the enzyme within plant organs has been more difficult to establish. A good example demonstrating tissues rich in lipoxygenase was provided by Vernooy-Gerritsen et al. (7). Through immunofluorescent staining, they showed that the predominant location of lipoxygenase in soybean cotyledons after 20 hours of germination was the cytoplasm of storage parenchyma cells. Later in germination the enzyme was confined primarily to the epidermis, hypodermis, and cells surrounding the vascular bundle.

The subcellular location of lipoxygenase, which might give a clue to its function, has been varied. Vacuoles (8), cytoplasm (9), and protein bodies (9) have been reported as sites of lipoxygenase in some tissues. There is also substantial evidence that the chloroplasts of photosynthetic leaf tissue possess high lipoxygenase activity (10). Unfortunately, the inconsistency among plants with regard to lipoxygenase location in organs and subcellular organelles has not provided a clear indication of a physiological role for lipoxygenase in plants.

METABOLISM OF FATTY ACID HYDROPEROXIDES

Hydroperoxide Lyase

Many species of plants have hydroperoxide lyase enzymes that cleave fatty acid hydroperoxides into two fragments (11,12). If the substrate is a 13-hydroperoxy fatty acid, then the products are 12-oxo-cis-9-dodecenoic acid and either hexanal or cis-3-hexenal, depending on whether the hydroperoxide was derived from linoleic or linolenic acid, respectively (Fig. 2). For 9-hydroperoxy fatty acids the products of hydroperoxide lyase action are

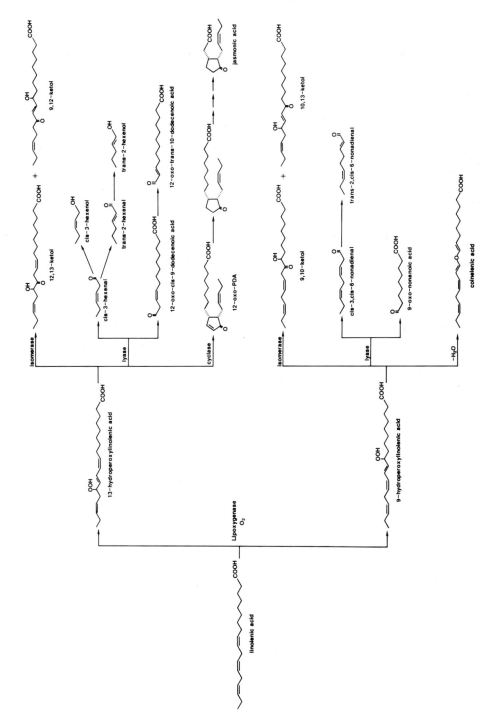

Figure 2. The linolenic acid cascade in plants.

nonanoic acid and cis-3-nonenal (from linoleic) or cis,cis-3,6-nonadienal (from linolenic). Most plants have active isomerases that quickly convert the cis-3-enal products to trans-2-enals. Both aldehyde isomers have distinctive aromas and usually contribute to the odor of the plant tissue. In certain plant tissues alcohol dehydrogenase actively converts the aldehyde products to the corresponding alcohols, which also have characteristic odors. Mushrooms have an unusual hydroperoxide lyase that uses the 10(S)-hydroperoxide of linoleic acid as substrate to produce eight- and ten-carbon fragments, 1-octen-3-ol (the characteristic flavor of mushrooms) and 10-oxo-trans-8-decenoic acid (13).

Hydroperoxide Isomerase

Hydroperoxide isomerase was the first enzyme discovered that used the hydroperoxide product of lipoxygenase as its substrate, the products being α- and γ-ketols (14,15). The 13-hydroperoxide isomer of linoleic or linolenic acid is converted to both 13-hydroxy-12-oxo and 9-hydroxy-12-oxo products (Fig. 2), in approximately a 4:1 ratio, respectively. From the 9-hydroperoxide isomer, 9-hydroxy-10-oxo and 13-hydroxy-10-oxo products are formed.

Hydroperoxide Cyclase

Only fatty acid hydroperoxides with ω-3 unsaturation serve as substrates for hydroperoxide cyclase. In plants, therefore, 13-hydroperoxy-linolenic acid is the only likely substrate for this enzyme. The product is an eighteen-carbon cyclic fatty acid containing a cyclopentenone ring (16) (Fig. 2). It has been given the trivial name of 12-oxo-phytodienoic acid (12-oxo-PDA). The two side chains are known to have a cis configuration with respect to the plane of the cyclopentenone ring. Although the absolute stereoconfiguration has not been established unequivocally, it is probably 9S,13S, as shown in Figure 2. So far, hydroperoxide cyclase activity has always been found in conjunction with hydroperoxide isomerase activity; the two activities have never been separated. It is quite likely that both activities are associated with the same protein. The only plant tissue in which the subcellular location of hydroperoxide cyclase is known with certainty is spinach leaves, where the enzyme is exclusively in the membranes of the chloroplasts (Vick and Zimmerman, unpublished observations).

The product of hydroperoxide cyclase, 12-oxo-PDA, is a key intermediate in the biosynthesis of jasmonic acid (17) (Fig. 2). First, the ring double bond of 12-oxo-PDA is reduced by NADPH in the presence of 12-oxo-PDA reductase (18). Then, after three cycles of β-oxidation of the carboxyl side chain, jasmonic acid is formed.

Divinyl Ether Formation

Up to now this reaction has only been observed with extracts of potato tubers. The 9-hydroperoxy isomers of linoleic and linolenic acids appear to be the exclusive substrates for the reaction. During the reaction one of the hydroperoxy oxygen atoms is inserted into the fatty acid chain to form a divinyl ether linkage. The products from linoleic and linolenic acids have been termed colneleic and colnelenic acids, respectively (5) (Fig. 2).

PHYSIOLOGICAL ROLES OF LIPOXYGENASE PATHWAY METABOLITES

Fatty Acid Hydroperoxides

It is unfortunate that the discussion of physiological roles for any of the lipoxygenase pathway metabolites is based more on speculation than on experimental evidence. In those cases where proposed functions have an

experimental basis, there almost always exist contradictory data to make
the suggestion ambiguous. Currently, proposals concerning the function of
fatty acid hydroperoxides in plants revolve around three main areas of
discussion: ethylene biosynthesis, senescence, and enzyme regulation.

Ethylene biosynthesis can be classified into two categories: (1)
normal formation of ethylene at certain stages of plant growth, and (2) the
formation of ethylene that results from physical, chemical, or biological
stress to the plant. Most of the ethylene biosynthetic pathway is now
understood, except for the final step which involves the conversion of
1-aminocyclopropane-1-carboxylic acid (ACC) to ethylene. There is general
agreement that the conversion requires oxygen and that it probably involves
the participation of free radicals. Because the lipoxygenase reaction
requires oxygen and has free radical intermediates, its participation in
ethylene formation has been suggested. Several investigators have provided
evidence that lipoxygenase is either directly or indirectly responsible for
ethylene production in vitro (19-21), but other experiments indicate that
properties of in vitro systems do not correspond favorably with the
properties of in vivo systems (22). The available evidence seems to rule
out lipoxygenase in ethylene biosynthesis via the normal, in vivo pathway,
but does not exclude its participation in the production of stress ethylene.

Plant senescence is another area of investigation in which lipoxygenase
has been implicated. Lipoxygenase activity increases in senescing pea
leaves, and the senescence can be retarded by exposure of the leaves to
lipoxygenase inhibitors (23). The senescence is thought to result from
increased lipid peroxidation products in cellular membranes, which then
acquire more rigid gel phase properties, become permeable, and eventually
disintegrate (24). However, there are some observations that are not consistent with a role for lipoxygenase in senescence. For example, lipoxygenase
in senescing wheat and rye leaves declined, not increased, during senescence
(25). In fact, high lipoxygenase activity is often observed in rapidly
growing tissues such as young seedlings. Moreover, the fatty acid hydroperoxides produced by lipoxygenase would most likely be quickly metabolized by
other enzymes such as hydroperoxide lyase, isomerase, or cyclase, which also
have high activity in young seedlings.

Enzyme activity regulation has also been proposed as a function for
fatty acid hydroperoxides, which are highly reactive with sulfhydryl groups
of proteins. Douillard (26) has proposed that hydroperoxides may regulate
the activity of several Calvin cycle enzymes in the chloroplast. According
to the hypothesis, the sulfhydryl groups of thioredoxin are oxidized by
fatty acid hydroperoxides. In turn, the oxidized thioredoxin would then be
unable to supply the necessary reducing power to keep the sulfhydryl groups
of the enzymes in their reduced, active states.

Aldehydes and ω-Oxoacids

The products of hydroperoxide lyase are ω-oxoacids and aldehydes (Fig.
2). Neither product has a clear function in plant metabolism. The 12-carbon oxoacid, 12-oxo-trans-10-dodecenoic acid, mimics the effects of wounding
in certain bioassays by stimulating the formation of callous tissue, and
therefore it is most likely the active component of the wound hormone,
traumatin (27). It is easily oxidized to trans-2-dodecenedioic acid,
commonly called traumatic acid. Traumatic acid was originally thought to be
the active component of traumatin (28), but now it appears that it was an
oxidation product of the oxoacid during purification. No metabolic role has
yet been suggested for 9-oxo-nonanoic acid which results from lyase action
on the 9-hydroperoxide isomer of linoleic or linolenic acid.

The physiological function of the aldehyde products is also enigmatic. It is known that the synthesis of hexanal and hexenal in tea leaves varies with the season, being higher in summer than winter, and this could possibly be related to temperature acclimation of the leaves (29). Defense mechanisms have also been suggested as a role for aldehydes in plants. Fungal growth (30) and bacterial growth (31) are both inhibited by trans-2-hexenal. The toxicity exhibited by α,β-unsaturated aldehydes is due to the high reactivity of the enal function with sulfhydryl, amino, and hydroxyl groups through addition reactions or formation of Schiff bases (32). It is possible that these products regulate certain metabolic pathways by inhibiting critical sulfhydryl-containing enzymes.

Ketols

The formation of α- and γ-ketols appear to be a dead-end pathway in plants. Neither compound has yet been shown to be further metabolized by tissue slices or cell-free homogenates. In young, developing seedlings the concentration of α-ketols is very low and increases only slightly during germination (33). The concentration changes do not parallel the dramatic changes in activity of the enzymes responsible for their synthesis, lipoxygenase and hydroperoxide isomerase. This probably results from an intracellular compartmentation of the enzymes separate from their substrates. This compartmentation would be lost during maceration of the tissue, as was demonstrated in corn seedlings by the five-fold increase in α-ketol concentration within ten minutes after mincing the tissue (33). It could be that isomerases function to convert excess toxic fatty acid hydroperoxides to innocuous ketol products as a protective mechanism.

The γ-ketols are structurally similar to 4-hydroxyalkenals, which are extremely reactive with sulfhydryl groups and lead to the inactivation of enzymes (32). The chemical reactivity of γ-ketols with glutathione has been investigated and it was found that a glutathione conjugate was formed, although the exact structure of the product was not characterized (Feng and Zimmerman, unpublished observations). In addition to chemical reactions, it is possible that enzymes, such as glutathione-S-transferases, are present to catalyze such reactions.

12-Oxo-PDA and Jasmonic Acid

The cyclopentenone structure of 12-oxo-PDA is similar to that of prostaglandin A_1 of animals. Because the prostaglandins have powerful regulating activity in animals, it is tempting to speculate that 12-oxo-PDA could have a regulatory role in plant metabolism. Although 12-oxo-PDA may have physiological activity of its own, most likely its role in plant metabolism is to serve as a precursor to jasmonic acid. Jasmonic acid and its methyl ester have been detected in many plant species. Both have been of considerable research interest in recent years because they affect plants by retarding growth or promoting senescence (34-40).

The biosynthesis of jasmonic acid is probably the best example available of a metabolic function for plant lipoxygenase. Many investigators are now suggesting that jasmonic acid may be a previously unrecognized plant hormone. Its presence has been detected in phloem exudates (41), indicating that it is translocated in the phloem. Although many reports implicate jasmonic acid involvement in growth retardation and senescence, it is usually detected in young, actively growing plant tissues (42). Consequently, jasmonic acid may be involved in other phases of plant development which are not yet recognized. In support of this view, both jasmonic acid and 12-oxo-PDA stimulated adventitious root formation in mung bean seedlings pretreated with 5 μM indolebutyric acid (43). Jasmonic acid also promoted synthesis of chlorophyll in photoheterotrophic suspension cultures of soybean and increased the synthesis of several polypeptides, one of which is regulated by cytokinins. The results suggested that jasmonic acid may act by interfering with the manner in which cytokinins stimulate cellular activity (44).

Oxidized Pigments

The lipoxygenase-catalyzed cooxidation reactions discussed in "The Lipoxygenase Reaction" section offer yet another possible role for lipoxygenase. In a cooxidation reaction with linoleic acid that was catalyzed by lipoxygenase, violaxanthin was cleaved to xanthoxin (45). Xanthoxin has been shown to be a precursor to abscisic acid in tomato shoots and dwarf bean (46). Although there is good evidence that abscisic acid is synthesized from mevalonic acid in the isoprenoid biosynthetic pathway, it is possible that an alternative pathway involving lipoxygenase is also operative in some plants.

COMPARISON WITH MAMMALIAN SYSTEMS

Plants possess two substrates, linoleic and linolenic acids, that are candidates for conversion to oxygenated fatty acids by the lipoxygenase pathway. In animals, arachidonic acid is the likely substrate for oxygenation, and two major enzyme systems are active: the lipoxygenase pathway and the cyclooxygenase pathway. The cyclooxygenase pathway, which is not known to operate in plants, leads to the formation of the prostaglandins, including thromboxane and prostacyclin. There are three major lipoxygenase pathways in animals, resulting in oxygen incorporation at carbon 5, 12, or 15 of arachidonic acid. The 12- and 15-lipoxygenase pathways result in mono- and trihydroxylated arachidonic acid metabolites whereas the 5-lipoxygenase pathway leads to the synthesis of the leukotrienes.

Collectively, the products of arachidonic acid oxygenation are known as the icosanoids. These metabolites are now well known as important regulators of mammalian metabolism. Part of the reason for the rapid advances in mammalian lipoxygenase research compared with plant lipoxygenase research can be understood by considering the chronology of mammalian icosanoid research. Historically, all of the important icosanoids were first detected by bioassay; their chemical structures were later elucidated by physicochemical methods. In contrast, many "octadecanoids" of the plant lipoxygenase pathway have been characterized, but no satisfactory bioassays are available to provide important clues as to the function of the octadecanoids.

REFERENCES

1. E. Andre and K. -W. Hou, C. R. Hebd. Seances Acad. Sci. 194:645 (1932).
2. R. T. Holman and S. Bergström, in: "The Enzymes," J. B. Sumner and K. Myrback, eds., Vol. 2, Pt.1, p. 559 (1951).
3. B. Axelrod, in: "Advances in Chemistry Series: Food Related Enzymes", No. 136, p. 324 (1974).
4. G. A. Veldink, J. F. G. Vliegenthart, and J. Boldingh, Prog. Chem. Fats other Lipids 15:131 (1977).
5. T. Galliard and H. W. -S. Chan, in: "The Biochemistry of Plants. A Comprehensive Treatise," P. K. Stumpf and E. E. Conn, eds., Vol. 4, p. 131, Academic Press, New York (1980).
6. B. A. Vick and D. C. Zimmerman, in: "The Biochemistry of Plants. A Comprehensive Treatise," P. K. Stumpf, ed., Vol. 9, Academic Press, New York, in press (1986).
7. M. Vernooy-Gerritsen, A. L. M. Bos, G. A. Veldink, and J. F. G. Vliegenthart, Plant Physiol. 73:262 (1983).

8. D. A. Wardale and E. A. Lambert, Phytochemistry 19:1013 (1980).
9. M. Vernooy-Gerritsen, J. L. M. Leunissen, G. A. Veldink, and J. F. G. Vliegenthart, Plant Physiol. 76:1070 (1984).
10. R. Douillard and E. Bergeron, C. R. Hebd. Seances Acad. Sci. Ser. D 286:753 (1978).
11. B. A. Vick and D. C. Zimmerman, Plant Physiol. 57:780 (1976).
12. T. Galliard and D. R. Phillips, Biochim. Biophys. Acta 431:278 (1976).
13. M. Wurzenberger and W. Grosch, Biochim. Biophys. Acta 794:25 (1984).
14. D. C. Zimmerman, Biochem. Biophys. Res. Commun. 23:398 (1966).
15. H. W. Gardner, J. Lipid Res. 11:311 (1970).
16. D. C. Zimmerman and P. Feng, Lipids 13:313 (1978).
17. B. A. Vick and D. C. Zimmerman, Biochem. Biophys. Res. Commun. 111:470 (1983).
18. B. A. Vick and D. C. Zimmerman, Plant Physiol. 80:202 (1986).
19. D. V. Lynch and J. E. Thompson, FEBS Lett. 173:251 (1984).
20. J.-F. Bousquet and K. V. Thimann, Proc. Natl. Acad. Sci. USA 81:1724 (1984).
21. A. Kacperska and M. Kubacka-Zebalska, Physiol. Plant. 64:333 (1985).
22. D. V. Lynch, S. Sridhara, and J. E. Thompson, Planta 164:121 (1985).
23. Y. Y. Leshem, S. Grossman, A. Frimer, and J. Ziv, in: "Advances in the Biochemistry and Physiology of Plant Lipids," L. A. Appelqvist and C. Liljenberg, eds., p. 193, Elsevier/North Holland Biomedical Press, Amsterdam (1979).
24. K. P. Pauls and J. E. Thompson, Plant Physiol. 75:1152 (1984).
25. M. Kar and J. Feierabend, Planta 160:385 (1984).
26. R. Douillard, Physiol. Veg. 19:533 (1981).
27. D. C. Zimmerman and C. A. Coudron, Plant Physiol. 63:536 (1979).
28. J. English Jr., J. Bonner, and A. J. Haagen-Smit, Science 90:329 (1939).
29. J. Sekiya, T. Kajiwara, and A. Hatanaka, Plant Cell Physiol. 25:269 (1984).
30. R. T. Major, P. Marchini, and T. Sproston, J. Biol. Chem. 235:3298 (1960).
31. H. Schildknecht and G. Rauch, Z. Naturforsch. 16b:422 (1961).
32. E. Schauenstein, H. Esterbauer, and H. Zollner, "Aldehydes in Biological Systems," Pion Limited, London (1977).
33. B. A. Vick and D. C. Zimmerman, Plant Physiol. 69:1103 (1982).
34. J. Ueda and J. Kato, Plant Physiol. 66:246 (1980).
35. J. Ueda and J. Kato, Z. Pflanzenphysiol. 103:357 (1981).
36. S. O. Satler and K. V. Thimann, C. R. Hebd. Seances Acad. Sci. Ser. D 293:735 (1981).
37. W. Dathe, H. Rönsch, A. Preiss, W. Schade, G. Sembdner, and K. Schreiber, Planta 153:530 (1981).
38. J. Ueda and J. Kato, Agric. Biol. Chem. 46:1975 (1982).
39. H. Yamane, H. Abe, and N. Takahashi, Plant Cell Physiol. 23:1125 (1982).
40. S. Tsurumi and Y. Asahi, Physiol. Plant. 64:207 (1985).
41. J. M. Anderson, Plant Physiol. 77S:75 (1985).
42. A. Meyer, O. Miersch, C. Büttner, W. Dathe, and G. Sembdner, J. Plant Growth Regul. 3:1 (1984).
43. D. C. Zimmerman and B. A. Vick, Plant Physiol. 72S:108 (1983).
44. J. M. Anderson, Plant Physiol. 80S:34 (1986).
45. R. D. Firn and J. Friend, Planta 103:263 (1972).
46. H. F. Taylor and R. S. Burden, J. Exp. Bot. 24:873 (1973).

ENZYMIC OXYGENATIVE-CLEAVAGE REACTION OF LINOLENIC ACID IN LEAVES
-CHLOROPLASTIC LIPOXYGENASE AND FATTY ACID HYDROPEROXIDE LYASE
IN TEA LEAVES

Akikazu Hatanaka, Tadahiko Kajiwara and Jiro Sekiya

Department of Agricultural Chemistry, Faculty of Agriculture
Yamaguchi University
Yamaguchi 753, Japan

INTRODUCTION

Since (2E)-hexenal (leaf aldehyde) and (3Z)-hexenol (leaf alcohol) were first found in some bushes and tea leaves(1-3), many early works have been reported that these C_6-volatile compounds are originating from plant tissues and largely responsible for the characteristic odor of various fruits and green leaves of vegetables and trees. Linoleic acid and linolenic acid having (1Z,4Z)-pentadiene moiety were reported to be precursors of hexanal and hexenals, respectively, in 1966.(4) The reactions involved in the major biosynthetic pathway for C_6-aldehydes consist of four sequential steps; acyl hydrolysis of lipids, hydroperoxidation of linoleic acid and linolenic acid, cleavage of the fatty acid hydroperoxides and isomerization of (3Z)-hexenal to (2E)-hexenal (Fig. 1).(5-9) The three enzymes and one non-enzymic factor involved in the pathway are lipolytic acyl hydrolase(LAH), lipoxygenase, fatty acid hydroperoxide lyase (hydroperoxide lyase) and an isomerization factor (Fig. 1). Volatile C_9-aldehydes are formed by a similar pathway.(7, 10,11) Among these enzymes lipoxygenase and hydroperoxide lyase are the most important. The substrate specificities and the product specificities of these enzymes determine the composition of volatile aldehydes formed from linoleic acid and linolenic acid. The present report describes occurrence of, and substrate and product specificities of, lipoxygenase and hydroperoxide lyase in leaves, particularly in tea leaves.

MATERIALS AND METHODS

Plant materials and enzyme solution. Fresh leaves of tea plants (Camellia sinensis cv. Yabukita) were harvested in August at a local farm. The chloroplasts were prepared as a 4,000 g precipitate and stored in 2.1 M sucrose at -20°C, as described previously(12); these were type II, broken chloroplasts. One gram of wet chloroplasts corresponded to 3.5 g chloroplast-sucrose suspension. Before use, chloroplasts(3.5 g) in 2.1 M sucrose were washed with 10 ml of McIlvaine's buffer, pH 6.3, by centrifugation at 10,000 g for 10 min. The washed chloroplast fragments(1.0 g) were suspended in 10 ml of McIlvaine's buffer, pH 6.3, were used to measure lipoxygenase activity. One gram of washed chloroplasts (wet weight) contained 3 to 5 mg of chlorophyll and 100 to 150 mg of protein.

Preparation of chemicals. Linoleic acid, linolenic acid, γ-linolenic acid and

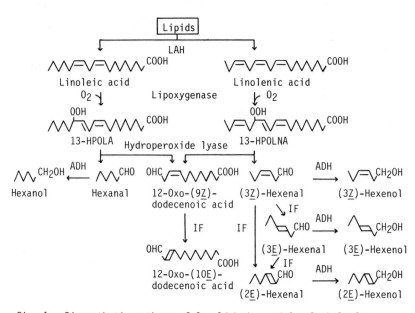

Fig. 1 Biosynthetic pathway of C_6-aldehydes and C_6-alcohols from linoleic acid and linolenic acid in plant tissues.
LAH, lipolytic acyl hydrolase. IF, isomerization factor.
ADH, alcohol dehydrogenase.

($9\underline{E},12\underline{E}$)-octadecadienoic acid (linoelaidic acid) used were 99% pure, and arachidonic acid was over 90% pure. Linoleyl alcohol and linoleyl aldehyde were prepared as described previously.(13) Positional isomers and geometrical isomers of linoleic acid were synthesized as described previously.(14) Fatty acid hydroperoxides were prepared using lipoxygenases as described previously.(15)

<u>Lipoxygenase assay</u>. Oxygen uptake by lipoxygenase reaction was measured at 25°C with a Clark type oxygen electrode, on assumption that there was 0.24 mM dissolved oxygen in an air-saturated solution at 25°C.(13) The reaction mixture contained 0.33 mM substrate (0.1 ml of substrate dispersed in 0.2% Tween 20) and 0.1 ml of the chloroplast suspension (0.1 mg/ml) in 3 ml of McIlvaine's buffer, pH 6.3. Oxygen uptake was represented as the initial rate per min (0.5 to 1.5 min).

<u>Hydroperoxide lyase assay</u>. Hydroperoxide lyase was determined by measurement of the formation of hexanal from 13-hydroperoxide of linoleic acid at pH 6.3.(13) The substrate (6 μmol) dissolved in diethyl ether was pipetted into a 50-ml flask and the solvent was evaporated <u>in vacuo</u>. Then, 10 ml of chloroplast suspension or leaf homogenate was added. The mixture was sealed in the flask with a rubber stopper and incubated at 35°C for 10 min. The C_6-aldehydes formed were measured by the headspace method with GLC.

C_6-Aldehyde forming-activity. C_6-Aldehyde formation from C_{18}-fatty acid was determined by the procedure used for hydroperoxide lyase activity except for the addition of C_{18}-fatty acids instead of the 13-hydroperoxides.

Preparation of fatty acid hydroperoxides with chloroplasts. Chloroplast fragments (1 g wet weight) and linoleic acid (200 mg; 714 μmol) were suspended in McIlvain's buffer, pH 6.3, and incubated for 20 min at 35°C. After pH was adjusted to 2.5 with 2 N HCl to stop the reaction, the reaction products were extracted with diethyl ether. The ether extract was concentrated in vacuo. Based on the absorption at 234 nm, about 20% of linoleic acid was found to be converted to the hydroperoxides. The obtained hydroperoxides of linoleic acid were purified by column chromatography (SiO_2 for dry column).(16)

Analyses of the fatty acid hydroperoxides. The separated fatty acid hydroperoxides were reduced with $NaBH_4$ and subsequently esterified with CH_2N_2 as described previously.(16) Geometrical isomers of hydroxy fatty acid methyl ester derivatives were analyzed directly with HPLC and optical isomers were analyzed with GLC after esterification of the hydroxy fatty acid methyl ester with R-(+)-MTPA followed by oxidative ozonolysis as described previously.(16)

RESULTS AND DISCUSSION

Occurrence of lipoxygenase in leaves.
 There were some early works which suggested that green leaves involve lipoxygenase activity.(17-21) In tea plant most of the lipoxygenase activity was found in chloroplast fragments (thylakoid-rich fraction).(12,13) However, the solubilization of the enzyme from chloroplast thylakoids was difficult and a solubilized enzyme has not been obtained. Therefore, lipoxygenase bound to the thylakoids was used in the experiments below. The chloroplast lipoxygenase of tea leaves showed maximal activity at pH 6.3. The enzyme activity was very stable for a long term (more than 4 years) when the chloroplast thylakoids suspended in a 2.1 M sucrose solution were stored at -20°C or less.

Substrate and product specificities of chloroplast lipoxygenase.
 Substrate specificity of tea chloroplast lipoxygenase was examined. Saturated and monoenoic fatty acids did not act as substrates for tea chloroplast lipoxygenase. Linoleic acid and linolenic acid, which are possible potent natural substrates for the enzyme in the plant tissues, acted as the substrates and linolenic acid was better substrate among them (Table 1). γ-Linolenic acid and arachidonic acid were also good substrates. All the C_{18}-fatty acids acting as the substrates had (1Z,4Z)-pentadiene moiety at C-9 and C-12. (11Z,14Z)-Double bond in arachidonic acid correspond to (9Z,12Z)-double bond in linoleic acid when they are numbered from the ω-terminal. Geometrical isomer of linoleic acid, (9E,12E)-octadecadienoic acid (linoelaidic acid), did not act as the substrate (Table 1). Among the positional isomers of linoleic acid, (8Z,11Z)-, (10Z,13Z)- and (11Z,14Z)-octadecadienoic acids showed reactivities for tea chloroplast lipoxygenase, but the reactivities were about a half of that of linoleic acid (Table 1). Linoleic acid derivatives including methyl ester, alcohol and aldehyde acted as the substrate, but to a lesser extent. From these results, the fatty acids having (1Z,4Z)-pentadiene moiety at C-9 and C-12 in C_{18}-fatty acid and at C-11 and C-14 in arachidonic acid act as good substrates for tea chloroplast lipoxygenase. Further introduction of double bonds into C_{18}-fatty acids may increase the reactivity as the substarate. Carboxy group at C-1 is important but not essential for the substrate.
 Product specificity of tea chloroplast lipoxygenase was examined using linoleic acid, methyl linoleate and linoleyl alcohol (as the substrate). Major products from linoleic acid were 13-hydroperoxides (96%); particularly 13-(S)-hydroperoxy-(9Z,11E)-octadecadienoic acid (more than 60% of all pro-

Table 1 Substrate specificity of lipoxygenase and C_6-aldehyde-forming activity

Substrate	Lipoxygenase	C_6-aldehyde-forming activity	
	Relative activity (%)		Product
Linoleic acid	100	100	Hexanal
Methyl linoleate	75*,*2	14*,*2	Hexanal
Linoleyl alcohol	84*,*2	16*,*2	Hexanal
Linoleyl aldehyde	68*,*2	8*,*2	Hexanal
Linolenic acid	181*,	114*	Hexenals
γ-Linolenic acid	147*	18*	Hexanal
Arachidonic acid	108*	0*	
(6$\underline{Z}$,9$\underline{Z}$)-octadecadienoic acid	27	0*3	
(7$\underline{Z}$,10$\underline{Z}$)-	25	0*3	
(8$\underline{Z}$,11$\underline{Z}$)-	56*3	0*3	
(10$\underline{Z}$,13$\underline{Z}$)-	44*3	tr*3	
(11$\underline{Z}$,14$\underline{Z}$)-	60*3	0*3	
(12$\underline{Z}$,15$\underline{Z}$)-	31	0*3	
(9$\underline{Z}$,12$\underline{E}$)-	nd*4	tr*3	Hexanal
(9$\underline{E}$,12$\underline{Z}$)-	nd*4	tr*3	Hexanal
(9$\underline{E}$,12$\underline{E}$)-	0	0*3	

* cited from ref. 13. *2 assayed at pH 4.5. *3 cited from ref. 33. *4 not determined.

ducts) (Table 2). 9-Hydroperoxides were minor products (4%). It was the case with methyl linoleate as the substrate. However, when linoleyl alcohol was used as the substrate, 9-hydroperoxides and 13-hydroperoxides accounted for 41 and 59%, respectively, and the both positional specificity and stereospecificity of the products were decreased (Table 2). From these results functional group at C-1 affected the product specificity. The product specificity obtained here was similar to that with soybean lipoxygenase when substrate was linoleic acid.(16) However, product composition with methyl linoleate was different slightly.(16)

Occurrence of hydroperoxide lyase in plant tissues.
   Hydroperoxide lyase utilized 9- and/or 13-hydroperoxides of linoleic acid and linolenic acid as the substrate to give C_6- and/or C_9-aldehydes and counterparts of oxo-acids. Since hydroperoxide lyase was first identified in the non-green tissues(5,11,22), the occurrence of the enzyme has been reported in various non-green and green plant tissues. Non-green tissues in which hydroperoxide lyase occurs include etiolated seedlings of watermelon (5), cucumber(23) and alfalfa(23), soybean seed(21,24,25), fruits of cucumber (11,23,26), tomato(6,27), pear(28) and apple(27), and cultured tobacco cell (29). Hydroperoxide lyases also occur in the green tissues including leaves or kidney bean(30), tea(15) and others(21), peel tissue or cucumber fruit

Table 2 Compositions of the positional, geometrical and optical isomers of the hydroperoxides formed by incubation with tea chloroplasts

Substrate	Geometrical				Optical			
	13-HPO		9-HPO		13-HPO		9-HPO	
	(Z,E)	(E,E)	(E,Z)	(E,E)	R	S	R	S
Linoleic acid	80	16	3	1	20	76	2	2
Methyl linoleate	96	1	1	2	9	88	1	2
Linoleyl alcohol	39	20	5	36	28	31	nd*	nd*

* not determined.

(31), and cultured tobacco green cells.(29) Hydroperoxide lyase activity in the green tissues consists of at least two enzymes, of chloroplastic and non-chloroplastic type.(21,31)

Most of non-chloroplastic type of hydroperoxide lyase has been reported to be membrane-bound one. Non-chloroplastic hydroperoxide lyase was solubilized from fruits of cucumber(26), pear(28) and tomato(27) with Triton X-100 and then purified partially. Watermelon hydroperoxide lyase was purified 42-fold without any particular solubilization treatment.(5) Chloroplastic hydroperoxide lyase was solubilized from tea leaves with 0.2% Tween 20 and partially purified.(15) Although attempts at purification of hydroperoxide lyase have been made, the enzyme has not been purified to homogeneity thus far.

Optimal pH of the hydroperoxide lyase was 5.5-8.0. In tea plant, the solubilized enzyme showed optimal activity at pH 7-8(15), whereas optimal pH of chloroplast-bound enzyme was 7.0.(13)

Substrate and product specificities of chloroplast hydroperoxide lyase.

Hydroperoxide lyase is classified into three types based on the substrate specificity; 9-hydroperoxide-specific, 13-hydroperoxide-specific and non-specific. The first 9-hydroperoxide-specific enzyme cleaves exclusively the 9-hydroperoxides to give C_9-aldehydes and a C_9-oxo-acid. Pear fruit hydroperoxide lyase belongs to this type.(28) The 13-hydroperoxide-specific enzyme cleaves the 13-hydroperoxides to give C_6-aldehydes and C_{12}-oxo-acid. Watermelon seedling(5), tea leaf(15), cultured tobacco cell(29), tomato fruit (30), alfalfa seedling(23) and soybean seed(24) contain this type of the enzyme. Kidney bean leaf(30), and cucumber fruit(11,22) and seedling(23) contain the non-specific enzyme which can cleave both the 9- and 13-hydroperoxides to give the corresponding products.

13-Hydroperoxides of linoleic acid and linolenic acid were the best substrates both for chloroplast-bound hydroperoxide lyase and the enzyme solubilized from chloroplast in tea plant (Table 3). 9-Hydroperoxides of linoleic acid and linolenic acid did not act as the substrates for C_6-aldehyde and C_9-aldehyde formation. When 13-hydroperoxy-(6Z,9Z,11E)-octadecatrienoic acid (13-HPO-γ-linolenic acid) was used, the reactivity was 18% of that of 13-HPOLA. 15-Hydroperoxy-(5Z,8Z,11Z,13E)-eicosatetraenoic acid (15-HPO-arachidonic acid) did not act as a substrate for hexanal formation. Conversion of the carboxy group to a methyl ester or alcohol function greatly reduced reactivity. The 12-keto-hydroxy derivative(α-ketol) and 13-hydroxy-linoleic acid did not act as substrates. As to configuration of hydroperoxy group of the 13-hydroperoxides of linoleic acid, the (S)-configuration is

Table 3 Substrate specificity of tea chloroplast hydroperoxide lyase

Substrate	Relative activity (%)	Product
13-Hydroperoxy-(9Z,11E)-octadecadienoic acid (13-HPOLA)	100*	Hexanal
Methyl 13-hydroperoxy-(9Z,11E)-octadecadienoate (13-HPOLMe)	27*,*2	Hexanal
13-Hydroperoxy-(9Z,11E)-octadecadienol (13-HPOLAL)	53*,*2	Hexanal
13-Hydroperoxy-(9Z,11E,15Z)-octadecatrienoic acid (13-HPOLNA)	102*	(3Z)- & (2E)-Hexenal
13-Hydroperoxy-(6Z,9Z,11E)-octadecatrienoic acid (13-HPO-γ-LNA)	22*	Hexanal
15-Hydroperoxy-(5Z,8Z,11Z,13E)-eicosatetraenoic acid (15-HPO-arachidonic acid)	0*	(Hexanal)
9-Hydroperoxy-(10E,12Z)-octadecadienoic acid (9-HPOLA)	0*3	((3Z)- & (2E)-Nonadienal)
9-Hydroperoxy-(10E,12Z,15Z)-octadecatrienoic acid (9-HPOLNA)	0*3	((3Z,6Z)- & (2E,6Z)-Nonadienal)
12-Oxo-13-hydroxy-(9Z)-octadecenoic acid (α-Ketol)	0*3	(Hexanal)
13-Hydroxy-(9Z,11E)-octadecadienoic acid (13-hydroxy-LA)	0	(Hexanal)

* cited from ref. 13. *2 assayed at pH 6.3. *3 cited from ref. 15.
( ); expected product.

favored by the tea chloroplast enzyme.(32) These results thus indicate the optimum requirements for the substrate for tea chloroplast hydroperoxide lyase; C_{18}-straight chain fatty acid with free carboxy group, attachment of hydroperoxy group to C-13 with (S)-configuration, and (Z)-double bond at C-9 and (E)-double bond at C-11 in the chain. Further introduction of (Z)-double bond at C-15 increased reactivity by 27% and at C-6 decreased it by 45%, respectively.

Overall reaction of lipoxygenase and hydroperoxide lyase in tea chloroplast.
    Overall reaction of enzyme system consisting of lipoxygenase and hydroperoxide lyase in tea chloroplasts was investigated. As described earlier, tea chloroplast thylakoids contained lipoxygenase and hydroperoxide lyase. (13) This enzyme system catalyzes formation of C_6-aldehydes including hexanal, (3Z)-hexenal and (2E)-hexenal from linoleic acid and linolenic acid but not C_9-aldehydes from these fatty acids.
    Substrate specificity of the enzyme system is illustrated in Table 1 as C_6-aldehyde-forming activity. Linoleic acid and linolenic acid were the good substrates for overall reaction or the tea chloroplast enzyme system. As expected from the structure, the products were hexanal from linoleic acid and (3Z)-hexenal and (2E)-hexenal from linolenic acid. Methyl linoleate, linoleyl alcohol and γ-linolenic acid did not act as the substrate although

they were good substrates for lipoxygenase reaction to 13-hydroperoxides.(13) This indicates that a rate-determining step of the overall reaction is hydroperoxide lyase rather than lipoxygenase. This conclusion is consisted with the results shown in Table 3 which showed the strict substrate specificity of hydroperoxide lyase. Positional isomers of linoleic acid may not act as substrates for the enzyme system because, for example, neither pentanal nor heptenal as expected products were found in incubation of tea chloroplast with (10$\underline{Z}$,13$\underline{Z}$)-octadecadienoic acid (Table 1).

Thus, good substrates for the tea chloroplast enzyme system consisting of lipoxygenase and hydroperoxide lyase were linoleic acid and linolenic acid which are commonly occurring fatty acid in plant tissues particularly after acyl hydrolysis of lipids on aging, injury and processing etc. The resultant products are volatiles including hexanal, (3$\underline{Z}$)-hexenal and (2$\underline{E}$)-hexenal which are responsible for characteristic odor of plant products. In addition 12-oxo-(9$\underline{Z}$)-dodecenoic acid and 12-oxo-(10$\underline{E}$)-dodecenoic acid reported as plant wound hormone are formed as a counterpart of volatile C_6-aldehydes.

ACKNOWLEDGEMENTS

This work was supported in part by grants-in-aid from the Ministry of Education, Science and Culture of Japan (A. H. and T. K.) and from Agricultural Chemical Foundation of Japan (J. S.).

REFERENCES

1. T. Curtius and H. Frunzen, Justus Liebigs Ann. Chem., 390, 89 (1912).
2. P. von Romburgh, Chem. Zentralbl., 1920,I: 83.
3. S. Takei, Y. Sakato, M. Ohno and Y. Kuroiwa, Bull. Agric. Chem. Soc. Japan, 14, 709-715 (1938).
4. F. Drawert, W. Heimann, R. Emberger and R. Tressl, Justus Liebigs Ann. Chem., 694, 200-208 (1966).
5. B. A. Vick and D. C. Zimmerman, Plant Physiol., 57, 780-788 (1976).
6. T. Galliard and J. A. Matthew., Phytochemistry, 16, 339-343 (1977).
7. T. Galliard, in "Biochemistry of Wounded Plant Tissues"; G. Kahl, ed., Walterde, Gruyter, Berlin, West Germany, 1978, 155.
8. J. Sekiya, T. Kajiwara and A. Hatanaka, Plant Cell Physiol., 25, 269-280 (1984).
9. A. Hatanaka, T. Kajiwara and J. Sekiya, Phytochemistry, 15, 1125-1126 (1976).
10. A. Hatanaka, T. Kajiwara and T. Harada, Phytochemistry, 14, 2589-2592 (1975).
11. T. Galliard, D. R. Phillips and J. Reynolds, Biochim. Biophys. Acta, 441, 181-192 (1976).
12. J. Sekiya, S. Numa, T. Kajiwara and A. Hatanaka, Agric. Biol. Chem., 40, 185-190 (1976).
13. A. Hatanaka, T. Kajiwara, J. Sekiya, M. Imoto and S. Inouye, Plant Cell Physiol., 23, 91-99 (1982).
14. T. Kajiwara, T. Koda and A. Hatanaka, Agric. Biol. Chem., 43, 1781-1782 (1979).
15. A. Hatanaka, T. Kajiwara, J. Sekiya and S. Inouye, Phytochemistry, 21, 13-17 (1982).
16. A. Hatanaka, T. Kajiwara, J. Sekiya and M. Asano, Z. Naturforsch., 39c, 171-173 (1984).
17. M. Holden, Phytochemistry, 9, 507-512 (1970).
18. A. Pinsky, S. Grossman and M. Trop, J. Food Sci., 36, 571-572 (1971).
19. A. Ben-Aziz, S. Grossman, P. Budowski and I. Ascarelli, Phytochemistry, 10, 1823-1830 (1971).
20. S. Grossman, A. Ben-Aziz, I. Ascarelli and P. Budowski, Phytochemistry, 11, 509-514 (1972).

21. J. Sekiya, T. Monma, T. Kajiwara and A. Hatanaka, Agric. Biol. Chem., 50, 521-522 (1986).
22. T. Galliard and D. R. Phillips, Biochim. Biophys. Acta, 431, 278-287 (1976).
23. J. Sekiya, T. Kajiwara and A. Hatanaka, Agric. Biol. Chem., 43, 969-980 (1979).
24. T. Matoba, H. Hidaka, H. Narita, K. Kitamura, N. Kaizuma and M. Kito, J. Agric. Food Chem., 33, 852-855 (1985).
25. T. Matoba, H. Hidaka, K. Kitamura, N. Kaizuma and M. Kito, J. Agric. Food Chem., 33, 856-858 (1985).
26. D. R. Phillips and T. Galliard, Phytochemistry, 17, 355-358 (1978).
27. P. Schreier and G. Z. Lorenz, Z. Naturforsch., 37c, 165-173 (1982).
28. I. -S. Kim and W. Grosch, J. Agric. Food Chem., 29, 1220-1225 (1981).
29. J. Sekiya, S. Tanigawa, T. Kajiwara and A. Hatanaka, Phytochemistry, 23, 2439-2443 (1984).
30. J. A. Matthew and T. Galliard, Phytochemistry, 17, 1043-1044 (1978).
31. D. A. Wardale, E. A. Lambert and T. Galliard, Phytochemistry, 17, 205-212 (1978).
32. T. Kajiwara, J. Sekiya, M. Asano and A. Hatanaka, Agric. Biol. Chem., 46, 3087-3088 (1982).
33. A. Hatanaka, T. Kajiwara and T. Koda, Agric. Biol. Chem., 43, 2115-2117 (1979).

FATTY ACID ß-OXIDATION IN HIGHER PLANTS

Bernt Gerhardt

Botanisches Institut, Universität Münster
D-4400 Münster, Federal Republic of Germany

INTRODUCTION

Thirty years ago, Stumpf and Barber[1] studying fatty acid oxidation of in vitro systems from plants suggested that plant cells possess the ß-oxidation sequence known for mammalian and microbial cells at that time. Subsequent studies, mainly from Stumpf's laboratory, in principle established the existence of the ß-oxidation pathway in plant cells.

Until recently, fatty tissues of germinating oilseeds have mostly been used in order to study fatty acid ß-oxidation of higher plant cells. For these tissues, subcellular location of the ß-oxidation pathway in the glyoxysomes has been demonstrated[2,3] and is well established[4]. The glyoxysomal ß-oxidation system differs in some respect from the classic mitochondrial ß-oxidation system of mammalian cells (see below). A ß-oxidation system as it occurs in glyoxysomes has also been discovered in the peroxisomes of mammalian systems[5,6,7] and those of eucaryotic microorganisms[8]. In mammalian cells, the peroxisomal ß-oxidation system coexists and appears to cooperate with the mitochondrial ß-oxidation system[7,9].

Glyoxysomes belong to the class of peroxisomes and are characterized by having the glyoxylate cycle. They only occur in fatty tissues of germinating seeds. Glyoxysomes are considered to be the exclusive site of ß-oxidation in these tissues[4]. Concerning all other tissues of higher plants, ß-oxidation was still considered to be a mitochondrial process. But there was little experimental support for this assumption. Recent reports on the occurrence of a mitochondrial ß-oxidation system in pea cotyledons[10,11] do not unequivocally demonstrate this. At present, there is no convincing evidence that plant mitochondria possess ß-oxidation activity. Among other arguments[12], all attempts to demonstrate activity of acyl-CoA dehydrogenase, the initial enzyme of the classic mitochondrial ß-oxidation system, in mitochondria from plant tissues have been unsuccessful[13,14,15,16].

During the last years, it has been established that not only fatty tissues of germinating seeds but also nonfatty plant tissues possess a peroxisomal ß-oxidation system[12]. The current informations with respect to the peroxisomal ß-oxidation system of higher plant cells will be reviewed in this article. A comparison between this system and the peroxisomal ß-oxidation systems of mammalian cells and eucaryotic microorganisms is not intended here.

ß-OXIDATION ENZYMES

Up to now, ß-oxidation enzymes have only been purified from cucumber cotyledons. The enzymes appear to be characterized by having a very alkaline isoelectric point. Using peroxisomes as an enzyme source, characteristics of single ß-oxidation reactions have also been studied.

Fatty acid activation has been investigated using glyoxysomes from castor bean endosperm[17] and peroxisomes from mung bean hypocotyls[18]. Acyl-CoA synthetase as fatty acid activating enzyme was established by the dependence of acyl-CoA formation from ATP and CoA and the generation of acyl-CoA, AMP and pyrophosphate as reaction products in a 1:1:1 stoichiometry[18]. The acyl-CoA synthetase has not yet been purified. At 100 µM substrate concentration, the enzyme activates long-chain fatty acids 5- to 10-times faster than short-chain fatty acids. The enzyme of the castor bean glyoxysomes is also active towards ricinoleic acid, the major fatty acid in castor bean endosperm. The enzyme of the mung bean hypocotyl peroxisomes utilizes most effectively linolenic, linoleic and oleic acid, the predominent fatty acids of membrane lipids. Half- maximal reaction rate with palmitic acid as substrate was obtained at a fatty acid concentration of 33 µM[18].

The first oxidative reaction of the peroxisomal ß-oxidation pathway is catalyzed by an acyl-CoA oxidase which transfers the electrons directly to oxygen, producing hydrogen peroxide[2,15]. In contrast, the acyl-CoA dehydrogenase of the mammalian mitochondrial ß-oxidation pathway is tightly coupled to the respiratory chain. Therefore, an acyl-CoA dependent oxygen uptake, which is catalyzed by an acyl-CoA dehydrogenase, and consequently the mitochondrial ß-oxidation system itself are sensitive to inhibitors of the respiratory chain. The peroxisomal ß-oxidation system is insensitive to these inhibitors due to its acyl-CoA oxidase reaction. The expected stoichiometry, oxygen uptake : hydrogen formation : enoyl-CoA formation = 1:1:1, of the acyl-CoA oxidase reaction has been demonstrated[15].

The acyl-CoA oxidase, a flavoprotein, has recently been purified[19]. The protein has an alkaline isoelectric point and consists of two identical subunits. The activity of the purified acyl-CoA oxidase was low towards short-chain acyl-CoAs (25 µM) and increased with long-chain acyl-CoAs. Using peroxisomes as enzyme source, activities of the acyl-CoA oxidase with short-chain acyl-CoAs (50 µM) have been reported which ranged from half to higher thant the activity obtained with palmitoyl-CoA[13,14,16,20]. Medium-chain and unsaturated C_{18} acyl-CoAs were oxidized most effectively[16]. Half-maximal reaction rates were obtained with short-chain acyl-CoAs at substrate concentrations approximately 3-times higher thant those (10-20 µM) required with medium- and long-chain acyl-CoAs[21].

The enoyl-CoA hydratase and 3-hydroxyacyl-CoA dehydrogenase reactions are catalyzed by a bifunctional protein[22]. This is characteristic of the peroxisomal ß-oxidation sytem. The bifunctional protein of cucumber glyoxysomes has an isoelectric point of pH 9.8 and consists of a single polypeptide chain with a molecular weight of 75000. Concerning the substrate specificity, it has been reported[23], that the maximum enoyl-CoA hydratase activities decreased 100-fold as the acyl-chain length increased from C_4 to C_{16}. Half-maximal reaction rates with long-chain enoyl-CoAs were obtained at substrate concentrations 10-fold higher than those required with short-chain enoyl-CoAs.

The 3-oxoacyl-CoA thiolase of cucumber glyoxysomes[24] consists of two identical subunits with a molecular weight of 45000. The protein has an isoelectric point of pH 8.5. Data on the substrate specificity of the enzyme have not yet been reported.

OPERATION OF THE ß-OXIDATION PATHWAY

It has been demonstrated[2,3,14,20,25] that the single peroxisomal ß-oxidation enzyme activities are linked to form a ß-oxidation capacity. When provided with a fatty acid (plus ATP and CoA) or an acyl-CoA (plus CoA) and NAD^+, the peroxisomes catalyze the formation of NADH (product of the 3-hydroxyacyl-CoA dehydrogenase reaction) and acetyl-CoA (or labeled acetyl-CoA in those cases where a labeled substrate was used). The reaction is insensitive to inhibitors of the respiratory chains of plant mitochondria. The stoichiometry of the overall ß-oxidation was determined to be 1:1:1 with respect to oxygen uptake, NADH and acetyl-CoA formation[2]. At low substrate concentrations (about 10 μM) the ß-oxidation rates with long-chain acyl-CoAs were 2- to 4-fold higher than those with medium- and short-chain acyl-CoAs[2,25]. No dependence on the chain-lenght was observed at high acyl-CoA concentrations (200 μM)[20].

Peroxisomes appear to be capable of degrading long-chain, saturated, even-numbered fatty acids completely to their constituent acetyl units. Medium- and short-chain acyl-CoAs, intermediates in long-chain acyl-CoA oxidation, are oxidized by the acyl-CoA oxidase and serve as substrates for the overall in vitro ß-oxidation by peroxisomes. However, storage triacylglycerols, the substrate source for ß-oxidation in fatty tissues, and membrane lipids, a possible substrate source for ß-oxidation in nonfatty tissues, contain unusual and unsaturated fatty acids. Additional enzyme reactions are required to link the catabolism of these fatty acids to the ß-oxidation sequence.

The completeness of fatty acid or acyl-CoA degradation by the peroxisomal ß-oxidation system has not yet thoroughly been investigated. From the data reported by Cooper and Beevers[2] it can be calculated that the palmitoyl-CoA went through three cycles of ß-oxidation. Studies on the degradation of palmitic acid, labeled along the hydrocarbon chain, by cell-free systems from fatty tissues indicated that complete oxidation of the fatty acid occurred[1,26]. Hutton and Stumpf[27] who studied the pathway of ricinoleic acid catabolism concluded that the castor bean glyoxysomes appear to possess the addional enzymes required to catabolize ricinoleic acid to the level of propionate. The proposed metabolic pathway for ricinoleic acid degradation includes an α-oxidation step. 3-cis-2-trans-enoyl-CoA isomerase, an enzyme required for the catabolism of unsaturated fatty acids, has been demonstrated in cotton glyoxysoms[20].

The peroxisomal ß-oxidation system appears to be carnitine-independent. Carnitine acyltransferase and carnitine acetyltransferase activities have not been detected in peroxisomes[18,20]. In addition, palmitoylcarnitine is not oxidized by peroxisomes[2,16,20]. The acyl-CoA synthetase is tightly bound to the peroxisome membrane. Latency and resistance to protease treatment indicate that the enzyme is located on the inner side of the organelle membrane[18]. This topology of the acyl-CoA synthetase implies that not acyl-CoAs but fatty acids have to cross the peroxisomal membrane in order to initiate fatty acid ß-oxidation. At present, it is unknown as to how the peroxisomes are provided with both ATP and CoA required for the acyl-CoA synthetase reaction and how an accumulation of the reaction products AMP and pyrophosphate is prevented within the organelles.

Continuous operation of the ß-oxidation requires a reoxidation of the NADH produced by the 3-hydroxyacyl-CoA dehydrogenase reaction. Concerning this NADH reoxidation there exist different models[28,29]. They are based on studies on glyoxysomes and may also be applicable to the peroxisomes of nonfatty tissues[12,25].

The fate of the acetyl-CoA generated during ß-oxidation is only known for the glyoxysomes in which the acetyl-CoA enters the glyoxylate cycle. Activities of isocitrate lyase and malate synthase, the key enzymes of the glyoxylate cycle, have previously not been detected in the peroxisomes from nonfatty tissues[4,13,14,15]. Very recently, the occurrence of these activities in spinach leaf peroxisomes has been reported as well as the presence, in spinach leaf tissue, of powerful inhibitors of these activities[30]. Thus, it may turn out that the fate of the acetyl-CoA produced by fatty acid ß-oxidation is identical in all plant tissues.

REFERENCES

1. P.K. Stumpf and G.A. Barber, Fat metabolism in higher plants. VII. ß-Oxidation of fatty acids by peanut mitochondria. Plant Physiol. 31:304 (1956).

2. T.G. Cooper and H. Beevers, ß-Oxidation in glyoxysomes from castor bean endosperm. J. Biol. Chem. 244:3514 (1969).

3. D. Hutton and P.K. Stumpf, Fat metabolism in higher plants. XXXVII. Characterization of the ß-oxidation system from maturing and germinating castor bean seeds. Plant Physiol. 44:508 (1969).

4. A.H.C. Huang, R.N. Trelease, and T.S. Moore, Plant Peroxisomes. Academic Press, New York (1983).

5. P.B. Lazarow and C. de Duve, A fatty acyl-CoA oxidizing system in rat liver peroxisomes; enhancement by clofibrate, a hypolipidemic drug. Proc. Nat. Acad. Sci. 73:2043 (1976).

6. P.B. Lazarow, Rat liver peroxisomes catalyze the ß-oxidation of fatty acids. J. Biol. Chem. 253:1522 (1978).

7. N.E. Tolbert, Peroxisomes and glyoxysomes. Ann. Rev. Biochem. 50:133 (1981).

8. A. Tanaka, M. Osumi, and S. Fukui, Peroxisomes of alkane-grown yeast: fundamental and practical aspects. In: Peroxisomes and Glyoxysomes, H. Kindl and P.B. Lazarow, eds., Ann. New York Acad. Sci., New York (1982).

9. G.P. Mannaerts and L.J. Debeer, Mitochondrial and peroxisomal ß-oxidation of fatty acids in rat liver. In: Peroxisomes and Glyoxysomes, H. Kindl and P.B.Lazarow, eds., Ann. New York Acad. Sci., New York (1982).

10. C. Wood, M.N.H. Jalil, J. McLaren, D.C.S. Yong, A. Ariffin, P.H. McNeil, N. Burgess, and D.R. Thomas, Carnitine long-chain acyltransferase and oxidation of palmitate, palmitoyl coenzyme A and palmitoyl-

carnitine by pea mitochondria preparations. Planta 161:255 (1984).

11. C. Wood, N. Burgess, and D.R. Thomas, The dual location of ß-oxidation enzymes in germinating pea cotyledons. Planta 167:54 (1986).

12. B. Gerhardt, Basic metabolic function of the higher plant peroxisome. Physiol. Vég. 24:397 (1986).

13. M. Macey and P.K. Stumpf, ß-Oxidation enzymes in microbodies from tubers of Helianthus tuberosus. Plant Sci. Lett. 28:207 (1982).

14. M. Macey, ß-Oxidation and associated enzyme activities in microbodies from germinating peas. Plant Sci. Lett. 30:53 (1983).

15. B. Gerhardt, Localization of ß-oxidation enzymes in peroxisomes isolated from nonfatty plant tissues. Planta 159:238 (1983).

16. B. Gerhardt, Peroxisomes - site of ß-oxidation in plant cells. In: Structure, Function and Metabolism of Plant Lipids, P.A. Siegenthaler and Eichenberger, eds, Elsevier Science Publishers B.V., Amsterdam (1984).

17. T.G. Cooper, The activation of fatty acids in castor bean endosperm. J. Biol. Chem. 246:3451 (1971).

18. H. Gerbling and B. Gerhardt, in preparation.

19. H. Kindl, personal communication.

20. J.A. Miernyk and R.N. Trelease, Control of enzyme activities in cotton cotyledons during maturation and germination. IV. ß-Oxidation. Plant Physiol. 67:341 (1981).

21. B. Gerhardt, Substrate specificity of peroxisomal acyl-CoA oxidases. Phytochemistry 24:351 (1985).

22. J. Frevert and H. Kindl, A bifunctional enzyme from glyoxysomes. Purification of a protein possessing enoyl-CoA hydratase and 3-hydroxyacyl-CoA dehydrogenase activities. Eur. J. Biochem. 107:79 (1980).

23. J.A. Miernyk and R.N. Trelease, Substrate specificity of cotton glyoxysomal enoyl-CoA hydratase. FEBS Lett. 129:139 (1981).

24. J. Frevert and H. Kindl, Purification of glyoxysomal acetyl-CoA acyltransferase. Hoppe-Seyler's Z. Physiol. Chem. 361:537 (1980).

25. B. Gerhardt, Higher plant peroxisomes and fatty acid degradation. In: Peroxisomes in Biology and Medicine, D. Fahimi and H. Sies, eds., Springer Verlag, Heidelberg (in press).

26. C. Rebeiz and P. Castelfranco, An extra-mitochondrial enzyme system from peanuts catalyzing the ß-oxidation of fatty acids. Plant Physiol. 39:932 (1964).

27. D. Hutton and P.K. Stumpf, Fat metabolism in higher plants. LXII. The pathway of ricinoleic acid catabolism in the germinating castor bean (Ricinus communis L.) and pea (Pisum sativum L.). Arch. Biochem. Biophys. 142:48 (1971).

28. J.J. Mettler and H. Beevers, Oxidation of NADH in glyoxysomes by a malate aspartate shuttle. Plant Physiol. 66:555 (1980).

29. D.B. Hicks and R.P. Donaldson, Electron transport in glyoxysomal membranes. Arch. Biochem. Biophys. 215:280 (1982).

30. N.E. Tolbert, personal communication.

DESATURATION OF FATTY ACIDS ON COMPLEX-LIPID SUBSTRATES

Sten Stymne, Gareth Griffiths and Keith Stobart*

Department of Food Hygiene, Swedish University of
Agricultural Sciences, Box 7009, S-750 07 Uppsala, Sweden
*Botany Dept., Univ. of Bristol, U.K.

INTRODUCTION

Almost 20 years ago Gurr et al. (1), working with Chlorella, suggested that complex lipids in plants may serve as substrates for desaturase enzymes, particularly those that catalysed the synthesis of C18-polyunsaturated fatty acids. This proposal was contrary to our understanding of the desaturase enzymes in animals where acyl-CoA substrates were directly involved. Hence it was considered that perhaps in plants the polyunsaturated fatty acid products were being rapidly transferred from CoA to oxygen esters. More recently, with the advent of reliable techniques, our knowledge of the substrates involved in plants has increased considerably and the evidence, for instance, that the oleate in microsomal phosphatidylcholine (PC) is the substrate for a $\Delta 12$-desaturase is well documented (2-4). There are also reliable data to show that α-linolenic acid (α-18:3) is synthesised in leaves from linoleate (18:2) associated with the chloroplast lipid, MGDG (5-7). On the other hand, although the evidence is of a less satisfactory nature, the α-18:3, that is synthesised in certain oil-seeds occurs in microsomal PC (8,9). Plants therefore differ from animals in that the synthesis of polyunsaturated fatty acids is characterised by complex lipid substrates. This raises further important problems, particularly those concerned with the mode(s) of entry of the fatty acid substrate into the complex lipid and the mechanism(s) by which the polyunsaturated products are then made available for the assembly of other lipids. Our work is largely concerned with the synthesis of C18-polyunsaturated fatty acids in oil-seeds and its relationship to the production of triacylglycerol (TAG). Although this research is part of a programme concerned with the acyl quality of oil-crops, it has, in our opinion, many fundamental aspects that may be relevant to the understanding of lipid synthesis and organelle development in other tissues. Developing oil-seeds offer many advantages in that they usually possess exagerated enzyme activities, often uninterupted by competing reactions, which makes it easier to interpret many of the basic events concerned with polyunsaturated fatty acid synthesis and lipid assembly. Here we describe aspects of our work concerned with 1. the influx and efflux of acyl groups in PC for desaturation 2. the biosynthesis of gamma(γ)-18:3 and 3. the specificity of desaturase enzymes.

RESULTS AND DISCUSSION

1(a) Acyl exchange

We have previously shown that microsomal membranes prepared from developing oil-seeds catalyse the transfer of oleate in acyl-CoA to position sn-2 of PC and that concomitant with this is the the return of fatty acids from the complex lipid to the acyl-CoA (see 10). This movement, coupled to desaturation in PC, regulates the quality of the acyl-CoA pool from which polyunsaturated fatty acids are selected for in the acylation of glycerol 3-phosphate (G3P) and subsequent TAG synthesis. The movement of acyl groups between PC and acyl-CoA has been found in microsomes from the maturating oil-seeds of many spp. and can range in activity from 1-2 nmol/ mg protein x min in rape and linseed to over 18 nmol/mg protein x min for safflower. The mechanism for this bidirectional movement appears to involve the forward and back reactions of an acyl-CoA:lysoPC acyltransferase (10). It is through the reverse reaction of this enzyme that the acyl acceptor (1-acyl-lysoPC) is regenerated for further acylation from the acyl-CoA pool. Similar properties of the acyltransferases in some animal systems may also serve in the acyl remodelling of complex lipids (11-16). In-vivo in oil-seed tissue a similar result would occur through the action of a phospholipase A_2 and the subsequent esterification of the fatty acids to CoA by an ATP dependent acyl-CoA synthetase. In seeds however there is no evidence for A_2 activity and the microsomal membranes do not require ATP to catalyse the acyl transfer.

1(b) Diacylglycerol(DAG) and PC interconversion

There is now good evidence to show that oil-seeds generally possess a CDP-choline:diacylglycerol cholinephosphotransferase (17,18). The enzyme

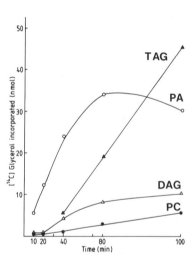

Fig. 1. TAG synthesis in microsomes from developing seeds of rape Microsomes (100 nmol PC) were incubated with $[^{14}C]$G3P (400 nmol), 16:0-CoA (100 nmol) and 18:2-CoA (100 nmol).

appears to be rather novel and catalyses the interconversion of DAG with PC during the movement of glycerol backbone from phosphatidate (PA) to TAG (19). The activity in microsomes of this mode of entry into PC again can vary between spp. (10). Highly polyunsaturated plants, such as safflower and sunflower, have good activity whereas in rape microsomes (fig. 1), that still have high rates of TAG formation, it is relatively low. For comparison, active TAG synthesising microsomes from avocado mesocarp had no DAG - PC interconversion (20), a result now confirmed in in-vivo studies (21).

There are therefore two possible modes of entry/exit of substrate/product from the PC molecule (Fig. 2). In many spp. rich in linoleate both reactions are generally high and will result in the continuous enrichment of the glycerol backbone with polyunsaturated fatty acids. In those species (e.g. rape) with lower polyunsaturated oils these reactions appear to operate at a lower magnitude.

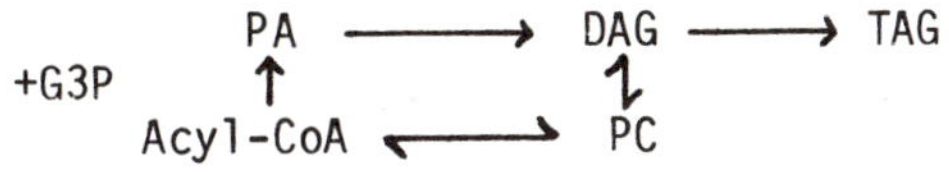

$$+G3P \quad \begin{array}{c} PA \longrightarrow DAG \longrightarrow TAG \\ \uparrow \qquad \updownarrow \\ Acyl\text{-}CoA \longleftrightarrow PC \end{array}$$

Fig. 2. Entry of substrate into PC and its relationship to TAG formation

2. Synthesis of γ-18:3

Understanding γ-18:3 formation in plants is particularly important, not only because of the therapeutic value of the acid, but also because it is the only C18-polyunsaturated fatty acid whose synthesis can be compared directly to an animal counterpart. We have speculated (10), for experimental purposes, that the efficient insertion of a double bond between the methyl-end of a C18-fatty acid and an already established centre of unsaturation, as occurs in plants, may require the intramolecular environment of a complex lipid. On the other hand the introduction of double bonds between an existing double bond and the carboxyl-end may necessitate, as in animals, the involvement of acyl-CoA. It has also been suggested (22) that plant desaturases which catalyse the synthesis of 18:2 from 18:1 may utilise both acyl-CoA and PC substrates. These proposals were tested in studies on γ-18:3 synthesis in the developing seeds of borage (Borago officinalis). Preliminary in-vivo studies showed that γ-18:3 was synthesised from 18:1 via 18:2. Microsomal preparations also proved to be highly active in γ-18:3 formation. Microsomal PC was labelled in-situ with [^{14}C]linoleate under non-desaturating conditions (minus NADH). After washing, the membranes were incubated with NADH. The results (table 1) show that the microsomes possessed a Δ6-desaturase system that catalysed the formation of γ-18:3 from the 18:2 in PC.

Table 1. Desaturation of in-situ labelled [^{14}C]18:2-PC in microsomes of Borago

Incubation time (min)	[^{14}C]radioactivity in PC (nmol)	
	18:2	γ-18:3
0 (-NADH)	13	0
80 (-NADH)	10	0
80 (+NADH)	8	3

Microsomes were pre-labelled with [^{14}C]18:2 by incubation with [^{14}C]18:2-CoA. Washed microsomes were then incubated with (2µmol) or without NADH and the radioactivity determined in the 18:2 and γ-18:3 of the PC

This was confirmed in further experiments in which microsomes were incubated with [^{14}C]18:2-CoA and non-radioactive γ-18:3-CoA in the presence of NADH. Here (table 2) the sp. radioactivity of the synthesised γ-18:3 was greater in PC than in the acyl-CoA pool and is confirmation that the Δ6-desaturase enzyme utilises the acyl substrate in the microsomal complex lipid, PC.

Table 2. Desaturation of [^{14}C]18:2 in microsomes incubated with [^{14}C]18:2-CoA in the presence of non-radioactive γ-18:3-CoA

Lipid	Incubation time (min)	18:2 [^{14}C]nmol	SA*	γ-18:3 [^{14}C]nmol	SA*
Acyl-CoA	0	74	29	0	0
	90	16	15	1	0.6
PC	0	0	0	0	0
	90	19	3	6	1.5

Microsomes (equiv. 192 nmol PC) were incubated with [^{14}C]18:2-CoA (65 nmol), γ-18:3-CoA (100 nmol) and NADH (2 μmol)
*SA, d.p.m. × 10^{-2}/nmol

Hence, it appears that perhaps all the desaturase enzymes in plants that utilise C18-unsaturated fatty acid substrates do so on complex lipids. The reasons for this are probably not directly mechanistic and may have evolved as a rapid adaptive response of key membranes to changes in the environment. Desaturation on PC would be advantageous since the enzyme does not initially have to contend with the competing reactions inherent with an acyl-CoA substrate.

3(a) Positional specificity

A positional analysis of the fatty acids in microsomal PC from borage showed that whilst both sn-1 and 2 carbon atoms contained linoleate the majority of the γ-18:3 was at position 2 (table 3).

Table 3. Fatty acid composition of microsomal PC

	Fatty acids (%)					
	16:0	18:0	18:1	18:2	γ-18:3	20:1
Position sn-1	29	11	15	39	2	4
Position sn-2	tr.	tr.	21	47	32	tr.

Purified PC was treated with phospholipase A_2 and the products separated by TLC.

This was confirmed with PC obtained from cotyledons incubated with [^{14}C]18:2. Here [^{14}C]18:2 was present at both positions but only position 2 contained radioactive γ-18:3. These results indicate that the Δ6-desaturase in borage is positionally specific for the acyl substrate in PC. If this is so then it raises important questions regarding all the desaturase enzymes that utilise complex lipids. For instance, is the desaturation of 18:1 to 18:2, that is known to occur at the sn-1 and 2 positions in PC (2, 20), catalysed by different Δ12-desaturase enzymes? In further investigations borage seed microsomes were incubated with NADH and the endogenous fatty acids in the sn-1 and 2 positions of PC ana-

lysed. The results (Fig. 3) show that the mass of 18:1 at both positions
rapidly declined. Concomitant with this was a dramatic increase in the
18:2 at sn-1 and very little change in this acid at sn-2. Gamma-18:3 only
increased at the sn-2 position of PC. The results illustrate that whilst
Δ12-desaturase activity occurs at both positions the Δ6-desaturase is specific for position 2. Also noteworthy is the observation that during γ-18:3
formation little change occurred in the mass of the original endogenous
pool of 18:2 at sn-2 whilst most of the oleate was consumed (also noted
in incubations up to 360 min). If the majority of the 18:2 at sn-2 of PC
is available to the Δ6-desaturase then the results would indicate that
the enzyme has a particularly low affinity for its acyl-substrate. The
control of the quality and quantity of the polyunsaturated fatty acids in
PC is therefore complex and appears to involve many interacting factors.

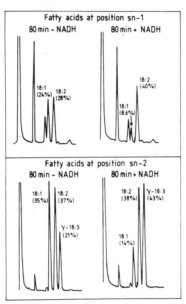

Fig. 3. Desaturation of the endogenous acyl-substrate in microsomal PC.
Borago microsomes were incubated with NADH (2μmol). Purified PC
was treated with phospholipase A_2 and the products separated by
TLC. After methylation the fatty acids from sn-1 and 2 were analysed by GLC.

3(b) Complex lipid specificity

During our work with borage it was observed that the leaf lipids
contained appreciable amounts of the tetraenoic fatty acid, 18:4
(Δ6,9,12,15), where it was particularly abundant in MGDG (37%) (Fig. 4).
Such leaf lipids also contained α-18:3 and smaller quantities of γ-18:3.
A survey of a number of spp. in the family Boraginaceae has shown that if
the leaf contains both α and γ-18:3, then 18:4 is always present. It
seems, therefore, that there is a close correlation between the presence
of 18:4 and the activity of the Δ6 and Δ15-desaturase enzymes. These
observations could be useful in understanding the relationship between
the synthesis of unsaturated C18-fatty acids in the plant cell and the
assembly of complex lipids for specific membrane (chloroplast) formation.
Certainly such tissue provides a rather unique system with which it
should be easier to interpret the movement of acyl species between the
cell compartments and to assess more precisely the regulatory mechanisms
involved. To this end we speculate here on the relationship between Δ6

and Δ15-desaturase activity and the synthesis of 18:4 (fig. 5). In the borage leaf we assume that the translation/transcription machinery for the microsomal Δ6 2-linoleoyl-PC desaturase enzyme-protein, that is found in the seed, still operates in the leaf and hence that the gene expression for this particular protein is here non-organ specific. The leaves, however, possess chloroplasts which in its turn necessitates the induction of Δ15-desaturase enzyme(s) for the synthesis of α-18:3 and plastid lipid production. Chloroplasts are considered to be the site of a Δ15-desaturase that utilises the 18:2 in MGDG. The source of this 18:2 is most probably microsomal and comes from PC by mechanisms that are still unclear. In leaves of borage the 18:2 in microsomal PC is most likely transferred to the chloroplast together with γ-18:3. The Δ15-desaturase in the chloroplast, that normally uses 18:2, is non-specific and also introduces a further double bond in γ-18:3 to yield 18:4. A positional analysis of the fatty acids in MGDG showed that the 18:4 was present only at the sn-2 position where it amounted to over 70% (fig. 4). Since the γ-18:3 is only synthesised at the sn-2 position of microsomal PC this observation could indicate that the PC backbone is transferred intact to the chloroplast and there used for MGDG formation (23). It is still possible that a similar result may arise if the enzymes that acylate G3P select γ-18:3-CoA exclusively for position sn-2 and the resultant species of phosphatidate is transferred to the chloroplast for MGDG formation (24).

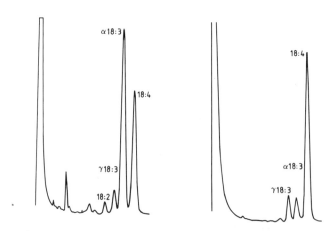

Fig. 4. Fatty acid composition of MGDG from Borago leaves. Left: Total MGDG; Right: Position sn-2 of MGDG.

To add further support to the above speculations we have now found that some spp. in the Boraginaceae (e.g. Cynoglossum officinalis) produce seed lipids that contain both α and γ-18:3 and also 18:4. The observation that some spp. produce seeds which contain α-18:3 and yet do not possess chloroplasts at any stage during maturation is interesting in itself. We had previously considered, as had others, that perhaps α-18:3 was only found in seeds with cotyledons that were of a chlorophyllous nature.

Some in-situ desaturase was also found with phospholipids other than

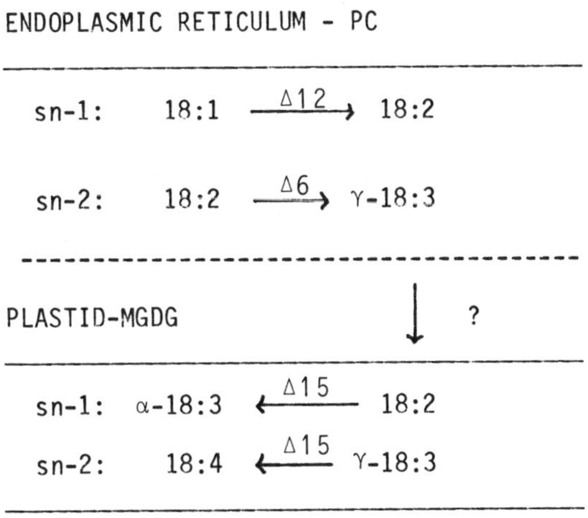

ENDOPLASMIC RETICULUM - PC

---

sn-1:   18:1 $\xrightarrow{\Delta 12}$ 18:2

sn-2:   18:2 $\xrightarrow{\Delta 6}$ γ-18:3

- -

PLASTID-MGDG   ↓ ?

---

sn-1:   α-18:3 $\xleftarrow{\Delta 15}$ 18:2
sn-2:   18:4 $\xleftarrow{\Delta 15}$ γ-18:3

Fig. 5. Scheme for the synthesis of 18:4 in Borago-leaves

PC, in microsomes from developing borage-seeds. This was noticable in PE in membranes incubated with NADH (Table 4). After incubation the endogenous 18:1 in PE had decreased with a concomitant increase in γ-18:3. Little change occurred in the 18:2. Again, this raises further questions about the specificity of desaturase enzymes for complex lipid substrate. Are the same desaturases involved irrespective of the headgroup composition of the complex lipid? It should be noted, however, that PE in microsomes from oil-seeds has never, in our experience, been associated with the desaturation of exogenously supplied acyl-CoA and appears therefore not to be a prime substrate in the production of polyunsaturated fatty acids for other lipids.

Table 4. Desaturation of endogenous acyl-substrate in microsomal phosphatidylethanolamine (PE) from the developing seeds of borage.

Incubation time (min)	% fatty acids (of total)		
	18:1	18:2	γ-18:3
0	11	54	9
360	5	53	15

Microsomes were incubated as in Fig. 3.

GENERAL REMARKS

We have endeavoured here to give a wider appreciation of the problems associated with the desaturation of C18-unsaturated fatty acids in plants. Where appropriate we have felt entitled to speculate in the hope of stimulating experimentation on this important aspect of lipid metabolism. In summary we would stress the importance to understand in more detail 1. the modes of entry/exit of substrate/product in PC 2. the positional specificity of the desaturase enzymes for the acyl substrate in complex lipids and 3. the regulation of the desaturase activity by the concentration of acyl-substrate within the complex lipid. A thorough knowledge of all these factors is essential before it will become feasible to manipulate successfully, through the techniques of genetic engineering, the polyunsaturated fatty acid constituents of plant lipids.

ACKNOWLEDGEMENT

The authors acknowledge financial support from the Swedish Natural Science Research Council, Swedish Council for Forestry and Agricultural Research and the National Swedish Board for Technical Development. G.G. is in receipt of an SERC(U.K.)-overseas fellowship.

REFERENCES

1. M. I. Gurr, M. P. Robinson, and A. T. James, Eur. J. Biochem. 9:70 (1969).
2. C. R. Slack, P. G. Roughan, and J. A. Browse, Biochem. J. 179:649 (1979).
3. S. Stymne, A. K. Stobart, and G. Glad, Biochim. Biophys. Acta 752:198 (1983).
4. D. J. Murphy, I. E. Woodrow, and K. D. Mukherjee, Biochem. J. 225:267 (1985)
5. P. G. Roughan, J. B. Mudd, T. T. McManus, and C.R. Slack, Biochem. J. 184:571 (1979).
6. J. C. Hawke, and P. K. Stumpf, Arch. Biochem. Biophys. 203:296 (1980).
7. A. V. M. Jones, and J. L. Harwood, Biochem. J. 190:851 (1980).
8. J. A. Browse, and C. R. Slack, FEBS-Lett. 131:111 (1981).
9. S. Stymne, and L-Å. Appelqvist, Plant Sci. Lett. 17:287 (1980).
10. S. Stymne and A. K. Stobart, Triacylglycerol biosynthesis, in: "The Biochemistry of Plants: a Comprehensive Treatise, Vol. 10," P. K. Stumpf, ed., Acad. Press, N. Y. (1986).
11. R. F. Irvine, and M. Dawson, Biochem. Biophys. Res. Commun. 91:1399 (1979).
12. J. Trotter, and E. Ferber, FEBS-Lett. 128:237 (1981).
13. I. Flesch, B. Ecker, and E. Ferber, Eur. J. Biochem. 139:431 (1984).
14. R. M. Kramer, M. Pritzka, and G. Bereziat, J. Biol. Chem. 259:2403 (1984).
15. S. Stymne, and A. K. Stobart, Biochim. Biophys. Acta 837:239 (1985).
16. J. G. Nijssen, and H. van den Bosch, Biochim. Biophys. Acta 875:458 (1986).
17. C. R. Slack, L. C. Campbell, J. A. Browse, and P. G. Roughan, Biochim. Biophys. Acta 754:10 (1983).
18. C. R. Slack, P. G. Roughan, J. A. Browse, and S. E. Gardiner, Biochim. Biophys. Acta 833:438 (1985).
19. A. K. Stobart, and S. Stymne, Biochem. J. 232:217 (1985).
20. A. K. Stobart, and S. Stymne, Planta 163:119 (1985).
21. G. Griffiths, Ph.D. Thesis, Univ. of Bristol, U.K. (1986).
22. D. Howling, L. J. Morris, M. I. Gurr, and A. T. James, Biochim. Biophys. Acta 260:10 (1972).
23. C. R. Slack, P. G. Roughan, and N. Balasingham, Biochem. J. 162:289 (1977).
24. G. Griffiths, S. Stymne, A. Beckett, and A. K. Stobart, in "Regulation of Chloroplast Differentiation," A. Akoyunoglou, ed., Liss, Inc., N. Y. (1986).

RAPID ENZYMIC PEROXIDATION OF POLYUNSATURATED FATTY

ACIDS ON HYDRATION OF WHEAT MILLING PRODUCTS

T. Galliard, S.P.C. Tait and D.M. Gallagher

R.H.M. Research Ltd., Lord Rank Research Centre
High Wycombe, England, U.K.

INTRODUCTION

As reported elsewhere in this volume, a lipase, specifically located in the bran tissues of wheat grain, is responsible for the hydrolysis of endogenous triacylglycerols (TG) in wholemeal to free (unesterified) fatty acids, 60-65% of which are polyunsaturated fatty acids (F-puFA). This process occurs in wholemeal flour during storage under ambient conditions.

When wholemeal flour is mixed with water, dissolved oxygen is consumed due to a lipoxygenase-catalysed peroxidation of F-puFA[1,2]. The fatty acid oxidation products that are formed in low amounts in aqueous suspensions of white flour have been described previously[3]. This paper illustrates the more substantial lipid peroxidation reactions that occur in aqueous suspensions and doughs of wholemeal flour and points out several important practical implications of this process.

MATERIALS AND METHODS

Materials. Wholemeal flour was obtained from commercial flour mills or by laboratory milling of wheat and was stored either at -20°C within 12h of milling or maintained at $+20^\circ$C, 65% relative humidity.

Methods were as described elsewhere[1,2] for measuring a) the O_2-uptake of aqueous suspensions of milling products; b) total lipid and free fatty acids; c) distribution of ^{14}C-label in oxidation products from 1-^{14}C linoleic acid; d) conjugated oxidation products (COP). Peroxide values were obtained by a standard method[4].

RESULTS

F-puFA accumulate in wheat milling products under normal moisture (10-14%) and ambient temperature conditions. In wholemeal, there is no substantial loss of total puFA over many months of ambient storage. However, when mixed with water for 15 min, wholemeal that had been stored for 19 weeks at 20°C with no loss of puFA, showed a 28% loss in total puFA and a corresponding increase in COP (from 0.27 to 1.80mg COP/g wholemeal) in the 15 min.

The oxidation of F-puFA on hydration of milling products is accompanied by consumption of dissolved O_2. The rate and extent of oxygen uptake increases linearly with the time of storage and correlates closely ($r \geqslant 0.96$; $P < 0.001$) with FFA content of the material. The oxygen consumption is due to peroxidation of F-puFA, catalysed by lipoxygenase that is located predominantly in the germ of wheat grain.

The primary oxidation products are fatty acid hydroperoxides (FAHPO); secondary products are mainly mono- and tri-hydroxy fatty acids. The amounts and proportions of oxidation products vary, depending upon the storage time of the milling products. Table 1 shows that hydrated stored wholemeal (in this case dough) contains more oxidation products (mainly FAHPO) than a dough from freshly-milled wholemeal.

Table 1. Comparison of lipid oxidation products in doughs from freshly-milled and stored wholemeal.

Wholemeal used	COP content (mg/g wholemeal)	Products from ^{14}C-linoleic acid (% total counts)		
		residual F-puFA	FAHPO	secondary oxidation products
freshly-milled	0.50	12.0	4.7	83.3
stored 9 months at $20^\circ C$	2.13	15.1	62.6	22.3

The amounts and the proportions of secondary oxidation products depend upon the storage history of the material. For example, when doughs containing ^{14}C-linoleic acid were prepared from freshly-milled wholemeal and from wholemeal that had been stored for 9 months at $20^\circ C$, the distribution of ^{14}C-labelled products differed markedly (Table 1). Hydroperoxides (FAHPO) predominated in the dough from stored wholemeal, whereas secondary oxidation products contained most of the ^{14}C-counts when freshly-milled wholemeal was used.

Almost all of the available F-puFA is oxidised within 15 min of hydration and in doughs from stored wholemeal, the FAHPO are stable, showing no substantial further conversion over at least 2h.

DISCUSSION

Fig. 1 summarises the processes by which endogenous TG of wheat grain are degraded enzymically during storage and hydration of milling products. The bran lipase is described elsewhere in this volume. The extent of lipid oxidation is determined by the amount of substrate (F-puFA) that accumulates in the stored material. The oxidation products can represent at least 25% of the total acyl lipid in the original material. In the case of wheat bran, this represents 1-2% of the dry weight.

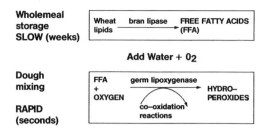

Figure 1

There are several practical implications of these results:-
a) <u>Measurement of oxidative rancidity</u>, e.g. peroxide value, of stored milling products, gives much lower values if performed on the dry, rather than hydrated, material. For example, a wholemeal stored for 25 weeks at $20^\circ C$ gave peroxide values before and after hydration of 0.025 and 0.134 mEq.O_2/kg respectively.

b) <u>The baking performance</u> of stored wholemeal is adversely affected by increased FFA levels and by removal of essential O_2 from dough by lipid peroxidation. Co-oxidation reactions (Fig. 1) cause pigment bleaching and, probably, have important effects on protein functionality.

c) <u>The nutritional value</u> of hydrated wheat milling products is affected by loss, due to oxidation, of essential puFA. The nutritional effects of the resulting oxidation products are not known, although the structural similarities between some of these and ricinoleic acid (the major component of castor oil) is worth noting !

d) <u>Inhibition of enzymic degradation</u> can be achieved for wheat bran and germ by appropriate heat/moisture treatment. This is not possible for wholemeal without denaturing the important gluten proteins, but the degradation can be limited by avoidance of high-lipase wheat and by appropriate milling and storage procedures.

ACKNOWLEDGEMENTS

This work, reported in detail elsewhere[1,2,5], was supported by the U.K. Ministry of Agriculture, Fisheries and Food and is Crown Copyright © 1986.

REFERENCES

1. Galliard,T. J.Cereal Sci. 4 (1986) 33-50.
2. Galliard,T. J.Cereal Sci. 4 (1986) 179-192.
3. Graveland,A. Lipids, 8 (1973) 599-605.
4. I.S.O. Standard 3976 (1977).
5. Galliard,T. in "Chemistry and Physics of Baking" (J.M.V.Blanshard, P.J.Frazier and T.Galliard, eds.) Royal Society of Chemistry, London (1986) 199-215.

ACTION OF BORON ON ETHYLENE PRODUCTION AND LIPOXYGENASE ACTIVITY IN MICROSOMES FROM SUNFLOWER COTYLEDONS

Andrés Belver, Pilar Rodriguez, Marta Roldán
and Juan P. Donaire

Department of Biochemistry
Estación Experimental del Zaidín, (C.S.I.C.)
Granada (Spain)

ABSTRACT

Conversion of ACC to ethylene by microsomes isolated from sunflower cotyledons grown in a medium with 50 ppm of boron was strongly increased relative to membranes developed under 10 ppm or without boron, and SHAM, BHT and SOD provoked a high inhibition. A strong increase in ethylene takes place upon incubating ACC and linolenic acid with crude microsomes or with a protein fraction which includes lipoxygenase activity.

INTRODUCTION

The conversion of ACC to ethylene is believed to be a free radical associated process, which in microsomes is attributed to hydroperoxides generated by membrane-associated lipoxygenase[1]. In this work, we will attempt to establish the hypothesis that the growth of sunflower cotyledons could be linked to a boron mediated lipid catabolic process which is geared to an ethylene evolving system.

MATERIALS AND METHODS

Seeds of Helianthus annuus L. cv. HK were grown for 6 days in the dark at 26° C in a medium containing 0.2 mM $CaSO_4$ and no, 10 or 50 ppm of boron. Crude microsomes prepared from the cotyledons were suspended in 50 mM Hepes, pH 8.5, containg 0.5 mM $MnCl_2$, solubilized with 0.1 % Brij W-1 and the protein membrane fraction passed on Sephadex G-50 fine. Ethylene and lipoxygenase were measured as shown by Donaire and Belver[2] and proteins according to Markwell et al.[3].

RESULTS AND DISCUSSION

The addition of ACC to the incubation medium stimulated less the ethylene production in microsomes from cotyledons grown in media with 10 ppm or without boron than in 50 ppm. SHAM, BHT and SOD induced a high inhibition of ethylene, mainly in microsomes from cotyledons grown in 50 ppm of boron (Table 1). The high content of free fatty-acids previously detected by us in cotyledons developed in 50 ppm of boron[4] could justify the increased ethylene rate that we have observed here at microsomal level, an autooxidation or enzymatic lipid peroxidation being the cause.

Table 1. Ethylene formation by microsomes obtained from cotyledons grown in no, 10 or 50 ppm of boron

Treatments	Ethylene (nl/ h x mg of proteins)		
	no boron	10 ppm	50 ppm
Standard reaction	9.3	7.8	16.3
- ACC	0.9	0.9	1.1
+ SHAM (1.3 mM)	2.5	1.7	0.4
+ BHT (0.02 %)	1.6	1.6	1.4
+ SOD (50 U.I)	2.2	3.6	2.5

Standard reaction = Microsome incubation (1 h) in 1.0 ml of 50 mM Hepes, pH 8.5, containing 0.5 mM $MnCl_2$, 2.0 mM ACC and 2.0 mg of proteins. The values are the average of three independent experiments not differing by more than 9 % from the mean. ACC = 1-aminocyclopropane-1-carboxylic acid; SHAM = salicylhydroxamic acid; BHT = 2,6-dietert-butyl-p-cresol; SOD = bovine superoxide dismutase.

A high increase of ethylene takes place upon incubation of crude microsomes with ACC and linolenic acid. Furthermore, a split of lipoxygenase and ethylene-forming enzyme, separated on Sephadex G-50 from solubilized microsome membranes, reflects that linolenic acid was required for the ethylene formation by the proteins of the first peak (lipoxygenase activity) but not by those of the second one (ethylene-forming enzyme) (Fig. 1 and Table 2).

Table 2. Ethylene-forming enzyme (nl/ h x mg of proteins) and lipoxygenase (μmol O_2/ min x mg of proteins) activities in microsomes from cotyledons grown in no, 10 or 50 ppm of B

Treatments	Ethylene						Lipoxygenase		
	no boron		10 ppm		50 ppm		no	10	50
	+Ln	-Ln	+Ln	-Ln	+Ln	-Ln			
Crude Micr.	168.2	9.3	153.6	7.8	189.0	16.3	0.51	0.48	0.41
Sol. micr. mem.	141.4	8.9	136.8	6.1	152.8	13.6	0.82	0.96	0.84
Fract. Sephadex									
Fraction A	121.2	2.7	114.3	2.9	109.9	3.1	2.14	2.67	2.07
Fraction B	12.2	11.8	16.0	15.7	15.9	14.6	n.d.	n.d.	n.d.
Fractions A+B	151.0	11.0	140.1	13.5	143.7	13.3	1.02	0.96	0.71

The values are the average of two independent experiments not differing by more than 6 % (ethylene) or 10 % (lipoxygenase) from the mean; n. d. = not detected; Ln = linolenic acid.

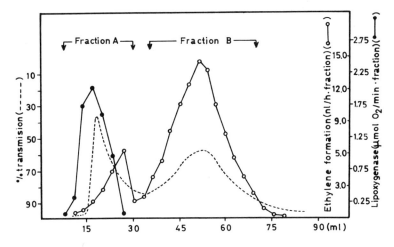

Fig. 1. Fractionation pattern of solubilized crude microsomes by Sephadex G-50. Cotyledons were previously grown in 10 ppm of boron. A similar fractionation pattern was obtained when microsomes from tissues grown in media with no or with 50 ppm of boron were used (not shown).

From the present results, it is possible to suggest that boron may have an important role in the growth of sunflower seedlings by regulating the ethylene formation in cotyledon tissues associated with free radicals derived from lipid membrane degradation.

ACKNOWLEDGEMENTS

This research was supported by financial assistance from the Consejo Superior de Investigaciones Científicas (Spain). Proyecto nº 217-1985.

REFERENCES

1. D. V. Lynch, S. Sridhara, and J. E. Thompson, Lipoxygenase-generated hydroperoxides account for the nonphysiological features of ethylene formation from 1-aminocyclopropane-1-carboxylic acid by microsomal membranes of carnation, Planta 164: 121 (1985).
2. J. P. Donaire and A. Belver, Effect of boron on metabolic changes in sunflower cotyledons grown under dark and light conditions. J. Plant Physiol., 120: 389 (1985).
3. M. A. Markwell, S. M. Hassa, L. Bieber, and N. E. Tolbert, A modification of the Lowry procedure to simplify protein determination in membrane and lipoprotein samples, Anal. Biochem., 87: 206 (1978).
4. A. Belver and J. P. Donaire, Partial purification of soluble lipoxygenase of sunflower cotyledons: action of boron on the enzyme and lipid constituents, Z. Pflanzenphysiol., 109: 309 (1983).

PRELIMINARY CHARACTERIZATION OF LIPOXYGENASE FROM THE ENTOMOPATHOGENIC

FUNGUS LAGENIDIUM GIGANTEUM

Christopher A. Simmons, James L. Kerwin and Robert K. Washino

Department of Entomology
University of California
Davis, CA 95616 USA

INTRODUCTION

In many biological systems cell proliferation and differentiation are inhibited by polyunsaturated fatty acids and their oxidative metabolites (Gavino et al., 1981; Morisaki et al., 1984). Using developmentally synchronized cultures of Lagenidium giganteum (Oomycetes:Lagenidiales), a facultative parasite of mosquito larvae, it has been documented that oxidative lipid metabolism is necessary for the induction and maturation of its sexual stage, the oospore (Kerwin et al., 1986). These initial investigations, which used compounds known to selectively inhibit mammalian lipoxygenases and cyclooxygenases and partly purified eicosanoid extracts from L. giganteum cultures, demonstrated that oxygenase activity is associated with discrete developmental events and different oxidative metabolites appear to be produced during specific stages of fungal morphogenesis.

Verification of the presence of lipoxygenase enzyme(s) in L. giganteum and preliminary characterization of their substrate specificities are presented in the following report.

MATERIALS AND METHODS

The California strain of L. giganteum (ATCC 52675) maintained as previously described (Kerwin and Washino, 1986a) was used in these investigations. Oosporogenesis was induced in liquid shake culture using 1 liter of medium in 2800 ml fernbach flasks. Media consisted of (gram/liter of deionized water): 1 glucose, 1 yeast extract, 1 hydrolyzed lactalbumin, 0.5 dehydrated egg yolk, 0.02 tyrosine, 2 mM calcium, 0.5 mM magnesium, pH 6.5 . Cultures of the fungus were grown at 120 rpm on a rotary shaker at 23 ± 3 °C. Mycelium was collected on nylon filters ca. 40 h after inoculation with 7 ml stock culture at the developmental stage corresponding to the primordial oospore previously described (Kerwin and Washino, 1986b) and stored at -70°C.

Five grams of frozen material were homogenized for 2 min under CO_2 in 100 ml 0.2 M phosphate buffer, pH 7.5, using a Braun homogenizer. The homogenate was fractionated using differential centrifugation at 4°C and the following supernatant fractions collected: 10 K - 10000 x g for 15 min; 18

K - 18000 x g for 30 min; 100 K - 100000 x g for 60 min; and a 100000 x g for 60 min pellet - the microsomal fraction. The microsomal fraction was resuspended in 1 ml phosphate buffer and all material used immediately or stored at -70°C.

Lipoxygenase activity was assayed by monitoring absorbance at 234 nm due to diene conjugation. Fatty acid substrates (Sigma) were prepared by mixing 0.05 ml fatty acid with 0.05 ml denatured 95% EtOH followed by gradual dilution with distilled water to a final volume of 50 ml. The reaction mixture consisted of 2.7 ml 0.2 M phosphate buffer, 300 microliter substrate and 50 microliter enzyme homogenate. The reaction was monitored for 10 min at 23°C using a Cary 219 spectrophotometer interfaced with an Apple IIe, Master Kinetics software.

RESULTS AND DISCUSSION

Using the incubation conditions described, lipoxygenase activity was not detectable in any of the particulate fractions of the L. giganteum homogenate, including the microsomal fraction. This is in agreement with the cytosolic nature of most plant (Veldink et al., 1977) and animal (Bailey and Chakrin, 1981) lipoxygenases.

Of the 8 fatty acids tested as L. giganteum lipoxygenase substrates (delta 9,12-octadecadienoic, delta 9,12,15-octadecatrienoic, delta 6,9,12-octadecatrienoic, delta 11,14-eicosadienoic, delta 8,11,14-eicosatrienoic, delta 5,8,11,14-eicosatetraenoic, delta 5,8,11,14,17-eicosapentaenoic, and delta 4,7,10,13,16,19-docosahexaenoic acids, all-cis), only arachidonic (C-20:4), eicosapentaenoic (C-20:5) and docosahexaenoic (C-22:6) acids were consistently metabolized (Table 1). It is noteworthy that the C-18 compounds apparently are not suitable substrates for the L. giganteum enzyme under the incubation conditions used. If this initial observation is verified by work in progress, this will be the first documented instance of such selective lipoxygenase substrate specificity.

A previously described fungal lipoxygenase from Fusarium oxysporum readily metabolized 18-carbon fatty acids (Matsuda et al., 1978), as does a recently described lipoxygenase from Saprolegnia parasitica (Hamberg et al., 1986), an oomycetous fungus more closely related to L. giganteum. The

Table 1. Substrate Specificity of Lagenidium giganteum Lipoxygenase

Substrate[a]	Specific Activity (Δ abs 234 nm/min/microgram protein) x 10^3		
	Fraction		
	10 K	18 K	100 K
C-20:3	1.1	0	0
C-20:4	1.6	1.5	2.0
C-20:5	3.2	2.9	2.4
C-22:6	0.4	0.4	0.6

[a] Positions of the double bonds are given in the text. Compounds showing no diene conjugation: Δ9,12-C-18:2; Δ9,12,15-C-18:3; Δ6,9,12-C-18:3; Δ11,14-C-20:2. Preincubation of the 100 K fraction with alpha-naphthol (0.1 mM) inhibited enzyme activity by ca. 95%.

Saprolegnia enzyme is of interest since it appears to be tightly bound to hydroperoxide isomerase which rapidly converts a 15-hydroperoxy eicosatetraenoic acid intermediate into a nonconjugated epoxy alcohol (Hamberg et al., 1986; R. P. Herman, pers. commun.). If a comparable enzyme complex exists in L. giganteum, it could explain why no diene conjugation is observed upon incubation of C-18 polyunsaturated fatty acids; however, it does not explain why conjugation occurs upon incubation with e.g. arachidonic acid under identical conditions. It must be postulated that selective substrate specificity is exhibited by either the lipoxygenase or the isomerase.

Evidence for cyclooxygenase activity associated with particulate fractions of L. giganteum homogenates has recently been obtained (Kerwin et al., unpubl.). Characterization of this type of oxygenase and confirmation of our preliminary observations with the lipoxygenase(s) are in progress.

REFERENCES

Bailey, D. M. and Chakrin, L. W., 1981, Arachidonate lipoxygenase, Annu. Rep. Med. Chem.,16:213.
Gavino, U. C., Miller, J. S., Ikhareba, S. O., Milo, G. E., and Cornwell, D. G., 1981, Effect of polyunsaturated fatty acids and antioxidants on lipid peroxidation in tissue cultures, J. Lipid Res.,22:763.
Hamberg, M., Herman, C. A., and Herman, R. P., 1986, Novel biological transformations of 15-L$_s$-hydroperoxy-5,8,11,13-eicosatetraenoic acid, Biochim. Biophys. Acta,877:447.
Kerwin, J. L. and Washino, R. K., 1986a,Regulation of oosporogenesis by Lagenidium giganteum: promotion of sexual reproduction by unsaturated fatty acids and sterol availability, Can. J. Microbiol.,32:294.
Kerwin, J. L. and Washino, R. K. 1986b, Oosporogenesis by Lagenidium giganteum: induction and maturation are regulated by calcium and calmodulin, Can. J. Microbiol.,32. In press.
Kerwin, J. L., Simmons, C. A. and Washino, R. K., 1986, Eicosanoid regulation of oosporogenesis by Lagenidium giganteum, Prostaglandins Leukotrienes Med.,21. In press.
Matsuda, Y., Beppu, T. and Arima, K., 1978, Crystallization and positional specificity of hydroperoxidation of Fusarium lipoxygenase, Biochim. Biophys. Acta,530:439.
Morisaki, N., Lindsey, J. A., Stitts, J. M., Zhang, H. and Cornwell, D. G., 1984, Fatty acid metabolism and cell proliferation. V. Evaluation of pathways for the generation of lipid peroxides, Lipids, 19:381
Veldink, G. A., Vliegenthart, J. F. G. and Boldingh, J., 1977, Plant lipoxygenases, Prog. Chem. Fats Other Lipids, 15:131.

CHARACTERIZATION OF TOMATO LIPOXYGENASE

Peter A. Biacs and Hussin Daood

Central Food Research Institute
Budapest /Hungary/

INTRODUCTION

Lipoxygenase enzyme is responsible for the oxidation of fatty acids and their glycerides containing methylene interrupted system of double bonds such as linoleic, linolenic and arachidonic acids. The presence of lipoxygenase in the economically most important tomato variety in Hungary /ventura/ has been proved. In this paper the extraction, the purification steps and the characterization of the enzyme are described.

MATERIALS AND METHODS

Fruits of Lycopersicon esculentum variety ventura were diced in Tris buffer at pH = 7 for 15 min, filtered, centrifuged and the supernatant subsequently fractionated with $/NH_4/_2SO_4$. The active supernatant was dialyzed overnight against redistilled water. /Table 1./

Table 1. FRACTIONATION of TOMATO LIPOXYGENASE by AMMONIUM SULFATE

Amm.sulf. concentration	Total unit	Total protein /mg/	Yield %	Sp.activity	Fold purification
Crude extract	5600	460	100	12.2	1
0-0.3 saturat.	850	43.5	15	19.5	1.6
0.3-0.6 saturat. Pellet	200	62	3.5	3.2	0.26
Supernatant before dialysis	9900	181	176	55	4.5
Supernatant after dialysis	4400	132	78.5	34	2.8

Sp. activity = $Unit.ml^{-1}$/mg $protein.ml^{-1}$

For the enzyme assay the method of Al-Obaidy and Siddiqi /1981/ was used. One enzyme unit was defined as the amount of the enzyme producing 0.001 change in the absorbance at 234 nm in a minute.

RESULTS AND DISCUSSIONS

Results of the fractionation are summarized in Table 1. The active supernatant of the 0.6 saturation fraction was used for further investigation.

Using linoleic acid substrate the absorbance at 234 nm showed proportional increase with the time /Figure 1/ allowing initial reaction rate determination.

The linear relationship between the initial reaction rate and the concentration of the enzyme is maintained over a wide range of enzyme concentration /Figure 2/.

The enzyme activity showed remarkable changes during the ripening process of the fruit. Lipoxygenase activity changes are represented on Figure 3 at different stages of ripening.

Pellet of the second fractionation was further separated by aceton. The precipitated proteins /A/ redissolved in Tris buffer and added to the active supernatant fraction brought about 20% increase in the enzyme activity, while the supernatant colour liquor /B/ caused 90% enzyme inhibition indicating the separation of some natural inhibitor /Figure 4/.

The effect of EDTA on the undialyzed active fraction of 0.6 saturation suggests strong binding of the activating metal and the prosthetic group of the enzyme. The resistance of the enzyme to high concentrations of EDTA is in favour of a true lipoxygenase and not a hematin compound /Figure 5/.

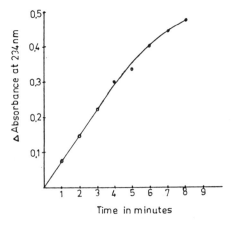

Fig.1.

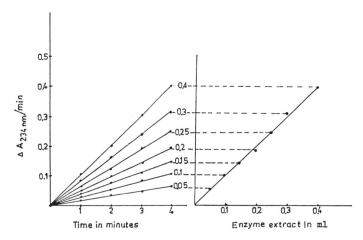

Fig.2.

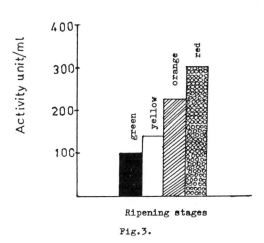

Fig.3.

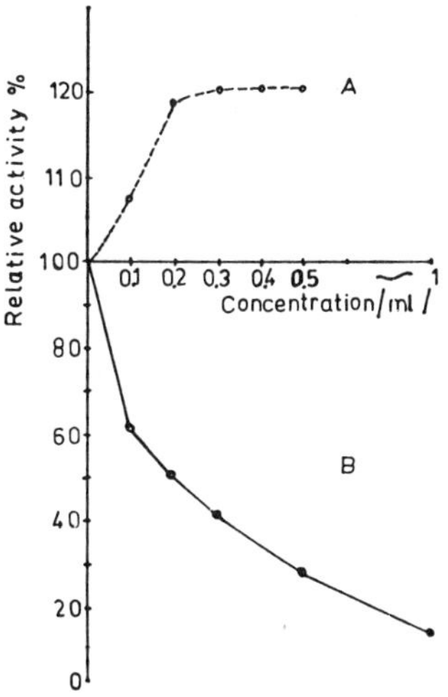

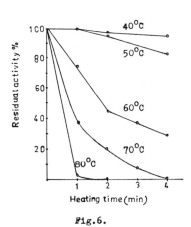

Fig.4.

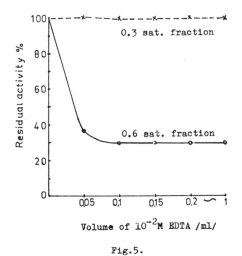

Fig.5.

Fig.6.

427

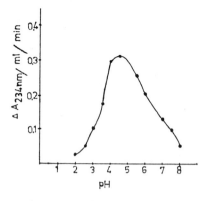

Fig. 7.

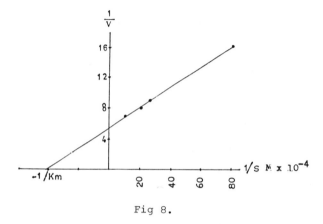

Fig 8.

Heat stability of the partially purified tomato lipoxygenase measured at different temperatures is illustrated on Figure 6. Moderate heat stability also seems to exclude the participation of hematin compounds, since they are stable to heat.

The enzyme has a rather low pH optimum at pH = 4.5 as represented by Figure 7, which is important for the processing of this tomato variety.

The K_m value calculated from the Lineweaver-Burk plot of the partially purified tomato lipoxygenase /Figure 8/ was found to be $K_m = 0.027 \times 10^{-4}$ M which is higher than that of the tomato lipoxygenase present in the ACE 55 FV variety $Km = 0.0015 \times 10^{-4}$ M published by Bonnet and Crouzet /1977/.

REFERENCES

Al-Obaidy, H.M. and Siddiqi, A.M., 1981, Properties of broad bean lipoxygenase, J. Food Sci., 46:622.
Bonnet, J.L. and Crouzet, J., 1977, Lipoxygenase from tomato fruit: Partial purification and study of some properties, J. Food Sci., 42:625.

MULTIPLE PATHWAYS OF LINOLENIC ACID SYNTHESIS OPERATE AND INTERACT IN LEAF TISSUE

Helen A. Norman and Judith B. St. John

USDA, ARS, Weed Science Laboratory
Beltsville, Maryland 20705

INTRODUCTION

Different biosynthetic pathways have been proposed for trienoic fatty acids(FA) in 18:3/16:3 and 18:3/18:3 monogalactosyldiacylglycerol (MGDG) molecular species[1]. In the 'prokaryotic' pathway, 18:1/16:0 MGDG is synthesized within the chloroplast and desaturated *in situ* to form 18:3/16:3 MGDG. A 'eukaryotic' pathway involving microsomal phosphatidylcholine (PC) as substrate for desaturation of 18:1 to 18:2 provides the diglyceride precursor for 18:3/18:3 MGDG. After galactosylation, the MGDG is further desaturated in the chloroplast to form 18:3/18:3. We have used a recently identified mutant of Arabidopsis thaliana (L.) ('JB1') deficient in 18:3,[2] and secondly 4-chloro-5-dimethylamino-2-phenyl-3(2H)-pyridazinone (BASF 13-338, Sandoz 9785) which inhibits 18:2 desaturation[3] to obtain definitive evidence for the synthesis of 18:3 via these different pathways.

MATERIALS AND METHODS

Lipids were labeled with radioactivity *in vivo* by incubating detached leaves with [^{14}C] FA. BASF 13-338, in acetone solution, was added to petri dishes and the acetone evaporated off before adding incubation medium to provide a 10^{-4} molar concentration. Details of the HPLC methods used to quantify lipid molecular species and the determination of their radioactivity have been provided elsewhere.[4,5]

RESULTS AND DISCUSSION

The major MGDG molecular species in wild-type ('WT') Arabidopsis was 18:3/16:3 (Table I). This was reduced in the 'JB1' mutant, and 18:3/16:2 and 18:2/16:2 MGDG increased. There was more eukaryotic MGDG in 'JB1', and in contrast to the decline of 18:3 in prokaryotic MGDG, the level of 18:3/18:3 increased, suggesting that the eukaryotic path was enhanced.

We found a very selective labeling of MGDG molecular species after incorporation of [^{14}C]18:1 (Fig.1). Detached leaves were incubated with [^{14}C]18:1 for 1 h, washed, and further incubated in nonradioactive medium. [^{14}C]18:3 accumulated only in the sn-2 position of 18/18C species, and prokaryotic MGDG was unlabeled. Therefore the 18:3 at sn-2 in

18:3/18:3 MGDG is formed by a different pathway (accessible to [^{14}C]18:1) than both 18:3 at sn-1, and also 18:3 at sn-1 of the 18:3/16:3 prokaryotic MGDG. In 'JB1', labeling of 18:3/18:3 was slow compared to 'WT', but the difference between the 18:2/18:3 species was less, suggesting that 18:2/18:3 is the precursor of 18:3/18:3 MGDG, and desaturation at sn-1 (occurring on MGDG) is more inhibited by the mutation.

Table I. Major MGDG Molecular Species in 'WT' and 'JB1' Arabidopsis

Molecular Species	μmol/g dry wt	
	'WT'	'JB1'
18:3/16:3	169.6 ± 5.8	61.0 ± 4.9
18:3/16:2	1.9 ± 0.3	68.9 ± 6.2
18:2/16:2	0.9 ± 0.2	52.8 ± 5.0
18:3/18:3	6.3 ± 0.9	18.8 ± 1.7
18:2/18:3	2.9 ± 0.4	24.1 ± 3.2
18:2/18:2	2.8 ± 0.5	11.2 ± 1.4

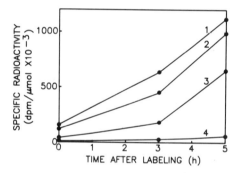

Fig. 1. Radioactivity of 18:2/18:3 in 'WT' (1) and 'JB1'(2), and of 18:3/18:3 in 'WT' (3) and 'JB1'(4) after labeling with [^{14}C] 18:1.

Previous work suggested PC provides diglyceride precursors for 18/18C MGDG.[1] In 'WT' leaves the specific activity of 18:3 in 18:2/18:3 PC was higher than in 18:2/18:3 MGDG (Table II). This difference was even greater in 'JB1', where [^{14}C]18:3 increased more rapidly in 18:2/18:3 PC. The labeling patterns suggested desaturation of 18:2/18:2 PC provides 18:2/18:3 PC as precursor for eukaryotic MGDG.

When leaves were incubated with [^{14}C] 12:0, the C16 and C18 FA at both positions of eukaryotic and prokaryotic MGDG were labeled indicating access to the FA elongation system. In 'WT', declining radioactivity in 18:2/16:2 and 18:3/16:2 coincided with a rise in 18:3/16:3 showing progressive desaturation in the prokaryotic pathway (Fig. 2). In 'JB1', the 18/16C species were labeled more slowly and radioactivity accumulated in 18:2/16:2, indicating inhibition of these chloroplast desaturase(s).

Table II. Labeling of PC Molecular Species Following Incubation with [^{14}C]18:1

Molecular Species	Time (h) After Labeling					
	'WT'			'JB1'		
	0	3	5	0	3	5
	dpm/µmol (x10^{-3})					
18:2/18:2	2600	4504	5867	3200	6037	4092
18:2/18:3	1200	2204	5336	1295	2492	6162
18:3/18:3	40	79	159	38	77	150

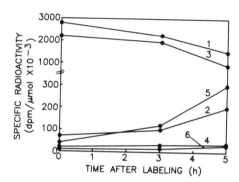

Fig. 2. Radioactivity of 18:2/16:2 in 'WT' (1) and 'JB1'(2), 18:3/16:2 in 'WT'(3) and 'JB1'(4), and of 18:3/16:3 in 'WT'(5) and 'JB1'(6) after labeling with [^{14}C]12:0.

Further evidence for different pathways of 18:3 formation was obtained with BASF 13-338. When leaves prelabeled with [^{14}C]12:0 for 1 h were incubated with BASF 13-338 for 16 h the level of [^{14}C]18:3 at sn-1 of MGDG was decreased by 75%, and at sn-2 by only 30%. Desaturation at sn-1 (occurring in the chloroplast) was thus more sensitive to BASF 13-338.

REFERENCES

1. P. G. Roughan and C. R. Slack, Trends Biochem Sci. 9:383 (1984).
2. J. A. Browse, C. R. Somerville, and P. J. McCourt, in: W. Siegenthaler, W. Eichenberger, eds., Proc. 6th International Symposium on the Structure, Function and Metabolism of Plant Lipids. Elsevier/North Holland, Amsterdam, pp. 167-170 (1983).
3. J. B. St. John, Plant Physiol. 57:38 (1976).
4. H. A. Norman and J. B. St. John, Plant Physiol. 81: in press (1986).
5. H. A. Norman and J. B. St. John, J. Lipid Res., in press (1986).

THE EFFECT OF TEMPERATURE ON DESATURATION OF GALACTOLIPID FATTY ACIDS IN BRASSICA NAPUS

John P. Williams, K. Mitchell and M. Khan

Botany Department
University of Toronto
Toronto, Ontario M5S 1A1

INTRODUCTION

Plants, animals and bacteria exposed to low growth temperatures often exhibit higher levels of unsaturated fatty acids in the glycerolipids of their cellular membranes. This general phenomenon has been known for some time and several theories have been put forward to explain it. No theory, however, has received complete acceptance and the reason for an apparently higher level of desaturase enzyme activity at low temperatures remains to be explained.

Fatty acids in many leaves are desaturated in conjunction with phospholipids before incorporation into galactolipid (the cytosolic pathway). In 16:3-plants, the situation is further complicated by the presence of a chloroplastic pathway in addition to the cytosolic pathway of glycerolipid biosynthesis and fatty acid desaturation.

We have been able to use the presence of the chloroplastic pathway (involving the biosynthesis of 16/18C molecular species of glycerolipids) to study the rates of desaturation of fatty acids in monogalactosyldiacylglycerol (MGDG) and digalactosyldiacylglycerol (DGDG) under different environmental conditions. These show that immediately after $^{14}CO_2$-feeding radioactivity is incorporated almost exclusively into the 16/18C molecular species of galactolipids allowing us to trace the rate of desaturation in these molecular species.

MATERIALS AND METHODS

Brassica napus (var. Tower) plants were grown from seed in growth chambers under 16h day/8h night regimes with a light intensity of 15 klux. Three growth temperatures (10, 20 and 30°C) and young leaves of similar physiological age were used for the experiments.

Leaves were harvested, allowed to assimilate $^{14}CO_2$ for 5 min and the incubations timed from the completion of this feeding period. The leaves were fed in the same chambers in which they were grown, under essentially identical conditions. The same three temperatures (10, 20 and 30°C) were used for feeding and subsequent incubations.

RESULTS

Fatty Acid Composition

The lipids of these leaves were extracted, separated and analyzed as previously described (1, 2, 3). Fatty acids were estimated quantitatively and for radioactivity.

Table I. Fatty Acid Composition of Leaf Lipids in Plants Grown at 10, 20 and 30°C

Lipid	Temp	Fatty acid (mole %)							
	°C	16:0	16:1	16:2	16:3	18:0	18:1	18:2	18:3
MGDG	10	2(1)	tr	2(1)	41(1)	0(0)	1(1)	2(1)	53(1)
	20	3(2)	1(1)	3(1)	31(1)	tr	1(1)	6(2)	55(3)
	30	4(1)	2(1)	2(2)	26(1)	1(1)	3(1)	7(1)	56(1)
DGDG	10	12(1)	2(1)	tr	9(1)	1(1)	1(0)	4(1)	72(1)
	20	13(1)	2(1)	0(0)	4(1)	2(0)	1(0)	7(1)	70(3)
	30	22(0)	1(0)	0(0)	2(0)	5(0)	3(0)	8(1)	59(1)
PG	10	27(1)	24(2)	tr	2(1)	1(1)	2(1)	9(1)	35(1)
	20	24(2)	23(3)	0(0)	5(3)	2(1)	3(1)	15(2)	28(2)
	30	28(2)	27(2)	tr	1(0)	4(1)	7(1)	19(1)	15(1)
PC	10	18(1)	2(1)	0(0)	1(1)	3(1)	4(1)	23(1)	50(1)
	20	18(1)	2(1)	0(0)	1(1)	5(1)	7(2)	28(2)	38(2)
	30	23(2)	2(1)	tr	tr	7(0)	11(1)	30(2)	26(1)

Figures in parentheses are standard deviations.
n=5 for 10 and 20°C and n=3 for 30°C samples.

The results in Table 1 show that at lower temperatures in MGDG and DGDG there is an increase in the level of 16:3 but little change in the level of unsaturation in the 18C fatty acids particularly MGDG. In contrast, in PG and PC, which lack 16:3, there is a significant increase in 18:3 at lower temperatures. At lower temperatures there is also an increase in the ratio of 16/18C:18/18C molecular species in MGDG. At 10°C the MGDG consists almost entirely of 16/18C molecular species (90% at 10°C, 76% at 20°C and 68% at 30°C).

The results, therefore, confirm the trend to higher levels of unsaturation at lower temperatures but also suggest that the chloroplastic pathway is favoured, almost to the exclusion of the cytosolic pathway at temperatures of 10°C or lower. This means that $^{14}CO_2$ tracer techniques may be used in plants grown at 10°C to determine rates of biosynthesis and fatty acid desaturation of 16/18C molecular species (through the chloroplastic pathway) with little interference from accumulation of radioactivity through the cytosolic pathway.

^{14}C-Labelling of Fatty Acid Plants Grown at 10, 20 and 30°C

The ^{14}C-labelling pattern of the fatty acids from MGDG of leaves grown at 10 and 30°C are shown in Table II. The leaves were fed and incubated at 20°C to remove the effects of temperature on photosynthesis and any direct effect on the activity of fatty acid synthetase and desaturase.

Table II. Distribution of Radioactivity in MGDG Fatty Acids in Plants Grown at 10 and 30°C

Growth Temp °C	Incubation Temp °C	Time min	% Radioactivity (fatty acids)							
			16:0	16:1	16:2	16:3	18:0	18:1	18:2	18:3
10	10	0	19	7	22	6	0	8	29	10
10	20	30	6	5	23	14	0	4	26	22
10	20	60	5	4	20	18	0	3	22	27
30	10	0	30	11	8	2	0	24	18	5
30	20	30	42	2	2	1	3	36	13	2
30	20	60	35	4	4	1	2	27	20	7

The leaves sampled at 0 time were fed at 10°C in order to limit the amount of desaturation during the 5 min feeding period.

Levels of radioactivity appearing in 16C and 18C fatty acids are approximately equal; confirming that the radioactivity is predominantly a measure of 16/18C molecular species biosynthesis and desaturation. Plants grown at 10°C incorporate significant levels of radioactivity into 16:2 and 18:2 immediately following feeding; some radioactivity is found in 16:0 and 18:1. At 30°C virtually all radioactivity is located in 16:0 and 18:1. Incubation for 30 and 60 min results in desaturation from 16:2 and 18:2 to 16:3 and 18:3 respectively in 10°C grown plants but not in plants grown at 30°C.

These plants are, therefore, preconditioned by growth temperature and exhibit high and low rates of desaturation regardless of incubation temperature.

Desaturation takes place in two stages: one from 16:0 and 18:1 to 16:2 and 18:2, the other from 16:2 and 18:2 to 16:3 and 18:3, respectively.

In plants grown at 10°C the first stage occurs in two phases: a rapid phase to 16:2 and 18:2 that occurs in the first 5 min during feeding and a slower phase (as indicated by little further desaturation of 16:0 even after 60 min). In plants grown at 30°C, there is little or no indication of the existence of the rapid phase. The rapid phase in plants grown at 10°C, is less apparent for the second stage of desaturation to 16:3 and 18:3 and results in the accumulation of radioactivity in 16:2 and 18:2.

Labelling of fatty acids in plants grown at 10°C and incubated at 10, 20 and 30°C

Plants grown at 10°C exhibit both a rapid and slow phase of desaturation in MGDG. In order to determine the effect of temperature on the rate of desaturation of these plants, they were grown at 10°C and then fed and incubated at 10, 20 and 30°C for 30 and 60 min. The results are contained in Table III.

Table III. Distribution of Radioactivity in Fatty Acids of MGDG.

Growth Temp °C	Incubation Temp °C	Time min	% Radioactivity (fatty acids)							
			16:0	16:1	16:2	16:3	18:0	18:1	18:2	18:3
10	10	0	19	7	22	6	0	8	29	10
10	10	30	18	7	20	5	0	9	33	9
10	10	60	20	6	23	4	0	4	32	11
10	20	30	6	5	23	14	0	4	26	22
10	20	60	5	4	20	18	0	3	22	27
10	30	30	16	1	22	13	0	5	22	21
10	30	60	13	0	19	21	0	3	16	29

At all three incubation temperatures the phenomenon of a very rapid phase of desaturation is evidenced by high levels of radioactivity in 16:2 and 18:2. Subsequent desaturation of 16:2 and 18:2 to 16:3 and 18:3 occurs slowly at 10°C and more rapidly at 20 and 30°C respectively.

Therefore, high levels of unsaturated fatty acids in MGDG are due primarily to the rapid phase of desaturation. The desaturase shows normal kinetics, indicating higher rates of desaturation at higher temperatures. This suggests that the high levels of unsaturated fatty acids found at low temperatures are not due to increased enzyme activities because of an increase in O_2 concentration nor abnormal enzyme kinetics.

One possible explanation for the occurrence of the rapid desaturation phase may be localisation and/or complexing of the galactosyltransferase and desaturase enzymes in the membranes of plants grown at low temperatures. Other explanations may be possible and these are presently under investigation.

BIBLIOGRAPHY

1. J.P. Williams, M. Khan, and K. Mitchell, "Structure, Function and Metabolism of Plant-lipids. P.-A. Siegenthaler and W. Eichenberger, eds., Elsevier Publisher, B.V. (1984).

2. J.P. Williams, G.R. Watson, M. Khan, S.P.K. Leung, Plant Physiol. 55:1038-1042 (1975).

3. J.P. Williams, P.A. Merrilees, Lipids, 5:367-370 (1970).

4. M. Khan, J.P. Williams, J. Chromatogr., 140:179-185 (1977).

ACYL LIPID METABOLISM IN RHODOTORULA GRACILIS (CBS 3043) AND THE EFFECTS OF METHYL STERCULATE ON FATTY ACID DESATURATION

C.E. Rolph[1], R.S. Moreton[2], I.S. Small[2] and J.L. Harwood[1]

[1]Dept. of Biochemistry
University College
Cardiff, U.K.

[2]Cadbury-Schweppes PLC
The University
Reading, U.K.

INTRODUCTION

The oleaginous yeast, Rhodotorula gracilis (CBS 3043) is capable of accumulating large amounts of storage lipid when grown under nitrogen-limiting conditions. The storage lipids produced under these conditions are located in intracellular storage vesicles, and are primarily triacylglycerols. The major fatty acids present in the acyl lipids have been shown to be palmitic, stearic, oleic, linoleic and α-linolenic acids; indicating the presence of Δ9, Δ12 and Δ15 desaturase enzymes.

In order to define these desaturation systems further, we have carried out experiments using different radiolabelled precursors with both whole cells and cell-free extracts. In addition, we have utilised sterculate (a known inhibitor of Δ9 desaturation) to alter the synthesis of unsaturated fatty acids in R. gracilis.

MATERIALS AND METHODS

Growth of organisms and lipid analysis

R. gracilis was grown in N-limited media (C:N ratio = 34.5), in baffled shake flasks in an orbital incubator at 150 r.p.m. at 30°C. Cells were harvested and lipids extracted[1] and analysed by t.l.c. and g.l.c.[2], except that the solvent system for neutral lipids was petroleum ether/diethyl ether/ acetic acid (80:20:2, by vol). Acyl-CoA's were analysed by the method of Stymne and Stobart[3].

Pulse-chase experiments

Log-phase cells were subjected to a 350nM pulse of [1-^{14}C]acetate, (for acyl-CoA pool analyses [^{14}C-acetate] = 700nM) for 30 mins, after which the pulse was diluted 100 fold with unlabelled acetate. Samples were removed from the vessel, the cells freeze-dried and lipids extracted and analysed as above.

Assay of the Δ12 desaturase in cell free extracts

0.1ml, (0.2mg protein) aliquots of supernatants from French pressed log phase cells in 0.1M Tris-HCl (pH 7.2) were assayed for Δ12 desaturase activity in an incubation mixture containing 1.25mM ATP, 2.0mM NADH, 2.0mM NADPH and 0.1μCi [1-^{14}C]stearoyl-CoA in 0.1M Tris-HCl (pH 7.2) (total volume, 2.0mls) at 30°C. Reactions were stopped by the addition of KOH, and fatty acids extracted and analysed[2].

RESULTS AND DISCUSSION

The incorporation of [1-^{14}C]acetate into the major acyl lipids of R. gracilis was studied in a series of pulse chase experiments followed by analysis of the fatty acid composition of individual phospholipids and triacylglycerol and acyl-CoA pools. As reported earlier[4], labelled linoleic acid was observed to accumulate first in phosphatidylcholine, whilst its appearance in the other phospholipids and the triacylglycerol pool did not occur until later in the time-course. These results implied that phosphatidylcholine could play a role in Δ12 desaturation in vivo. Further analysis of the labelling pattern in the remaining acyl lipids showed a simultaneous accumulation of labelled linoleate in the acyl-CoA pool (Fig. 1).

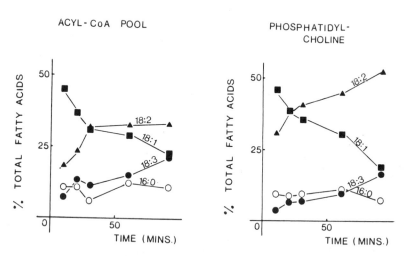

Fig. 1 Time-Course of the Relative Labelling of Different Fatty Acids from [1-^{14}C]Acetate in Phosphatidylcholine and the Acyl-CoA Pool in R. gracilis

However, the amounts of labelled linoleate in the acyl-CoA pool, in terms of relative proportions or total quantities were considerably less than for phosphatidylcholine. It is possible that R. gracilis may contain two Δ12 desaturase enzymes, one using oleoyl-phosphatidylcholine as substrate and the other using oleoyl-CoA. Such a situation has been reported for Candida lipolytica[5].

The activity of the Δ12 desaturase(s) was also followed in cell-free extracts using [1-^{14}C]stearoyl-CoA as substrate. Label in the acyl-CoA pool was rapidly depleted and labelled oleic acid was observed to accumulate in both phosphatidylcholine and phosphatidylethanolamine. No Δ12 desaturation was observed until after 300 minutes had elapsed, and at this point labelled linoleate appeared in both phosphatidylcholine and phosphatidylethanolamine. This evidence implies that in vitro,

Table 1 Effect of Methyl Sterculate on the Fatty Acid Composition of the Major Acyl Lipids in R. gracilis

	Fatty Acid	Fatty Acid Composition (% total)			
		TG	PC	PE	PS+PI
Control	16:0	17.8	14.9	22.9	33.9
	18:0	8.9	5.9	8.9	8.6
	18:1	45.1	32.8	39.8	33.9
	18:2	15.7	31.7	18.6	11.3
	18:3	6.0	11.9	4.0	1.5
+ Sterculate	16:0	12.4	16.0	21.2	32.2
	18:0	39.4	14.4	15.9	19.2
	18:1	19.6	11.2	17.4	16.5
	18:2	12.8	21.9	22.4	12.3
	18:3	10.6	35.1	18.6	8.6

R. gracilis was grown for 24hrs (stationary phase) in N-limited media ± methyl sterculate (1μl/ml of culture, Methyl sterculate = 84% pure). Results are means (n=3).

Table 2 The Effect of Methyl Sterculate on the Incorporation of Labelled Acetate and Ammonium Stearate into C_{18} Unsaturated Acids of R. gracilis

		Distribution of counts (% total [^{14}C]acids)	
Precursor	Fatty Acid	Control	+Methyl sterculate
[$1-^{14}C$]Acetate	16:0	9.3±1.1	10.6±0.6
	18:0	9.5±0.6	25.3±2.6
	18:1	70.1±1.2	54.3±1.6
	18:2	9.2±2.3	5.4±0.4
	18:3	n.d.	n.d.
NH_4^+[$1-^{14}C$]Stearate	16:0	n.d.	n.d.
	18:0	58.9±3.3	95.8±2.1
	18:1	37.1±3.6	4.2±2.1
	18:2	4.3±0.4	n.d.
	18:3	n.d.	n.d.

R.gracilis was grown for 15hrs (log-phase) in a N-limited media ± methyl sterculate (1μl/ml of culture. Purity of methyl sterculate = 84%). Cells were incubated with [$1-^{14}C$]acetate or [$1-^{14}C$]Stearate for 1h before analysis. Results are means ±S.D. (n=2), n.d. = none detected.

phosphatidylethanolamine can also act as a substrate for the Δ12 desaturase, agreeing with the proposals of Sanchez et al[6] for Δ12 desaturation in safflower microsomal preparations.

The acyl compositions of the phospholipids and of the triacylglycerol pool were determined for R. gracilis grown in a N-limited media containing methyl sterculate. In all of the acyl lipids studied (Table 1), the oleate content was observed to fall and that of stearate rise, indicating inhibition of the Δ9 desaturase. In addition, the relative proportion of linolenate in these acyl lipids was increased. This increase in linolenate was most pronounced in the phospholipids, and in phosphatidylcholine, linolenic acid was found to be the major fatty acid present.

The effects of methyl sterculate on R. gracilis were also studied via the use of [1-^{14}C]acetate and ammonium [1-^{14}C]stearate as possible precursors for oleic acid synthesis (Table 2). When labelled stearate was the precursor molecule, methyl sterculate was able to decrease the conversion of stearate to oleate from approximately 40% to 4%, whilst in samples in which [1-^{14}C]acetate was the precursor, very little inhibition was observed. Such evidence implies that the conversion of acetate to oleate is by a route which is not susceptible to sterculate inhibition. Similar observations have been reported for both plant[7] and mammalian[8] desaturase systems.

In work performed by Jeffcoat and Pollard[9], the inhibitory form of sterculic acid in mammalian tissue was reported to be sterculoyl-CoA. If this is also true for R. gracilis, we may tentatively propose that the alternative route of oleic acid synthesis in the presence of methyl sterculate is via a Δ9 desaturase system that is capable of desaturating stearate when attached to a complex lipid. Further desaturation of phospholipid substrate to linoleate and then linolenate would explain the lack of sterculate inhibition of the latter's synthesis. These possibilities are being investigated.

Acknowledgement: C.E.R. was supported by a CASE studentship from the S.E.R.C. and Cadbury-Schweppes PLC.

REFERENCES

1. R.S. Moreton, Modification of fatty acid composition of lipid accumulating yeast with cyclopropene fatty acid desaturase inhibitors, Appl.Microbiol.Biotechnol. 22:41-45 (1985).

2. P. Bolton and J.L. Harwood, Fatty acid biosynthesis by a particulate preparation from germinating pea, Biochem.J. 168:261-269 (1977).

3. S. Stymne and A.K. Stobart, Involvement of acyl exchange between acyl-CoA and phosphatidylcholine in the remodelling of phosphatidylcholine in microsomal preparations of rat lung, Biochem.Biophys.Acta. 837:239-250 (1985).

4. C.E. Rolph, R.S. Moreton, I.S. Small and J.L. Harwood, Acyl lipid metabolism and fatty acid desaturation in the yeast Rhodotorula gracilis (CBS 3043), Biochem.Soc.Trans. (in press).

5. G. Ferrante and M. Kates, Pathways for desaturation of oleoyl chains in Candida lipolytica, Can.J.Biochem.Cell Biol. 61:1191-1196 (1983).

6. J. Sanchez, V.P. Agrawal and P.K. Stumpf, Metabolism of ^{14}C-oleoyl-CoA by Safflower microsomes: The effect of ATP and CoA, in "Structure, Function and Metabolism of Plant Lipids" Siegenthaler, P-A and Eichenberger, W. eds. pp 77-80, Elsevier, Amsterdam (1984).

7. A.T. James, P. Harris and J. Bezard, The inhibition of unsaturated fatty acid biosynthesis in plants by sterculic acid, European J.Biochem. 3:318-325 (1968).

8. P.K. Raju and R. Reiser, The alternate route for the biosynthesis of oleic acid in the rat, Biochem.Biophys.Acta. 176:48-53 (1969).

9. R. Jeffcoat and M.R. Pollard, Studies on the inhibition of the desaturases by cyclopropenoid fatty acids, Lipids. 12:480-485 (1977).

METABOLISM OF EICOSAPOLYENOIC ACID LIPIDS IN RACE SPECIFIC INTERACTIONS

BETWEEN PHYTOPHTHORA INFESTANS AND POTATO

R. M. Bostock.

Department of Plant Pathology
University of California, Davis

INTRODUCTION

Arachidonic (AA) and eicosapentaenoic (EPA) acids are efficient elicitors of the metabolic cascade occurring in potato tuber during hypersensitivity expression.[1,2,3] Important components of this cascade include the accumulation of sesquiterpene phytoalexins, deposition of lignin and hydroxyproline in tuber cell walls, evolution of ethylene and ethane, and stimulation of oxidative enzymes. Induction of this response following inoculation with incompatible races of the fungal pathogen, Phytophthora infestans, in which EPA and AA occur primarily as esters in membrane and storage lipids, restricts fungal ingress and prevents colonization of host tissue.[4,5] Compatible isolates of the pathogen colonize potato tissue and defense reactions are not induced. The interaction between P. infestans and potato provides a unique opportunity for the study of biologically active lipids in recognition phenomena in plant-pathogen interaction.

Structure-activity studies indicate that free eicosapolyenoic acids with unsaturation at carbon 5 in the chain are most active in the induction of sesquiterpene phytoalexins.[1,2,6] Potato lipoxygenase, which is most reactive with free unsaturated fatty acids, has been implicated to participate in the hypersensitive response since inhibitors of this enzyme prevent AA elicitor activity.[7] In light of these observations, I proposed that release of eicosapolyenoic acids from fungal lipid during ingress by incompatible races of P. infestans would facilitate induction of the hypersensitive response by presenting the most biologically active forms of these elicitors to the plant cell. To test this hypothesis, I compared the metabolism of lipids containing esterified eicosapolyenoic acids and acyl esterase activity in potato disks after inoculation with incompatible and compatible isolates of P. infestans.

METHODS

Fungal lipids containing ^{14}C-radiolabelled EPA and AA were applied at concentrations below the threshold of induction of the tuber response to potato disks (cv. Kennebec) which were subsequently inoculated with spores of an incompatible (race 0) or compatible (race 1.4) isolate of P. infestans. Lipids were extracted at various times after inoculation and analyzed by TLC and liquid scintillation counting. Acyl hydrolase activity was evaluated in the top mm of potato disks extracted in a buffer containing 0.3 M sucrose, 0.1 M tricine-NaOH, pH 7, and 5 mM β-mercaptoethanol. Activities were determined with p-nitrophenyl palmitate as substrate at pH 8.[8]

RESULTS AND DISCUSSION

Inoculation of potato disks treated with P. infestans lipids containing radiolabelled EPA and AA resulted in a decline in the proportion of radioactivity recovered in the triglyceride fraction (TG) and increases in the free fatty acid (FFA) and polar lipid fractions (Fig. 1). The reactions proceeded similarly in both incompatible and compatible interactions. Analysis of the fatty acid methyl esters obtained by methanolysis of lipids recovered from potato disks indicated that the treatments had no effect on the levels of elicitor fatty acids relative to the other fatty acids (Table 1). This suggests that the eicosapolyenoic acids were not preferentially metabolized. Recoveries of radioactivity in the lipids were similar for all three treatments (data not shown). Acyl esterase activities were similar in both interactions and declined during the course of the experiments (Fig. 2).

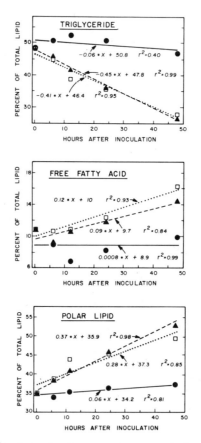

Fig. 1. The effect of inoculation on the proportion of radioactivity recovered in the triglyceride (TG), free fatty acid (FFA) and polar lipid fractions from potato disks treated with fungal lipids containing radiolabelled AA and EPA. -●-, control; -□-, race 0; -▲-, race 1.4.

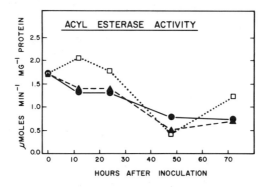

Fig. 2. Acyl esterase activity in the upper 1 mm of potato disks uninoculated or inoculated with P. infestans. - □ -, control; -●-, race 0; -▲-, race 1.4.

Table 1. The proportion of EPA and AA in the fatty acids obtained by methanolysis of lipids recovered from potato disks treated with radiolabelled fungal lipids and inoculated with P. infestans.

PERCENT EPA AND AA

HOURS AFTER TREATMENT

	0	6	12	24	48
CONTROL	33	29	31	29	34
RACE 0	33	32	30	31	34
RACE 1.4	33	32	27	35	32

The data do not support the hypothesis that hydrolysis of eicosapolyenoic acid lipids occurs more rapidly in the incompatible interaction. Evidence for deacylation of TG, the most abundant elicitor lipid in the fungus, was not obtained. Disks treated with purified TG and then inoculated showed a similar increase in polar lipid but not in FFA (data not shown). What appeared on thin-layer chromatograms to be oxidation products accumulated and probably account for most of the increase in radioactivity observed in the polar lipid fraction. In preliminary experiments, significant lipase activity was not detected in any treatment. Although differences in acyl esterase activity were not detected, the slight increase in FFA observed after inoculation probably derives from polar lipids. Lipolytic acyl hydrolase, the most abundant acyl hydrolase in potato tuber, does not hydrolyze TG but readily attacks a variety of phospholipids and mono- and di-acyl glycerols.[6]

If lipolytic activities are affected similarly and free EPA and AA are available in both interactions, perhaps a reaction subsequent to lipolysis (eg. lipoxygenase) is activated during the incompatible interaction or, alternatively, is inhibited in the compatible interaction. It is also possible that lipids containing eicosapolyenoic acids induce the response without prior release of free EPA and AA.

REFERENCES

1. R. M. Bostock, J. A. Kuc' and R. A. Laine. Eicosapentaenoic and arachidonic acids from Phytophthora infestans elicit fungitoxic sesquiterpenes in the potato. Science 212:67 (1981).
2. R. M. Bostock, R. A. Laine and J. A. Kuc'. Factors affecting the elicitation of sesquiterpenoid phytoalexin accumulation by eicosapentaenoic and arachidonic acids in potato. Plant Physiol. 70:1417 (1982).
3. R. M. Bostock, D. A. Schaeffer and R. Hammerschmidt. Comparison of elicitor activities of arachidonic acid, fatty acids and glucans from Phytophthora infestans in hypersensitivity expression in potato tuber. Physiol. Molecular Plant Pathol. 28:(in press) (1986).
4. J. R. Creamer and R. M. Bostock. Characterization and biological activity of Phytophthora infestans phospholipids in the hypersensitive response of potato tuber. Physiol. Molecular Plant Pathol. 28:215 (1986).
5. M. J. Kurantz and S. F. Osman. Class distribution, fatty acid composition and elicitor activity of Phytophthora infestans mycelial lipids. Physiol. Plant Pathol. 22:363 (1983).
6. C. L. Preisig and J. A. Kuc'. Arachidonic acid-related elicitors of the hypersensitive response in potato and enhancement of their activities by glucans from Phytophthora infestans (Mont.) de Bary. Arch. Biochem. Biophys. 236:379 (1985).
7. D. A. Stelzig, R. D. Allen, and S. K. Bhatia. Inhibition of phytoalexin synthesis in arachidonic acid-stressed potato tissue by inhibitors of lipoxygenase and cyanide-resistant respiration. Plant Physiol. 72:746 (1983).
8. T. Galliard. The enzymic deacylation of phospholipids and galactolipids in plants. Purification and properties of a lipolytic acyl-hydrolase from potato tubers. Biochem. J. 121:379 (1971).

MEDIUM AND LONG CHAIN BIOSYNTHESIS

MODULATION OF FATTY ACID SYNTHESIS IN PLANTS BY THIOLACTOMYCIN

Mitsuhiro Yamada, Misako Kato, Ikuo Nishida, Kazuo Kawano, Akihiko Kawaguchi and Tomoko Ehara*

Department of Biology, The University of Tokyo, Tokyo 153 and *Department of Microbiology, Tokyo Medical College Tokyo 160, JAPAN

INTRODUCTION

The synthesis of fatty acids and lipids in plants are complicated because of involvement in both the chloroplast and the endoplastic reticulum for de novo synthesis, elongation and desaturation of fatty acids, accompanied by their esterification to glycerides. Nevertheless, a controlled flow of the metabolic path proceeds in a plant cell, as shown by a given ratio of lipid and fatty acid composition in a organelle membrane. Administration of the inhibitor for the synthesis of fatty acids and lipids is used as a tool to elucidate the mechanism of the controlled flow[1]. However, even when an inhibitor is specific for a definite reaction of fatty acid synthesis, the effect of the inhibition on overall synthesis of fatty acids are different in different lipids and different plants. For example, cerulenin inhibits 3-ketoacyl-ACP synthase in de novo synthesis of fatty acids and not that in elongation of fatty acids[2], but the effect of cerulenin on lipid synthesis in greening barley leaves indicates an increase of 16:0 and a marked decrease of 18:3 in MGDG, and conversely a decrease of 16:0 and an increase of 18:3 in PC, although the synthesis of both glycolipids and phospholipids are considerably inhibited[3].

Thiolactomycin (TLM) selectively inhibits type II fatty acid synthases from E. coli and higher plants, but does not type I fatty acid synthases from fungi and mammals[4]. In this paper, the effect of TLM on fatty acid synthesis is confirmed in enzyme and cell levels. Pulse- and chase-experiments with plant tissues in the presence of TLM accounts for the controlled synthesis of lipids in higher plants. Finally the modulated lipid composition of plant tissues by continuous administration of TLM is examined in relation to the chloroplast structure and the photosynthetic activity of the tissues.

MATERIALS AND METHODS

TLM, a gift from Chugai Pharmaceutical Company Ltd, Takata 3-41-8, Toshima-ku, Tokyo 171, Japan, were solubilized in methanol and added to the incubation medium. The concentration of TLM was determined by the absorbance at 238 nm using HPLC. E. coli fatty acid synthase activities

Table 1. Effect of TLM on the individual enzymes of the fatty acid synthase system in E. coli

Enzyme	TLM(µM)	Specific activity (units/mg protein)	
Overall fatty acid synthase	0	1.15	(100)
	3.5	0.27	(23)
	35	0	(0)
ACP acetyltransferase	0	0.037	(100)
	3.5	0.006	(16)
	35	0	(0)
ACP malonyltransferase	0	225	(100)
	35	301	(131)
3-Ketoacyl-ACP synthase	0	80.2	(100)
	3.5	39.9	(50)
	35	1.9	(2)
3-Ketoacyl-ACP reductase	0	177	(100)
	3.5	170	(96)
	35	91.7	(52)
3-Hydroxybutyryl-ACP dehydrase	0	2.05	(100)
	35	2.32	(113)
Enoyl-ACP reductase	0	21.1	(100)
	35	18.9	(90)

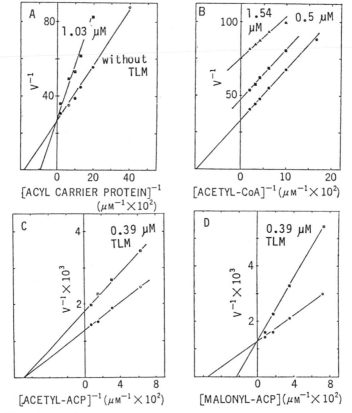

Fig. 1. Kinetics of E. coli ACP acetyltransferase and 3-ketoacyl-ACP synthase activities in the presence of TLM. (A) ACP acetyltransferase, 30 µM acetyl-CoA; (B) ACP acetyltransferase, 44.2 µM ACP; (C) 3-ketoacyl-ACP synthase, 55.3 µM malonyl-ACP; (D) 3-ketoacyk-ACP synthase, 63.3 µM acetyl-ACP

were determined by Reference 5. Fatty acid synthesis by spinach chloroplasts was measured by Reference 4 and anaerobic condition was provided by exchange of O_2 to N_2 in a Thunberg tube. Pseudomonas aeruginosa M-507 cells were grown with TLM at 32°C for 24 hr and fatty acids from the dried cells were analyzed. Chase experiments after pulse-labeling with [1-^{14}C]acetate in greening Avena leaves were carried out in the presence and absence of TLM according to Reference 6. Etiolated leaves from Avena and Brassica seedlings were greened by illumination with continuous absorption of TLM (100 μg/ml) and subjected to lipid analysis and electron microscopic observation. Oxygen evolution from the leaves in the light was determined by O_2 electrode.

RESULTS AND DISCUSSION

Inhibition of Fatty Acid Synthase System by TLM

As shown in Table 1, of individual six enzymes from E. coli ACP:acetyltransferase and 3-ketoacyl-ACP synthase were inhibited more than 50 % at 3.5 μM TLM, although the other enzymes retained full activities at this concentration. TLM reversibly inhibited fatty acid synthase. The inhibition of acetyltransferase was competitive for ACP, but uncompetitive for acetyl-CoA. The inhibition of 3-ketoacyl-ACP synthase was also competitive for malonyl-ACP but noncompetitive for acetyl-ACP (Fig. 1).

In spinach leaves cerulenin is inhibitory to 3-ketoacyl-ACP synthase I which is responsible for de novo synthesis of 16:0, but inert to 3-ketoacyl-ACP synthase II which is responsible for the elongation of 16:0 to 18:0. Table 2 shows the effect of TLM on fatty acid synthesis from [1-^{14}C]acetate by spinach chloroplasts under anaerobic condition. Under this condition 18:0 was accumulated by blocking the desaturation from 18:0 to 18:1 and addition of TLM resulted in more decreases of 18:0 than 16:0 without any effect in 18:1 synthesis. Pseudomonas aeruginosa synthesized 16:1(9) and 18:1(11) in addition to the synthesis of 16:0 and 18:0 by anaerobic path. Addition of TLM to the culture medium resulted in more decrease of 18:0 and 18:1 rather than 16:0 and 16:1 (Table 3). These results indicate that TLM is more inhibitory to the elongation of C16 to C18 than de novo synthesis of C16, in contrast to the effect of cerulenin.

Effect of TLM on Lipid Synthesis in Leaves

Based on different responses of TLM to C16 and C18 syntheses, we

Table 2 Effect of TLM on incorporation of [1-^{14}C]acetate into fatty acid by spinach chloroplasts under anaerobic condition

TLM (μg/ml)	Incorporation (dpm)	Distribution (%)					C18/C16
		12:0	14:0	16:0	18:0	18:1	
0	10,800	5.6	4.5	15.0	56.0	19.0	5.0
0.13	12,250	4.2	4.2	22.4	53.6	15.7	3.5
0.26	8,800	3.4	4.8	15.7	57.0	19.0	4.8
0.52	5,570	5.1	5.7	18.8	51.9	18.6	3.8

Reaction mixture (1 ml): 200 μM acetate (5 μCi), 10 mM $NaHCO_3$, 0.4 mM G3P 3 mM ATP, 0.2 mM CoA, 4 mM DTT, 1 mM $MnCl_2$, 25 mM Hepes buffer (pH 7.9) and chloroplasts (80 μg chlorophyll). Incubated at 25°C and 20,000 lux for 1 hr.

Table 3 Effect of TLM on fatty acid composition of Pseudomonas aeruginosa

TLM (µg/ml)	Dry weight of cells (mg)	Fatty acid composition (mol %)				
		14:0	16:0	16:1	18:0	18:1
0	51.8	0.67	40.0	12.7	0.83	45.8
0.069	46.2	0.85	36.7	17.8	0.63	44.0
0.49	43.8	3.29	40.1	36.9	0.43	19.3
1.02	15.8	4.70	48.1	35.9	t	11.4
1.29	10.5	3.47	49.2	35.2	t	12.1

t: trace

plan to explore the metabolic regulation of the eukaryotic path for 18:3/18:3-MGDG synthesis, ranging between the chloroplast and the endoplasmic reticulum. Fig. 2 shows the chase experiment after pulse-labeled with [1-^{14}C]acetate for greening Avena leaves. Continuous exposure of the leaf segments to TLM blocked the flow of the conversion of PC to MGDG in the eukaryotic path (Fig. 2). When leaves were pulse-labeled with [1-^{14}C]acetate in the presence of TLM and chased without TLM, the conversion pattern of PC to MGDG was not affected, suggesting that a temporary cease of fatty acid supply does not modulate

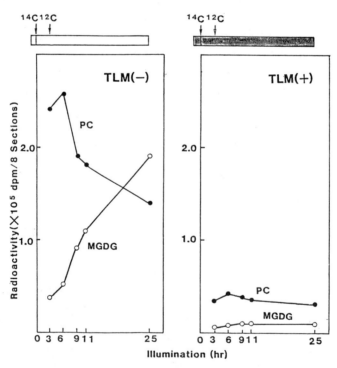

Fig. 2 Effect of TLM on the synthesis of PC and MGDG in greening Avena leaves. Etiolated leaf sections (4 cm length from the top) were fed with [1-^{14}C]acetate (40 µCi/0.2 ml) and TLM (100 µg/ml) in the light (2,400 lux, 25°C) for 25 hr.

the flow of the eukaryotic path (data not shown). More suppressed supply of 18:1 than that of 16:0 by TLM inhibition resulted in the increase of 16:0 and the decrease of C18 in PC at earlier stage of greening, but fatty acid composition in PC was recovered to the normal ratio at later stage of greeening (Fig. 3). In MGDG, fatty acid composition was little affected by TLM throughout greening stages. These results indicate the presence of compensative mechanism in the flow of the eukaryotic path.

Modulation of Lipid Composition by TLM in Avena and Brassica Leaves

Fig. 4 shows changes of lipid composition in Avena leaves greened from their etiolated leaves in the presence of TLM for 24 hr. In the total lipid composition PC remained unchanged, but MGDG markedly decreased by TLM treatment. Since Avena is a typical 18:3-plant in which the eukaryotic path from 18:2/18:2-PC to 18:3/18:3-MGDG is actively operated, the effect of TLM seems to appear in MGDG as end product, but not in PC as intermediate. In spite of the unchanged content of PC, there were a marked decrease of 18:1 and a compensating increase of 18:3 in its fatty acid composition (Table 4). The reduced

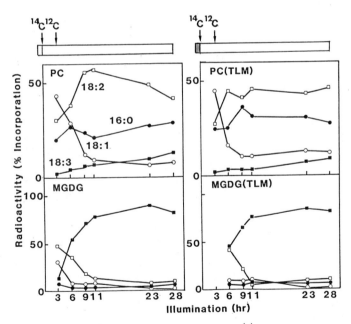

Fig. 3 Chase after pulse-labeling with [1-^{14}C]acetate and TLM in greening Avena leaves. Etiolated leaf sections (4 cm length from the top) were preincubated with TLM (100 μg/ml) for 1 hr in the dark and fed with [1-^{14}C]acetate (40 μCi/0.2 ml), 3.6 mM for 3 hr in the light (2,400 lux, 25°C), then chased without TLM for 28 hr in the light (2,400 lux, 25°C).

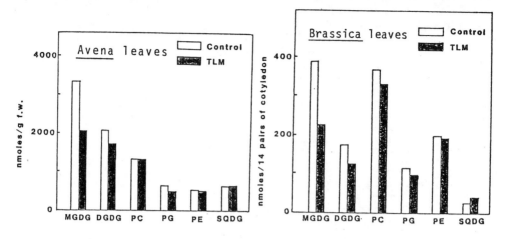

Fig. 4 Lipid contents of Avena and Brassica leaves greened in the presence of TLM. Etiolated leaves were greened at 2400 lux and 25°C in the presence of 5 mM acetate and TLM (100 μg/ml) for 24 hr.

content of MGDG ascribes to the decrease in both C16 and C18, whereas the reduced content of PG is due to the decrease in C18. The latter seems to be caused by more suppressed supply of 18:1 than that of 16:0 by TLM. Of interest is that the desaturation of 16:0 to 16:1(3t) in PG was markedly inhibited by TLM. The results of a TLM-treated experiment with Brassica leaves was shown in Table 5. In a reduced content of total lipids, PC was little reduced, but MGDG was to some extent, as in the case of Avena leaves. Fatty acid composition of PC was little affected in Brassica leaves. Since Brassica is 16:3-plant in which MGDG synthesis is divided by both prokaryotic and eukaryotic paths and the supply of fatty acids to the eukaryotic path through PC is reduced, less effect of TLM on fatty acid composition of PC would be expected. Marked decreases of 18:2 and 18:3 in MGDG account for suppressed supply of 18:1

Table 4 Effect of TLM on the fatty acid composition of greened Avena leaves. The experimental condition was the same as in Fig. 4.

Lipid	TLM	Fatty acid (nmoles/g f.w.)						
		16:0	16:1	18:0	18:1	18:2	18:3	Total
MGDG	−	215.3	−	85.8	+	44.2	3030.8	3376.1
	+	119.6	−	12.0	+	30.9	1888.9	2051.4
DGDG	−	376.4	−	15.2	19.5	105.2	1575.7	2092.0
	+	322.4	−	15.1	15.9	128.7	1241.3	1723.4
PC	−	613.1	−	38.4	69.1	442.7	211.9	1375.2
	+	621.7	−	38.3	20.9	412.6	278.6	1372.1
PG	−	214.5	174.1	38.3	32.7	42.5	158.2	660.3
	+	282.3	87.9	14.5	+	26.4	104.1	515.2
PE	−	231.7	−	24.5	20.2	212.8	76.8	566.0
	+	202.0	−	77.4	+	183.7	49.1	512.2
SQDG	−	312.5	−	27.6	39.1	31.1	218.5	628.8
	+	320.8	−	72.6	+	24.7	235.1	653.2

Table 5 Effect of TLM on fatty acid composition of greened Brassica leaves. The experimental condition was the same as in Table 4, except for the use of etiolated Brassica leaves.

Lipid	TLM	Fatty acid (nmoles/14 pairs of cotyledon)							
		16:0	16:1	16:3	18:0	18:1	18:2	18:3	Total
MGDG	−	10.7	−	42.8	3.8	4.1	16.7	311.4	389.5
	+	12.1	−	25.7	2.2	2.5	5.2	177.4	225.1
DGDG	−	27.3	−	−	4.9	3.8	5.2	132.8	174.7
	+	33.9	−	−	4.6	6.3	2.7	79.3	126.8
PC	−	95.9	−	−	+	51.0	163.4	58.6	368.9
	+	88.1	−	−	+	50.0	135.5	55.8	329.4
PG	−	63.8	21.5	−	1.8	2.5	10.0	15.2	114.8
	+	56.1	14.7	−	+	11.9	5.2	8.5	96.4
PE	−	69.1	−	−	5.7	12.8	92.9	20.2	200.7
	+	66.5	−	−	8.4	12.8	80.0	27.0	194.7
SQDG	−	16.4	−	−	1.4	2.0	3.4	4.0	27.2
	+	26.0	−	−	1.8	+	3.1	8.9	39.8

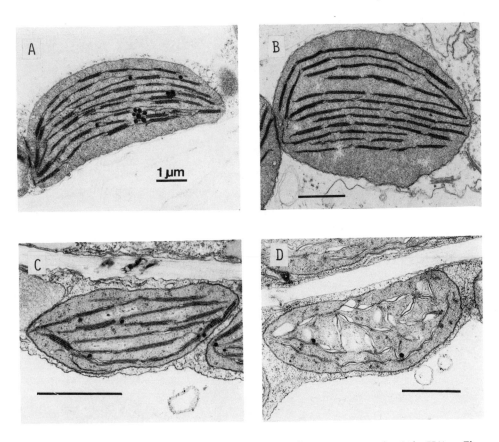

Fig. 5 Chloroplasts of Avena and Brassica leaves greened with TLM. The experimental conditions were the same as in Tables 4 and 5. A black bar in a photograph indicates 1 μm. (A) Avena, control; (B) Avena, presence of TLM; (C) Brassica, control and (D) Brassica, presence of TLM.

by TLM inhibition to the prokaryotic synthesis of this lipid. The desaturation of 16:0 to 16:1(3t) and of 18:1 to 18:2 in PG was suppressed by TLM treatment.

Changes in Structure and Photosynthetic Activity of the Chloroplast by Modification of Lipid Content

Fig. 5 shows the structure of Avena and Brassica chloroplasts formed from etioplasts in the presence of TLM. Avena chloroplasts turned from elipse into round shape, suggesting a change in permeability of the envelope membrane. The thylakoid membranes of Brassica chloroplasts were swollen everywhere, leading to loose stacking of thylakoids. Thus, insufficiency of MGDG and/or PG induced the membrane transformation of the chloroplast. However, the photosynthetic activites in TLM-treated Avena and Brassica leaves as measured by O_2 evolution decreased only a little, in spite of visible changes of the chloroplast structure (data not shown).

REFERENCES

1. S. E. Gardiner, P. G. Roughan and C. P. Slack, Plant Physiol. 70: 1316 (1982)
2. T. Shimakata, and P. K. Stumpf, Proc. Natl. Acad. Sci. USA 79: 5808 (1982)
3. G. Laskay, T. Farkas and E. Lehoczi, J. Plant Physiol. 118: 267 (1985)
4. I. Nishida, A. Kawaguchi, and M. Yamada, Plant & Cell Physiol. 25: 265 (1984)
5. I. Nishida, A. Kawaguchi and M. Yamada, J. Biochem. 99: 1447 (1986)
6. J. Ohnishi and M. Yamada, Plant & Cell Physiol 21:1607 (1980)

FATTY ACID SYNTHESIS IN DEVELOPING OILSEEDS

Michael R. Pollard and Sheo S. Singh

ARCO Plant Cell Research Institute

6560 Trinity Court, Dublin, CA 94568, USA

INTRODUCTION

De novo fatty acid synthesis (FAS) supplies fatty acids for a large variety of lipids, including membrane lipids, epicuticular waxes and cutin, neutral storage lipids, and many natural products, such as acyl salicylates, polyacetylenes or macrocyclic lactones. Many of these specific lipids are found in highly differentiated plant cells, where they can often represent a major part of cell dry weight. The regulation of FAS, which in most cases supplies the same fatty acids (palmitic and oleic acids) for different end products, is therefore a key question.

We will review de novo FAS and its regulation in the maturing oilseed, where most of the fatty acids are channelled into neutral storage lipids, usually triacylglycerols. This is a topical subject, given recent speculation about the possibility of manipulating oil content and composition of oilseed crops using the tools of molecular biology. The pathway of FAS in the seed is reasonably well defined, but the regulation of this pathway is poorly understood. Regulation can be either at the metabolic level, being concerned with measured enzyme activities, the result of effectors on this activity, the availability of substrates, and the interactions between organelles; or at the genetic level, being concerned with the number of genes involved, and the regulation of gene expression, resulting in new mRNAs and ultimately in the synthesis of new enzymes. In this review we will concentrate largely on regulation at the metabolic level, and we will also take the reviewer's prerogative and describe current projects in our laboratory on malonyl-CoA decarboxylation and medium-chain FAS in developing seeds.

THE LOCALISATION AND PATHWAY OF DE NOVO FATTY ACID SYNTHESIS

De novo FAS is the sequence of reactions utilising acetyl-CoA and malonyl-CoA to produce fatty acids via acyl-ACP thioesters for export from the plastid.[1,2] The salient features of this pathway are as follows:

1) In a variety of oil rich tissues, including avocado mesocarp,[3] castor bean endosperm,[4] and cotyledons of linseed and safflower,[5] plastids have been shown to have high rates of FAS activity. The assumption of exclusivity of fatty acid synthesis to the plastid seems reasonable, though not proven to the high degree of exactness for leaf tissue.[6]

2) FAS activity, as assayed in vitro in soluble extracts from plants, has long been known to require exogenous, reduced acyl carrier protein (ACP). More recently,

purification of the individual activities of FAS, namely acetyl-CoA:ACP transacylase, malonyl-CoA:ACP transacylase, ß-ketoacyl-ACP synthetase, ß-ketoacyl-ACP reductase, ß-hydroxyacyl-ACP dehydrase, and enoyl-ACP reductase, has shown that FAS has a "procaryotic" or "non-associated" organisation. That is, each enzyme activity is catalysed by a separate polypeptide. This holds true for the oilseed tissues of avocado mesocarp[7] and safflower.[8]

3) The most complete examination of the relative rates for the individual activities listed above, including Cuphea, safflower and rapeseeds, suggests acetyl-CoA:ACP transacylase and ß-ketoacyl-ACP synthetase as rate-limiting.[9]

4) Once palmitoyl-ACP has been produced, a second condensing enzyme, ß-ketoacyl-ACP synthetase II, elongates it to stearoyl-ACP. $\Delta 9$ Desaturation of this intermediate produces oleoyl-ACP. This desaturase has been purified from developing safflower seeds.[10]

5) The terminal reaction in the pathway requires the hydrolysis of the long-chain acyl-ACPs to free fatty acids. This activity shows a high degree of specificity for oleoyl-ACP.[10,11] Free fatty acids have been demonstrated to be the major end product of FAS when safflower plastids are incubated with [1-^{14}C] acetate in the presence of ATP.[5] Whether the addition of CoA would result in the accumulation of a labelled acyl-CoA pool remins to be tested, but can be expected by analogy with chloroplasts, where the export of fatty acids is facilitated by a free fatty acid: CoA ligase on the cytoplasmic side of the chloroplast envelope. It is a good assumption that in maturing oilseed tissue plastids probably export >95% of the fatty acids produced. There is no evidence for a "procaryotic" utilisation of acyl-ACPs by a plastid glycerol-3-phosphate acyl transferase, as is the case in leaves.[12] The glycerolipids of seeds have a "eucaryotic" acyl distribution, with no C_{16} acids at the sn-2 position.[1] The subsequent activities of acyl chain modification and lipid assembly all appear to be 'microsomal', that is, associated with the endoplasmic reticulum. The precise mechanism of free fatty acid movement across the plastid envelope, and the transfer of acyl-CoA from the plastid, remains to be explored. Also, there is an interesting report of a long-chain acyl-carnitine synthesizing system in etioplasts of barley leaves, a possibility that could also be tested in seed plastids as a mechanism for acyl transfer.[13]

CARBON SUPPLY FOR FATTY ACID SYNTHESIS

Carbon supply for FAS begins with the import of sucrose from the apoplast, and ends with the conversion of acetyl-CoA to malonyl-CoA, the step immediately preceding ACP-dependent FAS. The subject has been most extensively studied in the castor bean endosperm, and was reviewed in 1982 by Dennis and Miernyk.[14] Briefly, the plastid contains a complete set of isozymes for the glycolytic pathway from hexose to pyruvate, which is distinct from the cytosolic set. Stoichiometrically, acetyl-CoA, ATP and reducing equivalents produced by glycolysis are almost exactly balanced by the needs of FAS, but the operation of an oxidative pentose-phosphate pathway to produce extra reducing equivalents has also been shown. The plastid, at least in the case of the castor bean, does not appear to possess invertase, while there are conflicting reports as to whether the first enzyme in the pentose phosphate shunt, glucose-6-phosphate dehydrogenase, is also found in the plastid.[14,15] Although there is a difference in the literature as to whether glucose-6-phosphate can support FAS in plastids of the developing castor bean, both Yamada and Usami[16] and Miernyk and Dennis[4] agree that pyruvate is a better substrate than acetate, confirming the observation of an active pyruvate dehydrogenase complex in these organelles.[17] 3-Phosphoglyceric acid was also observed to support FAS, and is likely to be transported into the plastid via the phosphate translocator.[4] Hexoses give lower but measurable rates of FAS.[4,16]

While plastids can undoubtedly utilise acetate effectively for FAS it is very unlikely that free acetate is a significant carbon source in vivo. Apart from the studies cited above showing a complete glycolytic pathway operating within the plastid, acetate feeding studies with intact tissue of oilseeds producing large amounts of C_{20} and C_{22} fatty acids show clearly a two pool phenomenon.[1] The specific activity of acetate utilised for chain elongation is over an order of magnitude greater than that utilised for de novo FAS, implying poor utilisation of free acetate by the plastid. Even in seeds where there is no competition by the elongases, acetate cannot sustain more than a tiny fraction of the endogenous rate of lipid synthesis, even at saturating concentations.[18] The acetate:CoA ligase present in plastids probably functions as a scavenger system. Free acetate can be produced in the plastid as a consequence of cysteine biosynthesis, being released in the conversion of O-acetyl-L-serine to cysteine.

Carbon for FAS could also be produced via ATP-citrate lyase. This activity, which will convert citrate to oxaloacetate plus acetyl-CoA, has been demonstrated in soybean extracts.[19] But there is no current evidence for citrate transport into the plastid nor of localisation of this activity in the plastid to support citrate cleaving enzyme as a source of carbon for FAS, at least for avocado mesocarp plastids.[4] Extracts of developing soybean also contain a $NADP^+$-dependent malic enzyme.[20] NAD^+-dependent malic enzyme, which produces pyruvate and carbon dioxide from malate, is an enzyme specific to the mitochondrial matrix in higher plants.[21] The localisation of $NADP^+$-malic enzyme in immature soybeans, and the possibility of pyruvate production other than by pyruvate kinase, and the utilisation of this pyruvate in FAS, remain to be determined.

The carboxylation of acetyl-CoA to give malonyl-CoA requires a biotin carrier protein and two activities; an ATP-dependent biotin carboxylase activity utilising bicarbonate, and a transcarboxylase activity to transfer the carboxylate group to acetyl-CoA. The overall sequence is catalysed by the acetyl-CoA carboxylase (ACC), which is often termed "a key enzyme" or "the first committed step" in FAS. ACC has recently been purified from two oilseed tissues, rapeseed,[22] castor bean,[23] and from avocado mesocarp.[24] Using a rapid purification procedure designed to minimize proteolysis, ACC in rapeseed was purified to homogeneity and shown to have a single 220 kD polypeptide subunit, indicating a multienzyme ("eucaryotic") organisation. Although the castor bean ACC was not purified to homogeneity, a 528 kD complex was isolated and detailed kinetic analysis showed that it was unlikely to be regulated by most cellular metabolites, but that dead-end and product inhibition by ADP and P_i could occur. Thus, regulation of ACC by plastid energy charge is possible, as proposed by Eastwell and Stumpf for wheat germ ACC.[25]

CONTROL OF SEED OIL CONTENT

Some general comments on the factors that could regulate the lipid content of seeds are appropriate. Lipid content will be defined as the percentage oil in the mature embryo expressed on a dry weight basis. This precision is necessary, as it is possible to affect seed oil content simply by breeding for thinner seed coats.[26] Unlike fatty acid composition, where classical plant breeding has revealed simple genetic relationships, breeding for oil content has revealed much more complexity. The case of rapeseed, reviewed by Stefanson,[27] is typical of many oilseeds in that oil content is controlled in an additive manner by many genes, in that maternal genotype rather than embryo genotype influences oil content, in that there is an inverse relationship between oil and protein content, and in that environmental factors such as temperature or nitrogen availability have a considerable impact. This complexity is not surprising given that the triacylglycerols are the end product of an extended pathway, with competition between various intracellular metabolic sinks for carbon supplied by sucrose.

It is unclear whether the rate of FAS within the seed plastid is enzyme limited, or metabolite limited, or both. Unlike the chloroplast, where transport mechanisms and stromal metabolite concentrations are known, there is almost a complete dearth of similar information for developing seeds. This is an obstacle to understanding the regulation of FAS in seeds. It would be helpful to know the steady state levels of ATP, ADP, NADH, NADPH, CoA, acetyl- and malonyl-CoA, ACP, acyl-ACPs and glycolytic intermediates in the plastid. As the seed during its maturation phase is a 'one-way' metabolic system, there may not be a fine control, as exemplified by carbon metabolism in leaf mesophyll cells.[28] Interactions between cytoplasmic and plastid compartments could also be significant in the regulation of the rate of FAS. At this time, no mechanism for feedback control on FAS in the plastid by lipid assembly in the endoplasmic reticulum is known. Nor are the precise interactions between cytoplasmic and plastid glycolytic pools understood.

Information is beginning to accumulate to suggest a concerted induction of all the activities associated with lipid synthesis during seed maturation, with a corresponding decline these activities during the seed desiccation period. This has been noted for ACC activity in rapeseed,[29] for malonyl-CoA incorporation into fatty acids in Cuphea seeds,[30] and for ACP levels in soybean.[31] In the case of ACP, assay by activity correlated well with an immunochemical assay, and is probably an indication of true induction of protein synthesis during development rather than activation. Likewise, the ACC activity surge in maturing rapeseed was shown to parallel the appearance of a 225 kD peptide that, from its binding to avidin-agarose is likely to be the ACC peptide.[29] Similar developmental surges have been shown for glycolytic activities.[32] During this conference, the first in vitro evidence for a developmental surge and decline in the post-plastid activities of lipid assembly will be presented. The data comes from studies on diacylglycerol acyl transferase activity in cocoa beans (Fritz and McHenry, this volume). Given that the concerted induction of all activities involved in lipid biosynthesis is likely, the importance of any one enzyme as a key activity in the pathway will be lessened. In certain cases, namely seeds containing unusual storage lipids, we can speculate that tissue/development specific genes must be expressed in order to explain the novel metabolism. What is not clear is how many of the remaining activities of lipid biosynthesis are represented by such specific genes, or instead by constitutive genes with the appropriate regulatory elements. The underlying mechanism for concerted induction is also unknown. We can expect to see much emphasis on these topics in coming years. Only in the case of ACP, where two isozymes have been isolated, do we have a partial answer. One ACP form is found in the seed, but both are found in the leaf.[33] This does not necessarily imply that the ACP isoform found in both seed and leaf tissue is the same gene product in both tissues.

We have seen that there is considerable scope for metabolic regulation of FAS. Key branch points early in the pathway would involve partitioning of carbon between lipid and amino-acid synthesis. The enzymes involved could then be considered "regulatory". Adams et al.[34] have pointed out regulation at the level of PEP utilisation by competition between PEP carboxylase and pyruvate kinase. Additional competition in the plastid for PEP by DHAP synthase, for aromatic amino-acid synthesis, can also be envisaged, as can a further branch point at pyruvate, where pyruvate dehydrogenase and acetolactate synthase (to branched amino-acids) compete. Given the absence of large "metabolic sinks" for acetyl-CoA and malonyl-CoA other than FAS, the pyruvate dehydrogenase complex might just as reasonably be considered as the first committed step of neutral lipid biosynthesis as ACC. And finally in this section of speculation, the procaryotic nature of plant ACP-dependent FAS should not blind us to the fact that in vivo FAS may have a high degree of organisation, and act as a channelled-metabolic system. Certainly there is precedence for procaryotic FAS to assemble into high MW aggregates that can be purified,[35,36] as there is for association of ACC with plastid membranes.[23,24]

Obviously, our understanding of the regulation of FAS in oilseeds is still in its infancy.

MALONYL-COENZYME A DECARBOXYLATION

High levels of malonyl-CoA incorporation into fatty acids can be sustained in vitro without the need for acetyl-CoA as demonstrated for cell free homogenates of Cuphea seeds.[30] This observation has been extended to safflower seed extracts, and, more specifically, to extracts obtained by ammonium sulphate precipitation of 20,000 xg supernatant followed by Sephadex G25 chromatography of the resolubilised protein. With this procedure, it was possible to discount low MW precursors giving rise to primer for FAS. Malonyl-CoA decarboxylation was then examined as a possible source of acetyl-CoA in these incubations. Decarboxylation activity was identified by four independent assay methods: trapping of the acetate released after alkaline hydrolysis of the incubation mixture; HPLC; $^{14}CO_2$ trapping from [1,3-^{14}C] malonyl-CoA;[37] and by a coupled spectrophotometric method.[37] A literature search revealed that malonyl-CoA decarboxylation had been described in several plant tissues in 1961 by Hatch and Stumpf,[38] but appears to have received scant attention since. We were surprised to find high levels of decarboxylase activity in the developing seed. The K_m for malonyl-CoA in safflower extracts was 75 μM. At 250 μM malonyl-CoA the rate of decarboxylation was calculated as ca. 7 nmoles/min/embryo. This compares to a rate of fatty acid synthesis in the same safflower extracts of ca. 4 nmoles/min/embryo, measured at 250 μM malonyl-CoA and 0.4 mg ACP/ml.

We are currently characterising the malonyl-CoA decarboxylase activity and investigating its role in vivo. It does not appear to be implicated in de novo FAS in vivo. Evidence for this comes largely from two experiments. Firstly, the development profile for malonyl-CoA decarboxylase does not coincide with that for de novo FAS activity. In the dessicating safflower seeds FAS activity returns to barely detectable levels, whereas malonyl-CoA decarboxylase activity remains at maximum levels. Secondly, localisation experiments point to an extra-plastid localisation of malonyl-CoA decarboxylase. If malonyl-CoA decarboxylase does not play a role in de novo FAS, it might function instead as a cytoplasmic malonyl-CoA scavenger. There is precedence in the animal literature for a scavenging role.[39] An interesting point then arises as to the presence of the activity in plant tissues that produce long-chain fatty acids. The elongation reactions occur on the endoplasmic reticulum and utilise a cytoplasmic pool of malonyl-CoA. Also, researchers need to be aware of malonyl-CoA decarboxylation as a contaminating activity in the in vitro study of ACC and de novo FAS.

MEDIUM CHAIN FATTY ACID SYNTHESIS IN SEEDS

In cases where novel fatty acids are found in seed storage lipids, their biosynthesis is usually understood in terms of novel, membrane-bound activities in the cytoplasmic compartment acting to modify the acyl chain. However, in the case of medium chain fatty acids (MCFA; C_8 to C_{14}), a modified de novo FAS system is implicated, but as yet no satisfactory explanation of their biosynthesis has been found.[11,18,40,41] MCFA, and particularly lauric acid, have considerable commercial importance, particularly in the soap and detergent industry. Their natural sources are coconut oil and palm kernel oil, both of which are tropical products. Interest in a source of MCFA which can be grown as an annual oilseed crop in temperate regions has been a driving force for the research effort into the domestication and cultivation of the genus Cuphea.[42] The lipids of Cuphea seeds have a variety of well-defined MCFA distributions, including some species with a bimodal chain length specificity,[30] and it is for this precise specificity that we must look in vitro. No genetic analysis of the trait of MCFA biosynthesis is currently available.

It is a well-documented phenomenon that, in the assay of de novo FAS in vitro, the distribution of chain length of the acyl products can be manipulated to shorter chain lengths by altering assay parameters. This can be achieved in a number of ways, all of which increase chain-initiation events at the expense of chain elongation events. For example, in extracts of Cuphea lutea ($C_{10:0}$+$C_{12:0}$), increasing acetyl-CoA and ACP concentrations results in a build up of MCFA products, presumably by increasing the formation of acetyl-ACP and by having excess ACP to ensure that a large intermediate acyl-ACP pool can build up.[30] A similar type of control can be found in extracts of plant tissues that do not contain MCFA, as for potato tubers.[43] When the B-keotacyl-ACP synthetase is preferentially inhibited with cerulenin,[44] or when the level of acetyl-CoA:ACP transacylase is increased or the levels of malonyl-CoA:ACP transacylase or B-ketoacyl-ACP synthetase are decreased in reconstitution experiments with purified FAS enzymes,[9] a shorter chain acyl distribution is observed. All these experiments lead to interesting speculation, in that an increase in chain initiation events (ie., acetyl-ACP production and utilisation) relative to chain elongation events (ie., malonyl-ACP production and utilisation) could have a role in MCFA biosynthesis in vivo, but they ignore the most important facet, chain termination, where specificity must ultimately reside. All the experiments described above show a broad acyl chain length distribution for the total acyl pool. Previous workers have been unable to pinpoint a specific chain termination mechanism, including the most likely candidate, the acyl-ACP thioesterase.[11,40]

In order to reexamine the mechanism of MCFA biosynthesis in seeds we have run a series of comparative assays at high rates of FAS, utilising [2-^{14}C] malonyl-CoA as substrate, with soluble extracts of safflower and Cuphea inflata seeds. A particular advantage of Cuphea inflata is the very high specificity of chain length control it exhibits, as 86% of its fatty acid content in the seed oil is capric acid ($C_{10:0}$). The results of a typical experiment are shown in Table 1. The safflower extract does not respond to factors that promote MCFA production, whereas the Cuphea inflata extract, under identical conditions, does. More importantly, the data show a bimodal specificity in the end product (free fatty acid) for Cuphea inflata. The fatty acid synthetase of Cuphea clearly has the capacity to synthesise normal chain fatty acids (C_{16}+C_{18}), and has an active thioesterase mechanism to release C_{16} and C_{18} fatty acids. But also, under conditions favouring the build-up of MCFA intermediates there is a preferential release of $C_{10:0}$ into the free fatty acid fraction. This result was repeatable over several experiments, which consistently showed a strong preference for release of $C_{10:0}$ over $C_{8:0}$ or $C_{12:0}$ when compared to the composition of the acyl-thioester pool. Thus we have a favourable point to start a study on the mechanism of MCFA in seeds, as we are observing the correct specificity in vitro. The study is currently in progress.

Table 1. Fatty Acid Synthesis in Cuphea and Safflower Seed Extracts: Chain Length Distribution

Assay Conditions[a] Fraction	Label Incorporated (%)	[^{14}C] Fatty Acid Chain Length Distribution					
		8	10	12	14	16	18
Safflower, 0μM acetyl-CoA, 40 μg/ml ACP							
Acyl-ACP	0.5						
Free fatty acids	14	-	-	-	-	23	77
Safflower, 50 μM acetyl-CoA, 360 μg/ml ACP							
Acyl-ACP	0.5	-	15	10	15	15	45
Free Fatty Acids	36	-	-	-	-	19	81
Cuphea inflata, 0μM acetyl-CoA, 40 μg/ml ACP							
Acyl-ACP	1	-	15	28	7	20	30
Free fatty acids	17	3	4	1	1	15	76
Cuphea inflata, 50 μM acetyl-CoA, 360 μg/ml ACP							
Acyl-ACP	3	5	32	51	5	4	3
Free fatty acids	22	3	16	4	2	24	51

[a]Each assay also contains 250 μM [2-^{14}C] malonyl-CoA, and 1 mM each NADH, NADPH. Enzyme extracts are obtained by ammonium sulphate precipitation (0-70%) of a 20,000 xg supernatant, with resolubilisation and G-25 chromatography of the pellet.

REFERENCES

1. C. R. Slack and J. A. Browse, Synthesis of storage lipids in developing seeds, in: "Seed Physiology, Volume 1. Development", D. R. Murray, ed., Academic Press, Sydney (1984).
2. P. K. Stumpf, Biosynthesis of saturated and unsaturated fatty acids, in: "The Biochemistry of Plants: A Comprehensive Treatise. Volume 4. Lipids: Structure and Function", P. K. Stumpf. ed., Academic Press, New York (1980).
3. P. J. Weaire and R. G. O. Kekwick, The synthesis of fatty acids in avocado mesocarp and cauliflower bud tissue, Biochem. J. 146:425 (1975).
4. J. A. Miernyk and D. T. Dennis, The incorporation of glycolytic intermediates into lipids by plastids isolated from the developing endosperm of castor oil seeds (Ricinus communis L.), J. Expt. Bot. 34:712 (1983).
5. J. Browse and C. R. Slack, Fatty-acid synthesis in plastids from maturing safflower and linseed cotyledons, Planta 166:74 (1985).
6. J. B. Ohlrogge, D. N. Kuhn and P. K. Stumpf, Subcellular localisation of acyl carrier protein in leaf protoplasts of Spinacia oleracea, Proc. Natl. Acad. Sci. USA 76:1194 (1979).
7. I. Caughey and R. G. O. Kekwick, Characteristics of some components of the fatty acid synthetase system of plastids from the mesocarp of avocado (Persea americana) fruit, Eur. J. Biochem. 123:553 (1982).
8. T. Shimakata and P. K. Stumpf, The procaryotic nature of the fatty acid synthase of developing Carthamus tinctorium L. (Safflower) seeds, Arch. Biochem. Biophys. 217:144 (1982).
9. T. Shimakata and P. K. Stumpf, The purification and function of acetyl coenzyme A:acyl carrier protein transacylase, J. Biol. Chem. 258:3592 (1983).
10. T. A. McKeon and P. K. Stumpf, Purification and characterization of the stearoyl-ACP desaturase and the acyl-ACP thioesterase from maturing seeds of safflower, J. Biol. Chem. 257:12141 (1982).
11. J. B. Ohlrogge, W. E. Shine and P. K. Stumpf, Fat metabolism in higher plants: characterization of plant acyl-ACP and acyl-CoA hydrolases, Arch. Biochem. Biophys. 189:382 (1978).
12. G. Roughan and R. Slack, Glycerolipid synthesis in leaves, Trends Biol. Sci. 9:383 (1984).
13. D. R. Thomas, M. Noh Hj Jalil, A. Ariffin, R. J. Cooke, I. McLaren, B. C. S. Yong and C. Wood, The synthesis of short- and long-chain acylcarnitine by etiochloroplasts of greening barley leaves, Planta 158:259 (1983).
14. D. T. Dennis and J. A. Miernyk, Compartmentation of nonphotosynthetic carbohydrate metabolism, Ann. Rev. Plant Physiol. 33:27 (1982).
15. Y. Satoh, Q. Usami and M. Yamada, Glucose-6-phosphate dehydrogenase in plastids from developing castor bean seeds, Plant Cell Physiol. 24:527 (1983)
16. M. Yamada and Q. Usami, Long chain fatty acid synthesis in developing castor bean seeds. IV. The synthetic system in protoplastids, Plant Cell Physiol. 16:879 (1975).
17. E. E. Reid, P. Thompson, C. R. Lyttle and D. T. Dennis, Pyruvate dehydrogenase complex from higher plant mitochondria and proplastids. Plant Physiol. 59:842 (1977).
18. S. S. Singh, T. Y. Nee and M. R. Pollard, Acetate and mevalonate labeling studies with developing Cuphea lutea seeds, Lipids 21:143 (1986).
19. D. R. Nelson and R. W. Rinne, Citrate cleavage enzyme from developing soybean cotyledons. Incorporation of citrate carbon into fatty acids. Plant Physiol. 55:69 (1975).
20. C. A. Adams and R. W. Rinne, Interactions of phosphoenolpyruvate carboxylase and pyruvic kinase in developing soybean seeds, Plant Cell Physiol. 22:1011 (1981).
21. R. Douce, in "Mitochondria in Higher Plants. Structure, Function, and Biogenesis," Academic Press, Orlando (1985).
22. A. R. Slabas and A. Hellyer, Rapid purification of a high molecular weight subunit polypeptide form of rapeseed acetyl-CoA carboxylase. Plant Sci. 39:177 (1985).

23. S. A. Finlayson and D. T. Dennis, Acetyl-coenzyme A carboxylase from the developing endosperm of Ricinus communis. Isolation and characterization, Arch. Biochem. Biophys. 225:576 (1983).
24. S. B. Mohan and R.G.O. Kekwick, Acetyl-coenzyme A carboxylase from avocado (Persea americana) plastids and spinach (Spinacia oleracea) chloroplasts. Biochem. J. 187:667, (1980).
25. K. C. Eastwell and P. K. Stumpf, Regulation of plant acetyl-CoA carboxylase by adenylate nucleotides, Plant Physiol. 72:50 (1983).
26. A. L. Urie, Inheritance of partial hull in safflower, Crop Sci. 26:493 (1986).
27. B. R. Stefansson, The development of improved rapeseed cultivars, in: "High and low erucic acid rapeseed oils. Production, usage, chemistry, and toxicological evaluation," J. K. G. Kramer, F. D. Sauer and W. J. Pigden, eds., Academic Press, Toronto (1983).
28. M. Stitt, Fine control of sucrose synthesis by fructose-2,6-bisphosphate, in: "Regulation of carbon partitioning in photosynthetic tissue", R. L. Heath and J. Preiss, eds., ASPP monograph (1985).
29. E. Turnham and D. H. Northcote, Changes in the activity of acetyl-CoA carboxylase during rapeseed formation, Biochem. J. 212:223 (1983).
30. S. S. Singh, T. Nee and M. R. Pollard, Neutral lipid biosynthesis in developing Cuphea seeds, in: "Structure, function and metabolism of plant lipids," P. A. Siegenthaler and W. Eichenberger, eds., Elsevier Science Publishers, Amsterdam (1984).
31. J. B. Ohlrogge and T. M. Kuo, Control of lipid synthesis during soybean seed development:enzymic and immunochemical assay of acyl carrier protein, Plant Physiol. 74:622 (1984).
32. R. J. Ireland and D. T. Dennis, Isoenzymes of the glycolytic and pentose-phosphate pathways during the development of the castor oil seed, Can. J. Bot. 59:1423 (1981).
33. J. B. Ohlrogge and T. M. Kuo, Plants have isoforms for acyl carrier protein that are expressed differently in different tissues, J. Biol. Chem. 260:8032 (1985).
34. C. A. Adams, T. H. Broman and R. W. Rinne, Use of [$3,4-^{14}C$] glucose to assess in vivo competition for phosphoenolpyruvate between phosphoenolypyruvate carboxylase and pyruvate kinase in developing soybean seeds, Plant Cell Physiol. 23:959 (1982).
35. M. L. Ernst-Fonberg, Fatty acid synthetase activity in Euglena gracilis variety bacillarius. Characterisation of an acyl carrier protein dependent system, Biochemistry 12:2449 (1973).
36. R. W. Hendren and K. Bloch, Fatty acid synthesis from Euglena gracilis. Separation of component activities of the ACP-dependent fatty acid syntehtase and partial purification of the β-ketoacyl-ACP synthetase, J. Biol. Chem. 255:1504 (1980).
37. P. E. Kolattukudy, A. J. Poulose and Y. S. Kim, Malonyl-CoA decarboxylase from avian, mammalian and microbial sources, Methods Enzymol. 71:150 (1981).
38. M. D. Hatch and P. K. Stumpf, Fat metabolism in higher plants. XVII. Metabolism of malonic acid and its α-substituted derivatives in plants, Plant Physiol. 36:121 (1961).
39. E. J. Mitzen, A. A. Ammouni and N. H. Koeppen, Developmental changes in malonate-related enzymes of rat brain, Arch. Biochem. Biophys. 175:436 (1976).
40. K. C. Oo and P. K. Stumpf, Fatty acid biosynthesis in the developing endosperm of Cocos nucifera, Lipids 14:132 (1979).
41. A. R. Slabas, J. Harding, A. Hellyer, C. Sidebottom, H. Gwynne, R. Kessell and M. P. Tombs, Enzymology of plant fatty acid biosynthesis, in: "Structure, Function and Metabolism of Plant Lipids", P. A. Siegenthaler and W. Eichenberger, eds., Elsevier Science Publishers, Amsterdam (1984).
42. F. Hirsinger, Agronomic potential and seed composition of Cuphea, an annual crop for lauric and capric seed oils, J. Am. Oil Chem. Soc. 62:76 (1985).
43. K. P. Huang and P. K. Stumpf, Fat metabolism in higher plants. XLIV. Fatty acid synthesis by a soluble fatty acid synthetase from Solanum tuberosum, Arch. Biochem. Biophys. 143:412 (1971).

44. N. M. Packter and P. K. Stumpf, Fat metabolism in higher plants. The effect of cerulenin on the synthesis of medium- and long-chain acids in leaf tissue, Arch. Biochem. Biophys. 167:655 (1975).

MEDIUM AND LONG-CHAIN FATTY ACID SYNTHESIS

John L. Harwood

Department of Biochemistry
University College
Cardiff CF1 1XL, U.K.

INTRODUCTION

De novo fatty acid synthesis in plant tissues appears to be catalysed by the operation of a high molecular weight (multifunctional protein) acetyl-CoA carboxylase and a Type II fatty acid synthetase. Elongation of the product of the synthetase, or of endogenous acids, is catalysed by Type III fatty acid synthetases with malonyl-CoA as the source of the additional carbons.

ACETYL-CoA CARBOXYLASE

Acetyl-CoA carboxylase catalyses the ATP-dependent formation of malonyl-CoA from acetyl-CoA and bicarbonate. This reaction is the first committed step for fatty acid synthesis since, as noted above, malonyl-CoA is required for elongation as well as de novo synthesis of fatty acids. In bacterial systems the enzyme consists of three separable proteins - biotin carboxylase, biotin carboxyl carrier protein (BCCP) and BCCP:acetyl-CoA transcarboxylase - while in mammals three functional domains are present on a single polypeptide chain. Early purification attempts on plant acetyl-CoA carboxylase led to results which suggested that the enzyme was either similar to the bacterial one or, possibly, represented an intermediate form. However, it is now clear that these results were influenced by the high activity of endogenous proteinases and it is generally agreed that seeds contain a single polypeptide with functional domains. Leaf tissues may also contain a multifunctional protein, although the situation there is less clear.

Plant acetyl-CoA carboxylase is localised in the chloroplasts of leaf tissues[1,4]. Its activity, which is low in the dark, is raised considerably by the illumination of leaves[5]. On illumination it is known that stromal concentrations of ATP and Mg^{++} increase while that of ADP decreases[6]. From kinetic data it can be estimated that nucleotide and Mg^{++} changes would each result in a doubling of acetyl-CoA carboxylase activity. This would be in line with the observation of Eastwell and Stumpf[7] that wheatgerm acetyl-CoA carboxylase can be tightly controlled in vitro through its requirement for ATP and its inhibition by ADP and AMP. In addition, during active photosynthesis, light-driven removal of protons from the stroma raises its pH from about 7 to over 8. This change can bring about a three-fold increase in acetyl-CoA carboxylase which together

with the nucleotide and Mg^{++} effects would raise activity up to 24-fold[8].
Such an increase in fatty acid synthesis rates are also seen when
incubations are carried out in light versus dark conditions. Moreover,
fatty acid biosynthesis and acetate activation in spinach chloroplasts has
also been suggested to be mediated by ATP/ADP ratios, pH and Mg^{++}
concentration[5,9]. A caveat should be added that generalisations about
acetyl-CoA carboxylase regulation in different plant tissues should only
be made with extreme caution. For example, recent experiments with the
soya bean enzyme[10] showed that, while ATP protected the crude enzyme from
loss of activity during dilution, pre-incubation of the purified enzyme
with ATP had no effect on activity. In this case, also, ADP and AMP were
inhibitory and it was reported that the soya bean carboxylase was under
the control of relative nucleotide concentrations.

The above results imply that acetyl-CoA carboxylase may be the
rate-limiting enzyme for de novo fatty acid synthesis. Indeed, in seeds
the level of acetyl-CoA carboxylase activity correlate with the
accumulation of lipid in developing castor bean[11] and rape[12] seeds.
However, unlike the mammalian enzyme, plant acetyl-CoA carboxylase does
not seem to be consistently stimulated by tricarboxylic acids c.f.[2,10,13].

Acetyl CoA-carboxylase has been purified from a number of plant
sources. As mentioned above, early purifications from avocado[2], barley
embryo[14] and wheat germ[15] gave preparations which yielded 3-6 subunits on
PAGE under denaturing conditions. Rapid purification of the enzyme from
parsley using an avidin-affinity column resulted in a preparation which
had a molecular weight of 210,000 on SDS-PAGE[16]. This result has been
confirmed and the same workers also purified the wheat germ enzyme in the
presence of proteinase inhibitors to yield a peptide with a 240,000
molecular weight[17].

However, the situation concerning the molecular weight of plant
acetyl-CoA carboxylases does not seem to be getting any simpler with time.
Thus, a preparation from maize leaves which seemed to contain all
functional activity of the carboxylase showed a single peptide of 60,000
on SDS-PAGE[6]. The native enzyme had a molecular weight of about 500,000
by gel filtration. Nikolau et al.[18] examined biotin-containing enzymes by
using Western blotting following SDS-PAGE and the use of (^{125}I) -
Streptavidin to probe for biotin in proteins from two C_3 plant leaves and
two C_4 plant leaves. No high molecular weight biotin-proteins were
detected but all species showed bands of 62,000 molecular weight in
addition to one or two other biotin-proteins. It was suggested that the
62,000 molecular weight band corresponded to acetyl-CoA carboxylase[18].

Recently, acetyl-CoA carboxylase has been purified from two cultivars
of soya beans[10]. Interestingly, while both showed a band of 240kDa on
SDS-PAGE one cultivar also showed additional bands of 65 and 58 kDa. The
origin (or functions) of these additional bands was not commented on.

In summary, the plant acetyl-CoA carboxylase appears to be a multifunctional protein which has a subunit (molecular weight of about 240,000
when purified from seed tissues such as rape, castor bean or soya bean.
The enzyme from leaf tissues appears to have a subunit molecular weight of
about 60,000 but isoenzymes are found which have locations in different
leaf cell types c.f.[18].

FATTY ACID SYNTHETASE

Results from three laboratories in 1982 firmly established that the
plant fatty acid synthetase is a Type II enzyme. In fact, prior to that
time, the isolation of ACP from three plants[19] implied that the plant

synthetase had activities present on individual polypeptide chains. Partial separation of β-ketoacyl-ACP synthetase, β-ketoacyl-ACP reductase, acetyl CoA:ACP transacylase and malonyl-CoA:ACP transacylase was achieved from barley chloroplasts[20]. From avocado fruit, Caughey and Kekwick[21] purified the β-ketoacyl-ACP reductase and malonyl-CoA:ACP acyltransferase to homogeneity and also purified the enoyl-ACP reductase. However, the most thorough study was that by Shimakata and Stumpf mainly with spinach leaves. Purifications of acetyl CoA:ACP transacylase[22], β-ketoacyl-ACP synthetase I[23], β-ketoacyl-ACP synthetase II[24], β-ketoacyl-ACP reductase β-hydroxyacyl-ACP dehydrase and enoyl-ACP reductase[25] have been reported, the last three to homogeneity.

Several interesting features of the plant fatty acid synthetase systems can be highlighted. Firstly, several species of ACP have been found in different plants e.g.[26,27] which may carry out different functions in vivo. Similarily, the malonyl-CoA:ACP transacylase from soybean leaves has also been shown recently to exist in at least two isoforms[28]. Only one of these forms was found in soybean seeds. The activity of β-ketoacyl-ACP synthetase II which shows a high substrate specificity is vital in determining the balance of 16C/18C products[24]. In fact, this enzyme could be regarded as part of the Type III synthetase, palmitate elongase. The β-ketoacyl-ACP synthetase I is sensitive to cerulenin while synthetase II is sensitive to arsenite[24] thus accounting for the actions of these inhibitors on fatty acid synthesis c.f.[29]. The spinach leaf acetyl-CoA:ACP transacylase is also inhibited by arsenite and this enzyme may catalyse the rate limiting reaction in the fatty acid synthetase of this tissue[22]. The β-ketoacyl-ACP reductase has been reported in two forms and, in the case of avocado, the NADPH form has been resolved from that using NADH[21]. In other tissues such as spinach or safflower, NADPH is a more effective substrate than NADH[25,30]. Likewise, two forms of the enoyl-ACP reductase have been found. Type I utilises crotonyl-ACP and is NADH-specific[25]. Type II uses 2-decenoyl-ACP as substrate and NADPH in preference to NADH. Both types are present in safflower, castor bean and rape seeds[31] although only Type I seems to be present in leaf tissue. The Type I enoyl-ACP reductase has been purified to homogenity from spinach leaves[25] and rape seed[32] where they have native molecular weights of 115,000 and 140,000 respectively. With the spinach enzyme a subunit molecular weight of 32,500 was found suggesting the existance of a tetramer[25]. The rape seed enzyme also seems to be tetramer but in this case with two different subunits, of 33,600 and 34,800 respectively[32]. Interestingly, in rape seeds the Type I enoyl-ACP reductase had no activity towards a C_{18} substrate suggesting that the Type II enzymes had to be used for stearate formation[31]. Since the Type I enzyme is the only one contained in spinach leaves, one presumes that it must have been active with C_{18} substrate, in contrast to the rape seed Type I enzyme - although the spinach enoyl reductase was only tested with substrates up to C_{16}[25].

There is general agreement that, in photosynthetic tissues, the plastid is the major (perhaps exclusive) site of de novo fatty acid synthesis c.f.[33]. However, it has also been shown that various membrane fractions can catalyse the synthesis of various fatty acids by de novo processes as well as by elongation e.g.[34]. In microsomal fractions from germinating pea, partial-proteolysis experiments indicated that two sorts of membrane fragments were active[35]. Because of the reported high activity of chloroplast preparations, including thylakoid fractions[36] it was suggested that one of the active membranes could have been derived from plastids.

Indeed, if chloroplasts are purified by Percoll gradients from lettuce or pea leaves, broken by osmotic shock and the Class II chloroplasts assayed for fatty acid synthesis with [^{14}C]malonyl-CoA, then the majority of activity is still associated with the chloroplasts rather than being released (Table 1). However, it will be noted that stromal marker enzymes are also poorly released indicating that osmotic shock is not a good way of releasing lettuce stromal contents. Nevertheless, successive washing

Table 1. Distribution of Fatty Acid Synthesis and Marker Enzyme Activity in Fractions from Lettuce Chloroplasts

Activity	Distribution (% total chloroplast activity)			
	Particulate (Class II chloroplasts)	Soluble (Stroma)	Yeda-press fractions Particulate	Soluble
Fatty Acid Synthesis (4)	67+7	30+11	3+1	63+6
RuBPCase (3)	83+15	8+1	6+1	79+15
(NADP)GAPDH (2)	84+3	10+3	5+4	82+tr.
Ca^{++}-ATPase (2)	95+3	2+1	90+6	4+1
Chlorophyll (1)	92	8	92	tr.
Protein (4)	90+8	12+3	56+5	35+3

Fatty acid synthesis was measured using [^{14}C]malonyl-CoA under the conditions described in Walker and Harwood[37]. Results show means +S.D. for the number of experiments shown in parentheses.

Table 2. Effect of Successive Washing on Protein Release from Chloroplasts

	% Activity in total chloroplast fraction			
	Supernatant			Particulate
Lettuce	I	II	III	
Fatty Acid Synthesis	51+2	29+3	3+1	11+1
(NADP)GAPDH	53+4	32+2	4+2	3+3
Protein	22+4	9+2	9+tr.	56+8
Pea				
Fatty Acid Synthesis	86+2	9+6	3+1	6+1
(NADP)GAPDH	99+2	9+6	n.d.	n.d.

Fatty acid synthesis was measured with [^{14}C]malonyl-CoA under the conditions described in Harwood and Walker[37] where further details can be found. Means +S.D. for 3 experiments are shown.

Table 3. Fatty Acid Products Synthesised by Chloroplast Fractions from [^{14}C]Malonyl-CoA

Distribution of radioactivity (% total ^{14}C-acids)

		12:0	14:0	16:0	16:1	18:0	18:1
Pea	Total chloroplast fraction	4	11	28	3	51	6
	Stroma	3	11	25	3	55	5
Lettuce	Total chloroplast fraction	n.d.	6	37	4	48	7
	Class II chloroplasts	1	8	56	6	23	6
	Stroma	tr.	3	15	tr.	81	1
	Yeda-press supernatant	n.d.	n.d.	11	n.d.	76	14

Data from Walker and Harwood[37]

of the chloroplasts could remove most of the fatty acid synthesising activity from both pea and lettuce (Table 2). A small but significant amount of fatty acid synthetase remained associated with membranes even when all (pea) or most (lettuce) of the NADP-GAPDH had been removed. Thus, although the chloroplast fatty acid synthetase is essentially soluble, it associates with the membrane in vitro and is likely to also in vivo.

Since chloroplast fatty acid synthetase could associate with thylakoid membranes, especially in lettuce, and because fatty acid synthesis is likely to integrate with membrane lipid formation in vivo, we tested the effect of different chloroplast fractions on the pattern of products (Table 3). The results showed some differences in the patterns produced and, in particular, that the β-ketoacyl-ACP synthetase II of pea may be less tightly associated with membranes than synthetase I.

ELONGATION

Recently, we have studied the Type III Synthetases (elongases) of ageing potato tubers and asked the question - is more than one enzyme responsible for the synthesis of fatty acids greater than 18:0? Ageing potato discs were used because they show high rates of fatty acid synthesis and increasing synthesis of very long chain fatty acids (20C-24C) with time (Table 4).

The increased synthesis of VLCFA's with time was shown to be due to enzyme induction and could be blocked by cycloheximide (Table 5) or by puromycin when these were added to the ageing medium. The inhibitors had little effect when added to the assay medium.

Since a sequential production of 20:0 then 22:0 then 24:0 acids was seen with time and because this could be blocked by cycloheximide, we added the inhibitor after different periods of ageing to see if the synthesis of particular VLCFA's could be blocked by this means. Representative data, with appropriate controls, are shown in Table 6 and they showed clearly that individual elongation systems are present for the synthesis of arachidic, behenic and lignoceric acids respectively[38].

These conclusions are in agreement with indirect evidence from leek[39], genetic studies with barley[40] and the solubilisation of an enzyme specific for stearate elongation from leek[41].

Table 4. Effect of Ageing on the Products of Fatty Acid Synthesis by Potato Discs

Ageing time(h)	Distribution (% total [^{14}C]fatty acids)				
	16C	18C	20C	22C	24C
0	44±1	51±1	4±1	n.d.	n.d.
2	50±3	42±4	5±1	1±tr.	n.d.
4	56±2	35±2	6±1	2±1	n.d.
8	50±2	41±2	5±1	3±1	1±tr.
24	46±1	37±5	5±tr.	6±3	5±2

For details see Walker and Harwood[38]

Table 5. Effects of Cycloheximide Additions upon Assay of Fatty Acid Synthesis by Aged Potato Discs

Ageing (h)	Cycloheximide Addition		Pattern (% total [^{14}C]fatty acids)				
	Ageing	Assay	16C	18C	20C	22C	24C
0	−	−	27	71	2	n.d.	n.d.
0	−	+	26	72	1	n.d.	n.d.
8	−	−	33	56	4	4	2
8	−	+	29	56	4	5	2
8	+	−	19	79	1	n.d.	n.d.
8	+	+	20	80	n.d.	n.d.	n.d.

For further details see Walker and Harwood[38]

Table 6. Effect of Sequential Treatment of Potato Discs during Ageing with Cycloheximide

Treatment	Ageing (h)	Products (% total [^{14}C]fatty acids)				
		16C	18C	20C	22C	24C
Control	0	24±5	69±5	tr.	n.d.	n.d.
	4	25±3	56±4	9±tr.	8±1	n.d.
	8	23±tr.	48±3	10±2	11±1	5±tr.
+ Cycloheximide	8	24±1	70±4	1±tr.	n.d.	n.d.
Control then + Cycloheximide	4+4	28±tr.	59±1	5±tr.	5±1	1±tr.
Hybrid sample*	8	25±3	56±4	7±1	6±1	3±tr.

For further results and details see Walker and Harwood[38]

SUMMARY

1. Acetyl-CoA carboxylase has a high molecular weight in its native form. The enzyme from seeds has a subunit molecular weight of about 240K but that from leaves of 60K.

2. Plant fatty acid synthetase is concentrated in plastids and is a soluble Type II enzyme.

3. Stearate synthesis is dependent on β-ketoacyl-ACP synthetase II.

4. Further elongation of stearate relies on chain-length specific elongases.

ACKNOWLEDGEMENT

The financial support of the S.E.R.C. is gratefully acknowledged.

REFERENCES

1. C.G. Kannangara and C.J. Jenson, Eur.J.Biochem. 54:25 (1975)

2. S.B. Mohan and R.G.O. Kekwick, Biochem.J. 187:667 (1980)

3. B.J. Nikolau, J.C. Hawke and C.R. Slack, Arch.Biochem.Biophys. 211:605 (1981)

4. E. Thomson and S. Zalik, Plant Physiol. 67:655 (1981)

5. Y. Nakamura and M. Yamada, Plant Sci.Lett. 14:291 (1979)

6. B.J. Nikolau and J.C. Hawke, Arch.Biochem.Biophys. 228:86 (1984)

7. K.C. Eastwell and P.K. Stumpf, Plant Physiol. 72:50 (1983)

8. A. Hellyer, H.E. Bambridge and A.R. Slabas, Biochem.Soc.Trans. 14:in press (1986)

9. A. Sauer and K. Heise, Plant Physiol. 73:11 (1983)

10. D.J. Charles and J.H. Cherry, Phytochemistry 25:1067 (1986)

11. P.D. Simcox, W. Garland, V. DeLuca, D.T. Canvin and D.T. Dennis, Can.J.Bot. 57:1008 (1979)

12. E. Turnham and D.N. Northcote, Biochem.J. 212:223 (1983)

13. S.A. Finlayson and D.T. Dennis, Arch.Biochem.Biophys. 225:576 (1983)

14. K. Brock and C.G. Kannangara, Carlsberg Res.Commun. 41:121 (1976)

15. N.C. Nielsen, A. Adee and P.K. Stumpf, Arch.Biochem.Biophys. 192:446 (1979)

16. B. Egin-Buhler and J. Ebel, Eur.J.Biochem. 133:335 (1983)

17. A.R. Slabas and A. Hellyer, Plant Sci. 39:177 (1985)

18. B.J. Nikolau, E.S. Wurtele and P.K. Stumpf, Plant Physiol. 75:895 (1984)

19. R.D. Simoni, R.S. Criddle and P.K. Stumpf, J.Biol.Chem. 242:573 (1967)

20. P.B. Hoj and J.D. Mikkelsen, Carlsberg Res.Commun. 47:119 (1982)

21. I. Caughey and R.G.O. Kekwick, Eur.J.Biochem. 123:552 (1982)

22. T. Shimakata and P.K. Stumpf, J.Biol.Chem. 258:3592 (1983)

23. T. Shimakata and P.K. Stumpf, Arch.Biochem.Biophys. 220:45 (1983)

24. T. Shimakata and P.K. Stumpf, Proc.Nat.Acad.Sci. (U.S.) 79:5808 (1982)

25. T. Shimakata and P.K. Stumpf, Arch.Biochem.Biophys. 218:77 (1982)

26. J.B. Ohlrogge and T-M. Kuo, Plant Physiol. 74:622 (1984)

27. P.B. Hoj and I. Svendsen, Carlsberg Res.Commun. 48:285 (1983)

28. D.J. Guerra and J.B. Ohlrogge, Arch.Biochem.Biophys. 246:274 (1986)

29. B.R. Jordan and J.L. Harwood, Biochem.J. 191:791 (1980)

30. T. Shimakata and P.K. Stumpf, Arch.Biochem.Biophys. 217:144 (1982)

31. A.R. Slabas, J. Harding, A. Hellyer, C. Sidebottom, H. Gwynne, R. Kessel and M.P. Tombs, Dev.Plant Biol. 9:3 (1984)

32. A.R. Slabas, C. Sidebottom, R. Kessell, A. Hellyer and M.P. Tombs, Biochem.Soc.Trans. 14:in press (1986)

33. J.B. Ohlrogge, D. Kuhn and P.K. Stumpf, Proc.Nat.Acad.Sci. (U.S.) 76:1194 (1979)

34. P. Bolton and J.L. Harwood, Biochem.J. 168:261 (1977)

35. J. Sanchez, B.R. Jordan, J. Kay and J.L. Harwood, Biochem.J. 204:463 (1982)

36. P.J. Weaire and R.G.O. Kekwick, Biochem.J. 146:439 (1975)

37. K.A. Walker and J.L. Harwood, Biochem.J. 226:551 (1985)

38. K.A. Walker and J.L. Harwood, Biochem.J. 237:41 (1986)

39. C. Cassagne and R. Lessire, Arch.Biochem.Biophys. 191:146 (1978)

40. P. von Wettstein-Knowles, Dev.Plant Biol. 3:1 (1979)

41. V.P. Agrawal and P.K. Stumpf, Arch.Biochem.Biophys. 240:154 (1985)

CUTICULAR LIPIDS IN PLANT-MICROBE INTERACTIONS

P.E. Kolattukudy, William F. Ettinger and Joseph Sebastian

Biotechnology Center
The Ohio State University
Columbus, Ohio

INTRODUCTION

Cuticle constitutes the boundary between higher plants and their environment. Therefore, this layer might be expected to play an important role in the interaction of the plant with environmental factors. The plant cuticle is composed almost entirely of lipids and the role of some of these lipids in the interaction between plants and microbes has become clear in the recent years. In this brief review, we shall confine our discussion to two specific examples of such interactions: a detrimental one with pathogenic fungi and a beneficial one with phyllospheric bacteria which might provide fixed nitrogen in return for the use of some of the cuticular components as the carbon source. In this context, we will deal only with the role of the insoluble lipid-derived polymer, cutin, but not the role of soluble waxes that are always constituents of the cuticle.

COMPOSITION OF CUTIN

This insoluble polymer, which is the major structural component of plant cuticle, is composed of hydroxy and hydroxyepoxy acids. The most common major components are 16-hydroxy-C_{16}, 18-hydroxy-C_{18}, 18-hydroxy-9,10-epoxy-C_{18} and 9,10,18-trihydroxy-C_{18} fatty acids (Fig. 1). The less common and minor components found are derived from or analogous to these acids. Indirect chemical studies showed that there are no free primary alcohol groups, showing that they must all be involved in ester linkages resulting in the linear polymer structure. About one half of the hydroxyl groups in the middle of the chains are also esterified indicating the presence of considerable amounts of branching and/or cross linking in this polymer.

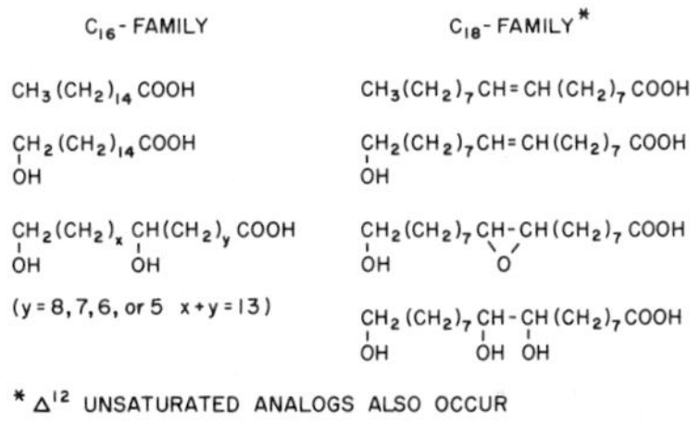

C_{16}-FAMILY	C_{18}-FAMILY*
$CH_3(CH_2)_{14}COOH$	$CH_3(CH_2)_7CH=CH(CH_2)_7COOH$
$\underset{OH}{CH_2}(CH_2)_{14}COOH$	$\underset{OH}{CH_2}(CH_2)_7CH=CH(CH_2)_7COOH$
$\underset{OH}{CH_2}(CH_2)_x\underset{OH}{CH}(CH_2)_yCOOH$	$\underset{OH}{CH_2}(CH_2)_7\underset{\diagdown O \diagup}{CH-CH}(CH_2)_7COOH$
(y=8,7,6, or 5 x+y=13)	$\underset{OH}{CH_2}(CH_2)_7\underset{OH}{CH}-\underset{OH}{CH}(CH_2)_7COOH$

*Δ^{12} UNSATURATED ANALOGS ALSO OCCUR

Figure 1. Cutin Monomers

HOW DO FUNGI BREACH THE CUTIN BARRIER?

<u>Isolation of Cutinase</u>: Plant pathologists debated for the better part of a century about the mechanism by which fungi penetrate the cuticular barrier (1). The view that mere physical force of growth of the infection structure is adequate to achieve penetration was countered by the argument that enzymatic breaching of the polyester barrier is essential for the penetration. The postulated enzyme, named cutinase, was isolated only in the mid-seventies when it was discovered that several pathogenic fungi secreted cutinase when grown on cutin as the sole carbon source (2). Cutinase was purified and characterized from the extracellular fluid of cutin-grown <u>Fusarium solani pisi</u> (3,4). Since then, cutinase has been purified from a variety of pathogenic fungi (5,6).

<u>Properties of Cutinase</u>: All the fungal cutinases are quite similar in size at about 25 kDa (5,6). They are all glycoproteins containing a few percent O-glycosidically attached carbohydrates. They all have very similar amino acid composition. Some noteworthy features of the composition include the presence of two disulphide bridges with no free SH groups, one Trp and one or two His. Immunological cross-reactivity is observed only with cutinases from closely related species of fungi. Catalysis by fungal cutinases probably involves the active serine catalytic triad (7). The involvement of active serine was shown by titration of the serine residue with tritiated diisopropylfluorophosphate. The evidence for participation of histidine in catalysis was provided by the finding that the enzyme activity was lost when the imidazol residue was carbethoxylated by treatment of the enzyme with diethylpyrocarbonate. Treatment of the enzyme with carbodiimide resulted in the modification of three carboxyl residues without affecting the enzymatic activity. Carbodiimide treatment in the presence of SDS resulted in the modification of an additional carboxyl group and this caused the loss of enzyme activity. Even though such rigorous evidence for participation of the catalytic triad has not been provided for cutinases from other sources, inhibition by active serine-directed reagents indicate that all fungal cutinases catalyze the hydrolysis of cutin using the catalytic triad.

<u>Evidence for involvement of cutinase in fungal penetration</u>: To determine whether fungal pathogens secrete cutinase during the penetration into their hosts, ferritin conjugated antibodies prepared against cutinase and electron microscopy were used. By this method, it was shown that <u>F. solani pisi</u> secreted cutinase during penetration into pea hypocotyl with intact cuticular barrier (8). This infection could be completely prevented by specific inhibition of cutinase either with chemicals or antibodies (9). This protection was not observed when inoculation was done on hypocotyl with mechanically breached cuticle/wall barrier showing that protection resides in the penetration phase of infection. Similar protection against fungal attack was subsequently observed with other host-pathogen systems (Table 1).

A test was performed to determine whether cutinase inhibitors can protect crop plants in the field. A biweekly spray of papaya fields with a potent cutinase inhibitor resulted in protection of the fruits against lesion formation by <u>C. gloeosporioides</u> (Fig. 2). The degree of protection achieved by this antipenetrant treatment was not as complete as that obtained by a fungicide. Such a result might be expected as the antipenetrant is only a protective agent which cannot affect fungi that already penetrated the defensive barrier.

CLONING AND SEQUENCING OF cDNA FOR CUTINASE

Induction of cutinase by the hydroxy acids of cutin hydrolysate in

TABLE I: Demonstrated cases of plant protection against fungi by cutinase inhibitors.

Pathogen	Host
Fusarium solani f. pisi	Pea hypocotyl
Colletotrichum gloeosporioides	Papaya fruits
Venturia inequalis	Apple seedling leaves
Colletotrichum capsici	Peper fruits
Colletotrichum graminicola	Corn seedling leaves

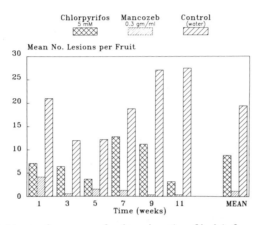

Figure 2: Protection of papaya fruits in the field from Colletotrichum gloeosporioides by a cutinase inhibitor. Chlorpyrifos, an organic phosphate inhibitor of cutinase, a fungicide (Mancozeb) or water was sprayed on the growing fruits on a biweekly interval and after the fruits reached maturity, the lesions observed were recorded (W. Nishijima, M.B. Dickman, S. Patil and P.E. Kolattukudy, unpublished).

glucose-grown F. solani pisi allowed us to isolate the mRNA and develop a cell-free translation system (10). A cDNA library was prepared and the clones unique to the induced culture was screened by hybrid-selected translation and cutinase cDNA was sequenced (11) (Fig. 3). The amino acid sequence of nearly 40% of the protein was determined by sequencing peptides derived from the enzyme including those derived from the active sites. The nucleotide sequence agreed completely with the amino acid sequence. More recently, the cDNA for cutinase from Colletotrichum capsici was cloned and sequenced, thus allowing us to get the first glimpse of how cutinases from different genera of fungi might be related (W.F. Ettinger and P.E. Kolattukudy, manuscript in preparation). As seen in Figure 3, there are regions of homology between the two enzymes but there are considerable differences. The active serine containing region is highly conserved. The aspartic acid residues which contribute the COOH group to the catalytic triad are located in nearly the same place in both enzymes. The position of the histidine residue involved in catalysis is conserved. The positions of the two disulphide bridges, that are crucial in keeping the catalytic triad residues in proper juxtaposition, are also

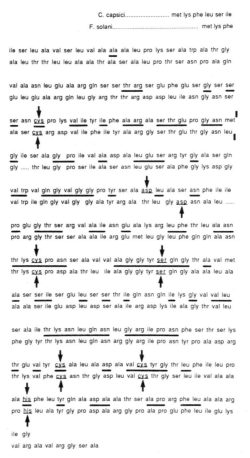

Figure 3: Amino acid sequences of cutinases from a Colletotrichum and a Fusarium species deduced from the nucleotide sequence of cloned cDNA. Regions of complete homology are shown by the lines. The aspartic acid, serine and histidine residues indicated by the arrows are the members of the catalytic triad and the position of the cys residues are also indicated by arrows. The two cys residues closest to the carboxyl terminus are in disulphide bridge with each other and so are the other two cys residues. The positions of the intron which interrupts the coding sequences in the genes are indicated by the vertical lines.

highly conserved. The other regions that are presumably not crucial for the function of the enzyme have diverged considerably.

CLONING OF CUTINASE GENE

Using the cDNA as a probe libraries of genomic DNA from Fusarium solani pisi have been screened and the cutinase gene has been isolated and sequenced (C.L. Soliday and P.E. Kolattukudy, manuscript in preparation). The open reading frame completely corresponded to the cDNA and was interrupted by a single 51 bp intron (Fig. 4). This intron has a considerable degree of homology with introns in other fungal genes. More recently, the cutinase gene from Colletotrichum capsici was cloned and sequenced (W.F. Ettinger and P.E. Kolattukudy, unpublished). This gene

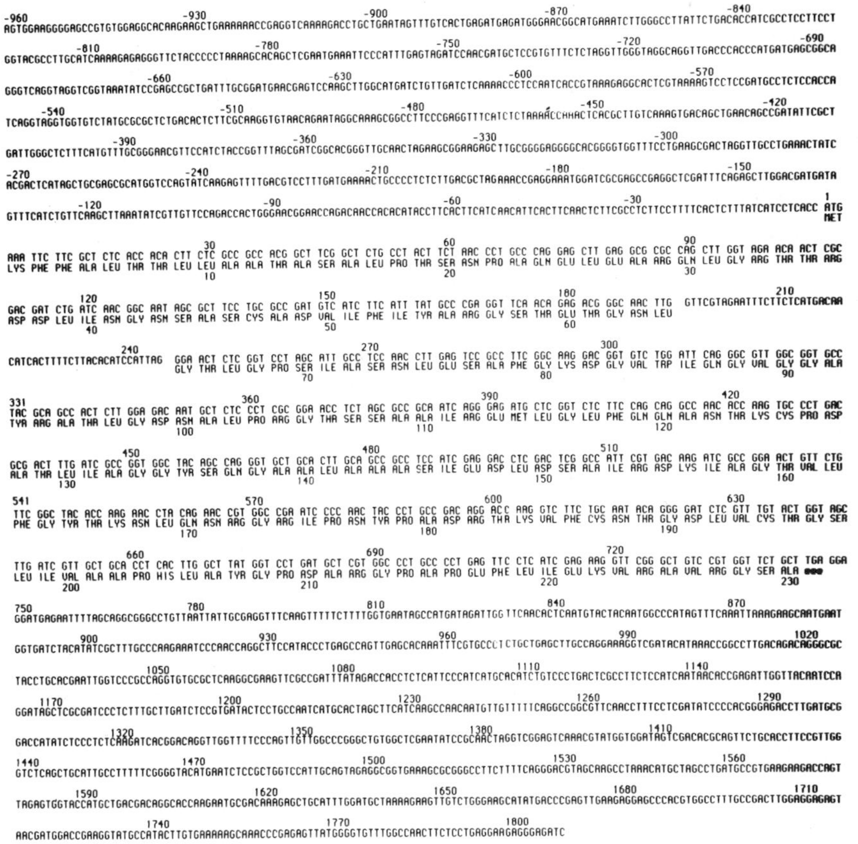

Figure 4: The nucleotide sequence of cutinase gene from Fusarium solani pisi. Southern blots show that this isolate contains another copy of the gene. However, the copy shown above is completely homologous with the cloned cDNA which would code for a protein with the primary structure that matches the major cutinase isozyme secreted by the fungus.

also contained one intron of about the same size and at about the same position as the intron found in the Fusarium gene. The putative promoter segment from the 5'-region of the cutinase gene from F. solani pisi has been used to transform F. solani pisi and Colletotrichum capsici with a hygromycin resistance marker gene (M. Dickman and P.E. Kolattukudy, unpublished). With this promoter assay, the regulatory region of the gene is being defined.

HYDROXY-FATTY ACIDS TRIGGER THE EXPRESSION OF CUTINASE GENE IN FUNGAL SPORES

Cutinase is induced in the spores of F. solani pisi by cutin, the insoluble polymer (12). This observation, together with the finding that spores of highly virulent isolates of F. solani pisi contain small amounts of cutinase (13), suggested that the actual inducer of cutinase might be the small amounts of cutin monomers generated by the enzyme carried by the spores. In fact, chemically prepared cutin hydrolysate and monomers isolated from it induced cutinase in Fusarium spores (Fig. 5). The most unique monomers of cutin, 10,16-dihydroxy-C_{16} acid and 9,10,18-trihydroxy-C_{18} acid, that are found nowhere else in nature, were found to be the most potent inducers of cutinase. The enzyme began to appear in the medium

60 minutes after the spores contacted cutin or 45 minutes after the addition of the isolated monomer (Fig. 6). With ^{32}P-labeled cutinase cDNA as a hybridization probe cutinase gene transcripts could be detected within 15 minutes after the spores contacted cutin. Nuclear runoff experiments showed that the hydroxy fatty acids unique to cutin dramatically increased the transcription rate of cutinase gene (M. Dickman, C.X. Huang and P.E. Kolattukudy, unpublished). These results clearly show that the hydroxy-fatty acids play a crucial role in the interaction between pathogenic fungi and their hosts and these lipids trigger the expression of specific genes involved in such interactions.

CUTINASE IN PHYLLOSPHERIC BACTERIA

The report that phyllospheric bacteria can at least partially substitute for nitrogen furtilizers (14) prompted us to test whether these bacteria can use cuticular components as carbon source. When such a

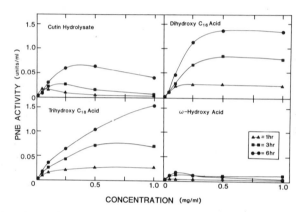

Figure 5: Induction of cutinase in the spores of <u>Fusarium solani pisi</u> by cutin hydrolysate and monomer fractions isolated from it by thin-layer chromatorgraphy. Paranitrophenyl butyrate (PNB) hydrolyzing activity of the medium was measured after 1, 3 or 6 hours of incubation.

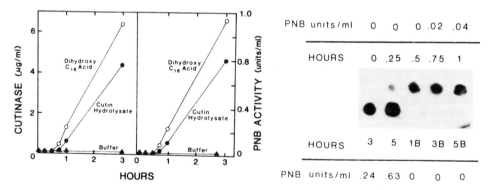

Figure 6: Induction of cutinase in <u>Fusarium solani pisi</u> by cutin hydrolysate or 10,16-dihydroxy-C_{16} acid isolated from it. On the left, the levels of immunologically measured cutinase protein and the spectrophotometrically measured catalytic activity are shown. On the right, the levels of cutinase transcripts as measured by dot blot assays with ^{32}P-labeled cutinase cDNA are shown. B indicates buffer containing no cutin.

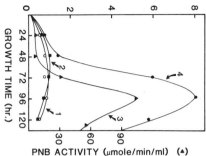

Figure 7: Induction of cutinase activity in Pseudomonas putida cultures in nutrient broth-yeast extract with or without cutin hydrolysate (1 and 2) and with cutin (3). Tracing 4 indicates release of labeled monomers from ^{3}H-labeled cutin.

culture was examined, two morphologically distinguishable bacterial species were found (Joseph Sebastian and P.E. Kolattukudy, manuscript submitted). One was a Corynebacterium sp. that could grow in a nitrogen-free medium but could not produce cutinase and the other was a strain of Pseudomonas putida which could not grow on nitrogen-free medium but could be induced to generate cutinase. Both together could grow without exogenous fixed nitrogen using cutin as the carbon source. The two bacterial species cohabit the phyllosphere and presumably, one provides fixed nitrogen while the other generates the carbon source from the plant cuticle for the two to grow together while providing fixed nitrogen to the host plants.

Induction of cutinase in P. putida requires cutin (Fig. 7) and is not under catabolite repression, unlike the fungal induction which is catabolite repressed (15). The bacterial cutinase was purified to homogeniety (Joseph Sebastian and P.E. Kolattukudy, unpublished). It has a molecular weight of about 30,000 and is immunologically unrelated to the fungal enzyme. However, the catalytic properties such as the basic pH optimum appear to be similar to those of the fungal enzyme. The bacterial cutinase gene has been cloned and preliminary results suggest that certain key regions such as that around the active serine might have homology with the fungal enzyme.

CONCLUSION

Since the outer layer of plants is derived entirely from lipids, it would not be surprising to find that these boundary lipids play significant roles in the interaction of plants with microbes. As indicated in this paper, we are beginning to get concrete examples of such roles. The composition of the complex mixture of cuticular lipids and the lipid-derived polymers show a high degree of species specificity and these components might play a role in recognition between plants and microbes. Since enzymatic release of polymer-bound components can take place during the early phases of interaction between plants and microbes, such released components might be used as agents to trigger other events in the microbes such as triggering of genes involved in penetration in fungi, virulence genes in bacteria, etc. or defense genes in the host plants. The recent report that flavanoids trigger bacterial nodulation genes (16) and our observation that cutin monomers trigger induction of genes involved in fungal penetration into their hosts (12) might be the first demonstrated examples of such interactions among plants and microbes and such information would enable us to manage such interactions for the benefit of man.

Acknowledgement

The work described in this review was supported in part by a grant, DMB 8306835, from the National Science Foundation and conducted at the Institute of Biological Chemistry, Washington State University, Pullman, WA.

References

1. van den Ende, G., and Linskens, H.F. (1974) Cutinolytic enzymes in relation to pathogenesis. Ann. Rev. Phytopathol. 12, 247-258.
2. Purdy, R.E., and Kolattukudy, P.E. (1973) Depolymerization of a hydroxy-fatty acid biopolymer, cutin, by an extracellular enzyme from Fusarium solani f. pisi: Isolation and some properties of the enzyme. Arch. Biochem. Biophys. 159, 61-69.
3. Purdy, R.E. and Kolattukudy, P.E. (1975) Hydrolysis of plant cuticle by plant pathogens. Purification, amino acid composition and molecular weight of two isozymes of cutinase and a non-specific esterase from Fusarium solani f. pisi. Biochemistry. 14, 2824-2831.
4. Purdy, R.E., and Kolattukudy, P.E. (1975) Hydrolysis of plant cuticle by plant pathogens. Properties of cutinase I, cutinase II, and a non-specific esterase isolated from Fusarium solani pisi. Biochemistry. 14, 2832-2840.
5. Kolattukudy, P.E. (1984) Cutinases from fungi and pollen. In: "Lipolytic Enzymes," ed. B. Borgstrom, H. Brockman. pp. 471-504. Elsevier/North Holland, Amsterdam.
6. Kolattukudy, P.E. (1980) Cutin, suberin and waxes. In: "The Biochemistry of Plants," ed. P.K. Stumpf. 4, 571-645. Academic Press, New York.
7. Koller, W., and Kolattukudy, P.E. (1982) Mechanism of action of cutinase: Chemical modification of the catalytic triad characteristic for serine hydrolases. Biochemistry. 21, 3083-3090.
8. Shaykh, M., Soliday, C.L., and Kolattukudy, P.E. (1977) Proof for the production of cutinase by Fusarium solani f. pisi during penetration into its host, Pisum sativum. Plant Physiol. 60, 170-172.
9. Maiti, I.B., and Kolattukudy, P.E. (1979) Prevention of fungal infection of plants by specific inhibition of cutinase. Science. 205, 507-508.
10. Flurkey, W.H., and Kolattukudy, P.E. (1981) In vitro translation of cutinase mRNA: Evidence for a precursor form of an extracellular fungal enzyme. Arch. Biochem. Biophys. 212, 154-161.
11. Soliday, C.L., Flurkey, W.H., Okita, T.W., and Kolattukudy, P.E. (1984) Cloning and structure determination of cDNA for cutinase, an enzyme involved in fungal penetration of plants. Proc. Natl. Acad. Sci. USA. 81, 3939-3943.
12. Woloshuk, C.P., and Kolattukudy, P.E. (1986) Mechanism by which contact with plant cuticle triggers cutinase gene expression in the spores of Fusarium solani f. sp. pisi. Proc. Natl. Acad. Sci. USA. 83, 1704-1708.
13. Koller, W., Allan, C.R., and Kolattukudy, P.E. (1982) Role of cutinase and cell wall degrading enzymes in infection of Pisum sativum by Fusarium solani f. sp. pisi. Physiol. Plant Pathol. 20, 47-60.
14. Pati, B.R., and Chandra, A.K. (1981) Effect of spraying nitrogen-fixing phyllospheric bacterial isolates on wheat plants. Plant and Soil. 61, 419-427.
15. Lin, T.S., and Kolattukudy, P.E. (1978) Induction of a polyester hydrolase (cutinase) by low levels of cutin monomers in Fusarium solani f. sp. pisi. J. Bacteriol. 133, 942-951.
16. Peters, N.K., Frost, J.W. and Long, S.R. (1986) Leuteolin, a plant flavone induces expression of Rhizobium meliloti nodulation genes. Science. 233, 977-980.

PLANT ELONGASES

Claude Cassagne, René Lessire, Jean-Jacques Bessoule and Patrick Moreau

Institut de Biochimie Cellulaire et Neurochimie du CNRS
1 rue Camille Saint-Saëns
33077 Bordeaux cedex, France

INTRODUCTION

In the early sixties, several investigators demonstrated the elongation of liver microsomal fatty acids in the presence of malonyl-CoA, ATP and reducing equivalents in the form of NADPH and/or NADH (see for example, Lorch et al., 1963). Furthermore, elongation was shown to occur in the absence of ATP, provided that a fatty acyl-CoA was added to the reaction mixture (Nugteren, 1965). The general mechanism of microsomal fatty acid elongation was supposed to be similar to that of de novo synthesis : condensation of the preferred substrate to form a beta-ketone, reduction of the beta-ketone to a secondary alcohol, dehydration of the alcohol to form a trans-alpha-beta-double-bond and reduction of the double-bond. Besides this microsomal elongation, also observed in Brain (Bourre, 1980), a mitochondrial elongation has also been demonstrated (Wakil, 1961).

In higher plants, the in vivo biosynthesis of the aliphatic chains present in the epicuticular waxes has been largely investigated and the incorporation of exogenous C20-C24 acids into VLCFA has been observed in several tissues, leading to the conclusion that, as in mammals, VLCFA are synthesized from an as yet undefined precursor (for review, Kolattukudy et al., 1976, 1980; Von Wettstein-Knowles, 1979, 1982; Lessire et al., 1982). The synthesis was observed both in germinating cotyledons and in leaf epidermis and has been demonstrated in cell-free preparations of Pea cotyledons (Macey and Stumpf, 1968), excised epidermis of Pea leaves (Kolattukudy and Buckner, 1972) and in leek epidermis (Cassagne and Lessire, 1974). Since preparations catalyzed the elongation of endogenous acyl-moieties, the nature of the acyl-substrate remained obscure as did the nature of the elongation product. Owing to the membrane nature of the elongating systems, and because of the fact that the elongation is a multi-step process, the resolution of the questions concerning the elongation mechanism, the localization of the elongases, their number and properties as well as the intracellular fate of the VLCFA have been relatively neglected until recently. The aim of this review is to summarize the major recent break-throughs which have led to a greatly improved knowledge of the VLCFA metabolism and to a renewal of the approaches concerning the elongases. The different elongases will be discriminated successively from a metabolic point of view (nature of the substrates, nature of the products), by their intracellular localization and, finally, by their physical and chemical properties after solubilization.

SUBSTRATES OF THE ELONGASES

The VLCFA synthesis is strictly membrane-bound, uses malonyl-CoA and NADPH, or NADPH + NADH. If there is general agreement on these points, discrepancies concerning the nature of the substrate of the elongases can be observed in the literature (Table 1).

Table 1 : In vitro study of VLCFA formation in higher plants

Plant tissue	Substrate	VLCFA synthesis (nmol/mg prot/h)	Subcellular location	References
Germinating Pea cotyledons	endogenous	0.8	microsomes	11
	endogenous	0.2	microsomes	3
	stearic acid	0.3	microsomes	7
	DSPC	n.d.		
Pea epidermis	endogenous	< 0.1	microsomes	4
Leek epidermis	endogenous	0.15	20,000g S	8 & 5
	stearoyl-CoA	6	microsomes	6
	stearoyl-CoA	13.4	ER	6
Etiolated Maize coleoptiles	stearoyl-CoA	12.1	ER	9
Leek epidermis	C12-C22-CoA	1.3-4.6	microsomes	1
	endogenous	3	microsomes	1
Etiolated leek seedlings	C12-C24-CoA	2.6-9.8	microsomes	10
	endogenous	5.2	microsomes	10
Developing B.juncea seeds	oleoyl-CoA	n.d.		2
Developing meadow foam seeds (L.alba)	free fatty acid, acyl-CoAs	n.d.	homogenate	14
Developing Jojoba seeds (S.chinensis)	C18-CoA	9.5*	homogenate,	13
	C18:1-CoA	9*	12,000g	13
	C18-ACP	34*	wax pad	13
Etiolated leek seedlings	C18-CoA	12.5	ER	
	C20-CoA	3.8	Golgi	
	endogenous	12	"heavy membranes"	12

n.d. : not determined; ER : endoplasmic reticulum; *percentage of elongation of the substrate; 1. Agrawal et al., 1984; 2. Agrawal & Stumpf, 1985; 3. Bolton & Harwood, 1977; 4. Buckner & Kolattukudy, 1972; 5. Cassagne & Lessire, 1974; 6. Cassagne & Lessire, 1978; 7. Jordan & Harwood, 1980; 8. Lessire, 1973; 9. Lessire et al., 1982 a; 10. Lessire et al., 1985; 11. Macey & Stumpf, 1968; 12. Moreau, 1986; 13. Pollard et al., 1979; 14. Pollard & Stumpf, 1980.

Bolton and Harwood (1977), studying the fatty acid biosynthesis by a particulate preparation from germinating Pea, observed that the acyl-CoA thioesters apparently played no role in the C20 to C24 fatty acid formation, the addition of CoA decreasing the overall synthesis which was stimulated by ACP. However, the low level of C20-C24 fatty acid synthesis (Table 1) indicated either that the experimental conditions were far from optimal, or that germinating pea cotyledons are a poor material for the study of this synthesis. The addition of phospholipids increased the elongation (Jordan and Harwood, 1980), but to only a limited extent (less than 20% of the total fatty acid synthesis), so that the role of phospholipids as acyl-donors in elongation should not be over-estimated, at least until the modulation of the activity of the membranous elongases by the exogenous phospholipids has been investigated.

Another interesting case is that of the seeds containing erucic acid in which the waxes are stored internally (Brassica sinapsis, Crambe abyssinica, Tropaeolum majus, Limnanthes alba and Simmondsia chinensis.

Extracts of jojoba seeds (Simmondsia chinensis) readily utilized oleoyl-CoA as the precursor of the four principal unsaturated components that make up the wax esters in this tissue (Pollard et al., 1979). Stearoyl-CoA was also elongated, as was stearoyl-ACP which, in terms of percentage elongation, seemed to be a more active substrate than stearoyl-CoA (Table 1). As the jojoba seed does not contain saturated VLCFAs or alcohols, it was hypothesized that the stearoyl-CoA and the C18-ACP were not available in this tissue due to the efficiency of the stearoyl-ACP desaturase (Stumpf, 1980). In the microsomes from leek epidermal cells, it was shown that stearoyl-CoA, rather than free stearic acid or stearoyl-ACP, was a substrate of the elongase(s) (Cassagne and Lessire, 1978). The fact that free stearic acid elongation was stimulated by the addition of ATP-Mg^{2+} and CoASH, suggested that the substrate was in fact the stearoyl-CoA formed by the acyl-CoA synthetase that is present in this membrane fraction (Lessire and Cassagne, 1979). On the other hand, the replacement of CoASH by ACP from E.coli resulted in a drastic decrease of the elongation. These observations were recently confirmed and completed by an analysis of the substrate specificity of leek elongases (Agrawal et al., 1984; Agrawal and Stumpf, 1985; Lessire et al., 1985a). These authors noticed that, despite the presence of large amounts of unsaturated C18 fatty acids in the microsomes (C18:1, 21%, C18:2 50%), only the saturated fatty acids were selected for elongation in the leek system, and it was shown that this is due to the fact that the microsomal leek elongases accept saturated acyl-CoAs, but not oleoyl-CoA, C18:2-CoA or C18:3-CoA. Surprisingly, elaidoyl-CoA, as well as C20:1 (11 cis)CoA and C22:1 (13 cis)CoA, were accepted by the leek elongases, indicating that the elongation system leading from C20 to C22-C26 fatty acids, did not have any preference between saturated and unsaturated acyl-CoAs. In this latter case, the specificity observed in vivo for saturated fatty acids, is simply due to the absence of unsaturated C20 and C22 fatty acids in leek.

NATURE OF THE PRODUCT OF THE ACYL-CoA ELONGASES

Indirect evidence has suggested that the products of the acyl-CoA elongase were also acyl-CoAs (Abdulkarim et al., 1982; Cassagne and Lessire, 1982; Pollard et al., 1979). Using a new separation technique (Juguelin and Cassagne, 1984), Lessire et al. (1985 b) demonstrated that, in the early stages of the elongation process by leek seedling microsomes (5 or 10 min), all the neosynthesized VLCFA were found in the acyl-CoAs (Table 2) and appeared only later in the other lipid classes (phosphatidylcholine -PC- and neutral lipids). It was also demonstrated that there is a net acyl-transfer from acyl-CoA to PC and not an acyl-exchange as shown, for example, in the microsomes from safflower seeds (Stymne et al., 1983). Thus, although one cannot entirely rule out the possibility that elongation takes place with undetected short-lived intermediates (possibly phospholipids), it is highly likely that the products of the elongase are acyl-CoAs. This result is to be compared with those already reported for Pea cotyledons (Bolton and Harwood, 1977; Sanchez and Harwood, 1981) where the VLCFA were found in the acyl-lipids. In the latter case, no firm conclusion should be drawn as regards the nature of the product of the elongase, as the determination was made after a 4 hour incubation.

Table 2 : Products of leek elongases

Elongases	Preferred substrate	Products	C_2 unit addition	References
Acyl-CoA elongases	C18-CoA C20-CoA	C20-C22-CoA C22-C26-CoA	multiple multiple	3 & 4
ATP-dependent elongases	endogenous	FFA ?	unique	1 & 2

FFA : free fatty acids; 1. Agrawal et al., 1984; 2. Lessire et al., 1985; 3. Lessire et al., 1985 a; 4. Lessire et al., 1985 b.

INTRACELLULAR LOCALIZATION OF THE ELONGASES

A significant amount of indirect evidence has led to the hypothesis that at least two elongases are responsible for VLCFA formation (see for review : Kolattukudy, 1976; Von Wettstein-Knowles, 1982; Lessire et al., 1982). This was particularly clear in the case of leek where, in addition to the microsomal acyl-CoA elongases already demonstrated, an ATP-dependent elongase, using an endogenous substrate, was evidenced (Agrawal et al., 1984). It was convincingly shown that the role of ATP was to form an acyl-CoA via the acyl-CoA synthetase present in leek microsomes (Lessire and Cassagne, 1979). However, the products, either free fatty acids or acyl-CoAs (Lessire, unpublished data), resulted from the addition of single C2-unit, contrasting with the multiple C2-unit additions observed with true acyl-CoA elongases (Lessire et al., 1985 a). Additional evidence was gained from the study or the localization of the various elongases.

Cassagne and Lessire (1978), and Lessire et al. (1982 b) fractionated the microsomes from leek epidermis and from etiolated Maize coleoptiles and showed that most of the C18CoA elongating activity was located in an endoplasmic reticulum-enriched fraction, whereas the plasmalemma-enriched fraction, which contained most of the VLCFA, had a low elongating activity which could be attributed to a contamination by endoplasmic reticulum (ER), or to an intrinsic plasmalemmal elongation.

Recently (Moreau, 1986; see also Moreau et al., this volume), plasmalemma was purified from etiolated leek seedlings by sucrose gradient fractionation followed by a two-phase polymer partition. The purified plasmalemma (upper phase) was practically devoid of acyl-CoA elongase and ATP-dependent elongase activities which were 150 to 600 times more active in the other fractions. The C18-CoA elongase was chiefly associated with the CDP-choline-diacylglycerol phosphotransferase (ER), whereas the C20-CoA elongase was located in the fraction exhibiting the maximal latent IDPase activity (Golgi). The ATP-dependent elongase, present all along the gradient, but chiefly in the heavier fraction containing the plasmalemma, disappeared from this membrane fraction after phase partition. This demonstrates a different localization for the different elongases. In addition, the ATP-dependent elongase does not elongate exogenous C18-CoA or C20-CoA (Table 3). Thus, if plasma membrane is devoid of elongating activity, its VLCFA have to be brought to this membrane, as hypothesized by Lessire et al. (1982).

Table 3 : Differential localization of the leek microsomal elongases

	(1) Golgi	(2)	(3) PM upper phase	(3) lower phase
C18-CoA elongase	+++	+	-	+
C20-CoA elongase	+	+++	-	+
ATP-dependent elongase	+	+	-	+++
CDP-choline-DG transferase	+++	+	-	+
Latent IDPase	+	+++	-	+
Glucane synthetase II	-	-	+++	++

(1) Membrane banding at 1.12 $g.cm^{-3}$; (2) Membrane banding at 1.14 $g.cm^{-3}$; (3) Membrane banding at 1.17 $g.cm^{-3}$.

The first demonstration of an in vivo transfer of the VLCFA came from work using etiolated leek seedlings pulse-labelled with $(1-^{14}C)$ acetate and chased for different times. By fractionation of the microsomes into 4 bands, it was shown that 60% of the total fatty acids (and 100% of the VLCFA) were transferred during the chase from the lightest to the heavier membranes (Moreau et al., 1986). Finally, an increase in the label of the VCLFA of the purified plasmalemma was observed during the chase following an in vivo labeling of the etiolated leek seedlings by (^{14}C) acetate; since the plasmalemma was devoid of any elongating activity, it follows that, unless the elongases were present in this membrane but were degraded or inhibited during the purification process, these results constitute a proof of the transfer of the VLCFA to the plasmalemma (Moreau, 1986).

SOLUBILIZATION OF THE PLANT ELONGASES (see also Lessire et al., this volume).

Very few -and unsuccessful- studies have been so far devoted to the solubilization of the microsomal acyl-CoA elongases from mammals. The addition of deoxycholate (DOC) to microsomes of rat liver resulted in the inactivation of the acyl-CoA elongases and it was postulated that separate enzymes, rather than a single multifunctional enzyme, were operative in this material (Bernert and Sprecher, 1977; 1979). The further purification of beta-OH acyl-CoA dehydrase, as well as other indirect evidence (Bernert and Sprecher, 1979) supported this view. From Table 1, it is not surprising that the first and, up to now, the only attempts in higher plants have been carried out on leek which, of all plants so far studied, exhibits the highest elongating activity.

Agrawal and Stumpf (1985), and Lessire et al. (1985), reported that DOC will at least partially inactivate the microsomal elongase, even at low concentrations when the detergent acts not as solubilizing agent, but as a membrane perturbant. However, after Triton X-100 or beta-D-octylglucoside treatment, the elongases were probably not dissociated into discrete multiple enzymes, as the activity, in both cases, was retained and led to the formation of saturated VLCFA. The elongases were partly purified by gel filtration and DEAE sephadex chromatography.

According to Hjelmeland and Chrambach (1984), two criteria have to be met before the solubilization of an enzyme may be considered as being certain :

1. retention of function in the supernatant after centrifugation at 150000g for 1 hour; obviously, this depends both on the density and on the temperature of the medium. Thus, a medium containing, for example, an excess of glycerol may render some species "soluble" by reason of "non sedimentability".

2. after gel filtration, proteins that elute in the void volume are to be considered as not certainly solubilized, whereas proteins with larger elution volumes are considered to be soluble, or solubilized.

If one accepts these criteria, the elongases from leek microsomes were clearly solubilized by Lessire et al. (1985 c). These authors studied the elongation of C18-CoA and C20-CoA. By all the criteria studied, there were two elongases separated from each other by ultracentrifugation on sucrose gradients, gel filtration, and DEAE chromatography (Table 4).

Table 4 : Properties for the solubilized acyl-CoA elongases

	C18-CoA elongase	C20-CoA elongase
Effect of cerulenin 140 µM	no effect	inhibition
Effect of Triton X-100	stimulation	stimulation
Effect of DOC	no effect	inhibition
Ultracentrifugation on sucrose gradient	0.51 M	0.62 M
Apparent molecular weight after gel filtration	350 kD	650 kD
Elution from DEAE Sephadex (NaCl, M)	0.25	0.40

As pointed out by Agrawal and Stumpf (1985), the elution of the activity as a single sharp peak on gel filtration and ion exchange chromatography, suggested, but in no way established, that all of the partial activities involved in the elongation of acyl-CoAs have to be located in a single protein, like fatty acid synthetase from Eukaryotes (Wakil and Stoops, 1983). Agrawal and Stumpf estimated an apparent molecular weight of several million kD while Lessire et al. estimated it at 350 kD and 650 kD. This discrepancy is probably due to the fact that Agrawal and Stumpf used a high percentage of glycerol, which is known to favour the aggregation of proteins. In any case, the data concerning the molecular weight should be analyzed with care since they do not account for the true value of the weight of the elongase because of the presence of detergents and the existence of mixed micelles. Bearing these restrictions in mind, the data so far available are compatible with the existence of two complexes containing all the activities required for C18-CoA and C20-CoA elongation.

PERSPECTIVES

Important break-throughs have been realized during the last years with regard to the progress of our knowledge of VLCFA. These molecules are now considered as "event markers" during the biogenesis of the plasma membrane, and the question is no longer to decide whether the intermembrane transfer takes place, but to study how it takes place. New systems have been designed allowing this study, and the progress of the techniques renders possible a real in vivo study of the intracellular pathway followed by these molecules. On the other hand, the situation of the elongases, for a long time obscured by the fact that there are numerous types within a cell, is growing more and more clear. The purification of these enzymes, which is now at hand, will allow, for the first time, the study of the various steps of the elongating process which, up to now, have only been postulated. Antibodies raised against elongases should be prepared in a near future, opening the way to a study of this multi-functional enzyme by the means of molecular biology.

REFERENCES

Abdulkarim, T., Lessire, R., and Cassagne, C., 1982, Physiol. Vég., 20:679.
Agrawal, V.P., Lessire, R., and Stumpf, P.K., 1984, Arch. Biochem. Biophys., 230:580.
Agrawal, V.P., and Stumpf, P.K., 1985 a, Lipids, 20:361.
Agrawal, V.P., and Stumpf, P.K., 1985 b, Arch. Biochem. Biophys., 240:154.
Bernert, J.T., and Sprecher, H., 1977, J. Biol. Chem., 252:6736.
Bernert, J.T., and Sprecher, H., 1979, Biochim. Biophys. Acta, 573:436.
Bolton, P., and Harwood, J.L., 1977, Biochem. J., 168:261.
Bourre, J.M., 1980, in : "Neurological mutations affecting myelination", N. Baumann, ed., Elsevier, Amsterdam, p. 187.
Cassagne, C., and Lessire, R., 1974, Physiol. Vég., 12:149.
Cassagne, C., and Lessire, R., 1978, Arch. Biochem. Biophys. 191:146.
Cassagne, C. and Lessire, R., 1982, in : "Biochemistry and metabolism of plant lipids", J.F.G.M. Wintermans and P.J.C. Kuiper, eds., Elsevier, Amsterdam, p. 79.
Hjelmeland, L.M., and Chrambach, A., 1984, in : "Methods in Enzymology", W.B. Jakoby, ed., Academic Press, New York, 104:305.
Jordan, B.R., and Harwood, J.L., 1980, Biochem. J., 191:791.
Juguelin, H., and Cassagne, C., 1984, Anal. Biochem., 142:335.
Kolattukudy, P.E., and Buckner, J.S., 1972, Biochem. Biophys. Res. Commun., 46:801.
Kolattukudy, P.E., 1976, in : "Chemistry and Biochemistry of natural waxes, P.E. Kolattukudy, ed., Am. Elsevier, New York, pp. 1-15
Kolattukudy, P.E., 1980, in : "The biochemistry of plants", P.K. Stumpf and E.E. Conn, eds., Academic Press, New York, 4:571.
Lessire, R., 1973, Thèse de 3e Cycle, Bordeaux.
Lessire, R., and Cassagne, C., 1979, Plant Sci. Lett., 16:31.
Lessire, R., Abdulkarim, T., and Cassagne, C., 1982, in : "The plant cuticle", D.F. Cutler, K.L. Alvin and C.E. Price, eds., Academic Press, London, p. 167.

Lessire, R., Hartmann-Bouillon, M.A., and Cassagne, C., 1982, Phytochemistry, 21:55.
Lessire, R., Juguelin, H., Moreau, P., and Cassagne, C., 1985 a, Phytochemistry, 24:1187.
Lessire, R., Juguelin, H., Moreau, P., and Cassagne, C., 1985 b, Arch. Biochem. Biophys., 239:260.
Lessire, R., Bessoule, J.J., and Cassagne, C., 1985 c, FEBS Lett., 187:314.
Lorch, E., Abraham, S., and Chaikoff, I.L., 1963, Biochim. Biophys. Acta, 70:627.
Macey, M.J.K., and Stumpf, P.K., 1968, Plant Physiol., 43:1637.
Moreau, P., Juguelin, H., Lessire, R., and Cassagne, C., 1986, Phytochemistry, 25:387.
Moreau, P., 1986, Ph.D. Thesis, Bordeaux.
Nugteren, D.H., 1965, Biochim. Biophys. Acta, 106:280.
Pollard, M.R., Mc Keon, T., Gupta, L.M., and Stumpf, P.K., 1979, Lipids, 14:651.
Sanchez, J., and Harwood, J.L., 1981, Biochem. J., 199:221.
Stumpf, P.K., in : "The biochemistry of plants", P.K. Stumpf, and E.E. Conn, eds., Academic Press, New York, 4:177.
Stymne, S., Stobart, A.K., and Glad, G., 1983, Biochim. Biophys. Acta, 752:198.
Wakil, S.G., 1961, J. Lipid Res., 2:1.
Wettstein-Knowles, P. von, 1979, in "Advances in the biochemistry and physiology of plant lipids", L.A. Appleqvist and C. Liljenberg, Elsevier/North-Holland Biomedical Press, Amsterdam, 1.
Wettstein-Knowles, P. von, 1982, Physiol. Vég., 20:797.

GENES, ELONGASES AND ASSOCIATED ENZYME SYSTEMS IN EPICUTICULAR WAX SYNTHESIS

Penny von Wettstein-Knowles

Institute of Genetics, University of Copenhagen, and
Department of Physiology, Carlsberg Laboratory,
Gamle Carlsberg Vej 10, DK-2500 Copenhagen, Denmark

INTRODUCTION

Elongases are enzyme complexes which condense short carbon chains to a primer and prepare the growing chain for the next addition. Soluble plastid fatty acid synthetase (FAS) in higher plants is a special example in which the initial primer is acetyl-acyl carrier protein (ACP), the donor of C_2-units is malonyl-ACP and the product palmityl-ACP. Addition of another C_2-unit to give stearoyl-ACP is not accomplished by FAS but by the soluble plastid palmityl elongase[1,2]. Epidermal cells of leek appear to lack the latter complex[3]. Other elongases are generally believed to be located within the epidermal cells wherein they are affiliated with or are part of the microsomal membranes[4,5,6,7]. Such complexes carry out the elongation steps required to synthesize the very long chains characteristic of the epidermal waxes. Coenzyme A (CoA) rather than ACP derivatives are thought to serve as substrates for the elongases[4,5,7,8]. The latter fall into two groups depending on whether they use acyl-CoA or β-ketoacyl-CoA chains as primers[9,10,11]. Before arriving on the outermost surface of the cuticle wall, the long acyl-CoA chains normally enter an associated enzyme system. Attention has been focused on two such complexes using acyl elongase products[4]. The reductive system yields aldehydes plus free and esterified primary alcohols. The decarboxylative system, named for the apparent decarboxylation that occurs, gives rise to the hydrocarbons, secondary alcohols and ketones. Elongated acyl-CoA compounds not entering one of the associated pathways may appear in the epicuticular wax as free acids. While these are interpreted as left-overs[4], they are occassionally the major lipid class in a wax as, for example, on leaf sheaths of sorghum[12]. By contrast, such left-overs have not been detected coming from β-ketoacyl elongase products[13]. All the latter enter one of two associated enzyme systems which will be detailed below.

Mutations in genes potentially affecting wax synthesis and/or its deposition have been isolated in a number of species because of an alteration in the phenotype of the cuticle surface[4]. I do not intend to catalogue these genes herein. Rather, I will focus on how useful they have been or might be in i) elucidating wax synthesizing mechanisms and subcellular sites, ii) providing tools for isolating the various enzymes or complexes thereof and iii) revealing the relationships among the elongases and associated enzyme systems. To accomplish this I will discuss the oor-cqu gene in barley (Hordeum vulgare), the glossy genes in maize (Zea mays) and the gl_2 gene in brussel sprouts (Brassica oleracea var. gemmifera) together with the wsp gene in peas (Pisum sativum).

THE cer-cqu GENE AND THE β-KETOACYL ELONGASE

Genetic analysis of mutations modifying the phenotype of barley spike waxes revealed the presence of three very closely linked complementation (functional) groups, cer-c, -q and -u on chromosome 4. Among the 522 mutants mapped to these three groups, 13 belonged to more than one complementation group with all possible combinations being identified. Revertants to wild type from all 13 multiple mutants have been induced with NaN_3 in frequencies from 1.3-15.1 x 10^{-5}. These results demonstrate that cer-c, -q and -u correspond to three functional domains of a polypeptide which is determined by the gene designated cer-cqu. Furthermore, this polypeptide functions as a multimer with a minimum of two units. Thus far, nine of the 36 wild type revertants have been shown to arise from mutational events within the cer-cqu gene and one to mutation of a non-linked gene to a dominant suppressor[11,14,15].

Comparison of the wax composition on various organs of wild type and cer barleys including a mutant belonging to each of the specified complementation groups led to two deductions[10,16]. Firstly, barley wax lipids are synthesized by two parallel elongation systems. The acyl elongase system(s) produce hydrocarbons, aldehydes, primary alcohols, fatty acids and esters[4]; the β-ketoacyl elongase system, the β-diketones, hydroxy-β-diketones, alkan-2-ol esters and 7-oxoalkan-2-ol esters (Fig. 1). Secondly, cer-c^{36}, -q^{42} and -u^{69} created blocks at three sites; cer-q^{42} and -c^{36} very early in the elongation and cer-u^{69} in the associated enzyme system inserting a hydroxyl group into the completed β-diketone molecule (Fig. 1). Subsequent analyses reveal that cer-cqu mutants can be divided into ten distinct phenotypic classes on the basis of the composition of their spike waxes[11,13]. Moreover, mutants belonging to different complementation groups can give rise to the same phenotypic class, for example, cer-q^{1459} and -u^{472}, or cer-u^{69} and -q^{678}. While these results are in accord with cer-c, -q and -u being part of a single gene, they fail to disclose any additional sites of action of this gene in the biosynthetic pathway.

The condensing activity of the β-ketoacyl elongase yields 3-oxoacyl compounds which may have one of two fates[8,10,11,17]. As shown on the left side of Fig. 1 if a C_2-unit is added to a C_{12} or C_{14}-CoA primer, the β-carbonyl is not immediately reduced. Instead an apparent decarboxylation to a methyl ketone occurs before reduction of the oxo to a hydroxyl group. The resulting C_{13} and C_{15} alkan-2-ols are invariably esterified. These three reactions appear to be tightly coupled. Thereafter an oxo group may be inserted into the ester onto carbon 7 of the C_{15}-2-ol moieties only. The right side of Fig. 1 shows that in two successive rounds of elongation when primarily C_{16}-CoA serves as the primer, reduction of the β-carbonyls does not occur in either case. During the subsequent six or five C_2-unit additions the β-carbonyls are removed presumably in a manner analogous to that in saturated fatty acid synthesis. An apparent decarboxylation yields a completed β-diketone (96% hentriacontan-14,16-dione) into which a hydroxyl group may be inserted onto carbon 25. How protection of one or two β-carbonyl groups is accomplished by the β-ketoacyl elongase is unknown. One can envisage either interaction with a prosthetic group such as copper[9] and/or the absence or inability of the requisite enzyme activities to act on the specified chain lengths even though the condensing activity is unimpeded, as explified by chalcone synthase[18] and 6-methylsalicylic acid synthetase[19], respectively. This conservation of specific β-carbonyl groups is one of the unique features of the β-ketoacyl elongase which distinguishes it from the other elongases participating in epicuticular wax biosynthesis and from plastid FAS and palmityl elongase.

Another interesting facet of the β-ketoacyl vs acyl elongases is its marked sensitivity to arsenite as illustrated in Table 1 for the esterified alkan-2-ols, and in Fig. 2 of reference 9 for the β-diketones. This difference is parallel to that between palmityl elongase and fatty acid synthetase, respectively (see reference 11). The cause of this difference in

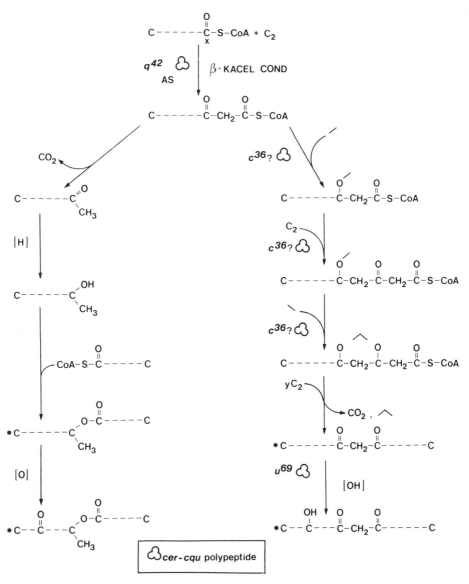

Fig. 1. Scheme for synthesis of barley 3-oxoacyl-CoA derived epicuticular lipids containing sites of action of the cer-cqu polypeptide. In alkan-2-ol ester synthesis (left) x is primarily 12 and 14, whereas in β-diketone synthesis (right) x is predominantly 16 and y is generally 6 although sometimes 5. Arsenite (AS) is presumed to inhibit the condensing enzyme of the β-ketoacyl elongase (β-KACEL COND). Retention of 3-oxo groups in two successive rounds of elongation is shown by ∕ and ∧. A * indicates that the lipid occurs in the epicuticular wax.

Table 1. Effect of Pretreatments with Arsenite on Incorporation of $(1-^{14}C)$-Myristate into Five Epicuticular Lipid Fractions by Tissue Slices from cer-c^{36} and -q^{42} Spikes Minus the Awns (cpm × 10^{-4})[20]

Arsenite (mM)	Aldehyde	Ester			Hydrocarbon
		Acid	Alkan-1-ol	Alkan-2-ol	
cer-c^{36}					
0	10.8	19.4	1.2	3.5	1.0
0.1	37.9	21.5	1.2		0.4
10	13.7	15.8	0.7		0.1
cer-q^{42}					
0	12.8	18.9	2.1		0.6
1.0	3.5	17.3	1.0		0.04

spinach is the presence in these two complexes of different condensing enzymes β-ketoacyl-ACP synthethases II and I, respectively, with the appropriate characteristics[1,2]. The β-ketoacyl elongase uses primarily C_{12}, C_{14} and C_{16} chains as primers as does synthethase II, although the former are CoA and the latter ACP derivatives. This is probably serendipitous as a comparison of characterized β-diketones from different plant species reveals that β-ketoacyl elongases can use primers of very varied chain length[11]. An analogous deduction can be drawn from the few studied alkan-2-ols. For example, in sorghum seedling leaves C_{10}-CoA ostensibly serves as a substrate for esterified nonan-2-ols in the epicuticular wax[17].

In the course of demonstrating that a 3-oxoacyl compound was the key intermediate in synthesis of β-diketone and esterified alkan-2-ol lipids (Fig. 1), the mutant cer-q^{42} was shown to be defective in the condensing activity of the β-ketoacyl elongase[8]. That is, radioactively labelled C_{16} acyl-, 3-oxoacyl- and 3-hydroxyacyl-CoAs were fed to tissue slices prepared from spikes minus awns of cer-u^{69}, -c^{36} and -q^{42} plants. The acyl- and 3-oxoacyl-CoAs served as precursors for esterified alkan-2-ols in cer-c^{36} tissue and for both esterified alkan-2-ols and β-diketones in cer-u^{69} tissue. While neither of these two potential substrates yielded labelled β-diketones with -q^{42} tissue, the 3-oxoacyl-CoA but not the acyl-CoA proved to be a good substrate for esterified alkan-2-ols. The 3-hydroxyacyl-CoA was ineffective as a precursor in all cases. That the 3-oxoacyl-CoA, which is beyond the enzymatic defect caused by cer-q^{42}, was unable to serve as a β-diketone precursor is presumably due to the fact that, in contrast to alkan-2-ol synthesis, two successive rounds of 3-oxoacyl conservation are required in β-diketone synthesis[11] (Fig. 1).

On the basis of experiments with protoplasts from epidermal cells of barley leaf sheaths, the proposal was made that the plasmalemma and/or cell wall was the site of the epicuticular wax synthesizing machinery[6]. The observations summarized above are pertinent in this respect. The cer-cqu determined polypeptide contains the enzymatic activity for the final step in the associated pathway consisting of at least the apparent decarboxylation and hydroxylation reactions. The latter would be expected to occur close to the plant surface. This is where the decarbonylation activity which is the final step in hydrocarbon synthesis in peas has been located[21]; more specifically in a cutin containing fraction of a microsomal preparation. That the condensing activity of the β-ketoacyl elongase, which is the first step in this elongation sequence, is also present in the cer-cqu polypeptide infers that acyl elongases in parallel elongation systems are similarly located. From whence come the C_{12}, C_{14} and C_{16}-CoA primers used by the β-ketoacyl elongase? Are they intermediates produced by the soluble ACP-dependent FAS studied in epidermal cells[3], products of an endoplasmic reticulum elongase[4,5] (see Fig. 6 in reference 7) or of a plasmalemma located elongase which would have the advantage of a close proximity to the other wax synthesizing enzymes?

TEN glossy GENES IN MAIZE

The composition of the wax on the first, glaucous leaves of a maize plant differs from that on the latter appearing non-glaucous leaves; designated young and mature, respectively. Young leaf wax is almost entirely composed of lipids arising from an associated reductive pathway (63% primary alcohol, 20% aldehyde and 16% ester)[22]. In mature leaf wax these account for 14, 9 and 42%, respectively, while the hydrocarbons from the decarboxylative pathway amount to 17% and free acids, presumably leftovers, to 14%[23]. Lack of data precludes speculation upon the actual or relative efficiencies of the two associated enzyme systems in the two types of leaves. The prominent hydrocarbons C_{31} and C_{29} in both leaf waxes can be accompanied by large amounts of C_{33}, but this is not consistently observed[22,24]. Almost all aldehydes and primary alcohols on young leaves have 32 carbons, whereas on mature leaves those with 26-30 carbons account for 65 and 57%, respectively. A more striking difference in the two waxes, however, is in their ester alcohol compositions (Table 2), regardless of which set of data is used for the young leaves. To account for the summarized observations, I propose that in mature leaves the reductive pathway (R_2) prefers shorter chains especially for esterification, than does the analogous system (R_1) in young leaves. The substantial proportion of free acids in the mature leaf wax suggests that R_2 is less efficient than R_1; while the long lengths of the free acids (69% C_{26}-C_{32}) and hydrocarbons (78% C_{29}-C_{33}) implies that the elongating capacity of the two leaf types is comparable.

In maize at least 10 genes are known to reduce glaucousness of young leaves[25,26]. The appropriate wax from a number of mutants has been studied in detail[22,27,28]. The potential role of the five genes gl1, gl7, gl8, gl15 and gl18 is of interest with respect to the just detailed differences in young and mature leaf waxes. The extent of wax reduction in these mutants which varies from 24-73% is not directly correlated with the large variation in the relative amounts of the reductive pathway derived lipids. Decreases in primary alcohol and aldehyde, however, are compensated for by ester increases. Comparison of the ester alcohol distributions with those of the wild type (Table 2) suggests that the reductive pathway has been altered in these five mutants. The accumulation of shorter chains cannot be attributed to a feedback from an elongation impediment since the chain length distributions of the aldehydes, primary alcohols and hydrocarbons are similar to those of the wild type. Rather, the ester alcohols on the young leaves of gl1, gl8 and gl18 appear to have been produced via an R_2-like reductive system or by a combination of the latter and R_1. A parallel can be drawn between the just described observations and the mode of action of Cer-yy in barley which turns spike wax into leaf wax[29]. In gl7 a third system (R_3) appears to function which selects short chains and in contrast to R_1 and R_2, 28 and 30 but not 32 chains. The esters of gl15 contain prominent amounts of C_{16}, C_{18} and C_{32} alcohols inferring the presence of still another system, designated R_4. Whether or not the observed differences in the ester alcohol distributions attributed to R_1-R_4 arise from changes in components of the reductive system itself and/or are the result of changes in the abutting membrane or cell wall components is of course an open question. They certainly cannot be due to production, transportation or use of substrates by the elongases as has been suggested for four of these mutants[30].

Table 2. Percent of Important Ester Alcohols from Leaf Wax of Six Maize Genotypes[a]

	Number of Carbons									References
	16	18	20	22	24	26	28	30	32	
Mature plant Wild type (R_2)				55	27			3		23
Young plant Wild type (R_1)									100	22
			13	7	13				53	24
gl1			20	11	20	26			11	22
gl8			17	19	36				9	22
gl18			16	18	33				19	22
gl7 (R_3)			12	12	26		31	11	0	22
gl15 (R_4)	14	38							39	27

[a] Four reductive pathways (R_1-R_4) are presumed to give rise to different types of homologue distributions (see text). The wild type is WF-9.

The primary defects resulting from mutations in gl2, gl3 and gl4 leading to 20-44% as much wax as on the wild type young leaves have been assigned to the terminal elongation steps[22]. Within both decarboxylative and reductive derived lipids, the increased importance of the shorter homologues can be attributed to feed-back effects within the existing pathways. But how can the increase of hydrocarbons to 12% of gl2 wax be accounted for by this hypothesis? Certainly a simple block confined within an elongase[28] fails to to explain why mutation at the gl11 locus results in such marked compositional changes of the wax which amounts to 70% of that in the wild type. Instead, I suggest that in gl11 an alteration in the terminal elongase occurs at the site normally interacting with both associated pathways in such a manner as to modify their properties also (Fig. 2). This more pliable hypothesis can explain why i) a decrease in C_{31} alkanes is accompanied by an increase of C_{21}-C_{25} homologues, ii) no aldehydes are present in the wax, iii) the 27% primary alcohols are almost all C_{32} and iv) the esters, accounting for 66% of the wax, have increased amounts of C_{20}-C_{24} and C_{30} alcohol moieties at the expense of the C_{32} ones. Which and how many of the terminal elongase components do the genes gl2, gl3, gl4 and gl11 represent? Transposon tagging[31] could be used to isolate them. Nucleotide sequencing thereafter would reveal whether any duplicated genes were included. This is possible as gl3 and gl4 are on chromosome 4 and the other two genes on chromosome 2.

In contrast to the hitherto mentioned maize glossy genes, gl5 is unique in that its precise role in wax synthesis is clearly revealed from wax compositional analyses[22]. Mutation of gl5 inhibits reduction of aldehydes to primary alcohols (Fig. 2), and thereby is a potential tool for identifying the deduced aldehyde reductase in protein extracts.

ELONGASES AND ACYL-CoA REDUCTASES

The pathway(s) by which elongated chains lose a carbon and enter the decarboxylative pathway has proved extraordinarily obdurate to experimentation. In 1984, however, a cutin containing fraction from a pea microsomal preparation was shown to convert a C_{18} aldehyde into a C_{17} alkane apparently via a decarbonylation reaction[21]. Furthermore, the original microsomal preparation and the first particle fraction prepared therefrom catalyzed the reduction of acyl CoAs to the appropriate aldehyde. Given this acid to aldehyde to alkane sequence defects in the decarbonylation reaction might result in accumulation of aldehydes in a wax. While this has been asserted to happen in a number of instances where waxes with dominating amounts of the decarboxylative lipids have been studied[32,33,34,35], I believe this to be true in only one case. In the wax of the gl$_2$ mutant in brussel sprout[33] C_{30} aldehydes account for ≃ 14.5 µg/cm^2 and the 29 carbon alkanes, secondary alcohols and ketones for 4.30 µg/cm^2, whereas in the wild type the corresponding values are trace and 45. If a cutin containing microsomal fraction from the wild type but not gl$_2$ was able to carry out the decarbonylation reaction, this might prove useful in identifying the decarbonylase protein. In peas a related situation deserves exploration for the same reason. Namely, while 10.7 µg of alkanes are present/cm^2 of lower epidermis, only 1.3 µg occur/cm^2 of upper epidermis. The two epidermal cell layers can be readily separated from one another almost free of mesophyll tissue in the pea Arg mutant[36]. At the lipid class level the Arg mutant does not induce any significant changes in wax composition[13]. In addition on the respective leaf surfaces of the pea mutant wsp, the alkanes are reduced to 0.2 and 0.04 µg/cm^2 in the absence of an accumulation of aldehydes[35] suggesting that mutation of this gene produces a defective acyl-CoA reductase but normal decarbonylase. This hypothesis could be readily tested using lower epider-

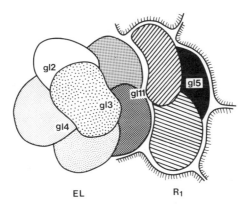

Fig. 2. Schematic presentation of the terminal elongase (EL) components with the associated reductive pathway (R_1) components within the epidermal plasmalemma and/or cell walls of young maize leaves. The studied mutations at gl2, gl3, and gl4 modify EL components. Although all three are positioned within EL as if their effects were entirely confined therein, this is indubitably an oversimplification in the case of gl2 (see text). The mutation at gl11 differs from the others in occurring at the EL site normally interacting with associated enzyme systems so as to alter their properties also. The aldehyde reductase of R_1 is affected by the gl5 mutation.

mal peels from Arg wsp plants, whose wax at the lipid class level is also apparently the same as on wsp plants[13].

Over the years observations have accrued indicating that an acyl-CoA reductase participates in synthesis of the aldehyde precursors of the primary alcohols arising via the reductive pathway[4,22,37]. This enzyme cannot be the one participating in alkane formation as on the lower and upper leaf surfaces of wsp pea plant leaves, primary alcohols are present (0.3 and 9.0 µg/cm^2) in wild type amounts (0.7 and 7.1 µg/cm^2), respectively. A second line of evidence supporting the notion of two acyl-CoA reductases comes from experiments with inhibitors and radiotracers. Early studies with pea tissue slices led to the deduction that the tested thiols inhibited the apparent decarboxylation step[32]. Since the decarbonylase reaction is reported to be insensitive to dithioerythritol[21], one is forced to conclude that this thiol inhibits the alkane yielding acyl-CoA reductase. The analogous primary alcohol yielding enzyme is not affected by dithiotreitol as the following data illustrate. Incorporation of label from ^{14}C-acetate into alkanes was reduced from 352 to 32 x 10^3 cpm with a simultaneous increase from 262 to 457 x 10^3 cpm split between the aldehydes and primary alcohols when barley spikes were pretreated with dithiothreitol[9,38]. The thiol 2-mercaptoethanol is now known to act at the C_{20-22} elongation step and not at the apparent decarboxylation step as originally believed[38].

Decarboxylation reactions have also been invoked in the synthesis of two other wax lipids, the β-diketones and esterified alkan-2-ols (Fig. 1). Their synthesis as that of the primary alcohols is not inhibited by dithiothreitol[9,13,38]. Thus we have the interesting possibility that if a decarbonylase reaction is involved as in alkane synthesis, the required acyl-CoA reductase must be insensitive to dithiothreitol as is that in primary alcohol synthesis. On the other hand, perhaps true decarboxylations take place as has been shown for hydrocarbon synthesis in termites[39].

REFERENCES

1. T. Shimakata and P.K. Stumpf, Fatty acid synthetase of Spinacia oleracea leaves, Plant Physiol. 69:1257 (1982).
2. T. Shimakata and P.K. Stumpf, Purification and characterization of β-ketoacyl-ACP synthetase 1 from Spinacia oleracea leaves, Arch. Biochem. Biophys. 220:39 (1983).
3. R. Lessire and P.K. Stumpf, Nature of the fatty acid synthetase systems in parenchymal and epidermal cells of Allium porrum L. leaves, Plant Physiol. 73:614 (1983).
4. P. von Wettstein-Knowles, Genetics and biosynthesis of plant epicuticular waxes, in: "Advances in the Biochemistry and Physiology of Plant Lipids," L.-Å. Appelquist and C. Liljenberg, eds., Elsevier/North-Holland Biomedical Press, Amsterdam (1979).
5. C. Cassagne and R. Lessire, Biosynthesis of saturated very long chain fatty acids by purified membrane fractions from leek epidermal cells, Arch. Biochem. Biophys. 191:146 (1978).
6. J.D. Mikkelsen, Synthesis of lipids by epidermal and mesophyll protoplasts isolated from barley leaf sheaths, in: "Biogenesis and Function of Plant Lipids," P. Mazliak, P. Benveniste, C. Costes, and R. Douce, eds., Elsevier/North-Holland Biomedical Press, Amsterdam (1980).
7. V.P. Agrawal, R. Lessire, and P.K. Stumpf, Biosynthesis of very long chain fatty acids in microsomes from epidermal cells of Allium porrum L., Arch. Biochem. Biophys. 230:580 (1984).
8. J.D. Mikkelsen, Biosynthesis of esterified alkan-2-ols and β-diketones in barley spike epicuticular wax: synthesis of radioactive intermediates, Carlsberg Res. Commun. 49:391 (1984).

9. J.D. Mikkelsen and P. von Wettstein-Knowles, Biosynthesis of β-diketones and hydrocarbons in barley spike epicuticular wax, Arch. Biochem. Biophys. 188:172 (1978).
10. P. von Wettstein-Knowles, Biosynthetic relationships between β-diketones and esterified alkan-2-ols deduced from epicuticular wax of barley mutants, Molec. gen. Genet. 144:43 (1976).
11. P. von Wettstein-Knowles, Role of cer-cqu in epicuticular wax biosynthesis, Biochem. Soc. Tran. (in press) (1986).
12. P. Avato, G. Bianchi, and G. Mariani, Epicuticular waxes of Sorghum and some compositional changes with plant age. Phytochemistry 23:2843 (1984).
13. P. von Wettstein-Knowles, unpublished.
14. P. von Wettstein-Knowles and B. Søgaard, Genetic evidence that cer-cqu is a cluster-gene, in: "Barley Genetics IV," Edinburgh University Press, Edinburgh (1981).
15. B. Søgaard and P. von Wettstein-Knowles, unpublished.
16. P. von Wettstein-Knowles, Genetic control of β-diketone and hydroxy-β-diketone synthesis in epicuticular waxes of barley, Planta 106:113 (1972).
17. P. von Wettstein-Knowles, J.D. Mikkelsen, and J.Ø. Madsen, Nonan-2-ol esters in sorghum leaf epicuticular wax and their collection by preparative gas chromatography, Carlsberg Res. Commun. 49:611 (1984).
18. R. Schüz, W. Heller, and K. Hahlbrock, Substrate specificity of chalcone synthase from Petroselinum hortense, J. Biol. Chem. 258:6730 (1983).
19. F. Lynen, H. Engeser, J. Friedrich, W. Schindlbeck, R. Seyffert, and F. Wieland, Fatty acid synthetase of yeast and 6-methylsalicylate synthetase of Penicillium patulum - two multienzyme complexes, in: "Microenvironments and Metabolic Compartmentation," P.A. Srere and R.W. Estabrook, eds., Academic Press, New York (1978).
20. P. von Wettstein-Knowles, Effects of inhibitors on synthesis of esterified alkan-2-ols in barley spike epicuticular wax, Carlsberg Res. Commun 50:239 (1985).
21. T.M. Cheesbrough and P.E. Kolattukudy, Alkane biosynthesis by decarbonylation of aldehydes catalyzed by a particulate preparation from Pisum sativum, Proc. Natl. Acad. Sci. USA 81:6613 (1984).
22. G. Bianchi, Genetic control of composition of epicuticular waxes of maize: a survey, Genet. Agr. 33:75 (1979).
23. G. Bianchi, P. Avato, and F. Salamini, Surface waxes from grain, leaves, and husks of maize (Zea mays L.), Cereal Chem. 61:45 (1984).
24. P. Avato, J.D. Mikkelsen, and P. von Wettstein-Knowles, Effect of inhibitors on synthesis of fatty acyl chains present in waxes on developing maize leaves, Carlsberg Res. Commun. 45:329 (1980).
25. Maize Genetics Cooperation Newsletter, 60:150-169 (1986).
26. M.G. Neuffer, L. Jones, and M.S. Zuber, "The Mutants of Maize," Crop Science Society of America, Madison (1968).
27. G. Bianchi, P. Avato, and F. Salamini, Glossy mutants of maize IX. Chemistry of glossy 4, glossy 8, glossy 15 and glossy 18 surface waxes, Heredity 42:391 (1979).
28. P. Avato, G. Bianchi, and F. Salamini, Absence of long chain aldehydes in the wax of the glossy 11 mutant of maize, Phytochemistry 24:1995 (1985).
29. U. Lundqvist and P. von Wettstein-Knowles, Dominant mutations at Cer-yy change barley spike wax into leaf blade wax, Carlsberg Res. Commun. 47:29 (1982).
30. A. Bianchi, G. Bianchi, P. Avato, and F. Salamini, Biosynthetic pathways of epicuticular wax of maize as assessed by muation, light, plant age and inhibitor studies, Maydica 30:179 (1985).
31. N.V. Fedoroff, D.B. Furtek, and O.E. Nelson, Cloning of the bronze locus in maize by a simple and generalizable procedure using the transposable controlling element Activator (AC), Proc. Natl. Acad. Sci. USA 81:3825 (1984).

32. J.S. Buckner and P.E. Kolattukudy, Specific inhibition of alkane synthesis with accumulation of very long chain compounds by dithioerythritol, dithiothreitol, and mercaptoethanol in Pisum sativum, Arch. Biochem. Biophys. 156:34 (1973).
33. E.A. Baker, The influence of environment on leaf wax development in Brassica oleracea var. gemmifera, New Phytol. 73:955 (1974).
34. P.J. Holloway, G.A. Brown, E.A. Baker, and M.J.K. Macey, Chemical composition and ultrastructure of the epicuticular wax in three lines of Brassica napus (L), Chem. Phys. Lipids 19:114 (1977).
35. P.J. Holloway, G.A. Brown, E.A. Baker, and M.J.K. Macey, Chemical composition and ultrastructure of the epicuticular wax in four mutants of Pisum sativum (L), Chem. Phys. Lipids 20:141 (1977).
36. H.C. Hoch, C. Pratt, and G.A. Marx, Subepidermal air spaces: basis for the phenotypic expression of the Argentum mutant of Pisum, Amer. J. Bot. 67:905 (1980).
37. P.E. Kolattukudy, Enzymatic synthesis of fatty alcohols in Brassica oleracea, Arch. Biochem. Biophys. 142:701 (1971).
38. J.D. Mikkelsen, The effects of inhibitors on the biosynthesis of the long chain lipids with even carbon numbers in barley spike epicuticular wax, Carlsberg Res. Commun. 43:15 (1978).
39. A.J. Chu and G.J. Blomquist, Decarboxylation of the tetracosanoic acid to n-tricosane in the termite Zootermopsis angusticollis, Comp. Biochem. Physiol. 66B:313 (1980).

THE PURIFICATION OF ACETYL-CoA:ACYL CARRIER PROTEIN TRANSACYLASE

FROM BRASSICA CAMPESTRIS LEAVES

Ann-Marie A. Wolf[1] and John T. Perchorowicz[2]

Calgene, Inc.
1920 Fifth Street
Davis, CA 95616 USA

INTRODUCTION

In recent years, the molecular organization of the plant fatty acid synthetases has been examined in extracts from various plant tissues, and in all cases was found to be nonassociated and similar to the prokaryotic type of E. coli (1,2,3). Many of the individual component enzymes have been partially purified including malonyl-CoA:acyl carrier protein (ACP) transacylase (4), β-ketoacyl-ACP reductase (5,6), β-hydroxylacyl-ACP dehydrase (6), enoyl ACP reductase (5,6) and acetyl-CoA:ACP transacylase (ATA) (1). ATA has the lowest specific activity in comparison with the other enzymes in the system and has been reported as catalyzing the rate-limiting step in the plant fatty acid synthetase (FAS) system: the thioester transfer reaction between acetyl-CoA and acyl carrier protein to form acetyl-ACP.

ATA has previously been partially purified from spinach leaf tissue (1). In the current study, ATA was purified to homogeneity from Brassica campestris using $(NH_4)_2SO_4$ fractionation, gel filtration and chromatofocusing. During the course of the purification, two distinct forms of ATA were found: ATA_1 which elutes at a pH of 7.2 from the chromatofocusing column and ATA_2 which elutes at a pH of 6.8. Amino acid composition has been determined for ATA_2 giving an approximate molecular weight of 45,000. Two isoenzymes of malonyl-CoA:ACP transacylase (MTA) (7), as well as two forms of ACP (8,9) have been reported in the literature, adding credibility to the possibility of two isoenzymes of ATA. Investigation of the chemical and physical properties of the two forms of ATA are underway. This paper describes the purification scheme of the two forms of ATA.

MATERIALS AND METHODS

Materials. $[1-^{14}C]$Acetyl-CoA (55.6 mCi/mmol) and $[2-^{14}C]$Malonyl-CoA (48.6 mCi/mmol) were purchased from New England Nuclear and Amersham, respectively. Leupeptin and ACP were obtained from Sigma. PBE, polybuffer 74, polybuffer 96 and Sephadex G-75 were obtained from Pharmacia Fine Chemicals, and dithiothreitol (DTT) was purchased from Boehringer Mannheim Biochemical. All other chemicals were of reagent grade.

---

[1]Current address: Genelabs, Inc., 8/1 Industrial Rd., Building J, San Carlos, CA 94070.
[2]To whom correspondence should be addressed.

Enzyme Assays. ATA activity was assayed as follows: The reaction mixture (50µl) contained 0.1 M Tris-HCl, pH 8.1, 50µM ACP, 5mM DTT, 30µM [1-^{14}C] acetyl-CoA and ATA (1). The ATA was preincubated with the ACP, buffer, and DTT for 15 min. at 30°C, before the reaction was started with the addition of [1-^{14}C] acetyl-CoA. The reaction was stopped after 1 or 5 min. with the addition of 250µl of ice-cold perchloric acid, and the samples were placed in an ice bath for 15 min. to ensure complete precipitation of the product. Control reactions lacking ACP but with the enzyme preparation, as well as reactions with ACP but lacking the enzyme preparation were included to check for nonspecific precipitation of the radioactive substrates. The samples were filtered through 0.45µm Millipore filters (HAWP 025) and each rinsed with approximately 100 ml of ice-cold 5% perchloric acid. ^{14}C trapped on the filters was determined by liquid scintillation spectroscopy. MTA assays were carried out in a similar fashion except that 50µM [2-^{14}C] malonyl-CoA and MTA were added in place of the [1-^{14}C] acetyl-CoA and MTA, and the final reaction time was 2 min.

Protein Determination. Protein was monitored by reading the absorbance at 280nm or by using the Bradford protein assay.

Preparation of Crude Extracts. All of the material above the cotyledons was harvested from greenhouse-grown Brassica campestris seedlings, when the plants were about 15cm tall. The material was weighed and homogenized at 4°C in a Waring blender with a minimum volume of phosphate buffer (pH 7.5) containing 5mM DTT and 10µm leupeptin. The homogenate was then frozen and could be stored at -20C° indefinitely. All the subsequent steps were carried out at 4C°. After thawing overnight at 4C°, the homogenate was centrifuged for 30 min. at 10,000 g, and filtered through several layers of cheesecloth. The filtrate was centrifuged again, filtered through Miracloth, made 20% in glycerol and stirred for 30 min.

Ammonium Sulfate Fractionation. The filtrate was brought to 50% $(NH_4)_2SO_4$ staturation and centrifuged at 10,000g for 40 min. The supernatant was filtered through Miracloth, and the pellet was discarded. The filtrate was then brought to 85% $(NH_4)_2SO_4$ staturation and centrifuged as above. The pellet obtained from the 85% cut was dissolved in 0.05 M NDG buffer and dialyzed against the same buffer overnight at 4°C. NDG buffer consists of 0.05M sodium phosphate pH 7.5, 5 mM DTT, 10µM leupeptin and 20% glycerol.

Chromatography. The dialyzed solution was concentrated with an Amicon Ultrafiltration Unit using a PM10 membrane (molecular weight cut off of 10,000). The concentrated solution was applied to a Sephadex G-75 column (5cm x 55cm) and eluted with 0.05M NDG buffer, pH 7.5, at a flow rate of 20.5 ml/hr. Active fractions were pooled and concentrated with the Amicon Ultrafiltration Unit, again with the PM10 membrane.

The concentrated fractions were dialyzed overnight against 0.025M Tris-CH_3COOH pH 8.3 containing 5mM DTT and 20% glycerol, and then applied to a polybuffer exchange column (1cm x 50cm) equilibrated with the same buffer. A linear pH gradient from 8.3 to 5.0 was used to elute the ATA, with the elutant consisting of a concentrated polybuffer 96 (30%) and polybuffer 74 (70%) pH 5.0 solution, which was diluted 1:10 with water and glycerol to make a final solution containing 5mM DTT and 20% glycerol. The column was washed with 10 column volumes of the elutant at a flow rate of 15.3 ml/hr.

Two ATA activity peaks were found: ATA_1 at pH 7.2 and ATA_2 at pH 6.8. Active fractions of each were pooled and applied to a second polybuffer exchange column (0.7 cm x 20 cm) and eluted using the same conditions as described above except with a flow rate of 7.0 ml/hr. The active fractions were concentrated using a Millipore Immersible CX-10 filter (molecular weight cut off of 10,000).

SDS Polyacrylamide Gel Electrophoresis. SDS-PAGE was carried out according to the methods of Laemmli (10) and silver stained.

RESULTS AND DISCUSSION

Purification of ATA. ATA was purified over 1000-fold from Brassica campestris crude extract with an activity recovery of about 25% (Table I). The freeze/thaw at 4°C caused a large amount of contaminating protein to drop out without a loss of activity. SDS polyacrylamide gel electrophoresis of the extract before and after the freeze/thaw revealed the loss of protein bands of molecular weight of 50,000.

Table I. Purification of ATA_1 and ATA_2.

	Protein (mg)	Total Activity (nmol/min)	Specific Activity (nmol/min/mg)	Purification (-fold)
Crude Extract	2008	4.02	.002	
$(NH_4)_2SO_4$	595	13.1	.022	11
SEPHA	132	6.36	.048	24
PBE ATA_1	2.35	7.81	3.32	1660
ATA_2	1.90	1.56	0.82	410
PBE ATA_1	0.60	1.44	2.40	1220
ATA_2	0.48	1.05	2.16	1080

Figure 1 shows the results of the Sephadex G-75 column of the dialyzed 50-85% $(NH_4)_2SO_4$ pellet. One main peak of ATA activity was found, along with a much smaller activity peak. Comparing it to the earlier calibration of the column with a BioRad Protein Molecular Weight Standards kit, the much smaller peak eluted in the area of ovalbumin (MW 43,000), while the main peak eluted with the void volume. The main peak of activity probably consists of aggregates of ATA, since the molecular weight of ATA was found to be around 45,000. Both peaks were saved and further purified.

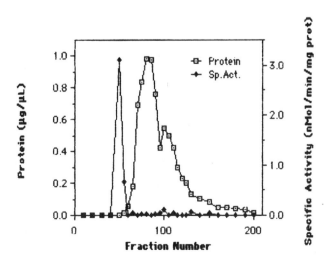

Figure 1. Sephadex G-75 gel filtration chromatography.

Figure 2 shows the results from the polybuffer exchange column of fractions eluting with the void volume from the gel filtration volumn. Two activity peaks were found: one at a pH of 6.77 and the other at pH 7.18. MTA was one of the main contaminants of ATA. Two peaks of MTA activity were seen: one at pH 5.1 and the other at a pH of approximately 7.5 (data not shown). No MTA activity was found in the ATA peaks.

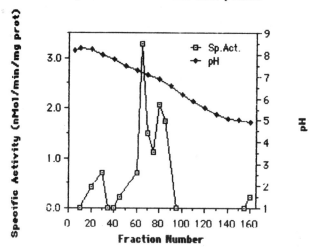

Figure 2. Polybuffer exchange chromatography.

Both peaks had two main protein bands on SDS-PAGE: one at approximately 43,000 and the other at 92,000 (data not shown). The peak at pH 6.77 was also combined and applied to a second polybuffer exchange column (data not shown). One peak of activity was found, and the appropriate fractions were pooled, concentrated and electrophoresed on an SDS polyacrylamide gel. One protein band was seen (Figure 3). This sample, ATA_2, was used for amino acid analysis. When the activity peak at pH 7.18 (ATA_1) was rechromatofocused, considerable protein was lost and specific activity declined 35% (Table I).

Assay for ATA Activity. The assay for spinach ATA was reported as being linear for at least 5 min. (1). For Brassica ATA the assay was only linear for approximately 1 min. after which the activity dropped off rapidly (data not shown). This loss of activity was no longer seen after the ATA had been applied to the first polybuffer exchange column. Active fractions from the polybuffer exchange column gave a time course that was linear for about five min. with no loss of incorporated label. The rapid loss of incorporated label in partially purified samples could be due to the presence of interfering acyl-ACP hydrolases (11,12). As the enzyme is purified the interfering or competing enzyme is removed or inactivated. Both ATA_1 and ATA_2 gave similar time courses.

The overall recovery for ATA activity from the crude extract was 25% with a greater than 1000-fold purification (Table I). As shown in Table I, a significant amount of protein was lost from each chromatofocusing column. ATA activity was also rapidly lost when glycerol was not included in the buffer or when stored at 4°C. The reducing agent DTT was included in the buffer throughout the purification to stabilize the ATA activity. When DTT was not included or the concentration lowered, the ATA activity was adversely affected.

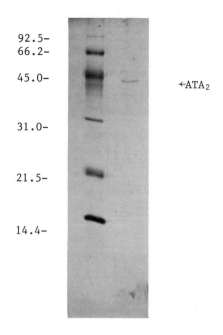

Figure 3. SDS polyacrylamide electrophoresis of
ATA$_2$ following the second polybuffer exchange column.

REFERENCES

1. Shimakata, T. and Stumpf, P.K. 1983. The purification and function of acetyl coenzyme A:acyl carrier protein transactylase. J.Biol.Chem., 258: 3592-3598.
2. Shimakata, T. and Stumpf, P.K. 1982. Molecular structure of fatty acid synthetase (FAS) in plants. Fed. Proc., 41: 1192.
3. Shimakata, T. and Stumpf, P.K. 1982. The procaryotic nature of the fatty acid synthetase of developing Carthanus tinctorius L. (safflower) seeds. Arch.Biochem.Biophys., 217: 144-154.
4. Stapleton, S.R. and Jaworski, J.G. 1982. Malonyl-CoA:acyl carrier protein transacylase from spinach. Fed.Proc., 41: 1192.
5. Caughey, L. and Kekwick, R.G.O. 1982. The characteristics of some components of fatty acid synthetase system in the plastids from the mesocarp of avocado (Persea Americana) fruit. Eur.J.Biochem., 123: 553-561.
6. Shimakata, T. and Stumpf, P.K. 1982. Purification and characterization of β-ketacyl-[acyl-carrier-protein] reductase, β-hydroxyacyl-[acyl-carrier-protein] dehydrase, and enoyl-[acyl-carrier-protein] reductase from Spinacea oleracea leaves. Arch.Biochem.Biophys., 218: 77-91.
7. Guerra, D.J. and Ohlrogge, J.B. 1986. Partial purification and characterization of two forms of malonyl-coenzyme A-acyl carrier protein transacylase from soybean leaf tissue. Arch.Biochem.Biophys., 246: 274-283.
8. Hoj, P.B. and Suendsen, I.B. 1984. Barley chloroplasts contain two acyl carrier proteins coded for by different genes. Carlsberg.Res.Commun., 49: 483-492.
9. Ohlrogge, J.B. and Kuo, T.M. 1985. Plants have isoforms for acyl carrier protein that are expressed differently in different tissues. J.Biol.Chem., 260, 8032-8037.
10. Laemmli, U.K. 1970. Cleavage of structural proteins during assembly of the head of bacteriophage T4. Nature, 227: 680-685.

11. Ohlrogge, J.B., Shine, W.E. and Stumpf, P.K. 1978. Fat metabolism in higher plants, characterization of plant acyl-ACP and acyl-CoA hydrolases. Arch.Biochem.Biophys., 189: 382-391.
12. Shine, W.E., Mancha, M. and Stumpf, P.K. 1976. Fat metabolism in higher plants. The function of acyl thioesterases in the metabolism of acyl-coenzymes A and acyl-acyl carrier proteins. Arch.Biochem.Biophys., 172: 110-116.

REGULATION OF ACETYL COENZYME A SYNTHESIS IN CHLOROPLASTS

Klaus-Peter Heise and Hans-Jürgen Treede

Institut für Biochemie der Pflanze
University of Göttingen
D-3400 Göttingen, FRG

INTRODUCTION

Photosynthetically active chloroplasts are able to synthesize acetyl-CoA either from acetate via acetyl-CoA synthetase (ACS) (1) or from pyruvate via the pyruvate dehydrogenase complex (PDC) (2).
Evidence for the coexistence of two alternative sources of acetyl-CoA in chloroplasts has been provided by
1. the demonstration of both enzyme systems in stromal extracts of chloroplasts from peas, spinach and maize mesophyll (2-6).
2. the partial purification of chloroplast PDCs from the same leaf tissues (7,8).

A common characteristic of both enzyme systems in green (2-8) as well as colorless plastids (9), is an alkaline pH-optimum (pH 8.0) and a high Mg requirement ($\geq$ 5mM). These properties distinguish them from corresponding ones of other cellular compartments (e.g. mitochondrial PDC) (10). The ACS with its additional requirement for Mg ATP (3) appears better adapted to photosynthetically active chloroplasts than the PDC which, in contrast, seems to be rather inhibited by free ATP (e.g. by Mg-complexation) (5). It is questionable that this ATP-dependent inactivation of chloroplast PDC occurs in vivo because the chloroplast stroma is apparently lacking in either $\overline{\text{ATP-favored}}$ PDC kinase activities (7) or in not-complexed ATP.

As possible mechanisms controlling the relative involvement of both acetyl-CoA synthesizing systems within chloroplasts, the stromal availability of acetate and pyruvate (6) and feedback control of chloroplast PDC by acetyl-CoA or NADH (5) in accordance with the mitochondrial complex (10) are being discussed below.

All methods and materials have been described elsewhere in detail (3-8).

RESULTS AND DISCUSSION

In investigations of the endogenous acetate and pyruvate levels in leaf tissues, significant species-specific differences in the proportion of both

Table I. Acetate and pyruvate content of different leaf tissues (A) and the estimated level of both metabolites in corresponding chloroplasts (B)

	A leaves[a]		B chloroplasts[b]	
	(nmol × mg^{-1} Chl)		(μM)	
	pyruvate	acetate	pyruvate	acetate
spinach	20 + 10	59 + 20	80 + 40	433 + 199
peas	66 + 19	38 + 10	264 + 76	276 + 108
maize	185 + 42	39 + 23	740 + 168	296 + 200

[a] Without middle rib
[b] Chloroplast metabolite levels have been estimated by assuming a general stromal volume of 25 µl/mg Chl and stromal portions of acetate with about 15 - 20% and of pyruvate with about 10% of their overall concentration (6).

metabolites have been found albeit with relatively high fluctuations (Table I, A). This points to variations in their availability for acetyl-CoA formation.

The relative cellular distribution probability of both precursors has been derived from nonaqueous fractionation studies with spinach leaves (6) which, related to the overall concentration, suggest an enrichment of pyruvate in the cytosol (about 50%) and of acetate in the mitochondria (about 30%) but only lower amounts of acetate (15 - 20%) and pyruvate (10%) in the chloroplast compartment.

Based on this information and assuming a general stromal volume of 25 µl/mg Chl (6) the chloroplast levels of both metabolites have been estimated (Table I, B) and compared with the substrate requirements of ACS and PDC from kinetic measurements in rapidly prepared stromal extracts (Table II, 1a). These calculations point to a predominant role of acetate in acetyl-CoA synthesis of spinach chloroplasts and to an increasing involvement of pyruvate in pea and maize mesophyll chloroplasts (6). The recently observed, light-driven pyruvate uptake by maize mesophyll chloroplasts (11) even favors PDC activities within these organelles.

Besides this stromal precursor availability, the highly labile PDC (8) has been considered to play an important role in controlling the relative involvement of both pathways in acetyl-CoA formation since the individual subcomplexes are able to function either together or independently of each other (5) and since the end products acetyl-CoA and NADH strongly inhibit the PDC by feedback control (5) (Table II, b).

The observed analogies in kinetic properties (Table II) and in size (7,8) with the E. coli complex give rise to speculations on the homology between the prokaryotic and the chloroplast PDC.

Table II. Some analogies between kinetic constants of acetyl-CoA synthesizing enzymes in Higher Plant chloroplasts (A) with corresponding ones from Escherichia coli (B) (12,13).

Enzyme	kinetic constant		A in chloroplasts from			B in Escherichia coli
			spinach	maize (mesophyll)	peas	
ACS	a) app. K_m acetate	(mM)	0.1	0.1–0.15	0.1–0.15	–
PDC	app. K_m pyruvate	(mM)	0.2–0.6	0.1–0.15	0.1–0.30	0.2–0.6
	b) app. K_m NAD$^+$	(μM)	7.7	10	n.d.	10
	app. K_m CoA	(μM)	2.9	0.8	n.d.	2
	app. K_i NADH	(μM)^c	1.3	2.4	n.d.	n.d.
	app. K_i acetyl-CoA	(μM)^d	8.6	6.5	n.d.	6

Substrate requirements of the chloroplast enzymes (a) have been determined in stromal extracts (5). For residual kinetic constants (b) chloroplast PDC was pelleted at 140 000 x g (8).
n.d. = not determined

^{c} versus NAD$^+$
^{d} versus CoA

REFERENCES

1. Kuhn, D.N., Knauf, M. and Stumpf, P.K. (1981) Arch. Biochem. Biophys. 209, 441-450
2. Williams, M. and Randall, D.D. (1979) Plant Physiol. 64, 1099-1103
3. Sauer, A. and Heise, K.-P. (1984) Z. Naturforsch. 39, 268-275
4. Liedvogel, B. (1985) Anal. Biochem. 148, 182-189
5. Treede, H.-J. and Heise, K.-P. (1985) Z. Naturforsch. 40, 496-502
6. Treede, H.-J., Riens, B. and Heise, K.-P. (1986) Z. Naturforsch. in press
7. Camp, C.J. and Randall, D.D. (1985) Plant. Physiol. 77, 571-577
8. Treede, H.-J. and Heise, K.-P. (1986) submitted to Z. Naturforsch.
9. Reid, E.E., Thompson, P., Little, C.R. and Dennis, D.T. (1977) Plant. Physiol. 59, 842-853
10. Randall, D.D., Rubin, P.M. and Fenko, M. (1977) Biochim. Biophys. Acta 485, 336-349
11. Flügge, U.I., Stitt, M. and Heldt, H.W. (1985) FEBS Lett. 183, 335-339
12. Visser, J., Kester, H., Jeyaseelan, K. and Topp, R. (1982) Methods Enzymol. 89, 399-408
13. Schwartz, E.R. and Reed, L.J. (1970) Biochemistry 9, 1434-1439

LIPID PRECURSORS IN PLANT CELLS: THE PROBLEM OF ACETYL COA GENERATION FOR PLASTID FATTY ACID SYNTHESIS

Bodo Liedvogel

Institut für Biologie II, Zellbiologie
Albert-Ludwigs-Universität
D-7800 Freiburg i.Br. F.R.G.

INTRODUCTORY SURVEY

In terms of compartmentation the generation of acetyl CoA and isopentenyl pyrophosphate which are the precursors of the lipophilic domains of lipid molecules in plants is not finally understood. For this reason this topic is debated rather controversially in the literature (for review see ref. 1). The problem of generation and compartmentation of isopentenyl pyrophosphate will be dealt with in a separate contribution (H. Kleinig, this vol.). The point in question here is the intraplastid acetyl CoA formation for fatty acid synthesis.

Different hypotheses have been argued in order to explain the in vivo formation of acetyl CoA in the plastid stroma the major ideas being: (i) activation of the acetate anion derived from acetyl CoA hydrolysis in the mitochondrion ('acetate pathway'); (ii) acetyl CoA generation by a plastid pyruvate dehydrogenase complex*; in addition (iii) ATP citrate lyase and (iv) a carnitine acetyltransferase are thought to act as acetyl group donors (ref. 1). The acetate pathways' plausibility summarized for the spinach system by Stumpf's group (ref. 2) was due to a number of supporting data, e.g. the acetate anion freely permeates biomembranes, is a well incorporated precursor of all plastid FAS* systems; its activation is restricted to the stroma compartment where acetyl CoA synthetase is localized, and acetate is present in sufficient amounts in the tissue. In addition, PDHC which may serve as a most direct source for acetyl CoA was thought to be absent from the chloroplasts. However, the latter two assumptions have met with a differing valuation in the meantime (ref. 3).

On the other hand, a series of non-green plastids (leucoplasts, chromoplasts) has been described which clearly possess an active PDHC and additionally a full set of glycolytic enzymes for providing the complex with its substrate pyruvic acid. Also for chloroplasts of different species evidence could be presented for the existence of actively working PDHCs; but an uninterrupted glycolytic pathway from the photosynthate 3-PGA to pyruvate seems to be absent at least from mature chloroplasts and the lack of phosphoglyceromutase has been affirmed in particular by different authors (cf. ref. 1). In chloroplasts - unlike in photosynthetically inactive types of plastids - an unprevented carbon flow to pyruvate via glycolysis would of course be of grave consequence and a strict metabolic control at the level of 3-PGA is

---
* PDHC = pyruvate dehydrogenase complex; FAS = fatty acid synthesis

to postulate. For tissues bearing mature chloroplasts, e.g. the spinach leaf, reliable turnover measurements of membrane lipids and of their acyl groups in particular are lacking. For this reason, estimations on the in vivo rates of FAS as well as of intraplastid activities of pyruvate producing enzymes are hardly to specify and may be rather low and, therefore, not easy to assay.

The very intention of the work to present here was to take up again the problem of carbon supply for chloroplast FAS using a young plant system still on the point of development and growth. For this purpose the white mustard (Sinapis alba L.) seedling has been chosen. At the stage the seedlings were used for chloroplast isolation their storage fat was almost completely catabolized but the cotyledons are still expanding and are depending on newly synthesized membrane lipids or on their acyl moieties, respectively.

These cotyledon chloroplasts were examined for their capability to synthesize long-chain fatty acids starting from acetate and pyruvate as labeled substrates. The enzymatic activities converting these compounds to the general precursor acetyl CoA - acetyl CoA synthetase and PDHC - were characterized in terms of specific activities and cofactor requirements. Finally, a point of high interest was to go in search of the glycolytic enzymes leading from the primary photosynthate 3-PGA to pyruvate within these chloroplasts' stroma.

MATERIALS AND METHODS

The details of the procedures employed in this investigation will be found in references 3, 4 and 5.

RESULTS

Cotyledon chloroplasts from mustard seedlings were purified on linear Percoll gradients and checked for purity by EM and marker enzyme analyses. The EM evaluation showed the organelles' structural intactness (90 %), neither other cell organelles or exogenous contaminants could be observed. Various marker enzymes of cytoplasm, mitochondria and microbodies were measured in the stroma fraction. The attached contamination was calculated not to exceed 0.4 % on the basis of specific activities.

Fatty acid synthesis. These chloroplasts incorporate labeled acetate as well as pyruvate with considerable rates in long-chain fatty acids (Tab. 1). Incorporation rates are linear with time (up to 20 min) and show proportionality over a wide chlorophyll content (up to 200 µg per assay). Both of the precursors are utilized at comparable rates at about saturating conditions (240 µM), but administering rather high substrate amounts the pyruvate incorporation was twice as much in comparison with acetate. The incorporation of both substrates shows a strict requirement of light and a slight dependence on exogenously applied CoA. While acetate utilization was completely independent on exogenous ATP, those of pyruvate could be increased by its application.

PDHC and acetyl CoA synthetase. The respective activities have been assayed with the aid of a radiochemical method basing on acetyl CoA trapping with dithioerythritol (ref. 5). The specific activities of the respective enzymes are: PDHC at 0.25 and 0.65 mM pyruvate, 6.2 and 18.2 nmol/mg protein/min; acetyl CoA synthetase at 0.25 and 0.65 mM acetate, 6.1 and 14.8 nmol/mg protein/min. The dependences of acetyl CoA synthetase on its cofactors CoA and ATP is an absolute one, while that of PDHC differs for the various cofactors: the marked requirement for TPP as known from plant mitochondrial

Tab. 1. [2-^{14}C]Pyruvate and [1-^{14}C]Acetate Incorporation into Long-Chain Fatty Acids and Distribution Pattern of the Acids Synthesized by Isolated Cotyledon Chloroplasts from Mustard Seedlings

Substrate	Incorporation Rate; nmol Substrate/mg Chl/min		Distribution of Labeled Acids %			
	Concentration Range		Phospho-lipids	Galacto-lipids	FFA*	DAG*
	0.24 mM	2 mM				
[2-^{14}C]Pyruvate	0.9	3.6	5	2	6	87
[1-^{14}C]Acetate	0.96	1.6	11	3	4	72

* FFA = Free Fatty Acids; DAG = Diacylglycerol

PDHCs could not be observed (minus TPP 69 % of control). With NAD$^+$ being absent the activity decreased to 31 % and without CoA to 2.8 % of the control. ATP (1 mM) stimulated the complex by 53 %, and 3-PGA by 30 %.

Plastid glycolytic enzymes. The cotyledon chloroplasts' capability was tested to produce pyruvate from 3-PGA. For this purpose the presence of phosphoglyceromutase, enolase and pyruvate kinase has been examined in the stroma fraction and - for comparison - in a membrane-free 100,000 x g supernatant which contained also cytoplasmic constituents. These results are expressed in nmol/mg protein/min (values for 100,000 x g fraction in brackets) phosphoglyceromutase 12.1 (102); enolase 38 (78); pyruvate kinase 11.9 (61).

CONCLUSIONS

Cotyledon chloroplasts from developing mustard seedlings have given proof to incorporate both substrates, radiolabeled acetate and pyruvate, into fatty acids with high rates. The enzymes responsible for acetyl CoA formation, acetyl CoA synthetase and PDHC, were measured and found to exhibit similar high activities. Furthermore, a metabolic sequence leading from the photosynthate 3-PGA to pyruvate is localized within the plastid stroma, their in vitro measured activities being sufficient to supply the PDHC with pyruvate. From the author's point of view the generation of intraplastid acetyl CoA via PDHC which is produced by plastid glycolytic enzymes would represent a most concise way to supply the FAS with its substrate. However, the mode to regulate the acetyl CoA generating steps may be different in mature chloroplasts and/or the in vivo rates of FAS may be rather low in such plastids of fully differentiated tissues.

References. 1. B. Liedvogel, Acetyl coenzyme A and isopentenylpyrophosphate as lipid precursors in plant cells - biosynthesis and compartmentation, J. Plant Physiol. 214:211 (1986). 2. P.K. Stumpf, Fatty acid biosynthesis in higher plants, in: "Fatty Acid Metabolism and Its Regulation", S. Numa, ed., Elsevier, Amsterdam (1984). 3. B. Liedvogel, Acetate concentration and chloroplast pyruvate dehydrogenase complex in Spinacia oleracea leaf cells, Z. Naturforsch. 40c:182 (1985). 4. B. Liedvogel and R. Bäuerle, Fatty acid synthesis in chloroplasts from mustard (Sinapis alba L.) cotyledons: Formation of acetyl coenzyme A by intraplastid glycolytic enzymes and pyruvate dehydrogenase complex, Planta (submitted). 5. B. Liedvogel, A new radiochemical method for determination of pyruvate dehydrogenase complex and acetyl-coenzyme A synthetase, Anal. Biochem. 148:182 (1985).

PARTIAL PURIFICATION AND CHARACTERIZATION OF ACETYL-CoA SYNTHETASE FROM

MATURE SPINACH LEAVES

        C. A. Zeiher and D. D. Randall

        Dept. of Biochemistry
        University of Missouri-Columbia
        Columbia, Missouri 65211

INTRODUCTION

Acetyl-CoA synthetase (ACS) catalyzes the conversion of acetate to acetyl-CoA. In photosynthetic tissue, this enzyme is localized in the chloroplast (1) where it potentially provides a key source of acetyl-CoA for fatty acid, isoprenoid, and branch-chain amino acid biosynthesis. Acetyl-CoA synthetase's contribution to chloroplast acetyl-CoA is presently controversial because of the identification of alternative sources of acetyl-CoA (i. e. pyruvate dehydrogenase complex (2,3), and carnitine acyltransferase (4)), in the chloroplast. To further elucidate the role of ACS in chloroplast acetyl-CoA metabolism we have partially purified and characterized ACS from mature spinach leaves.

MATERIALS AND METHODS

ACS activity was measured by the radioassay described by Roughan et al (5). The reaction mixture contained in 350 ul: 50 mM Hepes-KOH (pH 8.0), 5 mM $MgCl_2$, 2 mM ATP, 0.5 mM CoA, 0.5 mM DTT, and 1 mM [^{14}C]-acetate (1 uCi/umole). The reaction was initiated by the addition of [^{14}C]-acetate unless otherwise stated and assayed for 30 min at 30°C. One unit of ACS activity was defined as 1 nmole acetyl-CoA produced/min.

ACS was extracted from mature spinach leaves purchased from a local market. The extraction media contained 100 mM Tris pH 7.8, 5 mM 2-mercaptoethanol, 5 uM leupeptin, 1 mM benzamidine, 1 mM ϵ-aminocaproic acid, and 1 mM EDTA. DEAE Sephacel, Red A (Amicon) and Sephadex G-100 columns were equilibrated with 20 mM Tris pH 7.8, 5 mM 2-mercaptoethanol, and 20% glycerol (Buffer A).

RESULTS AND DISCUSSION

Several different plant species (i.e. spinach, pea, soybean, corn, wheat, and rye) were initially screened for ACS activity. Leaves (5 g) were homogenized, centrifuged at 17300g and the supernatant desalted on a Sephadex G-25 column. From the desalted leaf extracts ACS was found to be highest in spinach (7.7 units/mg protein) followed by peas (4.01 units/mg protein) and corn (3.79 units/mg protein). Rye, soybeans,

Table I: Summary of Purification of Acetyl-CoA Synthetase from Mature Spinach Leaves.

Purification Step	Total Protein (mg)	Total Activity (Units)	Yield (%)	Enrichment (fold)
Crude	1715	11250	100	1.0
$(NH_4)_2SO_4$ 40-60%	950	8284	74	1.3
DEAE Sephacel	417	5325	47	1.9
Red A	5	2105	19	64.2
Sephadex G-100	1	1320	12	201.1

and wheat had significantly lower activity of 1.87, 1.17, 1.09 units/mg protein, respectively. Because of the high ACS specific activity observed, spinach leaves were chosen as the source of ACS for further purification.

For purification 175 g of spinach leaves were homogenized with a polytron in 525 ml of extraction media and centrifuged at 14600g. The enzyme in the supernatant was precipitated using 40-60% $(NH_4)_2SO_4$ fractionation. The pellet was dissolved in buffer A, desalted, and applied to DEAE Sephacel. The enzyme was eluted with a linear gradient of 0 to 500 mM KCl in buffer A with the enzyme eluting at approximately 200 mM KCl. The most active fractions were pooled and directly applied to a Red A column. The column was washed with buffer A followed by 500 mM KCl in buffer A. ACS was eluted with 1 M KCl. The active fractions were pooled, concentrated, and applied to Sephadex G-100 column. Table I summarizes the results of this purification. Using this procedure, ACS was purified 200 fold to a specific activity of 1.3 umole/min/mg protein, the highest specific activity reported to date for ACS from photosynthetic tissue. Both affintiy (AG-ATP, AG-CoA, AG-AMP) and hydrophobic chromatography have been tried at various points during purification but attempts to date have been unsuccessful in binding ACS to these columns.

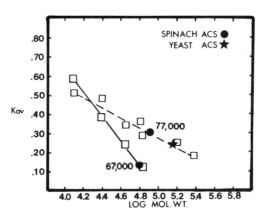

Fig.1: Molecular weight determination of acetyl-CoA synthetase using gel filtration chromatography. Partially purified spinach ACS was applied to Sephadex G-100-120 column (———) and FPLC (Pharmacia) Sepharose 12 (-- -- --) column. Sephadex G-100-120 and FPLC Sepharose 12 columns were calibrated with molecular weight markers ranging in size from 12.5 to 68 KD and 12.5 to 240 KD, respectfully. The molecular weight markers were cytochrome C (12.5 KD), chymotrypsinogen A (25 KD), hen egg albumin (45 KD), hemoglobin (64.5 KD), bovine serum albumin (68 KD), aldolase (158 KD), yeast acetyl-CoA synthetase (151 KD) and catalase (240 KD).

The molecular weight of ACS estimated by a calibrated Sephadex G-100 column was 67,000 and 77,000 by a FPLC Sepharose 12 column (figure 1). This was similar in size to the potato tuber ACS (59,500; 6), and ACS isolated from mitochondria of bovine mammary gland (63,000; 7); but considerably smaller in size than ACS isolated from baker's yeast (151,000; 8). The partially purified spinach ACS had a pH optimum of 7.8 with 90% of maximum activity observed between pH 7.6-8.2 and 50% of maximum activity observed as low as pH 6.6 (figure 2). The temperature optimum of partially purified ACS was 50°C. Above 50°C spinach ACS was unstable with complete loss of activity observed if ACS was preincubated at 60°C for 30 min prior to assaying at 30°C. Arrhenius plots of ACS activity gave an Ea value of 10329 cal/mol and a Q_{10} of 1.54.

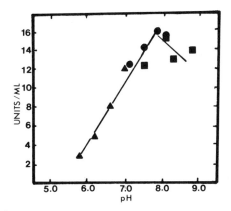

Figure 2: The pH dependency of partially purified acetyl-CoA synthetase. Three different buffers were used at 50 mM concentration: Mes (▲), Hepes (●) and Bicine (■). The reaction was initiated by addition of enzyme and pH determined at completion of assay.

REFERENCES

1. D. N. Kuhn, M. Knauf, and P. K. Stumpf, Subcellular localization of acetyl-CoA synthetase in leaf protoplasts of Spinacia oleracia, Arch. Biochem. Biophys. 212:441 (1981).
2. B. A. Elias and C. V. Givan, Localization of pyruvate dehydrogenase complex from chloroplasts of Pisum sativum chloroplasts, Plant Sci. Lett. 17:115 (1979).
3. M. Williams and D. D. Randall, Pyruvate dehydrogenase complex from chloroplasts of Pisum sativum L., Plant Physiol. 64:1099 (1979).
4. I. Mclaren, C. Wood, M. N. A. Jalil, B. C. S. Yong, and D. R. Thomas, Carnitine acyltransferase in chloroplasts of Pisum sativum, Planta 163:197 (1985).
5. P. G. Roughan, R. Holland, and C. R. Slack, On the control of long-chain fatty acid synthesis in isolated spinach (Spinacia oleracia) chloroplasts, Biochem. J. 184:193 (1979).
6. K. P. Huang and P. K. Stumpf, Fat metabolism in higher plants. XLI: Properties of potato acetyl-coenzyme A synthetase. Arch. Biochem. Biophys. 140:158 (1970).
7. S. Quershi and R. M. Cook, Utilization of volatile fatty acids in ruminants. V. Purification of acetyl-CoA synthetase from mitochondria of lactating bovine mammary gland, J. Agric. Food Chem. 23:555 (1975).
8. E. P. Frenkel and R. L. Kitchens, Acetyl-CoA synthetase from Baker's yeast (Saccharomyces cereviseae), Meth. Enz. 71:317 (1981).

ACETYL-CoA CARBOXYLASE AND BIOTIN-CONTAINING PROTEINS IN CARROT SOMATIC EMBRYOGENESIS

Basil J. Nikolau, J. Croxdale*, T.H. Ulrich and E.S. Wurtele

NPI, 417 Wakara Way, Salt Lake City, UT 84108 and
*Botany Dept., University of Wisconsin, Madison, WI 53706

Somatic embryogenesis in carrots is induced by transferring carrot cultures from a maintenance medium (containing the auxin 2,4-dichlorophenoxyacetic acid (2,4-D)) into an inductive medium lacking the auxin 2,4-D[1]. In maintenance medium, the culture contains embryogenically incompetent cells and embryogenic cell clusters. Embryogenically incompetent cells are typically large and highly vacuolated and divide but do not undergo somatic embryogenesis in our inductive medium. The embryogenic clusters are composed of 5-15 small, cytoplasmically rich cells which, upon induction, undergo the process of somatic embryogenesis, developing sequentially through globular, heart, torpedo and germinating embryo stages.

We have developed fractionation procedures, based upon previous methods[2,3,4], to purify from maintenance cultures embryogenically incompetent cells and embryogenic cell clusters; and from induced cultures, embryos at various stages of development and non-embryogenic cells.

Examination of embryos by scanning electron microscopy indicates a heavy deposition of cuticle which was absent from the non-embryogenic and embryogenically incompetent cells (data not shown). Therefore, we examined the activity of acetyl-CoA carboxylase, a biotin-containing protein, catalyzing the synthesis of malonyl-CoA required for fatty acid biosynthesis.

Acetyl-CoA carboxylase activity (on the basis of protein or fresh weight) was similar in embryogenically incompetent cells and embryogenic cell clusters in maintenance medium (Table 1). However, the activity of this enzyme, per mg protein, was 2- to 15-fold higher in cells and embryos from an induced culture. More importantly, acetyl-CoA carboxylase activity increases as the embryos develop from embryogenic cell clusters to form globular, heart and torpedo embryos. These data are consistent with the requirement for higher acetyl-CoA carboxylase activity in embryos for cuticle biosynthesis.

Since acetyl-CoA carboxylase is a biotin-containing protein, we analyzed the biotinyl polypeptides in carrot cultures by a Western procedure, with ^{125}I-labelled streptavidin as a probe[5]. Analysis of cells in maintenance medium showed a major biotinyl polypeptide in the embryogenically incompetent cells, with a molecular weight of 70,000, although less abundant biotin-containing polypeptides of approximately 200kDa, 140kDa,

Table 1. Acetyl-CoA carboxylase activity in carrot culture fractions.

Culture fractions	Acetyl-CoA Carboxylase Activity (nmol/min)	
	per mg protein	per g fresh weight
Maintenance culture		
Embryogenic cell clusters	0.3	1.3
Embryogenically incompetent cells	0.4	1.3
Induced culture		
Non-embryogenic cells	0.8	1.8
Globular embryos	2.6	23.4
Globular and heart embryos	5.0	70.0
Torpedo embryos	5.3	88.5

50kDa, 35kDa and 32kDa were also detected (Figure 1, lane 1). In contrast, the embryogenic cell clusters had a major biotinyl polypeptide of 32kDa in size and the relative abundance of the 70kDa biotinyl protein was reduced compared to that of the incompetent cells (Figure 1, lane 2); the abundance of the other biotin-containing proteins were unaltered.

Analysis of cultures after induction of embryogenesis, showed that developing embryos through the globular, heart and torpedo stages of development accumulated increasing levels of biotinyl proteins of 50kDa, 35kDa and 32kDa, whereas the levels of the 200kDa, 140kDa and 70kDa biotin-containing proteins were similar (Figure 1, lanes 4-6). The non-embryogenic cells in inductive medium contained only the 70kDa biotinyl protein (Figure 1, lane 3).

The presence of multiple biotinyl proteins in plants has been documented before [5,6], however some plants, for example barley, have a single biotinyl protein correlated with acetyl-CoA carboxylase[7]. The agreement in the molecular weights between the biotinyl protein found in barley and the sub-

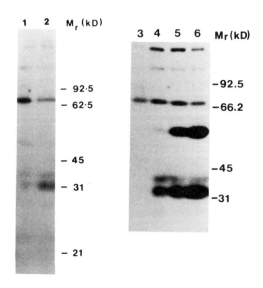

Figure 1. Biotinyl proteins in carrot culture fractions. Proteins from crude extracts of embryogenically incompetent cells (lane 1), embryogenic cell clusters (lane 2) from a maintenance culture; and non-embryogenic cells (lane 3), small globular embryos (lane 4), large globular and heart embryos (lane 5), and torpedo embryos (lane 6) where fractionated by SDS-PAGE, transferred to nitrocellulose and probed with ^{125}I-labelled streptavidin[6].

units of the purified maize leaf[8] and soybean[9] acetyl-CoA carboxylase to the 70kDa biotinyl protein found in carrot culture cells indicates that this polypeptide may originate from acetyl-CoA carboxylase. Similarly, a 200kDa subunit of acetyl-CoA carboxylase has been identified in the purified enzyme from non-photosynthetic tissue of several species[9,10,11], indicating that this polypeptide is also related to acetyl-CoA carboxylase. The functions of the other biotinyl proteins are unclear. One possibility is that they represent further isozymes or processing of acetyl-CoA carboxylase. Consistent with this possibility is that finding that free Mg^{2+} inhibits acetyl-CoA carboxylase activity from an induced culture by 50%, whereas it causes complete inhibition of the enzyme from a maintenance culture (data not shown), indicative of multiple forms of this enzyme in these two culture conditions.

In order to identify the function and characterize the regulation of the accumulation of these biotinyl polypeptides, we are purifying the individual proteins using affinity chromatography with immobilized avidin. Furthermore, we have synthesized an olignucleotide whose sequence codes for the known amino acid sequences of biotinyl-proteins[12]. This oligonucleotide detects two transcripts on Northern analysis of polyA+ RNA from carrots. We are currently cloning these transcripts.

REFERENCES

1. Halperin, W., and D.F. Wetherell, Adventive embryony in tissue cultures of the wild carrot Daucus carota, Amer. J. Bot. 51, 274 (1964).
2. Fujimura, T., and A. Komamine, Synchronization of somatic embryogenesis in a carrot cell suspension culture, Plant Physiol. 64, 162 (1979).
3. Giulino, G., D. Rosellini, and M. Terzi, A new method for the purification of the different stages of carrot embryoids, Plant Cell Rep. 2, 216 (1983).
4. Warren, G.S., and M. Fowler, Physical method for the separation of various stages in the embryogenesis of carrot cell cultures, Plant Sci. Lett. 9, 71 (1977).
5. Nikolau, B.J., E.S. Wurtele, and P.K. Stumpf, Use of streptavidin to detect biotin-containing proteins in plants, Anal. Biochem. 149, 488 (1985).
6. Nikolau, B.J., E.S. Wurtele, and P.K. Stumpf, Tissue distribution of acetyl-Coenzyme A carboxylase in leaves, Plant Physiol. 75, 895 (1984).
7. Nikolau, B.J., E.S. Wurtele, and P.K. Stumpf, Subcellular distribution of acetyl-Coenzyme A carboxylase in mesophyll cells of barley and sorghum leaves, Arch. Biochem. Biophys. 235, 555 (1984).
8. Nikolau, B.J., and J.C. Hawke, Purification and characterization of maize leaf acetyl-Coenzyme A carboxylase, Arch. Biochem. Biophys. 228, 86 (1984).
9. Charles, D.J., P.M. Hasegawa, and J.H. Cherry, Characterization of acetyl-CoA carboxylase in the seed of two soypean genotypes, Phytochem. 25, 55 (1986).
10. Slabas, A.R., and A. Hellyer, Rapid purification of a higher molecular wieght subunit polypeptide form of rape seed acetyl-CoA carboxylase Plant Sci. Lett. 39, 177 (1985).
11. Egin-Buhler, B., and J. Ebel, Improved purification and further characterization of acetyl-CoA carboxylase from cultured cells of parsley (Petroselinum hortense), Eur. J. Biochem. 133, 335 (1983).
12. Wood, H.G., and R.E. Barden, Biotin enzymes, Ann. Rev. Biochem. 46, 385 (1977).

PYRUVATE REVERSAL OF S-ETHYL DIPROPYLCARBAMOTHIOATE (EPTC) INHIBITION OF PYRUVATE DEHYDROGENASE COMPLEX

R. E. Wilkinson and T. H. Oswald
University of Georgia
Experiment, GA 30212-5099

INTRODUCTION

EPTC decreased fatty acid synthesis and desaturation (1-3), ent-kaurene synthesis and oxidation (4, 5) and total isoprenoid pigment quinone, chlorophyll, and contents (6, 7). Commonality of substrates and cofactors in these diversified biosynthetic processes includes Ac^-, O_2, NADPH and/or NADH, and ATP. EPTC did not inhibit photosynthesis (8); thus, it is presumed that O_2, NADPH and/or NADH, and ATP are probably not limiting for these biosynthetic processes.

Acetate is produced by seed triglyceride fatty acid β-oxidation but this supply of Ac^- is exhausted in about 14 days. Activity of chloroplastic PDC is sufficient to supply Ac^- for fatty acid and isoprenoid syntheses (9) but chloroplasts also contain Ac-CoA synthetase (10); thus, Ac^- derived from mitochondrial PDC could also be utilized for fatty acid and isoprenoid syntheses. The Ac^- concentration in _Spinacia oleracea_ leaves approximated <100 µM (10). Consequently, an inhibition of chloroplastic and mitochondrial PDC by EPTC could explain the observed physiological responses.

METHODS AND MATERIALS

^{14}C Incorporation into Fatty Acids. Wheat (_Triticum aestivum_ L. cv Stacy) was grown in the greenhouse. Chloroplasts isolation (9), chlorophyll detemination (11), and fatty acid synthesis (3) followed published procedures. Assays contained EPTC (0, 10, or 100 nM), dichlormid (2,2-dichloro-N,N-di-2-propylenylacetamide) (0, 1, 10, or 100 M), and $[2-^{14}C]$ acetate (15.8 µCi/µmol) or $(2-^{14}C)$ pyruvate (1 µCi/µmol). After 30 min light exposure (45 µein/m^2/sec) and 1 h dark, reactions were terminated by addition of 1 ml acetone. Fatty acid methyl esters (3) were extracted in n-pentane and quantified by liquid scintilation spectrometry. Double reciprocal plots of *Ac and *Pyr incorporation were subjected to linear regression analyses. Incubation proceeded for 30 min in the light (45 µein/m^2/sec) + 60 min in the dark (25°C) and were terminated with 1 ml acetone.

Mitochondrial PDC. Mitochondria were isolated from greenhouse grown wheat (cv. Stacy) (21 day) by minor modifications of published procedures (12). Protein determination (13) and assay condition follow published procedures (14, 15) Mitochondria (Class I) had an Acceptor

Control Ratio = 4.33. Activity was monitored as the increase in E_{340} using a Varian Techtron model 635 UV-visible dual beam recording spectrometer. Data were analyzed by standard errors.

RESULTS AND DISCUSSION

EPTC v *Ac⁻ Incorporation. [2-^{14}C] Acetate was incorporated into chloroplast fatty acids. The incorporation was inhibited by EPTC and the inhibition was reversed by 0.16 mM Ac⁻ (Fig. 1).

EPTC v *Pyruvate Incorporation. ^{14}C from [2-^{14}C] pyruvate was incorporated into chloroplast fatty acids and the incorporation was inhibited by EPTC (Fig. 2). The inhibition was reversed by pyruvate (90 nM). However, this inhibition of incorporation of [2-^{14}C] pyruvate into chloroplast fatty acids might have resulted from an inhibition of the fatty acid synthesis complex rather than an inhibition of PDC.

EPTC v Mitochondrial PDC. Activity of wheat mitochondrial PDC was measured by the increase in NADH spectrophotometrically. Pyruvate + NAD⁺ + CoA → Ac-CoA + NADH. EPTC inhibited wheat mitochondrial PDC activity 10% at 3.3×10^{-14} M and >90% at 3.3×10^{-7} M EPTC (Fig. 3). Additionally, 2-oxoglutorate dehydrogenase was inhibited 70% by 3.3×10^{-8} M and 85% by 3.3×10^{-6} M EPTC (Fig. 3).

Dichlormid v. Ac⁻ Incorporation: Dichlormid inhibited [2-^{14}C] acetate incorporation into chloroplast fatty acids and the inhibition was reversed by 90 nM acetate (Fig. 4).

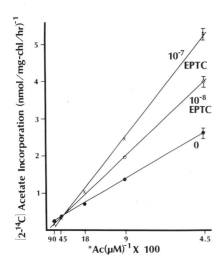

Fig. 1. Influence of EPTC on the incorporation of [2-^{14}C] acetate into wheat chloroplast fatty acids. ●-● = no EPTC, o-o = 10^{-8}M EPTC, x-x = 10^{-7}M EPTC. Slopes significantly different <1%

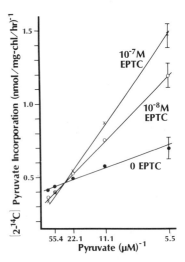

Fig. 2. Influence of EPTC on the incorporation of (2-^{14}C) pyruvate (●-● = no EPTC) into wheat chloroplast fatty acids. ●-● = no EPTC, o-o = 10 nM EPTC, x-x = 100 nM EPTC. Slopes significantly different <1%

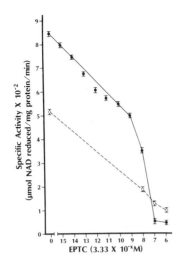
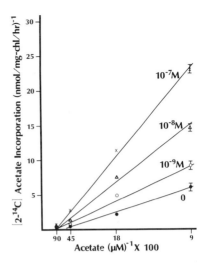

Fig. 3. EPTC inhibition of wheat mitochondrial pyruvate dehydrogenase complex (o-o) and 2-oxoglutarate dehydrogenase complex (●-●).

Fig. 4. Influence of dichlormid on the incorporation of [$2-^{14}C$] acetate into wheat chloroplast fatty acids. ●-● - no dichlormid, o-o = 10^{-9}M dichlormid, Δ-Δ = 10^{-8}M dichlormid, x-x = 10^{-7}M dichlormid. Slopes significantly different <1%

ACKNOWLEDGEMENTS

This work was conducted with funds from the Hatch Project 1306, Department of Agronomy, University of Georgia Agricultural Experiment Station, Georgia Station, Experiment, Georgia. Technical support by Mrs. Lois Wilson, Mrs. Sharon Wu, and Mrs. Cynthia McCormick is gratefully acknowledged.

REFERENCES

1. R. E. Wilkinson and A. E. Smith, Weed Sci. 23:100 (1975).
2. R. E. Wilkinson and A. E. Smith, Weed Sci. 24:235 (1976).
3. P. Karunen and L. Eronen, Physiol. Plant. 40:101 (1977).
4. R. E. Wilkinson and D. Ashley, Weed Sci. 27:270 (1979).
5. R. E. Wilkinson, Pest. Biochem. Physiol. 19:321 (1983).
6. R. E. Wilkinson, Pest. Biochem. Physiol. 8:208 (1978).
7. R. E. Wilkinson, Bot. Gaz. 138:270 (1977).
8. F. M. Ashton, Weeds 11:295 (1963).
9. P. J. Camp and D. D. Randall, Plant Physiol. 77:571 (1985).
10. B. Loedvogel, Z. Naturforsch. 40C:182 (1984).
11. D. I. Arnon, Plant Physiol., 24:1 (1949).
12. W. D. Bonner, Methods Enzymol., 10:126 (1967).
13. O. H. Lowry, N. J. Rosenbrough, A. L. Farr, and R. J. Randall, J. Biol. Chem., 193:265 (1951).
14. L. J. Reed and B. B. McKherigee, Meth. Enzymol. 13:55 (1969).
15. C. J. Stanley and R. N. Perham, Biochem. J. 191:147 (1980).

ACYL-CoA ELONGATION SYSTEMS IN Allium porrum MICROSOMES

René Lessire, Jean-Jacques Bessoule and Claude Cassagne

Institut de Biochimie Cellulaire et Neurochimie du CNRS
1 rue Camille Saint-Saëns
33077 Bordeaux cedex, France

INTRODUCTION

The synthesis of very long chain fatty acids (VLCFA) has been demonstrated in cell-free preparations from higher plants[1,2,3]. The leek epidermal cell microsome system is the best documented [4,5,6] and it is in this system that the existence of an endogenous precursor ATP-dependent elongation system was established[4]. The elongation of exogenous acyl-CoAs in the absence of ATP has also been demonstrated[5] and it was shown that acyl-CoAs were the reaction products of the acyl-CoA elongating system[6]. TCA and cerulenin inhibitory effects and in direct evidence suggest the presence of distinct C18-CoA and C20-CoA elongating systems. In order to test this hypothesis, the elongation activities were solubilized.

RESULTS AND DISCUSSION

Solubilization of C18-COA and C20-CoA elongation systems

The elongation of C18-CoA and C20-CoA by $(2-^{14}C)$malonyl-CoA was tested at different concentrations of Triton X-100, n-octyl-beta-D-glucopyranoside and deoxycholate. The results showed that deoxycholate completely inhibited the formation of C22-C24 and C26 acids while Triton X-100, at a concentration of 0.3 mM, increased by more than three times the VLCFA synthesis. Consequently, Triton X-100 was used for the solubilization of the C18-CoA and C20-CoA elongation systems.
Leek epidermal cell microsomes, prepared as described earlier[4], were incubated for 1 hour at 4°C in the presence of different concentrations of Triton X-100 and then centrifuged at 100,000g for 1 hour. The elongation activities were measured using the supernatant as the enzyme source. The results reported in Table 1 show that, even in the presence of a low detergent/protein ratio, the solubilization of membranous proteins occurred and reached more than 87 % of the total amount of microsomal proteins for a ratio of 1.5. For both C18-CoA and C20-CoA elongation activities, the specific activities recovered in the supernatant increased when the detergent/protein ratio was lower than 1 and then decreased for the higher ratios. Figure 1 shows the same phenomenon for the total activity, the optimum conditions for the solubilization required a Triton X-100/protein ratio of 1. As a control

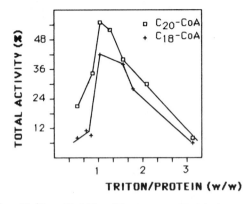

Figure 1 : Triton X-100 effect on solubilized C18-CoA and C20-CoA elongation activities

of solubilization, the supernatant was subjected to Sephacryl filtration chromatography and sucrose density gradient centrifugation [7]. In both experiments, no membrane fragments were detected, proving that the soluble fraction was uniquely constituted of solubilized proteins.

Table 1 : Effect of Triton X-100 on specific activities of solubilized C18-CoA and C20-CoA elongases

Triton/ proteins (w/w)	Specific activity in the solubilized fraction (% of specific acitvity of microsomes)				Solubilized proteins (% of microsomes)
	C18-CoA elongation		C20-CoA elongation		
	Exp. 1	Exp. 2	Exp. 1	Exp. 2	
0	0	0	0	0	0
0.5	-	15.1	34.2	40.3	56.3
0.7	29.4	-	-	-	-
0.8	-	13.8	50.4	57.5	62.5
1	88.2	51.4	82.0	90.9	65
1.4	39.2	-	-	-	-
1.5	-	42.4	56.4	62.9	87.5
1.7	32.2	-	-	-	-

Separation of C18-CoA and C20 elongases

The solubilized proteins were loaded onto a DEAE 23 chromatography column (bed volume = 5 ml). After washing with 2 volumes of 0.08 M

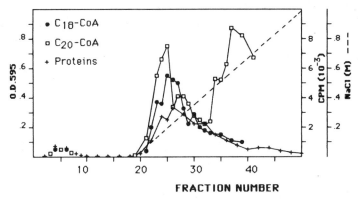

Figure 2 : DEAE-cellulose column chromatography of solubilized microsomal proteins

Hepes buffer (pH 7.0) containing 2 % glycerol, 10 mM beta-mercaptoethanol and 0.02 % triton X-100, the proteins were eluted by using a 0 to 1 M NaCl gradient. The elution profile is reported in Figure 2. The distribution of the C18-CoA elongation activity appeared as a single peak eluted at 0.25 NaCl while the C20-CoA elongase presented two different peaks of activities corresponding to 0.25 M and 0.5 M NaCl, respectively. These results are in very good agreement with our previous findings which showed that C18-CoA and C20-CoA elongation activities were separated by gel filtration chormatography and by sucrose density gradient centrifugation. All these data demonstrate the presence in leek epidermal microsomes of two distinct acyl-CoA elongases using C18-CoA and C20-CoA respectively as substrates. The purification of these two enzymes is under-way in our laboratory using filtration and anion exchange chromatographies. The partially purified enzymes have been subjected to SDS-page electrophoresis and the preliminary results showed the absence of very high molecular weight proteins. However, we observed six or seven greatly enriched protein bands suggesting a procaryotic nature for the structure of the elongases.

REFERENCES

1. Macey, M.J.K., and Stumpf, P.K., Plant Physiol. 13:1637, (1968).
2. Kolattukudy, P.E., and Buckner, J.S., Biochem. Biophys. Res. Commun. 46:801, (1972).
3. Cassagne, C., and Lessire, R., Physiol. Vég. 12:149, (1974).
4. Agrawal, V.P., Lessire, R., and Stumpf, P.K., Arch. Biochem. Biophys. 230:580, (1984).
5. Lessire, R., Juguelin, H., Moreau, P., and Cassagne, C., Phytochem. 24:1187, (1985).
6. Lessire, R., Juguelin, H., Moreau, P., and Cassagne, C., Arch. Biochem. Biophys. 239:260, (1985).
7. Lessire, R., Bessoule, J.J., and Cassagne, C., FEBS Lett. 187:314, (1985).

LIPID BIOSYNTHESIS IN OIL PALM PROTOPLASTS

Ravigadevi Sambanthamurthi[1], Khaik-Cheang Oo[2] and Augustine Soon-Hock Ong[1]

[1]Palm Oil Research Institute of Malaysia (PORIM), P O Box 10620, 50720 Kuala Lumpur, Malaysia, [2]Department of Biochemistry, University of Malaya, 59100 Kuala Lumpur, Malaysia

INTRODUCTION

Protoplasts offer several advantages for metabolic studies. They provide a homogenous single-celled system and ensure the even distribution of precursors into the cell. The reaction products can be isolated more quickly and at higher purity. One of the main attractions of employing protoplasts for studying lipid metabolism is their ease of lysis. Organelles can be separated with minimal injury. The study of lipid metabolism in mesocarp and embryoid protoplasts of the oil palm is described.

MATERIALS AND METHODS

Oil palm fruits (*E. guineensis* Jacq var. tenera and *E. oleifera*) were obtained fresh from the PORIM Research Station at Serdang. *E. guineensis* embryoids clone P9 were kindly supplied by Dr K Paranjothy, PORIM.

Protoplasts were isolated from oil palm mesocarp 16 - 20 weeks after anthesis, and embryoids at three stages of development : a) Embryoids which had just differentiated from callus; b) Embryoids just prior to shoot formation; c) Very young shoots which had differentiated from embryoids.

Two ml of protoplast suspension were incubated with 15 μCI ^{14}C-acetic acid, sodium salt for 6 hr and the products of the incubation extracted as described by Oo *et al* (1). Separation of the lipid classes by thin layer chromatography employed the hexane: diethyl ether: acetic acid (80:20:2, v/v/v) solvent system. The incorporation of radioactivity into each component lipid class was determined by liquid scintillation counting. Fatty acid composition was analysed on a gas liquid chromatograph (GLC) attached to a proportional counter as described by Oo *et al* (1).

RESULTS AND DISCUSSION

Incorporation of ^{14}C- acetate into lipids increased with the age of the mesocarp from which the protoplasts were derived and was maximum at 20 weeks. Embryoid protoplasts were more active than mesocarp protoplasts as seen by the higher incorporation of radioactive acetate into lipids at all three stages of embryoid development. (Table 1.)

Table 1. Incorporation of Radioactivity by Protoplasts

Protoplast Source	Total Cell Count	% of Substrate Radioactivity		
		CO_2	Lipid Extract	Aqueous Extract
E. guineensis Mesocarp				
16 weeks	4×10^5	0.2	1.8	18.8
17 weeks	4×10^5	0.4	2.7	6.8
18 weeks	4×10^5	0.4	3.6	6.9
20 weeks	2×10^3	0.4	5.0	3.1
E. oleifera Mesocarp				
18 weeks	4×10^4	0.6	4.4	3.2
20 weeks	2×10^3	0.3	5.2	3.1
Embryoids				
a	1×10^3	1.1	8.6	2.5
b	$< 10^3$	2.7	11.1	3.7
c	$< 10^3$	1.8	9.8	4.1

Table 2. Lipid Classes of Oil Palm Protoplasts

	Percentage Composition							
	E. guineensis Mesocarp			*E. oleifera* Mesocarp		Embryoids		
	16 WK	17 WK	20 WK	18 WK	20 WK	a	b	c
SE	0.1	0.2	0.5	0.9	0.3	1.0	0.6	0.3
TG	4.9	9.4	23.5	7.4	14.9	59.0	13.3	11.3
FFA	7.9	20.7	51.1	23.3	37.8	24.1	50.0	25.6
S	0.04	0.3	0.1	1.5	2.9	0.3	0.1	0.2
DG	7.6	12.8	9.3	41.2	29.2	7.0	10.2	11.4
MG	3.8	13.9	4.7	5.6	5.3	3.9	4.6	8.4
PL	75.7	42.8	8.9	20.4	9.9	5.2	17.4	42.2

SE - Sterol Ester; TG - Triacylglycerol; FFA - Free Fatty Acids;
S - Sterol; DG - Diacylglycerol; PL - Polar Lipids; MG - Monoacylglycerol

Table 3. Total Fatty Acid Composition of Protoplast Lipids

Protoplast Source	Percentage of Fatty Acids					
	C14:0	C16:0	C16:1	C18:0	C18:1	C18:2
E. guineensis Mesocarp						
16 weeks	3.3	42.8	18.4	5.5	28.5	1.6
17 weeks	8.0	43.5	17.1	23.0	7.5	1.2
18 weeks	6.4	42.0	13.0	28.7	8.6	1.4
20 weeks	3.4	29.6	29.7	8.4	29.0	-
E. oleifera Mesocarp						
18 weeks	5.1	44.5	7.9	20.0	14.4	8.2
20 weeks	5.6	28.9	39.2	2.5	23.8	-
Embryoids						
a	2.0	29.6	29.5	8.4	30.6	-
b	2.5	28.1	25.0	17.0	26.8	1.2
c	2.4	29.6	32.5	7.0	28.6	-

Incorporation into triacylglycerols (TG) by mesocarp protoplasts increased with the age of the fruit and was maximum at 20 weeks. (Table 2). Incorporation into polar lipids (PL) however was maximum in protoplasts from the young fruit and fell sharply as the fruit ripened. Thus at 20 weeks the mesocarp protoplasts are involved in the synthesis of storage oils while the young fruit is involved in cellular and membrane synthesis. Embryoid protoplasts showed the reverse trend with TG levels decreasing and PL levels increasing as embryogenesis proceeded. This reflects an increasing synthesis of cellular and membrane lipids during embryogenesis. Turnham and Northcote (2) studying the incorporation of ^{14}C-acetate into lipids during embryogenesis in oil palm tissue cultures however, reported increasing TG levels.

The fatty acid composition of protoplast lipids is tabulated (Table 3). An interesting feature common to all the protoplasts was the synthesis of C16:1 as a major fatty acid (8-39%). The proportion of C16:1 synthesized remained approximately constant (13-18%) from weeks 16-18 but increased sharply at week 20 (29.7%) in tenera mesocarp protoplasts. *E. oleifera* protoplasts followed a similar trend. The increase in C16:1 with the concomitant decrease in C16:0 synthesis at week 20 suggests that C16:1 is formed by the desaturation of C16:0. In embryoid protoplasts the levels of both C16:0 and C16:1 remained approximately constant at all three stages. Another interesting feature is that the ratio of C18:1 in mesocarp protoplasts varied with the age of the fruit and appeared to follow a cycle. At week 16, C18:0 was only a minor fatty acid and C18:1 present at high levels. This was reversed at weeks 17-18 with the ratio in favour of C18:0. At week 20, the ratio was again reversed in favour of C18:1. The embryoid protoplasts however, showed no significant change in fatty acid composition.

CONCLUSION

The results show that lipid metabolism is altered in *E. guineensis* and *E. oleifera* protoplasts. The conditions of isolation probably play a major role in determining the metabolic activity of protoplasts. Webb and Williams (3) reported a change in lipid metabolism of *Vicia faba* mesophyll protoplasts. However, Sato (4) also working on *V. faba* found no such alteration in the lipid metabolism of mesophyll cell protoplasts. The synthesis of C16:1 as a major fatty acid is interesting because this fatty acid is only a trace component (0.1-0.3%) in palm mesocarp oil (5). The findings that both C16:1 and C18:1 are present in increased amounts in 20 week mesocarp protoplasts suggest that both fatty acids are produced by a common desaturase whose specificity is altered during protoplast isolation.

ACKNOWLEDGEMENT

We thank the Director-General of PORIM for his support and permission to present this paper.

REFERENCES

1. Oo, K.C., Teh, S.K., Khor, H.T. and Ong, A.S.H. (1985), Lipids 20. 205 - 210.
2. Turnham, E. and Northcote, D.H. (1984) Phytochemistry 23, 35 - 39.
3. Webb, M.S. and Williams, J.P. (1984) Plant and Cell Physiol 25. 1551 - 1559.
4. Sato, N. (1985) Plant and Cell Physiol 26. 805 - 811.
5. Tan, B.K. and Oh, F.C.H. (1981) PORIM Technology 3. 1 - 5.

IDENTIFICATION OF PROTEINS ASSOCIATED WITH CHANGES IN THE LINOLENATE CONTENT OF SOYBEAN COTYLEDONS

Xuemin Wang, David F. Hildebrand, and Glenn B. Collins

Department of Agronomy
University of Kentucky
Lexington, Kentucky 40546

INTRODUCTION

The relatively high content of linolenate in soybean seeds has been considered as a major factor resulting in oxidative and flavor instability of soybean oil. Various approaches have been undertaken in attempts to manipulate the linolenate content in soybean seeds. Reduction of linolenate biosynthesis due to a substituted pyridazinone (4-chloro, 5-dimethylamino, 2-phenyl-3(2H) pyridazinone, also San 9785) was reported in soybean cotyledons developing in vitro.[1] Mutants with a lower linolenate content have been identified in soybeans. Of these, the mutant C1640 has about a 50% reduction of linolenate content in mature seeds and this low linolenate mutation is controlled by one nuclear locus.[2] This report presents a comparison of the effects of San 9785 and the mutation (C1640) on the reduction of linolenate content and compares the chemical and mutation effects on protein synthesis in order to identify the gene products associated with the regulation of linolenate production in developing soybean cotyledons.

MATERIALS AND METHODS

Developing cotyledons were excised from greenhouse-grown soybean plants. One cotyledon was placed in the medium containing San 9785 (100 um) and the other cultured in the control solution[3] without San 9785. These cotyledons were collected after two weeks in culture and then lyophilized. Neutral lipids from the cotyledons were extracted with petroleum ether. Fatty acid metyl esters were prepared and analyzed by gas chromatography.[4]

Cotyledons (7mm in length), after incubation in the solution with San 9785, were radioactively labelled with ^{35}S methionine (100 uci/ml) for one hour. Proteins were extracted as previously described.[5] Equal amounts of radioactivety based on counts per minute were applied to each gel. Two dimensional gel electrophoresis was conducted according to O'Farrell's[6] procedure except 8 M urea was used in the isoelectrical focusing gels.

RESULTS AND DISCUSSION

The low linolenate mutant C1640 exhibited decreased levels of linolenate in neutral lipids of the cotyledons at various developmental stages when compared to cv. Century, with the difference being largest toward maturity. There were no significant changes in other major fatty acids except a corresponding increase in the linoleate content. When the cotyledons were cultured in the presence of San 9785, the content of linolenate was significantly reduced in both genotypes, Century and C1640. A greater decrease of linolenate formation due to San 9785 was seen in the cotyledons excised at early stages of development in both genotypes, suggesting a more active synthesis of linolenate at the early stages. In addition, a similar reduction of linolenate content in Century and C1640 occured at the 5 and 7 mm length stages. However, at the later stages (9 and 11 mm) San 9785 was much less effective in decreasing the linolenate content of the mutant. This result is consistent with the above observation that the activity for linolenate synthesis in the low linolenate mutant is greatly decreased in the later stages of cotyledon development. Examination of phosphatidylcholine (PC) molecular species revealed that the predominant alteration such as a buildup of 18:2/18:3 and a decrease in 16:0/18:3 PC species in the mutant resembles the changes in Century cotyledons treated with San 9785.[4]

The changes of protein synthesis in the low linolenate mutant and San 9785 treated cotyledons of Century were compared with untreated Century cotyledons. Distinct alterations of two protein spots observed in the mutant also occurred in the cotyledons of Century treated with San 9785. The formation of one protein was substantially reduced in the mutant and its synthesis was also greatly inhibited with San 9785 treatment (Fig.1). This protein appears to have a molecular weight of 100,000 Da and an isoelectric point of 5.5. Another protein increased in both the low linolenate mutant and San 9785 treated cotyledons (Fig.2). The higher concentration of San 9785 enhanced the synthesis of this protein (Fig.2,D.E).

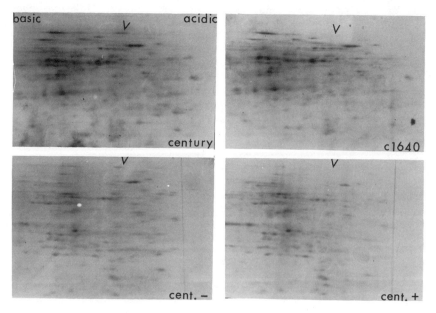

Fig. 1. Autoradiograms of 2-D electrophoretic analysis. Cent.+ & - were from Century cotyledons labelled after preincubation in the media with and without San 9785 for 3hs. The pH gradient in the first (horizontal) dimension was from 3 to 10. A 10-20% SDS polyacrylamide gel was run in the second (vertical) dimension. Marker (V) indicates the positions of the changed protein.

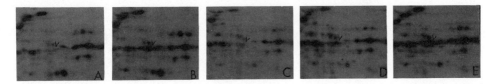

Fig. 2. Portions of the fluorograms of 2-D electrophoretic analysis. A and B were from Century and C1640, respectively. C,D, and E were from Century cotyledons incubated in the media with 0, 100, and 500 um San 9785 for 20 hs. . A 10% SDS polyacrylamide gel was run in the second dimension.

These results demonstrate the similarity of the effects of the low linolenate mutation on the reduction of linolenate content to that of San 9785 modulation and also show that the changes in the proteins in the mutant cotyledons parallel those resulting from San 9785 treatment. Though detailed studies are needed to prove the involvement of these altered proteins in the process of linolenate biosynthesis, the current investigation provides a suitable system to study the association of these proteins in the regulation of linolenate content. Studies on the relation of these polypepetides to low linolenate content have been undertaken by examining their cosegregation with low linolenate level in F2 progeny. Further efforts are now in progress to isolate and biochemically characterize these proteins.

ACKNOWLEDGEMENTS

The research is supported by a grant from the American Soybean Association. We particularly thank Drs. J.R. Wilcox and J.B. St.John for providing us the mutant C1640 and the compound San 9785, respectively.

REFERENCES

1. J. B. St John and M.N.Christiansen, Chemical manipulation of soybean (Glycine max L. Merr.) oil quality, in: "Bioregulator: Biochemistry and uses." R. L. Ory and F. R. Ritting, eds. ACS Symposium Series N. 257 (1984).
2. J. R. Wilcox and J. F. Cavins, Inheritance of low linolenic acid content of the seed oil of a mutant in Glycine max, Theor. Appl. Genet., 71:74 (1985).
3. F. C. Hsu and R. L. Obendorf, Compositional analysis of in vitro matured soybean seeds, Plant Sci. Letters. 27:129 (1982)
4. X. M. Wang, D. F. Hildebrand, H. D. Norman, M. L. Dahmer, J. B. St John, and G. B. Collins, Reduction of linolenate content in soybean cotyledons by a substituted pyridazinone, Phytochemistry (in review).
5. C. C. Des Francs, H. Thiellement, and D. De Vienne, Analysis of leaf proteins by two dimensional gel electrophoresis, Plant Physiol. 78: 178 (1985).
6. P. H. O'Farrell, High resolution two dimensional electrophoresis of proteins, J. Biol. Chem. 250: 4007 (1975).

ACYLTRANSFERASES IN DEVELOPING SEEDS OF OILSEED RAPE

Denis J. Murphy

Deparment of Botany
University of Durham, Durham, DH1 3LE
United Kingdom

INTRODUCTION

The oilseed rape Brassica napus is the major oil-bearing crop grown in temperate regions of Canada, Northern Europe and Asia. Despite greatly increased production, together with recent demands for higher oil quality, surprisingly little is known about the mechanism of oil biosynthesis in this important crop. The principal route of storage triacylglycerol (TG) biosynthesis in oilseeds is believed to be via the Kennedy pathway. In addition, high polyunsaturate seeds like sunflower (1), safflower (2) and linseed (3) are able to exchange acyl moieties between a phosphatidylcholine (PC) pool and the Kennedy pathway. This exchange was not found in the low polyunsaturate tissue, avocado mesocarp (4), but was recently found in another low polyunsaturate tissue, oilseed rape (5).

While the "microsomal" fraction of most oilseeds is highly active with respect to TG biosynthesis, the role of other subcellular fractions is less well documented. In this report the properties and subcellular distribution of the acyltransferases responsible for channelling acyl moieties either to PC or to the Kennedy pathway in oilseed rape are described.

MATERIALS AND METHODS

Microsomal, oilbody, and other subcellular fractions were prepared from developing rapeseed, harvested at about 4 weeks after flowering, by standard procedures (5). Incubations with radiolabelled substrates and analytical procedures have been described elsewhere (6).

RESULTS AND DISCUSSION

Microsomal fractions of oilseed rape readily incorporated $(1-^{14}C)$ oleoyl CoA into PC, PA and TG, indicating the presence of active Kennedy pathway and PC-acyltransferase activities. Acyl transfer to PC was hugely stimulated in the presence of exogenous LPC but, in the presence of exogenous LPA the vast majority of the acyl transfer was to PA (Table 1). In each case the acylation of the added lysophospholipid was virtually the only route for oleate metabolisation, resulting in the accumulation of labelled PC or PA over the short term. By means of pulse-chase studies, it was possible to observe some flow of label from PA towards TG and PC, which is consistent with the proposed operation of an equilibrium between DG and PC (1-4). It was more difficult to chase label out of the PC pool and this may in part be due to the relatively large DG pool in rapeseed microsomes, which would favour the DG → PC reaction, but not the reverse reaction. The rapeseed microsomes used here also had a very low oleate desaturate activity and it is possible that, if the purpose of the DG ⇌ PC equilibrium is indeed to enrich the DG pool in polyunsaturates (4,5) then there would be a selection against the labelled oleoyl-PC for DG formation.

Table 1

Effect of exogenous acyl acceptors upon oleoyl CoA incorporation by rapeseed microsomes

Addition (nmol ml^{-1})	Time (min)	PC	PA % total ^{14}C	DG	TG
none	20	35	20	1	6
LPC (27)	20	77	9	tr	tr
LPC (110)	20	93	2	tr	1
LPC (240)	20	88	4	tr	tr
LPA (40)	20	10	64	1	2
LPA (125)	20	6	75	2	3
LPA (200)	20	3	80	2	4
LPC (110)	20 + 240*	88	7	tr	tr
LPA (125)	20 + 240*	8	45	10	13
LPA (125) + LPC (110)	20		69	1	2

*pulse chase. See ref 5 for incubation conditions

The possibility that the LPA- and LPC- dependent acyltransferases were due to a single enzymic activity was investigated by following their kinetic properties under various conditions. The two activities exhibited dramatically different responses to a range of mild detergents, with the LPC-acyltransferase invariably proving more susceptible to such treatment. The LPC-acyltransferase was also more susceptible to inhibition by BSA. Finally, while both activities exhibited Michelis-Menten kinetics with respect to oleoyl CoA, their kinetic constants and their behavious in the presence of their lysophospholipid substrates were very different. The rapeseed LPC-acyltransferase had an apparent K_m for oleoyl CoA of 15.3uM (in the presence of 110 nmoles ml^{-1} LPC), which is slightly more than the value found in the safflower enzyme, i.e. 9.5 uM (7). The kinetics of the rapeseed enzyme with respect to LPC concentration were not straightforward, showing an increase in activity with LPC amounts up to 110 nmoles.. ml^{-1} but thereafter a rapid decline with higher LPC concentrations. The LPA-acyltransferase had an apparent K_m for oleoyl CoA of 12.5uM. It also gave straightforward kinetics with respect to LPA and an apparent K_m for LPA of 31uM was derived. The conclusion from these studies is that there are probably at least two separate enzymes responsible for acyltransfer to LPA and LPC respectively in rapeseed. Acyltransfer to other lysophospholipids such as LPE or LPI occurred at only very low rates in this tissue.

Table 2

Subcellular distribution of acyltransferase activities in rapeseed

Fraction	Additions	Incorporation of oleate nmol.mg^{-1} protein 20 min^{-1}					
		PC	PA	MG	DG	TG	Total
Microsomes	none	5.6	1.2	2.0	2.3	1.0	12.1
	LPC	24.2	0.3	0.3	0.4	0.4	25.6
	LPA	4.7	11.8	2.1	3.3	0.5	22.4
Oil bodies	none	5.5	2.7	2.0	1.9	1.4	13.5
	LPC	17.4	8.5	1.7	2.0	2.2	31.8
	LPA	2.6	12.3	1.8	3.0	2.5	22.2
Soluble Fraction	none	0.3	0.9	3.5	4.8	tr	9.5
	LPC	0.6	0.5	4.7	6.3	tr	12.1
	LPA	0.2	0.8	5.7	5.6	tr	12.3

The results of subcellular localisation studies are summarised in Table 2. The microsomal and oil body fractions both contained all the activities of the Kennedy pathway and had especially high LPC- and LPA- acyltransferase activities. The specific activities of the microsomal enzymes were generally higher than those of the oil bodies, but not dramatically so. The soluble fraction had high MG and DG-forming activities, but was unable to form PC or TG from oleoyl CoA. It is possible that the soluble fraction contains G3P and LPA acyltransferases but that the products are rapidly dephosphorylated to form MG and DG respectively (8). Soluble (9) and microsomal (10) G3P acyltransferases have been found in cocoa and safflower seeds respectively.

ACKNOWLEDGEMENTS

These studies were financed by the Agriculture and Food Research Council, U.K. and by the Nuffield Foundation. The expert technical assistance of Duncan Libby is greatly appreciated.

REFERENCES

1. Stobart, A.K. and Stymne, S. (1985) Biochem. J. 232, 217-221
2. Slack, C.R., Roughan, P.G., Browse, J.A. and Gardiner, S.E. (1985) Biochim. Biophys. Acta 833, 438-448
3. Stymne, S. and Stobart, A.K. (1985) Planta 164, 101-104
4. Stobart, A.K. and Stymne, S. (1985) Planta 163, 114-125
5. Murphy, D.J. (1986) J. Plant Physiol., in press
6. Murphy, D.J., Mukherjee, K.D. and Latzko, E.T. (1983) Biochem. J. 213, 249-252
7. Moreau, R.A. and Stumpf, P.K. (1982) Plant Physiol. 69, 1293-1297
8. Murphy, D.J. and Mukherjee, K.D. (1986) Biochem. J., submitted
9. Fritz, P.J. Kauffman, J.M., Robertson, C.A. and Wilson, M.R. (1986) J. Biol. Chem 261. 194-199
10. Ichihara, K. (1984) Arch. Biochem. Biophys. 232, 686-698

PROPERTIES OF ACYL-(ACYL-CARRIER PROTEIN):GLYCEROL-3-PHOSPHATE ACYLTRANSFERASE FROM GREENING SQUASH COTYLEDONS

Ikuo Nishida[1], Margrit Frentzen[2] and Norio Murata[1]

[1]National Institute for Basic Biology, Okazaki, Japan
[2]Institut für Allgemeine Botanik, Universität Hamburg
Hamburg, FRG

INTRODUCTION

Fatty acid and molecular composition analyses of the lipids from higher plants indicate that a proportion of saturated molecular species of phosphatidylglycerol (PG) from plastids are well correlated with the chilling sensitivity of plants[1-3]. The thermotrophic phase behavior of aqueous dispersions of leaf lipids suggests that only PG from chilling-sensitive plants, but no other lipid from either chilling-sensitive or chilling-resistant plants, undergoes phase transition at room temperature or above[4]. The major saturated molecular species of PG are sn-1,2-dipalmitoyl and 1-palmitoyl-2-(trans-3)hexadecenoyl. Since in plastid PG either palmitic acid (16:0) or trans-3-hexadecenoic acid (16:1t) are esterified to the C-2 position of glycerol backbone, the proportion of 16:0 at the C-1 position is equal to that of the saturated molecular species. Acyl-(acyl-carrier protein):glycerol-3-phosphate acyltransferase (hereinafter referred to as acyltransferase) in plastids carries out the first step of lipid biosynthesis in the chloroplasts i.e. esterification at the C-1 position of glycerol-3-phosphate[5,6]. We speculated that the substrate selectivity of this enzyme directs the fatty acid composition at the C-1 position and therefore determines the molecular species composition. To test this hypothesis, we purified the enzyme from greening cotyledons of squash, a chilling-sensitive plant, and examined its substrate selectivity.

MATERIALS AND METHODS

Four hundred g dry weight of squash (Cucurbita moschata, cv. Shirakikuza) seeds were soaked for 3 h, then germinated and grown on moist vermiculite in darkness for 4-5 days at 30°C. The etiolated cotyledons were illuminated for 9 h with fluorescent light at an intensity of 20 µE/m^2 sec at 25°C. About 400-g fresh weight of the greening cotyledons were homogenized in 800 ml of 50 mM Tris-HCl (pH 7.4), and the acyltransferase was purified from the homogenate as summarized in Table 1. The enzyme was assayed according to Bertrams and Heinz[7] with palmitoyl-CoA as the acyl donor. The substrate selectivity of the enzyme was studied according to Frentzen et al.[8] with acyl-ACP as the acyl donor.

RESULTS AND DISCUSSION

The acyltransferase was purified from the homogenate of greening squash cotyledons by acid treatment, hydrophobic chromatography with butyl-Toyopearl, gel filtration chromatography with Sephacryl S-300, and anion exchange chromatography with DEAE-Toyopearl (Table 1). In the last step, the enzyme activity was separated into two fractions; one type of enzyme (designated as AT1) was not adsorbed, on the DEAE-Toyopearl column equilibrated with 50 mM Tris-HCl (pH 7.4), while the other (designated as AT2) was adsorbed on the column. The two types of acyltransferase were detected in the stroma fraction of intact chloroplasts from the greening squash cotyledons (data not shown). The physicochemical properties of the isomeric acyltransferase from squash cotyledons are summarized, together with those from spinach and pea (chilling-resistant plants) in Table 2. Isomeric acyltransferases in pea chloroplasts having different isoelectric points of 6.3 and 6.6 have been reported[7].

Kinetic parameters of the two types of acyltransferase from squash were compared with those from spinach (Table 3). In both AT1 and AT2 from squash the K_m for palmitoyl-ACP (16:0-ACP) was twice as large as that for oleoyl-ACP (18:1-ACP), whereas in the spinach acyltransferase the former was ten times as large as the latter. These observations suggest that both AT1 and AT2 from squash have no strict selectivity to 16:0-ACP and 18:1-ACP, whereas the spinach enzyme has a much higher affinity to 18:1-ACP than to 16:0-ACP. The K_m for glycerol-3-phosphate greatly depended on the acyl donors. The K_m measured with 16:0-ACP as the acyl donor, was 50-100 times as large as that with 18:1-ACP as the acyl donor. The K_m of spinach acyltransferase for glycerol-3-phosphate with 16:0-ACP was ten times as high as the glycerol-3-phosphate concentration (100-200 μM) in spinach chloroplasts[9]. If the concentration of glycerol-3-phosphate in squash chloroplasts is the same as in spinach chloroplasts, the squash enzyme should esterify 16:0 to the C-1 position much more than the spinach enzyme.

Table 3 shows that the squash acyltransferase can also efficiently use 18:0-ACP. The low level of 18:0 at the C-1 position of chloroplast lipids may result from a low concentration of 18:0-ACP in the chloroplasts.

Table 1. Purification of Acyl-(Acyl-Carrier Protein):Glycerol-3-phosphate Acyltransferase from Greening Squash Cotyledons

Purification step	Total Activity nmol/min (%)	Protein mg	Specific Activity nmol/min mg protein	Purification factor
Homogenate	7,500 (100)	25,000	0.3	1
pH 5.4 supernatant	7,200 (96)	9,000	0.8	3
Butyl-Toyopearl	5,600 (75)	3,500	1.6	5
Sephacryl S-300	4,200 (56)	950	4.4	15
DEAE-Toyopearl				
not absorbed (AT1)	1,200 (16)	390	3.1	10
absorbed (AT2)	2,000 (27)	80	25	83
AT1	1,200 (16)	390	3.1	10
CM-Toyopearl	980 (13)	54	18	60
Sephacryl S-300	600 (8)	9.2	65	217
ACP-Affinity	440 (6)	0.06	7,200	24,000
AT2	620 (8)	25	25	83
Butyl-Toyopearl	350 (5)	4.8	73	240
Sephacryl S-300	310 (4)	2.9	110	370

Table 2. Physicochemical Properties of Acyl-(Acyl-Carrier Protein):
Glycerol-3-Phosphate Acyltransferase from Higher Plants

Plant	Molecular weight	Isoelectric point[a]
Squash		
AT1	27,000[b]	6.6
AT2	38,000[b]	5.5
Spinach[c]	42,000[d]	5.2
Pea[c]	42,000[d]	6.3 & 6.6

[a]Determined by isoelectric focusing. [b]Determined by HPLC on TSKgel G3000SW. [c]Data from Ref. 7. [d]Determined by chromatography on Sephadex G-100.

Table 3. Substrate Selectivity of Acyl-(Acyl-Carrier Protein):Glycerol-3-Phosphate Acyltransferases from Squash (Chilling-Sensitive) and Spinach (Chilling-Resistant)

Plant	Acyl-ACP	K_m (µM) for		V_{max} nmol/min mg protein (rel.[a])
		acyl-ACP	glycerol-3-phosphate	
Squash				
AT1	16:0	0.72	300	3800 (1.7)
	18:0	0.60	30	6700 (2.9)
	18:1	0.36	4	2300 (1.0)
AT2	16:0	4.3	310	140 (3.2)
	18:0	1.1	500	130 (3.0)
	18:1	2.2	6	44 (1.0)
Spinach[b]	16:0	3.2	3150	320 (3.6)
	18:0	3.3	3160	260 (2.9)
	18:1	0.3	31	90 (1.0)

[a]Relative value. [b]Data from Ref. 8.

ACKNOWLEDGMENT

This work was supported by a Grant-in-Aid for Scientific Research (61440002) from the Ministry of Education, Science and Culture, Japan.

REFERENCES

1. N. Murata, N. Sato, N. Takahashi and Y. Hamazaki, Plant Cell Physiol. 23:1071-1079 (1982).
2. N. Murata, Plant Cell Physiol. 24:81-86 (1983).
3. N. Murata and K. Kurisu, in: "Structure, Function and Metabolism of Plant Lipids", P.-A. Siegenthaler and W. Eichenberger, eds., Elsevier Science Publishers, Amsterdam (1984).
4. N. Murata and J. Yamaya, Plant Physiol. 74:1016-1024 (1984).
5. M. Bertrams and E. Heinz, Planta 132:161-168 (1976).
6. J. Joyard and R. Douce, Biochim. Biophys. Acta 486:273-285 (1977).
7. M. Bertrams and E. Heinz, Plant Physiol. 68:653-657 (1981).
8. M. Frentzen, E. Heinz, T. A. McKeon and P. K. Stumpf, Eur. J. Biochem. 129:629-636 (1983).
9. A. Sauer and K.-P. Heise, Z. Naturforsch. 38c:399-404 (1983).

LIPID BIOSYNTHESIS IN EPIDERMAL, GUARD AND MESOPHYLL CELL PROTOPLASTS
FROM LEAVES OF *VICIA FABA* L

Naoki Sato

Department of Botany, Faculty of Science
University of Tokyo
Hongo, Bunkyo-ku, Tokyo 113, Japan

INTRODUCTION

Epidermis is known to be the site of synthesis of long-chain fatty acids which are supposed to be precursors to components of epicuticular wax[1]. Epidermis of leaf tissue is composed of epidermal cells and guard cells, which are different in cellular architecture, including plastid morphology: plastid development is generally repressed in epidermal cells while guard cell chloroplasts are active in photosynthesis and starch synthesis. In the light of the current view that the plastid is the sole site of synthesis of fatty acids, I suspected that epidermal and guard cells are quite different in ability to synthesize long-chain fatty acids.

MATERIALS AND METHODS

Seeds of *Vicia faba* L. cv. Uchikoshi No. 1 were germinated and grown at $25°C$ under a 14-h photoperiod for 4 weeks. Laminae of the 3rd to 6th leaf pairs were cut with razor blade, soaked in 0.8% Macerozyme solution in 550 mM sorbitol/5 mM MES-KOH (pH 5.8), vacuum-infiltrated, and incubated at $25°C$ for 30 min. Then abaxial (lower) epidermis which loosely adhered to mesophyll tissue was peeled manually. Epidermal peels and remaining leaf tissue were further digested in Macerozyme and Cellulase solutions. Liberated protoplasts were fractionated and purified by step density gradient centrifugation in Percoll.
The protoplasts (about 10^6 cells) suspended in 200 µl of 20 mM HEPES-KOH (pH 7.0), 550 mM sorbitol, 0.5 mM $CaCl_2$ were incubated with 10 µCi sodium [2-^{14}C]acetate (0.8 mM) at $25°C$ for 60 min under incandescent light at 800 W/m^2. Lipids were extracted by the Bligh-Dyer method and separated by two dimensional TLC. Fatty acids were analyzed by radio gas chromatography. Experimental details were described in reference 2.

RESULTS AND DISCUSSION

Protoplast preparation

Mesophyll cell protoplasts (MCP), guard cell protoplasts (GCP) were more than 90% pure upon microscopic inspection. Two fractions of epidermal cell protoplasts (ECP) were obtained: heavy and light ECPs. Heavy ECP

Table 1. Distribution of radioactivity in lipid classes after labelling with [2-^{14}C]acetate for 60 min (%)

Protoplast	P C	P E	P I	P G	P A	S G	MGDG	F A	T G
MCP fr.	34	5	1	14	2	0	2	32	9
GCP fr.	44	6	5	14	0	2	1	2	17
H-ECP fr.	37	7	4	8	0	2	1	2	25
L-ECP fr.	34	15	1	9	0	2	1	4	24

Abbreviations: MCP fr., mesophyll cell protoplast fraction; GCP fr., guard cell protoplast fraction; H-ECP fr., heavy epidermal cell protoplast fraction; L-ECP fr., light epidermal cell protoplast fraction.
DGDG and SQDG were not labelled.

contained 23% GCP but light ECP was 99.9% pure. Chlorophyll content in pg/cell was 184 in MCP, 2.2 in GCP, and 0.3 in both fractions of ECP (the values presented were corrected for the contamination by other types of cells). The presence of chlorophyll in epidermal cells was verified by fluorescence microscopy: epidermal cells in epidermal peels as well as isolated ECP contained plastids which emitted bright red fluorescence. Electron microscopic observations showed that the plastids in ECP possessed thylakoid membranes which formed grana stacking consisting of 2-3 lamellae. ECP fractions, however, did not show any measurable activity of photosynthetic carbon fixation attributable to ECP. These results demonstrate that epidermal cells of Vicia faba leaves contained chlorophyllous plastids which are clearly distinguished from proplastids.

Lipid content in MCP

Content of major lipid classes in pmol fatty acids per cell was determined. MGDG, 276; DGDG, 254; TG, 67; PC, 65; PG, 56; FFA, 23; PE, 21. The results showed no sign of extensive degradation of lipids which was reported to occur during protoplast isolation^2.

Incorporation of [^{14}C]acetate

Activity of acetate incorporation into lipids in terms of fmol/cell/h was 5.0 in MCP, 3.0 in GCP, 4.4 in heavy ECP and 2.2 in light ECP. Methanolysis of lipids showed that more than 94% of the label in lipids in MCP, GCP and heavy ECP was localyzed in acyl moiety. In light ECP, 80% of the label was found in acyl moiety. These results indicate that all types of cells possess comparable activities of fatty acid synthesis.
Analysis of distribution of radioactivity in lipid classes (Table 1) showed that PC contained 34-44% of radioactivity in all protoplast preparations. FA contained 32% of radioactivity in MCP, whereas, in GCP and ECP, TG contained 17-25% of radioactivity. The distribution of radioactivity was similar in GCP and ECP but different in MCP.
Analysis of distribution of radioactivity in fatty acids (Table 2) showed that 16:0 and 18:1 were major labelled fatty acids in all protoplast preparations. GCP and ECP also contained radioactive 20:0 and 22:0. Careful measurement of radioactivity in fatty acids of MCP showed less than 0.5% of the total radioactivity was present in 20:0 and 22:0. These results showed that both guard and epidermal cells possessed the ability to synthesize very long chain fatty acids, but mesophyll cells did not.

Table 2. Distribution of radioactivity in fatty acids after labelling with [2-^{14}C]acetate for 60 min (%)

Protoplast	16:0	17:0	18:0	18:1	20:0	22:0
MCP fr.	19	0	Tr	81	0	0
GCP fr.	42	0	5	37	6	10
H-ECP fr.	38	2	5	37	8	10
L-ECP fr.	46	3	8	27	10	6

In conclusion, epidermal cells as well as guard cells and mesophyll cells have comparable activities of fatty acid synthesis, though they contain chloroplasts of different size and different morphology. The ability to synthesize fatty acids longer than C_{18} was present in both guard and epidermal cells but not in mesophyll cells.

References

1. P. E. Kolattukudy, Further evidence for an elongation-decarboxylation mechanism in the biosynthesis of paraffins in leaves, Plant Physiol, 43: 375 (1968).
2. N. Sato, Lipid biosynthesis in epidermal, guard and mesophyll cell protoplasts from leaves of Vicia faba L., Plant Cell Physiol., 26: 805 (1985).
3. M. S. Webb and J. P. Williams, Changes in the lipid and fatty acid composition of Vicia faba mesophyll protoplasts induced by isolation, Plant Cell Physiol., 25: 1541 (1984).

ONTOGENETIC VARIATIONS IN THE CHEMICAL COMPOSITION OF MAIZE SURFACE LIPIDS

P. Avato, G. Bianchi and *F. Salamini

Dipartimento di Chimica Organica, Università, I-27100 Pavia (Italy) and *Max Planck Institut für Züchtungsforschung D-5000 Köln 30 (FRD)

INTRODUCTION

Morphological variations of wax surface structures during the ontogeny of many plants have been often observed[1]. Seedlings of maize are typically covered by a wax layer up to the fifth-sixth leaf stage of growth. From there on the leaf surface assumes a _glossy_ appearance, which is maintained throughout the plant life[2]. Ultrastructural modifications of waxes induced by age are usually associated with changes in the chemical composition[3]. Thus, comparison of waxes from young seedlings and adult plants of maize revealed well defined differences in their chemistry. The results from the compositional analyses of the epicuticular waxes from maize at the two different stages of growth are the subject of the present contribution. Experimental data are also correlated with earlier observations on the biochemical pathways involved in the synthesis of maize surface lipids. Details on the experimental procedures are given in reference 4.

RESULTS

Wax isolated from maize seedlings consist of alcohols (63%), aldehydes (20%) and esters (16%) plus traces of acids[5], whereas that from mature leaves is characterized by esters (42%) as dominant class of compounds. The percentages of the other wax classes are: alkanes, 17%; acids, 14%; alcohols, 14% and aldehydes, 9%. Besides, sterols which have not been detected in the seedlings, account for 4% in the surface waxes from mature leaves.

Table 1 shows the homologue composition of four of wax fractions. Three chains, C_{27}, C_{29} and C_{31} prevail in the alkanes. Mature leaves are characterized also by high amounts of the C_{33}, C_{35} alkanes. Distributions of alcohols and aldehydes from seedlings are almost indistinguishable, being the 32 carbon chain by far the most prominent (99, 96%). The same chain only accounts for 28 and 18% in the alcohols and aldehydes from mature leaves, respectively. A noticeable difference is also observable

549

Table 1. Composition (%) of wax fractions from seedlings (S) and mature leaves (L) of maize

Chain length	Alkanes S	Alkanes L	Aldehydes S	Aldehydes L	Alcohols S	Alcohols L	Acids S	Acids L
16							25	2
18							13	2
20							4	3
22						1	6	5
23	2	1		2		t	-	1
24	t	1		1		3	14	8
25	5	3		2		1	-	t
26	t	1		9		14	22	10
27	13	9		3		3	-	2
28	t	2	1	24	t	18	12	19
29	29	30	-	4	-	t	-	2
30	t	2	3	32	1	25	t	24
31	49	30	-	3	-	3	-	2
32	t	t	96	18	99	28	4	16
33-35	2	21		2		4		4

in the free acids patterns. Homologues in the range C_{26}-C_{32} amount to 69% in the adult plants. By contrast, in the seedlings 48% of the total is due to C_{24}-C_{28} chains with high amounts of C_{16} (25%) and C_{18} (13%) acids.

GC analysis of intact esters (Fig. 1) from wax seedlings(S) and adult plants (L) shows differences in their distributions[4], as also confirmed by the analysis of their alcohol and acid moieties. Unlike seedlings, where the C_{32} is practically the only esterified alcohol, in the adult leaves a wide range of chains contributes to the alcohol moieties, C_{24} (55%) and C_{26} (27%) being dominant. Longer chains than in wax from mature leaves

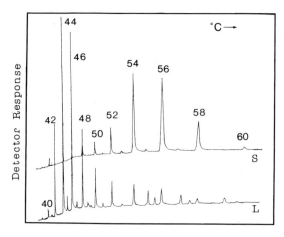

Fig. 1. GC analysis of intact esters (See text).

accumulate in the esterified acids of seedlings (C_{22}-C_{26}; C_{24} = 49% vs C_{20}-C_{24}; C_{20} = 42% in the adult plants). The three phytosterols, campesterol (14%), stigmasterol (16%) and ß-sitosterol (41%) have been so far identified in the sterol fraction from mature leaves wax.

DISCUSSION

The biosynthesis of the wax coat present on young maize plants can be genetically dissected taking advantage of several available glossy (waxyless) mutants. In general, all the mutants so far analyzed accumulate in their surface lipids larger amounts of esters than the wildtype[5]. Synthesis of this class of compounds has been interpreted as being unaffected by mutations and taking place independently from that of the other wax components. Thus, two distinct elongation-decarboxylation (ED) enzymic complexes have been proposed for maize surface lipids formation. ED-I is responsible for the synthesis of the long chain alkanes, aldehydes and alcohols. ED-II mainly controls the production of the shorter chain moieties of the esters.

From data here reported it is apparent how wax composition of mature plants of maize resembles that of several glossy mutants affecting the juvenile wax layer[5]. The reductive processes controlled by ED-I giving rise to aldehydes and alcohols appear to be not very active and show a loose-chain length specificity in synthesizing the C_{32} homologues typical of the seedling waxes (Table 1). Thus, accumulation of underused elongated acyl chain as free fatty acids is observed. By contrast, synthesis of esters is not negatively affected indicating that it occurs independently from the synthesis of the long chain components of the juvenile wax, that is in the ED-II system. In conclusion, the differences found in the wax chemical composition during the ontogenesis of maize prove that in both adult and young plants the ED-II pathway is effective; in the young plant the other pathway, ED-I, superimposed on ED-II, is also active. A most interesting question is to determine the physiological role of the predominance of given wax chemical components at different growth stages.

REFERENCES

1. E.A. Baker, Chemistry and morphology of plant epicuticular waxes, in: "The Plant Cuticle", D.F. Cutler, K.L. Alvin, and C.E. Price, eds., Academic Press, Inc., New York (1982).
2. A. Bianchi and G. Marchesi, The surface of the leaf in normal and glossy maize seedlings, Z. Vererbungsl. 91: 214 (1960).
3. P. Avato, G. Bianchi, and G. Mariani, Epicuticular waxes of Sorghum and some compositional changes with plant age, Phytochem. 23: 2843 (1984).
4. G. Bianchi, P. Avato, and F. Salamini, Surface waxes from grain, leaves, and husks of maize (Zea Mays L.), Cereal Chem. 61: 45 (1984).
5. A. Bianchi, G. Bianchi, P. Avato, and F. Salamini, Biosynthetic pathways of epicuticular wax of maize as assessed by mutation, light, plant age and inhibitor studies, Maydica 30: 179 (1985).

MECHANISM OF BIOSYNTHESIS OF ß-DIKETONES AND ALKAN-2-OL ESTERS FROM EPICUTICULAR WAXES

Giorgio Bianchi

Dipartimento di Chimica Organica, Università, I-27100 Pavia

Long chain aliphatic ß-diketones and alkan-2-ol esters are common constituents of plant surface lipids[1]. The ß-dicarbonyl pattern, two carbonyl groups 1,3 one another, has suggested that the biosynthesis of ß-diketones involves condensation of appropriate ß-ketoesters and esters, <u>via</u> a 'biological Claisen reaction'[2]. An alternative route has been advanced according to which ß-diketones are generated by an elongation-decarboxylation reaction on a protected ß-keto acid precursor[3].

The reaction mechanisms advanced are discussed in the light of a new hypothesis and in relationship to feeding experiment data, chemical genetic evidence and an abiological biomimetic synthesis of ß-diketones. As hentriacontan-14,16-dione **6** is the commonest ß-diketone found in cereal plant waxes we shall use it as model for our discussion (Scheme).

Possible biosynthetic pathways to ß-diketones and alkan-2-ol esters

The ß-ketopalmitate anion **1** that exists as two resonance hybrids, is attacked in a Claisen-type reaction by an activated palmitate **2** at the oxygen (O acylation) or at the carbon atom (C acylation) to yield intermediates **3** and **4**, respectively (Scheme). Diacylacetate **4**, a very reactive compound, readily extrudes CO_2 on ester hydrolysis to give hentriacontan-14,16-dione **6**. The latter may arise also from the acyloxy-α,ß-unsaturated derivative **5** through 1,3-sigmatropic rearrangement, both thermally and photochemically allowed,[4a] or also through a free radical process[4b].

As shown in the Scheme, ß-acyloxy-α,ß-unsaturated acid derivatives **3** may enter alternative reactions: they either rearrange to **4** or undergo reduction to alkan-2-ol ester **7** preceeded by decarboxylation to **5**; the latter vinylester may transform into hentriacontan-14,16-dione **6**.

<u>Discussion</u>. Labeling experiments in which a number of ^{14}C labeled acids were administrated to barley organ slices gave ß-diketones containing the isotopic marker[3]. Diketones were isolated and degradated by basic hydrolysis to fatty acids whose radioactivity was measured. Some data most pertinent to our discussion are reported in the Table.

Table. Distribution of label among fatty acids obtained by cleavage of hentriacontane-14,16-dione

Substrate	Radioactivity (%) of fatty acids	
	C_{14}	C_{16}
$\|1-{}^{14}C\|$ -myristic acid	50	50
$\|1-{}^{14}C\|$ -palmitic acid	–	100
$\|1-{}^{14}C\|$ -stearic acid	–	–

In the experiments with myristic acid the radioactivity was evenly distributed between the C_{14} and C_{16} acid moieties, while with palmitic acid 100% of the label was in the C_{16} chain; stearic acid was not incorporated. These feeding experiment data do not reporesent evidence supporting univocally the protection-elongation-decarboxylation mechanism of ß-diketones formation. On the contrary, in our opinion, the data fit better the alternative pathways presented in the Scheme. We intend to test the soundness of our mechanistic analysis planning feeding experiments with the hypothesized immediate precursors of ß-diketones via the route advanced by the cited Author[3], the diketodotriacontanoic acids whose synthesis has been accomplished in our laboratory[5]. We expect the diketoacids not to be metabolized.

We know of a demonstrated example "in vitro" of the 'biological Claisen-type' reaction to ß-diketones, that of the condensation of dihydroferuloylacetate with an hexanoate residue to give [6]-gingerdione[6].

Furthermore, route via C acylation presents a clear chemical homology with a published 'biogenetic-type synthesis' of hentriacontane-14,16--dione accomplished from appropriate derivatives of ß-ketostearate and myristic acid[7].

SCHEME

$$CH_3(CH_2)_{12} \overset{O^-}{\underset{\|}{C}}-CHCOX \longleftrightarrow CH_3(CH)_{12} \overset{O^-}{\underset{|}{C}}=CHCOX \quad \mathbf{1}$$

$$+ \quad CH_3(CH_2)_{14} COX \quad \mathbf{2}$$

O-acylation →

$CH_3(CH_2)_{12} C=CHCOX$ (with $O-C(CH_2)_{14}CH_3$ at (a)) $\xrightarrow[\text{rearrangement (a)}]{[1,3]}$ $CH_3(CH_2)_{12} \overset{O}{\underset{\|}{C}} \underset{|}{CHC}(CH_2)_{14} CH_3$ with COX **4**

[**3**]

$CH_3(CH_2)_{12} C=CH_2$ (with $O-C(CH_2)_{14}CH_3$ at (a)) $\xrightarrow[\text{rearrangement (a)}]{[1,3]}$ $CH_3(CH_2)_{12} \overset{O}{\underset{\|}{C}} CH_2 \overset{O}{\underset{\|}{C}} (CH_2)_{14} CH_3$ **6**

[H] ↓

$CH_3(CH_2)_{12} \underset{|}{CH} \underset{OCO(CH_2)_{14}CH_3}{CH_3}$ **7**

C-acylation →

X = CoA, ACP, OH

Compositional[8] data of waxes from wheat varieties and related species, barley[3], and mutants of these cereals provide useful biosynthetic pieces of information regarding ß-diketones and alkan-2-ol esters synthesis.

A characteristic of the class of ß-diketones from waxes is their peculiar response to the genetic asset of the plant. Among the numerous barley and wheat mutations affecting the production of ß-diketones, none was found in which these molecules were altered in chain length and structure.

The homologue and isomer compositions remain the same with hentriacontane-14,16-dione dominant. This unchangeability of the enzyme system governing the production of ß-diketones is in marked contrast to the frequency by which mutations alter the chain patterns of the other wax classes, particularly the alkanes which are proved to arise through elongation-decarboxylation.

Thus, the complete lack of sensitivity to mutations and the failure of detecting any free long chain diketoacids in ß-diketones containing waxes are against to the suggestion that ß-dicarbonyl compounds arise <u>via</u> decarboxylation.

Of interest, recently a tetraploid wheat, <u>cv</u>. Trinakria, has been found inhibiting the production of ß-dicarbonyl compounds[1]. This line and the crosses with ß-diketones producing wheat varieties would be suitable material for studying the manner of action of biosynthetic and inhibitor biosynthetic genes responsible for the production of ß-diketones and alkan-2-ol esters.

A GM study of the esters of this kind from rye[7], and previous related data, revealed that they contain the sole tridecan-2-ol and pentadecan-2-ol esterified with the even homologous acids in the strict range C_{16}-C_{22}. These findings contrast with alkan-1-ol esters that comprise numerous primary alcohols and wider ranges of acids. Furthermore, while alkan-1-ols occur also free and in high amount in the waxes, the alkan-2-ols have never been isolated in the free state.

On the basis of the data presented and discussed hereto, the conclusion should be drawn that ß-diketones and alkan-2-ol esters arise from "head-to head" type reactions of normal to medium acyl- and ß-ketoacyl chains. The protection-elongation-decarboxylation reaction for the formation of the former class of compounds, and the intermediacy of alkan-2-ols for the latter are hypotheses hardly tenable.

REFERENCES
1. G. Bianchi and M.L. Figini, <u>J.Agric.Food Chemistry</u> 34: 429(1986).
2. G. Bianchi and M. Corbellini, <u>Phytochemistry</u> 16: 943 (1977).
3. P. von Wettstein-Knowles in "Biochemistry and Metabolism of Plant Lipids". J.F.G.M. Wintermans and P.J.C. Kuiper, eds. <u>Elsevier Biomedical Press B.V.</u> and references therein cited, 69 (1982).
4. J. March, in "Advanced Organic Chemistry" (Mc Graw-Hill Kogakusha) (a) pp. 1037-1044; (b) p. 506 (1977).
5. G. Bianchi, unpublished results.
6. P. Dennif and D.A. Whiting, <u>J.C.S. Chem.Comm.</u> 711 (1976).
7. G. Bianchi and M. Grugni, <u>Gazz. Chim. Ital.</u> 115: 633 (1985).
8. G. Bianchi, <u>Genet. Agr.</u> 39: 471 (1985).

<u>Acknowledgement</u>   I thank CNR (Rome) for financial aid.

EPICUTICULAR WAX FORMATION ON NEEDLES OF *PICEA ABIES* AND *PINUS CEMBRA*

M.S. Günthardt-Goerg

Institute of Plant Biology, University of Zurich
Zollikerstrasse 107
CH-8008 Zürich / Switzerland

INTRODUCTION

Needles (plucked from trees at the alpine timberline) still enclosed in the buds and after flushing were examined by scanning electron microscopy. First wax protrusions emerge from the cuticle on the rim of the subsidiary cells and cover the slightly sunken guard cell mother cells. The stomata are fully developed and covered by the first wax layer before bud break. Crystallization of chromatographically isolated wax components show the wax structure to vary with its composition. Wax secretions of different structure seem to follow each other (Günthardt-Goerg 1985). This presentation intends to show the wax composition during the first months after flushing and when the different wax components are formed. At the alpine timberline the wax layer has to build up in a short vegetation period (July - October) to protect the young needles in the coming winter. The wax layer is poorly developed and/or has another composition under unfavourable conditions. This fact constitutes one reason for the needles' death through winter desiccation.

METHODS

Wax extracted from entire needles with chloroform, was analysed by means of TLC and GLC. Unknown compounds were isolated and identified by TLC, GLC, IR, MS, NMR, reference substances, and formation of derivatives. Details: Günthardt-Goerg 1986. Plant material: Monthly (first day of month) taken youngest needles of 2 healthy trees (*Picea abies (L.) Karst., Pinus cembra L.*) at the timberline, 2000 m a.s.l., Davos, Switzerland.

RESULTS

1. Composition of epicuticular waxes: Figure 1

Development of wax stops during winter. The wax samples taken from November to March had nearly the same composition. Figure 1 shows the percentage of the fractions constituting the March-sample. Total amounts for each species are taken as 100 %, small quantities < 1 % are not mentioned. The following fractions were found: HFA = free ω-hydroxy fatty acids, homologue

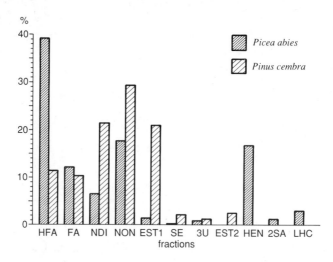

percentage = $C_{12}:C_{14}:C_{16}$ = 21:47:32 (*Picea abies*), $C_{12}:C_{14}:C_{16}$ = 29:49:22 (*Pinus cembra*). FA = free fatty acids, homologue percentage: *Picea abies* = $C_{14}:C_{16}:C_{20}:C_{22}:C_{24}:C_{26}:C_{28}:C_{30}$ = 3:2:14:26:28:15:5:7, homologue percentage: *Pinus cembra* = $C_{12}:C_{14}:C_{16}:C_{18}:C_{20}:C_{22}:C_{24}:C_{26}:C_{28}:C_{30}:C_{32}:C_{34}$ = 38:3:5:4 :13:8:11:3:3:3:7:2. NDI = nonacosane-5,10-diol, a secondary diol. NON = nonacosan-10-ol, a secondary alcohol. EST1 = estolides, *Picea abies*: estolides up to C_{44} with irregular percentage, saponification products of this fraction = C_{12} primary alcohol, $C_{14,16,20,22}$ FA and $C_{12,14}$ ω-HFA, *Pinus cembra*: estolide fraction contains estolides No. 2 (C_{36}), No. 5 (C_{38}) and No. 7 (C_{40}) with the following saponification products: C_{12} primary alcohol, $C_{12,14}$ FA, $C_{12,14,16}$ ω-HFA, $C_{12,14,16}$ α-ω-diols. SE = probably short chain esters (< C_{36}). 3U = 3 unknown compounds which resist saponification and give some evidence to be ketones C_{29}, C_{31} and C_{33}. EST2 = estolides No. 1, 3, 4, 6 and 8 (only *Pinus cembra*). HEN = hentriacontan-4-ol, a secondary alcohol (only *Picea abies*). 2SA = two additional secondary alcohols (only *Picea abies*). LHC = longchain hydrocarbons with retention times near alcanes $C_{40} - C_{46}$.

2. Development of wax fractions: Figure 2

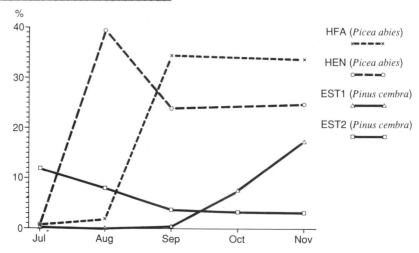

The first wax layer present on the needles two weeks after flushing was constituted by fatty acids, nonacosane-5,10-diol, nonacosan-10-ol, shortchain esters and 3 ketones (?) equal in *Picea abies* and *Pinus cembra* but in different amounts. In addition, first wax of *Picea abies* contained estolides and unknown longchain hydrocarbons, the wax of *Pinus cembra* on the other hand estolides No. 1,3,4,6,8. Figure 2 shows the development of some interesting fractions (total amount of all fractions per monthly analysed samples is taken as 100 %). Free HFA appeared in very small amounts in August and were abundant only as late as in September and following months on *Picea abies*-needles, whereas on *Pinus cembra*-needles the amount of HFA and of FA was increasing continuously from July to November. The secondary alcohol HEN *(Picea abies)* was present in August and following months. EST2 *(Pinus cembra)* were decreasing in favour of EST1. In the wax extract of *Picea abies*-needles considerable amounts (105 - 281 ppm) of a substance, ressembling phenols rather than wax components were found in September and in the following months. The substance was identified as p-hydroxy acetophenone (p-HAP). Homologue percentage in the fractions comprising several homologues did generally not change in the wax of *Picea abies* with the exception of the first sample taken in July and the fractions estolides as well as longchain hydrocarbons, which had fluctuating homologue composition. Most fractions of *Pinus cembra* indicated also a development at the level of the homologue composition.

CONCLUSIONS

Monthly analyses of the epicuticular wax of youngest needles from the timberline support the observation by SEM, which suggested, that the epicuticular wax layer is formed by layers following each other and not by one layer of a lipid mixture exuded all at once. Some components appear 1 - 3 months later (Fig. 2), others are decreasing, increasing or fluctuating. p-hydroxy acetophenone is an unexpected substance in the wax extracts of the youngest needles of a healthy, 80 year old *Picea abies*-tree with amounts as high as 1.3 - 4.8 mg/g needle dryweight (September - March). It has been determined in entire needles of damaged trees. They contained 2.7 - 14.4 mg p-HAP/g dw. p-HAP showed inhibiting properties in biotests and was therefore thought to be involved in stress phenomena (Hoque 1985). In addition to the genetical variation, the above shown facts of the wax formation contribute to the considerable variation found in randomly taken wax samples. If seasonal or human made disturbances occur, formation of certain wax components might be retarded or repressed, so that the resulting wax composition could be different.

LITERATURE

Günthardt, M.S., 1985, Entwicklung der Spaltöffnungen und der epicuticulären Wachsschicht bei *Pinus cembra* und *Picea abies*. Botanica Helvetica 95, 5-12.
Günthardt-Goerg, M.S., 1986, Epicuticular Wax of needles of *Pinus cembra*, *Pinus sylvestris* and *Picea abies*. Eur. J. For. Path. (in Press).
Hoque, E., 1985, Norway spruce dieback: Occurrence, isolation and biological activity of p-hydroxy acetophenone and p-hydroxy acetophenone-O-glucoside and their possible roles during stress phenomena. Eur. J. For. Path. 15, 129-145.

ACKNOWLEDGEMENT: I am grateful to Dr. Dr. h. c. Th. Keller (Swiss Federal Institute for Forestry Research, Birmensdorf) for valuable advice and to Miss M. Bättig and Miss V. Michellod for technical assistance. We thank the Swiss National Science Foundation and the Swiss Federal Institute for Forestry Research for financial support.

INTERACTION OF FEDERAL, INDUSTRIAL AND ACADEMIC RESEARCH

INTER-ORGANIZATIONAL COLLABORATION: A KEY TO SUCCESS IN TECHNOLOGICAL INNOVATION

Ronald L. Sampson

Queensland Innovation Centre Limited
46 Charlotte Street
Brisbane, Queensland 4000
Australia

Every country, industry, organization and researcher faces enormous challenges in today's rapidly changing world. The stakes are high for everyone in this increasingly competitive environment as economic growth and prosperity, standards of living, and even the survival of entire social systems are at risk. The challenges of change and competition we face are global in nature, as are the opportunities they create. Collaboration between individuals and organizations of all types is becoming a universal response to these challenges, and researchers in every field are at the forefront of the effort to turn challenges into opportunities. By way of survey and illustration, this paper is intended to review the situation with particular emphasis on why: (1) technological innovation is so important in today's world; and, (2) inter-organizational collaboration is seen as a key to successful innovation. To provide examples of such collaboration in action, some of the worldwide initiatives to foster collaboration are cited. Particular emphasis is given to those initiatives which help bridge between the very different sectors of industry, government and academe.

On The Challenges of Change

As pointed out by Merrifield (1983) and other, a worldwide explosion of technology and information is occurring, and the rate of change in every field of endeavor is enormous and, from some viewpoints, even alarming. It has been reported that over 90 percent of all mankind's scientific knowledge has been generated within only the last 30 years - with 75 percent developed within the last 20 years. The rate of change is accelerating, and the total body of knowledge is now doubling every ten years. Information is merely the reporting of knowledge, so it should come as no surprise that there are now over 100,000 different technical journals published worldwide - reporting over 72 billion pieces of new information each year (Buggie, 1982).

The explosion in technology is creating a more competitive world, and much effort has previously been devoted to interpretation and analysis of the challenges we now face. In their report on global competition, the President's Commission on Industrial Competitiveness (Young, 1985) stated that the challenge for America is "to maintain our high standard of living in an increasingly competitive world environment".

Given the economic changes that have evolved since that report, one might be tempted to embellish that statement today in far more foreboding terms.

On an individual basis, the technology explosion and the information overload it creates do impose inordinate challenges for each researcher to stay at the forefront of his or her respective field of technology. The National Science Foundation reported that the knowledge acquired by a graduate chemical engineer in 1960 was obsolete within five years. It is likely that an analysis today would reveal a far more frightening rate of obsolesence for any scientist or engineer - particularly those in "high technology" fields, such as biotechnology. Being up-to-date in one's discipline and avoiding obsolesence is, of course, essential for any researcher to be able to generate original new work of merit, and even to earn a livelihood in one's occupation of choice. The old adage that "what you don't know won't hurt you" is certainly not true today - if indeed it ever was.

On The Need for Technological Innovation

Great challenges and particularly great changes invariably create great opportunities, and it pays to look upon problems and challenges as merely "opportunities in disguise". Indeed one can frequently find a solution to a problem in a definition of the problem.

In 1966, Schmooker correlated technological innovation with economic activity. In 1982, Abernathy suggested that declining innovation in the United States had set the stage for the current state of difficulty that country faces. The literature is full of citations which document the obvious - that a decline in innovative capabilities presages a decline in individual businesses, industries, or even the economies of whole countries. Does this not suggest that if we can trace much of our current economic woes back to a decline in technological innovation, would not an increase in technological innovation provide an answer to these same problems?

Many prominent researchers and authorities have come to just that conclusion. Among others, Drucker (1985) sees innovation as a way for industry to respond to the challenges of change. He sees technological innovation as a key to growth and the development of "entrepreneurial economies" which avoid the economic stagnations which otherwise result from the decline of mature industries. Obviously, technological innovation is not in itself a panacea for all ills. There are many social, political and environmental factors which also impact on business and economic growth. Nonetheless, if we look carefully at any of these factors, we quickly find that they also impact on the effectiveness of the innovation process in very direct ways.

But how could we have a decline in innovation at the same time as we witness an explosion in knowledge and information? To understand this apparent dichotomy, reflection on the basic definitions involved might be useful. Sometimes even restating the obvious can be helpful.

In at least one of the Webster Dictionary editions, innovation is defined as "the process of creating change by introducing something new". Innovation of course involves change, and is based on something new - be it a new idea, new invention, new product, or new process. Similarly, we remember that technology is "applied science" and science, of course, is knowledge derived from research. Therefore "technological innovation" could simply be viewed as "the process of using research-based knowledge to change things". In this light, deficiencies in innovation may be seen as deficiencies in the application of research - that is, its reduction to practice. The evidence does suggest that although we might be generating a lot of new information, we are not using it all that effectively. In other words, we have room for improvement in the practice of innovation. Consideration of improving a process must necessarily begin with consideration of the process itself.

On the Process of Innovation

The process of innovation is actually a very complex and iterative one full of many interactions, feedback loops and recycles (Kline, 1985). However it is sometimes useful to think of it as a simple linear process of three primary steps (Felker, 1984).

Under this latter view, the first step can be considered as that of invention. This is the research or discovery phase wherein a fundamental idea or concept is developed and reduced to the simple practice which demonstrates its technical validity, or at least plausibility. Importantly, research is seen as a part of the entire innovation process, and not something separate from it.

Concepts which appear to offer sufficient merit, progress to the second step - that of translation. In this step, the concept is translated into a viable product form and tested. Technological capability is married with marketability, and the product and the process to make it are developed to the point that commercial viability can be meaningfully assessed.

Favorable results of that assessment lead to the third step - commercialization. This is the moment of truth - the time when the market casts the only meaningful vote. Success is rewarded with financial returns and business prosperity and, we might add, increased funding for new developments. Failure at this point is costly, and continued business survival depends on the returns of successes being sufficient to pay back the costs of all the failures as well.

On the Opportunities Change Creates

There is ample evidence that innovation - particularly as practiced by small entrepreneurial organizations - can be a primary contributor to economic growth and prosperity. It has been widely reported that during a period significant contraction in mature industries in the U.S., major new growth in employment was realized through the creation of new businesses. From 1972 to 1982, 21 million new jobs were created by new businesses which were developing at an average rate of about 600,000 per year. During that period, high technology industries increased their employment at faster rates than other sectors - albeit from a small base (OTA, 1984). In addition, there is a largely unmeasured multiplier effect for new high technology jobs. The Department of Commerce has estimated that for every new "high tech" job created, nine other jobs in support or peripheral areas are created. Application of this multiplier to the available data on "high tech" employment growth does indeed indicate that "high tech" industry development is a major employment creator. Regardless of the actual statistics, one can find adequate basis for the premise that technological innovation is a vehicle for growth - and the means to turn the challenges of change into opportunities.

On the Need for Collaboration

Much attention has been paid to the question of how we do technological innovation better. Among others, Schmitt (1984) recognized that successful innovation requires a close and intimate coupling between technologists and the business people who will bring technologies to market. Importantly, this close coupling must be sustained throughout the entire innovation process. In other words, coupling must begin in the research stage and cannot wait until the development one. Experience amply demonstrates that technology transfer undertaken as a separate, sequential process to be undertaken after research is completed is seldom effective, and never efficient.

Particularly the Need for Inter-organizational Collaboration

To maintain worldwide competitiveness, the need for collaboration beyond organizational or institutional boundaries is widely recognized. To remain competitive in either research or commercial arenas, immense resources and investments are required. In an age when the resources and funding available to any organization are limited, inter-organization collaboration provides a means to achieve progress which otherwise might not be attainable.

There is ample evidence that cooperation between companies to solve modern technological problems and compete in international markets increases effectiveness (Felker, 1984). This realization has brought about changes in U.S. anti-trust laws which promote collaboration in research - even between competitors. In their report to the President of the United States, the Business-Higher Education Forum (Anderson, 1983) stated that collaboration between all sectors would be required to restore America's competitive capacity. Repeatedly, this same point has been restated and emphasized by other (e.g., Young, 1985).

In Australia, a national system of innovation centres has been formed which is evolving into a network to promote collaboration between individuals, companies and public institutions. Although each centre is organized somewhat differently, the centres are jointly funded and supported by State and Federal Governments. They are intended to facilitate the process of innovation, that is the development and commercialization of new products and technologies, and early results are encouraging. Although there have been innovation centers in the U.S. for several years, they are not linked into the kind of network that is evolving in Australia, and their record of success has been quite variable.

Recently, there has been a new emphasis on collaboration between academia and industry. Keyworth (1985) pointed out that there has been a striking decline in industry's dependence on government to stimulate the development of new commercial technology. This in turn has resulted in an increasing reliance of industry on universities - for both fundamental knowledge (that is the results of research) and, of course, new technical talent. Many others have recognized the need for improving the interaction between universities and industry (e.g., Ross, 1985). Recognition of this need has resulted in efforts to stimulate partnerships between scientists and engineers in academic and governmental laboratories and their counterparts in industry. In fact, such an objective has been adopted as a goal of national science policy (Keyworth, 1984).

Focus on the academic/industrial interface has led to initiatives such as the National Science Foundation's program to create Centers for Engineering Research. This latest NSF initiative emphasizes multi-investigator, inter-disciplinary research that is more problem-oriented than discipline-oriented (Schmitt, 1985). The program is intended to foster cooperation between industry and academia and increase the rates of commercialization of technologies spawned in the university environments (Coulton, 1986). Under the program, an NSF grant funds establishment of each center. The university involved then seeks to establish a consortium of supporting companies – with each company making an annual monetary contribution to the center. In return, the companies review and approve research projects supported by each center and have preferential access and pre-emptive rights to research results. Although it is too early to measure tangible results from this program, the initial interest in it from the academic sector has been enthusiastic. In the first round of proposals submitted for the centers, NSF was inundated by 142 proposals from 106 different institutions requesting a total of $2 billion under a program budgeted at only $10 million. Initially, ten centers were planned and several are already operational.

Similarly, in other countries there is increasing emphasis on building the bridges between academe and industry to foster technological innovation. In Australia, for example, a new program known as the Teaching Company Scheme has been introduced for that purpose. Under this scheme, a company works in partnership with a tertiary institution on an R&D project that will "substantially improve the company's technology and performance". Academic staff from the tertiary institution collaborate in the planning and execution of a Teaching Company Project with graduate engineers or scientists appointed as Teaching Company Associates. These Associates work full-time under the supervision of both company and academic staff. The Federal Government in Australia provides a grant to the tertiary institution to cover up to one-half of the Associate's salary and administrative costs for the institution. These grants are competitive and projects must be approved in advance on the basis of merit. The program can apply to both proprietary and non-proprietary developments with appropriate agreements negotiated as necessary to protect intellectual property rights (and potential commercial value).

Increasingly, joint ventures and direct cooperative relationships are being established between industrial companies and universities. In Australia, such relationships are commonly facilitated by establishment of distinct organizational entities to effectively "broker" the services or resources of a given institution. Many of these appear to function in a manner similar to certain programs in the United States – like MIT's Industrial Liaison Program. Others are organized as more autonomous, non-profit companies. One of the more successful of these organizations, Uniquest Ltd., provides services for the University of Queensland. Uniquest not only brokers consulting and other research services, but also organizes company startups and secures venture finance. In several instances, Uniquest has even managed new public offerings of equity in new ventures, and taken a direct equity position in them.

These evolving relationships between the academic and industrial sectors do face potential conflicts of interest and may involve some level of compromise. Superficially, the objectives of each type of organization appear almost diametrically opposed. Apparent differences on issues such as secrecy versus publication, and knowledge versus profit and return on investment must be reconciled. However, increasingly these issues have been found to be not as contradictory as they appear, and ample room is being found for collaboration to occur without compromise of either organization's objectives. Indeed, combinations of the resources of both academic and industrial organizations are being found to yield complementary, and even synergistic benefits and facilitate the achievement of significant intellectual and commercial results (Gray, 1985).

The interface between government researchers and their industrial counterparts is also receiving attention. In the U.S., other initiatives being taken to spawn greater collaboration between the Federal Laboratories and the private sector to increase the return on investments in these laboratories and also to help create new industries or expand old ones (Brody, 1985). Some of these initiatives involve changes in public law to increase the autonomy of governmental labs and enable royalty returns can flow back to the labs and the individual researchers. Indeed, such innovative concepts as rewarding governmental researchers with profit royalties would have been unheard of a few years ago; yet today they are moving toward reality.

In the United Kingdom, the British Technology Group (BTG) appears to have been quite successful in facilitating the commercialization of research from public institutions. BTG was formed about 25 years ago to broker technologies developed in governmental and academic laboratories. Having long ago paid back the ₤50 million government investment in BTG, that group today generates revenues of nearly ₤25 million from royalties alone.

On Relevance to Research

Even if the global problems our communities face seem remote from a researcher's laboratory, the changes in the competitive environment do directly affect individual researchers. The competition for resources, and particularly funding, is keener today than ever before. Despite an upturn in national expenditures on research in the U.S., a high degree of selectivity is being imposed on any research proposal from any source, and each proposal is undergoing intense scrutiny in a highly competitive climate. (This is encouraged by the general decline in funding available to support public sector institutions and the competitive pressures felt in the private sector.) Research funding in the future will increasingly flow to those areas which yield the most productive results. This means that research proposals today must be sold on the basis of relevance and the promise of not only increasing knowledge, but yielding a return that justifies the investment.

To do relevant work, researchers cannot be isolated from those who would build on the research and translate it into benefits for the community. This means new bridges and partnerships must be formed at earlier stages in the process of R&D or, in today's terms, in the process of innovation. The faster a field of technology is changing, the more important these new network relationships will be.

And Particularly Research in Biotechnology

Like other fields of high technology, biotechnology is changing and advancing at an enormous and ever-increasing rate. This high rate of change is creating new vistas of opportunities for discoveries of immense potential value from both academic and commercial viewpoints. Of particular note, research relating to plants, and even specifically plant lipids, appear to be on the threshold of major developments. Developments in plant lipids do offer the promise for providing new materials of value which are readily, or even possibly, available from any other source. In addition, they offer propsects for providing competitive alternatives to certain petrochemicals for that inevitable (but now seemingly more distant) time when petroleum resumes its upward cost spiral. It seems that never before have the prospects for fundamental research in plant lipids offered as much potential for industrial application as they do now. In these context, it may be worthwhile to consider two inter-organization initiatives which have some relevance to plant lipid research.

Several years ago, a collaborative program was established between the government, academia and industry to support basic research work to develop a new crop source of plant lipids. In the context of this review, the subject of the research is not as important as the organizational model and precedents established by the program. That exploratory program was focused on the potential development of a new oilseed crop source of mid-chain triglycerides. Specifically, the program was intended to assess the prospects for domesticating and developing a wild plant, known as Cuphea, into a domestic crop. Under this program, matching funds from each the Federal Government (through the U.S.D.A.), the State of Oregon (through Oregon State University), and industry (through the Soap and Detergent Association) was leveraged to support fundamental exploratory work at the university. Although the jury is still out on whether or not Cuphea could ever be a viable new crop source of selected triglycerides, the program does demonstrate that collaborative private/public sector programs can be established at even very early stages of research. Not only do such programs provide support for specific research, they also act as a catalyst for other directly or indirectly related programs as well.

Another initiative being now proposed on a national basis would involve the establishment of an organizational entity to catalyze new partnerships between industry, government and academe to develop new agricultural products. Specifically, the New Farm and Forest Products Task Force is addressing the needs for diversification and better utilization of U.S. agriculture and developing actionable proposals. The Task Force is in itself a collaborative effort between the three sectors. One of the proposals it is generating involves establishment of a national Institute or Center for New Farm and Forest Products to foster the development of new products including new (or modified) crops and new uses for existing crops. As proposed, the Institute would be a non-profit, autonomous organization which would act as an advocate, network connector, coordinator and even funding source for collaborative inter-organizational efforts. One could anticipate that the Institute would be significantly involved with innovation in biotechnology including, of course, plant lipid developments, and thereby be of direct relevance and interest to researchers and others with interests in these fields.

In Summary

The survey presented above is intended to make the following points. First, the challenges researchers and all others now face are great. The competition for funds - be they business profits, government revenues or research grants - are greater than ever before. Second, technological innovation is providing a key to meet these challenges and realize the opportunities those challenges create. Thirs, inter-organizational collaboration is a key to the successful practice of technological innovation, and, finally, recognition of the value of collaboration is spawning new relationships between academia, government and industry. These relationships are in themselves stimulating and rewarding and offer much promise for a brighter future for us all.

REFERENCES

Abernathy, W.J. Competitive Decline in U.S. Innovation: The Management Factor. Research Management 25-5: 34-41. 1982.

American Oil Chemists' Society. Cuphea: Diverse fatty acid composition may yield oleochemical feedstock. JAOCS 62-1: 6-12, 1985.

Anderson, R. et al., America's Competitive Challenge: The Need for a National Response. Washington, D.C.: Business-Higher Education Forum; 1983.

Brody, H. At Your Service. High Technology 39-44; July 1985.

Buggie, F.D. New Product Development Strategies. New York; AMACOM; 1981.

Coulton, R. Private Communication, National Science Foundation, Washington, D.C., Jan. 1986.

Drucker, P.F. Innovation and Entrepreneurship - Practice and Principles. New York: Harper & Row; 1985.

Gray, P.E. A Case of Synergism. ChemTech. 15-9; 519-521, 1985.

Keyworth, G.A. Four Years of Reagan Science Policy: Notable Shifts in Priorities. Science 224-2644: 9-13, 1984.

Keyworth, G.A. Science and Technology Policy: The Next Four Years. Technology Review 45-53; Feb-Mar 1985.

Kline, S.J. Innovation Is Not A Linear Process. Res. Man. 37-45; Jul-Aug 1985.

Merrifield, D.B. Forces of Change Affecting High Technology Industries, Nat. Jour. 253-6; Jan 29 1983.

Office of Technology Assessment, Congress of the United States. Technology, Innovation and Regional Economic Development. Washington, D.C. 1984.

Ross, I.M. The Global Contest in Industrial Competitiveness Has Just Begun. Res. Man.: 10-14; May-Jun 1985.

Schmitt, R.W. National R&D Policy: An Industrial Perspective. Science 224: 1206-9, 1984.

Schmitt, R.W. and Brooks, H. Current Science and Technology Policy Issues: Two Perspectives. Washington, D.C.; The George Washington University; 1985.

Schmookler, J. Invention and Economic Growth. Cambridge, Mass.; Harvard Univ. Press; 1966.

Young, J.A. et al., Global Competition: The New Reality, The Report of the President's Commission on Industrial Competitiveness; Washington, D.C.; U.S. Gov. Print Off., 1985.

THE PALM OIL RESEARCH INSTITUTE OF MALAYSIA - A UNIQUE RESEARCH UNIT

Augustine S.H. Ong

PORIM
P.O. Box 10620
50720 Kuala Lumpur

INTRODUCTION

PORIM was established in 1979 out of the need to provide R&D support for the rapidly growing palm oil industry. Research in palm oil has been carried out both privately and in government agencies; firstly in the Department of Agriculture and later in the Malaysian Agricultural Research and Development Institute (MARDI). However the tremendous growth of the industry in the seventies necessitates the establishment of a fully committed single crop Research and Development Institute (RDI).

When PORIM was established most of the research and supportive staff from Palm Oil Sub-Division of MARDI chose to be transferred to PORIM and they formed the core of the pioneers of the new Institute.

FUNCTIONS AND RESEARCH OBJECTIVES OF PORIM

Before considering the structure and mechanism of PORIM it is thought appropriate to outline the functions and objectives of the Institute.

The main functions of the Institute as stipulated by an Act of Parliament, the Palm Oil Research and Development Board Act 1979, are:-

(a) To conduct and promote research into the production, extraction, processing, storage, transportation, marketing, consumption and uses of palm oil and oil palm products;

(b) To secure, where the public interest or the interest of the oil palm industry so requires, the development and exploitation of any result of research, as to when it is not being developed and exploited or sufficiently developed or exploited;

(c) To collect, collate and disseminate information relating to oil palm, palm oil, oil palm products and other vegetable and animal oils and fats, and to promote the use of palm oil and palm oil products in competition with other materials or as complementary thereto.

With these stipulated functions in mind the broad research objectives are:

(a) To reduce unit cost of production

(b) To find more uses for palm oil and palm kernel oil in both the edible and non-edible areas.

(c) To establish quality assurance for Malaysian palm oil

(d) To introduce mechanisation in harvesting and in field transportation

(e) To undertake and promote nutritional studies of palm oil and to project a good image for palm oil.

STRUCTURE AND MECHANISM OF PORIM

The structure of PORIM is outlined in Figure 1. The management policies are made by the PORIM Board which is responsible to the Minister of Primary Industries. The Chairman of the Board is appointed by the Honourable Minister of Primary Industries. The Board has three committees: the Establishment Committee to look into new appointment and promotion, the Tender and Development Committee to look into finance, budget, tender and development, and the Programme Advisory Committee (PAC) to scrutinise research projects. The overall running of the Institute is under the care of the Director-General who is assisted by a Deputy Director-General and four Directors. There is also a Technical Advisory Committee (TAC) similar to PAC to advise the Director-General on research projects.

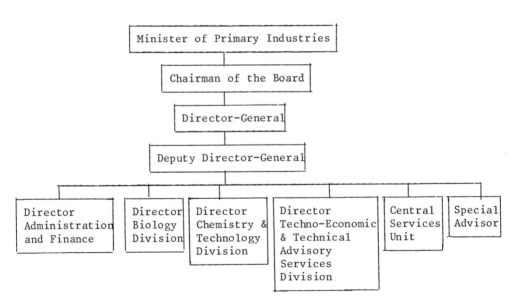

Figure 1 · PORIM Organisational Structure

After outlining the structure of PORIM the various activities and relationships are discussed in detail below.

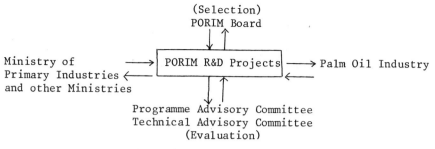

Figure 2 : PORIM in Perspective

A Management Board Where the End-Users or Clients are Represented

The PORIM Board consists of 13 members including the Chairman of the Board and the Director-General of the Institute. Six out of 11 other members are from the palm oil industry with the remaining five from the Government. The industry as end-users of research results and clients is very well represented in the management board.

Permanent Committees

As mentioned earlier the Board has three committees viz Establishment Committee, Tender and Development Committee, and Programme Advisory Committee (PAC). The PAC is further divided into three Sub-Committees namely Biology, Technology and End-Use, and Technical Promotion and Commercial Sub-Committee. Each Sub-Committee consists of both local and foreign experts. The PAC meets yearly and it scrutinises and advises the PORIM Board on the research programmes.

A Feedback Mechanism

Workshops, seminars and courses are organised by PORIM not only as a means of transfer of technology but also as a mechanism for feedback from the industry in particular and the public in general. In addition to these regular visits both advisory and trouble shooting in nature are also organised regularly (both local and overseas) and these also function as a feedback mechanism. PORIM is represented in about 40 committees where interaction with the Industry and Government is realised.

OBJECTIVES, CONCEPTS AND STRATEGIES

With the context of PORIM, the three research divisions namely Biology, Chemistry & Technology, and Techno-Economic and Technical Advisory Services, operate. The objectives, concepts and strategies of the research programme undertaken by these divisions are outlined below:

Biology

Objectives. To direct biological research for optimum yield and quality of palm oil

To reduce production cost

To undertake studies in order to obtain zero-waste concept.

To organise the resource output and personnel for extension services.

Concepts and Strategies

Basic and applied research projects are undertaken with a multi-disciplinary approach involving several disciplines such as agronomy and nutrition, genetic and breeding, vegetative propagation through tissue culture, crop physiology, crop protection, by-product utilisation technology, farm mechanisation and biotechnology.

Maximum yield and reduction in cost of production are expected to be achieved through improvement in plant breeding, agronomy and physiology, improved planting materials via tissue culture and biotechnology, crop protection, labour saving, farm mechanisation and effective transfer of technology to the industry.

Efforts will be concentrated on optimum utilisation of oil palm by-products so as to maximise income to the industry, thus reducing the unit cost of production.

The Division is also involved in yield prediction, germplasm prospection, conservation and evaluation, nation wide urea trial and serveillance studies for disease and pest control.

It cooperates with the other extension agents in the country in formulating and undertaking effective extension services specific to the oil palm industry.

Chemistry and Technology

Objectives. To conduct and promote research of the palm oil from the point of chemistry, technology and applications of palm oil.

To improve the efficiency of present technology and to promote new technology in milling, processing and effluent treatment.

To enhance the quality of palm oil.

To increase the usage of palm oil in edible and non-edible applications.

To conduct research on optimum utilisation of palm oil by-products.

Concepts and Strategies

Research efforts in chemistry and technology of palm oil are focussed on the expansion and improving the current uses of palm oil and palm kernel oil and to find new uses for both edible and inedible purposes. In this aspect effective communication with the industry at various stages of research and development would be maintained from laboratory to pilot plant scales.

Concentrated efforts are being made towards the enhancement of quality in palm oil through integrated projects on quality improvement, quality control and quality assurance. These include factors affecting quality at the mill and refinery. Every effort is being made to find new uses for palm oil and palm kernel oil. The emphasis is on high value added products with large marketing potential. Attention is also being given to areas in energy conservation, automation, enzymology and exploitation of high value minor components from palm oil and its products including waste resource utilisation through appropriate treatment technology, to obtain no waste technology. There is need to look into the diversification of industrial applications of palm-based oleochemicals.

Nutrition and biochemistry research would continue to examine parameters which are presently controversial in the edible oils and fats industry and on techniques to increase efficiency of production and extraction of new products.

Where possible research findings should be translated into commercial applications. Thus laboratory findings should be tested in pilot plants and then introduced to the industry.

An important strategy is to have vision and think ahead of the industry. Efforts are being stepped up to identify future trends and to carry out research in promising areas so that PORIM would be ready to provide new technology to the industry when required in the future.

Techno-Economic

Objectives. To provide quantitative and evaluated economic information to the oil palm industry.

To provide economic advisory services to the industry.

Evaluation of techno-economic feasibility of new technologies, products etc for the oil palm industry.

Concepts and Strategies

Research efforts in techno-economics are designed to obtain information in order to assist the industry, the government and others in their planning and decision making on oil palm, palm oil and their products.

The main research areas are in agricultural economics, market and price economics, econometrics and process economics. The ever changing and dynamic situations prevailing in the oil and fat industry, particularly on palm oil, require close monitoring and research into aspects mentioned.

In collaboration with other Divisions of PORIM and outside organisations, appropriate projects are linked together for better evaluation.

Technical Advisory Services

Objectives. To publicise the performance characteristics of palm oil.

To be the main channel for identifying product opportunities or problems and methods in dealing with them.

To identify present and potential problems and constraints to the utilisation of palm oil by user or potential user.

Concepts and Strategies

First hand information is collected from consumer or potential consumer countries by regular round of visits. Suitable opportunities are identified and reported.

Information on the performance characteristics of the palm oil disseminated to the consumers and potential consumers by lectures, meetings, exhibitions and publications.

Research projects are formulated, usually in collaboration with other organisations, when a product development opportunity or technical problem area is appropriately dealt with in that way.

RESEARCH PROGRAMMES

The research programmes of PORIM are scrutinised by two committees of local and international experts. The Technical Advisory Committee advises the Director-General while the Programme Advisory Committee reports to PORIM's Board on the Division research programmes and activities.

There are over 140 on going projects in the Institute and in this paper, the research programme of Chemistry and Technology Division will be elaborated. Research projects can be grouped into seven research areas viz, food uses, non-food uses, analytical, milling and processing, nutrition and biochemistry, effluent treatment technology and engineering services. A description of the current research programme is given below:

Food Uses

Palm oil is a semi-solid product with a wide range of applications in food uses. The most important uses are in margarines, shortenings, vanaspati and frying fats but it is also used as an ingredient in ice cream, biscuits and other foods. The Division is actively involved in increasing the proportion and improving the performance of palm oil in existing edible products and at finding new uses. The following topics are under investigation:

o Development of vanaspati and bakery formulations. Maximum use of palm oil for these products is investigated.

o Utilization of palm mid-fractions. Refiners in Malaysia are producing further downstream products such as palm mid-fractions. Such products have been surveyed for their composition and properties and their potential uses in cocoa butter extenders and other applications are being evaluated.

o Margarine and shortening formulations. Margarines and shortenings from various countries have been evaluated. The substitution of other fats in these products with palm oil is being studied in relation to their properties and usage.

o Palm oil in frying. Refined palm oil has been shown to be an excellent medium for frying. However, it tends to darken rapidly and the cause for this colour development is under investigation. The use of crude and partially processed palm oils have been applied to frying and the products evaluated for their taste, stability and keepability. Study on blending of palm olein with other liquid oils is being carried out and results are promising.

Non Food Uses

Oils and fats are converted into fatty acids and other derivatives for use in lubricants, plastics, resins, soap, surfactants, cosmetics, toiletries and textile chemicals. The Division has started research in this area to optimize and enhance the usage of palm oil in oleochemicals for the above applications. Among the topics being studied are:-

- o Characterization of by-products from palm oil refining. Palm fatty acid distillates are produced from the physical refining of palm oil. These are used as the feedstock for the production of fatty acids. The properties and composition of the distillates produced by Malaysian refiners have been documented.

- o Palm oil as a diesel substitute. Palm oil and its fractions are being converted into alkyl esters for use as a diesel substitute. The kinetics of the esterification reactions are being studied together with the isolation of useful by-products. Engine performance based on the use of the alkyl esters is also evaluated.

- o Application of palm fatty acids and their derivatives. Information is being gathered with regard to the present and future volume of fatty acids going into various inedible uses and the specifications involved. The suitability of palm fatty acids for substitution will be considered. Studies are also in progress on the production of epoxidized palm oil and monobasic and dibasic acids for use in plasticizers, stabilizers, lubricants and synthetic fibres. Colour reversion and odour development in palm fatty acids will be investigated.

- o Soap from palm oil. Toilet soap has traditionally been made from tallow or tallow based fatty acids. The substitution of tallow by palm oil in soap manufacture has been successfully done in Malaysia and will be extended to other countries by the use of suitable blends and formulations.

Analytical

The Division has trained staff for research into lipid methodology and analysis using modern equipment. Analyses into the quality, composition and properties of palm oil are also conducted. Among the topics covered are:

- o Survey on palm oil, its fractions and palm kernel oil. Comprehensive data on the chemical and physical characteristics of these oils and fractions have been established, leading to greater consumer awareness on end-use applications of palm oil. Specifications for sludge palm oil have been proposed.

- o Factors affecting the hydrolytic and oxidative stability of palm oil. The effects of moisture, iron and added anti-oxidants are monitored. Such studies will lead to improved procedures for enhancing the stability and keepability of palm oil.

- o Residual colour of refined palm oils. Studies to isolate the compounds responsible for the residual colour in refined palm oil have been initiated. Attempts will be made to determine the mechanisms for its formation and to suggest possible solutions for its prevention.

o Crystallization of palm oil. One of the limitations encountered in the use of palm oil is its unusually slow rate of crystallization. The cause for the slow crystallization and attempts to shorten it are being studied. Possible ways to modify the crystal size are also investigated.

o Analytical methods and collaborative trials. The Division is concerned in the development of analytical methods for research and quality control. Standardization of methods for quality evaluation are being done through collaborative trials with the industry while methods for the detection of adulteration of palm oil products are also pursued.

Milling and Processing

The Division is concerned with all the processes used from the point at which the fresh fruit enters the oil mill, through the refining process and including the storage of the end products. The main concern is to establish the effect of each unit process on quality and to investigate possible improvements in the technology and process efficiency. The following topics may be mentioned:

o Seasonal variations in oil quality. It has been confirmed that the free fatty acids of crude palm oil increase during the peak production season. Bleachability decreases at this time. Oxidation may not be the only factor involved.

o Manual on milling technology and code of practice. The accumulated technical experience of the industry has been compiled into a manual on milling technology while a code of practice has been published for the edible oil refineries to ensure hygienic and safe end-products.

o Pre-treatment practices in physical refining. The use of different degumming agents and bleaching earth on crude oils of varying quality is being investigated. The reduction of phospholipid content to a low level appears to be most important, particularly when physical refining is used. The original phospholipids and phosphorus compounds remaining in refined oils are being isolated and identified.

o Pre-treatment practices in physical refining. The use of different degumming agents and bleaching earths on crude oils of varying quality is being investigated. The reduction of phospholipid content to a low level appears to be most important, particularly when physical refining is used. The original phospholipids and phosphorus compounds remaining in refined oils are being isolated and identified.

o Effects of physical refining. The high temperatures in physical refining may affect the glycerides of the oils and hence their physical properties. Such changes are likely to be small in palm oil because of its relatively low level of polyunsaturation but they may still be significant in some applications.

o Modification processes. Palm oil and palm kernel oil properties can be modified by the physical process of fractionation or the chemical processes of hydrogenation and interesterification.

These processes may be used in combination. Significant changes
in properties can be effected leading to wider applications.
Laboratory work on the processes has been proceeding
and the work is now entering the pilot plant phase to produce
larger amounts of material for commercial evaluation.

Biochemistry and Nutrition

The Division is involved in studies to understand the biochemistry
of the oil palm and the nutritional values of palm oil. The production of
new palm oils of specialized composition through plant breeding are also
considered. Among the topics being studied are:

- Biochemical changes in the oil palm fruit. The synthesis of
  oil in the fruit and the biochemical and morphological changes
  accompanying it are under investigation. Attempts are also
  made to determine the cause for the rapid hydrolysis of
  triglycerides in bruised fruits.

- Isolation of non-glyceride components. These are being isolated from oil palm tissues for use in commercial and pharmaceutical applications.

- New oil palms. The Division is actively engaged in conjunction with plant breeders from the Biology Division to produce
  oil palms which will yield oils of specific fatty acid types
  so as to achieve the desired properties for the oils. Among
  these are oils which are high in oleic and linoleic acids for
  use as salad oils and oils high in stearic acid for confectionery
  fats.

- Nutritional research on palm oil. In research done in Holland,
  palm oil has been found to be less thrombogenic than has been
  suggested by its composition. Research efforts will attempt
  to confirm the lack of thrombosis observed for palm oil and
  to try to identify the exact component(s) responsible for the
  anti-thrombotic effect. The effects of palm oil on the lipoprotein structure have been compared with those of other oils
  where like the polyunsaturated oils, HDL level has been increased.
  The relationship between coronary heart diseases (CHD) and
  dietary fat intake using numerous edible oils including palm oil
  will be studied to establish a possible mechanism on heart
  diseases.

Effluent Treatment Technology

Environmental legislation requires that all effluents from the oil
mills and refineries should be treated prior to its discharge so as not to
pollute the natural waterways. The Division conducts research to assess
and improve current effluent treatment systems and to propagate new
treatment technology which are cost effective and efficient. Research
activities include:

- Survey on current treatment systems for palm oil mill effluent
  (POME). Five treatment systems were considered. All could
  meet the limits of most of the parameters set by the Department
  of Environment with regard to effluent discharge.

- Survey on treatment systems for refinery effluent. The costs
  and efficiency of imported technology will be evaluated with
  regard to local conditions. A pilot plant study on the biological

treatment of waste from physical refining plant has resulted in the development of a new treatment system known as Sequential Batch Reactor (SBR).

o Land application of palm oil mill effluent. Palm oil mill effluent is high in nutrient content. When applied to land in controlled amounts, it improves crop yield and soil properties. The long term effects on water quality are being monitored. Results obtained so far indicate no contamination.

o Pilot plant studies on the thermophilic anaerobic digestion of POME. Anaerobic digestion at high temperatures has been shown to be more efficient for the breakdown of organic matter in addition to a reduction in the retention time. More biogas is also generated.

o Pilot plant studies on ponding systems and high rate ponds. Such systems have been shown to be efficient in reducing nitrogen and phosphorus by algae action through photosynthesis.

o Microbial ecology of aerobic and anaerobic treatment systems. Studies are done to enumerate, isolate and characterize some bacteria involved in the biological treatment systems. The objective is to maximise the efficiency of treatment systems.

o Production of microbial protein and exo-enzymes. The use of sterilizer condensate as a resource is looked into. The feasibility of using enzymic methods for recovering oil from press cake fibres will also be investigated.

o Improvement in effluent testing procedures. Attempts are made to develop methods which are time and cost-saving with no loss in efficiency and accuracy.

Engineering Services

The engineering services group was recently set up to provide a forum for exchange of ideas among engineers of the whole institute. Ad hoc engineering problems related to research projects, PORIM physical development and the oil palm industry are brought for discussion and subsequent action.

Philosophy and Strategy for Implementation

It is in the philosophy of PORIM that it is not possible to be self-sufficient in technical expertise all the time since science and technology are continuously changing with unexpected quantum advance from time to time. Thus it has adopted a strategy whereby PORIM will be staffed with a core of scientists and technologists to deal with industrial problems requiring immediate or mid-term solutions and an ability to forge linkages with various centres in the world including Malaysia through collaborative research projects. The list is already extensive and continues to grow (see Appendix 1).

PORIM also believes that there needs to be continual interaction with fresh minds or hybridisation of ideas. This could be achieved through vacation student training scheme and the newly established research fellowship scheme whereby University dons, scientists and technologists could spend a stipulated period of time to address themselves to specific problems based on their expertise.

It is also the intention of the Division to translate all laboratory findings into commercialisation. Thus economic evaluation inputs and pilot plant study are applied to any new finding with the potential for industry.

Services to Industry

It is recognised that a career in an R&D institute on an industrial crop like oil palm involves providing services and training to the industry.

These services include analytical work related to the quality of palm oil, palm kernel oil and their products. The Analytical Services Group which has been set up has analysed 2,894 samples in 1985. The trend is increasing and thus computerisation has been resorted to.

Besides analytical services, the Division provides consultancy services on food and non-food applications of palm oil and palm kernel, palm oil milling technology, effluent treatment technology for both mills and refineries. Work on formulations of products for potential consumers in several overseas countries has been successfully conducted and recommendations made. Many of these have contributed to increased trading in palm oil and its products.

The transfer of knowledge and technology from research findings is best done through courses, seminars and workshops. These functions provide the opportunity for feedback from the industry. The Division has been organising regularly several courses including the following:

o   Chemistry of Fatty Acids and Lipids Course

o   Palm and Palm Kernel Oil Processing Course

o   Course on Product Technology

o   Palm Oil Mill Engineers

The contribution of the staff to training would be significant and this is considered relevant and important.

Evaluation

A system of evaluation on the quality and quantity of work output is a must. This is carried out through research group discussion, monthly research progress and time utilisation reports, half-yearly and annual research reports. The Technical Advisory Committee and the Programme Advisory Committee do play an important role in the evaluation of research progress and new research proposals.

In the final analysis, the contribution of R&D efforts should be gauged in the effects they have on the industry and the number of patents, articles in International Journals, conferences papers, project reports etc. The output from PORIM is given in Table 1. However, greater efforts will be put for higher contribution both in terms of quality and quantity.

Achievements

In a short span of seven years, the division has recruited and trained an important core staff and acquired a good range of modern instrumentation and pilot plant facilities to deal with all problems arising from the palm oil industry. Some of the highlights of its achievements are described as follows:

o It has obtained baseline data for palm oil, palm kernel oil and their products and the information forms the basis for standard specifications for the products. Through cross-checks with quality control laboratories and method development work it contributes to quality assurance to the industry. Further, the schemes for the certificate of competency for mills and refineries have been implemented with the resultant improvement in the quality of palm oil and processed palm oil.

o Since 90% of palm oil is used as food, much effort has been directed at utilisation of palm oil in the existing food systems. This can be exemplified in the formulation of vanaspati for India, Pakistan and the Middle East and margarine for different countries such as South Korea, Egypt, Russia, East Germany and the Scandinavian countries. It has also been demonstrated that instant noodles fried in palm oil and palm stearin are endowed with longer shelf life. Thus, palm oil is popular in South Korea for such application. Work on blending palm olein with other liquid oils shows that palm olein confers improved properties to some liquid oils e.g. rapeseed oil and rice bran oil, thus opening greater utilisation of palm olein for frying in countries where such liquid oils are produced.

o The PORIM effluent laboratory has been accepted as the referral laboratory for effluent testing. A multidisciplinary approach has been applied to the effluent problem in particular a study of the anaerobic digestion. Better understanding has been obtained with regard to the efficiency of biogas production with reduction in the percentage of hydrogen sulphide as a by-product. Besides contributing to effluent abatement at the mills it is also active in assisting the refining industry in tackling its effluent problems.

o The Division has built up a number of laboratory pilot plant facilities for refining and processing of palm oil as well as for formulation of products such as margarine and soap. These facilities are an asset to end use research as well as a valuable service to the refining industry.

o Several findings have developed to a stage where pilot plant studies would be appropriate. Applications for patents have been filed for some of these. These can be illustrated with the pilot plant for the production of palm diesel, with a capacity of 3,000 tonnes per year. This plant has been operational since November 1985 and the exhaustive field trial for vehicles commenced in March 1986. The other plants are: Vitamin E pilot plant for recovery of vitamin E from palm fatty acid distillate and soap and detergent production. Finally, fundamental work on biosynthesis has resulted in the isolation of protoplast in oil palm fruits of varying ages varying from 16 to 18 weeks after anthesis. This finding could lead to better understanding of new biosynthetic pathways and also to protoplast fusion in genetic engineering work.

Future Prospects

The production of palm oil is expected to increase to 6.5 million tonnes in 1990 and may exceed 10 million tonnes in the year 2000 when the area under oil palm may exceed 2 million hectares. Hence palm oil will be an even more important commodity. Thus, it is important to project the research trend in the next decades.

With increasing production, the efficiency and modernisation of the palm oil mills will be critical. An important step towards these objectives is the establishment of an experimental research mill where developmental study can be conducted. This includes application of microprocessor and automation. It is hoped that by the end of the next decade all the problems on quality will be resolved so that greater attention could be devoted to research on biotechnology and the application of such findings to industrial applications for example the enzymatic fat splitting process where considerable energy can be saved.

R&D inputs are also expected to upgrade our commodity industry to speciality products industry where there will be added value. Finally, efforts on nutritional research will be increased so that the full implications of palm oil in terms of nutritional values could be determined. To implement this programme effectively there will be a modest increase of staff as anticipated in Table 2 and hopefully the necessary financial support would be forthcoming.

CONCLUSION

Work on the chemistry and technology of palm oil in Malaysia was limited until the establishment of PORIM in 1970. In the short span of seven years, PORIM has made a good start inspite of the handicaps of space, facilities and pre-occupation with the construction of the present headquarters. PORIM requires to be nurtured into full development with continuous financial and moral support and maintaining a high morale among the staff.

PORIM has been well accepted by the palm oil industry and by overseas consumers for information on palm oil and palm kernel oil. Its analytical and effluent laboratories have been used as referral laboratories. It has provided consultancy service in product formulation and effluent treatment technology. Its instrumental and pilot plant facilities have been made available to the industry. It is gradually gaining good reputation for research and development on palm oil at both local and International levels.

ACKNOWLEDGEMENT

The author would like to express his gratitude to the Director-General of PORIM for his advice, encouragement and approval for this article.

Table 1. List of Patents, Articles, PORIM Reports, Conference Papers and PORIM Publication as at June 1986

	Bio Div	Chem & Tech Div	TE & TAS Div	DG/CSU Admin Div	Grand Total
Patents	-	5	-	-	5
Articles	38	51	25	23	137
Conference Papers	113	196	70	59	438
Reports	50	151	163	36	400
Sub-total	201 (20.5%)	403 (41.1%)	258 (26.3%)	118 (12.0%)	980 (100%)

Table 2. Staff Projection

| Year | Senior Staff | | | | | Sub Total | Junior Staff | | | | Sub Total | Total |
	AF	B	CT	TT	CSU		AF	B	CT	TT	CSU		
1984	11	24	30	18	9	92	101	103	84	18	24	330	422
1985	10	22	31	17	8	88	157	117	99	25	33	431	519
1986	12	36	35	20	9	112	230	129	136	27	45	567	679
1987	13	37	36	20	10	116	236	138	146	28	49	597	713
1988	13	37	36	20	12	118	243	144	155	28	51	621	739
1989	14	39	37	20	12	122	247	148	165	28	55	643	765
1990	14	39	37	20	12	122	255	152	169	28	57	661	783

AF : Administration and Finance
(including staff required by the Development and Maintenance Unit)
B : Biology
CT : Chemistry and Technology
TT : Techno-Economic and Technical Advisory Services
CSU: Central Services Unit

RESEARCH COLLABORATION WITH OTHER INSTITUTIONS   Appendix 1

Institution		Research Project
1. Limburg University, The Netherlands	(a)	Effect of Palm Oil on Arterial Thrombosis and Related Phenomena
	(b)	Comparison of Palm Oil with another Saturated Vegetable Oil
2. National Chemical Laboratory for Industry, Japan	(a)	Study on a Novel Utilization of Palm Oil and Its Fatty Acids and the Utilization of Carotenes in Palm Oil.
3. Japanese Association of Industrial Fermentation, Japan	(a)	Technology for the concentration of Vitamin E from Palm Fatty Acid Distillate
4. Korea Advance Institute of Science and Technology, South Korea	(a)	Quality Deterioration of Malaysian RBD Palm Oil during Shipping - Malaysia/South Korea
	(b)	Margarine Formulation for Korean Market.
	(c)	Blending of Cooking Oil with Palm Oil and Other Oils for Korean Food System
	(d)	An Epidemiological Study on Palm Oil
	(e)	Utilisation of Oil Palm By-Products for Food Use
	(f)	Acetone-Butanol Fermentation of Palm Oil Sludge and Palm Kernel Meal
5. University of Masschusetts, U.S.A	(a)	Carotenoids in Palm Oil
6. The North East Wales Institute of Higher Education, U.K.	(a)	Photodegradation of Palm Oil
7. University of Calcutta, India	(a)	Studies on Better Utilization of Palm Oil and Palm Oil Fractions in Products.
8. Institute of Medical Research, Malaysia	(a)	Nutritional Studies on Malaysian Processed Palm Oil
9. Universiti Malaya, Kuala Lumpur	(a)	Biochemical Causes for Increased Free Fatty Acids
	(b)	Palm Oil Splitting by Immobilised Lipase
	(c)	Biosynthesis of Fatty Acids and Glycerides in the Oil Palm
	(d)	Trace Organic Constitutents in Palm Oil
10. Universiti Sains, Malaysia, Penang	(a)	Fractionation by Density Gradients
	(b)	Fundamental Studies on the Absorption of Beta Carotenes by Bleaching Earths.

11.	Universiti Pertanian Malaysia	(a)	An Investigation of Characteristics of RBD Olein during Frying
12.	Universiti Kebangsaan Malaysia	(a)	Assessment of Biological Activity of Vitamin A (Carotene) from Palm Oil and Its Effect on the Lipid Level in Blood, Heart and Adipose Tissue.
		(b)	Biochemical Aspects of the Physiology of Fruit Ripening in Oil Palm
		(c)	The Use of Crude and Semi-Processed Palm Oil in Selected Malaysian Food Products.
13.	Standards and Industrial Research Institute of Malaysia and Rubber Research Institute of Malaysia	(a)	Pilot Plant Study on Biogas Production
14.	Rubber Research Institute of Malaysia	(a)	Palm Oil Mill Effluent Treatment Technology
15.	Commonwealth Scientific and Industrial Research Organisation, Glenthorne Laboratories, Australia	(a)	The Effect of Palm Oil on Cardiac Arrhythmia
16.	National Institute of Nutrition, Hyderabad, India	(a)	Palm Oil, Acceptability and Its Metabolic Influences.

BIOTECHNOLOGY OF LIPIDS IN INDUSTRY

Rolf D. Schmid

Dep. of Biotechnology

Henkel KGaA, Düsseldorf, W. Germany

I. Introduction

In industrialized countries, renewable resources are increasingly used as starting materials for chemistry. Thus already in 1980, 9% of the chemical feedstock in W. Germany was based on agricultural resources (1) (fig. 1).

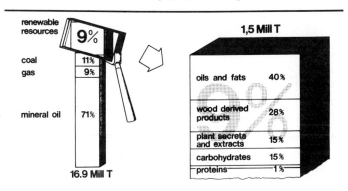

Fig 1.

Fats and oils hold a large share of renewable raw materials. This is due to the fact that

1. different from fossil feedstocks which predominantly are based on C_1-C_6 chemicals, fats and oils provide access to $C_{10}-C_{20}$ base chemicals by applying simple unit processes like hydrolysis, esterification and hydrogenation (2) (fig. 2).

587

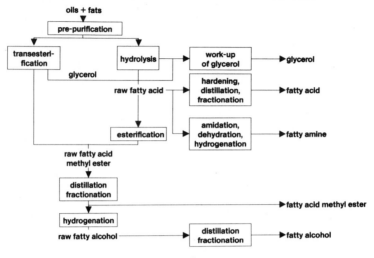

Fig. 2

2. contrary to petrochemistry, feedstock prices in oleochemistry have not undergone strong fluctuations. This is attributable to nearly 90% of the world harvest in fats and oils being absorbed by the food industry, a field in which fats and oils are largely interchangeable; since oil crops are grown in many parts of the world, prices for this commodity are controlled by strong international competition (3).

There is one notable exception to this general pattern. Medium-chain fatty acids in the range of C_{10} to C_{14} - an important raw material to the oleochemical industry - are available from but a few oil crops, especially from coconut oil. Since the large majority of this oil is coming from the Philippines, the price of these "laurics" is highly dependent on events in this country, e.g. its annual harvest.

Within this framework, the imminent technological changes, both in plant breeding and bioreactor technology, are providing for new opportunities to secure raw material supply and add new specialities to the oleochemical industry (4-10).

II. Plant cultivation and biotechnology

Over several decades, oil crop breeding has resulted in major improvements (11, 12). As an example, rapeseed was developed into a new crop low in erucic acid and glucosinolates - undesirable components for the use of this oil in foods. A more recent example is the development of sunflower oil with an oleic acid content exceeding 80%.

Tissue culture techniques have been increasingly used in plant breeding. An outstanding success utilizing this method was achieved by improving palm oil production (13). Similar techniques are now being applied to the coconut palm (14).

An ambitious program has been reported: the application of rDNA-related techniques to vegetable oil synthesis in plants. First promising results from this project include the cloning of ACP from spinach into bacteria (15). An important prerequisite for such endeavors consists in a rather detailed knowledge of the biosynthesis of fatty acids in plants on an enzymatic level including aspects of its regulation and compartimentation. Encouraging results have been obtained in this field (16), though many questions still remain unresolved.

For a company producing specialty chemicals from fats and oils, a major concern must necessarily be the safe supply of its raw material at a reasonable price. However, this need does not necessarily imply sponsor breeding projects, too. From a list of competing pathways to C_{12}-fatty alcohols (Table 1), it becomes apparent that many options are available to secure this important intermediate including petrochemistry-based reactions. On the basis of a risk-benefit analysis it thus becomes obvious that companies owning plantations or active in the seed business may have a strong argument to invest in plant breeding projects, while such oleochemical industries emphasizing derivatization and marketing instead may often confine themselves to influence supply on a political level (17).

Table 1. Routes to C_{12} alcohols

status
o hydrogenation of coconut-/palmkernel fatty acid methyl esters
o paraffin oxidation to oxo-alcohols
o ethylene oligomerisation by Ziegler-/SHOP-process

ongoing/future projects
o more supply of coconut-/palmkernel oil
o Cuphea breeding
o breeding for petroselenic acid production/ozonolysis
o genetic engineering of C_{16}-/C_{18}-plants
o cloning of C_{12}-specific thioesterase in plants
o genetic engineering of microbial fatty acid synthetase/single cell oil
o $\Delta 6$-desaturation/oxidative fission of stearic acid in bioreactor
o monoterminal oxidation of dodecane
o enzymatic hydration of C_{12}-α-olefin

The Cuphea project is a good example of this attitude. Among the wealth of information provided through the NRRC screening program in Peoria, Illinois (18), the Cuphea family was recognized as producer of C_8-C_{14}-fatty acids (fig. 3)

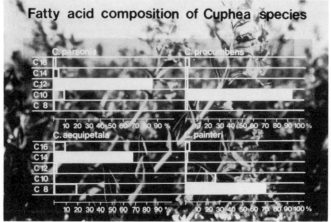

Fig. 3

As early as 1974, work was initiated at Göttingen University with the perspective to examine the potential of Cuphea as an agricultural crop (19). After first encouraging results became available in the early 1980ies (20), an USDA/SDA project was set up in 1978 and carried out by the USDA and university researchers. Although there are still major drawbacks in applying Cuphea to farming, recent progress makes it look feasible that within the next decades medium-chain fatty acids might become commercially available from Cuphea plants grown in the United States and Southern Europe (21).

III. Bioreactors

In recent years, the attempts to apply immobilized enzymes or microorganisms to the production of commodity chemicals have been steadily increasing. High-fructose corn-syrup (22), acrylamide (23), ethanol (24) and soap (25) are notable examples of this tendency. With respect to oleochemistry, four types of reactions are worth mentioning:
a) enzymatic hydrolysis of triglycerides,
b) enzymatic interesterification of triglycerides or ester synthesis,
c) enzymatic reduction of fatty acids to fatty alcohols,
d) functionalization of the fatty acid hydrophobic side chain

The enzymatic hydrolysis of oils and fats by lipase from castor seed was described as early as 1902 (26). In spite of a host of information on the action of lipases (27, 28), however, the first commercial application of such a process was described as late as 1981 for the production of soaps from plant oils (25).
Since then, this approach has been vigorously studied by a number of laboratories (29-32). Most recently, the application of lipases to the enzymatic hydrolysis of triglycerides led to the construction of a semicommercial plant for the manufacture of ultrapure unsaturated fatty acids (33).

We have been developing a membrane-type bioreactor for lipase-catalysed fat splitting (32). Using a stirred reactor, triglyceride hydrolysis was nearly complete within three hours (fig. 4).

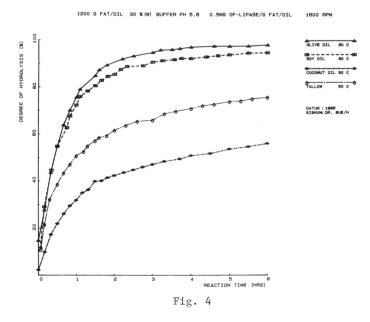

Fig. 4

However, industrially important oils and fats such as coconut oil or beef tallow, were less susceptible to enzymatic hydrolysis. By increasing the concentration of lipase, these triglyceride mixtures could be quantitatively split within 3 hours too. Cost calculations showed, however, that enzymatic fat splitting will be significantly more expensive than the present standard steam splitting process, even if present days' high prices of lipases could be reduced by a factor of 10, e.g. through cloning procedures.

The free energy of triglyceride hydrolysis is close to 0 kcal/mole (34); it is thus quite easy to direct lipase-catalysed reactions towards ester synthesis as long as reactant concentrations are selected appropriately (35-39). The potential of lipases to catalyse synthetic reactions is obviously most promising where little competition to traditional chemistry prevails. Thus, reactions requiring mild conditions and high selectivity hold promise to this type of reaction. As an example, we studied the lipase-catalysed synthesis of 2-methyl-pentanoyl prenylester, a component of blackberry flavor. This ester can be chemically prepared by base-catalysis from its components, but Wagner-Meerwein rearrangements tend to impair the fragrance characteristics of the resulting mixture. By applying Candida cylindracea lipase adsorbed on celite it proved possible to set up a bioreactor which yielded the desired product in good yields over 30 weeks indicating high stability of the adsorbed lipase (fig. 5) (35, 36).
As another example, we tested lipase-catalysed interesterification of trioleate with oleyl alcohol producing a mixture of oleyl oleate and monooleyl glycerol. After 500 days of continuous operation, lipase activity had decreased to 44% of the original activity (fig. 5).

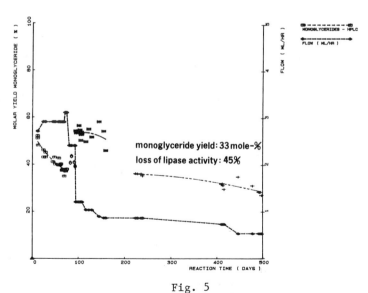

Fig. 5

Further industrial applications of the synthetic potential of lipases have been reported in the area of foods. Thus, immobilized lipase was efficiently used in a solvent-free reaction system to produce cocoabutter from palm oil and stearic acid (40). This process includes intermittent removal of reaction water by periodical drying under vacuum. A related example is the use of immobilized lipase for the interesterification of plant oils with tallow leading to spreadable butter-like fats without resorting to any catalytic hydrogenation (41).

It has been suggested that a bioreduction process could be developed to convert fatty acid esters into fatty alcohols at the expense of the glycerol contained in triglycerides. This approach does not take into account, however, that glycerol is a major factor in the economics of commercial triglyceride hydrolysis. In an endeavor to investigate the potential of this reaction from an analysis of the enzymes in question, we initiated research at some W. German universities, where the enzymes under question were purified from several microorganisms and characterized. The turnover numbers from our experiments and from literature support the view that the enzymatic steps involved in the bioreduction of fatty acids to fatty alcohols are too slow to the extent of several orders of magnitudes for being applied to a commercial process (42).

With very few exceptions, the chemistry of fats and fatty acids is so far limited to the hydrophilic part of the fatty acid molecule (fig. 6) (43).

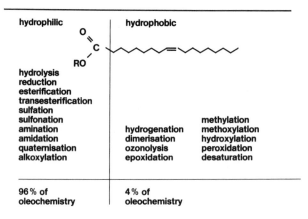

Fig. 6

The number of naturally occurring fatty acids modified in the hydrophobic side chain, however, exceeds 1,000. It may thus be stipulated that the potential of biochemistry for the biotransformation of fatty acids is very high indeed. The on-going projects in this field include terminal oxidation of fatty acids to dicarboxylic acids. Dicarboxylic acids can be produced both from petrochemical and oleochemical raw materials (44) and have conquered significant markets (Table 2).

Table 2. Technically important dicarboxylic acids

Dicarboxylic acid	chain-length [C]	application	raw material	price (DM/kg)	market size (Tt)
Adipic acid	6	polyester plasticizer	butadiene	2.20	2000
Azelaic acid	9	polyester plasticizer	oleic acid	8	28
Sebacic acid	9	polyester plasticizer	ricinoleic acid	9	9
Dodecanoic acid	12	polyester bleach	butadiene	11	5
Brassylic acid	13	polyester fragrance	eruic	15	<1

Quite recently, brassylic acid produced from the corresponding alkane by fermentation has been introduced into the Japanese market (45), and fermentative productions of adipic acid was reported (46). Whereas high yields for diterminal alkane oxidation by block mutants of microorganisms have been reported in the patent literature (44), the potential of microorganisms to oxidize the terminal methyl group of fatty acids so far appears to be limited. This may be due to the better susceptibility of alkanes to biochemical transport as well as to toxic side-effects of fatty acids. It seems quite feasible, however, that the selection of appropriate strains and the modification of biochemical pathways for fatty transport, synthesis, degradation and regulation eventually will yield powerful biocatalysts for the terminal oxidation of fatty acids.

IV. Conclusion

New methods in biotechnology such as plant tissue culture, the application of recombinant DNA-techniques to plants and the biotransformation of oleochemicals by immobilized enzymes and microorganisms hold promise to a) secure the raw material basis of the fats and oils industry, and b) to lend themselves to the development of new processes and products.
Projects in oil-crop breeding as well as bioreactor development projects have to face competition from petrochemical intermediates and increasing sophistication of traditional chemistry. However, in some areas of biotechnological research on lipids it has become visible that biotransformation reactions will lead to products hitherto unavailable from industrial chemistry. In order to further develop this field, the accumulation of more basic knowledge of molecular biochemistry of fatty acids in plants will be as important as more efforts to master the commercial biotransformation of sparingly water-soluble compounds.

V. References

(1) J. Semel and R. Steiner, Nachwachsende Rohstoffe in der chemischen Industrie, Chem. Ind. 9:489 (1983)

(2) J. Knaut and H. J. Richtler, Trends in industrial uses of palm and lauric oils, JAOCS, 62:317 (1985)

(3) H. Fochem, Der Weltmarkt der Pflanzenöle, ihre Produktion, Verwendung und Verwertung, Fette, Seifen, Anstrichm., 87:6 (1985)

(4) F. Falbe and R. D. Schmid, Biotechnology and the surfactants and detergents industry, Fette, Seifen, Anstrichm. 88:203 (1986)

(5) B. W. Werdelmann and R. D. Schmid, The biotechnology of fats, Fette, Seifen, Anstrichm., 84:436 (1982)

(6) G. Allen, The needs for the 1990s: biotechnology, Chem. and Ind. 784, Dec. 1985

(7) Y. Tsujisaka and M. Iwai, The application of biotechnology to lipid-related Industry, Yukagaku 31:772 (1982)

(8) C. Ratledge, Biotechnology as applied to the fats and oils industries, Fette, Seifen, Anstrichm., 86:379 (1984)

(9) P. Darbon, Biotechnologie des corps gras, Biofutur 9, Feb. 1986

(10) C. Ratledge, P. Dawson and J. Rattray, "Biotechnology for oil and fats industry," American Oil Chemists Society, Champaign, Il. (1984)

(11) G. Röbbelen, Biogenese und Verfügbarkeit pflanzlicher Fettrohstoffe, Fette, Seifen, Anstrichm. 86:373 (1984)

(12) A. T. James, The biotechnology of oilseed crops, JAOCS, 62:204 (1985)

(13) L. H. Jones, The oil palm and its clonal propagation by tissue culture, Biologist 30:181 (1983)

(14) Kao Corp., Tokyo, Annual Report 1984

(15) Anon., Calgene moves to apply rDNA to vegetable oils, Genetic Technol., 6:1 (1986)

(16) P. K. Stumpf, Fatty acid biosynthesis in higher plants, in: "Fatty acid metabolism and its regulation," S. Numa, ed., Elsevier Science Publ., Amsterdam (1984)

(17) W. Stein, Improvement of oil-seeds from an industrial point of view, in "Improvement of oil-seed and industrial crops by induced mutations", pp. 233-242, Internat. Atomic Energy Agency, Wien (1982)

(18) L. H. Princen, Development of new crops for industrial raw materials, JAOCS, 61:281 (1984)

(19) F. Hirsinger, Untersuchungen zur Beurteilung der Anbauwürdigkeit einer neuen MCT-Ölpflanze Cuphea, Pflanzenzüchtung 85:157 (1980)

(20) F. Hirsinger, Agronomic potential and seed composition of Cuphea, an annual crop for lauric and capric seed Oils, JAOCS, 62:76 (1985)

(21) Anon., Cuphea, JAOCS, 62:7 (1985)

(22) R. van Tilburg, Isomerization of corn-starch based glucose syrups, Food Sci. Technol., 14:175 (1985)

(23) Anon., Nitto will start 1st microbe acrylamide plant, Japan Chem. Week 4, Feb. 1985

(24) H. Tadashi, Production process by fermentation, Kagaku Kogaku, 48:29 (1984)

(25) Anon., Miyoshi establishes lipase fat splitting, Japan Chem. Week 2, May 1981

(26) W. Connstein, E. Hoyer and H. Wartenberg, Über fermentative Fettspaltung, Berichte d.D. chem. Gesellschaft 35:3988 (1902)

(27) A. R. Macrae, Extracelluar microbial lipases, in: "Microbial enzymes and biotechnology", W.M. Fogarty, ed., Applied Sciene Publishers, Barking (1983)

(28) Y. Tsujisaka and M. Iwai, Comperative study on microbial lipases, Chemistry and Chem. Ind., 58:60 (1984)

(29) Y. Kimura, A. Tanaka, K. Sonomoto, T. Nihira and S. Fukui, Application of immobilized lipase to hydrolysis of triacylglyceride, European J Appl. Microbiol. Biotechnol. 1:321 (1983)

(30) S. Fukui and A. Tanaka, Enzymatic reactions in organic solvents, Endeavour, New Series, 9:10 (1985)

(31) M. M. Hoq, T. Yamane and S. Shimizu, Continuous hydrolysis of olive oil by lipase, JAOCS, 62:1016 (1985)

(32) M. Bühler and C. Wandrey, Fette, Seifen, Anstrichm., in preparation

(33) Anon., Nippon oil and fats develops highly-purified fatty acids, Japan Chem. Week, 6:6 (1984)

(34) H. P. Kaufmann and M. C. Keller, Zur Thermodynamik der Fettspaltung, 34. Mtgl.,Fette, Seifen, Anstrichm., 44:105 (1937)

(35) G. Lazar, Estersynthesen mit Lipasen, Fette, Seifen, Anstrichm., 87:394 (1985)

(36) G. Lazar, A. Weiss and R. D. Schmid, Synthesis of esters by lipases, JAOCS, in press

(37) A. R. Macrae and R. C. Hammond, Present and future applications of lipases, Biotechnology and Genetic Engineering Reviews, 3:193 (1985)

(38) A. R. Macrae, Enzyme-catalysed modification of oils and fats, Philos. Trans. R. Soc. London, 310:227 (1985)

(39) A. R. Macrae, Interesterification of fats and oils, Studies in Organic Chemistry, 22:195 (1985)

(40) T. T. Hansen and P. Eigtved, A new immobilized lipase for interesterification and ester synthesis, JAOCS, in press

(41) P. Eigtved and T. T. Hansen, Characteristics of immobilized lipase in ester synthesis and effects of water and temperature in various reactions, JAOCS, in press

(42) H. Simon and H. Grisebach, personal communication

(43) W. Stein, Fatty chemical and petrochemical raw materials - contrast or complementation, Fette, Seifen, Anstrichm., 84:45 (1982)

(44) M. Bühler and J. Schindler, Aliphatic hydrocarbons, in: "Biotechnology," H. -J. Rehm and G. Reed, ed. Verlag Chemie, Weinheim (1984)

(45) Anon.,Nippon Mining develops brassylic acid by fermentation CEER, Chem. Econ. Eng. Rev., 17:37 (1985)

(46) Anon., Synthesis of nylon raw materials using biotechnology Chemical Economy & Engineering Review, 18:35 (1986)

ALGAL LIPIDS

LITTLE KNOWN FACTS OF PLANT LIPID METABOLISM

A. A. Benson

Scripps Institution of Oceanography
La Jolla, California 92093

Discovery of the galactosylglycerols in wheat flour lipids by Herb Carter and his colleagues (Carter et al., 1956, 1961) engendered interest in the membrane lipid components of our diets. Being major components of chloroplasts (Benson et al., 1959) their remarkable linolenate component stimulated concern for their membrane-lipoprotein interactions and for the evolutionary relationships of such structures. Most of these questions remain unexplored.

Three little-known-facts about chloroplast lipids should be familiar to all plant lipidologists. First is the role of galactolipids in the historical development of molecular biology. When Monod and Jacob (Monod et al., 1952) discovered induction of enzyme synthesis (Pardee et al., 1959) and the E. coli lac operon they assayed de novo synthesized beta-galactosidase (Cohn and Monod, 1951) using lactose or beta methylthiogalactoside as inducers. They did not yet realize that E. coli already had been trained for millions of years by an inducer, a thousand times more active. Beta galactosylglycerol is the natural inducer for the lac operon. The cow kept her secret all the time until Melvin Cohn (Borstein et al., 1965) recognized beta galactosylglycerol in the E. coli culture. Remember that, when you order a salad with a glass of milk!

The second LKF relates to discovery, here at the University of California, Davis which explained why we can be here to think about plant lipids at all. In 1951 Paul Stumpf and Eldon Newcomb (1952) discovered in peanut cotyledons a novel enzyme activity capable of degrading fatty acids, one carbon atom at a time. This alpha oxidase enzyme system was interesting but made only minimal impact on humanity until Jim Mead (Levis and Mead, 1964; Mead et al., 1986) at UCLA discerned it in brain with relatively high activity. For once plant biochemists had led the animal biochemists with a critical enzyme. Mead's discovery provided Dan Steinberg (1972) with the clue to revelation of the genetic block in patients with Refsum's Disease, a disastrous hereditary disorder in which phytanic acid esters accumulate in the body because of its inability to degrade this product of chlorophyll's phytol oxidation. Patients accumulated hundreds of grams of phytanic lipids in their spleens and livers. Their lives were nerve-wrecked and brief, clearly a neurological disorder. Unless phytanic acid can be oxidized once by an alpha oxidase, its side-chain methyl groups block beta oxidative degradation. Stable myelin structure requires long straight, preferably saturated, fatty chains.

When the four sidechain methyl groups of phytanic or dihydrophytanic acid are incorporated into lipid bilayer structures their stability is greatly reduced. The myelin is broken and no longer an effective insulator for the neuroelectrical activity of mammalian nerves. For this reason we have active alpha oxidase in our brain which we should use to express our appreciation to Paul Stumpf and his avocados and peanuts.

Now for a third LKF: One of the most actively metabolized plant lipids on earth is an arsenic compound! In most of the ocean's surface water, phosphate concentrations are so reduced that the ubiquitous isomorphous ion, arsenate, can exceed it in concentration. Algae absorb arsenate in their quest for phosphate and so must detoxicate it immediately or die. Phosphorous and arsenic differ in that phosphate esters are relatively stable compared to arsenate esters. Further, phosphate is not reduced by metabolic reductants while arsenate and arsenite are readily reduced to yet more toxic forms (Knowles and Benson, 1983; Knowles, 1982) which bind sulfhydryl groups. In its effort to release its critical sulfhydryl enzymes from their arsenic bondage the arsenic is freed as a methylated and 5-deoxyribosylated trialkylarsine oxide identified by Edmonds and Francesconi (1981; Edmonds et al., 1982).

This sulfate ester of arsenoribosylglycerol accumulates in all aquatic plants. It is a derivative of an equally universal arsenophospholipid, an arsenoribosylphosphatidylglycerol, of all such plants. The steady state concentration of this lipid is low and varies from one species to another depending on its rates of removal. It appears to mediate arsenic excretion by oxidative or hydrolytic cleavage from its phospholipid at the outside of the plasma membrane. Arsenical membrane lipids such as this also appear to play a role in the removal of arsenic from some animal systems via the gill membrane (Benson and Summons, 1981).

These arsenicals, all derived from arsenoribosylphosphatidylglycerol, are innocuous analogs of trimethylamine oxide and readily excreted by humans. One of their derivatives, arsenobetaine is found in shrimp, lobster, and some fish. It, too, is non-toxic, another of Nature's solutions to its arsenic problem.

To these little-known-facts we will add today the reports of four active laboratories on studies of algal lipids. Each will contribute to the impact of plant lipid studies upon the future of biology.

REFERENCES

Benson, A. A., and Summons, R. E., 1981, Arsenic accumulation in Great Barrier Reef invertebrates, Science, 211:482.

Benson, A. A., Wintermans, J.F.G.M., and Wiser, R., 1959, Chloroplast lipids as carbohydrate reservoirs, Plant Physiol., 34:315.

Burstein, C., Cohn, M., et Monod, J., 1965, Role du lactose et de ses produits metaboliques dans l'induction de l'operon lactose chez Escherichia coli, Biochim. Biophys. Acta, 95:634.

Carter, H. E., Hendry, R. A., and Stanacev, N. Z., 1961, Wheat flour lipids: III. Structure of the mono- and digalactosylglycerol lipids, Lipid Res., 2:223.

Carter, H. E., McCluer, R. H., and Slifer, E. D., 1956, Lipids of wheat flour. I. Characterization of galactosylglycerol components, J. Am. Chem. Soc., 78:3735.

Cohn, M., and Monod, J., 1951, Purification et proprietes de la β-galactosidase(lactase) de E. coli, Biochim. et Biophys. Acta, 7:153.

Edmonds, J. S., and Francesconi, K. A., 1981, Arseno-sugars from brown kelp (Ecklonia radiata) as intermediates in cycling of arsenic in a marine ecosystem, Nature, 289:602.

Edmonds, J. S., Francesconi, K. A., Healy, P. C., and White, A. H., 1982, Isolation and crystal structure of an arsenic-containing sugar sulphate from the kidney of the giant clam, Tridacna maxima. X-ray crystal structure of (2S)-3-[2-deoxy-5-(dimethylarsinoyl)-β-D-ribofuranosyloxy]-2-hydroxypropyl hydrogen sulphate, J. Chem. Soc. Perkin Trans. I:2989.

Knowles, F. C., 1982, The enzyme inhibitory form of inorganic arsenic, Biochem. Int., 4:647.

Knowles, F. C., and Benson, A. A., 1983, Mode of action of a herbicide. Johnsongrass and methanearsonic acid, Plant Physiol., 71:235.

Levis, G. M., and Mead, J. F., 1964, An alpha-hydroxyacid decarboxylase in brain microsomes, J. Biol. Chem., 239:77.

Mead, J. F., Alfin-Slater, R. B., Howton, D. R., and Popjak, G., 1986, Lipids, Chemistry, Biochemistry and Nutrition. 486pp. Plenum Publ. Corp. New York.

Monod, J., Pappenheimer Jr., A. M., and Cohen-Bazire, G., 1952, La cinétique de la biosynthése de la β-galactosidase chez E. coli consideré comme fonction de la croissence, Biochim. Biophys. Acta, 9:648.

Newcomb, E. H., and Stumpf, P. K., 1952, Fatty acid synthesis and oxidation in peanut cotyledons, in: "Phosphorus Metabolism, II," W. D. McElroy and B. Glass, eds., p. 291. Johns Hopkins Press, Baltimore. 930pp.

Pardee, A. B., Jacob, F., and Monod, J., 1959, The genetic control of and cytoplasmic expression of "inducibility" in the synthesis of β-galactosidase by E. coli, J. Molec. Biol., 1:165.

Steinberg, D., 1972, Phytanic acid storage disease: Refsum's Syndrome, in: "The Metabolic Basis of Inherited Disease," J. B. Stanbury, J. B. Wyngaarden, and D. S. Fredrickson, eds., p. 588. McGraw Hill, New York.

Stumpf, P. K., 1969, Metabolism of fatty acids, Ann. Rev. Biochem., 38:159.

UNIQUE CHARACTERISTICS OF CYANOBACTERIAL GLYCEROLIPIDS

Norio Murata

Department of Regulation Biology
National Institute for Basic Biology
Okazaki, 444 Japan

INTRODUCTION

The cyanobacteria (blue-green algae) are classified as Gram-negative bacteria (Stanier et al., 1976). Their cell envelope is composed of the outer membrane and plasma (or cytoplasmic) membrane separated by a peptidoglycan layer. In addition, they have intracellular photosynthetic membranes, i.e., the thylakoid membranes. Their membrane structure is similar to that of eukaryotic plant chloroplast, which contains outer and inner envelope membranes, surrounding the thylakoid membranes.

The composition of major lipids of the cyanobacteria is similar to that of higher plant chloroplasts, although a minor component, phosphatidylcholine, in the chloroplasts is replaced by monoglucosyl diacylglycerol in the cyanobacteria.

GLYCEROLIPIDS

The cyanobacteria contain four major glycerolipids, monogalactosyl diacylglycerol (MGDG), digalactosyl diacylglycerol (DGDG), sulfoquinovosyl diacylglycerol (SQDG) anda phosphatidylglycerol (PG) (Nichols et al., 1965; Hirayama, 1967; Stanier and Cohen-Bazire, 1977), and a minor component, monoglucosyl diacylglycerol (GlcDG) (Feige et al., 1980; Sato and Murata, 1982a). MGDG represents about 50% of the total glycerolipid content, whereas the GlcDG content apparently does not exceed 1% (Feige et al., 1980; Sato and Murata, 1982a).

Feige (1978) first discovered a glycerolipid (termed X-MGD and later identified as GlcDG), which migrated a little faster than MGDG in thin layer chromatography on silica gel, when he labeled lichens containing cyanobacteria with [^{14}C]bicarbonate. Feige et al. (1980) surveyed 30 species of cyanobacteria, and concluded that this lipid is ubiquitously present in these cyanobacteria but not in eukaryotic algae. The lipid was identified as β-glucosyl diacylglycerol (Feige et al., 1980; Sato and Murata, 1982a). The GlcDG is found also in the prokaryotic green alga, Prochloron (Murata and Sato, 1983). Fig. 1 compares chemical structures of GlcDG and MGDG.

The activities of glycerol-3-phosphate acyltransferase and 1-acylglycerol-3-phosphate acyltransferase, the enzymes involved in the first steps of glycerolipid synthesis, have been characterized in crude extracts of *Anabaena variabilis* (Lem and Stumpf, 1984b; Stapleton and Jaworski, 1984a,b). These enzymes use acyl-carrier protein (ACP), but not coenzyme

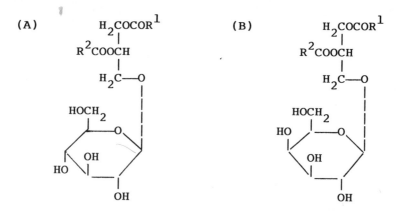

Fig. 1. Chemical structures of monoglucosyl diacylglycerol (A) and monogalactosyl diacylglycerol (B) from cyanobacteria.

A as the acyl carrier. These observations indicate that the precursor of the glycerolipids is phosphatidic acid which is converted to glyco-, sulfo- and phospholipids in this cyanobacterium.

Sato and Murata (1982c) observed in *A. variabilis* a membrane-associated activity of UDP-glucose:diacylglycerol glucosyltransferase which synthesizes GlcDG by transfer of glucose from UDP-glucose to diacylglycerol. This enzyme activity is located in both thylakoid and plasma membranes of *Anacystis nidulans* (Omata and Murata, 1986).

Feige et al. (1980) observed that in 30 species of cyanobacteria radioactivity from [^{14}C]bicarbonate was first incorporated into GlcDG, then shifted to MGDG. They proposed that GlcDG is a precursor of MGDG in the biosynthesis of glycolipids in the cyanobacteria. Sato and Murata (1982a) showed that in *A. variabilis* the conversion from GlcDG to MGDG results from epimerization of glucose to galactose, i. e., stereochemical isomerization at the C-4 of the glucose unit, but not by replacement of glucose by galactose. They further demonstrated that DGDG is produced by transfer of newly synthesized galactose to MGDG (Sato and Murata, 1982a). SQDG and PG are also rapidly labeled, suggesting that these lipids

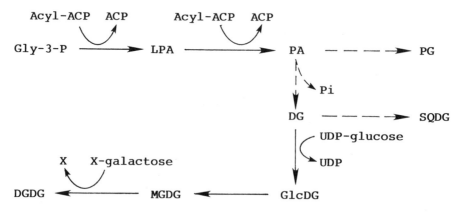

Fig. 2. A scheme for glycerolipid biosynthesis in *Anabaena variabilis*. The reactions indicated by broken arrows have not yet been experimentally demonstrated. Gly-3-P, glycerol-3-phosphate; LPA, 1-acylglycerol-3-phosphate; PA, phosphatidic acid; DG, 1,2-diacylglycerol; ACP, acyl-carrier protein; X, unidentified galactose carrier.

are directly synthesized, but not via GlcDG (Sato and Murata, 1982a).
A tentative scheme for glycerolipid biosynthesis is proposed in Fig. 2.

FATTY ACIDS

The fatty acid composition of glycerolipids from cyanobacteria was studied first by Holton et al. (1964) in *A. nidulans*, and by Levin et al. (1964) in *A. variabilis*. Subsequent analyses have characterized the fatty acids from a number of cyanobacteria. The fatty acids thus far known to be present in cyanobacteria are hexadecanoic acid or palmitic acid (16:0), Δ^9-cis-hexadecenoic acid or palmitoleic acid (16:1), hexadecadienoic acid (double bond positions undetermined) (16:2), octadecanoic acid or stearic acid (18:0), Δ^9-cis-octadecenoic acid or oleic acid (18:1), $\Delta^{9,12}$-cis-octadecadienoic acid or linoleic acid (18:2), $\Delta^{9,12,15}$-cis-octadecatrienoic acid or α-linolenic acid (α-18:3), $\Delta^{6,9,12}$-cis-octadecatrienoic acid or γ-linolenic acid (γ-18:3), and octadecatetraenoic acid (double bond positions undetermined) (18:4). Some species, such as *A. nidulans* (Allen et al., 1966; Sato et al., 1979), contain Δ^{11}-cis-octadecenoic acid (cis-vaccenic acid).

Kenyon (1972) and Kenyon et al. (1972) classified the cyanobacteria into four groups with respect to the composition and metabolism of fatty acids. Strains in the first group contain only saturated and monounsaturated fatty acids, whereas those in the other groups contain in addition, 18:2 fatty acid, as well as polyunsaturated fatty acids characteristic of each group. The second group is characterized by the presence of α-18:3, the third group by γ-18:3 and the fourth group by 18:4 fatty acids. Unicellular strains of cyanobacteria belong to the first and third groups (Kenyon, 1972). Filamentous cyanobacteria are distributed among all four groups (Kenyon et al., 1972; Oren et al., 1985). The prokaryotic green alga, Prochloron, contains only saturated and monounsaturated fatty acids as is the case in cyanobacteria of the first group (Perry et al. 1978; Johns et al. 1981; Murata and Sato 1983; Kenrick et al. 1984).

Distribution of specific fatty acids between the C-1 and C-2 positions of the sn-glycerol moiety has been studied in strains in the first and second groups. In the filamentous cyanobacteria such as *A. variabilis* (Sato et al., 1979), *Anabaena cylindrica*, *Oscillatoria chalybea*, *Nostoc calcicola* and *Tolypothrix tenuis* (Zepke et al., 1978), which contain approximately the same amounts of C_{16} and C_{18} acids, the C_{18} acids are esterified to the C-1 position, and the C_{16} acids to the C-2 position in all lipid classes (Fig. 3). In *A. nidulans*, most of the monounsaturated fatty acids, 14:1, 16:1 and 18:1, are esterified at the C-1 position and most of 16:0 is at the C-2 position in all lipid classes (Sato et al., 1979).

The fatty acid composition of individual lipids in the thylakoid membranes, plasma membrane and outer membrane prepared from *A. nidulans* is compared. A characteristic distribution is seen in MGDG, DGDG, SQDG and PG, regardless of the types of membrane from which they are derived (Murata et al., 1981; Omata and Murata, 1983).

Fatty acid biosynthesis has been extensively studied in *A. variabilis*. Sato and Murata (1982b) showed, by pulse-labeling for 6 min with [^{14}C]bi-

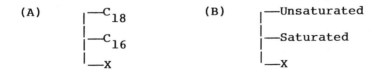

Fig. 3. Positional distribution of fatty acids in the glycerolipids in *Anabaena variabilis* (A) and *Anacystis nidulans* (B). X is a polar group.

carbonate, that radioactivity is incorporated into 16:0, 18:0 and 18:1 of GlcDG, SQDG and PG. An extension of labeling time up to 1 h resulted in an increase in radioactivity of 18:1 relative to that of the total fatty acids and corresponding decrease in that of 18:0. These changes in radioactivity of the 18:0 and 18:1 proceeded further during a subsequent chase for 10 h. It is suggested that, in *A. variabilis*, only saturated fatty acids, *i.e.*, 16:0 and 18:0, are first esterified to the lipids and then desaturated while they are bound to the glycerol moiety. The lipid-linked desaturation of 16:0 to 16:1 in MGDG was confirmed by a combination of [^{14}C]bicarbonate feeding and mass spectrometric analysis of the 2-acylglycerol moiety of ^{13}C-enriched MGDG (Sato et al., 1986).

The lack of fatty acid desaturation in the form of acyl-ACP or acyl-CoA was also suggested by *in vitro* experiments of Lem and Stumpf (1984a), and Stapleton and Jaworski (1984b), who studied fatty acid synthesis in crude extracts from *A. variabilis*. Under optimal conditions for fatty acid synthesis, only saturated fatty acids, 14:0, 16:0 and 18:0, were produced from [^{14}C]acetate. Neither added 18:0-ACP nor 18:0-CoA was desaturated. All of these observations suggest that saturated acyl-thioesters are the final products in fatty acid synthesis, and that all the desaturation reactions take place after the fatty acids are esterified to the glycerol moiety of the lipids.

The biochemical characteristics of the enzymes involved in fatty acid synthesis have been studied in crude extracts of *A. variabilis* by Lem and Stumpf (1984a), and Stapleton and Jaworski (1984a, b). The fatty acid synthetase in the crude extracts was found to be of a non-associated and ACP-dependent type, and therefore similar to the prokaryotic (or plant) type found in most bacteria and the chloroplasts of eukaryotic plants. Fatty acid elongation from 14:0 to 16:0 and from 16:0 to 18:0 requires ACP and NADPH, but not coenzyme A or NADH (Lem and Stumpf 1984a; Stapleton and Jaworski, 1984b).

CHANGES IN FATTY ACIDS IN RESPONSE TO TEMPERATURE

Isothermal Growth

Holton et al. (1964) first studied the effect of growth temperature on the fatty acid composition of the total lipids in *A. nidulans*. They found that with a lowering of growth temperature, 16:0 decreases and 16:1 increases, and that the average chain length of fatty acids decreases. Sato et al. (1979) and Murata et al. (1979) found that in *A. nidulans* the modes of changes with growth temperature in the fatty acid composition of each of all the major lipid classes are similar to that of the total lipids.

In contrast to the unicellular cyanobacteria, *A. variabilis* contains polyunsaturated fatty acids and responds to changes in growth temperature differently. When the growth temperature is lowered, 18:1 and 18:2 decrease and α-18:3 increases at the C-1 position of all the major lipid classes, while the levels of C_{16} acids remain nearly constant except for a minor decrease in 16:1 and increase in 16:2 at the C-2 positions of MGDG and DGDG (Sato et al., 1979; Sato and Murata, 1980b). The temperature dependence of the molecular species composition in *A. variabilis* (Table I) indicates that a decrease in growth temperature from 38°C to 22°C induces decreases in 18:1/16:0 and 18:2/16:0 species and increases in α-18:3/16:0 in all the major lipid classes, and decreases in 18:1/16:1 and 18:2/16:0 and increases in α-18:3/16:1 and α-18:3/16:2 in MGDG and DGDG.

An increase in unsaturation and decrease in the chain length of esterified fatty acids are known to lower the phase transition temperature of membrane lipids (Chapman et al., 1974). The changes in fatty acid composition with growth temperature can be regarded as an adaptive response to changes in the ambient temperature (Ono and Murata, 1982; Murata et al., 1984).

TABLE I. Molecular Species Composition of Glycerolipids in *Anabaena variabilis*[a]

Lipid	Growth temp. (°C)	C-1 C-2	18:0 16:0	18:1 16:0	18:2 16:0	18:3[b] 16:0	18:1 16:1	18:2 16:1	18:3[b] 16:1	18:3[b] 16:2
GlcDG	38		24	60	10	0	0	0	0	0
	22		22	40	24	6	0	0	0	0
MGDG	38		1	25	23	1	11	35	0	0
	22		2	2	12	34	0	3	32	12
DGDG	38		1	16	24	0	16	38	0	0
	22		1	4	20	19	1	9	37	4
PG	38		1	56	41	0	0	0	0	0
	22		0	10	26	61	0	0	0	0
SQDG[c]	38		10	48	26	0	0	0	0	0
	22		2	10	16	58	0	0	0	0

[a]Sato and Murata (1982b). [b]α-18:3. [c]Estimated from the positional distribution of fatty acids.

Temperature Shift

When the growth temperature is suddenly increased or decreased, the fatty acid composition of *A. variabilis* is rapidly altered (Sato and Murata, 1980a, b). For 10 h after the temperature shift from 38 to 28°C, the total amount of lipids stays at a constant level (Sato and Murata, 1980a), but a decrease in 16:0 and a concomitant increase in 16:1 take place at the C-2 position of MGDG. Then, as fatty acid and lipid synthesis resumes, the ratio of 16:0 to 16:1 is slowly restored to that seen prior to the temperature shift. This type of transient desaturation of 16:0 to 16:1 is not observed in the other major lipid classes such as DGDG. The rapid and transient introduction of a *cis* double bond into 16:0 which takes place at the C-2 position in MGDG and is regarded as an emergency acclimation to compensate for the decrease in membrane fluidity due to the decrease in temperature (Sato and Murata, 1980a, b). The rapid change in levels of unsaturation of the C_{16} acids after the temperature shift does not require *de novo* synthesis of fatty acids (Sato and Murata, 1981). The mass spectrometric analysis of ^{13}C-enriched MGDG indicates that during the change in unsaturation 16:0 is desaturated in the lipid-bound form and not replaced by previously synthesized 16:1 (Sato et al., 1986). Molecular oxygen, but not light, is necessary for the desaturation (Sato and Murata, 1981). The findings that desaturation is greatly suppressed by chloramphenicol and rifampicin suggest that a fatty acid desaturase, specific to 16:0 at the C-2 position of MGDG, is transiently synthesized after the downward shift in temperature (Sato and Murata, 1981).

Decreases in the contents of 18:1 and 18:2 and an increase in α-18:3 take place in MGDG, SQDG and PG, but these occur more slowly than the changes in C_{16} acids of MGDG. DGDG most slowly responds to the temperature shift. The desaturation of 18:1 and 18:2 after a temperature shift is also abolished by the inhibitors of protein synthesis or RNA synthesis (Sato and Murata, 1981).

An increase in growth temperature in *A. variabilis* transiently stimulates *de novo* synthesis of fatty acids but suppresses the desaturation of existing lipids to produce 16:1, 16:2 and α-18:3 (Sato and Murata, 1980a). As a result, the relative content of 16:0 rapidly increases and that of 16:1 rapidly decreases, but these changes are slowly reversed after *de novo* fatty acid synthesis and desaturation restore the original steady state levels. The rapid and transient changes in unsaturation level of the C_{16} acids occur exclusively in MGDG. Slower changes in the unsaturation level of the C_{18} acids take place in all the major lipid classes.

LIPID PHASE OF MEMBRANES

Phase transitions in the membrane lipids has long been proposed as a primary event in chilling injury of higher plants of tropical origin (Lyons, 1973; Raison, 1973). However, such transition has not been clearly demonstrated, probably because of the complicated membrane systems in higher-plant cells. Cyanobacteria are similar to chloroplasts of eukaryotic plants in their membrane structure, and therefore can be regarded as model systems for the study of molecular mechanisms of low temperature stress in the plants. The lipid phase of the membranes has been studied in the chilling-sensitive strain, *A. nidulans*, and a chilling-tolerant one, *A. variabilis*.

The temperatures critical for the onset of phase separation of the plasma membrane and thylakoid membranes of *A. nidulans* studied by various methods are summarized in Table II. The methods of freeze-fracture electron microscopy, carotenoid absorption and spin labeling provide nearly the same values for the temperature of the onset of phase separation in the plasma membranes from cells grown at $28°C$ and $38°C$. The four methods, chlorophyll fluorescence, spin labeling, X-ray diffraction and differential scanning calorimetry, all provide about the same values for the temperature of the onset of phase separation in the thylakoid membranes from cells grown at $28°C$ or $38°C$. According to the data summarized in Table II, the following facts are clearly suggested with respect to the lipid phase of the membranes in *A. nidulans*: (1) At growth temperatures both types of membranes are in the liquid crystalline state. (2) With decrease in temperature to below those which permit growth, the thylakoid membranes first enter the phase separation state. (3) About $10°C$ below the onset of phase separation in the thylakoid membranes, the plasma membrane enters the phase separation state. (4) The temperature for the onset of phase separation of both types of membrane depends on the growth temperature.

Mannock et al. (1985a, b) studied the phase behaviour of major lipids isolated from *A. nidulans* by differential scanning calorimetry and X-ray diffraction. In a cooling scan of aqueous dispersions of MGDG, DGDG, SQDG and PG from cells grown at $38°C$, the phase separation begins to appear at $17°C$, $11°C$, $15°C$ and $13°C$, respectively.

Irrespective of different features of the phase transition of the plasma and thylakoid membranes, there is no discernible difference between the plasma membranes and thylakoid membranes with respect to their polar lipid and fatty acid composition (Omata and Murata, 1983). The higher phase transition temperature of the thylakoid membrane with respect to the plasma membrane may be due to the high protein content of the thylakoid membranes, 70% of the dry weight, compared with the corresponding value of 40% found in the plasma membranes (Omata and Murata, 1983). The shift in phase transition temperature due to the growth temperature corresponds to the growth temperature-dependent alteration in the fatty acid composition (Holton et al., 1964; Sato et al., 1979).

A number of studies have demonstrated that *A. nidulans* is susceptible to low temperature. When the algal cells are exposed to temperatures near $0°C$, viability declines (Rao et al., 1977), and activities of overall photosynthesis, photosynthetic electron transport and phosphorylation all diminish (Forrest et al., 1957; Jansz and MacLean, 1973; Rao et al., 1977; Ono and Murata, 1981a; Vigh and Joó, 1983; Vigh et al., 1985). Also, potassium ions, glutamate and pteridines are released from the cells (Forrest et al., 1957; Jansz and MacLean 1973; Ono and Murata, 1981b; Vigh et al., 1985), and morphological alterations of the plasma and the

TABLE II. Temperatures for the Onset of Phase Separation of Plasma and Thylakoid Membranes and Those for the Characteristic Changes in Physiological Phenomena in *Anacystis nidulans* Grown at 28°C And 38°C, And *Anabaena variabilis* Grown at 22°C and 38°C.

Phenomenon	Sample	A. nidulans				A. variabilis			
		28°C		38°C		22°C		38°C	
		PM	TM	PM	TM	PM	TM	PM	TM
Onset of phase separation									
Freeze fracture electron microscopy[a]	C	5		16		<0		<0	
Carotenoid absorption[b]	C	5		15					
Chlorophyll a fluorescence[c]	C			13	25				
Spin probe[d]	M	5	14	13	23		7		17
X-ray diffraction[e]	M			16	26				
Differential scanning calorimetry[f]	L			13	25				
Critical for irreversible damage									
Photosynthesis[b,g]	C	5		16		<0		<0	
Hill reaction[b]	C	5		15					
Critical for leakage									
Electrolytes[g]	C	8		14		<0		<0	
Potassium ions[h]	C	7		17					
Amino acids[h]	C	7		17					
Break in Arrhenius plot									
Photosynthesis[g,i]	C			14	22		7		15
Phosphorylation[j]	M				24				

C, M and L stand for intact cells, isolated membranes and extracted lipids, respectively. PM and TM stand for plasma membrane and thylakoid membranes, respectively. [a]Ono and Murata (1982). [b]Ono and Murata (1981a), Vigh et al. (1985). [c]Murata and Fork (1975), Murata and Ono (1981), and Vigh and Joó (1983). [d]Wada et al. (1984), and Murata et al. (1975). [e]Tsukamoto et al. (1980). [f]Ono et al. (1983). [g]Murata et al. (1984). [h]Ono and Murata (1981b). [i]Murata et al. (1983). [j]Ono and Murata (1979).

thylakoid membranes are detectable (Brand et al., 1979). These low temperature-dependent phenomena are irreversible, and even when the cells are again warmed to growth temperature, only partial recoveries or none at all are observed. The critical temperatures for irreversible damage of the cells depend on growth temperature (Rao et al., 1977; Ono and Murata, 1981a, b; Vigh et al., 1985), suggesting that the temperature-dependent alteration of membrane lipids (Holton et al., 1964; Sato et al., 1979) may be involved in the susceptibility to low temperature.

Temperatures critical for irreversible damage and ion leakage, and for breaks in Arrhenius plots are listed in Table II. The critical temperatures for the irreversible damage of photosynthesis and the Hill reaction are 5°C and 15°C in cells grown at 28°C and 38°C, respectively, and correspond to the onset of phase separation in the plasma membranes. At about the same temperatures, potassium ions, free amino acids (Ono and Murata, 1981b) and electrolytes (Murata et al., 1984) begin to leak from the algal cells to the surrounding medium. In contrast, breaks in the Arrhenius plots of photosynthesis and phosphorylation appear at about 15°C and 25°C in cells grown at 28°C and 38°C, respectively (Murata et al., 1975, 1983; Ono and Murata, 1979), which correspond to the onset of phase separation in the thylakoid membranes.

Based on these observations, a mechanism can be proposed for low temperature-induced irreversible phenomena in *A. nidulans*. At the growth temperature, the plasma membrane and thylakoid membranes are both in the liquid crystalline state and are impermeable to ions and small molecules. With decrease in temperature, the thylakoid membranes go into the phase separation state and become permeable to ions and small molecules. Under these conditions, physiological activities such as photosynthesis and photosynthetic ATP formation are reversibly diminished as revealed by a break in the Arrhenius plot (Murata et al., 1975, 1983; Ono and Murata, 1979). With a further decrease in temperature, the plasma membrane enters the phase separation state and becomes permeable. Under these conditions, ions and small molecules in the cytoplasm leak out, and those in the surrounding medium leak in. This diminishes cellular metabolism leading to the death of the cell. Even when the temperature is raised to that which supports growth in unchilled cells, the concentrations of ions and small molecules are not recovered, thus resulting in irreversible damage of all physiological activities of the cells.

A. variabilis is tolerant of low temperatures. Freeze-fracture electron microscopy reveals that the plasma membrane of cells grown at 22°C and 38°C are both in the liquid crystalline state at 0°C (Ono and Murata, 1982). The electron paramagnetic resonance signal of a spin probe suggests that the thylakoid membranes from cells grown at 22°C and 38°C enter the phase separation state at about 5°C and about 15°C, respectively (Wada et al., 1984). Again in this alga, the temperature-dependent phase transition depends on the growth temperature. This is related to the growth temperature-dependent decrease in saturation level of their component fatty acids (Sato et al., 1979).

In *A. variabilis* no irreversible damage to photosynthesis nor electrolyte leakage occurs at 0°C (Murata et al., 1984). These results are related to the finding that the plasma membrane is in the liquid crystalline state above 0°C (Ono and Murata, 1982; Wada et al., 1984). In contrast, the thylakoid membranes enter the phase separation state above 0°C, and a clear break in the Arrhenius plot of photosynthesis appears (Wada et al., 1984; Murata et al., 1984). Therefore, in *A. variabilis* as in *A. nidulans*, phase transition of the thylakoid membranes does not induce irreversible damage in healthy cells, which is consistent with the mechanism for low temperature stress proposed for *A. nidulans*, *i.e.*, a phase transition in the plasma membrane is directly related to low-temperature damage.

ACKNOWLEDGEMENT

This work was supported in part by a Grant-in-Aid for Scientific Research (61440002) from the Japanese Ministry of Education, Science and Culture.

REFERENCES

Allen, C.F., Hirayama, O., and Good, P., 1966, in "Biochemistry of Chloroplasts," T.W. Goodwin, ed., Vol. 1, pp. 195-200. Academic Press, London.
Brand, J.J., Kirchanski, S.J., and Ramirez-Mitchell, R., 1979, Planta, 145:63-68.
Chapman, D., Urbina, J., and Keough, K.M., 1974, J. Biol. Chem., 249:2512-2521.
Feige, G.B., 1978, Ber. Deutsch. Bot. Ges., 91:595-602.
Feige, G.B., Heinz, E., Wrage, K., Cochems, N., and Ponzelar, E., 1980, in "Biogenesis and Function of Plant Lipids," P. Mazliak, P. Benveniste, C. Costes and R. Douce, eds., pp. 135-140.

Elsevier/North-Holland Biomedical Press, Amsterdam.
Forrest, H.S., Van Baalen, C., and Myers, J., 1957, Science, 125:699-700.
Hirayama, O., 1967, J. Biochem., 61:179-185.
Holton, R.W., Blecker, H.H., and Onore, M., 1964, Phytochemistry, 3:595-602.
Jansz, E.R., and MacLean F.I., 1973, Can. J. Microbiol., 19:381-387.
Johns, R.B., Nichols, P.D., Gillan, F.T., Perry, G.J., and Volkman, J.K., 1981, Comp. Biochem. Physiol., 69B:843-849.
Kenrick, J.R., Deane, E.M., and Bishop, D.G., 1984, Phycologia, 23:73-76.
Kenyon, C.N., 1972, J. Bacteriol., 109:827-834.
Kenyon, C.N., Rippka, R., and Stanier, R.Y., 1972, Arch. Microbiol., 83:216-236.
Lem, N.W., and Stumpf, P.K., 1984a, Plant Physiol., 74:134-138.
Lem, N.W., and Stumpf, P.K., 1984b, Plant Physiol., 75:700-704.
Levin, E., Lennarz, W.J., and Bloch, K., 1964, Biochim. Biophys. Acta, 84:471-474.
Lyons, J.M., 1973, Annu. Rev. Plant Physiol., 24:445-466.
Mannock, D.A., Brain, A.P.R., and Williams, W.P., 1985a, Biochim. Biophys. Acta, 817:289-298.
Mannock, D.A., Brain, A.P.R., and Williams, W.P., 1985b, Biochim. Biophys. Acta, 821:153-164.
Murata, N., and Fork, D.C., 1975, Plant Physiol., 56:791-796.
Murata, N., and Ono, T., 1981, in "Photosynthesis," G. Akoyunoglou, ed., Vol. 6, pp. 473-481. Balaban International Science Services, Philadelphia, Pa.
Murata, N., and Sato, N., 1983, Plant Cell Physiol., 24:133-138.
Murata, N., Troughton, J.H., and Fork, D.C., 1975, Plant Physiol., 56:508-517.
Murata, N., Ono, T., and Sato, N., 1979, in "Low Temperature Stress in Crop Plants: The Role of the Membrane," J.M. Lyons, D. Graham and J.K. Raison, eds., pp. 337-345. Academic Press, New York.
Murata, N., Sato, N., Omata, T., and Kuwabara, T., 1981, Plant Cell Physiol., 22:855-866.
Murata, N., Wada, H., Omata, T., and Ono, T., 1983, in "Effects of Stress on Photosynthesis," R. Marcelle, H. Clijsters and M. van Poucke, eds., pp. 193-199. Martinus Nijhoff/Dr W. Junk Publishers, The Hague.
Murata, N., Wada, H., and Hirasawa, R., 1984, Plant Cell Physiol., 25:1027-1032.
Nichols, B.W., Harris, R.V., and James, A.T., 1965, Biochem. Biophys. Res. Commun., 20:256-262.
Omata, T., and Murata, N., 1983, Plant Cell Physiol., 24:1101-1112.
Omata, T., and Murata, N., 1986, Plant Cell Physiol., 27:485-490.
Ono, T., and Murata, N., 1979, Biochim. Biophys. Acta, 545:69-76.
Ono, T., and Murata, N., 1981a, Plant Physiol., 67:176-181.
Ono, T., and Murata, N., 1981b, Plant Physiol., 67:182-187.
Ono, T., and Murata, N., 1982, Plant Physiol., 69:125-129.
Ono, T., Murata, N., and Fujita, T., 1983, Plant Cell Physiol., 24:635-639.
Oren, A., Fattom, A., Padan, E., and Tietz, A., 1985, Arch. Microbiol., 141:138-142.
Perry, G.J., Gillan, F.T., and Johns, R.B., 1978, J. Phycol., 14:369-371.
Raison, J.K., 1973, J. Bioenerg., 4:285-309.
Rao, V.S.K., Brand, J.J., and Myers, J., 1977, Plant Physiol., 59:965-969.
Sato, N., and Murata, N., 1980a, Biochim. Biophys. Acta, 619:353-366.
Sato, N., and Murata, N., 1980b, in "Biogenesis and Function of Plant Lipids," P. Mazliak, P. Benveniste, C. Costes and R. Douce, eds., pp. 207-210. Elsevier/North-Holland Biomedical Press, Amsterdam.
Sato, N., and Murata, N., 1981, Plant Cell Physiol., 22:1043-1050.
Sato, N., and Murata, N., 1982a, Biochim. Biophys. Acta, 710:271-278.
Sato, N., and Murata, N., 1982b, Biochim. Biophys. Acta, 710:279-289.
Sato, N., and Murata, N., 1982c, Plant Cell Physiol., 23:1115-1120.
Sato, N., Murata, N., Miura, Y., and Ueta, N., 1979, Biochim. Biophys.

Acta, 572:19-28.
Sato, N., Seyama, Y., and Murata, N., 1986, Plant Cell Physiol. 27:819-835.
Stanier, R.Y., and Cohen-Bazire, G., 1977, Annu. Rev. Microbiol., 31:225-274.
Stanier, R.Y., Adelberg, E.A., and Ingram, J.L., 1976, in: "The Microbial World," pp. 119-153. Prentice-Hall, Englewood Cliffs, N.J.
Stapleton, S.R., and Jaworski, J.G., 1984a, Biochim. Biophys. Acta, 794:240-248.
Stapleton, S.R., and Jaworski, J.G., 1984b, Biochim. Biophys. Acta, 794:249-255.
Tsukamoto, Y., Ueki, T., Mitsui, T., Ono, T., and Murata, N., 1980, Biochim. Biophys. Acta, 602:673-675.
Vigh, L., and Joo, F., 1983, FEBS Lett., 162:423-427.
Vigh, L., Gombos, Z., and Joo, F., 1985, FEBS Lett., 191:200-204.
Wada, H., Hirasawa, R., Omata, T., and Murata, N., 1984, Plant Cell Physiol., 25:907-911.
Wolk, C.P., 1973, Bacteriol. Rev., 37:32-101.
Zepke, H.D., Heinz, E., Radunz, A., Linscheid, M., and Pesch, R., 1978, Arch. Microbiol., 119:157-162.

LIPIDS OF DIATOMS AND OF HALOPHILIC DUNALIELLA SPECIES

Morris Kates

Department of Biochemistry
University of Ottawa
Ottawa, Ontario, Canada

INTRODUCTION

The major polar lipids of green algae, as of higher plants, generally reflect those that are characteristic of chloroplast membranes: mono- and digalactosyldiacylglycerol, sulfoquinovosyl diacylglycerol and phosphatidylglycerol. The long-chain fatty acid constituents, again as in higher plants, also reflect those characteristic of the chloroplast membrane and consist largely of C-18 polyunsaturated acids with lesser amounts of C-16 saturated and polyunsaturated acids.

However, there are several exceptions to this generalization. For example, certain green algae such as diatoms (Bacillariophyceae) and Dunaliella (Volvocales) contain novel polar lipids in addition to the above-mentioned lipids, and their fatty acyl chains have a wider range of chain lengths and degrees of unsaturation. In diatoms the novel polar lipids consist of three sulfolipids while Dunaliella species contain a novel zwitterionic non-phospholipid. Furthermore, these algae appear to lack phosphatidylserine and have only low amounts of phosphatidylethanolamine.

A brief review of the structure, properties and metabolism of these novel polar lipids will now be given.

LIPIDS OF DIATOMS

Overall Polar Lipid Composition

Of the five species of marine diatoms (Cylindrotheca fusiformis, Nitzschia angularis, Nitzschia thermalis, Cyclotella cryptica and Phaeodactylum tricornutum) and the one fresh-water diatom (Navicula pelliculosa) examined in an early study[1], all were found to contain major amounts of the four polar lipids usually associated with photosynthetic

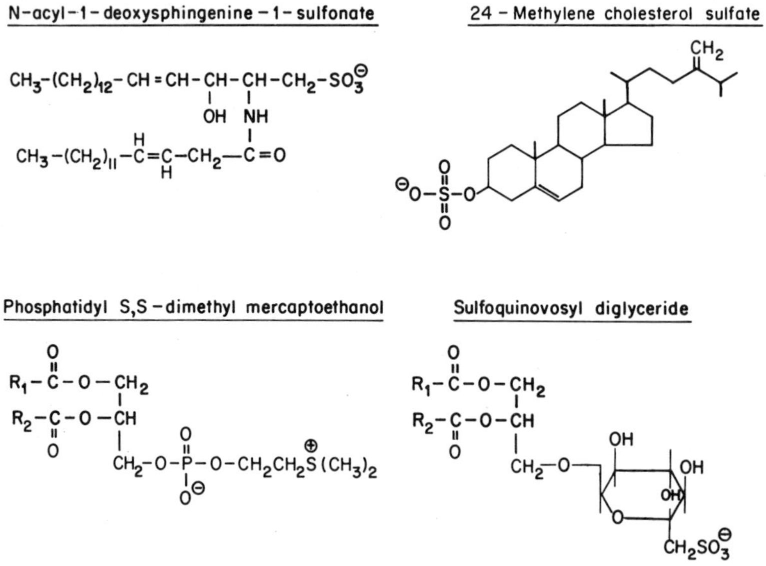

Fig. 1. Structures of sulfolipids in N. alba.

organisms: mono- and di-galactosyl diacylglycerols (MGD, DGD), sulfoquinovosyl diacylglycerol (SQD) and phosphatidylglycerol (PG). These same lipids were later also found present in other fresh-water (Nitzschia palea, Navicula muralis) and marine (Navicula incerta) diatoms[2] and in the non-photosynthetic diatom, Nitzschia alba[3]. A comparison of the polar lipid composition of N. alba with N. pelliculosa on a quantitative basis (Table 1) shows, however, that the non-photosynthetic diatom (N. alba), lacking a chloroplast structure, has much lower proportions of mono- and digalactosyl diglycerides than the photosynthetic diatom (N. pelliculosa) but the proportions of sulfoquinovosyl diglyceride and phosphatidylglycerol are about the same in both organisms.

Characteristically, all the diatom species examined[1,3] contained three novel sulfolipids, apart from sulfoquinovosyl diacylglycerol, which were detected by ^{35}S-labelling and identified in N. alba[3-5] as sterol sulfate (SS), deoxyceramide sulfonic acid (DCS) and the lecithin analogue phosphatidylsulfocholine (PSC) (see Fig. 1 for their structures). The structures of PSC[6] and DCS[7] have been confirmed by chemical synthesis. The sterol sulfate was shown to be the sulfate ester of 24-methylene cholesterol[5] and this was confirmed by chemical synthesis[4].

The two unidentified sulfolipids (U_1 and U_2) detected in N. palea, N. muralis and N. incerta[2] may well be sterol sulfate and the deoxyceramide

Table 1. Polar Lipid Composition of N. pelliculosa and N. alba

Lipid	% of total polar lipids	
	N. pelliculosa[a]	N. alba[b]
Phospholipids		
phosphatidylcholine (+ lyso derivative)	12	0
phosphatidylsulfocholine (+ lyso derivative)	2	41
phosphatidylethanolamine	<1	0
phosphatidylinositol	1	3
phosphatidylglycerol	5	4
cardiolipin	<1	1
Glycolipids		
monogalactosyl diglyceride	28	9
diglycosyl diglyceride	25	8
sulfoquinovosyl diglyceride (+ lyso derivative)	23	19
Sulfolipids		
sterol sulfate	1	8
deoxyceramide sulfonic acid	4	6

[a] Data from ref. 1.
[b] Data from ref. 3.

sulfonate, respectively, judging from their chromatographic mobilities; the presence of phosphatidylsulfocholine in these diatoms, however, is uncertain. In the non-photosynthetic diatom (N. alba)[3] phosphatidylsulfocholine completely replaced phosphatidylcholine (PC) but all of the photosynthetic diatoms contained both PC and PSC in varying ratios (Table 2). In all of the photosynthetic diatoms PC and PSC account for more than half of the phospholipids and about 10% of the total polar lipids, whereas in N. alba PSC alone accounts for more than 80% of the phospholipids and 40% of the total polar lipids (Table 1 & 2). The presence of both PC and PSC in the photosynthetic diatoms has recently been confirmed by NMR[8] and mass spectrometry[9]. Overall, sulfolipids are present in very high concentrations in diatoms particularly in N. alba where they account for 74% of total polar lipids; but even in N. pelliculose total sulfolipids account for at least 30% of the polar lipids (Table 1).

It should be noted that in all of the photosynthetic diatoms PG is a major phospholipid component and PI is a minor one, while in N. alba both PG and PI are minor components. Note also that nitrogenous base-containing phospholipids are completely absent in N. alba but small to trace amounts

Table 2. Phospholipid Composition of Diatoms

			% of total phospholipid				
Diatom	PC	PSC	PE	PE-ME	PI	PG	DPG
N. pelliculosa[a]	59	10	2	<1	5	24	<1
C. fusiformis[a]	21	25	<1	<1	28	25	<1
N. thermalis[a]	44	11	<1	<1	5	36	<1
N. angularis[a]	27	32	<1	<1	8	30	<1
C. cryptica[a]	30	22	<1	<1	5	40	<1
P. tricornutum[a]	36	18	<1	<1	7	36	<1
C. nana[b]	10	<1					
N. incerta[b]	3	nd					
N. alba[c]	0	83	0	0	7	8	2

[a] Data from ref. 1.
[b] Data from ref. 8, given as % of total lipids; nd, not detected.
[c] Data from ref. 3.

of PE and monomethyl-PE were detected in all of the photosynthetic diatoms[1]. Only small to trace amounts of cardiolipin (DPG) were detected in any of the diatoms.

Fatty Acid Composition

All the photosynthetic diatoms examined contained 16:0, 16:1 and 20:5 acids as major fatty acid components, with minor amounts of 14:0, 16:2, 16:3, 18:1, 18:2, 18:3 and 20:4 acids[1]. The fresh-water diatom N. pelliculosa, however, contained a lower proportion of 16:0 and a higher proportion of 16:3 than the marine diatoms[1]. In contrast, the major fatty acids in the non-photosynthetic diatom, N. alba[3,10], were 14:0, 16:0, 18:1 and 20:5 acids, which are the main fatty acids in the triglycerides constituting 86% of the total lipids[3]; the triglycerides (plus small amounts of diglycerides) are associated with cytoplasmic oil droplets. The photosynthetic diatoms examined have similar fatty acid compositions to those reported for other diatoms[2,11]. However, the diatom fatty acids are strikingly different in composition from those in leaves, green algae, blue green algae, euglenids and phytomonads which contain only minor amounts of 16:1 and 20:5 and major amounts of 18:3 (see ref. 11).

In regard to individual polar lipid components, fatty acid compositional data are available only for N. alba[3]. The two galactolipids MGD and DGD both contain major proportions of 14:0, 16:0 and 18:1 acids, while the SQD and its lyso derivative are rich in 20:5 and 22:6. In contrast, the glycolipids of higher plants and green algae contain major amounts of 18:3 acids. The main phospholipid, PSC, contained major amounts of 14:0, 18:1 and 20:5 acids corresponding to the major molecular species 14:0/14:0, 14:0/18:1 and 14:0/20:5[4,9]. An unusual feature of the deoxyceramide sulfonate is the presence of trans-3-hexadecenoic acid as the major N-acyl component[3,4].

Biosynthesis of Sulfolipids

As mentioned above, the three novel sulfolipids, as well as the sulfoquinovosyl diacylglycerol, are labelled with ^{35}S when diatoms are grown in the presence of [^{35}S]sulfate, but this raises the question of which pathways are used in their biosynthesis. In regard to the sterol sulfate, its biosynthesis most likely takes place by a sulfate transferase reaction from phospho-adenosylphosphosulfate (PAPS) to the 3-hydroxy group of 24-methylenecholesterol, analogous to that for biosynthesis of cholesterol sulfate[12]. Although this pathway appears plausible, it has not yet been tested experimentally in a cell-free system with [^{35}S]PAPS as sulfate donor.

The biosynthesis of the deoxyceramide sulfonic acid, in which the sphingosine base contains a sulfonate group at C-1 instead of the usual hydroxy group, might involve an analogous pathway to that for sphingosine[13] in which either cysteine or cysteic acid would substitute for serine in the condensation step with palmitoyl-CoA. With cysteine as precursor an oxidation step would be required after the condensation step. Cells of N. alba were found to incorporate ^{35}S into DCS when grown in the presence[14] of [^{35}S]cystine or [^{35}S]cysteine but not [^{35}S]methionine. DCS thus appeared to be synthesized from cysteine by the pathway mentioned above but the possibility existed that cysteine was first converted to cysteic acid which was the direct precursor in the biosynthesis. Recent studies (M. Kates and B.E. Volcani, unpublished results) using [^{35}S]cysteic acid indeed suggest that cysteic acid is the direct precursor of DCS synthesis, as has been found for biosynthesis of the analogous capnine or N-acyl capnine in gliding bacteria[15]. However, the pathway for conversion of cysteine to cysteic acid still remains to be elucidated.

In regard to the biosynthesis of phosphatidylsulfocholine, cells of N. alba incorporated ^{35}S into PSC and lyso-PSC when grown in the presence of either [^{35}S]methionine, [^{35}S]cysteine or [^{35}S]cystine[14]; however all of the ^{35}S incorporated from methionine appeared only in PSC and lyso-PSC. Methionine is thus the more direct precursor and labelling of PSC (+ lyso-PSC) with ^{35}S from cysteine (or cystine) results from incorporation of cysteine-derived sulfur into methionine via the cystathione intermediate, a pathway that is well-established in plants and microorganisms[16]. The reverse incorporation of methionine sulfur into cysteine clearly does not occur in N. alba since ^{35}S from methionine was not incorporated into deoxyceramide sulfonate. When N. alba cells were grown[14] in the presence of [Me-^{3}H]- and [^{35}S]methionine, PSC and lyso-PSC showed a higher ^{3}H/^{35}S ratio than that of the methionine substrate, showing that methionine supplies the sulfur atom and one of the sulfonium methyl groups and also functions as a methyl donor (as S-adenosyl methionine) to form S-methyl methionine. The latter may be converted to dimethyl-β-propiothetin, presumably by deamination and oxidative decarboxylation[17], and then to sulfocholine by further oxidative decarboxylation. Direct incorporation of sulfocholine into PSC (+ lyso-PSC) in N. alba has recently been demonstrated (P.-A. Tremblay, M. Kates and B.E. Volcani, unpublished results) by growing N. alba cells in medium containing [Me-^{3}H]- and [^{35}S]sulfocholine; both PSC and lyso-PSC had ^{3}H/^{35}S ratios close to that of the sulfocholine precursor. Incorporation of sulfocholine into PSC thus may occur by an analogous "Kennedy" pathway in which CDP-sulfocholine donates the phosphosulfocholine group to sn-1,2-diacylglycerol, catalyzed by the enzyme CDP-choline: diacylglycerol phosphocholine transferase, assuming this enzyme does not distinguish between choline and sulfocholine.

Evidence in favour of this suggestion was reported previously by Bjerve and Bremer[18] who found that radiolabel from both [^{35}S]sulfocholine and [Me-^{3}H]sulfocholine was incorporated into a phospholipid in rat heart and kidney that co-chromatographed with phosphatidylcholine. Anderson and Bilan[19] achieved virtually complete replacement of the choline of phosphatidylcholine or sphingomyelin by sulfocholine in mouse LM fibroblast cells grown in choline-free medium containing [^{35}S]sulfocholine. Recently, we have also succeeded in largely replacing choline of phosphatidylcholine by sulfocholine in the yeast Saccharomyces cerevisiae grown in a choline-free medium supplement with sulfocholine[20]. We have also shown (M. Kates and B.E. Volcani, unpublished results) that photosynthetic and non-photosynthetic diatoms can incorporate both choline and sulfocholine into the respective PC and PSC when grown in the presence of a mixture of these precursors.

LIPIDS OF HALOTOLERANT AND HALOPHILIC DUNALIELLA SPECIES

Overall Lipid Composition

Lipids of two halotolerant (D. parva Lerch and D. tertiolecta)[21] and six halophilic (D. viridis from the Dead Sea and unidentified species C_9, D_{11a}, D_{11b}, D_{13} and F_{20a} from the Sinai)[22] species of Dunaliella have been studied. Both the halotolerant and the halophilic species had characteristically high contents (40-55 mole %) of glycolipids, chiefly mono- and digalactosyl diacylglycerols and sulfoquinovosyldiacylglycerols, and low contents (6-19 mole %) of phospholipids (Table 3). The major phospholipids

Table 3. Lipid Composition (mol %) of Dunaliella Species[a]

Component	D. parva	D. tertiolecta	C_9	D_{11b}
phosphatidylcholine	9	4	1	2
phosphatidylethanolamine	nd	2	1	1
phosphatidylinositol	2	3	tr	tr
phosphatidylglycerol	6	8	4	3
phosphatidic acid	nd	2	tr	nd
total phospholipid	17	19	6	6
monogalactosyldiacylglycerol	21	22	22	24
digalactosyldiacylglycerol	11	21	19	15
sulfoquinovosyldiacylglycerol	7	10	14	14
unidentified glycolipids	1	tr	tr	tr
total glycolipid	40	53	55	53
diacylgyceryl-O-N,N,N-trimethylhomoserine	15	8	3	9
free fatty acid	13	7	14	7
neutral lipids	15	15	21	25

[a] Data from refs. 21 and 22.
Abbreviations: nd, not detected; tr, trace.

were phosphatidylglycerol and phosphatidylcholine, with lesser proportions of phosphatidylethanolamine (not detected in D. parva); phosphatidylinositol was present in the halotolerant species in appreciable proportions but was found in low amounts in the halophilic species; phosphatidic acid was present only in trace amounts or was not detected, except in D. tertiolecta (Table 3). In general, halophilic species had lower levels of phospholipids and higher levels of glycolipids, including several minor unidentified ones, than the halotolerant species. All Dunaliella species examined contained major amounts (15-25 mole %) of neutral lipids (chiefly triacylglycerols) and lesser amounts (7-14 mole %) of free fatty acids. All species also contained appreciable amounts (3-14 mole %) of a zwitterionic non-phospholipid which was identified by infrared, NMR and mass spectrometry as diacylglycerol-O-4'-(N,N,N-trimethyl homoserine)[23]:

$$\begin{array}{l} R-CO-O-CH_2 \\ | \\ R'-CO-O-C-H \quad\quad \overset{+}{N}(CH_3)_3 \\ | | \\ CH_2-OCH_2CH_2-CH-COO^- \end{array}$$

This 'betaine-type' lipid appears to be widespread in the plant kingdom having been reported present in: other members of the Volvocales family[24-27], other green algae[28,29], a chrysophyte[30], a human pathogenic fungus[31], and many species of fern (pteridophytes)[32]. The homoserine lipid is absent, however, in several algae species[28,29], and in all seed plants (angiosperms and gymnosperms)[32] examined.

Fatty Acid Composition

Total fatty acids of all Dunaliella species examined had very similar compositions: 16:0 was the major saturated fatty acid but low concentrations of 14:0, 15:0, 18:0 and 20:0 acids were also present; unsaturated fatty acids accounted for over 70% of the acids, 18:3n3 being the largest component and with lower levels of 18:2n6, 16:4, 18:1, 18:3n6 and 16:1 acids, in decreasing order.

Fatty acids of the individual lipid components had characteristic compositions for glycolipids and phospholipids: high levels of 16:0 in sulfoquinovosyldiacylglycerol but low levels in monogalactosyldiacylglycerol which was rich in 18:3n3 and 16:4 acids; digalactosyldiacylglycerol and the homoserine lipid both contained predominantly 16:0, 18:2n6 and 18:3n3 acids; phosphatidylglycerol was unique in containing high levels of 16:1 with lesser amounts of 16:0, 18:2 and 18:3 acids; phosphatidylcholine and phosphatidylethanolamine were similar in fatty acid composition, with 16:0 predominating and 16:1, 16:4, 18:0, 18:1, 18:2 and 18:3 acids also occurring in appreciable amounts.

REFERENCES

1. M. Kates and B.E. Volcani, Lipids of diatoms, Biochim. Biophys. Acta 116:264 (1966).
2. F.I. Opute, Lipid and fatty acid composition of diatoms. J. Exp. Bot. 25:823 (1974).
3. R. Anderson, B.P. Livermore, M. Kates and B.E. Volcani, The lipid composition of the non-photosynthetic diatom Nitzschia alba, Biochim. Biophys. Acta 528:77 (1978).
4. R. Anderson, M. Kates and B.E. Volcani, Identification of the sulfolipids in the non-photosynthetic diatom Nitzschia alba, Biochim. Biophys. Acta 528:89 (1978).
5. M. Kates, P. Tremblay, R. Anderson and B.E. Volcani, Identification of the free and conjugated sterol in non-photosynthetic diatom, Nitzschia alba, as 24-methylene cholesterol, Lipids 13:34 (1978).
6. P.-A. Tremblay and M. Kates, Chemical synthesis of sn-3-phosphatidyl sulfocholine, a sulfonium analogue of lecithin, Can. J. Biochem. 57: 595 (1979).
7. N.N. Karpyshev, A.S. Bushnev, E.N. Zoonkova and R.P. Evstigneeva, Synthesis of ceramide-1-deoxy-1-sulfonic acid, Bioorg. Chim. 3:1373 (1977).
8. P. Bisseret, S. Ito, P.-A. Tremblay, B.E. Volcani, D. Dessort and M. Kates, Occurrence of phosphatidylsulfocholine, the sulfonium analog of phosphatidylcholine, in some diatoms and algae. Biochim. Biophys. Acta 796:320 (1984).
9. P. Bisseret, D. Dessort, Y. Nakatani and M. Kates, Ammonia desorption chemical ionization mass spectrometry of phosphatidylsulfocholine-phosphatidylcholine mixtures. Chem. Phys. Lipids 36:309 (1985).
10. T.G. Tornabene, M. Kates and B.E. Volcani, Sterols aliphatic hydrocarbons and fatty acids of a non-photosynthetic diatom, Nitzschia alba, Lipids 9:279 (1974).
11. J.A. Erwin, Comparative biochemistry of fatty acids in eukaryotic microorganisms in "Lipids and Biomembranes of Eukaryotic Microorganisms", J.A. Erwin, ed., Academic Press, New York (1973).

12. R.K. Bannerjee, and A.B. Roy, The formation of cholesteryl sulfate by adrenostolane sulfotransferase, Biochim. Biophys. Acta 137:211 (1967).
13. W. Stoffel, Studies on the biosynthesis and degradation of sphingosine bases, Chem. Phys. Lipids 5:139 (1970).
14. R. Anderson, M. Kates and B.E. Volcani, Studies on the biosynthesis of sulfolipids in the diatom Nitzschia alba, Biochim. Biophys. Acta, 573:557 (1979).
15. D.R. Abbanat, W. Godchaux, G. Polychroniou and E.R. Leadbetter, Biosynthesis of a sulfonolipid in gliding bacteria, Biochem. Biophys. Res. Commun. 130:873 (1985).
16. G.A. Maw, in "Sulfur in Organic and Inorganic Chemistry", A. Senning ed., Marcel Dekker, New York (1972) pp. 113.
17. R.C. Greene, Biosynthesis of dimethyl-β-propiothetin, J. Biol. Chem. 237:2251 (1962).
18. K.S. Bjerve and J. Bremer, Sulfocholine (dimethylhydroxyethylsulfonium chloride) and choline metabolism in the rat, Biochem. Biophys. Acta 176:570 (1969).
19. R. Anderson and P. Bilan, Replacement of mouse LM fibroblast choline by a sulfonium analog, Biochim. Biophys. Acta 640:91 (1981).
20. M. Kates and P.-A. Tremblay, L'analogue sulfonium de la lecithine, le phosphatidylsulfocholine, peut-il-remplacer efficacement la lecithine dans les membranes naturelles? Rev. Can. Biol. 40:343 (1981).
21. R.W. Evans, M. Kates, M. Ginzburg and B.-Z. Ginzburg, Lipid composition of halotolerant algae, Dunaliella parva Lerche and Dunaliella tertiolecta, Biochim. Biophys. Acta 712:186 (1982).
22. R.W. Evans and M. Kates, Lipid composition of halophilic species of Dunaliella from the Sinai, Arch. Microbiol, 140:50 (1984).
23. R.W. Evans, M. Kates, G.W. Wood, Identification of diacylglycerol-O-(N,N,N-trimethyl)-homoserine in the halotolerant algae, Dunaliella parva, Chem. Phys. Lipids, 31:331 (1982).
24. A. Fried, A. Tietz, A. Ben-Amotz, W. Eichenberger, Lipid composition of the halotolerant alga, Dunaliella bardawil, Biochim. Biophys. Acta 713:419 (1982).
25. W. Eichenberger, A. Boschetti, Occurrence of 1(3), 2-diacylglyceryl-(3)-O-4'-(N,N,N-trimethyl)-homoserine in Chlamydomonas reinhardi FEBS Lett. 88:201 (1978).
26. D.R. Janero and R. Barnett, Isolation and characterization of an ether-linked homoserine lipid from the thylakoid membrane of Chlamydomonas reinhardtii 137[+]. J. Lipid Res. 23:307 (1982).
27. K.R. Moseley, G.A. Thompson Jr., Lipid composition and metabolism of Volvox carteri. Plant Physiol. 65:260 (1980).
28. W. Eichenberger, Distribution of diacylglycerol-O-4'-(N,N,N-trimethyl) homoserine, Plant Sci. Lett. 24:91 (1982).
29. N. Sato and M. Furuya, Distribution of diacylglyceryl-trimethylhomoserine and phosphatidylcholine in non-vascular green plants, Plant Sci. 38:81 (1985).
30. A.E. Brown, J. Elovson, Isolation and characterization of a novel lipid, 1(3),2-Diacylglyceryl-(3)-O-4'-(N,N,N-trimethyl)homoserine from Ochromonas danica. Biochemistry 13:3476 (1974).
31. T. Yamada and Y. Nozawa, An unusual lipid in the human pathogenic fungus Epidermophyton floccosum, Biohim. Biophys. Acta 574:433 (1979).
32. N. Sato and M. Furuya, Distribution of diacylglyceryl trimethylhomoserine in selected species of vascular plants, Phytochemistry, 23:1625 (1984).

METABOLISM OF GALACTOLIPIDS IN DUNALIELLA SALINA

Sung Ho Cho and Guy A. Thompson, Jr.

Department of Botany
University of Texas
Austin, TX 78713

INTRODUCTION

In spite of a relatively concerted effort recently directed to the study of plant lipid metabolism, many important and interesting questions remain unanswered. For a variety of reasons, plants often seem more intransigent than animals in yielding up their secrets. Our laboratory has initiated a study of membrane lipid synthesis and turnover in a simple eukaryotic green alga, Dunaliella salina, hoping that the ease of experimentation with it will lead to new and widely applicable knowledge.

So far, Dunaliella has indeed given us useful insights into the assembly and maintenance of membranes. Results published to date (e.g., 1-3) have dealt mainly with phospholipid metabolism. We would now like to report complementary findings involving the cell's galactolipids. Dunaliella seems well suited for studies of the metabolic interrelationships between one galactolipid and another and between the galactolipids and other lipid classes.

RESULTS

As in most higher plants, D. salina chloroplasts, which account for some three quarters of the cell's lipids, are composed mainly of monogalactosyldiacylglycerol (MGDG) and digalactosyldiacylglycerol (DGDG). Recent technical advances have made it feasible not only to purify MGDG and DGDG from other lipid classes but also to separate the individual molecular species of each galactolopid class. A comparison of the molecular species compositional differences between D. salina MGDG and DGDG (Table I) (4) reveals the same striking differences in fatty acid make-up that other workers have reported for higher plants. Findings of this type render it unlikely that DGDG is formed by a nonspecific placement of a second galactose onto a freshly made MGDG. In an effort to clarify the specificity of enzymes catalyzing these reactions and to pinpoint the site(s) and rates of galactolipid metabolism, we have carried out a number of pulse chase radioisotope labeling experiments. For the most part these involve short term labeling of the cells with ^{14}C-fatty acids, which enter D. salina with dispatch (1). Exogenous ^{14}C-fatty acids have also been shown to enter the prokaryotic galactolipid biosynthetic pathway of higher plants (5).

Table I

MOLECULAR SPECIES COMPOSITION OF D. SALINA GALACTOLIPIDS

Molecular Species	MGDG	DGDG
18:3/16:4	84	6
18:3/16:3 isomer	1	22
18:3/16:3	} 10	1
18:2/16:4		0
18:3/16:2 (+18:2/16:3 isomer)*	1	13*
18:3/18:3	0	8
18:2/16:2	tr	0
18:3/16:0	0	29
18:2/16:0	0	11
18:1/16:0	0	10
others	<4	tr

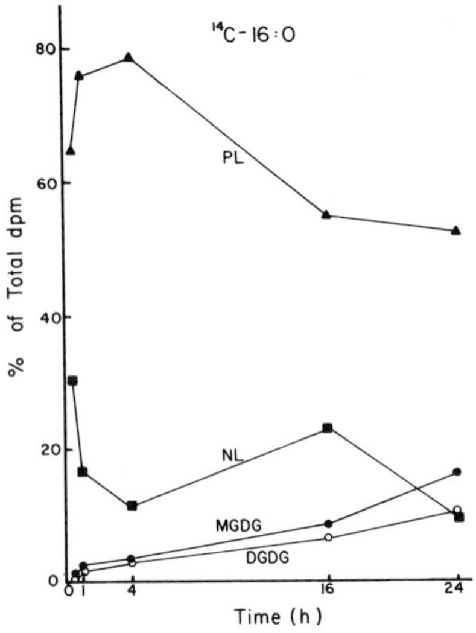

Figure 1

Table I. Molecular species composition of D. salina MGDG and DGDG. Figure preceeding the colon indicates the number of carbon atoms in the fatty acid while that following the colon represents the number of double bonds present. Pairs of numbers separated by a slash represent the components of a particular molecular species.

Figure 1. Incorporation of ^{14}C-16:0 into phospholipids (PL), neutral lipids (NL) and the galactolipids of D. salina following a 2 min pulse label.

D. salina cells exposed to tracer amounts of ^{14}C-16:0 or ^{14}C-18:1 for as little as 2 min. incorporate sizeable amounts of radioactivity into all classes of lipids (1). The radioactive fatty acids are initially concentrated primarily in neutral lipids and phospholipids, but with the passage of time the label is slowly transferred into glycolipids (Fig. 1). This pattern reflects an initial concentration of the radioactive fatty acid in lipids of extrachloroplastidal membranes, such as the endoplasmic reticulum, followed by a somewhat delayed movement of the isotope into the chloroplast, where glycolipid synthesis is confined.

The galactolipids of D. salina are almost exclusively of the "prokaryotic" type, meaning that they are assembled entirely within the chloroplast (6). Prokaryotic galactolipids can be conveniently identified by the presence in the sn-1 position of a C_{18} fatty acid and in the sn-2 position of a C_{16} fatty acid. "Eukaryotic" galactolipids, which are constructed using a diacylglycerol moiety donated from outside the chloroplast, characteristically have a C_{18} fatty acid bound at both the sn-1 and the sn-2 positions, and in D. salina are present in small amounts and only in DGDG (Table I). Thus D. salina provides an experimental system in which prokaryotic galactolipid biosynthesis can be examined with a virtual absence of eukaryotic galactolipid formation. A further experimental asset is that radioactivity administered in the form of ^{14}C-16:0 remains entirely within C_{16} fatty acids, saturated and unsaturated, and that fed as ^{14}C-18:1 is recovered only in

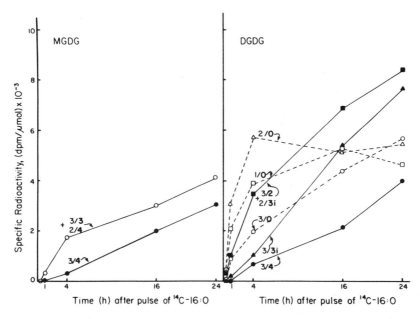

Figure 2. Labeling pattern of galactolipid molecular species. See Table I for complete identification of molecular species.

C_{18} fatty acids (1). (Conversely, added ^{14}C-12:0 is rapidly elongated to both C_{16} and C_{18} fatty acids).

Following a 2 min. pulse labeling of growing cells and a chase of different time periods, we have isolated MGDG and DGDG and measured the radioactivity of the major molecular species after separating them by HPLC. The specific radioactivities of these lipids at various times after labeling with ^{14}C-16:0 are shown in Fig. 2. Note that the one molecular species common to both MGDG and DGDG, namely, 18:3/16:4, has approximately the same specific radioactivity in both lipid classes. In DGDG, where a wider variety of molecular species is present in measurable amounts, there is a general tendency for the ones containing the more unsaturated C_{16} derivatives to become labeled relatively slowly in comparison to those having 16:0. Thus a distinctive pattern of labeling can be discerned even though the movement of ^{14}C-16:0 radioactivity into the galactolipids from other cellular lipids is continuous rather than as the sharply defined pulse of ^{14}C-16:0 into the cells as a whole (Fig. 1). Similar results were obtained in a separate series of experiments utilizing ^{14}C-18:1.

How can one use data of this type, which are basically similar to those reported by other workers, to decipher the pathways of galactolipid synthesis in chloroplasts? Let us begin by presenting our interpretation of current views on MGDG and DGDG biosynthesis (Fig. 3). It is generally agreed that the attachment of the first galactose, forming MGDG, takes place in the inner envelope of the chloroplast and that the conversion of MGDG to DGDG is also catalyzed by an envelope-associated enzyme (7). Although our findings (Fig. 2) are compatible with the notion that desaturation of galactolipids is a sequential process during which the fatty acids remain esterified to the

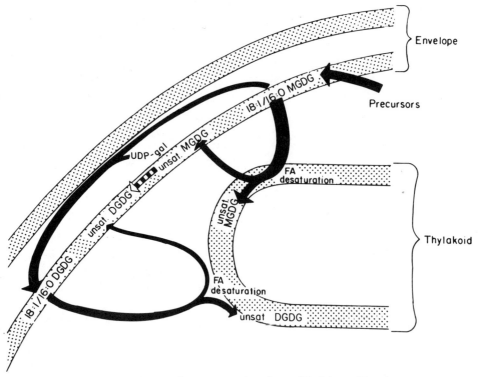

Figure 3. Probable sites of galactolipid synthesis.

parent lipid, it is unclear just where in the chloroplast this desaturation occurs. In Fig. 3 we have portrayed the insertion of double bonds as taking place in the thylakoid membranes since electron transport systems thought to be involved are situated there.

On the basis of our labeling data, we propose the following tentative conclusions regarding the options shown in Fig. 3. In view of the consistently similar specific radioactivities of 18:3/16:4 MGDG and 18:3/16:4 DGDG it would seem more likely that the 18:1/16:0 MGDG synthesized de novo in the chloroplast envelope and transported to the thylakoid membrane for stepwise desaturation would return in more unsaturated forms to the envelope membrane for conversion to DGDG there. The alternative pathway would involve galactosylation of 18:1/16:0 MGDG to yield an 18:1/16:0 DGDG that would be subsequently transported to the thylakoid for double bond insertion.

Either of the metabolic relationships between MGDG and DGDG outlined in Fig. 3 require extensive and probably reversible movement of galactolipids from envelope to thylakoid. Because of its shape, the MGDG molecule has a tendency to spontaneously assume non-bilayer structures (8), and these may enhance MGDG movement from its site of origin in the envelope to the thylakoid membrane and back again. The additional galactose moiety on DGDG prevents it from assuming the nonbilayer hexagonal II phase and in so doing should discourage its intramembraneous movement. This may account for the generally observed higher DGDG/MGDG ratio in chloroplast envelopes than in thylakoids (7) although the mobile MGDG inverted micelles are capable of accomodating a certain number of DGDG molecules (8).

We have begun to test these hypotheses by comparing the specific radioactivity of galactolipid molecular species from thylakoids of ^{3}H-16:0-labeled D. salina with that of equivalent molecular species from purified envelopes. At 16 hr after labeling, almost all the principal MGDG and DGDG molecular species of thylakoid membranes were virtually identical to their envelope counterparts with respect to specific radioactivity. Galactolipid movement between the two membrane types would therefore not seem to be a factor limiting the biosynthesis of any molecular species.

In Fig. 3 we did not consider certain interconversions that have been reported to modify the galactolipid composition of chloroplasts in vitro. In our labeling experiments we found no measurable amounts of 6-0-acyl-MGDG (9) or of trigalactosyldiacylglycerol (TGDG) or tetragalactosyldiacylglycerol (TeGDG) (10) by thin-layer chromatography of lipids from whole cells or cell fractions. Presumably these reactions are not of quantitative significance in intact D. salina cells.

Compared to phospholipids, the incorporation of radioactive fatty acids into galactolipids in vivo is usually quite slow. This is certainly the case in D. salina (Fig. 1). Our studies (1) suggest that the relatively rapid uptake of fatty acids into phospholipids is due not only to de novo synthesis but also to an active metabolic turnover of the acyl chains bound to existing phospholipids. In contrast, there is little evidence that the acyl chains of galactolipids are replaced in this way. For example, we have not detected the polyunsaturated C_{16}-fatty acids characteristic of D. salina galactolipids in the cell's free fatty acid pool, although all the fatty acids that typically occur bound to phospholipids are represented in that pool (Pike CS and Thompson GA Jr, unpublished observations).

As part of our systematic study of galactolipid metabolism in D. salina, we have looked for enzymes capable of participating in galactolipid fatty acid turnover. To date we have characterized two lipid acyl hydrolases of the alga. A microsomal enzyme is active in deacylating phosphatidylglycerol and phosphatidylethanolamine in vitro but does not act upon added MGDG (3). However, chloroplasts contain an enzyme which prefers MGDG as a substrate over other chloroplast lipids (11). When examined in chloroplast fragments, the acyl hydrolase requires high levels (10 mM) of Ca^{2+} for maximal activity and, unlike the microsomal acyl hydrolase, is not stimulated by calmodulin or inhibited by calmodulin antagonists. It also differs from the microsomal enzyme in not increasing its activity in low temperature-stressed cells.

Interestingly, the chloroplast enzyme specifically hydrolyzes the fatty acid bound at the sn-1 position of either endogenous or exogenous MGDG. DGDG is much less susceptible to attack. Although the enzyme readily deacylates all the MGDG molecular species common in D. salina, it is less active towards added eukaryotic molecular species and totally inactive towards catalytically hydrogenated D. salina MGDG.

Since there is no evidence for a metabolic turnover of D. salina MGDG in vivo, one may question the physiological significance of the acyl hydrolase present in homogenates. We suspect that this hydrolase activity may normally be latent, with significant action being restricted to periods of severe environmental stress. Support for this view comes from experiments in which cells prelabeled with ^{14}C-fatty acids were transferred to a medium lacking any source of nitrogen. Under these conditions there was over a period of 48 hrs a sharp drop in the appearance of radioactive fatty acids in galactolipids, and the triacylglycerols, which increased during this time, accumulated significant amounts of polyunsaturated C_{16}-fatty acids of the type normally occuring only in MGDG and DGDG. Despite these marked changes, the cells remained viable in the nitrogen-deficient medium.

DISCUSSION

Although galactolipids account for the bulk of all chloroplast membrane lipid components, we know relatively little about them in comparison with phospholipids. Recent studies of MGDG and DGDG physical properties (8) have shown that the second galactose moiety present in DGDG has a profound effect on the behavior of the lipids. MGDG but not DGDG easily forms non-bilayer structures which encourage intermembrane movement, promote membrane fusion, and generally favor the dynamic properties of the membrane lipids.

Because of this, regulation of the membrane physical state in plants subjected to environmental stress may involve a different strategy in chloroplasts than in other types of membrane. Most plant cell organelles have phospholipids as their primary lipid components, as do all animal membranes. Fluidity control here is effected, at least in its early stages, almost entirely by altering or replacing the phospholipid acyl chains (12). However, there appears to be little scope for metabolic turnover of galactolipid acyl groups, and the degree of acyl group unsaturation is characteristically so high that further desaturation would seem unlikely to alter the physical state significantly. Instead, chloroplasts of plants subjected to membrane fluidity-perturbing stress have been found to change their ratios of MGDG to DGDG (13, 14).

For scientists interested in the response of plants to stress as well as those simply involved in the study of chloroplast function, it is important to refine our knowledge of galactolipid metabolism. Our group has begun an analysis of MGDG and DGDG biochemistry in conjunction with an overall study of membrane lipid metabolism in D. salina. Galactolipid metabolism is somewhat simplified in this alga because almost all of the MGDG and DGDG arise through the prokaryotic pathway, which operates entirely within the chloroplast.

Our early findings show that the MGDG molecular species accumulating in the chloroplasts are all highly unsaturated. Labeling data suggest that MGDG is formed in the chloroplast envelope and moves very rapidly to the thylakoid membranes for desaturation of the bound 18:1 and 16:0 fatty acids. The ability of MGDG to form the highly mobile hexagonal-II phase continues to increase as its fatty acids become more unsaturated, and many of the molecules desaturated in the thylakoids quickly return to the envelope.

It appears that some of the MGDG molecules are rapidly converted into DGDG following their de novo synthesis. Other MGDG molecular species return to the envelope after being desaturated to varying degrees in the thylakoids, and these too may be transformed into DGDG. Perhaps due to their lower mobility, the newly made DGDG molecules seem much slower than MGDG in moving through subsequent desaturation steps. DGDG can itself ultimately be desaturated, as evidenced by an absolute decline in the radioactivity of the more saturated molecular species. But we feel that the close correspondence between the specific radioactivities of 18:3/16:4 MGDG and 18:3/16:4 DGDG throughout the time of our study indicate that the latter is formed largely through galactosylation of the former.

The fact that the average specific radioactivity of DGDG is always slightly higher than its presumed MGDG precursors during the time frame considered might be most easily explained by assuming that there is some "leakage" of ^{14}C-fatty acids into one or more DGDG molecular species without passing through the main MGDG pool. The most obvious possibility is that some of the more saturated DGDG molecular species, which our analyses have shown to contain C_{16} fatty acids only at their sn-2 positions, and to therefore be prokaryotic in origin, actually have at least traces of more highly radioactive fatty acids that entered them directly or indirectly via the eukaryotic pathways. However, evidence against this hypothesis comes from ^{14}C-12:0 labeling studies in which DGDG also has a higher specific radioactivity than MGDG despite the observation that ^{14}C-12:0 labels eukaryotic DGDG (as exemplified by the known eukaryotic species 18:3/18:3 DGDG) only very sparingly.

Other possibilities include the presence of two pools of MGDG in chloroplasts, one being more active than the other in accepting newly synthesized lipids and furnishing substrate for DGDG formation by the classical UDP-galactose galactosylation or by the galactolipid:galactolipid galactosyltransferase discovered by Wintermans and colleagues (10). Although we currently have no positive evidence for the presence of metabolically compartmentalized pools, we are initiating chloroplast fractionation experiments to look for the presence of distinct MGDG or DGDG subpopulation there.

REFERENCES

1. Norman HA, Smith LA, Lynch DV, Thompson GA Jr (1985) Arch Biochem Biophys 242: 157-167
2. Norman HA, Thompson GA Jr (1985) Arch Biochem Biophys 242: 168-175
3. Norman HA, Thompson GA Jr (1986) Biochim Biophys Acta 875: 262-269
4. Smith LA, Norman HA, Cho SH, Thompson GA Jr (1985) J Chromatog 346: 291-299
5. Thompson GA Jr, Roughan PG, Browse J, Slack CR, Gardiner SE Plant Physiol, in press
6. Roughan PG, Slack CR (1984) Trends Biochem Sci 9: 383-386
7. Douce R, Block MA, Dorne A-J, Joyard J (1984) in Subcellular Biochemistry, vol 10, Roodyn DB, ed, pp 1-84
8. Sen A, Williams WP, Brain APR, Quinn PJ (1982) Biochim Biophys Acta 685: 297-306
9. Heinz E, Rullkötter J, Budzikiewicz H (1974) Hoppe Seyler's Z Physiol Chem 355: 612-616
10. Van Besouw A, Wintermans JFGM (1978) Biochim Biophys Acta 529: 44-53
11. Cho SH, Thompson GA Jr Biochim Biophys Acta, in press
12. Lynch DV, Thompson GA Jr (1984) Trends Biochem Sci 9: 442-445
13. Lynch DV, Thompson GA Jr (1982) Plant Physiol 69: 1369-1375
14. Suss K-H, Yordanov IT (1986) Plant Physiol 81: 192-199

STEROL SYNTHESIS AND DISTRIBUTION AND ALGAL PHYLOGENY

Glenn W. Patterson
Department of Botany
University of Maryland
College Park, Maryland

ABSTRACT

The wide range of structural types of sterols in algae has been long recognized. Along with the possibility of several biosynthetic pathways, especially in the side chain biosynthesis, these different structures in the various algal taxa provide strong evidence for particular phylogenetic affiliations. As with other photosynthetic plants, all algae examined synthesize sterols through the cycloartenol pathway. During side chain alkylation, however, occurrence of both the 24(28) pathway and the 25(27) pathway have been demonstrated. Unlike in higher plants, 24-beta alkyl sterols are dominant in algae, although 24-alpha alkyl sterols are known in some diatoms. Most of the algal divisions can be easily characterized on the basis of their sterol composition. For instance, the red algae are unique in containing primarily cholesterol or related C-27 sterols. Blue green algae either contain no sterols or contain only small quantities of sterols. Brown algae contain almost exclusively fucosterol. Green algae are perhaps the most diverse with a very wide range of structural types that suggest phylogenetic relationships. Pyrrophyta contain sterols with an extra side chain methyl group not found in other algae. Chrysophyta appear to be a diverse group with many internal consistencies in sterol composition. As data become more accurate and more plentiful, it is clear that sterols can play an important role in determining the relationships between algae and in their evolutionary relationships with other organisms.

INTRODUCTION

It has been apparent for some time that variations in the sterol biosynthetic pathway and the resultant compositional differences between species provide excellent chemical characters for use in taxonomic considerations (1). Algae have no universal characteristic with respect to their sterol composition, although it is well known that nearly all algae with C-24 alkyl groups produce the 24- beta alkyl configuuration (2,3). All algae examined are consistent with other photosynthetic plants in that they utilize the cycloartenol rather than the lanosterol biosynthetic parthway (4). In some cases, great similarities are seen in the sterol compositions of species within the division, while in other divisions similarities exist only at or below the level of Order(3).

PHAEOPHYTA

Although data have been assembled for nearly 50 species of brown algae, only one of them, Halopteris scoparia, has cholesterol as the principal sterol(5). All other species contain fucosterol (24E-ethylidene-cholesterol) as the principal sterol(1), although cholesterol and 24-methylene cholesterol are frequently minor components in sterol mixtures from brown algae (1). Saringosterol is identified in many brown algae but is considered to be an oxidation product of fucosterol (6). Brown algae have not been demonstrated to contain significant quantities of sterols with saturated nine and ten-carbon side chains, suggesting that these algae cannot reduce the 24-methylene and 24-ethylidene double bonds. Fucosterol, found in all brown algae, has not been reported from any other algal division (1).

RHODOPHYTA

The sterols of red algae are markedly different from those of brown algae in that, for the most part, red algae are unable to synthesize sterols with alkyl substition at C-24 (3). Thus the principal sterol of red algae is cholesterol in nearly all species examined (Table 1). Several species contain desmosterol (24-dehydrocholesterol) or 22-dehydrocholesterol as the major sterol (1). In three species in the order Gelidiales, cholestanol is the principal sterol(7), and in Gastroclonium clavatum, 7-cholestenol is the major sterol (8). Each of the above sterols contains only 27 carbons with an unalkylated side chain. Recently a number of red algae have been discovered to contain liagosterol (25-hydroxy-23-dehydrocholesterol), a possible artifact, but in none of them is it the major sterol(9). Since the principal sterols of red algae have long been recognized to contain 27 carbons, it was suprising to note the reports on two red algae whose major sterols are C-28 sterols (8,10). Rytiphloea tinctoria and Vidalia volubilis contain as their principal sterols, 24-methylcholesterol and 24-methylenecholesterol, respectively. Cholesterol is clearly the principal sterol of most red algal species. Although cholesterol occurs in many algal species and is occasionally the principal sterol in several species of other algal taxa, only in the red algae is cholesterol dominant.

CHLOROPHYTA

The sterols of Chlorophyta are interesting from a taxonomic point of view. The simple and single-celled members frequently contain ergosterol or other sterols with a 5,7-double bond system (3). However, many single-celled Chlorophyta contain 5(6) or 7(8)-nuclear double bonds along with a nine or ten-carbon side chain which is either saturated

Table 1. Frequency of Various Sterol Structures in Algae

Algal Division	Sterol type and frequency			
Rhodophyta	27	- 5	- 0	(112)*
	27	- 5	- 24(25)	(5)
	27	- 5	- 22	(6)
	27	- 0	- 0	(4)
	27	- 7	- 0	(1)
	28	- 5	- 0	(1)
	28	- 5	- 24	(1)
Phaeophyta	29	- 5	- 24E	(50)
	27	- 5	- 0	(1)
Chlorophyta	27	- 5	- 0	(2)
	28	- 5	- 0	(3)
	28	- 5	- 24	(5)
	28	- 5,7	- 22	(9)
	28	- 5,8	- 22	(1)
	29	- 5	- 0	(11)
	29	- 5	- 22	(6)
	29	- 5	- 24Z	(10)
	29	- 5	- 25	(4)
	29	- 7	- 0	(1)
	29	- 7	- 22	(15)
	29	- 5,7	- 22	(1)
Chrysophyta	27	- 5	- 0	(2)
	27	- 5	- 22	(2)
	28	- 5	- 22	(11)
	28	- 5	- 24	(2)
	28	- 7	- 27	(1)
	28	- 8(9)	- 0	(1)
	29	- 5	- 0	(3)
	29	- 5	- 22	(6)
	30	- 5	- 0	(1)
Euglenophyta	27	- 5	- 0	(1)
	28	- 7	- 0	(1)
	28	- 5,7	- 22	(3)
	28	- 7	- 22	(1)
Cyanophyta	27	- 5	- 0	(5)
	28	- 5	- 22	(1)
	28	- 5,7	- 22	(1)
	29	- 5	- 0	(2)
	29	- 7	- 0	(1)
	29	- 7	- 22	(2)
Pyrrophyta	27	- 5	- 0	(3)
	27	- 0	- 0	(1)
	28	- 5	- 24	(1)
	29*	- 8(14)	- 24	(9)
	30*	- 0	- 22	(1)
	30*	- 0	- 17	(2)
	30*	- 0	- 0	(1)

* Times reported as principal sterol. Sterol structural types are represented by: first number = number of carbon atoms; second number, location of double bonds in nucleus; third number, location of double bonds in side chain; ex. cholesterol = 27-5-0. All double bonds at C-24 are 24(28) unless noted otherwise.

or contains a 22-double bond. None of the single-celled Chlorophyta contain cholesterol as a principal sterol. The larger, more complex members usually contain either cholesterol or sterols with 5,24(28) double bond systems. Many larger green algae contain 28-isofucosterol as their principal sterol. It is interesting to note that the 24(28) double bond here is always in the Z arrangement in contrast to the E arrangement found in brown algae (Table 1). With respect to evolution, it appears that the single-celled algae containing a 24-beta alkyl group are succeeded by the more advanced green algae which have dominant sterols with a 24(28) double bond (and thus no asymmetry at C-24) which in turn are followed by higher plants containing sterols mostly with 24-alpha alkyl groups.

CHRYSOPHYTA

Chrysophyta are important links in the marine food chain. Their sterols frequently include 5,22-sterols as principal components. In every case where modern methods have been used to determine the stereochemistry of these sterols at C-24, the twenty-eight carbon compound had an alpha methyl group (epibrassicasterol) while the twenty-nine carbon homolog had a beta ethyl at C-24 (poriferasterol)(11). These facts are unusual in two respects. Epibrassicasterol is the only sterol in algae to be clearly shown to contain a C-24 alpha alkyl group. Most algae have sterols with 100% of their C-24 alkyl groups in the beta configuration. The presence of 24-alpha methyl and 24-beta ethyl in the same alga is a demonstration of the presence of a biosynthetic mechanism producing opposite configurations in C-24 methyl vs ethyl sterols (11). Chrysophyta also frequently contain cholesterol and 24-methylene-cholesterol. The most abundant sterols of filter-feeding animals such as the oyster are cholesterol, 24-methylene-cholesterol, and epibrassicasterol. Since the oyster cannot synthesize sterols, it is not surprising that its sterol composition is so similar to that of its diet (12), which frequently includes members of the Chrysophyta. As with the Chlorophyceae, trends can be seen within the Division Chrysophyta. In the Class Chrysophyceae, poriferasterol is dominant, in the Class Xanthophyceae, clionasterol is the principal sterol in each species examined, while in the Bacillariophyceae (diatoms), epibrassicasterol is dominant, although not in every species(1). More data would be very useful in this area.

EUGLENOPHYTA

Only a few species in this Division have been examined for sterol composition. In most cases ergosterol is the principal sterol although in Astasia longa, cholesterol is the principal sterol(13).

CYANOPHYTA

The Cyanophyta (Cyanobacteria) were once thought to contain no sterols at all (1). The most significant characteristic about sterols in Cyanophyta is that they are usually present in much smaller amounts than in other photosynthetic plants. Many species are difficult to obtain in pure culture and analysis of non-axenically grown blue-greens is risky due to the ease of isolation of sterols from contaminant organisms. However, in recent years sterols have been identified from a dozen blue-greens but no pattern of structure has emerged (1).

PYRROPHYTA

The sterol composition of Pyrrophyta is the most unusual of any algae. Several of these algae contain cholesterol as the principal sterol and cholesterol is present in many other species (14). Most of the species examined thus far have contained dinosterol as the principal sterol. The structure of dinosterol (4,23,24-trimethylcholest-22-enol) is extremely unusual in that it contains an additional methyl group at C-4 on the nucleus and an additional methyl at C-23 in the side chain. Neither of these methyl groups is found on the principal sterol of any other algal species.

SUMMARY

Although more data are needed for most algal taxa the present data can be summarized in Table II.

Table II STEROLS CHARACTERISTIC OF ALGAL TAXA

ALGAL TAXA	CHARACTERISTIC STEROL
RHODOPHYCEAE	Cholesterol
PHAEOPHYTA	Fucosterol
CHLOROPHYTA	
single celled	No C-27, side chains saturated or with 22db
complex	Cholesterol, sterols with 24(28)db
CHRYSOPHYTA	
CHRYSOPHYCEAE	Poriferasterol
XANTHOPHYCEAE	Clionasterol
BACILLARIOPHYCEAE	Epibrassicasterol
EUGLENOPHYTA	Ergosterol?
CYANOPHYTA	Small quantities of sterols
PYRROPHYTA	Dinosterol

REFERENCES

1. G. W. Patterson, Sterols of Algae: Proceedings of International Symposium on Marine Algae of the Indian Ocean Region, Bhavnagar, India, 9-12 January, 1979, pp 37.

2. W. R. Nes, K. Krevitz, J. Joseph, W. D. Nes, B. Harris, G. F. Gibbons, and G. W. Patterson, The phylogenetic distribution of sterols in tracheophytes, Lipids 12: 511-527 (1977).

3. G. W. Patterson, The distribution of sterols in algae, Lipids 6: 120-127 (1970).

4. C. Anding, R. D. Brandt and G. Ourisson, Sterol biosynthesis in Euglena gracilis Z, European J. Biochem. 24: 259-263 (1971).

5. V. Amico, G. Oriente, M. Piatelli, C. Tringall, E. Fattorusso, S. Magno, L. Mayol, C. Santacroce, and D. Sica, Amino acids, sugars, and sterols of some Mediterranean brown algae, Biochem. Syst. Ecol. 3: 143-146 (1976).

6. B. A. Knights, Sterols in Ascophyllum nodosum, Phytochemistry 9: 903-905 (1970).

7. I. Chardon-Loriaux, M. Morisaki, and N. Ikekawa, Sterol profiles of red algae, Phytochemistry 15: 723-725 (1976).

8. E. Fattorusso, S. Magno, C. Santacroce, D. Sica, G. Impellizzeri, S. Mangiofico, M. Piatelli, and S. Sciuto, Sterols of Mediterranean Florideophyceae Biochem. Syst. Ecol. 4:135-138 (1976).

9. E. Fattorusso, S. Magno, C. Santacroce, D. Sica G. Impellizzeri, S. Mangiafico, G. Oriente, M. Piatelli, and S. Sciuto, Sterols of some red algae, Phytochemistry 14: 1579-1592 (1975).

10. A. Alcaide, M. Barbier, P. Potier, A. M. Magueur, and J. Teste. Nouveau resultats sur les sterols des algues rouges. Phytochemistry 8:2301-2303 (1969).

11. D. Raederstorff and M. Rohmer, Sterols of the unicellular algae Nematochrysopsis roscoffensis and Chrysotilla lamellosa: Isolation of (24E)-24- n-propylidinecholesterol and 24-n-propylcholesterol. Phytochemistry 23:2835-2838 (1984).

12. C. J. Berenberg and G. W. Patterson, The relationship between dietary phytosterols and the sterols of wild and cultivated oysters, Lipids 16:276-278 (1981).

13. M. Rohmer and R. D. Brandt, Les sterols it leurs precursors chez Astasia longa Pringsheim, Eur. J. Biochem. 36:446-454 (1973).

14. A. R. Loeblich, Dinoflagellate physiology and bio-chemistry In: Dinoflagellates, D. L. Spector, Ed. Academic Press, New York, (1984).

LIPIDS OF ACETABULARIA MEDITERRANEA. COMPOSITION, CELLULAR LOCALIZATION AND BIOSYNTHESIS

Waldemar Eichenberger and Annegret Gerber

Department of Biochemistry
University of Bern
Bern, Switzerland

INTRODUCTION

Many questions arise about the biosynthesis, the cellular localization and the function of the betaine lipid diacylglyceryltrimethylhomoserine (DGTS) which is widely distributed among cryptogamic green plants[1,2]. In Chlamydomonas, we previously found 15% of the toal DGTS to be localized in the thylakoid membranes[3], while other authors found this proportion to be 40% in the same organism[4] and even 90% in Dunaliella[5]. In the present experiments, we used Acetabularia because cell fractionation is easier with this alga, compared to other organisms.

MATERIAL AND METHODS

Acetabularia mediterranea was cultivated in Müller's medium[6] under 2000 lx light-dark (12:12) conditions at $22°C$. Lipids were extracted with methanol, separated by 2-dim. TLC and quantified as described before[3]. The positional distribution of fatty acids was determined using Rhizopus lipase. Chloroplast and microsomal fractions were isolated from homogenates of ^{14}C-labelled cells by differential centrifugation. NADH-cyt.c reductase and chlorophyll were used as markers for microsomes and chloroplasts, respectively.

RESULTS AND DISCUSSION

Besides MGDG and DGDG, Acetabularia cells contain relatively high proportions of SQDG and DGTS, each accounting for 20% of the polar lipid, as shown in Table 1. In contrast, total phospholipids account for only 5%, and PC for only 0.15% of total polar lipids. PC in Acetabularia was also reported by Moore and Tschismadia[7], though it could not be detected by Sato[1].

Table 1. Lipid composition of Acetabularia mediterranea

	µg per mg ether-soluble lipid ± standard error (% total polar lipid)									
MGDG	DGDG	SQDG	DGTS	PG	PE	PI	PC	Chl[a]	Car[b]
145 ± 23 (37)	83 ± 7 (20)	80 ± 2 (20)	81 ± 3 (20)	12 ± 1 (3)	5 ± 0.1 (1.2)	2.5 ±0.5 (0.6)	0.6 +0.05 (0.15)	86	38

[a]Chlorophylls [b]Carotenoids

We detected ca. 20 different fatty acids among which C_{16} and C_{18} fatty acids predominate, although C_{20} and C_{22} acids are also present, as shown in Table 2. MGDG, DGDG and SQDG contain almost exclusively C_{16} and C_{18} acids. For DGTS, a large amount of C_{20} and C_{22} acids is typical and they are mainly bound to the 2-position. It should be noted that the percentage of 2-positions occupied by C_{16} acids is 59 mol % in DGDG and SQDG, 22% in MGDG and 12% in DGTS. If we assume that this value reflects the origin of the diacylglycerol (DAG) portion[8], the results suggest that SQDG and DGDG are predominantly of chloroplastic origin, whereas the DAG portion of MGDG is mostly, and that of DGTS almost exclusively, synthesized in the cytoplasm.

Table 2. Fatty acid composition of total lipid and positional distribution of different chain lengths among the 1- and 2-position in glycerolipids of Acetabularia

		mol % fatty acid							
Fatty acid	total lipid	MGDG		DGDG		SQDG		DGTS	
		1-	2-	1-	2-	1-	2-	1-	2-
C-16	34	13	22	20	59	56	59	47	12
C-18	59	83	77	79	40	54	40	43	58
C-20	5	4	1	1		0.3	1	9	12
C-22	2							1	18

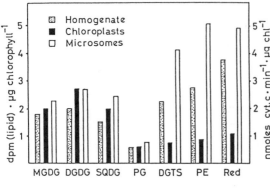

Fig. 1. Polar lipids and NADH-cyt.c reductase (Red) in cell homogenate, chloroplast and microsomal fractions of Acetabularia

In order to localize the different lipids within the cell, ^{14}C-labelled cells were homogenized and the chloroplast and the microsomal fractions isolated. Lipids were determined by counting their radioactive label. NADH-cyt.c reductase was used as a marker for microsomal membranes and chlorophyll for chloroplasts. As shown in Fig. 1, MGDG, DGDG, SQDG and PG each give very similar values in homogenate, chloroplasts and microsomes, as is expected for chloroplast constituents. In contrast, the distribution of DGTS is very similar to that of PE and reductase suggesting a co-purification of DGTS with the microsomal membranes. From the contents of reductase, PE and DGTS in homogenate, chloroplasts and microsomal fraction (figures not shown here) it can be calculated that the chloroplast fraction contains 33.5% of total DGTS. At the same time, chloroplasts contain also 29.6% of total PE and 27.6% of total reductase activity. Therefore, almost all of the DGTS found in the chloroplast fraction has to be ascribed to microsomal impurities.

Our results indicate that, in Acetabularia, DGTS is synthesized and mainly localized in extraplastidic membranes, although it is possible that minor amounts occur in the chloroplast. The origin of the other glycerolipids and the role of the very small amount of PC are under investigation.

The work has been supported by the Swiss National Science Foundation. We thank Prof. H.G. Schweiger and Dr. S. Berger (Max Planck-Institut für Zellbiologie, Ladenburg, FRG) for generously providing Acetabularia cells.

REFERENCES

1. N. Sato and M. Furuya, Distribution of diacylglyceryltrimethylhomoserine and phosphatidylcholine in non-vascular green plants, Plant Sci. 38: 81 (1985).
2. W. Eichenberger, Distribution of diacylglyceryltrimethylhomoserine in different algae, Plant Sci. Lett. 24: 91 (1982).
3. L. Mendiola-Morgenthaler, W, Eichenberger and A. Boschetti, Isolation of chloroplast envelopes from Chlamydomonas. Lipid and polypeptide composition, Plant Sci. 41: 97 (1985).
4. D. R. Janero and R. Barrnett, Isolation and characterization of an ether-linked homoserine lipid from the thylakoid membrane of Chlamydomonas reinhardtii 137^+, J. Lipid Res. 23: 307 (1982).
5. H. A. Norman and G.A. Thompson, Quantitative analysis of Dunaliella salina DGTS and its individual molecular species by HPLC, Plant Sci. 42: 83 (1985).
6. H. G. Schweiger, P. Dehm and S. Berger, Culture conditions for Acetabularia, in: "Progress in Acetabularia Research", C.L. Woodcock, ed., Academic Press, New York 1972, p. 319.
7. F. D. Moore and I. Tschismadia, Biosynthesis in isolated Acetabularia chloroplasts. III. Complex Lipids, in: "Progress in Acetabularia Research", C.L.F. Woodcock, ed., Academic Press, New York 1977, p. 159.
8. G. Roughan and R. Slack, Glycerolipid Synthesis in Leaves, Trends Biochem. Sci. 9: 383 (1984).

THE EFFECT OF ENVIRONMENTAL CONDITIONS ON FATTY ACID COMPOSITION OF THE RED ALGA PORPHYRIDIUM CRUENTUM

Zvi Cohen, Avigad Vonshak and Amos Richmond

Algal Biotechnology Laboratory
Jacob Blaustein Institute for Desert Research
Ben-Gurion University of the Negev
Sede-Boqer Campus 84990, Israel

INTRODUCTION

The unicellular red alga Porphyridium cruentum belongs to a selected group of algae that contains significant amounts of polyunsaturated fatty acids (PUFA), which are prostaglandin precursors. This group of fatty acids is of potential value as starting materials for prostanglandin synthesis and of potential clinical value for blood cholesterol reduction, prevention of blood platelet aggregation, and treatment of premenstrual syndrome (1). Porphyridium cultivation was suggested as a source for arachidonic acid (2,3). However, to be able to produce arachidonic acid or EPA on a large scale, the effects of environmental factors must be known and controlled. The fatty acid distribution in the exponential growth phase was reported to be erratic, and Porphyridum cultures cultivated under seemingly identical conditions produced fatty acids having very different fatty acid compositions (4,5). In an attempt to shed some light on this problem, we studied the effects of temperature, light and nitrogen content of the medium on the fatty acid distribution of Porphyridium.

MATERIALS AND METHODS

Porphyridium cruentum 13.80 was obtained from the Gottingen Algal Culture Collection. Cultures were grown on Jones' medium (6) in glass tubes (80x5 cm) incubated in a temperature-regulated water bath ($\pm 1°$), illuminated (unless otherwise stated) with four cool-white fluorescent lamps providing 170 $\mu E\ m^{-2}\ s^{-1}$ at the side of the bath. They were mixed by bubbling an air-CO_2 mixture (95:5) through a sintered glass tube placed in the bottom of each culture tube. For the nitrogen starvation experiments, cultures in the exponential phase of growth (28°) were centrifuged, washed and resuspended in a nitrogen-free medium. The cultures were kept for an additional 3 days under the same light and temperature conditions.

Lipids were separated by TLC according to Kates (7) and identified by comparison with standard lipids and by characteristic color reactions. Bands were scraped from TLC plates. For fatty acid analysis, samples from TLC plates or freeze-dried samples of Porphyridium were treated with methanol-acetyl chloride (95:5). Heptadecanoic acid was added and the mixture was handled up as previously described (8,9). Gas chromatographic analysis was performed on a SP-2330 fused silica capillary column at 200°. Fatty acids were identified by comparison with authentic samples and by GC-MS.

RESULTS AND DISCUSSION

Fractionation of the lipids of P. cruentum and fatty acid analysis of each lipid class showed (Table 1) that 20:4 acid (AA) is concentrated in the neutral lipids (NL) and in phosphatidyl choline (PC), while 20:5 acid (EPA) is mainly found in the glycolipids, monogalactosyl diglyceride (MG), digalactosyl diglyceride (DG) and sulfoquinovosyl diglyceride (SL). These data resemble those reported by Nichols (10) and by Nyberg (11, 12).

Effect of Light

Cultures of P. cruentum were cultivated at $28°$ under two light intensities. AA was the dominant PUFA under low light intensities, while EPA was preferred under high light intensities (Table 1). The AA/EPA ratio changed from 0.8 under high light to 3.2 under low light, and the total fatty acid content increased from 4.6% to 6.3% (percent of ash free dry weight).

A comparison of the fatty acid content and distribution in each of the major lipids (NL, MG, DG, SL, PC) showed that cultures grown under low light intensity had nearly twice the proportion of neutral lipids (mainly triglycerides) as the corresponding cultures grown under high light conditions. Under low light, the neutral lipids constituted up to 67.6% of total fatty acids. The quantities of other lipids were greatly reduced. Since neutral lipids, even under high light conditions, are rich in AA and relatively poor in EPA, doubling their weight in the mixture has a significant effect on the AA/EPA ratio of the total lipid mixture. Moreover, there is a shift in the AA/EPA ratio in each of the individual lipids, resulting from a decrease in EPA and an increase in AA. In an additional experiment, it was found that similar, yet not so pronounced results, could be obtained by comparing cultures grown under high light intensity at low high cell concentrations. The lipids of the high concentration culture resembled those of cultures grown under low light intensities.

Although Ahern et al (2) mentioned earlier that P. cruentum cultures grown under low light had twice the fatty acid content of corresponding high light cultures, the effect on the fatty acid composition was not observed, since EPA could not be chromatographically resolved from AA under the chosen analytical conditions. It seems that when growth is retarded due to light limitation, cell metabolism is shifted towards production of reserve materials i.e. neutral lipids, and fewer chloroplast components such as glycolipids are produced, resulting in a sharp increase in the AA/EPA ratio.

Temperature and Nitrogen Effects

A similar effect of increased AA/EPA ratio was observed when growth was retarded due to higher growth temperature or nitrogen starvation (Table 2). In both cases, a high AA/EPA ratio was obtained. Cultivation at $30°$ resulted in a ratio of 1.70 compared with 0.72 for cells grown at $28°$. The nitrogen effect was even more drastic, since nitrogen-starved cultures had a AA/EPA ratio of 2.95. Here again, the neutral lipid content doubled, yet the changes of fatty acid distribution within individual polar lipids had a different pattern than under light limited conditions (unpublished data).

The fatty acid profile of P. cruentun and especially the AA/EPA ratio are highly dependent on the growth rate. In fast growing cultures, EPA is the dominant fatty acid while AA prevails whenever growth is slowed down. However, it seems that growth rate is the main, although not the only factor, affecting the changes in these fatty acids.

This work was supported by a grant from the Israel National Council for Research and Development.

Table 1. The Effect of Light Intensity and Concentration Density on the Fatty Acids Content of P. cruentum[1] (weight percent)

	Total Lipids			Neutral Lipids			MG			DG			SL			PC	
	HL		LL	HL		LL	HL		LL	HL		LL	HL		LL	HL		LL
	LC	HC		LC	HC		LC	HC		LC	HC		LC	HC		LC	HC	
16:0	38.9	36.3	37.0	13.6	25.7	22.0	47.3	51.0	37.0	64.0	69.7	4.7	67.0	68.8	51.2	31.8	26.8	21.2
18:1	1.5	2.7	1.3	5.2	4.5	8.1	2.1	4.9	1.7	1.4	2.6	1.6	3.5	5.1	3.7	3.1	5.4	2.8
18:2	11.0	17.2	6.6	24.7	22.9	23.3	9.8	10.2	4.2	4.0	5.2	6.1	3.8	9.1	2.2	9.5	9.5	5.8
20:4	18.6	28.1	38.6	36.1	29.8	28.8	6.5	13.8	37.0	2.7	8.5	26.5	8.1	3.3	25.0	42.3	48.1	56.4
20:5	23.7	9.4	12.2	8.0	4.2	3.9	28.4	8.4	14.2	22.9	8.3	13.8	11.7	2.6	7.6	6.4	3.0	2.6
20:4 / 20:5	0.8	3.0	3.2	4.5	7.0	7.3	0.2	1.6	2.6	0.1	1.0	1.9	0.7	1.3	3.3	6.6	15.9	21.5
1% of FA	-	-	-	35.7	58.1	67.6	18.4	13.7	10.9	15.1	8.3	10.2	10.4	8.3	3.3	9.2	4.8	2.7

1. exponentially cultivated cultures at 28o.
HL - high light, 170 µE m^{-2} s^{-1}
LL - low light, 25 µE m^{-2} s^{-1}, 2.7 mg l^{-1} chlorophyll
LC - low cell concentration, 4.9 mg l^{-1} chlorophyll
HC - high cell concentration, 8.0 mg l^{-1} chlorophyll

Table 2. The Effects of Temperature and Nitrogen Starvation on the Fatty Acids Content of P. cruentum

Culture conditions	Temperature	16:0	16:1	18:0	18:1	18:2	18:3	20:2	20:3	20:4	20:5	20:4 / 20:5
+N	30o	34.1	1.9	1.8	2.1	15.4	1.0	-	2.3	25.9	15.2	1.70
+N	28o	34.4	2.6	1.1	0.8	10.4	0.8	0.1	0.5	20.6	28.6	0.72
-N, 3 days	28o	28.0	1.5	1.9	2.5	21.1	1.2	0.5	3.1	30.0	10.1	2.95

References

1. Horrobin, D.F., 1982, "Clinical Uses for Essential Fatty Acids", Eden Press Inc., Montreal.
2. Ahern, T.J., Katoh, S. and Sada, E., 1983, Arachidonic acid production by the red alga Porphyridium cruentum. Biotech. & Bioeng., 25:1057-70.
3. Vonshak, A., Cohen, Z. & Richmond, A., 1985, The feasibility of mass cultivation of Porphyridium, Biomass, 8:13-25.
4. Pohl, P. and Wagner, H., 1968, Inhaltsstoffe von algen – II. Uber die unterschiedliche fettsaurezusammensetzung von salz und susswasseralgen, Phytochem., 7:1565-72.
5. Kost, H.P., Senser, M. and Wanner, G., 1984, Effect of nitrate and sulphate starvation on Poprhyridium cruentum cells, Z. Pflanzenphysiol., 113.S:231-249.
6. Jones, R.E., Speer, L. and Kury, W., 1983, Studies on the growth of the red alga Porphyridium cruentum, Physiol. Plant, 16:636-43.
7. Kates, M., 1972, "Techniques of Lipidology", American Elsevier, NY.
8. Lepage, G. and Roy, C.C., 1984, Improved recovery of fatty acid through direct transesterification with prior extraction or purification, J. of Lipid Research, 25:1391-96.
9. Cohen, Z., Vonshak, A. and Richmond, A., Fatty acid composition of Spirulina strains under various environmental conditions, Phytochem., (submitted for publication).
10. Nichols, B.W. and Appelby, R.S., 1969, The distribution of arachidonic acid in algae, Phytochem., 8:1907-15.
11. Nyberg, H., 1985, The influence of ionic detergents on the phospholipid fatty acid compositions of Porphyridium purpureum, Phytochem., 24:435-40.
12. Nyberg, H. and Koskimies Soininen, 1985, The glycolipid fatty acids of Porphyridium purpureum cultured in the presence of detergents, Phytochem., 24:751-57.

NILE RED: A FLUOROPHORE USEFUL IN ASSESSING

THE RELATIVE LIPID CONTENT OF SINGLE CELLS

K. E. Cooksey[1], S. A. Williams[2] and P. R. Callis[2]

Departments of Microbiology[1] and Chemistry[2]
Montana State University
Bozeman, MT 59717

INTRODUCTION

The energy crisis in 1973 brought home to many people the fact that the world's supply of crude mineral oil was not unlimited. Since that time there have been various attempts to investigate alternate sources of energy and many of these have centered on the use of plants to convert solar radiation to a more conventional energy source. A particular example is the production of fuel-grade oil from microalgal biomass grown in brackish water ponds in the southwestern deserts of the USA. The algae used in this process must have a high lipid content as well as being able to grow at elevated temperatures and light levels. One of the objectives of our research was to design a method to select algal strains producing large quantities of neutral lipid. For the purposes of this study, neutral lipid was defined as the triacylglyceride and hydrocarbon fraction of the cell. Since neither of these groups of compounds have functional groups that are easy to assay chemically, bulk lipid is often weighed after extraction from tissues with non-polar solvents or solvent mixtures (e.g., Bligh and Dyer, 1959). The method is time consuming and requires a sample large enough to produce a weighing (for the extracted lipid) in the milligram range. Greenspan and Fowler (1985) drew attention to the fact that Nile Red was fluorescent in non-polar but not in polar environments and pointed out that the stain could serve as a fluorescent lipid probe (Greenspan et al., 1985). We have extended this idea into a semi-quantitative method for the determination of neutral lipid in small populations of single cells.

METHODS

Growth of Algae

Cells were grown axenically in appropriate media (Provasoli et al., 1957; Bold, 1942) at 100-200 μEinsteins m^{-2} sec^{-1} for the times indicated. Nitrate levels were reduced to near zero after 8 days.

Lipid Determination

Lipid was determined gravimetrically after $CHCl_3/CH_3OH$ extraction (Bligh and Dyer, 1959).

Nile Red Staining

Ten µl 250 µg ml^{-1} Nile Red (9-diethylamino-5H-benzo(α)phenoxazine-5-one) in acetone was added to 2.5 ml cell suspension followed by 90 µl acetone. After mixing vigorously, the fluorescence of the suspension was measured using a Becton-Dickinson FACS 440 flow cytometer (excitation 488 nm, emission 585±15 nm) or a Spex Fluorolog spectrofluorometer (excitation 525 nm, emission 585 nm). Samples for microscopy were prepared similarly.

RESULTS AND DISCUSSION

Experiments concerning the optimal Nile Red concentration for staining cells showed that there was little to choose in the range 1 to 10 µg ml^{-1}. Unstained cells did not autofluoresce between 540 and 640 nm (Fig. 1). Live stained cells faded after about 5 minutes. Formaldehyde-fixed cells did not stain, but it was possible to maintain for 2 hours by ethanol or formaldehyde fixation the fluorescence of previously stained cells. The fixed and stained cells had to be washed free of the fixative, however. Greenspan and Fowler (1985b) showed that above 600 nm, the more polar lipids of the cell contributed to the fluorescence signal. This was noticed for some large Nile Red-stained diatoms wherein the cell membrane had a reddish cast as opposed to bright yellow fluorescence of the lipid droplets. The fluorescence signal of all cells examined increased as the medium became depleted in nitrogen. This was borne out by lipid analyses.

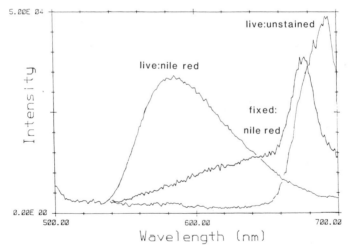

Fig. 1. Fluorescence of A. coffeaeformis (0.89 x 10^{-6} cells ml^{-1}) stained with Nile Red. The excitation wavelength was 525 nm. Spectra were obtained using a Spex Fluorolog spectrofluorometer (integration time 0.5 sec/nm, bandpass 4.5 nm).

We found that excitation and emission maxima were 525 nm and 585 nm respectively, although the peaks were broad. The peak widths suggest that an instrument with a filter-based optical system could be used to make fluorescence measurements with Nile Red. Using the protocol described above, we demonstrated that for A. coffeaeformis, lipid ml^{-1} (L) culture was linearly correlated (r = 0.950) to fluorescence (F) (flow cytometer data) by the relationship: F = 3.59 L - 1.89. Cultures of A. coffeaeformis were sampled with time over a period of 270 hr., when cell lysis was detected. A similar relationship was found when a fluorometer was used in place of the flow cytometer, however, in these experiments the independent variable was the number of cells x 10^{-6} (N) of a particular lipid composition. The relationships were

A. coffeaeformis   F = 4.14 N - 0.195   (r = 0.996)

Navicula sp. 9D   F = 0.557 N + 0.277   (r = 0.996)

Nile Red-stained cells remain viable.

ACKNOWLEDGMENTS

The authors acknowledge the financial support of the Solar Energy Research Institute and the help of D. Berglund, B. Cooksey, D. Davies and L. Priscu.

LITERATURE CITED

Bligh, E. G., and Dyer, W. J., 1959, A rapid method of total lipid extraction and purification, Can. J. Biochem. Physiol., 37:911-7.
Bold, H. C., 1942, The cultivation of algae, Bot. Rev., 8:69-138.
Greenspan, P., and Fowler, S. D., 1985, Spectrofluorometric studies of the lipid probe, nile red, J. Lipid Res. 26:787-789.
Greenspan, P., and Fowler, S. D., 1985, Nile red: a selective fluorescent stain for intracellular lipid droplets, J. Cell Biol. 100:965-973.
Provasoli, L., McLaughlin, J. J. A., and Droop, M. R., 1957, The development of artificial media for marine algae, Arch. Mikrobiol. 25:392-428.

LIPID ACCUMULATION IN SILICON-DEFICIENT DIATOMS

Paul G. Roessler

Biotechnology Research Branch
Solar Energy Research Institute
Golden, Colorado

INTRODUCTION

Previous studies have shown that the lipid contents of the the diatoms <u>Navicula pelliculosa</u> (2) and <u>Cyclotella cryptica</u> (3) increase significantly in response to silicon deficiency. The chemical nature of the lipids produced under these conditions was not determined, however. Furthermore, the biochemical mechanisms responsible for this switch to lipid accumulation have never been elucidated. The research described in this paper was therefore undertaken in an effort to further investigate the regulation of lipid biosynthesis in Si-deficient diatoms.

MATERIALS AND METHODS

<u>Organisms</u>: <u>Cyclotella cryptica</u> T13L and <u>Thalassiosira pseudonana</u> 3H were obtained from the Culture Collection of Marine Phytoplankton (Bigelow Lab., W. Boothbay Harbor, ME) and <u>Cylindrotheca fusiformis</u> was kindly provided by Dr. B.E. Volcani (Scripps Inst., La Jolla, CA).

<u>Lipid analyses</u>: Lipids were extracted from cells at 60°C for 1 h each with: 50% methanol in water, 2) methanol (twice), and 3) 50% methanol/50% chloroform (twice). Lipid mass was determined gravimetrically following phase separation by the method of Bligh and Dyer (1). Neutral lipids were separated from polar lipids by chromatography on silica gel-coated quartz rods (Chromarod Type S-II) or silica gel thin layer plates (for isotope experiments). After development in chloroform:acetic acid (100:0.5, v/v), the separated lipids were quantified by flame ionization detection (for Chromarods) and by liquid scintillation counting of plate scrapings (for isotope studies).

^{14}C <u>incorporation studies</u>: Exponential phase cultures were harvested by centrifugation and then resuspended in Si-free or Si-replete medium and returned to the incubator. Portions of the cultures were removed after 0,4,8,12, and 24 h and incubated in the light with $H^{14}CO_3^-$ (1 mCi/mmol, 0.06 mM). After 15, 30, and 45 min, 10 mL samples (x3) were removed, collected by filtration, and washed with isotope-free medium. The lipids from the cells on one set of filters were counted after extraction by the method of Bligh and Dyer (1). Chrysolaminarin was extracted from the cells on other filters with 0.1 N H_2SO_4, followed by Biogel P-2 chromatography and deproteinization prior to counting. Total ^{14}C incorporation was determined by direct counting of filters which had been digested in Beckman BTS-450.

RESULTS AND DISCUSSION

In order to identify a species suitable for biochemical studies, the first phase of this project examined the effects of Si-deficiency on several aspects of lipid metabolism in the diatoms C. fusiformis, C. cryptica, and T. pseudonana. In these experiments, cells were grown in both a Si-replete medium and in a medium containing a level of Si which became limiting while cultures were still in the exponential phase of growth. Significant interspecific differences were observed with respect to increases in lipid mass and total organic mass (i.e., ash-free dry mass, AFDM) which occurred after the onset of Si-limitation. Si-deficient and Si-replete C. fusiformis cultures exhibited similar gains in both total organic mass and lipid mass for at least 72 h after the onset of Si-deficiency. Lipid mass increased at a rate slightly greater than that of total organic mass in both cultures, however, leading to an increase in the lipid content of cells (as a percentage of the AFDM). After 72 h, the lipid content increased from 11% to 26% in Si-deficient cultures, and from 11% to 21% in Si-replete cultures.

C. cryptica cells responded quite differently to Si-deficiency. The exponential rate of increase for total organic mass during the first 12 h of Si-deficiency was reduced by 38% with respect to Si-replete cultures. Conversely, the synthesis of lipids in Si-limited cultures during this 12 h period was not diminished, leading to a substantial gain in lipid content (from 24% to 34% of the AFDM). After 12 h, there was little additional gain in lipid mass, AFDM, or lipid content in Si-deficient cultures.

The synthesis of lipids and total organic material was inhibited in Si-limited cultures of T. pseudonana, but not until the cells had been deprived of silicon for more than 36 h. The lipid content of Si-deficient cells remained between 15% and 20% during the 72 h starvation period, and between 14% and 17% in Si-replete cells.

All three species examined exhibited substantial increases in neutral lipid content in response to Si-deficiency. For C. fusiformis, the neutral lipid fraction (as a percentage of the total lipid mass) increased from 19% to 57% after 72 h of Si-deficiency. For Si-deficient C. cryptica cells, a maximum neutral lipid fraction of 64% was observed 36 h after the onset of Si-limitation, while the neutral lipid fraction of Si-replete cells never exceeded 32%. T. pseudonana cells had a neutral lipid fraction of 44% after 36 h of Si-deficiency, as compared to a maximum neutral lipid fraction of 22% in Si-replete cells.

Based on the rapid changes in lipid metabolism exhibited by Si-deficient C. cryptica cells, this species was chosen for additional studies. Since chrysolaminarin (a beta-1,3-linked glucan) and storage lipids are the primary energy reserves of diatoms, experiments were carried out which used $H^{14}CO_3^-$ as a tracer to determine the effects of Si-deficiency on the photosynthetic production of these compounds. As shown in Fig. 1, the percentage of ^{14}C partitioned into lipids nearly doubled (from 28% to 54%) after only 4 h of Si-deficiency, while the percentage of ^{14}C partitioned into chrysolaminarin decreased by 50%. These changes were followed by a gradual return to pre-limitation values. Control experiments with Si-replete cultures did not show these effects. The observed increase in the percentage of newly assimilated carbon partitioned into lipids was not due to a higher absolute rate of lipid synthesis, since the actual amount of ^{14}C incorporated into lipids decreased by 40% during the first 4 h of Si-deficiency. The amount of carbon assimilated into chrysolaminarin was reduced by an even greater amount (84%) during this period, however. Thus, differing amounts of decrease in the rates of synthesis of these compounds appear to be responsible for the changes in carbon partitioning observed in Si-limited C. cryptica cells.

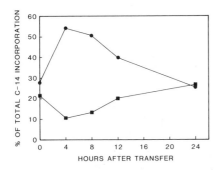
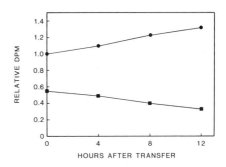

Fig. 1. Effect of Si-deficiency on partitioning of photoassimilated carbon in C. cryptica. (■) Chrysolaminarin; (●) Lipid.

Fig. 2. Redistribution of carbon into lipids in Si-deficient C. cryptica. (■) Chrysolaminarin; (●) Lipid. One unit = 22250 dpm.

Since earlier experiments with C. cryptica indicated that lipid mass increased at the same rate in Si-replete cultures and in cultures grown under Si-limiting conditions for 12 h or less, it seemed likely that Si-deficiency induced a slow conversion of cellular materials into lipids. To investigate this possibility, cells were incubated with $H^{14}CO_3^-$ for 1 h, harvested, and then resuspended in Si-free medium (isotope-free). The amount of radioactivity present in the lipid fraction was then determined after various periods of time. Fig. 2 indicates that there was, in fact, a slow redistribution of carbon from non-lipid materials into lipids, increasing the initial lipid mass by 32% during the first 12 h of Si-deficiency. This amount of carbon redistribution is sufficient to allow similar gains in lipid mass in Si-replete and Si-deficient cultures during this period, despite the large difference in photosynthetic lipid production rates. The amount of ^{14}C found in chrysolaminarin decreased during this time, although calculations suggested that a maximum of only 33% of the lipid could have originated from this source. Redistribution of carbon into lipids did not occur to a significant extent in cells transferred into a Si-replete medium. An additional finding from this pulse-chase experiment was that the ratio of ^{14}C-labeled triacylglycerols to total lipids increased with time during the first 12 h of Si-deficiency (from 13% to 42%). The amount of radioactivity in polar lipids decreased during this period, and calculations suggested that 25% of the triacylglycerols produced in this time period by means of this slow conversion process were derived from polar lipids, while the remainder originated from non-lipid sources.

The results of these experiments indicate that lipid accumulation in Si-deficient C. cryptica cells is due to two distinct processes: 1) an increase in the percentage of newly assimilated carbon partitioned into lipids, and 2) a redistribution of previously-fixed carbon from other cellular materials into lipids. Studies are currently underway to identify the enzymatic bases of these metabolic changes.

REFERENCES

1. Bligh, E.G., and W.J. Dyer. 1959. Can. J. Biochem. Physiol. 37:911-917.
2. Coombs, J., W.M. Darley, O. Holm-Hansen, and B.E. Volcani. 1967. Plant Physiol 42:1601-1606.
3. Werner, D. 1966. Arch. Mikrobiol. 55:278-308.

COMPOSITION AND POSITIONAL DISTRIBUTION OF FATTY ACIDS IN LIPIDS FROM THE DIATOM Phaeodactylum tricornutum

A. Kawaguchi, T. Arao, and M. Yamada

Department of Biology
The University of Tokyo
Tokyo, Japan

INTRODUCTION

Eicosapentaenoic acid (20:5) is generally considered the typical fatty acid of marine organisms. Various algae, the first link of the food chain, contribute probably to the marine ecosystem of this polyunsaturated fatty acid. The marine diatom, Phaeodactylum tricornutum, contains a considerable quantity of 20:5. In this paper we describe composition and positional distribution of fatty acids in lipid classes from this diatom.

MATERIALS AND METHODS

The stock cells of P. tricornutum were kindly gifted by Prof. Y. Isogai. P. tricornutum was grown at 20°C with aeration in a 2N3P medium (1) under the cycle of 16 hr light (6,000 lux) and 8 hr dark. Lipids were extracted from the cells by the procedure of Bligh and Dyer (2). The lipid classes were separated from each other by silicic acid column chromatography and thin-layer chromatography. The separated lipid classes were subjected to methanolysis to study their fatty acid compositions. The resulting methyl esters were analyzed by gas-liquid chromatography. The distribution of fatty acids at the sn-1 and sn-2 positions of glycerol backbone of glycolipids and phospholipids was investigated by means of enzymic hydrolysis with the lipase from Rhizopus delemar (3). Positional distribution of fatty acids in phospholipids was also confirmed by the enzymic hydrolysis with phospholipase A_2 (4).

RESULTS AND DISCUSSION

Fatty acid composition of P. tricornutum was different at different stages of growth (Table 1). At lag phase the major fatty acids were 16:0, 16:1 and 18:1. 20:5 accounted for only 6.4 % of the total fatty acids at this phase. At stationary phase the major fatty acids were 16:0, 16:1, 16:3 and 20:5. An increase in 20:5 and 16:3 and a decrease in 18:1 occurred at the transition from lag to stationary phases. The fatty acid composition of stationary phase cells are similar to the results reported by Kates and Volcani (5).

Table 1. Fatty Acid Composition of Total Lipids from P. tricornutum

Fatty acid	Fatty acid composition (molar %)		
	Lag phase*	Logarithmic phase**	Stationary phase***
14:0	5.8	8.8	6.8
16:0	26.7	20.0	14.5
16:1(9)+16:1(3)	34.2	24.5	27.2
16:2	1.3	3.2	3.8
16:3	2.9	6.1	13.4
18:0	2.7	2.7	0.7
18:1	20.0	8.8	2.8
18:2	0	3.6	3.0
20:5	6.4	22.3	27.8

*0.5×10^5 cells/ml, **5.0×10^6 cells/ml, ***2.5×10^7 cells/ml

The major lipid components identified were MGDG, DGDG, SQDG, PG and PC. In addition, minor amounts of PI and TG were identified. Table 2 shows fatty acid composition of the individual lipids isolated from stationary phase cells. All lipids showed high percentage of C16 acids. 20:5 was found in high content in MGDG, PG and PC. MGDG contained very low amounts of saturated fatty acid whereas SQDG contained relatively high percentage of saturated fatty acids. The high content of C18 acids was characteristic of PC. trans-3-Hexadecenoic acid [16:1(3)] exclusively occurred in PG and 16:3 was found predominantly in MGDG.

Table 2. Fatty Acid Composition of the Five Major Lipids from P. tricornutum

Fatty acid	Fatty acid composition (molar %)				
	MGDG	DGDG	SQDG	PG	PC
14:0	1.9	6.6	33.9	1.0	3.8
16:0	4.1	25.8	23.3	20.5	9.8
16:1(9)	22.2	27.6	30.7	15.6	13.8
16:1(3)	0	0	0	26.4	0
16:2	1.9	4.6	2.0	0	1.5
16:3	25.4	1.7	1.2	0	0.8
16:4	3.6	0	0	0	0
18:0	0	2.8	0	0	0
18:1	1.8	4.3	3.0	6.1	9.3
18:2	0.8	2.3	0.6	1.8	17.0
18:3	0.7	1.6	0.2	0.5	2.2
20:2	0.2	3.1	0	0	0
20:3	0.7	0.1	0	0.9	5.6
20:4*	0.6	0.3	0	0	2.8
20:4'**	2.1	0.5	0	0	2.5
20:5	34.0	18.7	5.2	27.2	30.9
Average number of double bonds per lipid molecule	3.05	1.59	0.70	1.95	2.61

*Arachidonic acid, **An isomer of arachidonic acid

Table 3. Distribution of Fatty Acids at the sn-1 and sn-2 Positions in the Five Major Lipids from P. tricornutum

Fatty acid	Fatty acid composition (molar %)									
	MGDG		DGDG		SQDG		PG		PC	
	1*	2*	1*	2*	1*	2*	1*	2*	1*	2*
14:0	3.2	0.6	11.9	0.3	62.2	5.6	1.8	0.2	6.8	0.8
16:0	0.1	8.1	31.4	17.3	1.5	45.1	10.6	30.4	6.5	13.1
16:1(9)	17.4	26.8	11.0	48.8	18.5	42.9	24.7	6.5	19.5	8.1
16:1(3)	0	0	0	0	0	0	12.2	40.6	0	0
16:2	1.5	2.3	1.6	7.5	4.0	0	0	0	0.6	2.4
16:3	3.6	47.2	2.1	1.1	1.4	1.0	0	0	0.3	1.3
16:4	1.2	6.0	0	0	0	0	0	0	0	0
18:0	0	0	3.9	1.4	0	0	0	0	0	0
18:1	0.9	2.7	1.4	7.2	1.0	5.0	0	12.0	1.6	17.0
18:2	0.4	1.2	0	4.7	0.8	0.4	0	3.6	4.1	29.9
18:3	0.5	0.9	0	6.3	0.4	0	0	1.3	0.1	4.3
20:2	0	1.7	4.2	1.6	0	0	0	0	0	0
20:3	0.5	0.9	0	0.2	0	0	0.8	1.0	3.5	7.7
20:4	1.2	0	0.6	0	0	0	0	0	4.3	1.3
20:4'	3.8	0.4	0.5	0.5	0	0	0	0	3.8	1.2
20:5	65.7	1.3	31.5	3.1	10.4	0	49.9	4.4	48.9	12.9

*The numbers 1 and 2 stand for sn-1 and sn-2 positions, respectively.

Table 3 shows the positional distribution of fatty acids in lipids. 20:5 was located almost exclusively at the sn-1 position of MGDG, DGDG, SQDG and PG. In PC 20:5 was located both positions but was predominant at the sn-1 position. 16:1(9) was dominant at the sn-2 positions of all lipids except for PC. 16:3 in MGDG was located almost exclusively at the sn-1 position. 16:1(3) in PG was mainly found at the sn-2 position and C18 acids in PC was located at the sn-2 position almost exclusively.

REFERENCES

1. Y. Komota, Y. Isogai, and K. Sato, Chem. Pharm. Bull. 31, 3771-3774 (1983)
2. E.G. Bligh, and W.J. Dyer, Can. J. Biochem. Physiol. 37, 911-917 (1959)
3. W. Fischer, E. Heinz, and M. Zeus, Hoppe-Seyler's Z. Physiol. Chem. 354, 1115-1123 (1973)
4. F. Haverkate and L.L.M. van Deenen, Biochim. Biophys. Acta 106, 78-92 (1965)
5. M. Kates and B.E. Volcani, Biochim. Biophys. Acta 116, 264-278 (1966)

LIPID METABOLISM IN TWO SPECIES OF RED MARINE ALGAE AS MODIFIED BY ENVIRONMENTAL FACTORS

Trevor R. Pettitt and John L. Harwood

Department of Biochemistry
University College
Cardiff CF1 1XL U.K.

INTRODUCTION

Little is known about lipid metabolism in marine algae and how it is affected by environmental factors. This is in spite of the fact that these organisms constitute a large part of the Plant Kingdom. The work up to 1979 has been summarized[1]. Recent work has studied the effect of light, temperature and heavy metals on lipid metabolism in the brown alga Fucus serratus[2].

The Rhodophyceae or red algae have high levels of 20C polyunsaturated fatty acids (PUFA's) but usually very low levels of 16C PUFA's[1]. There has been no comprehensive study on the acyl lipids of the Rhodophyceae although some aspects of the lipid composition in the unicellular red alga Porphyridium purpureum and how it is affected by detergents (common water pollutants) have been examined[3,4,5].

Changes in major environmental conditions such as light and temperature are known to cause changes in lipid composition and metabolism in higher plants and algae[1,6,7]. Here we describe time-course and pulse-chase experiments on the effects of these factors on lipid metabolism in two multicellular marine red algae, Chondrus crispus and Polysiphonia lanosa.

MATERIALS AND METHODS

Chondrus crispus and Polysiphonia lanosa (still attached to its brown alga host, Ascophyllum nodosum), collected from along the South Wales coast, were stored and incubated as described[2]. Lipids were extracted by the boiling isopropanol procedure[8]. Initial fractionation was on acid-washed Florosil columns[9] and further analysis was as before[2].

RESULTS AND DISCUSSION

Effect of Illumination

For photosynthetic organisms efficient utilization of light energy is of utmost importance so it is not surprising that light has been found to

Table 1. The effect of light on fatty acid labelling in Chondrus crispus

Treatment	Percentage of total [^{14}C] fatty acids						
	14:0	16:0	16:1	18:0	18:1	18:2	Others
Dark	6±1	31±1	2±1	7±3	40±6	3±tr	11±1
Light	1±1	19±2	3±2	tr	64±6	10±2	3±5

24h incubations, values are means ±S.D., tr = trace (<1%)

Table 2. The effect of light on acyl lipid labelling in Chondrus crispus

Treatment	Percentage of total [^{14}C] lipids				
	*MGDG	*DGDG	SQDG	PG	*PC
Dark	10.3±0.2	1.9±0.4	9.3±0.6	5.4±2.0	27.0±1.1
Light	14.3±2.6	6.9±0.6	13.0±5.0	18.5±1.7	17.6±2.9

24h incubations, values are means ±S.D. (n=3)
*MGDG = monoglycosyldiacylglycerol, DGDG = diglycosyldiacylglycerol
PC = phosphatidylcholine + phosphatidylsulphocholine

have a range of effects on lipid metabolism[10]. In the few algae which have been examined, light has been found to stimulate the production of unsaturated fatty acids[2,7,11]. Also the synthesis of those lipids typical of photosynthetic membranes (diglycosyldiacylglycerol [DGDG], monoglycosyldiacylglycerol [MGDG], sulphoquinovosyldiacylglycerol [SQDG], phosphatidylglycerol [PG]) appear to be stimulated in some algae such as the green Chlorella vulgaris and Euglena gracilis[7] but not in the brown alga Fucus serratus[2].

Light stimulated [^{14}C]-acetate incorporation into C. crispus about 5-fold and into P. lanosa about 3-fold. The distribution of radio-label between the C. crispus fatty acids with and without illumination is shown in Table 1. Illumination increased the relative rate of synthesis of 18:1w.9 and 18:2w.6 but decreased that for 14:0, 16:0 and 18:0 and is thus in agreement with other studies[6]. Only occasionally were trace amounts of label found in the 20C PUFA's after 24h incubations, indicating a slow turnover rate of these, quantitatively very important acids. P. lanosa showed a similar pattern.

Table 2 shows the effect of illumination on the distribution of activity between the major C. crispus lipids[12]. A sulphur-containing lipid which migrates with phosphatidylcholine (PC) was provisionally identified as phosphatidylsulphocholine[13]. Illumination stimulates the relative rate of labelling of MGDG, DGDG and PG while decreasing it for PC, triacylglycerol (TG), free sterol and free fatty acid. Except for a decrease in TG labelling, illumination caused no obvious changes in P. lanosa lipid synthesis. The increased labelling of the C. crispus MGDG, DGDG and PG, all considered typical chloroplast lipids, agrees well with other studies[7]. The absence of obvious changes in the P. lanosa "chloroplast" lipids may be due to their presence in large quantities outside the chloroplast and thus the effect of any changes within the chloroplast are diluted out by the large extrachloroplastic pools. This explanation has already been put forward for the lack of obvious changes in F. serratus[2]. PG is the most rapidly labelled lipid in the light for both algae and also in the dark for P. lanosa, indicating that it may be

Table 3. The effect of temperature on fatty acid labelling in C. crispus

Temp (°C)	Percentage of total [^{14}C] fatty acids					
	14:0	16:0	16:1	18:0	18:1	Others
4	1±2	28±5	6±2	12±4	53±10	1±1
15	10±1	37±6	5±2	11±2	34±6	3±1

24h incubations, values are means ±S.D. (n=3)

Table 4. The effect of temperature on acyl lipid labelling in C. crispus

Temp (°C)	Percentage of total [^{14}C] lipids				
	*MGDG	*DGDG	SQDG	PG	*PC
4	8.3±0.5	2.2±1.0	1.9±0.1	17.4±4.0	9.1±2.4
15	6.9±0.3	5.7±1.4	8.1±1.3	15.2±2.7	9.5±0.8

24h incubations, values are means ±S.D. (n=3), * = See Table 2.

involved in fatty acid metabolism before transfer to other lipids. Other authors have made similar observations in green algae[11].

Effect of Temperature

Lowering growth temperature has been widely shown to increase fatty acid desaturation in the membrane lipids of higher plant tissue and has been linked to the retention of membrane fluidity[14] although it may merely reflect increased oxygen solubility at lower temperatures[6]. The few studies on marine algae show similar effects[1,2].

Raising the temperature from 4°C to 15°C (the optimum temperature for these algae as determined by photosynthetic measurements) increased [^{14}C]-acetate incorporation for C. crispus about 3-fold. As temperature is raised from 4°C to 15°C the relative rate of synthesis of the saturated fatty acids 14:0 and 16:0 increased while that for 18:1w.9 decreased (Table 3) thus agreeing with other studies[1,2].

Frost hardening of plants is often accompanied by increased levels of PC and phosphatidylethanolamine (PE)[15,16] and may be explained by an increase in the levels of cholinephosphotransferase and ethanolaminephosphotransferase[17], however no such changes were found in the brown alga F. serratus[2]. Labelling of the major C. crispus lipids at different temperatures is shown in Table 4. No changes in PC or PE were observed but both DGDG and SQDG increased their relative rate of [^{14}C] incorporation at 15°C. This may indicate a role for the latter in the regulation of membrane fluidity or in the light harvesting assemblies (optimum photosynthetic rate at 15°C in this alga) as both lipids are considered typical of the chloroplasts.

In summary C. crispus and P. lanosa show similar alterations of fatty acid metabolism in response to illumination and temperature changes as do higher plants and other algae. The acyl lipids, particularly of C. crispus, also show changes in their metabolism but not always in the same way as has been observed in higher plants. It is unclear from this work whether these changes represent a difference in the rate of acyl lipid synthesis or modifications and/or replacement of fatty acids from pre-existing acyl lipids as the algae re-adjusts to changed environmental conditions.

The financial support of the S.E.R.C. is gratefully acknowledged.

REFERENCES

1. Pohl, P. and Zurheide, F. (1979), Fatty acids and lipids of marine algae and the control of their biosynthesis by environmental factors, in: "Marine Algae in Pharmaceutical Science" (Hoppe, H.A., Levring, T. and Tanaka, Y., eds.) pp. 473-523 Walter de Gruyter, Berlin.
2. Smith, K.L. and Harwood, J.L. (1984), Lipid Metabolism in Fucus serratus as modified by environmental factors, J.Exp.Bot. 35, 1359-1368.
3. Nyberg, H. (1985), The influence of ionic detergents on the phospholipid fatty acid composition of Porphyridium purpureum, Phytochem. 24, 435-440.
4. Nyberg, H. and Koskimies-Soininen, K. (1984), The glycolipid fatty acids of Porphyridium purpureum cultured in the presence of detergents, Phytochem. 23, 751-757.
5. Nyberg, H. and Koskimies-Soininen, K. (1984), The phospholipid fatty acids of Porphyridium purpureum cultured in the presence of Triton X-100 and sodium desoxycholate, Phytochem. 23, 2489-2495.
6. Harwood, J.L. (1983), Adaptive changes in the lipids of higher-plant membranes, Biochem.Soc.Trans. 11, 343-346.
7. Hitchcock, C. and Nichols, B.W. (1971) in: "Plant Lipid Biochemistry", Academic Press, London.
8. Harwood, J.L. (1975), Synthesis of sulphoquinovosyldiacylglycerol by higher plants, Biochim.Biophys.Acta 398, 224-230.
9. Christie, W.W. (1980) in: "Lipid Analysis", Pergamon press, Oxford.
10. Harwood, J.L. (1980), Plant acyl lipids: structure, distribution and analysis, in: "Biochemistry of Plants" (Stumpf, P.K. and Conn, E.E. eds.) Vol. 4, pp. 1-55, Academic Press, New York.
11. Nichols, B.W., James, A.T. and Breuer, J. (1967), Interrelationship between fatty acid biosynthesis and acyl-lipid synthesis in Chlorella vulgaris, Biochem.J. 486-496.
12. Pettitt, T.R. and Harwood, J.L. (1986), Lipid characterization and metabolism in two red marine algae, Biochem.Soc.trans. 14, 148-149.
13. Anderson, R., Kates, M. and Volcani, B.E. (1978), Identification of the sulfolipids in the non-photosynthetic diatom Nitzschia alba, Biochem.Biophys.Acta 528, 89-106.
14. Raison, J.K. (1980), Membrane lipids: structure and function, in: "Biochemistry of Plants" (Stumpf, P.K. and Conn, E.E., eds.) Vol. 4, pp. 57-83, Academic Press, New York.
15. Smolenska, G. and Kuiper, P.J.C. (1977), Effect of low temperature upon lipid and fatty acid composition of roots and leaves of winter rape plants, Physiol.Plant. 41, 29-35.
16. Clarkson, D.T., Hall, K.C. and Roberts, J.K.M. (1980), Phospholipid composition and fatty acid desaturation in the roots of rye during acclimatization of low temperature, Planta 149, 464-471.
17. Kinney, A.J., Clarkson, D.T. and Loughman, B.C. (1982), The effect of temperature on phospholipid biosynthesis in rye roots, in: "Biochemistry and Metabolism of Plant Lipids" (Wintermans, J.F.G.M. and Kuiper, P.J.C., eds.) PP. 437-440, Elsevier, Amsterdam.

LIPIDS OF *CHATTONELLA ANTIQUA* (RAPHIDOPHYCEAE)

N. Sato, Y. Nemoto and M. Furuya

Department of Botany, Faculty of Science
University of Tokyo
Hongo, Bunkyo-ku, Tokyo 113, Japan

INTRODUCTION

Chattonella antiqua is a unicellular marine species of raphidophyte, which is a member of Chromophyta. Chromophyta is a large group of algae consisting of brown algae, diatoms and other species that contain chlorophylls a and c. Although angiosperms have been the major plant materials in the study of lipid metabolism, lower plants and algae have unique fatty acids like arachidonic acid, eicosapentaenoic acid and an unusual class of lipid like diacylglyceryltrimethylhomoserine (DGTS).

As an attempt to use *C. antiqua* as a model organism in studying chromophytan lipid metabolism, we analyzed the composition of lipids and fatty acids. This is the first report of systematic analysis of lipids and fatty acids in Raphidophyceae.

MATERIALS AND METHODS

Cells of *C. antiqua* were grown photoautotrophically at $25°C$ under 12-h photoperiod as described previously[1]. Lipids were extracted by the Bligh-Dyer method and fractionated by two dimensional TLC using acetone/benzene/methanol/water (8:3:2:1) and chloroform/methanol/ammonia water (13:7:1) as developing solvents. After methanolysis, fatty acid methyl esters were analyzed by GLC using Thermon 1000 as a liquid phase. Fatty acid species were identified by GC-MS analysis of their methyl esters and pyrrolidides. Positional distribution of fatty acids in lipids were analyzed by limited hydrolysis with lipase from *Rhizopus delemar*.

RESULTS AND DISCUSSION

Lipid composition

TLC analysis of total lipids showed the presence of two unusual spots which were not found in the analysis of higher plant lipids. Based on the chromatographic mobilities and color reactions, one of them was identified as diacylglyceryltrimethylhomoserine (DGTS). The other polar lipid that showed no color reactions to ninhydrin, anthrono, molybdenum, Dragendorff, periodate-Schiff reagents remained unidentified. Quantitative analysis showed that major lipid classes were MGDG, SQDG and DGDG. Phospholipids

Table 1. Lipid content

	nmol/10^6 cells	%
Monogalactosyl diacylglycerol (MGDG)	129	29
Digalactosyl diacylglycerol (DGDG)	80	18
Sulfoquinovosyl diacylglycerol (SQDG)	128	29
Diacylglyceryltrimethylhomoserine (DGTS)	26	5.8
Phosphatidylglycerol (PG)	12	2.8
Phosphatidylcholine (PC)	23	5.2
Phosphatidylethanolamine (PE)	15	3.3
Phosphatidylinositol (PI)	6 *	1.4
Unknown	25	5.7
Total	445	100

* A diacyl structure is assumed.

and DGTS were less abundant (Table 1).

DGTS is a characteristic lipid class in lower green plants (Chlorophyta)[2]. At present, *Ochromonas* is the only chromophyte other than *Chattonella* that has been known to contain DGTS. It should be noted that brown algae[3,4] and diatoms[5] do not contain this lipid. In this sense, *C. antiqua* is exceptional among chromophytes. Whether the presence of DGTS is a common characteristic of raphidophyceae is to be investigated.

Table 2. Fatty acid composition in lipid classes

	MGDG	DGDG	SQDG	DGTS	P G	P C	P E	P I	Unknown
14:0	7	10	27	1	1	1	1	2	3
16:0	32	17	29	49	37	33	47	47	50
16:1(9)	14	13	4	0	2	1	1	3	4
(12)	}0	}0	8	}0	—	}0	}0	}1	}1
(3t)			—		9				
18:0	0	0	0	0	4	0	1	2	1
18:1	4	1	0	0	0	0	0	1	2
18:2	7	3	1	1	3	8	0	2	7
18:3ω3	6	12	5	0	1	2	2	1	3
18:4	12	16	5	0	2	2 *	0	2	8
20:4ω6	2	1	0	11	8	11	10	8	7
20:5	13	28	16	35	25	27	18	27	12
22:4?	0	0	0	0	2	5	9	2	1
22:6?	0	0	0	0	2	6	9	1	1

* 7% ω6 plus 4% ω3.

Fatty acid composition

Major fatty acids were identified as 16:0, 14:0 and 20:5 (5,8,11,14, 17). 16:1 (9+12+3t isomers), 18:0, 18:1 (9), 18:2 (9,12), 18:3 (6,9,12), 18:3 (9,12,15), 18:4 (6,9,12,15), 20:3, 20:4 (5,8,11,14), 20:4 (8,11,14, 17) and 22:0 were found as minor components. 22:4 and 22:6 were tentatively identified by retention time in GLC.

Analysis of fatty acids in different lipid classes showed specific esterification of fatty acids to lipid classes (Table 2): 18:4 was found mainly in the galactolipids; 14:0 in SQDG; C_{22} unsaturated acids in PE. It was especially interesting to note that different isomers of 16:1 were esterified to different classes of lipids: 9-isomer was found in the galactolipids; 3t-isomer was in PG as is the case in most other plant species, while a rare isomer, 16:1 (12), was found only in SQDG.

The fact that the major fatty acids of *C. antiqua* were 16:0 and 20:5 showed that this organism was quite different from the plants and algae belonging to other groups like Chlorophyta $(C_{16}+ C_{18})^4$, Phaeophyceae $(C_{16} + C_{18} + C_{20})^4$, and dinoflagellates $(C_{16} + C_{18} + C_{22})^6$. The low content of C_{18} acids noted in *C. antiqua* is a common character of Rhodophyta[4] and diatoms[5].

Positional distribution

16:0 was a major fatty acid in all classes of lipids and almost exclusively esterified at the C-1 position (data not shown). The location of 20:5 varied among lipid classes: it was found at the C-1 position in MGDG, DGDG and SQDG, but at the C-2 position in DGTS and PE. These results showed that the positional distribution of fatty acids in lipids of *C. antiqua* was distinct from other species of plants and algae including diatom[5]. Especially, the location of 16:0 and 20:5 at the C-1, and of 16:1 and 18:4 at the C-2 position of MGDG was not compatible with either "eukaryotic" or "prokaryotic" pathways of glycerolipid synthesis found in angiosperms[7].

In conclusion, *C. antiqua* is unique in lipid and fatty acid composition. Studies on chromophytan lipid metabolism will shed new light on the biosynthesis of molecular species of lipids especially those which contain C_{20} unsaturated acids.

References

1. Y. Nemoto and M. Furuya, Inductive and inhibitory effects of light on cell division in *Chattonella antiqua*, Plant Cell Physiol., 26: 669 (1985).
2. N. Sato and M. Furuya, Distribution of diacylglyceryltrimethylhomoserine and phosphatidylcholine in non-vascular green plants, Plant Sci., 38: 81 (1985).
3. K. L. Smith and J. L. Harwood, Lipids and lipid metabolism in the brown alga, *Fucus serratus*, Phytochemistry, 23: 2469 (1984).
4. G. R. Jamieson and E. H. Reid, The component fatty acids of some marine algal lipids, Phytochemistry, 11: 1423 (1972).
5. A. Kawaguchi, T. Arao and M. Yamada, Composition and positional distribution of fatty acids in lipids from the diatom *Phaeodactylum tricornutum*, this symposium.
6. G. W. Harrington, D. H. Beach, J. E. Dunham and G. G. Holz, Jr., The polyunsaturated fatty acids of marine dinoflagellates, J. Protozool., 17: 213 (1970).
7. E. Heinz and P. G. Roughan, Similarities and differences in lipid metabolism of chloroplasts isolated from 18:3 and 16:3 plants, Plant Physiol., 72: 273 (1983).

THE FUTURE - GENETICS/BIOTECHNOLOGY

THE FUTURE - GENETICS/BIOTECHNOLOGY

P.K. Stumpf

Department of Biochemistry & Biophysics
University of California
Davis, California 95616

Plant lipid biochemistry has made remarkable strides since the early 1950s when only primitive concepts and primitive techniques were available to the biochemist. As these concepts and techniques developed in the 1960s and 1970s rapid advances occurred along traditional lines.

Now in the 1980s the concepts and techniques of molecular biology are revolutionizing the whole science of biochemistry. However, plant lipid biochemistry has lagged behind for one reason or another.

Therefore, this section of the Symposium was specifically organized to examine (a) the classic procedures of plant breeding as it applies to the modification of oil crops, (b) the newer tissue culture techniques for clonal propagation of elite plants, (c) recent procedures for mutagenesis and their application to basic problems in plant lipid biochemistry and finally (d) the employment of molecular biology and recombinant DNA techniques in specific problems. Additional poster presentations further amplified the approaches of the new technique.

It is therefore entirely predictable that with these small beginnings, in the near future a very rapid expansion of plant molecular biology into the area of plant lipids will have major impacts not only in the basic knowledge necessary to understand more fully the biosynthesis, degradation and regulation of lipids in plant tissues but equally important on the agricultural and industrial side by which fatty acid composition, types of lipids, etc. can be designed for the societal needs of people

GENETIC MANIPULATION OF OILSEED QUALITY

R.K. Downey

Agriculture Canada, Research Station
107 Science Crescent
Saskatoon, Saskatchewan Canada S7N 0X2

INTRODUCTION

The nutritional and industrial value of a vegetable oil is largely determined by the presence and proportions of the individual fatty acids it contains. One of the main objectives of an oilseed plant breeder is to develop cultivars with fatty acid compositions which more closely meet market needs without adversely affecting seed or oil yield. Conventional oilseed breeding has been primarily concerned with the genetic expression of fatty acid composition. But the more that is known of fatty acid and lipid synthesis, the more effectively the breeder can assess the opportunities for oil quality improvement and design the appropriate breeding strategies.

Solar energy trapped by the plant in its photosynthetic organs is transported to the seed in the form of sucrose where it is converted into storage lipid in the endosperm and embryo. In general, fatty acid synthesis is under the genetic control of these lipid storage tissues and fatty acid composition of the stored lipid is quite different from that of other plant organs or tissues. Fatty acids are not transported from one plant tissue to another. The first product of the nonassociated plant fatty acid synthetase system, palmitic acid (16:0), is elongated to stearic acid (18:0) which in turn is desaturated to oleic acid (18:1) (Fig. 1). Palmitic and stearic acid do not usually accumulate to a large extent in oilseeds, although, as will be seen, genetic manipulation can have this result. Oleic acid can serve as a precursor to either an elongation pathway which gives rise to eicosenoic acid (20:1) and erucic acid (22:1) or the desaturation pathway producing linoleic (18:2) and linolenic (18:3) acids.

Plant breeders have not waited for the intricacies of fatty acid interconversions to be elucidated. Indeed, breeding for oil quality has contributed significantly to the understanding of the synthesis and accumulation of fatty acids in plants.

Prior to the development of gas-liquid chromatography in the early 1950's, methodology for determination of fatty acid composition was tedious. Usually it required over a kilogram of seed and at least two days to obtain an approximate composition. Gas-liquid chromatography on the other hand was capable of accurate analysis of a drop of oil in 20 to 30 minutes even in the early days when all peaks had to be quantitated by hand using triangulation.

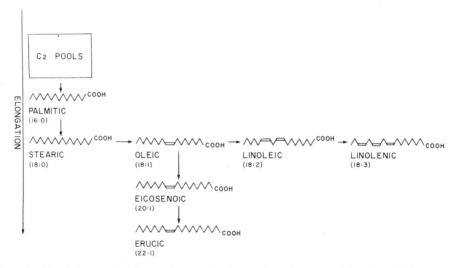

Fig. 1. The biosynthetic pathway of the major fatty acids in edible vegetable oilseeds. The chain elongation pathway from oleic to erucic acid is unique to the Brassica oil crops.

Among the first crops to be investigated using the newer technique of gas-liquid chromatography were the Brassica oilseed crops, rape (Brassica napus L.) and turnip rape (B. campestris L.). These crops which are now the world's fourth most important source of edible oil differed from other common vegetable oils in containing significant amounts of the monoenoic long chain fatty acids, eicosenoic and erucic (Table 1).

Table 1. Fatty acid composition of some principle vegetable oilseed crops

Crop	Fatty acid composition in %						
	16:0	18:0	18:1	18:2	18:3	20:1	22:1
Rapeseed	3	1	14	13	8	8	51
Canola[a]	4	1	62	20	10	2	<1
Soybean	11	4	25	52	8	–	–
Sunflower	7	4	16	73	–	–	–
Safflower	7	3	12	78	–	–	–
Flax	7	4	20	17	52	–	–

[a] low erucic acid rapeseed oil

FATTY ACID ELONGATION

In the 1950's and again in the early 1970's feeding experiments with laboratory animals indicated the nutritional value of rapeseed oil would be substantially improved if the erucic acid content could be reduced to <5% of the total fatty acids present. The identification of plants with essentially no erucic acid in their seed oil in Brassica napus L. (Stefansson et al, 1961) and B. campestris L. (Downey, 1964) not only resulted in a nutritionally superior new and natural vegetable oil, now called "canola", but offered an opportunity to learn more of the biosynthetic pathway of long chain fatty acids and its genetic control. First, it was demonstrated that the genetic make up of the developing embryo, rather than the maternal parent, determined the amount of the long chain fatty acids formed (Harvey and Downey, 1964). This form of inheritance has since been shown to apply to fatty acid syntheses in other seed oils. Second, it was demonstrated, using radioactive acetate, that eicosenoic acid was synthesized by the addition of an acetate to the carboxylic end of oleic acid and that the addition of a second acetate resulted in erucic acid (Downey and Craig, 1964). In radioactive feeding studies the distribution of radioactivity in oleic acid was found to be quite different from the elongation products, indicating that even though oleic acid is the precursor to eicosenoic and erucic acids the pathway for oleic acid synthesis is quite separate from the elongation pathway (Stumpf and Pollard, 1983).

It is still not clear whether one or two systems are involved in the conversion of 18:1 to 22:1. However, it is known that the chain elongation occurs in a separate compartment of the cell and that similar elongation systems operate in other oilseeds such as jojoba (Simmondsia californica), meadowfoam (Limnanthes alba), nasturtium (Tropaeodum majus), and other members of the Cruciferae (Stumpf and Pollard, 1983).

In rapeseed the amount of erucic acid formed in the developing embryo is controlled by a series of alleles which exhibit additive gene action. In B. napus and B. juncea two loci are involved while in B. campestris a single locus conditions erucic acid synthesis. By varying the alleles present at these loci it is now possible to fix the erucic level of a cultivar at almost any level from essentially zero to over 50%.

In the Brassica crops it has been found that the long chain fatty acids are located on the 1,3 position of the triglycerides. This distribution is of economic importance since there is a significant industrial market for high erucic acid oil. One of the major markets for such oils is in the plastics industry where erucamide is extensively used as a slip agent to prevent plastic products from sticking to each other and to the extruding machinery. Since the higher the erucic acid level the more valuable the oil is to the industry, several breeding and biotechnology programs are underway to increase erucic acid levels above the theoretical limit of 66.6%. That this may be possible is evidenced by the >70% erucic acid found in nasturtium seed (Swern, 1964).

FATTY ACID DESATURATION

Considerable differences occur between species in the level of polyunsaturated fatty acids they contain (Table 1). Some species, such as sunflower (Helianthus annus L.) and safflower (Carthamus tinctorius L.) contain essentially no linolenic acid but have high levels of linoleic acid while others, including soybean (Glycine max (L.) Merrill) rapeseed/canola (B. napus L. and B. campestris L.) and flax (Linium usitatissimum L.) all contain significant quantities of linolenic acid. Thies (1970) observed that linolenic acid occurs only in seeds which, during certain stages of their development, possess green photosynthetically active chloroplasts and doubted that linolenic acid could be significantly reduced or eliminated from such seed without severe adverse consequences.

However, recent developments in all three oilseed crops with significant linolenic acid content have demonstrated that the linolenic level of seed oils can be reduced to between 2 to 4% without seriously affecting oil content, seed germination or seedling vigor. The low linolenic acid mutants are not just biological curiousities but are of considerable economic importance since linolenic acid, with its three double bonds, is readily oxidized and the major cause of off flavors and reduced shelf life in edible oils. However, it should be noted that the presence of

8 to 10% linolenic acid (vitamin F) may be highly desirable from a nutritional point of view.

In rapeseed/canola the normal linolenic content is 8 to 10%. The breeding objective has been to reduce the level of linolenic to less than 3% while maintaining or increasing the level of the nutritionally desirable linoleic acid presently at 20-25%. Selection among the available germplasm has resulted in little gain since the level of both fatty acids are readily affected by environmental factors, particularly temperature, and a reduction or increase in one of the polyunsaturated acids has tended to result in a similar reduction or increase in the other. However, chemical mutation experiments by Rakow (1973) in B. napus were successful in reducing linolenic acid levels to 5.5% and subsequent selection produced material with linolenic levels as low as 3.2% (Röbbelen and Nitsch, 1975). Unfortunately these low linolenic mutants were agronomically poor. More recently, Stefansson (personal communication) using this material as a base, has developed agronomically acceptable summer B. napus strains containing less than 3% linolenic acid. The level of linoleic acid in these lines was maintained at about 23% (Table 2). The inheritance of the low linolenic characteristic has yet to be determined.

Table 2. Fatty acid composition of normal and modified rapeseed oil strains resulting from selection or mutation.

Oil Type	Fatty acid composition in % total fatty acids							
	16:0	16:1	18:0	18:1	18:2	18:3	20:1	22:1
Normal high erucic	3	<1	1	14	13	8	8	51
Low erucic (Canola)	4	<1	1	62	20	10	2	<1
Low 18:3 [a]	5	<1	2	64	24	<3	1	<1
High C16 [b]	10	4	1	51	19	13	1	<1

[a] Stefansson (personal communication)
[b] Jonsson (personal communication)

It is also of interest that work in Sweden (Jonsson and Persson, 1983) and elsewhere has resulted in selection strains with increased levels of the C16 fatty acids, palmitic and palmitoleic (Table 2). This composition could be highly desirable for use in the manufacture of 100% canola oil margarines. The present canola oils, which contain over 90% 18 carbon fatty acids tend to form undesirable β crystals on storage, unless other oils with significant amounts of non-C18 fatty acids are blended in. The presence of 14% or more C16 acids may provide sufficient variation in fatty acid carbon chain length to result in a stable hydrogenated oil crystal of the desired β-form. However, such an oil composition could undermine the present promotion of canola oil as containing only half the level of saturated fatty acids present in other common vegetable oils.

A considerable plant breeding effort has also been directed to improve the keeping quality of soybean oil by reducing the linolenic acid content from the present 8 to 9% to less than 3%. The most effective approach has been the use of chemical mutagens. To date the lowest linolenic acid mutant, identified as A5, has an average linolenic acid content of 3.8% and a linoleic level of about 43% (Table 3) (Hammond and Fehr, 1984). Wilcox et al (1984) have also produced a mutant line from the cultivar Century which contains 3.8% linolenic acid compared to 8% for Century. Crosses between these two mutants did not result in progeny with linolenic acid values below that of the lower linolenic parent, A5.

Table 3. Fatty acid compositions of normal and modified soybean and linseed oils resulting from mutation and selection

Oil Type	Fatty acid composition %					
	16:0	18:0	18:1	18:2	18:3	20:0
Soybean[a]						
Normal	15	4	24	49	8	0
Low linolenic[c]	9	4	40	43	<4	0
High stearic[b]	8	28	20	35	7	2
Flax[d]						
Normal parent	9	3	35	17	36	0
Low linolenic	9	5	36	48	2	0

[a] Hymowitz et al. (1972)
[b] Hammond and Fehr (1984)
[c] Hammond and Fehr (1983)
[d] Green (1986)

The mutagen treatments by Hammond and Fehr (1984) also produced a soybean line which contained a high level (28%) of stearic acid (Table 3). Although this novel fatty acid composition is of no immediate commercial interest because it is a solid at room temperature, it does suggest that the desaturation of stearic to oleic is open to genetic manipulation possibility resulting from a lower activity of stearoyl-ACP desaturase.

More recent mutagenic studies in flax have resulted in significantly lower linolenic levels in two mutants, each of which contained about 22% linolenic acid. The normal level for the parent cultivar, Glenelg, when grown under greenhouse conditions, was 34% linolenic acid. Crosses between the two mutants resulted in a line containing approximately 2% linolenic (Table 3) (Green, 1986). Inheritance studies indicated that two alleles are involved, with one allele contributing about 11% and the other approximately 0.5% linolenic acid. The alleles were found to occur at two independent loci and act in an additive manner. The fact that the low linolenic acid mutant in flax has an elevated linoleic content and the low linoleic acid canola selection contained the traditional level of linoleic is clear evidence that desaturation of oleic to linoleic and linoleic to linolenic acid are separate systems which can be genetically manipulated. Such mutant strains may also provide a means of isolating and constructing the major enzyme(s) and genes controlling this desaturation system.

The desaturation pathway from oleic to linoleic acid is also open to genetic manipulation as evidenced by the development of commercial cultivars of both safflower and sunflower which, depending upon the genotype, will yield either a high oleic or a high linoleic acid oil (Table 4).

Although low temperatures during seed development of these oilseed crops has tended to favor the synthesis of linoleic acid while high temperatures significantly increased the level of oleic acid, commercial cultivars have now been developed that are temperature stable for either high oleic or high linoleic fatty acid composition. Nutritionally the high linoleic acid strain may be more desirable as an edible oil, however industry has become increasingly interested in the very high oleic acid sunflower oil because of its greatly improved keeping qualities (Purdy, 1985).

Table 4. Fatty acid compositions of normal and modified safflower and sunflower oils resulting from selection or mutation

Oil Type	Fatty acid composition %			
	16:0	18:0	18:1	18:2
Safflower[a]				
Normal	8	2	18	72
High oleic	6	1	<u>79</u>	14
High stearic	7	<u>10</u>	14	69
Sunflower				
Normal northern[b]	6	5	18	69
High oleic[c]	4	4	<u>89</u>	<u>1</u>

[a] Knowles (1972)
[b] Earle et al. (1968)
[c] Purdy (1985)

In safflower, a high stearic acid strain, similar to that found in soybeans, was identified within the world germplasm collection. As with the soya mutant the level of stearic acid formed was found to be conditioned by two alleles at a single locus, with low stearic acid content being almost completely dominant over the high stearic acid characteristic (Ladd and Knowles, 1971; Hammond and Fehr, 1984). In safflower it was found that the level of stearic acid produced was inherited independently of the oleic-linoleic acid desaturase system (Ladd and Knowles, 1971). Although the level of stearic found in the safflower strain is less than that in the high stearic soybean mutant, similar processes appear to be in operation.

The proportions of oleic to linoleic acids synthesized in safflower are also under genetic control with three alleles, acting in an additive manner, conditioning the level from a single locus (Ladd and Knowles, 1971). The genetic block in the desaturation of oleic to linoleic acid in safflower is not complete, with approximately 14% linoleic acid being formed (Table 4). In sunflower, however, the development of a high oleic strain, by treatment of the Peredovick cultivar with chemical mutagens followed by selection for high oleic acid, resulted in an almost complete inhibition of linoleic acid synthesis (Table 4). Fick (1984) has reported the high oleic acid characteristic to be largely controlled by a single, partially dominant gene.

Recent studies on the incorporation of linoleic acid into triacylglycerols in developing sunflower and safflower seeds revealed a new cycle in which the transfer of oleic to linoleic acid between CoA and phosphatidylcholine, and the interconversion of phosphatidylcholine and diacylglycerol results in linoleic acid enrichment of the triacylglycerols (Stymne and Stobart, 1985) (Fig. 2). If this system were operational in sunflower the high oleic strains could result from either inefficient transfer of oleic and linoleic acids between CoA and phosphatidylcholine, or the interconversion of diacylglycerol with phosphatidylcholine.

Given this model for triacylglycerol synthesis for linoleic rich seeds a similar but separate exchange cycle might be postulated for linolenic acid accumulation. Certainly the evidence is now very strong that linoleic and linolenic acids are synthesized separately in the cell.

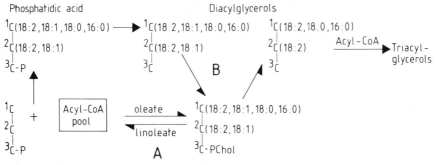

Fig. 2. Linoleic acid enrichment cycle. Oleic acid is transferred from CoA to the 2 position of phosphatidylcholine, coverted to linoleic acid and transferred back to CoA for incorporation into glycerol-3-phosphate. Glycerol-3-phosphate gives rise to phasphatidic acid, diacylglycoerol and triacylglycerol. Equilibriums between oleic and linoleic acid transfer, and diacylglycerol to phosphatidylcholine intercoversion, results in linoleic acid enrichment (from Stobart and Stymne, 1985).

From the array of fatty acid compositions which have been selected or induced in vegetable oil crops, it is evident that each step in the biosynthetic pathway of the major fatty acids is open to genetic manipulation in a significant way. This manipulation has beeen possible without serious side effects on the agronomic performance of the seed or the crop plant. Although the seed is clearly dependent upon the lipid content rather than the fatty acid composition for the energy of germination and early seedling development, the range of fatty acid composition which is tolerated is remarkable. The fact that the high stearic soybean mutant can synthesize and store an oil which would be solid or at least semi-solid during a large part of the oil accumulation period is a case is point.

The identification of normally functioning seeds of B. napus and flax with less than 3% linolenic acid has provided the biochemist with the means of clarifying and describing the separate biosynthetic steps in the synthesis of linolenic and linoleic acids. Given a clear understanding of how the developing embryo synthesizes and accumulates each of the major fatty acids, the conventional breeder and the biotechnologist can more effectively exploit additional opportunities for new and novel compositions. Eventually it should be possible to construct and insert genes to enhance or inhibit the formation of the fatty acid of interest.

REFERENCES

Downey, R. K., 1964, A selection of Brassica campestris L. containing no erucic acid in its seed oil, Can. J. Plant Sci., 44:295.
Downey, R. K. and Craig, B. M., 1964, Genetic control of fatty acid biosynthesis in rapeseed (Brassica napus L.), J. Amer. Oil Chem. Soc., 41:475.
Earle, F. R., van Etten, C. H., Clark, T. F., and Wolff, I. A., 1968, Compositional data on sunflower seed. J. Amer. Oil Chem. Soc., 45:876.
Fick, G. N., 1984, High oleic sunflower, Sunflower Research Workshop, Bismarck, N. D., Feb. 1.
Green, A. G., 1986, Genetic control of polyunsaturated fatty acid biosynthesis in flax (Linum usitatissimum), Theor. Appl. Genet. (in press).

Hammond, E. G., and Fehr, W. R., 1983, Registration of A5 germplasm line of soybean, Crop Sci., 23:192.

Hammond, E. G., and Fehr, W. R., 1984, Improving the fatty acid composition of soybean oil, J. Amer. Oil Chem. Soc., 61:1713.

Harvey, B. L., and Downey, R. K., 1964, The inheritance of erucic acid content in rapeseed (Brassica napus), Can. J. Plant Sci., 44:104.

Hymowitz, T., Palmer, R. G., and Hadley, H. H., 1972, Seed weight, protein, oil and fatty acid relationships within the genus Glycine, Trop. Agric., 49:245.

Jonsson, R., and Persson, C., 1983, Breeding for improved fatty acid composition in rapeseed, Proc. Intern. Rapeseed Congress, 6th, Paris, 311.

Knowles, P. F., 1972, The plant geneticist's contribution toward changing lipid and amino acid composition of safflower, J. Amer. Oil Chem. Soc., 49:27.

Ladd, S. L., and Knowles, P. F., 1971, Interactions of alleles at two loci regulating fatty acid composition of the seed oil of safflower (Carthamus tinctorius L.), Crop Sci., 11:681.

Purdy, R. H., 1985, Oxidative stability of high oleic sunflower and safflower oils, J. Amer. Oil Chem. Soc., 62:523.

Rakow, G., 1973, Selektion auf Linol- und Linolensäuregehalt in Rapssamen nach mutagener Behandlung, Z. Pflanzenzüchtg., 69:62.

Röbbelen, G., and Nitsch, A., 1975, Genetische und physiologische Untersuchungen neuer Mutanten, Z. Pflanzenzüchtg., 75:93.

Stefansson, B. R., 1985, Personal communication.

Stefansson, B. R., Hougen, F. W., and Downey, R. K., 1961, Note on the isolation of rape plants with seed oil free from erucic acid, Can. J. Plant Sci., 41:218.

Stobart, A. K., and Stymne, S., 1985, The regulation of the fatty acid composition of the triacylglycerols in microsomal preparations from avocado mesocarp and the developing cotyledons of safflower, Planta 163:119.

Stumpf, P. K., and Pollard, M. R., 1983, Pathways of fatty acid biosynthesis in higher plants with particular reference to developing rapeseed, in: "High and Low Erucic Acid Rapeseed Oil," J. K. G. Kramer, F. D. Sauer, and W. J. Pigden, ed., Academic Press, New York.

Swern, D., 1964, Structure and composition of fats and oils, in: "Bailey's Industrial Oil and Fat Products," D. Swern, ed., John Wiley and Sons, New York.

Thies, W., 1970, Chloroplast development and biogenesis of linolenic acid in ripening cotyledons of rapeseed. Proc. Intern. Rapeseed Congress, Ste. Adele., 348.

Wilcox, J. R., Cavens, J. F., and Nielsen, N. C., 1984, Genetic alteration of soybean oil composition by a chemical mutagen, J. Amer. Oil Chem. Soc., 61:97.

BIOTECHNOLOGY IN THE IMPROVEMENT OF THE OIL PALM
(Elaeis guineensis Jacq.)

By LH Jones
Unilever Research
Colworth Laboratory
Colworth House
Sharnbrook
Bedford MK44 1LQ

Abstract
The application of tissue culture methods for clonal propagation of the oil palm is summarised.

The benefits in terms of improved yield and quality are described and the potential application of recombinant DNA technology is discussed.

Introduction
During the 1960s the techniques of plant tissue culture developed rapidly. The ability to recover complete plants from aseptic culture quickly led to the realisation that this technique provided a rapid multiplication method for clonal propagation of selected individual plants with elite qualities.

Clonal propagation is widely used in horticulture to produce uniform varieties by multiplication of plants selected from segregating seedling populations. In the case of oil palm there is no conventional method, such as taking cuttings or bud grafting, available for conventional clonal propa- gation. Until the advent of tissue culture all oil palm plantations were planted with seedlings, with their attendant variability. Imagine an apple grower planting an orchard with seedling apples. Wide variation for yield[1] and oil composition[2] is observed in seedling populations of oil palm. Yield itself is subject to high environmentally induced variation but there are clearly highly heritable components of yield[3] and plant breeders have made steady progress in improving yield at a rate of about 10% per generation[4] by selection of elite individuals as parents for hybridisation. The variation seen in modern palm progenies suggests that major improvements are still possible by selection of the best individuals. Furthermore the availability of a clonal propagation method now makes it attractive to make broad based crosses to encourage new genetic combinations to segregate in highly variable progenies from which outstanding individuals can be selected. From physiological considerations Corley[5] has estimated that theoretically yields as high as 17 tonnes of oil per hectare per annum should be possible, against current yields in Malaysia of about 6 tonnes per hectare per annum.

Development of Tissue Culture Methods
Programmes to develop tissue culture methods for clonal propagation of oil palms started in the late nineteen sixties in the Unilever laboratories in England and in the IRHO laboratories in Paris. The first report of successful establishment of regenerated tissue culture plants in the field was made at the Malaysian International Oil Palm Conference in Kuala Lumpur in 1976.[6] This report was closely followed by a report of similar success by Rabechault and Martin.[7] In fact the first Unilever plants were sent to Malaysia from UK in late 1975, only one plant surviving the journey and subsequent transplanting shock. A second consignment in March 1976 proved more successful and during 1976 small numbers of several clones were sent to Malaysia for planting in trials. The first field planting of clonal palms was made in January 1977 at the Pamol plantation near Kluang.

In tissue culture of oil palm, as in most monocots, it is essential to use actively meristematic tissue as source material. Suitable actively dividing tissues are found in the root system, in developing leaf bases, in young leaf tissues and in immature inflorescences.
Conventional micropropagation from lateral buds is not possible in this species, and the first step is to induce a callus of undifferentiated cells by the use of media containing growth regulators of the auxin type such as 2,4-D or NAA.[8] The calluses may take several months to develop and initially grow rather slowly. The next step is to induce the formation of organised structures by modification of the growth regulator regime. (An appropriate protocol is reported by Nwankwo and Krikorian[9]). The developing structures are visible as small white globular bodies with a distinctive epidermal layer, and initially little internal organisation. They show many similarities to zygotic embryos in developing seeds and are termed embryoids. The embryoids occasionally germinate directly to form single plants with the characteristics of seedlings and we now have much biochemical evidence of their embryo-like nature.[10,11,12] More frequently, and more importantly for the multiplication processes they can be induced to proliferate to form a multinodular pro-embryonic mass,[13,14] when grown on appropriate media. Individual shoots can be induced to develop and can be removed for subsequent rooting, while the embryogenic material is recycled for further rounds of multiplication and shoot production. A single embryongenic event can thus give rise to a continuously multiplying production system. Plant production can continue in this way for several years from the same initial culture isolate. Recently, however, we have found abnormal flowering patterns developing in a few clones after several years in culture.[15] Although we do not yet know the cause of this abnormality, regular reisolation of fresh cultures from proven material is advisable to avoid the possibility of deleterious changes occurring in long-term cultures.

Fatty Acid Composition of Clonal Oils
The first clones were derived from unselected seedlings germinated aseptically and subsequently subdivided to provide source material for the early culture experiments. We did not therefore expect to obtain superior performance from these clones, but were more concerned with their uniformity and distinctness as clones and to establish that they followed a normal cycle of growth, flowering and fruiting. The early results from these clone trials[16] confirmed all these points and established that there were significant differences between clones in the fatty acid composition of the oil.[17] Indeed, surprisingly large differences were found in the first, very limited, range of clones examined. The high reproducibility within clone confirmed the strong genetic control of oil composition.

As more clonal plants became available in the late 1970's the same clones were also planted in small numbers in our estate at Lobe in Cameroon. When these plants came into bearing it was possible to compare oils from identical genetic material in the two different environments. The limited data obtained so far[18] indicated that oil from palms of the three clones examined showed a significantly higher level of longer chain C-18 fatty acids in Malaysia than in Cameroon. Since the unsaturated acids oleic and linoleic are in the C-18 group there was also a shift towards higher levels of unsaturated fatty acids and a higher iodine value in Malaysian compared with Cameronian oils. This bears out the trend usually seen in Malaysian and West African oils. Hitherto it has not been possible to distinguish the possible environmental effects from genetic differences in the palm populations in the two environments.

Recent surveys[19] have reinforced previous observations that considerable differences in fatty acid composition can be found within the existing populations of Elaeis guineensis. Still wider differences exist between this species and the South American Elaeis oleifera, with its more unsaturated and more liquid oil. In spite of good bunch yields the oil to bunch ratio in E. oleifera is very poor, and the species is not an economic plantation crop. It does, however, carry useful properties of disease resistance and low stature, and can be hybridised with E. guineensis. The hybrids have been widely planted and show characteristics intermediate in type between the two parents.[2] Some individual hybrid palms have both good yield and oil quality. The progenies from hybrid palms backcrossed with E.guineensis show great variability, segregating widely for various recombinations of the parental characters.[20] By using tissue culture it is possible to multiply individual hybrids, or selections from the F2 progeny to produce uniform clones, maintaining the unique genetic combinations which would be unable to breed true from seed. The Unilever-Harrisons & Crossfield partnership now has clonal plants derived from E.g X E.o hybrids and these will be tested in field trials over the next few years.

Selection Objectives

The primary objective of oil palm breeders is high oil yield. This is the most effective way of maximising the growers return. In areas such as parts of West Africa where Fusarium wilt is endemic, yields are only possible from wilt resistant varieties and for these areas selection for wilt resistance is a vital part of clonal selection. The first clones are now developing from wilt resistant palms in Zaire and are being tested for inherited wilt resistance, first at the nursery stage and subsequently in the field.

Other selection criteria can also be used to reduce plantation costs and hence combine with high oil yield to optimise growers returns. Such characters include ease of harvesting, represented by low stature and accessible bunches. Palms with uniform ripening bunches and low fruit abscission would also significantly reduce harvesting costs and contribute to high yield, and the ability to maintain high yields with low fertiliser inputs would also contribute to cost reduction. In a period of low oil prices such increases in efficiency are vital to maintain plantation profitability.

It is extremely unlikely that palms combining all these characters will be found by random selection. Many years of plant breeding will be needed to achieve the most favourable combinations of genetic characters, and this in turn is contingent upon identification of palms containing the desired features and demonstration of their heritability.

Selection for modified oil composition has so far been regarded as a very low priority in this list of selection objectives. This may seem surprising, since the high melting point of palm oil leads to a number of problems of recrystallisation in transport and storage. On fractionation palm oil yields an excess of relatively cheap stearin compared with the more valuable olein fraction. A lower melting point oil resulting from more unsaturated fatty acids should have a higher value than conventional palm oil, would have better nutritional value and would give palm oil more access to the large liquid oils market. On the other hand vegetable oils are closely linked to the soya bean oil price. The advances in vegetable oil refining, blending and interesterification enable manufacturers to produce oils and fats of any desired quality to meet market demand, which can change far more rapidly than plant breeding and the plantation crops industry can respond. In addition, if clones are selected for production of a specialist oil it will be necessary to grow relatively large areas of this type to feed dedicated mills with separate storage and transport systems. Unless several clones are produced with similar properties there will always be dangers from disease or other crop failures resulting from overplanting large single clone blocks. Plantation owners will only take the risk of planting clones with special oil composition if they can see an assured market for the oil for the productive lifetime of the plantation, say 25 years, giving them a better return than 'conventional' palm oil.

Application of Molecular Biology

Techniques in molecular biology are advancing rapidly and the ability to introduce functional foreign DNA into plants is now well established in a limited range of species. It is tempting to speculate on how this technology can be applied to oil palm breeding.

The present recombinant DNA technology is only possible under conditions where several stringent criteria are met. The desired gene for transfer has to be known, characterised and available, together with the appropriate promoters and regulatory and targeting sequences. It must be possible to transform the target plant cells and to recover the genetically modified cells using suitable selective marker genes such as antibiotic resistance, and finally it must be possible to recover complete plants in large numbers from the transformed cells. In monocots, such as oil palm we are as yet unable to use the Agrobacterium Ti or Ri plasmids as vectors, but it may be possible to use alternative methods. Transformation with foreign DNA is becoming possible even in monocots by direct DNA injection. Successful recovery of transformed cells would still be difficult in oil palm because the plating efficiency of cell cultures is low, relatively large inocula are required for successful establishment and most cell division potential resides in relatively large cell clusters. Although we have had limited success with colony formation from oil palm protoplasts[21] we have not obtained plants from the recovered colonies. Even if transformed cells can be recovered there are still major problems in achieving a high enough rate of plant regeneration to ensure recovery of rare transformed cells in selective media. We are not therefore yet in a position to contemplate using rNDA methods on oil palm breeding. Even more important, we have as yet no knowledge of the genes we might wish to transfer, and this is now the major priority. At present the technology is limited to the transfer of single genes coding for single proteins. Many of the characters we might wish to modify such as yield or drought resistance are not understood in terms of biochemical genetics and are certainly controlled by the interaction of many genes. Even where single genes might be involved, as in resistance to some diseases, the gene products are unidentified and when present are in very low abundance, so that a great deal of very refined biochemistry would be required to identify the protein concerned and construct probes which might be used to isolate the gene. No doubt such work will be done, but there is no expectation that it will be quick or easy. Considerable efforts are being directed towards the genetic modification of seed oils. In palm kernel oil we have a premium quality lauric oil which others seek to mimic in temperate crops. The techniques being developed require the fatty-acid modifying genes to be coupled to promoters activated by genes controlling seed development, so that the new genes are only expressed in developing seeds and do not interfere with biosynthesis of membrane and other metabolic lipids. If it were contemplated to modify palm mesocarp oil, similar regulatory sequences would be required, this time specific to the production of triglyceride in the fruit carpels. No work has been done on the biochemistry of palm oil biosynthesis and its tissue specific regulation. This would be essential groundwork to be done before any genetic manipulation can be contemplated.

In the shorter term, methods in molecular biology can prove of considerable value as analytical tools in more conventional oil palm breeding and in quality control of clonal propagation. DNA 'fingerprinting' methods are being developed for positive identification of clones and can be used in genetic analysis for the determination of the genetic diversity of potential parents. With time it will be possible to correlate specific DNA patterns with particular phenotypes and to build up genetic linkage maps, perhaps eventually to identify individual genes. These methods will greatly speed up evaluation of the progeny of specific crosses and simplify the selection of individuals for clonal propagation and eventually open the way to more direct genetic intervention.

We are also using cDNA probes prepared from mRNA of developing embryos at different stages of development to probe for stage specific gene expression in somatic embryos *in vitro*. This is an aid to providing culture conditions giving the right sequences of stimuli to ensure normal balanced development of the somatic embryos.

With further development of DNA fingerprinting methods it should be possible to develop quality control procedures that will enable us to identify cultures producing somaclonal variants, or other abnormal plants, and to have early warning of any genetic changes occurring in culture.

References
1) RHV Corley, 'Clonal planting material for the oil palm industry,' J. Perak Planters Assoc. for 1981, p.35 (1982).
2) RHV Corley. Palm Oil Composition and Oil Palm Breeding, The Planter, Kuala Lumpur, 55:467, (1979).
3) JJ Hardon, RHV Corely, and SC Ooi, Analysis of growth in oil palm. II. Estimation of genetic variances of growth parameters and yield of fruit bunches. Euphytica, 21:257 (1972)
4) JJ Hardon, RHV Corley and CH Lee, Breeding and Selection for Vegetative Propagation in the Oil Palm. in: 'Improvement of Vegetatively Propagated Plants,' Papers presented at Long Ashton Research Station Symposium, (1982). (in press).
5) RHV Corley, Potential Productivity of Tropical Perennial Crops., Expl. Agric., 19:217, (1983).
6) RHV Corley, JN Barrett, and LH Jones, Vegetative propagation of oil palm via tissue culture. in 'International Developments in Oil Palm,' DA Earp and W Newall, eds., Incorporated Society of Planters, Kuala Lumpur, (1977).
7) H Rabechault and JP Martin, Multiplication vegetative du Palmier a huile (Elaeis guineensis Jacq.) a l aide de cultures de tissus foliaires., C. R. Acad. Sci. Paris Ser. D. 283: 1735 (1976).
8) LH Jones, Propagation of clonal oil palms by tissue culture., Oil Palm News, 17:1 (1974).
9) BA Nwankwo and AD Krikorian, Morphogenetic Potential of Embryo-and Seedling-Derived Callus of Elaeis guineensis Jacq. var. pisifera Becc., Ann. Bot. 51:65, (1982).
10) LH Jones, Plant Cell Culture and Biochemistry: Studies for Improved Vegetable Oil Production, in 'Industrial Aspects of Biochemistry,' ed. B Spencer, Proceedings of FEBS Special meeting, Dublin, North Holland, Amsterdam, (1973).
11) E Turnham and DH Northcote, The incorporation of (1-14C) acetate into lipids during embryogenesis in oil palm tissue cultures., Phytochem. 23:35, (1984).

12) D Ferdinando, J Hulme and WA Hughes, Oil-Palm Embryogenesis : A Biochemical and Morphological Study in 'Experimental Manipulation of Ovule Tissues,' eds. GP Chapman, SH Mantell, and RW Daniels, Longman, New York, (1985).
13) C Lioret, in 'Oil Palm in Agriculture in the Eighties,' Incorporated Society of Planters, Kuala Lumpur, (1981).
14) K Paranjothy, A Review of Tissue Culture of Oil Palm and Other Palms, PORIM Occasional Paper No.3, Palm Oil Research Inst., Malaysia. (1982).
15) RHV Corley, CH Lee, IH Law and CY Wong, Abnormal flower development in oil palm clones. The Planter, Kuala Lumpur, 62:233 (1986).
16) RHV Corley, CY Wong, KC Wooi, and LH Jones, Early results from the first oil palm clone trials, in 'The Oil Palm in Agriculture in the Eighties,' Incorporated Society of Planters, Kuala Lumpur, (1981).
17) LH Jones, The Oil Palm and its clonal propagation by tissue culture, Biologist, 30 : 181, (1983).
18) LH Jones, Novel Palm Oils from Cloned Palms, JAOCS, 61:1717, (1984).
19) NT Arasu, Genetic Variation for Fatty Acid Composition in the Oil Palm. (Elaeis guineensis Jacq.), in Proceedings of Int. Soc. Oil Palm Breeders, PORIM, Malaysia, (1985).
20) J Meunier and JJ Hardon, Interspecific Hybrids Between Elaeis guineensis and Elaeis oleifera, in 'Oil Palm Research,' eds. RHV Corley, JJ Hardon and BJ Wood, Elsevier, Amsterdam, (1976).
21) A Bass and WA Hughes, Conditions for isolation and regeneration of viable protoplasts of oil palm (Elaeis guineensis), Plant Cell Reports, 3 : 169, (1984).

MUTANTS OF ARABIDOPSIS DEFICIENT IN FATTY ACID DESATURATION

C.R. Somerville[1], P. McCourt[1], L. Kunst[1] and J. Browse[2]

MSU-DOE Plant Research Laboratory[1]
Michigan State University
East Lansing, MI 48824

Plant Physiology Division, DSIR[2]
Palmerston North
New Zealand

INTRODUCTION

The isolation of a series of mutants with defects in lipid metabolism was originally motivated by the concept that, in order to study the functional significance of lipid unsaturation, it would be very useful to have a collection of otherwise isogenic mutant lines which differ only with respect to the activity of specific desaturases. We chose the small crucifer Arabidopsis thaliana (L.) as the experimental organism because it has a number of traits which render it particularly well-suited for physiological genetics (Estelle and Somerville 1986).

As a prelude to detailed genetic studies of lipid metabolism in Arabidopsis we have characterized the lipid composition and the pattern of lipid metabolism of the wild-type in some detail (Browse et al. 1986c). These studies established that Arabidopsis is a '16:3' plant in which approximately 38% of newly synthesized fatty acids enter the prokaryotic pathway of lipid biosynthesis (Roughan and Slack, 1982). Of the 62% which is exported as acyl-CoA species to enter the eukaryotic pathway, 56% (34% of the total) is ultimately reimported into the chloroplast. Thus, chloroplast lipids of Arabidopsis are about equally derived from the two pathways so that any mutation which affects only one pathway does not exert an absolute effect on membrane lipid composition.

MUTANT ISOLATION

Since there was no obvious way to identify mutants with altered lipid composition on the basis of gross phenotype, we screened for mutants by direct assay of fatty acid composition of leaf tissue. To facilitate this approach we devised a simple and rapid procedure for preparing fatty acid methyl esters from small samples of fresh leaf tissue (Browse et al., 1986a). Using this method it is possible to obtain quantitative information on the fatty acyl composition of the lipids from as little as 5 mg of leaf tissue within several hours. We then employed this method to measure the fatty acyl composition of total lipids from single leaves from randomly chosen plants in an ethylmethane sulfonate-mutagenized population of Arabidopsis. From among the first 2000 M2 plants examined in this way we identified 7 mutants with major changes in fatty acyl composition (Browse et al. 1985), and in subsequent searches we have identified several additional mutant lines. In all of these mutants the alteration in fatty acid composition has been found to be due to single, recessive nuclear mutations. The sites of the enzymatic lesions in four of these mutants (designated fadA, fadB, fadC, fadD) have been tentatively determined as illustrated in Figure 1.

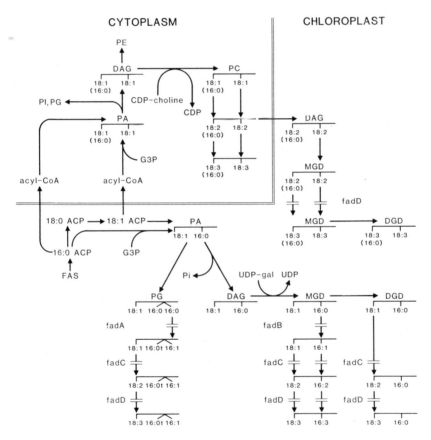

Figure 1. An abbreviated scheme for glycerolipid synthesis in Arabidopsis leaves showing the proposed locations of the enzymatic deficiencies in four mutants with altered fatty acid composition. The lesions are indicated by a break in the diagram adjacent to the gene symbol (ie., fadA, fadB, fadC and fadD).

THE fadA MUTATION

The first mutant to be characterized in detail was completely lacking the acyl group trans-hexadecenoic acid due to a mutation at a single nuclear locus designated fadA and (Browse et al. 1985). This acyl group normally occurs only on the sn-2 position of phosphatidyl glycerol (PG) in chloroplast membranes (Dubacq and Tremolieres, 1983). There were no other changes in the fatty acid composition of this mutant except for a corresponding increase in the amount of 16:0. Thus, it was inferred that the fadA locus encodes a desaturase which acts specifically on 16:0 at the sn-2 position of PG.

Since it is found in the chloroplast membranes of all higher plants, trans-16:1 has been thought to play an important role in some aspect of photosynthesis (reviewed in Dubacq and Tremolieres, 1983). The availability of a mutant line provided an excellent opportunity to test the various proposed roles. Therefore, we compared the mutant and the wild type by a number of criteria with the following results:

1. There was no apparent difference between the ultrastructural features of chloroplasts from the mutant and the wild type (Browse et al. 1985).

2. The photosynthetic characteristics of the mutant were not distinguishable from the wild-type when measured at either the whole-plant level by infrared gas analysis, or by photosynthetic electron transport assays with isolated chloroplast membranes (Browse et al. 1985).

3. The fluorescence induction kinetics and the low-temperature (77 K) fluorescence spectra of the mutant were indistinguishable from the wild type (McCourt et al., 1985).

4. On the basis of temperature-induced fluorescence enhancement yield measurements, the chlorophyll-protein complexes of the thylakoid membranes from the mutant were not more suceptible to thermal denaturation than those of the wild-type (McCourt et al. 1985).

The only functional difference that could be established between the mutant and the wild-type was that the dissociation of the putative oligomeric form of the chlorophyll a/b light harvesting complex (LHC-II) by SDS was slightly more sensitive to the NaCl concentration in the extraction buffer. This observation was consistent with previous suggestions that trans-16:1 participates in stabilizing the LHC-II oligomer (Remy et al., 1982; Tremolieres et al., 1981). It was also noted that the oligomeric form of the photosystem I (PSI) light harvesting complex (LHC-I) was also more suceptible to dissociation by SDS in membranes from the mutant.

The observation that the loss of trans-16:1 did not have a significant effect on chloroplast function provides a striking illustration of the principle that many components of organisms may not be absolutely required, but may serve subtle functions which, nevertheless, provide a selective advantage under some circumstances. Indeed, the fadA mutants do grow slightly more slowly than the wild-type. One possible explanation is that in the absence of trans-16:1 the insertion of LHC-II into the thylakoid membranes does not proceed as efficiently as in the wild-type so that chloroplast development is delayed. This possibility has been previously suggested on the basis of results from in vitro reconstitution experiments in which the uptake of LHC-II into artificial vesicles was stimulated by the presence of PG containing trans-16:1 (Remy et al. 1984).

THE fadD MUTATION

The distinguishing characteristic of a mutant with a lesion at the fadD locus is a substantial decrease in the amount of both 16:3 and 18:3 fatty acids in extracts of whole leaves and a corresponding increase in the amount of 16:2 and 18:2, respectively (Browse et al., 1986c). Thus, it was inferred that the fadD locus controls the activity of a desaturase which converts both 16- and 18-carbon dienoic to trienoic acyl groups. Analysis of the molecular species of MGD and DGD in the mutant and the wild-type indicated that the mutation specifically affects the desaturation of fatty acids in the prokaryotic pathway of lipid synthesis in the chloroplast and that both the sn-1 and sn-2 position of MGD was affected (Norman and St. John, 1986).

The effect of the fadD mutation was fully expressed only when plants were grown at temperatures above about 28°C. The simplest hypothesis to explain this is that the mutation renders the desaturase temperature sensitive. However, since there are other concievable explanations for this effect we are searching for additional mutant alleles of the fadD locus in order to determine if mutations at this locus always render the amount of trienoic acid responsive to temperature.

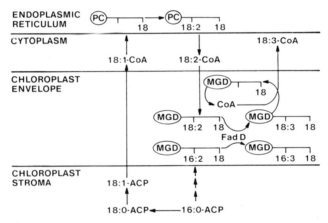

Figure 2. A scheme for control of 16:2 and 18:2 desaturation by the fadD gene product. We suggest that fadD controls a chloroplast desaturase which accepts both 16:2 and 18:2 as substrates. The specificity with respect to head group is not known but must include MGD since 16:3 is formed on MGD. In order to affect cytoplasmic lipid unsaturation either the acyl group must be exported (shown here) or the glycerolipid moiety must be transferred (not shown). In the latter case the transfer process would have to exclude MGD with 16-carbon acyl groups at the sn-2 position since such lipids are not found in cytoplasmic membranes (Roughan and Slack, 1982).

One of the unexpected and intriguing properties of the mutant is that the amount of 18:3 is reduced to a similar extent (about 45% reduction at temperatures above about 28°C) in all the major lipids (ie., MGD, DGD, SL, PG, PC, PE, PI). Since 16:3 is thought to be synthesized from 16:2 only in the chloroplast, the desaturase must be located within the chloroplast. However, there is no PE or PI in the chloroplast. Thus, it is apparent that either the same desaturase is located in both compartments, or that the 18:3 on PE and the other cytoplasmic lipids is synthesized in the chloroplast and transported to the cytoplasm. This could involve the transfer of 18:3 as either a CoA ester (Figure 2) or as a lipid. In view of the fact that transfer of acyl groups between compartments is already known to occur in plants (Dubacq et al., 1984; Kader et al., 1984), we favor this possibility. However, the recent observation that the same gene encodes both the cytoplasmic and mitochondrial histidine tRNA synthetase of yeast (Natsoulis et al., 1986) indicates that we cannot exclude this possibility.

The effects of the fadD mutation on photosynthesis have been examined in some detail. Our preliminary results indicate that the mutation has no apparent effect on the growth rate or the rate of photosynthesis (expressed on a chlorophyll basis) at any temperature in the range of $10°C$ to $30°C$ (McCourt et al., 1986). The mutation also has no effect on the fluorescence characteristics and has only a slight effect on membrane fluidity as measured by fluorescence polarization studies. Measurements of temperature-induced fluorescence yield enhancement also failed to reveal a difference between the mutant and the wild-type. Thus, we have been unable to detect major effects on the functional properties of chloroplast membranes associated with a major reduction in trienoic acids. This conclusion contrasts somewhat with the interpretation of other studies in which the amount of chloroplast membrane unsaturation was reduced by growth in chemical inhibitors (Laskay and Lehoczki, 1986; Leech et al., 1985) or by catalytic hydrogenation (Thomas et al., 1986). Therefore, we consider it possible that the effects of these experimental treatments may be due to non-specific effects.

OTHER MUTANTS

In addition to the fadA and fadD mutants we have isolated mutants (designated fadC in Fig. 1) which appear to be deficient in the chloroplast desaturase which converts 16:1 and 18:1 to 16:2 and 18:2, respectively. We have also isolated mutants (fadB) which accumulate high levels of 16:0 but are deficient in 16:1, 16:2 and 16:3 fatty acids. A third class of as yet uncharacterized mutants are deficient in 16:3 but do not have increased amounts of the more saturated 16-carbon fatty acids, and have slightly increased levels of all 18-carbon fatty acids. Thus, we consider it likely that this class of mutants is not deficient in a desaturase per se but affects the partitioning of fatty acids between the prokaryotic and eukaryotic pathways.

CONCLUSIONS

It appears that plant lipid metabolism may be amenable to experimental manipulation by the isolation of a wide spectrum of mutants. We anticipate that the mutants described here will provide useful tools for studies of the regulation and functional significance of lipid unsaturation. The anticipated development of more advanced methods of genetic analysis in Arabidopsis (Estelle and Somerville, 1986) may also facilitate the isolation of the genes which complement these mutations.

ACKNOWLEDGEMENTS

This work was supported in part by grants from the U.S. Department of Energy (#AC02-76ER01338), the McKnight Foundation and the US-NZ Agreement for Scientific and Technological Cooperation.

REFERENCES

Browse, J., McCourt, P., and Somerville, C.R., 1985, Science 227:763-765.
Browse, J., McCourt, P., and Somerville, C.R., 1986a, Anal. Biochem., 152:141-146.
Browse, J., McCourt, P., and Somerville, C.R., 1986b, Plant Physiol., 81:859-864.
Browse, J., Warwick, N., Somerville, C.R., and Slack, C.R., 1986, Biochem. J., 235:25-31.
Dubacq, J.P., Drapier, D., Tremolieres, A., and Kader, J.C., 1984, Plant Cell Physiol., 25:1197-1204.
Dubacq, J.P., and Tremolieres, A., 1983, Physiol. Veg., 2:293-312.
Estelle, M.A., and Somerville, C.R., 1986, Trends in Genet., 2:89-93.
Kader, J.C., Julienne, M., and Vergnolle, C., 1984, Eur. J. Biochem., 139:411-416.
Laskay, G., and Lehoczki, E., 1986, Biochim. Biophys. Acta 849:77-84.
Leech, R.M., Walton, C.A., and Baker, N.R., 1985, Planta 165:277-283.
McCourt, P., Browse, J., Watson, J., Arntzen, C.J., and Somerville, C.R., 1985, Plant Physiol., 78:853-858.
McCourt, P., Browse, J., and Somerville, C.R., 1986, Submitted.
Natsoulis, G., Hilger, F., and Fink, G.R., 1986, Cell 46:235-243.
Norman, H., and St. John, J.B., 1986, Plant Physiol., 81:731-736.
Roughan, P.G., and Slack, C.R., 1982, Ann. Rev. Plant. Physiol., 33:97-132.
Remy, R., Tremolieres, A., Ambard-Bretteville, F., 1984, Photobiochem. Photobiophys., 7:267-276.
Remy, R., Tremolieres, A., Duval, J.C., Ambard-Bretteville, F., and Dubacq, J.P.FEBS Lett., 137:271-275.
Thomas, P.G., Dominy, P.J., Vigh, L., Mansourian, A.R., Quinn, P.J., and Williams, W.P., 1986, Biochim. Biophys. Acta 849:131-140.
Tremolieres, A., Dubacq, J.P., Ambard-Bretteville, F., and Remy, R., 1981, FEBS Lett., 130:27-31.

ACYL CARRIER PROTEIN AS A PROBE OF THE MOLECULAR BIOLOGY OF PLANT FATTY ACID SYNTHESIS

J. B. Ohlrogge, P. D. Beremand, D. J. Hannapel, D. J. Guerra, D. E. Elmore, and D. N. Kuhn.*

Northern Regional Research Center, Agricultural Research Service, U.S. Department of Agriculture
1815 N. University St., Peoria, IL 61604; and
*Purdue University, West Lafayette, IN

Acyl carrier protein (ACP) is the best characterized protein in plant lipid metabolism. The stability and relative ease of purification of ACP have resulted in it being the first protein of plant fatty acid synthesis (FAS) to be purified to homogeneity (1), to have specific antiserum raised against it (2), and to have amino acid sequence data available (3-5). To date, ACP has been purified from spinach, avocado, castor bean, barley, rapeseed and soybean (6, and J. Ohlrogge, unpublished data.)

All forms of plant ACP that have been studied are common in their small size (9000 to 11,000 daltons) and in their acidic nature (pH = 4.0 - 4.2). Acyl carrier proteins have highly conserved structures based both on amino acid sequence homology and antibody cross-reactivity. In addition plant and bacterial acyl carrier proteins can be interchanged in several reactions of fatty acid synthesis and metabolism (1,7).

We are interested in the regulation of fatty acid biosynthesis and how the levels of proteins in this pathway are controlled. The central role of ACP in lipid metabolism and its well characterized structure led us to choose this protein as a representative marker of the fatty acid biosynthetic pathway. Our approach has been to examine the structure and expression of ACP in detail with the expectation that this will reveal some of the general control mechanisms which regulate plant fatty acid biosynthesis.

Subcellular localization of ACP

One of our first observations was on the subcellular distribution of ACP in spinach mesophyll cells. In 1979 we found that when protoplasts were gently lysed and their organelles seperated on sucrose gradients, essentially all of the ACP present could be attributed to the chloroplast fraction (2). This result revealed that plant mesophyll cells differ from animal and fungal cells by the absence of fatty acid synthesis in the cytoplasm. It is now known that plastids are also a major site of fatty acid synthesis in many non-green plant tissues although it has not been established whether other subcellular sites in these tissues also contribute to FAS.

The function of two or more forms of ACP may be related to their
participation in different reactions of plant lipid metabolism. As
discussed elsewhere in these proceedings, spinach leaf ACP-I and ACP-II
differ in activity in acyl transferase and thioesterase reactions (7).

Cloning of ACP:

The known amino acid sequences of spinach and barley ACP provide a
strategy for the isolation of cDNA and genomic clones that code for
this key protein. We are currently screening libraries with
oligonucleotide probes in an effort to obtain such clones. At least
two other laboratories have also taken this approach with reported
success. We have also constructed a synthetic gene which codes for
spinach ACP. This synthetic gene should prove valuable for a variety
of uses including expression of mature plant ACP in E. coli. Analysis
of ACP dependent reactions in plants has been hampered by a shortage of
plant ACP. Isolation of milligram quantities of this protein is
difficult but may be greatly aided if a synthetic ACP gene can be
effectively expressed in E. coli. In addition a synthetic gene may
prove useful for in vitro mutagenesis studies of ACP structure/function
relationships and as a probe of ACP nucleic acid sequences.

The design of the synthetic gene is shown below and is based on the
amino acid sequence of spinach ACP-I.

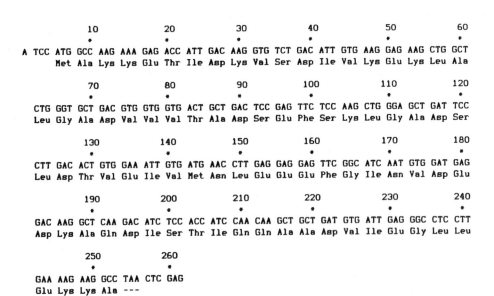

Figure 1. Sequence of spinach ACP-I synthetic gene. The DNA
sequence was designed based on the amino acid sequence of spinach
ACP-I (4) and codons were chosen using a plant codon usage table.

This synthetic gene has been constructed from 15 oligonucleotides coding for both strands and ranging in length from 11 to 44 bases. Our original attempts to assemble the entire gene in one construction led to a complete construct of the correct size (15) but we were unsuccessful in cloning this construct. We subsequently have succeeded in assembling the ACP gene by carrying out the assembly in two half gene segments, cloning these, and then combining the cloned halves. An outline of the construction of the first half of the ACP gene is shown in figure 2. Seven DNA fragments were annealed and ligated and the ligation products were analyzed by polyacrylamide gel electrophoresis. Figure 3 shows a densitometer scan of an autoradiogram of the ligation products. The desired 124 base construct is indicated by the arrow and is seen to represent only a minor proportion (1-2%) of the total ligation products. The other products evidently represent incompletely or incorrectly annealed and ligated assemblies. Nevertheless, when the entire ligation mixture is further ligated to Bam Hl cut vector, the cloning process selects for the correct construct. This is because incorrect assemblies are unlikely to be completely double stranded or contain two appropriate cohesive Bam Hl termini.

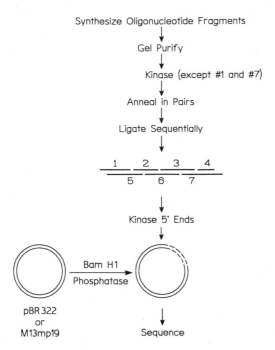

Figure 2. Scheme for construction of first half of ACP-I synthetic gene.

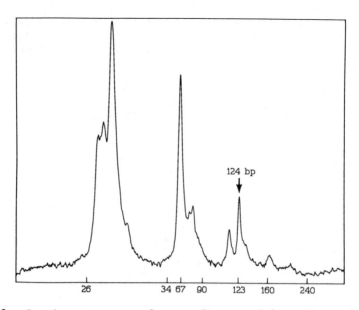

Figure 3. Densitometer scan of autoradiogram of ligation products of synthetic gene fragments. Seven oligonucleotides were ligated sequentially, kinased with ^{32}P and analyzed by electrophoresis in 8M urea, 12% polyacrylamide gels. The arrow indicates the ACP half gene construct of 124 base pairs. The X axis numbers represent the mobility of Msp fragments of pBR322 used as size markers.

After ligating the first half of the ACP gene into the phage vector M13mp19, E. coli cells were transformed and plaques with vector inserts were selected. Phage DNA from selected plaques was screened for ACP sequences by probing the DNA with a 32P-labelled oligonucleotide gene fragment. The positive clones were then sequenced by the dideoxy method. Sequencing is essential in the cloning of synthetic DNA in order to confirm not only that the correct fragments were ligated but also to assure that side reactions occurring during chemical DNA synthesis have not caused errors in the sequence. After identifying a completely correct first half clone, we constructed the second half in an analogous manner. We were able to join the two halves without the addition of any restriction sites in the middle of the ACP gene by taking advantage of the Hga 1 restriction enzyme. This enzyme cuts DNA five and ten bases away from its recognition sequence and thereby provided a method to generate cohesive termini in both halves which could be easily ligated together. This strategy is shown schematically in figure 4.

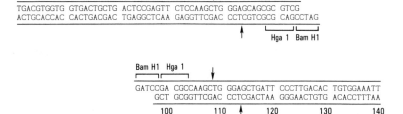

Figure 4. Hga 1 strategy for linking gene sections. Hga 1 cuts 5 and 10 bases away from its recognition sequence as indicated by the verticle arrows. The two half genes were constructed with the addition of linkers containing the Hga 1 site. After cloning the individual half genes, the sections were joined by mixing both clones, cutting with Hga and religating.

After assembly of the complete synthetic gene the junction of the two halves was also confirmed by sequencing. We then subcloned the ACP gene into the vector pTZ19 which contains a T7 RNA polymerase promoter. Using this vector we are able to synthesize RNA probes with high specific activity. Figure 5 presents preliminary data which demonstrate the ability of the ACP RNA probe to detect ACP messenger RNA in northern blots.

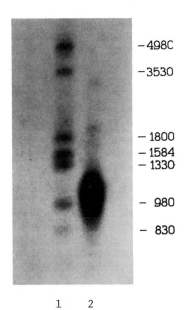

Figure 5. Northern blot of spinach leaf poly A^+ RNA probed with RNA transcribed from a synthetic spinach ACP-I gene. Two µg of RNA was separated on a 1.5% agarose gel including 8 mM methyl mercury hydroxide. Size standards (Lane 1) are ^{32}P labelled Hind III/Eco R1 fragments of lambda DNA. Hybridization was performed at $65°$ in 4.5 X SSC and 25% formamide.

Summary: By studying acyl carrier proteins in some detail, our lab and others have been able to gain several insights into the biochemistry and molecular biology of plant fatty acid synthesis. These "lessons from ACP" are summarized below. The recent isolation of synthetic and natural ACP genes should allow us to begin exploring the structure, organization and regulation of genes for plant lipid biosynthetic proteins.

LESSONS FROM ACP

1) Plant fatty acid synthetase is similar to the bacterial system and different from animals and fungi.

2) Essentially all fatty acid synthesis in spinach leaf mesophyll cells occurs in the chloroplasts.

3) During soybean seed development, turning on lipid synthesis is closely correlated with de novo synthesis of ACP.

4) Plants have at least two forms of ACP that are expressed differently in leaves and seeds.

5) Isoforms of ACP have different activity in acyl transferase and thioesterase reactions.

6) ACP is synthesized in vitro from mRNA as a larger precursor.

REFERENCES

1. Simoni, R. D., Criddle, R. S., and Stumpf, P. K. (1967). J. Biol. Chem. 242, 573-581

2. Ohlrogge, J. B., Kuhn, D.N., and Stumpf, P.K. (1979). Proc. Natl. Acad. Sci., USA, 76(3), 1194-1198

3. Matsamura, S., and Stumpf, P. K. (1968). Arch. Biochem. Biophys. 125, 932-941

4. Kuo, T. M., and Ohlrogge, J. B. (1984). Arch. Biochem. Biophys. 234, 290-296

5. Hoj, P. B., and Svendsen, I. (1983). Carlsberg Res. Commun. 48(4), 285-306

6. Ohlrogge, J. B., The Biochemistry of Plant Acyl Carrier Proteins. In: The Biochemistry of Plants: A Comprehensive Treatise, P. K. Stumpf, editor, In press

7. Guerra, D. J., Ohlrogge, J. B., and Frentzen, M. (1986). Plant Physiol., in press

8. Ohlrogge, J. B., and Kuo, T. M. (1984a). Plant Physiol. (Bethesda) 74(3), 622-625

9. Dorne, A. J., Corde, J. P., Joyard, J., Borner, T., and Douce, R. (1982). Plant Physiol. 69, 1467-1470.

10. Cashmore, A., et al. (1985). Biotechnology 3,803-808

11. Ohlrogge, J. B., and Kuo, T. M. (1984). Spinach Acyl Carrier Proteins: Primary Structure, mRNA Translation and Immunoelectrophoretic Analysis. In: Structure, Function, and Metabolism of Plant Lipids. P. A. Siegenthaler and W. Eichenberger, eds. Elsevier Science Publishers, pp. 63-67

12. Elhussein, S., Miernyk, J. A., and Ohlrogge, J. B. (1986). This Proceedings.

13. Hoj, P. B., and Svendsen, I. (1983). Carlsberg Res. Commun. 48(4),285-306

14. Ohlrogge, J. B., and Kuo, T. M. (1985). J. Biol. Chem. 260,8032-8037

15. Ohlrogge, J. B., Beremand, P. D., Kuhn, D. N., and Parker, P. E. (1986). Biochemical Society Trans. 14,579-581

OIL SEED RAPE ACYL CARRIER PROTEIN (ACP) PROTEIN AND GENE STRUCTURE

A.R. Slabas, J. Harding, P. Roberts, A Hellyer,
C. Sidebottom, C.G. Smith, R. Safford, J. de Silva,
C. Lucas, J. Windust, C.M. James and S.G. Hughes

Biosciences Division, Unilever Research, Colworth House
Sharnbrook, Bedford MK44 1LQ, U.K.

INTRODUCTION

Acyl carrier protein (ACP) plays a central rôle in lipid metabolism, serving as both a component of plant fatty acid synthetase (1) and as a substrate/cofactor for complex lipid biosynthesis (2). The protein has been purified from a number of plant sources and its amino acid sequence determined for the protein from both barley leaf (3) and spinach leaf (4) material. Both of these two previously mentioned sources of ACP have two detectable forms of the protein (5-6) whilst in seed material only one form has been detected (5). ACP has been shown, using immunological techniques, to be a developmentally regulated protein in maturing soy bean seeds. The activity of the protein appearing just prior to lipid accumulation (7). Despite the importance of this protein in lipid metabolism and the fact that seeds are a major site of lipid synthesis there is no reported literature on the characterization of ACP from seed material. The present study was aimed at a detailed characterization of ACP from rape (Brassica napus) seed:

1. Characterization of ACP Activity during Maturation of Rape Seed

   The ACP activity in maturing seeds of oil seed rape was determined using the malonyl-CoA exchange reaction (8). (Fig. 1.) It can be seen that the activity rises prior to the rise in in vivo fatty acid synthesis has ceased. This is similar to the situation reported for soy bean (7).

2. Isolation and Amino Acid Sequencing of ACP from Seed Material

   ACP from rape seed exhibits biological instability and we were unable to purify it using conventional methodology. The protein was purified by selective quantitative introduction of [^{3}H] palmitic acid into the pantetheine group using E. coli acyl ACP synthetase. The protein was then further purified to homogeneity taking advantage of the newly introduced hydrophobic domain. Fig. 2 shows an SDS gel of the purified product together with a fluorogram of the material. The slight contamination seen at higher molecular weight is probably deacylated acyl-ACP since no radioactivity is seen in this region of

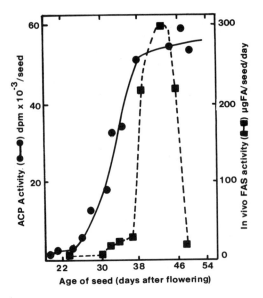

Figure 1. ACP Activity During Seed Development.

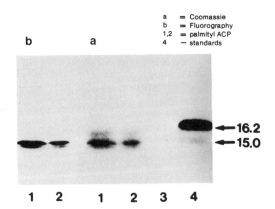

Figure 2. SDS Page and Purified Palmityl ACP.

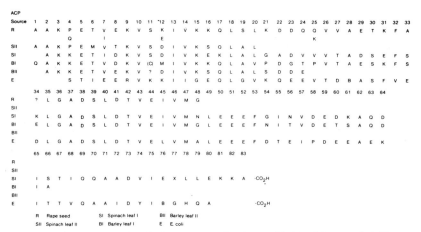

Figure 3. Amino Acid Sequence of First 48 Residues of ACP.

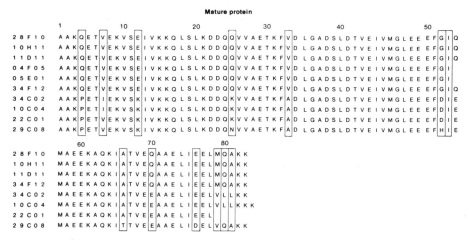

Figure 4. Amino Acid Sequences of ACP's Deduced from Gene Cloning.

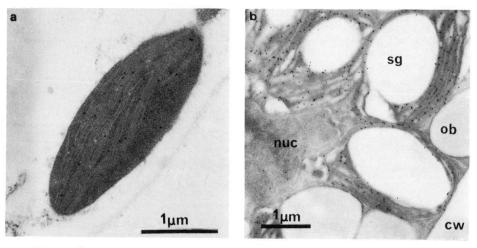

Figure 5. Immunological Localization of ACP in Chloroplasts (a) and Embryos (b).

the gel. The amino acid sequence of the first 48 residues is shown in Fig. 3. There is notably a high degree of homology with other ACP isolated so the reason for the lack of biological activity remains obscure.

3. Cloning of Rape Seed ACP Gene and Nucleotide Sequencing

The ACP gene was cloned from polyA mRNA which has been isolated from embryos 18-25 days post anthesis. The mRNA was size enriched for ACP prior to cDNA cloning. In vitro translational products precipitated by rabbit anti-spinach ACP antibodies were used to check for ACP messenger. The cDNA library constructed from such mRNA was probed with two heterologous oligonucleotide probes to ACP, (corresponding to LEEEF 64mer; VEIVM 96mer) under stringent hybridization conditions. The positive clones were sequenced. The gene transcript has a 51 amino acid N terminal extension which is not present in the mature protein. There are several variants in the sequence, the coding region of which is shown in Fig. 4.

4. Immunogold Localization of ACP in Leaf and Embryo Material

Using antibodies against spinach ACP we have localized ACP at the electron microscope level in both the chloroplasts of leaf (Fig. 5a) and proplastids of embryos (Fig. 5b) in both cases over 80% of the label located on membranes.

ACKNOWLEDGEMENTS:- We thank Dr. J. Ohlrogge for samples of rabbit anti-spinach ACP antibodies used in preliminary experiments.

REFERENCES

1. Stumpf, P.K. (1984) in "Fatty acid metabolism and its regulation". (Numa, S. Ed.) pp 155-179. Elsevier Sci. Pub. Amsterdam.
2. Frentzen, M., Heinz, E., McKeon, J. and Stumpf, P.K. (1983). Eur. J. Biochem. 129; 629-636.
3. Hoj, P.B. and Svendsen, I. (1983). Carl. Res. Commun. 48; 285-305.
4. Kuo, T.M. and Ohlrogge, J.B. (1984). Arch. Biochem. Biophys. 234; 290-296.
5. Hoj, P.B. and Svendsen, I. (1984). Carl. Res. Commun. 49; 483-492.
6. Ohlrogge, J.B., and Kuo, T.M. (1984). J. Biol. Chem. 260; 8032-8037.
7. Ohlrogge, J.B., and Kuo, T.M. (1984). Plant Physiol. 74; 622-625.
8. Majerus, P.W., Alberts, A.W. and Vagelos, P.R. (1969) in "Methods in Enzymology" XIV, 3450, (Lowerstein J.M., Ed.). Academic Press, New York and London.

COMPLETE AMINO ACID SEQUENCE OF NON-SPECIFIC LIPID TRANSFER PROTEIN FROM CASTOR BEAN SEEDS

Mitsuhiro Yamada[1], Shinichiro Watanabe[1], Kunio Takishima[2] and Gunji Mamiya[2]
[1]Department of Biology, The University of Tokyo, Tokyo 153 and
[2]Department of Biochemistry, National Defense Medical College
Saitama 359, Japan

INTRODUCTION

A lipid transfer protein (LTP) was purified from germinated castor bean seeds to homogeneity by column chromatographies. This protein, possessing about 9,000 of the molecular weight and higher than 10.5 of the isoelectric point, transferred PC, PI, PE, PG, PA, MGDG and DGDG, but not DG and TG, and the transfer activity remained unchanged after 30 min-heating at 90°C. This protein is one of non-specific LTPs as isolated from maize seedlings[1] and spinach leaves[2]. This paper presents the determination of complete amino acid sequence of castor bean non-specific LTP. Based on the amino acid sequence, structural aspect is discussed in relation to the interaction of the protein with the lipid molecule.

MATERIALS AND METHODS

The details of the procedures employed in these investigations will be found in references 3 and 4.

RESULTES AND DISCUSSION

The complete amino acid sequence of castor bean non-specific LTP are shown in Fig. 1. Sequence heterogeneity in the protein was found at the residue 42 (Ser) and 50 (Glu), being replaced by Thr and Asp, respectively. This fact suggests the occurrence of two genes for the LTP or the allelic variation of the same gene. The molecular weight of the LTP was calculated as 9,313 from the amino acid composition and the difference between (4 Asp + 1 Glu) and (6 Lys + 3 Arg) accounts for higher than 10.5 of the isoelectric point.

The sequence of the castor bean LTP can be divided into two regions from the sequence. Twelve of 14 acidic and basic amino acids were located on the latter half of the residue 44 to 92, whereas there were 2 ionic amino acids in the former region of the residue 1 to 43. The latter region is expected to interact with the polar head group of the lipid and the former region with acyl chains of the lipid.

In LTPs from plant sources[1,2], the part (the residue 1 to 30) of the amino acid sequence has so far been determined in non-specific LTP from

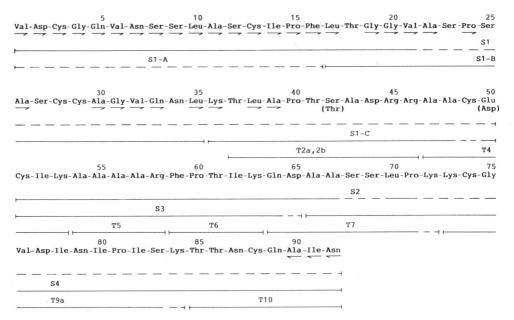

Fig. 1. The amino acid sequence of the non-specific LTP from germinated castor bean seeds. The solid lines indicate the sequenced part, whereas the broken lines indicate those part of the fragments in which the amino acid residues were not identified. S, Staphylococcus aureus protease peptide; S1-B and S1-C, chymotryptic peptides of S1 peptide; T, tryptic peptide. ⟶ Denotes sequence determined by Edman degradation of the intact protein and ⟵ denotes sequence determined by carboxypeptidase Y.

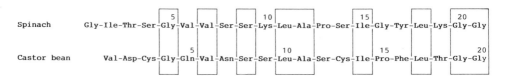

Fig. 2. Alignment of castor bean LTP and spinach LTP[5]. Identical residues are enclosed within boxes.

spinach[5]. Fig. 2 shows the comparison of this with the sequence of castor bean LTP. As enclosed within boxes, there was 45% homology between both LTPs, suggesting an importance of amino-terminal region in the sequence for lipid-protein interaction.

Fig. 3 shows predicted conformational profiles of PC-specific LTP from bovine liver[6] and non-specific LTP from germinated castor bean seeds according to Chou and Fasman[7]. The repeated pattern of β-structure in the latter region of the amino acid sequence is characteristic of bovine liver LTP, whereas there is no repeated β-structure in the latter region of castor bean LTP. This suggests difference in the interaction of the protein with the lipid between specific and non-specific LTPs, and therefore, difference in transport mechanism between both types of LTPs.

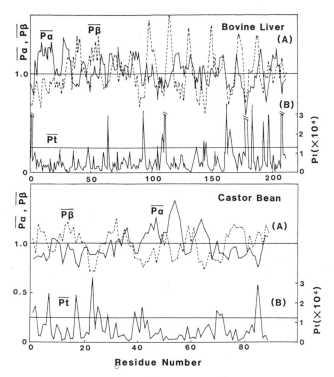

Fig. 3. Predicted conformational profile of the PC-specific LTP from bovine liver[6] and the non-specific LTP from germinated castor bean seeds[4] according to Chou and Fasman[7]. (A) The average helical potential ($\overline{P_\alpha}$) and the average β-sheet potential ($\overline{P_\beta}$) of tetrapeptides i to $i+3$ are shown. (B) Probability of β-turn occurrence (Pt).

REFERENCES

1. D. Douady, M. Grosbois, F. Guerbette and J.C. Kader, Biochim. Biophys. Acta 710:143 (1982)
2. J.C. Kader, M. Julienne and C. Vergnolle, Eur. J. Biochem. 139:411 (1984)
3. S. Watanabe and M. Yamada, Biochim. Biophys. Acta 876:116 (1986)
4. K. Takishima, S. Watanabe, M. Yamada and G. Mamiya, Biochim. Biophys. Acta 870:248 (1986)
5. P. Bouillon, C. Drischel, C. Vergnolle, J.C. Kader and H. Duranton, C.R. Acad. Sci. Paris, t. 300, Serie III, No 10, Physiol. Veg. 421 (1985)
6. R. Akeroyd, P. Moonen, J. Westerman, W. C. Puyk and K. W. A. Wirtz, Eur. J. Biochem. 114:385 (1981)
7. P. Y. Chou and G. D. Fasman, Adv. Enzymol. 47:45 (1978)

A POSSIBLE DIFFERENTIAL ROLE FOR PLANT ACYL

CARRIER PROTEIN ISOFORMS IN HIGHER PLANTS

D. J. Guerra, J. B. Ohlrogge and M. Frentzen

Northern Regional Research Center
Agricultural Research Service
U.S. Department of Agriculture
Peoria, IL 61604; and
Institut fur Allgemeine Botanik und Botanischer
Garten, Universitat Hamburg, West Germany

INTRODUCTION

Acyl carrier protein (ACP) dependent fatty acid synthesis (FAS) in plant leaf tissues is exclusively localized in the chloroplast (1). Two forms of ACP have been purified from spinach and barley leaf, but, only one ACP isoform predominates in spinach seed tissue (2, 3). Oleic acid is the major product of fatty acid synthesis by isolated chloroplasts (4). The regulation of plant lipid metabolism is believed to involve the export of oleic acid from its site of synthesis (plastid) to sites of complex lipid metabolism (eg. ER). This pathway requires the initial release of oleic acid from ACP by oleoyl-ACP thioesterase (5). In addition, the oleoyl moiety may also enter glycerolipid synthesis within the plastid through the action of acyl-ACP acyl-transferase (6). The preferred substrate for both the thioesterase and glycerol-3-phosphate acyl transferase reactions is 18:1-ACP (6,7).

The occurrence of two forms of ACP raises the question of their metabolic function. Because 18:1-ACP is the substrate for two competing reactions, ACP isoforms might facilitate acyl chain distribution within the cell. We have analyzed the reactivity and kinetics of oleoyl-ACP thioesterase and oleoyl-ACP glycerol-3-phosphate acyl transferase with oleoyl-ACP-I and oleoyl-ACP-II. Our results suggest that ACP isoforms may partially regulate the flow of oleic acid from its site of de novo synthesis in higher plants.

MATERIALS AND METHODS

Spinach (Spinacia oleracea) plants were grown as described (8). The enzymatic synthesis of [^{14}C]18:1-ACP isoforms was according to a published procedure (9), and was purified through the octyl sepharose step. Oleoyl-ACP thioesterase was partially purified from spinach tissue by methods which involved acid precipitation and either ion-exchange or affinity chromatography (7,8). Glycerol-3-phosphate acyl transferase was purified by ammonium sulfate fractionation, DEAE and

Sephadex chromatography (6). Plant acyl carrier proteins were purified as described (10), except that the heat step was omitted. Assays for the thioesterase and the acyl transferase were as described (8).

RESULTS AND DISCUSSION

Figure 1B shows the effect of [^{14}C]oleoyl-ACP isoforms on the reaction velocity of spinach oleoyl-ACP thioesterase. Oleoyl-ACP-I was a more efficient substrate for the reaction at all concentrations tested while oleoyl-ACP-II was a relatively poor substrate. The Km for oleoyl-ACP-I was 4.8 µM while that for oleoyl-ACP-II was 50 µM. For comparison, the Km for E. coli oleoyl-ACP was 1.4 µM.

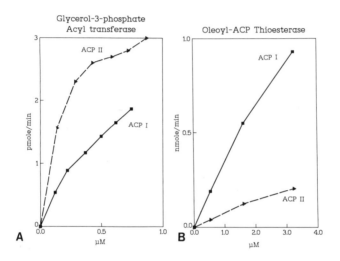

Figure 1A. sn-Glycerol-3-Phosphate Acyl Transferase Activity vs. Oleoyl-ACP Concentration. Reaction mixtures contained oleoyl-ACP (0-1.0 µM) from either spinach ACP I, or spinach ACP II, 0.6 mM glycerol-3-phosphate, acyltransferase from spinach leaves, 13.5 mM Mes, and 0.2% BSA in 0.25 mM Mops (pH 7.4). Values are averages of 3 determinations.

Figure 1B. Oleoyl-ACP Thioesterase Activity vs. Oleoyl-ACP Concentration. Reaction mixture contained oleoyl-ACP (0-3.24 µM) from either spinach ACP I, or spinach ACP II, and an appropriate amount of spinach achene oleoyl-ACP thioesterase in 0.1 M Tris pH 8.0. Values are averages of 2 determinations.

The glycerol-3-phosphate acyl transferase reaction preferred oleoyl-ACP-II over oleoyl-ACP-I at all substrate concentrations tested (Figure 1A). This is in direct contrast to the competing thioesterase reaction shown in Figure 1B. The Km for oleoyl-ACP-I by the acyl transferase was 0.75 µM while that for oleoyl-ACP-II was 0.16 µM. Based on Km values, the acyl transferase was 5X more reactive with ACP-II as cosubstrate. For comparison, the Km for E. coli oleoyl-ACP was 0.25 µM.

Our results lead us to suggest a possible model for the regulation of plant lipid metabolism in higher plants (Figure 2).

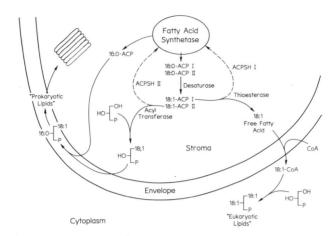

Figure 2. A proposed model for the acyl carrier protein directed apportionment of oleic acid between the eucaryotic and prokaryotic pathways of fatty acid metabolism.

Depending on the ACP isoform esterified to oleic acid within the plastid, the oleoyl moiety may enter into either plastid glycerolipid biosynthesis or may leave the plastid and enter lipid metabolism at sites such as the ER. The acyl transferase selectively transfers the oleoyl moiety from ACP-II and directs the synthesis of 1-oleoyl-lysophosphatidic acid. The thioesterase preferentially hydrolyzes the oleoyl moiety from ACP-I and directs free oleic acid to the outer envelope where it is subsequently esterified to CoA. The oleoyl-CoA thus formed is utilized outside the plastid for assembly of eukaryote type glycerolipids at the ER. The lysophosphatidic acid remains within the plastid as substrate for prokaryote-type glycerolipid synthesis. Our data suggest that the distribution of acyl chains between the prokaryotic and eukaryotic pathways may be controlled in part by the expression or acylation of ACP isoforms.

REFERENCES

1. J. B. Ohlrogge, D. N. Kuhn, and P. K. Stumpf, Subcellular localization of acyl carrier protein in leaf protoplasts of spinacia oleracea, Proc. Natl. Acad. Sci. USA 76:1194-1198 (1979).
2. P. B. Hoj, and I. Svendsen, Barley chloroplasts contain two acyl carrier proteins coded for by different genes, Carlsberg Res. Commun. 49:483-492 (1984).
3. J. B. Ohlrogge, and T. M. Kuo, Plants have isoforms for acyl carrier protein that are expressed differently in different tissues, J. Biol. Chem. 260:8032-8037 (1985).
4. P. K. Stumpf, D. N. Kuhn, D. J. Murphy, M. R. Pollard, T. McKeon, and J. MacCarthy, Oleic acid, the central substrate, in: "Biogenesis and Function of Plant Lipids" Mazliak P, Benviste P, Costes C, and Douce R, eds., pp 3-10, Elsevier/North-Holland (1980).

5. P. K. Stumpf, In Numa S, ed., Fatty Acid Metabolism and Its Regulation, Elsevier Science Publ., Amsterdam, pp 155-179 (1984).
6. M. Frentzen, E. Heinz, T. A. McKeon, and P. K. Stumpf, Specificities and selectivities of glycerol-3-phosphate acyl transferase and monoacylglycerol-3-phosphate acyl transferase from pea and spinach chloroplasts, Eur. J. Biochem. 129:629-636 (1983).
7. T. A. McKeon, and P. K. Stumpf, Purification and characterization of the stearoyl-acyl carrier protein desaturase and the acyl-acyl carrier protein thioesterase from maturing seeds of safflower, J. Biol. Chem. 257:12141-12147 (1982).
8. D. J. Guerra, J. B. Ohlrogge, and M. Frentzen, Activity of acyl carrier protein isoforms in reactions of plant fatty acid metabolism, Plant Physiol., In press (1986).
9. C. O. Rock, and J. L. Garwin, Preparative enzymatic synthesis and hydrophobic chromatography of acyl-acyl carrier protein, J. Biol. Chem. 254:7123-7128 (1979).
10. T. M. Kuo, and J. B. Ohlrogge, The primary structure of spinach acyl carrier protein, Arch. Biochem. Biophys. 234:290-296 (1984).

A PRELIMINARY CHARACTERIZATION OF PLANT HOLO-ACYL CARRIER PROTEIN SYNTHASE

Salah A. Elhussein,* Jan A. Miernyk and John B. Ohlrogge

Northern Regional Research Center, Agricultural Research Service, U.S. Department of Agriculture
1815 N. University Street, Peoria, IL 61604
*Agricultural Research Corporation, P.O. Box 126, Wad Medani, Sudan

INTRODUCTION

Acyl Carrier Protein (ACP) is a central cofactor in the biosynthesis of fatty acids and complex acyl lipids (1). ACP and CoA have a common prosthetic group, 4'-phosphopantetheine, to which fatty acids and intermediates are attached as thioesters. Using an E. coli auxotroph, Alberts and Vagelos (2) interpreted their in vivo experimental findings as suggesting that during de novo synthesis of ACP the prosthetic group is donated by CoA. The enzyme holo-ACP synthase (EC 2.7.8.7) was isolated from E. coli (3, 4) and shown to catalyze the transfer of 4'-phosphopantetheine from CoA to apo-ACP as follows:

$$CoA + apo\text{-}ACP \xrightarrow{Mg^{++}} holo\text{-}ACP + 3',5'\text{-}ADP$$

Herein we report the use of E. coli apo-ACP as a substrate to characterize the activity of holo-ACP synthase from higher plants.

MATERIALS AND METHODS

Holo-ACP synthase was prepared by homogenizing glasshouse-grown spinach (Spinacia oleracea L.) leaves or developing castor oil seed (Ricinus communis L., cv. Baker 296) endosperm with 0.1M Tris-HCl, pH 8.0. After centrifugation (27,000 g, 20 min.), supernatant proteins were precipitated with $(NH_4)_2SO_4$ to 60% of saturation, then redissolved and used to assay enzyme activity.

Holo-ACP was isolated from E. coli by the method of Rock and Cronan (5) and apo-ACP was prepared using 50% HF according to Prescott et al (4). Greater than 95% conversion of holo- to apo-ACP was achieved in 48 h.

Assay of holo-ACP synthase employed a 2-stage discontinuous reaction system. The first reaction contained the following components at the indicated final concentrations; holo-ACP synthase (ca. 200 μg protein), apo-ACP (6 μM), CoA (0.5 mM), $MgCl_2$ (10 mM) and DTT (2 mM) in a total volume of 60 μl. After incubation at 37° for 15 min (unless otherwise indicated), aliquots of 15 μl were taken and the holo-ACP formed was acylated with [^{3}H]-palmitate in the presence of E. coli acyl-ACP synthetase, ATP and other cofactors (in a total vol. of 50 µl), as described by Kuo and Ohlrogge (6). Aliquots (25 µl) from the second stage reaction were applied to DE-81 paper, washed, heated with 0.2 M NaOH and the radioactivity measured using a Beckman LS-9800 liquid scintillation spectrometer. The results in dpm were converted to moles of holo-ACP using standard amounts of ACP measured gravimetrically and enzymatically (2nd stage reaction). It was necessary to heat treat (70°, 10 min) the reaction mixture after the first stage in order to completely eliminate a very high background activity likely due to ATP-dependent, enzymatic acylation of CoA that occurs in the second reaction stage. This background activity, first thought to be due to the presence of holo-ACP in the apo-ACP preparation, was not reduced when NEM-alkylated apo-ACP was used nor when apo-ACP or acyl-ACP synthetase were completely omitted. Furthermore, HPLC/liquid scintillation analyses of a minus apo-ACP control reaction verified the absence of [^{3}H]-acyl-ACP.

Separations by HPLC were carried out using a Waters Associates Model 510 system with a Pharmacia Mono-Q anion-exchange column. Gradient elution (10 to 70% in 15 min, 1.5 ml/min, 8 min before run) was performed with 20 mM piperazine buffer, pH 6.0, containing 20% isopropanol (solvent A) and the same buffer containing 20% isopropanol and 1.0 mM NaCl (solvent B). Subcellular localization studies were accomplished using the methods previously described (7).

RESULTS AND DISCUSSION

Figure 1 shows reaction conditions for assay of holo-ACP synthase. In the first stage of the assay, holo-ACP is formed from apo-ACP and CoA by holo-ACP synthase. In the second stage, holo-ACP is quantitatively acylated with [^{3}H]-palmitate by E. coli acyl-ACP synthetase. Parallel tests of conditions which might affect acyl-ACP synthetase activity in the second stage were also performed. None of the conditions affected this second reaction. Relatively small aliquots of the first reaction (10-15 µl) were added to the second reaction mixture.

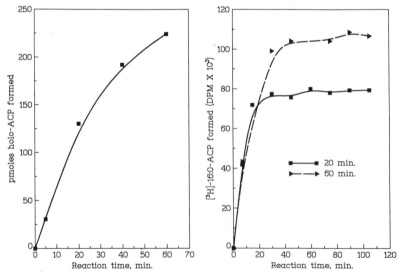

Fig. 1A. Holo-ACP formation by spinach holo-ACP synthase as a function of time. Spinach enzyme preparation (ca. 200 µg protein)

was incubated with apo-ACP (6 μM), CoA (0.5 mM), $MgCl_2$ (10 mM) and DTT (2 mM). The holo-ACP formed was acylated with [^3H]-16:0 in the presence of E. coli acyl-ACP synthetase and [^3H]-16:0-ACP then quantified by liquid scintillation spectrometry. Fig. 1B. Acylation of holo-ACP by E. coli acyl-ACP synthetase as a function of time. Holo-ACP synthesized in the first stage reaction by spinach holo-ACP synthase was transferred after 20 or 60 min to the second stage reaction. The [^3H]-16:0-ACP formed was assayed by liquid scintillation spectrometry.

Figure 2 shows that the reaction product, holo-ACP, after acylation with [^3H]-16:0 and E. coli acyl-ACP synthetase, had an HPLC mobility identical to authentic [^{14}C]-16:0-ACP. Similar results were obtained with the castor oil seed enzyme. Control incubations, without apo-ACP did not show significant amounts of [^3H]-16:0-ACP when similarly analyzed.

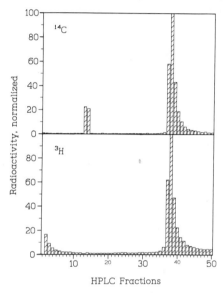

Fig. 2. HPLC identification of acyl-ACP by HPLC. The first and second stage reactions are described in the legend of Fig. 1. An aliquot from the second stage reaction was mixed with authentic [^{14}C]-16:0-ACP and analyzed by HPLC. Radioactivity in the HPLC fractions was measured by liquid scintillation spectrometry using a dual isotope counting mode.

Although the holo-ACP synthase reaction is linear with respect to time up to about 30 min (Fig. 1), initial rates were measured after 15 min. Incubation in this first reaction was followed by a 60 min incubation in the second reaction to transform all the holo-ACP to [^3H]-16:0-ACP.

The pH optimum for holo-ACP synthase was approx 8. The enzyme activity was absolutely dependent on apo-ACP and CoA. The Km values for spinach holo-ACP synthase were 1.82 ± 0.28 μM and 71.59 ± 9.90 μM for apo-ACP and CoA, respectively. Compared to the values obtained from E. coli (3), the spinach enzyme had an approx 5-fold higher Km value for apo-ACP but an approx 2-fold lower Km value for CoA. Increasing concentrations of $MgCl_2$ activated spinach holo-ACP synthase reaching a plateau around 10 mM. The Km value for $MgCl_2$ is 2.80±0.45 mM. The spinach enzyme was not activated by DTT. Spinach holo-ACP synthase is inhibited by the reaction product 3',5'-ADP with a Ki value of 0.52±0.12 mM.

In the purification protocol for holo-ACP synthase from E. coli (3), extracts were heated for 12 min at 54° without affecting enzyme activity. In contrast the spinach enzyme was extremely heat-labile. Only about 20% of the original activity remained after heating at 52° for 2.5 min.

In 1984, Ohlrogge and Kuo reported that spinach leaf ACP is synthesized in vitro as a larger precursor (8). These data suggested that ACP is nuclear encoded, synthesized in the cytosol, then translocated into the plastids for proteolytic processing. These results raised the question of when the pantetheine prosthetic group is attached during the processes of synthesis, transport and maturation. To address this question we have examined holo-ACP synthase activity in subcellular fractions from spinach leaves and developing castor oil seed endosperm. The data presented in Table 1 indicate that for both of these tissues holo-ACP synthase is localized in the cytosolic fraction. This suggests that the prosthetic group is added to preACP in the cytosol prior to uptake and proteolytic processing within the plastids.

TABLE 1. Subcellular localization of holo-ACP-synthase from spinach leaves and developing castor bean seeds.

Plant Tissue	Plastid	Subcellular fraction Mitochondria % of total activity in fractions	Microsomes	Cytosol
Spinach leaves				
Expt. 1	4.9	4.1	1.4	89.7
Expt. 2	0.03	0.6	0.0	99.4
Castor bean seed	6.3	14.9	1.3	77.5

Acknowledgement: S.A.E. acknowledges a fellowship award by the International Atomic Energy Agency.

REFERENCES

1. Ohlrogge, J. B. The Biochemistry of Plant Acyl Carrier Proteins, in "The Biochemistry of Plants: A Comprehensive Treatise." Vol. 10. P. K. Stumpf and E. E. Conn, eds. Academic Press. In press.

2. Alberts, A. W. and Vagelos, P. R. (1966). J. Biol. Chem. 241, 5201-5204.

3. Elovson, J. and Vagelos, P. R. (1968). J. Biol. Chem. 243, 3603-3611.

4. Prescott, D. J., Elovson, J. and Vagelos, P. R. (1969). J. Biol. Chem. 244, 4517-4521.

5. Rock, C. O. and Cronan, J. E., Jr. (1980). Analyt. Biochem. 102, 362-364.

6. Kuo, T. M. and Ohlrogge, J. B. (1984). Arch. Biochem. Biophys. 230, 110-116.

7. Miernyk, J. A. (1985). The Isolation and Characterization of Nongreen Plastids, in Modern Methods of Plant Analysis, New Series, Volume 1. Cell Components, HF Linskens and JF Jackson, eds., Springer-Verlag, Berlin, pp 259-295.

8. Ohlrogge, J. B. and Kuo, T. M. (1984). Spinach Acyl Carrier Proteins: Preliminary Structure, mRNA Translation and Immunoelectrophoretic Analysis, in "Structure, Function and Metabolism of Plant Lipids." P. A. Siegenthaler and W. Eichenberger, eds. Elsevier, pp. 63-67.

PHYSIOLOGICAL AND TRANSFORMATIONAL ANALYSES OF LIPOXYGENASES

D. F. Hildebrand, M. Altschuler, G. Bookjans, G. Benzion, T. R. Hamilton-Kemp, R. A Andersen, J. G. Rodriguez, J. C. Polacco*, M. L. Dahmer, A. G. Hunt, X. Wang and G. B. Collins

University of Kentucky   &   *University of Missouri
Lexington, KY 40546   Columbia, MO 65211

INTRODUCTION

Lipoxygenases (LOXs) represent a group of polyunsaturated fatty acid oxidases that are apparently ubiquitous in eukaryotic organisms. There are a growing number of reports of the presence of this enzyme in various human and other animal tissues. The list of plants containing active LOXs also continues to expand.

LOXs are of agricultural significance in so far as they apparently play an important role in the biogenesis of a number volatile flavor and aroma constituents that can affect food quality. Wheat leaves are particularly active in the generation of volatile compounds and were therefore investigated to determine the role of LOXs in the generation of these compounds.

Biochemical and genetic evidence indicates that there are three major LOXs in soybean seeds designated LOX 1, 2, and 3. Mutants have been found for all three LOXs. LOX 1 and 2 are tightly linked.[1] cDNAs have been cloned for the LOX mRNAs from developing soybean seeds.[2]

MATERIALS AND METHODS

LOX activity was determined by measuring the linoleate dependent production of conjugated dienes.[3] Western blot analysis was performed by electroblotting proteins separated on 10% LDS PAGE gels to a nylon membrane (Genescreen, NEN) and then incubating the membrane with rabbit monospecific LOX 1 antibodies followed by alkaline phosphatase linked goat anti-rabbit second antibodies and staining for alkaline phosphatase activity. Volatile compounds were determined as described by Andersen et al.[4]

The genomic DNAs were isolated from genomic DNA libraries in Charon 34 (soybean) and Charon 32 (tobacco). LOX cDNAs and genomic DNAs were cloned into a number of micro Ti plasmids particularly derivatives of pGA472[5] and then mobilized into Agrobacterium tumefaciens via standard triparental mating procedures.[6] Transformation of tobacco (Nicotiana tabacum) and soybeans (Glycine max) (and related species [G. canescens and G. clandestina]) was accomplished using various modifications of standard techniques[6] with the changes necessary for regeneration of transformed soybean plants[7].

RESULTS AND DISCUSSION

We find readily detectable LOX activity in virtually all parts of soybean and tobacco plants including seeds, pods, leaves, stems, roots and nodules. Moreover, these tissues all contain proteins that crossreact with monospecific LOX 1 antibodies (LOX 1 antibodies crossreact with LOX 2 and LOX 3 on immunoblots) on Western blots. A similar antigenic band is seen in extracts of soybean, red clover, alfalfa, tobacco and wheat leaves that crossreacts with antibodies to soybean seed LOX 1 (Fig. 1). That these crossreacting proteins are lipoxygenases is supported by the fact that monospecific antibodies were used and that they co-migrate with LOX active bands excised from non-denaturing PAGE gels and run along with the original extracts on Westerns blots (data not shown). The existence of LOXs in all or most all plant tissues might suggest a fundamental physiological role for LOXs, the possibility of which is made even more intriguing by the rapid expansion of literature indicating the importance of polyunsaturated fatty acid oxidases in many aspects of animal metabolism.[8,9]

Several inhibitors of LOX were also found to reduce the production of a number of volatile compounds in wheat leaf extracts. A particularly potent LOX inhibitor, acetonylacetone bisphenylhydrazone (AABPH), was also quite effective in reducing the production of a number of C6 volatiles including hexanal (a 93% reduction with 20 uM of AABPH), t-2-hexenal (89% reduction), 1-hexanol (80% reduction), t-2-hexen-1-ol (46% reduction) and c-3-hexen-1-ol (66% reduction). The addition of linoleate to the wheat leaf extracts markedly increased the production of saturated C6 volatiles; whereas, treatment of the wheat leaves with boiling water for 30 min. prior to extraction greatly reduced (>95%) the levels of nearly all volatiles generated (autoclaving soybean seeds prior to steam distillation of volatiles also greatly reduced the production of nearly all volatiles). These results indicate that the production of nearly all volatile compounds generated from wounding of whole plant tissues in addition to those generated via LOX action is enzymatically controlled.

Evidence is accumulating that the volatiles produced by LOXs in wounded plant tissues may play a role in induced pest resistance.[10] We are currently testing these compounds in terms of their induction of the transcripts and translation products in soybean leaves. The transcripts and translation products that are induced by products of LOX are being compared to those that are induced in soybean leaves by feeding of the two spotted spider mite.

These LOX cDNAs have been used to isolate genomic DNAs from soybean and tobacco genomic DNA libraries. The genomic DNA clones contain multiple introns and have exons with homologies to each other ranging from 50 to 90%. One 12.1 Kb soybean genomic DNA clone apparently contains all of LOX 1 and part of LOX 2. Transformed plantlets have been obtained from tobacco leaf disks and various G. clandestina tissues inoculated with Agrobacterium cells harboring micro-Ti's with LOX inserts. Transformation of G. max with LOX clones is in progress. Transformants with up and down regulated LOX activity, will be analyzed for alterations in pest resistance and volatile compound generation as well as for any other physiological changes in order to better understand the role(s) of polyunsaturated fatty acid oxidases in plant metabolism.

ACKNOWLEDGEMENTS

We thank Drs. Y. Ma and R. Goldberg for supplying the soybean and tobacco genomic DNA libraries, E. Nester for supplying many of the Ti plasmids and Agrobacterium strains and R. Yenofsky for soybean LOX 2 genomic DNA.

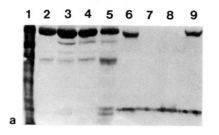
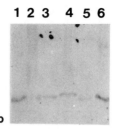

Figure 1. Western blot with monospecific LOX 1 antibodies (see Materials and Methods). a, lane 1: Sigma soybean LOX Type IV, 1 ug; lane 2: LOX 1 mutant soybean seed^1, 12 ug; lane 3: LOX 2 mutant seed, 12 ug; lane 4: LOX 3 mutant seed, 12 ug; lanes 5 & 6: wild type seeds; lane 6 soybean leaf extract, 25 ug; lane 7: tobacco leaf, 25 ug; lane 8: red clover leaf, 25 ug; lane 9: alfalfa leaf extract, 25 ug. b. lanes 1 & 2: wheat leaf extracts, 10 & 40 ug; lanes 3 & 4: tobacco leaf, 10 & 40 ug; lanes 5 & 6: soybean leaf, 10 & 40 ug.

REFERENCES

1. C. S. Davies and N. C. Nielsen, Genetic analysis of a nul-allele for lipoxygenase-2 in soybean, Crop Sci. 26:460 (1986).
2. W. G. Start, Y. Ma, J. C. Polacco, D. F. Hildebrand, G. A. Freyer, and M. Altschuler, Two soybean seed lipoxygenase nulls accumulate reduced levels of lipoxygenase transcripts, Plant Molec. Biol. (in press).
3. D. F. Hildebrand and M. Kito, Role of lipoxygenase in soybean seed protein quality, J. Agric. Food Chem. 32:815 (1984).
4. R. A. Andersen, T. R. Hamilton-Kemp, P. D. Fleming, and D. F. Hildebrand, Volatile compounds from vegetative tobacco and other plants obtained by steam distillation and headspace trapping, in "Biogenesis of aroma," T. H. Parliment and R. Croteau, eds., American Chemical Society Symposium Series, No. 317, ACS, Washington, D.C. (in press).
5. G. An, B. D. Watson, S. Stachel, M. P. Gordon, and E. W. Nester, New cloning vechicles for transformation of higher plants, EMBO J. 4:277 (1985).
6. C. Lichtenstein and J. Draper, Genetic engineering of plants, in "DNA cloning volume II," D. M. Glover, ed., IRL Press, Oxford (1985).
7. P. A. Lazzeri, D. F. Hildebrand, and G. B. Collins, A procedure for plant regeneration from immature cotyledon tissue of soybean, Plant Molec. Biol. Rep. 4:277 (1985).
8. E. L. Wheeler and D. Berry, In vitro inhibition of mouse epidermal cell lipoxygenase by flavonoids: structure-activity relationships, Carcinogenesis 7:33 (1986).
9. T. Saijo, H. Makino, S. Tamura, H. Kuriki, Y. Ashida, S. Terao, and Y. Maki, The antiallergic agent amoxanox supresses SRS-A generation by inhibiting lipoxygenase, Int. Archs. Allergy Appl. Immun. 79:231 (1986).
10. D. F. Hildebrand, J. G. Rodriguez, G. C. Brown, K. T. Luu and C. S. Volden, Peroxidative responses of leaves in two soybean genotypes injured by twospotted spider mites (Acari: Tetranychidae), J. Econ. Econ. Entomol. (in press).

INDEX

Abscisic acid, 205, 389
Absorption spectra, 65
Acetabularia mediterranea,
 lipids of, 637
Acetate content, different leaf tissues, 506
Acetyl CoA, 34
 carboxylase, 465, 517
 generation, for plastid fatty acid synthesis, 509
 synthesis, regulation, 505
 synthetase, 505, 510
  mature spinach leaves, 513
Acetyl CoA:acyl carrier protein transacylase, from Brassica campestris leaves, purification of, 499
Acholeplasma laidlawii, 146
Acid phosphatase, 371
Acyl-(acyl carrier protein): glycerol-3-phosphate acyltransferase
 from greening squash cotyledons, 541
 physicochemical properties from higher plants, 543
 purification, 542
 substrate selectivity, 543
Acyl-ACP, 253, 283, 284
Acylated steryl glucosides, 214
Acyl carrier protein (ACP), 248
Acyl CoA
 dehydrogenase, 400
 elongases, 483
 elongation systems, 525
 oxidase, 400, 401
 synthetase, 400, 403
1-Acyl-glycerol-3-phosphate acyltransferase, 603
Acyltransferase, 284
 in developing seeds of oilseed rape, 537
Adipic acid, 593
Alcohol dehydrogenase, 386
Alfalfa, 48
Algal phylogeny, 631

Alkaline phosphatase, 325
Alkan-2-OL esters, from epicuticular waxes, biosynthesis of, 553
Allium porrum, 221
 microsomes, 525
Amaranthus, 252
1-Aminocyclopropane-1-carboxylic acid (ACC), 387
Amyloplasts, 256
Anabaena variabilis, 299
 glycerolipid biosynthesis in, 604
Anthers, 345
Anthriscus, 249
Anti-endo conformations, 14
Antirrhinum majus, 197-199
Apium gravedens, 91
Arabidopsis, 252
Arabidopsis thaliana, 195
Arachidonic acid, 75, 76, 79
Arsenite, 492
Arsenoribosylglycerol, 600
Arsenoribosylphosphatidylglycerol, 600
ATPase, 194
Auxin, 229
25-Azasteroids, 103
Azelaic acid, 593

Bacillus cereus, 301
Bacteria, 146
Barley, 491
Bioactivities/gibberellins, 25
Biogenetic isoprene rule, 11
Bioreactors, 590
Bioregulators, 75
Biotechnology
 of lipids, in industry, 587
 research in, 568
Biotin-containing proteins, 517
Borago officinalis, 407
Borane, 17
Bradyrhizobium japonicum, 349
Branched chain fatty acids, 197
Brassica juncea, 341
Brassica napus, 205, 299, 341, 537
Brassylic acid, 593

Calcium, 77, 80, 111, 112
Calmodulin, 111, 112, 226, 230, 234, 321, 322
Camellia sinensis, 378, 391
Campesterol, 7, 15, 16, 92, 109, 206, 214
Camphor, 15, 16
Capsanthin, 138
Capsorubin, 138
Cardiolipin, 258
Carnitine acetyltransferase, 402
Carnitine acyltransferase, 402
β-Carotene biosynthesis, 34
Carotenoid, 37, 63, 64, 135
  biosynthesis, 40, 135
Castor bean, 329
Cerulenin, 449
Chervil, 252
Chilling, 252, 345
Chirality, 4
Chlamydomonas reinhardii, 298
Chlorella, 48
Chlorophyll, 63, 65, 135
Chlorophyta, sterol composition, 632
Chloroplast, 29, 34, 41, 250
  envelope, 293, 357
  in Dunaliella salina, 623
  membranes, 30, 31, 34
Chloropyrifos, 475
Cholesterol, 3-5, 7, 56, 59, 83-85, 91, 92, 95, 100, 109, 147, 214
Choline, 229
  kinase, 329, 330
CDP-Choline:diacylglycerol cholinephosphotransferase, 406
CTP-Cholinephosphatase cytidylyltransferase, 333
Cholinephosphate, 268, 325, 326, 330
CTP:Cholinephosphate
  cytidyltransferase, 325
  cytidylyltransferase, 266
Cholinephosphotransferase, 268, 270, 275, 326, 362
Chromoplast membranes, 41
Chrysophyta, sterol composition, 634
Cinnamomum comphora, 379
Clerodendrum splendens, 3
Cleyera japonica, 379
Cocoa butter, 337
Coenzyme A, 247
Cold-hardiness, 203
Compartmentation, 37
Critical mass, of sterol, 54
Cucurbita maxima, 7, 87
Cuticular lipids, 473
Cutin, 473

Cutinase, 474
  cDNA, 474
  gene
    cloning of, 476
    nucleotide sequence of, 477
    in phyllospheric bacteria, 478
Cyanobacteria, 249, 293
Cyanobacterial glycerolipids, 603
Cyanobacterium, 299
Cyanophyta, sterol composition, 634
Cycloartenol, 53, 56, 92, 98, 116
Cyclooxygenase, 389
Cyclotella cryptica, 649
Cylindrotheca fusiformis, 649
Cynoglossum officinalis, 410
Cytidylyltransferase, 268, 326, 333, 335
Cytochrome oxidase, 258
Cytosolic acetyl CoA, 108

Desaturation, 273, 303
Developmental change, 107
Developmental regulation, 53
Diacylglycerol, 258, 261, 275, 294, 406
1,2-Diacylglycerol, 248
Diacylglycerol acyltransferase, 337
Diatoms, lipid accumulation,
  in silicon-deficient, 649
Dibucaine, 321, 322
Digalactosyl diacylglycerol (DGDG), 293, 345, 603
Diglycerides, 229, 230
22-Dihydrobrassicasterol, 7
22,23-Dihydrobrassicasterol, 56
β-Diketone ester, from epicuticular waxes, biosynthesis of, 553
Diterpene, 16, 123
Docasonol, 108, 109
Dodecanoic acid, 593
Domain function, 59
Drought, 205

Early-13-hydroxylation pathway, 21, 22
Eelgrass (Zostera marina), 50
Eicosapentaenoic acid, 75, 76
Elongases, intracellular localization, 484
Elongation mechanism, 469
Embryoid protoplasts, 529
Enantiomeric cyclizations, 13
Endomembrane system, 266
Endoplasmic reticulum, 239, 257, 266
Enzyme segragation, 37
Epicuticular lipids, 492
  synthesis of, 491
Epicuticular wax formation
  in needles
    of Picea abies, 557
    of Pinus cembra, 557

Epicuticular wax synthesis, 489
25,25-Epiminolanosterol, 103
Ergosterol, 57, 58, 93, 147
Erucic acid, 341, 483
Esterase, 370, 371
Ethylene, 225, 227
  biosynthesis, 387
Etioplasts, 41, 42
Euglena, 297, 309
Euglena gracilis, 298
Euglenophyta, sterol composition, 634
Euphorbia lathyris, 111, 112, 115

Fatty acid, 140, 242
  biosynthesis, 34, 107
  branched chain, 197
  composition, 616
    in cyanobacterium
      isothermal growth, 606
      temperature shift, 607
    of Porphyridium cruentum, 641
  esters, 111, 133
  hydroperoxides, 256, 383, 386-388, 393
  synthesis, 510
    carbon supply, 456
    in Cuphea seed extracts, 460
    in developing oilseeds, 455
    in safflower seed extracts, 460
    medium chain in seeds, 459
  synthetase, 466
Fatty alcohols, 107
Floral initiation, 53, 83
Flowering, 53, 83, 107, 219
Fragrant compounds, 139
Free sterol, critical mass of, 59
Freezing, 279
Fungal gas, 21
Fungi, 5, 19, 75

Galactolipid
  metabolism in Dunaliella salina, 623
  synthesis, in Dunaliella salina, 626
Galactolipid:galactolipid galactosyl transferase, 257
Galactosylglycerols, 599
Gas chromatography, of gibberellins, 128
Gas chromatography-mass spectrometry, of gibberellins, 129
cer-cqu Gene, β-ketoacyl elongase, 490
Gene expression, 25
Geranyl pyrophosphate, 11-13, 15, 16
Gibberella fujikuroi, 19, 21, 103

Gibberellin (GAs), 19, 25, 124, 127
  bioactivities, 25
  biosynthesis, during development, 129
  gas chromatography of, 128
  high pressure liquid chromatography of, 128
  mutants, 23
  pathways, 21
ent-Gibberellins, 20
Ginko biloba, 5
Glomus fasciculatum, 349, 351
Glossy genes, in maize, 493
Glucans, 75-77
Glucocerebrosides, 214, 215
Glutathione, 388
Glutathione-S-transferases, 388
sn-Glycerol, 248
Glycerol phosphate acyltransferase, 253
sn-Glycerol-3-phosphate, 297
Glycerol-3-phosphate acyltransferase, 603
Glycerolipid, 131, 145
  biosynthesis, 63, 69, 71
  in Anabaena variabilis, 604
Glycine max, 229, 349
Glyoxylate cycle, 403
Glyoxysomes, 240
Golgi, 257, 266

Halosterol, 4, 5
Hexanal, 384, 388, 391, 397
cis-3-Hexanal, 384
Hexenol, 391
Hibiscus abelmoschus, 139
High pressure liquid chromatography, of gibberellins, 128
Hordeum vulgare, 83, 97
Horseradish peroxidase, 47, 49
Hydroperoxide, 396
  cyclase, 386
  isomerase, 386, 388
  lyase, 384, 386, 387, 391-395, 397
3-Hydroxyacyl-CoA dehydrogenase, 401, 403
Hydroxy-fatty acids, expression of cutinase gene, 477
Hypersensitive response (HR), 75, 77

Indol-3-yl acetic acid (IAA), 333
Indolebutyric acid, 388
Inositol phosphates, 229, 231
Inter-organizational collaboration, 563
Isocitrate lyase, 244, 403
Isopentenoid biosynthesis, 107
Isopentenyl pyrophosphate (ipp), 37
Isoprenoid synthesis, 34
Isothermal growth, fatty acid composition in cyanobacterium, 606

Japanese morning glory, 217
Jasmonic acid, 386, 388

ent-Kaurene, 20, 22-24, 123, 125
Ketols, 388
Kornberg-Pricer pathway, 261, 313-315

Lanosterol, 53, 115, 116
Linalyl pyrophosphate, 12, 14, 16
Linolenate content, of soybean cotyledons, 533
Linolenic acid, 383, 405
α-Linolenic acid, 345
γ-Linolenic acid, 393
Lipase, 242
Lipid
  biosynthesis
    in epidermal cell protoplasts, from leaves of Vicia faba, 545
    in guard cell protoplasts, from leaves of Vicia faba, 545
    in mesophyll cell protoplasts, from leaves of Vicia faba, 545
    in oil palm protoplasts, 529
  bodies, 241, 244, 246
  classes, in oil palm protoplasts, 530
  content of single cells, 645
  of Acetabularia mediterranea, 637
  of diatoms, 613
  of Halophilic dunaliella species, 613
    overall lipid composition, 618
  of Halotolerant species, overall lipids composition, 618
  phase, of membranes in cyano-cyanobacter, 608
Lipolytic acyl hydrolase, 255, 256, 369, 370, 391
Lipoxygenase, 77, 137, 383-389, 392, 393, 396, 397
Liquid-nitrogen homogenization, 49
Lolium temulentum, 83
Lowry-protein method, 46
Lubimin, 78-80
Luciferase, 330
Lyase, 396
Lysophosphatidic acid, 253
  acyltransferase, 538
  dependent acyltransferases, 538
Lysophosphatidyl choline-dependent acyltransferases, 538

Macrocyclic lactone
  biosynthesis, 142
  lactone musks, 139, 142
Main line sterols, 87

Maize, 48, 241, 244
  lipase, 243
  lipid bodies, 245
  surface lipids, chemical composition, 549
Malate synthase, 403
Malonyl Coenzyme A, decarboxylation, 458
Mamilactone A, 124, 125
Margaric acid, 349
Mass, of free sterol, 59
Mass spectrometry, of steryl esters, 96
Membranes, 3, 131, 145
  fluidity, 57, 85, 148
  models, 145
Mentha piperita (peppermint), 45
Methyl-branched acids, 198
Mevalonic acid (MVA), 20, 87, 109, 115, 141
Mevinolin, 37
Microsymbionts, 349, 351
Mirica rubra, 379
Monogalactosyl diacylglycerol (MGDG), 345, 603
  chemical structure of, 604
Monoglucosyl diacylglycerol, 603
  chemical structure of, 604
Monolayer films, 131, 145, 146
Monoterpenes, 11
Multiple functions, of sterols, 3, 148
Mung bean, 257
Mustard oils, 45

Narcissus pseudonarcissus, 297
Navicula pelliculosa, 649
Nicotiana tabacum, 197-199
Nile red, a fluorophore, 645
Nodulation, 351
cis-3-Nonenal, 386

Oil
  bodies, 239
  crop breeding, 588
Oleosomes, 239
Oleoyl acyl carrier protein, 253, 283, 285
Onion roots, 268
Ontogenetic changes, 7, 54, 107
  in sterols, 88
Oryza sativa, 345
Oryzalexin A, 124
3-Ooxoacyl-CoA thiolase, 401
β-Oxidation, 400
Oxysterols, 149

Paclobutrazol, 91
Palm Oil Research Institute of Malaysia, 571
Palmitoyl acyl carrier protein, 285

Palmitoyl CoA, 401
Pancreatic lipase, 343, 367
Parsley, 252
Patatin, 369
Pentacyclic triterpenoids, 53, 108, 111, 115
Peppermint (Mentha piperita), 45
Phaeodactylum tricornutum, fatty acids, 653
Phaeophyta, sterol composition, 632
Pharbitis nil, 217
Phosphatase, 330
3-Phosphate, 248
Phosphatidate, 248
Phosphatidate phosphatase, 252
Phosphatidylcholine, 145, 229, 274, 280, 325, 329, 333
Phosphatidylglycerol (PG), 603
Phosphatidylinositol, 207, 247
Phosphodiesterase, 325
Phospholipase
   A_2, 199, 226, 406
   C, 229, 258, 276
   D, 258
Phospholipid transfer proteins 353, 354
Phylloquinone (vitamin K_1), 29
Phytanic, 600
Phytoalexins, 75, 86, 123
Phytohormones, 19
Phytophthora cactorum, 103
Phytophthora infestans, 75, 76
Phytosterol distribution, 3, 120
Pinane, 17
Pinus pinea, 88
Pirodela oligorrhiza, 193
Pisum sativum, 88, 333
Plant
   elongases, 481
      solubilization of, 485
   enzyme isolation, 45
   phenolics, 46
Plasma membrane, 5, 75, 80, 108, 201, 207
Plastid glycolytic enzymes, 511
Plastome mutants, 197, 198
Plastoquinol biosynthesis, 32
Plastoquinone-9, 29, 34
Polar lipid composition, 613
Pollen, 345
Polyribosomes, 245, 246
Polyvinylpyrrolidone, 255
Porphyridium cruentum
   effect of light, 642
   fatty acid composition of, 641
   nitrogen effects, 642
   temperature effects, 642
Potato, 75, 77
Prenyllipid, 63, 69
   biosynthesis, 39, 70

Prenyllipid (continued)
   synthesis, 38
Prenylquinones, 29, 33, 63, 66
   distribution, 67
Protein kinase, 233, 234
Protozoa, 5
Pyrrophyta, sterol composition, 634
Pyruvate
   content, different leaf tissues, 506
   dehydrogenase complex, 505, 510
   inhibition of, 521

Rapeseed, 341
   lipases, 343
Refsum's disease, 599
Resistance, 75
Rhizopus arrhizus, 277, 315
Rhizopus delemar, 345
Rhodophyta, sterol composition, 632
Ribonucleic acid (RNA), 25
Ricinoleic acid, 341
Ricinus communis, 341
Rishitin, 78, 80
Ruzicka, 11, 12

Salvia officinalis, 14–16
Sebacic acid, 593
Secale cereale, 214
Seed
   maturation, 240
   oil content, control of, 456
Senescence, 225, 387, 388
Sesquiterpene, 16
   in disease resistance, 76
Sethoxydim, 63, 69
Side chain length, 4
Sinapis alba, 38, 42, 91
Sitosterol, 3, 4, 7, 92, 98, 109
β-Sitosterol, 206, 214
Solanum andigena, 85
Solanum tuberosum, 233, 321, 369
Somatic embryogenesis, 517
Sorghum bicolor, 107
Soybean, 230
   cotyledons, linolenate content of, 533
Sphaceloma manihoticola, 19
Spherosomes, 239
Spinach (Spinacia oleracea), 30, 48, 217, 249, 257
   chloroplasts, 248, 297
Spirodela oligorrhiza, 194
Squalene, 4, 88
Squalene-oxide, 53
Stearoyl acyl carrier protein, 283
Stereochemistry/sterols, 5, 6
Sterol, 53, 108
   biosynthesis, 34
   critical mass of, 54
   esterification, 57

Sterol (continued)
　main line, 87
　multiple functions, 53, 58, 93, 103, 148
　synthesis, in Alga, 631
Steryl
　esters, 95, 98, 206
　　mass spectrometry of, 96
　glucosides, 214
Stigmasterol, 91, 93, 109, 206, 214
Stroma, 30
Sulfolipids, biosynthesis of, 617
Sulfoquinovose, 309
Sulfoquinovosyl diacylglycerol (SQDG), 309, 313, 603
Sycamore, 257

Tanacetum vulgare, 14, 15
Tea, 388
　chloroplast, 397
Temperature shift, fatty acid composition in cyanobacterium, 607
Terpenes, 255
α-Terpineol, 13, 14
α-Terpinyl, 11
Thalassiosira pseudonana, 649
Theobroma cacao, 337
Thermolysin, 294, 295, 305
Thiolactomycin, 447
　effect on
　　ACP acetyltransferase, 448
　　ACP malonyltransferase, 448
　　enoyl-ACP reductase, 448
　　fatty acid composition of Pseudomonas aeroginosa, 450
　　3-hydroxybutyryl-ACP dehydrase, 448
　　3-ketoacyl-ACP reductase, 448
　　3-ketoacyl-ACP synthase, 448
　　lipid composition in Avena leaves, 451
　　lipid composition in Brassica leaves, 451
　　overall fatty acid synthase, 448
　lipid synthesis in leaves, 449
Thioredoxin, 387
Thylakoids, 30
α-Tocopherol, 29
　biosynthesis, 31
Tomato, 374
Tonoplast, 257
trans-2-hexenal, 388
trans-3-hexadecenoic acid, 345
Traumatic acid, 387
Traumatin, 387
Triacylglycerols, 206, 244, 337
Trierucin, 341

UDP gal:diacylglycerol galactosyltransferase, 293
UDP glucose:diacylglycerol glucosyltransferase, 604
Ulmus americana, 341
α,β-Unsaturated aldehydes, 388

Vascular plants, 5
Very long chain fatty acids, formation in higher plants, 482
Vicia faba, 299
Violaxanthin, 389
Vitamin K_1, biosynthesis, 32

Wax, 107

Xanthium strumarium, 83
Xanthoxin, 389

Yeast, 5

Zea mays (maize/corn), 22, 97, 99, 341
Zostera marina (eelgrass), 50